CHEMICAL ENGINEERING KINETICS

BUILDING THE LITERATURE OF A PROFESSION

Fifteen prominent chemical engineers first met in New York more than 50 years ago to plan a continuing literature for their rapidly growing profession. From industry came such pioneer practitioners as Leo H. Baekeland, Arthur D. Little, Charles L. Reese, John V. N. Dorr, M. C. Whitaker, and R. S. McBride. From the universities came such eminent educators as William H. Walker, Alfred H. White, D. D. Jackson, J. H. James, Warren K. Lewis, and Harry A. Curtis. H. C. Parmelee, then editor of *Chemical and Metallurgical Engineering*, served as chairman and was joined subsequently by S. D. Kirkpatrick as consulting editor.

After several meetings, this committee submitted its report to the McGraw-Hill Book Company in September 1925. In the report were detailed specifications for a correlated series of more than a dozen texts and reference books which have since become the McGraw-Hill Series in Chemical Engineering and which became the cornerstone of the chemical engineering curriculum.

From this beginning there has evolved a series of texts surpassing by far the scope and longevity envisioned by the founding Editorial Board. The McGraw-Hill Series in Chemical Engineering stands as a unique historical record of the development of chemical engineering education and practice. In the series one finds the milestones of the subject's evolution: industrial chemistry, stoichiometry, unit operations and processes, thermodynamics, kinetics, and transfer operations.

Chemical engineering is a dynamic profession, and its literature continues to evolve. McGraw-Hill and its consulting editors remain committed to a publishing policy that will serve, and indeed lead, the needs of the chemical engineering profession during the years to come.

THE SERIES

Bailey and Ollis: *Biochemical Engineering Fundamentals*
Bennett and Myers: *Momentum, Heat, and Mass Transfer*
Beveridge and Schechter: *Optimization: Theory and Practice*
Carberry: *Chemical and Catalytic Reaction Engineering*
Churchill: *The Interpretation and Use of Rate Data—The Rate Concept*
Clarke and Davidson: *Manual for Process Engineering Calculations*
Coughanowr and Koppel: *Process Systems Analysis and Control*
Danckwerts: *Gas Liquid Reactions*
Gates, Katzer, and Schuit: *Chemistry of Catalytic Processes*
Harriott: *Process Control*
Johnson: *Automatic Process Control*
Johnstone and Thring: *Pilot Plants, Models, and Scale-up Methods in Chemical Engineering*
Katz, Cornell, Kobayashi, Poettmann, Vary, Elenbaas, and Weinaug: *Handbook of Natural Gas Engineering*
King: *Separation Processes*
Knudsen and Katz: *Fluid Dynamics and Heat Transfer*
Lapidus: *Digital Computation for Chemical Engineers*
Luyben: *Process Modeling, Simulation, and Control for Chemical Engineers*
McCabe and Smith, J. C.: *Unit Operations of Chemical Engineering*
Mickley, Sherwood, and Reed: *Applied Mathematics in Chemical Engineering*
Nelson: *Petroleum Refinery Engineering*
Perry and Chilton (Editors): *Chemical Engineers' Handbook*
Peters: *Elementary Chemical Engineering*
Peters and Timmerhaus: *Plant Design and Economics for Chemical Engineers*
Ray: *Advanced Process Control*
Reed and Gubbins: *Applied Statistical Mechanics*
Reid, Prausnitz, and Sherwood: *The Properties of Gases and Liquids*
Resnick: *Process Analysis and Design for Chemical Engineering*
Satterfield: *Heterogeneous Catalysis in Practice*
Sherwood, Pigford, and Wilke: *Mass Transfer*
Slattery: *Momentum, Energy, and Mass Transfer in Continua*
Smith, B. D.: *Design of Equilibrium Stage Processes*
Smith, J. M.: *Chemical Engineering Kinetics*
Smith, J. M., and Van Ness: *Introduction to Chemical Engineering Thermodynamics*
Thompson and Ceckler: *Introduction to Chemical Engineering*
Treybal: *Mass Transfer Operations*
Van Winkle: *Distillation*
Volk: *Applied Statistics for Engineers*
Walas: *Reaction Kinetics for Chemical Engineers*
Wei, Russell, and Swartzlander: *The Structure of the Chemical Processing Industries*
Whitwell and Toner: *Conservation of Mass and Energy*

CHEMICAL ENGINEERING KINETICS

Third Edition

J. M. Smith

Professor of Chemical Engineering
University of California at Davis

McGraw-Hill Book Company

New York St. Louis San Francisco Auckland Bogotá Hamburg
Johannesburg London Madrid Mexico Montreal New Delhi
Panama Paris São Paulo Singapore Sydney Tokyo Toronto

This book was set in Times Roman.
The editor was Julienne V. Brown;
the production supervisor was Donna Piligra.
R. R. Donnelley & Sons Company was printer and binder.

CHEMICAL ENGINEERING KINETICS

4567890DODO898765432

Library of Congress Cataloging in Publication Data

Smith, Joe Mauk, date
    Chemical engineering kinetics.

    (McGraw-Hill chemical engineering series)
    Includes bibliographical references and index.
    1. Chemical engineering. 2. Chemical reaction,
Rate of. 3. Thermodynamics. I. Title.
TP149.S58   1981   660.2'994   80-16486
ISBN 0-07-058710-8

CONTENTS

PREFACE TO THE THIRD EDITION

In the decade since publication of the second edition, progress in reactor design again has necessitated rather complete revision of all chapters. The rewriting has provided an opportunity for improving the presentation of basic concepts. Major changes of this kind are in Chaps. 3 to 5. The mass conservation equations in Chap. 3 have been developed in a more general manner. The development is such that design expressions for specific reactor forms, as used in the remainder of the book, follow directly by simplification of the general equations. Analysis of recycle reactors, not included in the second edition, is presented first in Chap. 4. The material on semibatch reactors (Chap. 4) has been rewritten so that both this subject and recycle reactors provide additional illustrations of the general mass conservation equations developed in Chap. 3. For a clear presentation, mass and energy balances have been completely separated. Thus, the general equation for conservation of energy is developed in Ch. 5 where it is then used for nonisothermal reactor design.

A further change in the structure of the book has been made in introducing heterogeneous processes, catalysis, and catalyst properties. Adsorption and general aspects of catalysis are now included in Chap. 7, which in the second edition was but a short introductory treatment of heterogeneous processes. It is hoped that this change makes Chap. 7 a more meaningful introduction to the fluid-solid type of heterogeneous process that is emphasized in the remainder of the text. Then Chap. 8 is devoted mainly to physical properties and preparation of solid catalysts. A new section giving examples of catalysts is included in order to provide a guide for the kinds of catalysts that are known to be effective for different reactions.

A major objective of the third and earlier editions is to give examples of reactor design and interpretation of laboratory data for real chemical reactions. This often requires repetitive numerical calculations (for solution of differential equations). In the third edition these calculations are based primarily on a Runge-Kutta method. However, where it seems necessary to show more clearly the physical meaning of the calculations, the simpler, modified Euler method is used as well.

While the purposes of Chaps. 10 to 13 remain the same, the growing importance of three-phase reactors has required inclusion of design principles and applications for slurry and trickle-bed reactors. Also, some discussion of monolithic (continuous, solid catalyst phase) reactors is presented in Chap. 13. More complete treatment of fluidized-bed reactors, including a simple version of the bubbling-gas model has been given in Chaps. 10 and 13.

The subject material presented in each chapter remains generally the same as described in the Preface to the Second Edition, except for the changes already mentioned and other minor additions. The overall objective of the book is also unchanged—this is to provide a clear yet reasonably rigorous exposition of reactor design, with illustrations taken from practical and realistic chemical systems. The book should be readily understandable by students in the fourth, or possibly third, year of the undergraduate chemical engineering curriculum. The whole book can be comfortably covered in two semesters, and a bit less comfortably in two quarters.

Thanks are due to numerous students and colleagues for worthwhile discussions and suggestions, and Mr. C. S. Tan kindly assisted with many of the numerical calculations. Their contributions are sincerely acknowledged.

J. M. Smith

PREFACE TO THE SECOND EDITION

The first edition of *Chemical Engineering Kinetics* appeared when the rational design of chemical reactors, as opposed to empirical scaleup, was an emerging field. Since then, progress in kinetics, catalysis, and particularly in engineering aspects of design, has been so great that this second edition is a completely rewritten version. In view of present-day knowledge, the treatment in the first edition is inadequate with respect to kinetics of multiple-reaction systems, mixing in nonideal reactors, thermal effects, and global rates of heterogeneous reactions. Special attention has been devoted to these subjects in the second edition. What hasn't changed is the book's objective: the clear presentation and illustration of design procedures which are based upon scientific principles.

Successful design of chemical reactors requires understanding of chemical kinetics as well as such physical processes as mass and energy transport. Hence, the intrinsic rate of chemical reactions is accorded a good measure of attention: in a general way in the second chapter and then with specific reference to catalysis in the eight and ninth. A brief review of chemical thermodynamics is included in Chap. 1, but earlier study of the fundamentals of this subject would be beneficial. Introductory and theoretical material is given in Chap. 2, only in a manner that does not make prior study of kinetics mandatory.

The concepts of reactor design are presented in Chap. 3 from the viewpoint of the effect of reactor geometry and operating conditions on the form of mass and energy conservation equations. The assumptions associated with the extremes of plug-flow and stirred-tank behavior are emphasized. A brief introduction to deviations from these ideal forms is included in this chapter and is followed with a more detailed examination of the effects of mixing on conversion in Chap. 6. In Chaps. 4 and 5 design procedures are examined for ideal forms of homogenous reactors, with emphasis upon multiple-reaction systems. The latter chapter is concerned with nonisothermal behavior.

Chapter 7 is an introduction to heterogeneous systems. The concept of a global rate of reactions is interjected so as to relate the design of heterogeneous reactors to the previously studied concepts of homogeneous reactor design. A secondary objective here is to examine, in a preliminary way, the method of combining of chemical and physical processes so as to obtain a global rate of reaction.

Chapter 8 begins with a discussion of catalysis, particularly on solid surfaces, and this leads directly into adsorption and the physical properties of porous solids. The latter is treated in reasonable detail because of the importance of solid-catalyzed reactions and because of its significance with respect to intrapellet transport theory (considered in Chap. 11). With this background, the formulation of intrinsic rate equations at a catalyst site is taken up in Chap. 9.

The objective of Chaps. 10 and 11 is to combine intrinsic rate equations with intrapellet and fluid-to-pellet transport rates in order to obtain global rate equations useful for design. It is at this point that models of porous catalyst pellets and effectiveness factors are introduced. Slurry reactors offer an excellent example of the interrelation between chemical and physical processes, and such systems are used to illustrate the formulation of global rates of reaction.

The book has been written from the viewpoint that the design of a chemical reactor requires, first, a laboratory study to establish the intrinsic rate of reaction, and subsequently a combination of the rate expression with a model of the commercial-scale reactor to predict performance. In Chap. 12 types of laboratory reactors are analyzed, with special attention given to how data can be reduced so as to obtain global and intrinsic rate equations. Then the modeling problem is examined. Here it is assumed that a global rate equation is available, and the objective is to use it, and a model, to predict the performance of a large-scale unit. Several reactors are considered, but major attention is devoted to the fixed-bed type. Finally, in the last chapter gas-solid, noncatalytic reactions are analyzed, both from a single pellet (global rate) viewpoint, and in terms of reactor design. These systems offer examples of interaction of chemical and physical processes under *transient* conditions.

No effort has been made to include all types of kinetics or of reactors. Rather, the attempt has been to present, as clearly and simply as possible, all the aspects of process design for a few common types of reactors. The material should be readily understandable by students in the fourth undergraduate year. The whole book can be comfortably covered in two semesters, and perhaps in two quarters.

The suggestions and criticisms of numerous students and colleagues have been valuable in this revision, and all are sincerely acknowledged. The several stimulating discussions with Professor J. J. Carberry about teaching chemical reaction engineering were most helpful. To Mrs. Barbara Dierks and Mrs. Loretta Charles for their conscientious and interested efforts in typing the manuscript, I express my thanks. Finally, the book is dedicated to my wife, Essie, and to my students whose enthusiasm and research accomplishments have been a continuing inspiration.

J. M. Smith

LIST OF SYMBOLS

Included in the following list are common symbols employed throughout the book. Other, more specialized symbols are defined as used.

A	frequency factor in Arrhenius equation
A	area, $(\text{length})^2$
a	activity, or pore radius
a_m	external surface per unit mass, $(\text{length})^2/\text{mass}$
a_v	external surface per unit volume, $(\text{length})^{-1}$
C_i	concentration of component i, mol/vol.
C_e	concentration in exit stream
C_0, C_f	concentration in feed stream
C_b	concentration in bulk of fluid stream
$\bar{C}_i$	concentration of component i adsorbed on a catalyst surface, mol/(mass of catalyst)
C_s	concentration at catalyst surface
c_p	molal heat capacity or specific heat at constant pressure, energy/(temperature)(mol or per unit mass)
D	combined bulk and Knudsen diffusivity
$\mathscr{D}_{AB}$	bulk diffusivity of A or B in binary system, $(\text{length})^2/\text{time}$
$\mathscr{D}_K$	Knudsen diffusivity, $(\text{length})^2/\text{time}$
$\mathscr{D}_s$	surface diffusivity, $(\text{length})^2/\text{time}$
D_e	effective diffusivity (based upon total void plus nonvoid area), $(\text{length})^2/\text{time}$
D_L	axial dispersion coefficient, $(\text{length})^2/\text{time}$
d	tube diameter
d_p	particle or pellet diameter
E	activation energy, energy/mol
F	feed rate, mass or mol/time
ΔF	free-energy change for a reaction, energy/mol
f	fugacity
G	fluid mass velocity, mass/(area)(time)

H	enthalpy, energy/mass, or Henry's law constant
H'	enthalpy rate, energy/time
ΔH	enthalpy change for a reaction, energy/mol
h	heat-transfer coefficient, energy/(time)(area)(temperature difference)
$J(\theta)$	residence-time distribution function
K	equilibrium constant for a reaction
K_a	adsorption equilibrium constant
k	forward reaction-rate constant
k'	reverse reaction-rate constant
k_B	Boltzmann's constant, 1.3805×10^{-16} erg/K or 1.3805×10^{-23} J/K
k_e	effective thermal conductivity, energy/(time)(length)(temperature)
j	j-factor [see Eqs. (10-9) and (10-11)]
k_f	thermal conductivity, energy/(time)(length)(temperature)
k_m	mass transfer coefficient (particle-fluid), mass or mol/(time)(area) (concentration difference)
k_0	overall rate constant
L	length
M	molecular weight, mass/mol
m	mass
n_i	moles of component i
N_i	molal flow rate (or diffusion rate) of component i, mol/time
N_0	Avogadro's number, 6.023×10^{23} molecules/mol
p_i	partial pressure
p_t	total pressure
Pe	Peclet number, uL/D_L or ud_p/D_e
Pr	Prandtl number, $c_p \mu/k_f$
Q	volumetric flow rate, volume/time, or energy transfer as heat, energy
Q'	heat transfer rate, energy/time
q	heat flux, energy/(area)(time)
R	recycle ratio
R_g	gas constant, energy/(mol)(K)
Re	Reynolds number
r	radius, radial coordinate
$\mathbf{r}_i$	reaction rate of component i, mol/(vol.)(time)
$\bar{\mathbf{r}}_i$	average rate of reaction, mol/(vol.)(time)
$\mathbf{r}_p$	global reaction rate, mol/(mass catalyst) (time)
$\mathbf{r}_v$	global reaction rate, mol/(vol. of reactor)(time)
S	selectivity
S_o, S_p	overall selectivity, point selectivity
S	entropy, energy/(mol)(temperature)
ΔS	entropy change for a reaction, energy/(mol)(temperature)
S_g	pore surface area (of catalyst) per unit mass
Sc	Schmidt number, $\mu/\rho \mathscr{D}$

T	temperature, absolute
t	time
U	internal energy per mole or overall heat transfer coefficient, energy/(time)(area)(temperature difference)
u	superficial velocity, length/time
V_g	pore volume (of catalyst), vol./mass
V	reactor volume
v	velocity, length/time
v	specific or molal volume, vol./(mass) or vol./(mol)
W	mass of catalyst
w	weight fraction
X	active site on a catalyst surface
x	conversion, yield, or distance
y_i	mole fraction of component i
z	axial distance
γ	activity coefficient, or ratio of global rate to rate evaluated at bulk-fluid conditions
ε	void fraction
ε_S	solids fraction
η	effectiveness factor
θ	residence time
$\bar{\theta}$	mean residence time
λ	mean free path, length
μ	viscosity, mass/(length)(time)
ρ	density, mass/vol.
ρ_p	density of catalyst pellet, mass/vol.
Φ	Thiele-type modulus for a porous catalyst
ρ_B	density of bed of catalyst pellets, mass/vol.
δ	tortuosity factor (in a catalyst pellet)
ξ	extent of reaction
v_i	stoichiometric coefficient
ϕ_i	quantum yield

Subscripts

ave	average
b	bulk
c	catalyst or critical point
B	bed
g	gas
i	interface between phases
L	liquid
p	particle or pellet
s	surface, solid, or spherical

bold-face type denotes a vector (except for rate, **r**)

INTRODUCTION

The major objective of this book is to learn how to design equipment for carrying out desirable chemical reactions. The design and operation of such equipment, that is, reactors, requires rates of both physical and chemical processes. The principles governing such physical processes as energy transfer and mass transfer are often as important as those which govern chemical kinetics. This combination of physical and chemical operations is also a distinguishing feature of chemical engineering; the design of chemical reactors is an activity unique to the chemical engineer.

To design a reactor means to answer the following questions: What type and size of equipment are needed to accomplish the desired extent of reaction? What operating conditions (temperatures, pressures, flow rates) are required? What provisions are necessary for exchange of energy (usually as heat) with the surroundings? The answers to these questions constitute the *process design* of the reactor. A cost analysis to determine the most profitable design introduces further questions about construction materials, corrosion, utility requirements, and maintenance. In order to maximize profits, the instrumentation and a control policy (ranging from manual to closed-loop computer control) for optimum operation should be determined. Optimum design will also depend indirectly on estimates of such market conditions as the price-volume relationships for reactants and products. While these factors are important in overall design and operation of reactors, they will not be discussed in this book. Our use of the term *design* will be restricted to *process design*.

How do we go about combining rates of chemical and physical processes in order to design the reactor? The essential feature is to write conservation equations for mass and energy† for the chosen type of reactor. Solution of these equations, which can be either algebraic or differential equations, yields the extent of reaction and the operating conditions. There will be two kinds of terms in the conservation equations: (1) terms expressing physical processes, that is, rates of energy transfer and of mass transfer of specific chemical species, and (2) terms expressing rates of converting one chemical species to another species. This last quantity refers to a chemical process and, for each reaction involved, is termed the *intrinsic* rate of that chemical reaction. Such intrinsic rates, as yet, cannot be predicted with accuracy and so must be measured experimentally. However, there is a considerable body of knowledge about the variables that affect intrinsic rates and about equations which correlate rate data. This subject is known as *chemical kinetics*. It is discussed briefly in Sec. 1-2, and quantitatively and in detail in Chap. 2 for homogeneous reactions, and in Chaps. 8 and 9 for heterogeneous catalytic reactions. The purpose of these chapters is to provide expressions for the intrinsic rate that may be employed in the conservation equations.

The form of the conservation equations depends on the type of reactor but not on the specific chemical reactions involved. Further, the terms for mass and energy transfer in these equations are of the same form for a given type of reactor. Hence, the design problems are essentially the same for a given type of reactor; the only difference from one chemical reaction system to another is the intrinsic rate equation. Advantage is taken of this generalization in the organization of the book. Thus, in each chapter concerned with reactor design conservation equations are applied to various types of reactors. In Chap. 3, the two extreme classifications according to geometry, the stirred-tank and plug-flow reactors, are introduced. Then in Chaps. 4, 5, and 13, applications are considered for homogeneous and heterogeneous catalytic reactors. In Sec. 1-5, a qualitative introduction is given to various types of reactors. In contrast to intrinsic kinetics, reliable correlations are available for rates of many mass and energy transfer processes. Hence, experimental data are not required for evaluation of these terms in the conservation equations.

Although intrinsic reaction rates must be obtained from experimental data, such rates cannot always be obtained directly from such data. This is because concentrations and temperatures readily measured may not be those that exist where the reaction takes place. This is most likely to occur when more than one phase exists in the reactor. Consider the oxidation of sulfur dioxide gas to sulfur trioxide gas with air on a vanadium pentoxide catalyst (a porous solid). To supply SO_2 to the catalytically active solid surface, mass transfer must occur from the bulk SO_2 stream to the solid surface. Since such mass transfer requires a concentration difference, the concentration of SO_2 in the bulk gas mixture will be greater than the concentration at the catalyst surface. Hence, the measured rate of reaction will not be that equal to the intrinsic rate corresponding to the known, bulk

---

† The conservation-of-momentum principle is less used in reactor design. There are two reasons for this. The first is that pressure changes within a reactor usually are less important than composition and temperature changes. The second is that the geometry of many reactors is so complicated that, as yet, the momentum principle cannot be used to predict detailed velocity distributions, even though such information would be useful for design.

concentration of SO_2, but rather to the intrinsic rate corresponding to an unknown concentration at the catalyst surface. This is an illustration of coupling of chemical processes (intrinsic kinetics) and physical processes (in this example, mass transfer of SO_2) at the local rate level. In order to obtain an equation for the intrinsic rate from such data, the effect of mass transfer must be considered. We will use the term *global rate* to describe the measured rate, that is, the rate associated with bulk concentrations and temperatures. The relation between global and intrinsic reaction rates are considered in detail in Chaps. 10 to 12. An advantage of the global rate concept is that its use in reactor design means that the same form of conservation equations can be used for both homogeneous and heterogeneous systems.

In the next section, the effect of this interaction between chemical and physical processes on reactor design is discussed, qualitatively, in more detail.

1-1 Interpretation of Rate Data, Scale-up, and Design

The chemical engineer depends on data from the chemist's laboratory, the pilot plant, or a large-scale reactor for help in design work. As noted earlier, from this information expressions for the intrinsic rates of the chemical reactions involved, i.e., the chemical kinetics of the system, need to be extracted. To do this, the effects of physical processes must be separated from the observed data, leaving rate information for the chemical-transformation step alone. It will then be possible to reintroduce the influence of the physical steps for the particular reactor type and operating conditions chosen for the commercial plant. The interrelationship of the physical and chemical steps must be considered twice: once in obtaining intrinsic rate expressions from the available laboratory or pilot-plant data, and again in using these intrinsic rate equations to design the commercial-scale reactor. The first step, interpretation of the available data, is as important as the second, and entails generally the same type of analysis. Hence the interpretation of laboratory data will sometimes be discussed in parallel with the reactor-design problem in the chapters that follow. Also Chap. 12 is devoted solely to interpretation of laboratory data for catalytic reactions. Interpreting laboratory-reactor data does not always entail the same steps (in reverse order) as reactor design. Because there are fewer restraints (no economic ones, for example), there is more flexibility in choosing a laboratory reactor. It is common practice to design a laboratory reactor to minimize the significance of physical processes (see Chap. 12). This leads to more accurate results for the intrinsic rates of the chemical steps. For instance, a laboratory reactor may be operated at near isothermal conditions, eliminating heat transfer considerations, while such operation would be uneconomical in a commercial-scale system.

It is important to consider the relationship between reactor *scale-up* (projection of laboratory or pilot-plant data to the commercial reactor) and reactor design. In principle, if the intrinsic rates of the chemical reactions are known, any type of reactor may be designed by introducing the rates of the appropriate physical processes associated with that type of equipment. Scale-up is an abbreviated version of this design process. The physical resistances are not separated from the measured laboratory data but are instead projected directly to a large unit which presumably has the same interrelationship of chemical and physical processes. If the dimensions and operating conditions for the large-scale reactor can be determined to ensure that the interrelationships of chemical and physical processes are the same as in the laboratory unit, then the laboratory results may be used directly to predict the behavior of the large-scale reactor. In a scale-up process no attempt is made to determine the rates of the chemical steps, that is, to evaluate the chemical kinetics of the process. Scale-up may seldom be applicable, but when suitable, it provides a rapid means of obtaining approximate reactor sizes, and also indicates the important parameters in the interrelationship between physical and chemical processes (see Sec. 12-5).

Scale-up will have a much better chance for success if the laboratory and commercial operations are carried out in the same type of system. Suppose that laboratory data for the thermal cracking of hydrocarbons are obtained in a continuous tube through which the reaction mixture flows. If a tubular-flow reactor of this type is also proposed for the commercial plant, it may be possible to scale-up the pilot-plant operation in such a way that the temperature and concentration gradients within the tube will be the same in both reactors. Then performance of the large-scale reactor—for example, conversion of reactants to various products—can be predicted directly from the laboratory results. However, if the laboratory results were obtained from a batch reactor, a tank or vessel into which the reactants are initially charged (see Sec. 1-6), it is difficult to project them directly to a large-scale tubular reactor. In this case it would be necessary to analyze the laboratory data to obtain the rate equation for the chemical reactions and then use these results to design the commercial reactor. Our emphasis will be on this two-step process of determining the rates of reaction from laboratory data and then using these rates for design.

The foregoing comments do not imply that pilot-plant data from a small-scale replica of a proposed commercial unit are of no value. Such information provides an important evaluation of both the laboratory rate data and the procedures used to introduce the effects of physical processes on the performance of the pilot reactor and, presumably, on the final equipment.

Our discussion thus far of the interrelationship of chemical and physical processes in a reactor has been general. Let us look further into the problem by considering a very simple reaction system, the conversion of ortho hydrogen to the para form.† Owing to thermodynamic restrictions (see Sec. 1-3), this reaction must be carried out at a low temperature in order to obtain a large conversion to para hydrogen. At low temperatures it is necessary to use a catalyst to obtain a rapid rate of reaction. The preferred type of reactor is a continuous steady-state system in which hydrogen flows through a tube packed with pellets of the solid catalyst. Consider the interpretation of rate measurements made with a laboratory version of this type of reactor. The observed data would consist of measurements of hydrogen compositions in the inlet and exit streams of the reactor. Probable

---

† This reactor-design problem has some practical significance because of the superior storage properties of liquid hydrogen when it is in the para form. Noriaki Wakao and J. M. Smith, *AIChE J.*, **8**, 478 (1962).

variables would be the flow rate of hydrogen through the reactor, the mole fraction of para hydrogen in the feed to the reactor, and the temperature. The heat of reaction is negligible, so that the whole reactor system can easily be operated at isothermal conditions.

The first problem in designing a reactor for the production of para hydrogen is to obtain from the observed measurements a quantitative expression for the intrinsic rate of reaction at the surface of the catalyst. Specifically, we must separate from the observed data the diffusional resistances between the point at which the composition is measured—the exit of the reactor—and the point at which the chemical transformation occurs—the gas-solid interface at the catalyst surface. There are three diffusional effects that may cause a difference between the conversion measured in the exit of the reactor and that predicted from the rate at the catalyst interface. The first arises from the mixing characteristics of the fluid as it flows around the particles in the fixed bed. There may be some bypassing or short-circuiting, so that part of the flowstream does not come into contact with the catalyst; also, there may be diffusion or back-mixing of fluid as it flows through the bed. As a result, the observed amount of para hydrogen in the exit gas may be less than expected. The second factor is the tendency of the fluid to adhere to the catalyst pellet, so that the pellet is surrounded by a more or less stagnant layer of fluid which resists mass transfer. Thus a concentration gradient of para hydrogen must be established between the outer surface of the pellet and the bulk gas before the para hydrogen will move on into the gas stream. This reduces the amount of para hydrogen available to the bulk-gas phase. A third factor is that most of the active surface of the catalyst is in the pores within the pellet. The reactant must reach this interior pore surface by diffusing into the pellet, and the product must diffuse out. This process is impeded by intraparticle resistance, causing another reduction in the para-hydrogen content of the gas stream. Thus to determine the intrinsic rate of reaction on the surface of the catalyst (the chemical kinetics of the process) it is necessary to evaluate the concentration changes for each of these diffusional effects, ultimately arriving at the concentration of para hydrogen in the interior pore surface of the catalyst pellet. The interior concentration can then be used to establish a rate-of-reaction equation.

The second problem is use of the rate equation to design a commercial reactor. The individual diffusional resistances are now reintroduced so that we may determine the actual composition of para hydrogen in the exit stream from the reactor. Once the equation for the surface rate is known, it is possible, in principle, to predict the exit conversion for any type of reactor, any size catalyst pellet, any conditions of gas flow around the pellet, and any condition of mixing of fluid around the particles in the fixed bed.

If the same problem were approached from a scale-up standpoint, the procedure would be to attempt to choose the operating conditions and reactor size for the large-scale reactor such that the diffusional resistances were the same as in the laboratory equipment.

Rates are normally considered on a specific basis; that is, per unit volume of reaction mixture for a homogeneous reaction, or per unit mass of catalyst for a heterogeneous, fluid-solid catalytic reaction.

1-2 Chemical Kinetics

Chemical kinetics is the study of the rate and mechanism by which one chemical species is converted to another. The *rate* is the mass, in moles, of a product produced or reactant consumed per unit time. The *mechanism* is the sequence of individual chemical events whose overall result produce the observed reaction. Basolo and Pearson† have described the term "mechanism" as follows:

> By mechanism is meant all the individual collisional or elementary processes involving molecules (atoms, radicals and ions included) that take place simultaneously or consecutively in producing the observed overall rate. It is also understood that the mechanism of a reaction should give a detailed stereochemical picture of each step as it occurs. This implies a knowledge of the so-called activated complex or transition state, not only in terms of the constituent molecules but also in terms of the geometry, such as interatomic distances and angles. In most instances the postulated mechanism is a theory devised to explain the end results observed by experiments. Like other theories, mechanisms are subject to change over the years as new data is uncovered or as new concepts regarding chemical interreactions are developed.

It is not necessary to know the mechanism of a reaction in order to design a reactor. What is necessary is a satisfactory equation for the intrinsic rate. A knowledge of the mechanism is of great value, however, in extending the rate data beyond the original experiments and in generalizing or systematizing the kinetics of reactions. Determining the mechanism of a reaction is a very difficult task and may require the work of many investigators over a period of many years. Reaction mechanisms are reliably known for only a few systems. However, postulated theories for mechanisms are available for a wide variety of reactions, ranging from simple, gas-phase homogeneous systems to complicated polymerization reactions involving initiation, propagation, and termination steps.

Since reactor design necessitates a reliable intrinsic rate equation, this aspect of chemical kinetics is presented in detail in Chap. 2. Successful procedures for *predicting* rates of reactions will not be developed until reaction mechanisms are better understood. Nevertheless, it is important for those involved in reactor design to be aware of developments in this area so that they may take advantage of new principles of chemical kinetics as they are developed. A brief discussion of theories of reaction and mechanisms is included in Chap. 2.

The rate of a chemical reaction can vary from a value approaching infinity to essentially zero. In ionic reactions, such as those that occur on photographic film, or in high-temperature combustion reactions, the rate is extremely fast. The rate of combination of hydrogen and oxygen in the absence of a catalyst at room temperature is immeasurably slow. Most industrially important reactions occur at rates between these extremes, and it is in these cases that the designer must apply data on kinetics to determine finite sizes of reaction equipment. It is particularly important to know how the rate changes with operating parameters, the most important of which are temperature and composition of reaction mixture.

The first quantitative measurements of reaction rates were made in the middle

† F. Basolo and R. G. Pearson, "Mechanisms of Inorganic Reactions," John Wiley & Sons, Inc., New York, 1958.

of the nineteenth century by Wilhelmy,† Berthelot and St. Gilles,‡ and Harcourt and Esson.§ The first attempt to develop a theory explaining the manner in which molecules of a substance react was that of Arrhenius¶ in 1889. He postulated that the reactants had both inert and active molecules and that only the active ones possessed sufficient energy to take part in the reaction. Since these early developments there have been a great many experimental studies of reaction rates for a wide variety of reactions, but few noteworthy advances in theory were made until the work of Erying and Polanyi,†† beginning in 1920. Using only such fundamental information as the configurations, dimensions, and interatomic forces of the reacting molecules, they postulated an activated-complex theory for predicting the rate of reaction. Lack of exact knowledge of the interatomic forces, and hence of energy-position relations, for any but the most simple molecules has prevented the activated-complex theory from being useful for predicting reaction-rate data accurately enough for engineering work. While these theoretical developments have been of great value in the search for an understanding of how and why a chemical reaction takes place, the quantitative evaluation of the rate remains an experimental problem.

The large amount of experimental data on rates of chemical reactions have established reliable empirical forms for the mathematical expression of the effects of temperature and composition on the rate. These results are interpreted for various kinds of reactions in Chap. 2.

1-3 Kinetics and Thermodynamics

From the principles of thermodynamics and certain thermodynamic data the *maximum* extent to which a chemical reaction can proceed may be calculated. For example, at 1 atm pressure and a temperature of 680°C, starting with 1 mole of sulfur dioxide and $\frac{1}{2}$ mole of oxygen, 50% of the sulfur dioxide can be converted to sulfur trioxide. Such thermodynamic calculations result in maximum values for the conversion of a chemical reaction, since they are correct only for equilibrium conditions, conditions such that there is no further tendency for change with respect to time. It follows that the net rate of a chemical reaction must be zero at this equilibrium point. Thus a plot of reaction rate [for example, in units of g mol product/(s) (unit volume reaction mixture)] vs. time would always approach zero as the time approached infinity. Such a situation is depicted in curve *A* of Fig. 1-1, where the rate approaches zero asymptotically. Of course, for some cases equilibrium may be reached more rapidly, so that the rate becomes almost zero at a finite time, as illustrated by curve *B*.

Similarly, the *conversion* (fraction of reactant transformed or converted) calculated from thermodynamic data would be the end point on a curve of conversion vs. time such as that shown in Fig. 1-2. Again, curve *A* represents the case

† L. Wilhelmy, *Pogg. Ann.*, **81**, 413, 499 (1850).
‡ M. Berthelot and L. P. St. Gilles, *Ann. Chim. Phys.*, **63** (3), 385 (1862).
§ A. V. Harcourt and W. Esson, *Proc. Roy. Soc.* (*London*), **14**, 470 (1865).
¶ S. Arrhenius, *Z. Physik Chem.*, **4**, 226 (1889).
†† H. Eyring and M. Polanyi, *Z. Physik Chem. B*, **12**, 279 (1931).

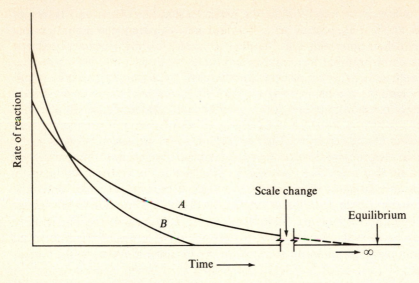

Figure 1-1 Rate of reaction vs. time.

where the time required to reach equilibrium conditions is great, while in case B the equilibrium conversion is approached more rapidly and is attained essentially at a finite time. Curves A and B could apply to the same reaction; the difference between them reflects the fact that in case B the rate has been increased, for example, by use of a catalyst. The rate of the reaction is initially increased over that for the uncatalyzed reaction, but the equilibrium conversion, as shown in Fig. 1-2, is the same for both cases.

The time available for carrying out a chemical reaction commercially is limited if the process is to be economically feasible. Hence the practical range of

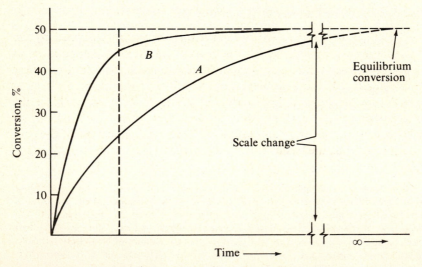

Figure 1-2 Conversion vs. time.

the curves in Figs. 1-1 and 1-2 is at the lower time values. However, the equilibrium conversion is important as a standard for evaluating the actual performance of the reaction equipment. Suppose a kinetics experiment is carried out with a time corresponding to the dashed vertical line shown in Fig. 1-2. At this point the conversion for the noncatalytic reaction is about 25% (curve *A*). A comparison with the equilibrium value of 50% indicates that the noncatalytic rate is rather low and that a search for a catalyst is advisable. Curve *B*, giving a conversion of 45%, shows the benefit of using a catalyst and also indicates that additional effort to find a more effective catalyst is unwarranted. Without prior knowledge of the equilibrium conversion, erroneous conclusions might well be drawn from the kinetic studies yielding curves *A* and *B*. For example, it might be reasoned that the catalyst giving curve *B* is only moderately effective, and considerable time might be spent in attempting to discover a catalyst which would give a conversion of 70 or 80%. Thermodynamic calculations are particularly valuable for such comparison of kinetic and equilibrium results. However, the actual design of reactors usually depends on the location of the curves shown in Figs. 1-1 and 1-2 and must therefore be determined by kinetic studies.

Prediction of the equilibrium conversion requires knowledge of the standard state free-energy changes for the reaction. Although the body of thermodynamic data is growing, it is still not possible to estimate the equilibrium conversion accurately for all reactions. The calculations and data available for gaseous systems are most reliable. The application of thermodynamics for such calculations is illustrated very briefly in the following section. More detailed treatment of the thermodynamics of chemical-reaction equilibrium is given in thermodynamics textbooks.†

The rate of energy transfer is important in determining the temperature distribution in reactors. Also, heats of reaction are significant in connection with equilibrium calculations. The following section deals with data and methods concerning heats of reaction, followed by a discussion of equilibrium conversion.

1-4 Thermodynamics of Chemical Reactions

Heat of reaction The *heat of reaction* is defined as the energy absorbed by the system when the products after reaction are restored to the same temperature as the reactants. The pressure must also be specified for a complete definition of the thermodynamic states of the products and reactants. If the same pressure is chosen for both, the heat of reaction is equal to the enthalpy change; this is the customary definition of the heat of reaction. The heat of any reaction can be calculated by combining heats of formation or heats of combustion of the products and reactants. Thus the basic information necessary for calculating heats of

† J. G. Kirkwood and Irwin Oppenheim, "Chemical Thermodynamics," McGraw-Hill Book Company, New York, 1961; G. N. Lewis and M. Randall, "Thermodynamics," 2d ed., rev. by K. S. Pitzer and Leo Brewer, McGraw-Hill Book Company, New York, 1961; J. M. Smith and H. C. Van Ness, "Introduction to Chemical Engineering Thermodynamics," 3d ed., McGraw-Hill Book Company, New York, 1975; H. C. Van Ness, "Classical Thermodynamics of Non-electrolyte Solutions," The Macmillan Company, Inc., New York, 1964; F. Van Zeggeren and S. H. Story, "The Computation of Chemical Equilibria," Cambridge University Press, New York, 1970.

Table 1-1 Standard heats of formation and combustion for reaction products $H_2O(l)$ and $CO_2(g)$ at 25°C, in calories per gram mole†

Substance	Formula	State	ΔH°_{f298}	$-\Delta H^\circ_{c298}$
		Normal paraffins		
Methane	CH_4	g	−17,889	212,800
Ethane	C_2H_6	g	−20,236	372,820
Propane	C_3H_8	g	−24,820	530,600
n-Butane	C_4H_{10}	g	−30,150	687,640
n-Pentane	C_5H_{12}	g	−35,000	845,160
n-Hexane	C_6H_{14}	g	−39,960	1,002,570
Increment per C atom above C_6	....	g	−4,925	157,440
		Normal monoolefins (1-alkenes)		
Ethylene	C_2H_4	g	12,496	337,150
Propylene	C_3H_6	g	4,879	491,990
1-Butene	C_4H_8	g	−30	649,380
1-Pentene	C_5H_{10}	g	−5,000	806,700
1-Hexene	C_6H_{12}	g	−9,960	964,260
Increment per C atom above C_6	....	g	−4,925	157,440
		Miscellaneous organic compounds		
Acetaldehyde	C_2H_4O	g	−39,760	
Acetic acid	$C_2H_4O_2$	l	−116,400	
Acetylene	C_2H_2	g	54,194	310,620
Benzene	C_6H_6	g	19,820	789,080
Benzene	C_6H_6	l	11,720	780,980
1,3-Butadiene	C_4H_6	g	26,330	607,490
Cyclohexane	C_6H_{12}	g	−29,430	944,790
Cyclohexane	C_6H_{12}	l	−37,340	936,880
Ethanol	C_2H_6O	g	−56,030	
Ethanol	C_2H_6O	l	−66,200	
Ethylbenzene	C_8H_{10}	g	7,120	1,101,120
Ethylene glycol	$C_2H_6O_2$	l	−108,580	
Ethylene oxide	C_2H_4O	g	−12,190	
Methanol	CH_4O	g	−48,100	
Methanol	CH_4O	l	−57,036	
Methylcyclohexane	C_7H_{14}	g	−36,990	1,099,590
Methylcyclohexane	C_7H_{14}	l	−45,450	1,091,130
Styrene	C_8H_8	g	35,220	1,060,900
Toluene	C_7H_8	g	11,950	943,580
Toluene	C_7H_8	l	2,870	934,500
		Miscellaneous inorganic compounds		
Ammonia	NH_3	g	−11,040	
Calcium carbide	CaC_2	s	−15,000	
Calcium carbonate	$CaCO_3$	s	−288,450	
Calcium chloride	$CaCl_2$	s	−190,000	
Calcium chloride	$CaCl_2 : 6H_2O$	s	−623,150	
Calcium hydroxide	$Ca(OH)_2$	s	−235,800	
Calcium oxide	CaO	s	−151,900	
Carbon	C	Graphite	....	94,052
Carbon dioxide	CO_2	g	−94,052	

Continued

Table 1-1 *(Continued)*

Substance	Formula	State	ΔH°_{f298}	$-\Delta H^{\circ}_{c298}$
	Miscellaneous inorganic compounds *(continued)*			
Carbon monoxide	CO	g	$-26,416$	67,636
Hydrochloric acid	HCl	g	$-22,063$	
Hydrogen	H_2	g	....	68,317
Hydrogen sulfide	H_2S	g	$-4,815$	
Iron oxide	FeO	s	$-64,300$	
Iron oxide	Fe_3O_4	s	$-267,000$	
Iron oxide	Fe_2O_3	s	$-196,500$	
Iron sulfide	FeS_2	s	$-42,520$	
Lithium chloride	LiCl	s	$-97,700$	
Lithium chloride	$LiCl \cdot H_2O$	s	$-170,310$	
Lithium chloride	$LiCl \cdot 2H_2O$	s	$-242,100$	
Lithium chloride	$LiCl \cdot 3H_2O$	s	$-313,500$	
Nitric acid	HNO_3	l	$-41,404$	
Nitrogen oxides	NO	g	21,570	
	NO_2	g	7,930	
	N_2O	g	19,513	
	N_2O_4	g	2,190	
Sodium carbonate	Na_2CO_3	s	$-270,300$	
Sodium carbonate	$Na_2CO_3 \cdot 10H_2O$	s	$-975,600$	
Sodium chloride	NaCl	s	$-98,232$	
Sodium hydroxide	NaOH	s	$-101,990$	
Sulfur dioxide	SO_2	g	$-70,960$	
Sulfur trioxide	SP_3	g	$-94,450$	
Sulfur trioxide	SO_3	l	$-104,800$	
Sulfuric acid	H_2SO_4	l	$-193,910$	
Water	H_2O	g	$-57,798$	
Water	H_2O	l	$-68,317$	

Source: Most values have been selected from Selected Values of Chemical Thermodynamic Properties, *Natl. Bur. Stds. Circ.* 500, 1952.

† Since most compilations of heats of reaction are given in cal/(g mol), the values in Tables 1-1 and 1-2 are in these units. To convert to SI units of joule/(kg mol) multiply cal/(g mol) by 4.1868×10^3. For example, the heat of formation of methane at 298 K is $-17,889$ $(4.1868 \times 10^3) = 74.898 \times 10^6$ J/(kg mol) or 74.898×10^3 kJ/(kg mol).

reaction are heats of formation and of combustion. Extensive tables of these data have been accumulated, and a few values for a temperature of 298 K (25°C) are given in Table 1-1.

In the absence of experimental data, there are procedures available for predicting heats of reaction. These are all based on predictions of the effects of differences in the chemical structure of the reactants and products. Reid, Prausnitz, and Sherwood† have described and evaluated the most reliable methods. This method is applicable to compounds involving carbon, hydrogen, oxygen, nitrogen, and the halogens.

† R. C. Reid, J. M. Prausnitz, and T. K. Sherwood, "The Properties of Gases and Liquids," pp. 223–277, McGraw-Hill Book Co., New York, 1977.

The variation of heat of reaction with temperature depends on the difference in molal heat capacities of the products and reactants. The following equation relates ΔH at any temperature T to the known value at the base temperature T_0:

$$\Delta H_T = \Delta H_{T_0} + \int_{T_0}^{T} \Delta c_p \, dT \qquad (1\text{-}1)$$

Here Δc_p is the difference in molal heat capacities,

$$\Delta c_P = \sum (N_i c_{P_i})_{\text{prod}} - \sum (N_i c_{P_i})_{\text{react}} \qquad (1\text{-}2)$$

If *mean* heat capacities $\bar{c}_p$ are known for the reactants and products over the temperature range T_0 to T, it is not necessary to integrate Eq. (1-1). Under these conditions the relationship of ΔH_T and ΔH_{T_0} is

$$\Delta H_T = \Delta H_{T_0} + \sum (N_i \bar{c}_{p_i})_{\text{prod}}(T - T_0) - \sum (N_i \bar{c}_{p_i})_{\text{react}}(T - T_0) \qquad (1\text{-}3)$$

When reactants and products enter and leave a reactor at different temperatures, it is usually simpler to bypass calculating ΔH_T and evaluate the desired energy quantity directly. This is illustrated in Example 1-1.

The effect of pressure on the heat of reaction for gaseous systems depends on the deviation of the components from ideal-gas behavior. If the reactants and products behave as ideal gases, there is no effect. Even for rather nonideal systems the effect of pressure is generally small. Details of the methods of calculating the effects of temperature and pressure are discussed in standard thermodynamics textbooks.

The application of heat-of-reaction information for calculating energy-transfer rates in reactors is illustrated in the following example.

Example 1-1 Ethylene oxide is produced by direct oxidation with air using a bed of catalyst particles (silver on a suitable carrier). Suppose that the stream enters the tubular flow reactor at 200°C and contains 5 mol % ethylene and 95% air. If the exit temperature does not exceed 260°C, it is possible to convert 50% of the ethylene to the oxide, although 40% is also completely burned to carbon dioxide. How much heat must be removed from the reactor, per mole of ethylene fed, in order not to exceed the limiting temperature? The average molal heat capacity of ethylene may be taken as 18 Btu/(lb mol)(°R) between 25 and 200°C and as 19 Btu/(lb mol)(°R) between 25 and 260°C. Similar values for ethylene oxide are 20 and 21 Btu/(lb mol)(°R). The pressure is essentially atmospheric.

SOLUTION Since heat effects at constant pressure are equal to enthalpy changes, the actual process may be replaced by one that utilizes the available heat-of-reaction data at 25°C (Table 1-1). The steps in this process are

1. Cool the reactants and air from 200 to 25°C.
2. Carry out the reactions at 25°C.
3. Heat the products and the air from 25 to 260°C.

The sum of the enthalpy changes for each step will be the total heat absorbed by the reaction system.

Step 1 With a basis of 1 mole of ethylene, there will be $\frac{95}{5}(1) = 19$ moles of air fed to the reactor. The mean heat capacity of air from 25 to 200°C is 7.0. Hence

$$\Delta H_1 = 1(18)(+77 - 392) + 19(7.0)(+77 - 392) = -5,700 - 41,900$$

$$= -47,600 \text{ Btu/lb mol}$$

Step 2 The only heat effect is due to the two reactions

$$C_2H_4 + \tfrac{1}{2}O_2 \rightarrow C_2H_4O(g)$$

$$C_2H_4 + 3O_2 \rightarrow 2CO_2 + 2H_2O(g)$$

Using the heat-of-formation data in Table 1-1, for the first reaction we obtain

$$\Delta H_{R_1} = -12,190 - 12,496 - 0$$

$$= -24,686 \text{ cal/g mol} \qquad \text{or} \quad -44,500 \text{ Btu/lb mol}$$

and for the second

$$\Delta H_{R_2} = 2(-57,798) + 2(-94,052) - 12,496 - 0$$

$$= -316,196 \text{ cal/g mol} \qquad \text{or} \quad -569,000 \text{ Btu/lb mol}$$

Since, per mole of ethylene, there will be 0.5 mole reacting to form ethylene oxide and 0.4 mole to be completely burned,

$$\Delta H_{T_0} = 0.5(-44,500) + 0.4(-569,000)$$

$$= -250,000 \text{ Btu/lb mol}$$

Step 3 The products will consist of the following quantities:

Ethylene $= 1 - 0.5 - 0.4 = 0.1$ mol
Ethylene oxide $= 0.5$ mol
Water vapor $= 2(0.4) = 0.8$ mol $(\bar{c}_p = 8.25)$
Carbon dioxide $= 2(0.4) = 0.8$ mol $(\bar{c}_p = 9.4)$
Nitrogen $= 19(0.79) = 15.0$ mol $(\bar{c}_p = 7.0)$
Oxygen $= 19(0.21) - \tfrac{1}{2}(0.5) - 3(0.4) = 2.6$ mol $(\bar{c}_p = 7.25)$

The values shown for $\bar{c}_p$ are mean values between 25 and 260°C:

$$\Delta H_3 = [0.1(19) + 0.5(21) + 0.8(8.25) + 0.8(9.4)$$

$$+ 15(7.0) + 2.6(7.25)](500 - 77)$$

$$= 150(500 - 77) = 63,500 \text{ Btu/lb mol}$$

Then the net heat absorbed will be

$$Q = -47,600 - 250,000 + 63,500$$

$$= -234,000 \text{ Btu/lb mol ethylene} \quad \text{or} \quad \frac{234,000}{1.8}(4.1868 \times 10^3)$$

$$= 544 \times 10^6 \text{ J/(kg mol)}$$

Hence the heat that must be removed is 234,000 Btu/lb mol of ethylene fed to the reactor.

Chemical equilibrium—Single reactions When a reaction occurs at equilibrium, the temperature and pressure in the system remain constant and the change in free energy is zero. These restraints can be used to develop the following relationship between the *standard* free-energy change $\Delta F°$ and the equilibrium constant K:

$$\Delta F° = -R_g T \ln K \tag{1-4}$$

The standard free-energy change $\Delta F°$ is the difference between the free energies of the products and reactants when each is in a chosen standard state. These standard states are chosen so as to make evaluation of the free energy simple and accurate. For example, for gases the standard state is normally that corresponding to unit fugacity at the temperature of the reaction. If the gas is ideal, this standard state reduces to 1 atm pressure.

The equilibrium constant K is defined in terms of the equilibrium activities a_i of the reactants and products. For a general reaction

$$aA + bB = cC + dD \tag{1-5}$$

the equilibrium constant is

$$K = \frac{a_C^c a_D^d}{a_A^a a_B^b} \tag{1-6}$$

The activities refer to equilibrium conditions in the reaction mixture and are defined as the ratio of the fugacity in the equilibrium mixture to that in the standard state; that is,

$$a_i = \frac{f_i}{f_i°} \tag{1-7}$$

For gaseous reactions with a standard state of unit fugacity the expression for the equilibrium constant becomes

$$K = \frac{f_C^c f_D^d}{f_A^a f_B^b} \tag{1-8}$$

If, in addition, the gases follow the ideal-gas law, the fugacity is equal to the pressure, and Eq. (1-8) reduces to

$$K = \frac{p_C^c p_D^d}{p_A^a p_B^b} \tag{1-9}$$

Here p (partial pressure) is the total pressure p_t times the mole fraction of the component in the mixture; for example,

$$p_A = p_t y_A \tag{1-10}$$

In many situations the assumption of ideal gases is not justified, and it is necessary to evaluate fugacities. This is the case for such reactions as the ammonia synthesis, where the operating pressure may be as high as 1500 atm. The fugacity in Eq. (1-8) is that of the component in the equilibrium mixture. However, the fugacity of only the pure component is usually known. To relate the two we must know something about how the fugacity depends on the composition. Normally this information is not available, so that it is necessary to make assumptions about the behavior of the reaction mixture. The simplest and most common assumption is that the mixture behaves as an ideal solution. Then the fugacity at equilibrium, f, is related to the fugacity of the pure component, f', at the same pressure and temperature, by

$$f_i = f'_i y_i \tag{1-11}$$

Substituting this expression in Eq. (1-8) leads to equations for the equilibrium constant in terms of pure-component fugacities and the composition of the equilibrium mixture,

$$K = \frac{(f'_C)^c (f'_D)^d}{(f'_A)^a (f'_B)^b} K_y \tag{1-12}$$

where

$$K_y = \frac{y_C^c y_D^d}{y_A^a y_B^b} \tag{1-13}$$

In gaseous reactions the quantity K_p is sometimes used. This is defined as

$$K_p = \frac{(y_C p_t)^c (y_D p_t)^d}{(y_A p_t)^a (y_B p_t)^b} = K_y p_t^{(c+d)-(a+b)} \tag{1-14}$$

From Eq. (1-9) it is clear that $K = K_p$ for an ideal-gas reaction mixture. For nonideal systems Eq. (1-14) may still be employed to calculate K_p from measured equilibrium compositions (K_y). However, then K_p is not equal to K determined from thermodynamic data, for example, from Eq. (1-4).

With Eq. (1-12), the composition ratio K_y can be determined from the equilibrium constant. This is a necessary step toward evaluating the equilibrium conversion from free-energy data. The steps in the process are as follows:

1. Evaluate $\Delta F°$.
2. Determine the equilibrium constant K, using Eq. (1-4).
3. Obtain K_y from Eq. (1-12).
4. Calculate the conversion from K_y.

The first and second steps require thermodynamic data. A brief tabulation of standard free-energy changes $\Delta F°$ at 298 K is given in Table 1-2. More extensive data have been assembled.[†] Also, estimation procedures have been developed for use when the data are unavailable.[‡]

[†] Selected Values of Chemical Thermodynamic Properties, *Natl. Bur. Stds. Circ.* 500 1952; Selected Values of Physical and Thermodynamic Properties of Hydrocarbons and Related Compounds, *Am. Petrol. Inst. Res. Proj.* 44, Carnegie Institute of Technology, Pittsburgh, 1953.

[‡] R. C. Reid, J. M. Prausnitz, and T. K. Sherwood, "The Properties of Gases and Liquids, pp. 223–277, McGraw-Hill Book Co., New York, 1977.

Table 1-2 Standard free energies of formation at 298 K, cal/(g mol)

Substance	Formula	State	ΔF°_{f298}
Normal paraffins			
Methane	CH_4	g	$-12,140$
Ethane	C_2H_6	g	$-7,860$
Propane	C_3H_8	g	$-5,614$
n-Butane	C_4H_{10}	g	$-4,100$
n-Pentane	C_5H_{12}	g	$-2,000$
n-Hexane	C_6H_{14}	g	-70
n-Heptane	C_7H_{16}	g	$1,920$
n-Octane	C_8H_{18}	g	$3,920$
Increment per C atom above C_8		g	$2,010$
Normal monoolefins (1-alkenes)			
Ethylene	C_2H_4	g	$16,282$
Propylene	C_3H_6	g	$14,990$
1-Butene	C_4H_8	g	$17,090$
1-Pentene	C_5H_{10}	g	$18,960$
1-Hexene	C_6H_{12}	g	$20,940$
Increment per C atom above C_6		g	$2,010$
Miscellaneous organic compounds			
Acetaldehyde	C_2H_4O	g	$-31,960$
Acetic acid	$C_2H_4O_2$	l	$-93,800$
Acetylene	C_2H_2	g	$50,000$
Benzene	C_6H_6	g	$30,989$
Benzene	C_6H_6	l	$29,756$
1,3-Butadiene	C_4H_6	g	$36,010$
Cyclohexane	C_6H_{12}	g	$7,590$
Cyclohexane	C_6H_{12}	l	$6,370$
Ethanol	C_2H_6O	g	$-40,130$
Ethanol	C_2H_6O	l	$-41,650$
Ethylbenzene	C_8H_{10}	g	$31,208$
Ethylene glycol	$C_2H_6O_2$	l	$-77,120$
Ethylene oxide	C_2H_4O	g	$-2,790$
Methanol	CH_4O	g	$-38,810$
Methanol	CH_4O	l	$-39,850$
Methylcyclohexane	C_7H_{14}	g	$6,520$
Methylcyclohexane	C_7H_{14}	l	$4,860$
Styrene	C_8H_8	g	$51,100$
Toluene	C_7H_8	g	$29,228$
Toluene	C_7H_8	l	$27,282$
Miscellaneous inorganic compounds			
Ammonia	NH_3	g	$-3,976$
Ammonia	NH_3	aq	$-6,370$
Calcium carbide	CaC_2	s	$-16,200$
Calcium carbonate	$CaCO_3$	s	$-269,780$

Continued

Table 1-2 *(continued)*

Substance	Formula	State	ΔF°_{f298}
Miscellaneous inorganic compounds *(Continued)*			
Calcium chloride	$CaCl_2$	s	−179,300
Calcium chloride	$CaCl_2$	aq	−194,880
Calcium hydroxide	$Ca(OH)_2$	s	−214,330
Calcium hydroxide	$Ca(OH)_2$	aq	−207,370
Calcium oxide	CaO	s	−144,400
Carbon dioxide	CO_2	g	−94,058
Carbon monoxide	CO	g	−32,781
Hydrochloric acid	HCl	g	−22,078
Hydrogen sulfide	H_2S	g	−7,892
Iron oxide	Fe_3O_4	s	−242,400
Iron oxide	Fe_2O_3	s	−177,100
Iron sulfide	FeS_2	s	−39,840
Nitric acid	HNO_3	l	−19,100
Nitric acid	HNO_3	aq	−26,410
Nitrogen oxides	NO	g	20,690
	NO_2	g	12,265
	N_2O	g	24,933
	N_2O_4	g	23,395
Sodium carbonate	Na_2CO_3	s	−250,400
Sodium chloride	$NaCl$	s	−91,785
Sodium chloride	$NaCl$	aq	−93,939
Sodium hydroxide	$NaOH$	s	−90,600
Sodium hydroxide	$NaOH$	aq	−100,184
Sulfur dioxide	SO_2	g	−71,790
Sulfur trioxide	SO_3	g	−88,520
Sulfuric acid	H_2SO_4	aq	−177,340
Water	H_2O	g	−54,635
Water	H_2O	l	−56,690

Notes: The standard free energy of formation ΔF°_{f298} is the change in free energy when the listed compound is formed from its elements with each substance in its standard state at 298 K (25°C). Standard states are:

1. Gases (*g*), the pure gas at unit fugacity and 25°C
2. Liquids (*l*) and solids (*s*), the pure substance at atmospheric pressure and 25°C
3. Solutes in aqueous solution (*aq*), the hypothetical 1-modal solution of the solute in water at atmospheric pressure and 25°C

The units of ΔF° are calories per gram mole of the listed substance.

Source: Selected mainly from F. D. Rossini et al., Selected Values of Properties of Hydrocarbons and Related Compounds, *Am. Petrol. Inst. Res. Proj.* **44**, Carnegie Institute of Technology, Pittsburgh, 1953, and loose-leaf supplements (by permission); F. D. Rossini et al., in D. D. Wagman (ed.), Selected Values of Chemical Thermodynamic Properties, *Natl. Bur. Stds. Circ.* 500, 1952, and loose-leaf supplements; G. N. Lewis and M. Randall, "Thermodynamics," 2d ed., revised by K. S. Pitzer and Leo Brewer, McGraw-Hill Book Company, New York, 1961.

Usually it is necessary to calculate the effect of temperature on $\Delta F°$ in order to obtain an equilibrium constant at reaction conditions. The *van't Hoff equation* expresses this relationship in differential form.

$$\frac{d(\ln K)}{dT} = \frac{\Delta H°}{R_g T^2} \tag{1-15}$$

where $\Delta H°$ is the standard-state enthalpy change for the reaction. Equation (1-15) has important implications in reactor design for reversible reactions. It shows that K will decrease with an increase in temperature for an exothermic reaction. Hence provisions must be made for removing the heat of reaction to avoid a thermo-dynamic limitation (decrease in K) to the potential conversion in exothermic systems. The oxidation of sulfur dioxide is a practical illustration. For endother-mic reversible reactions energy must be added to maintain the temperature if a decrease in K is to be avoided. Dehydrogenation of hydrocarbons, such as bu-tanes and butenes, is an example of a situation in which the addition of energy is important. If $\Delta H°$ is approximately independent of temperature, the integrated form of Eq. (1-15) between T_2 and T_1 is

$$\ln \frac{K_{T_2}}{K_{T_1}} = \frac{-\Delta H°}{R_g}\left(\frac{1}{T_2} - \frac{1}{T_1}\right) \tag{1-16}$$

If $\Delta H°$ is not constant, but can be expressed by Eq. (1-1), the integrated form is

$$\ln K_T = -\frac{\Delta H_0}{R_g T} + \frac{\Delta a}{R_g}\ln T + \frac{\Delta b}{2R_g}T + \frac{\Delta c}{6R_g}T^2 + C \tag{1-17}$$

where ΔH_0, C, and Δa, Δb, and Δc are constants and Δa, Δb, and Δc arise from the expression

$$\Delta c_p = \Delta a + \Delta b T + \Delta c T^2 \tag{1-18}$$

This equation implies $c_p = a + bT + cT^2$ for each component.

K can be determined from Eqs. (1-17) and (1-18) for a gaseous reaction at any temperature, provided the constants C and ΔH_0 can be evaluated. Experimental data for K at two temperatures is sufficient for this evaluation. Alternately, ΔH_0 can be found from the known heat of reaction at one temperature with Eq. (1-1). In this case only one experimental value of the equilibrium constant is needed in order to determine the constant C. Of course, in both methods it is necessary to have heat-capacity data for reactants and products in order to evaluate the coefficients Δa, Δb, and Δc.

A third method of calculating K at the reaction temperature is via Eq. (1-4). $\Delta F°$ at the desired temperature is obtained from the equation

$$\Delta F° = \Delta H° - T\,\Delta S° \tag{1-19}$$

To use Eq. (1-19) absolute entropies of reactants and products must be evaluated at the reaction temperature using heat-capacity data and the third law of thermody-namics. Such entropies are given in the same sources that were noted in Table 1-2. This method does not require any experimental values of K.

Application of the foregoing concepts in evaluating the equilibrium conversion from free-energy data is illustrated by the following examples.

Example 1-2 The following equilibrium data have been reported for the vapor-phase hydration of ethylene to ethanol:† at 145°C $K = 6.8 \times 10^{-2}$ and 320°C $K = 1.9 \times 10^{-3}$. From these data develop an expression for the equilibrium constant as a function of temperature.

SOLUTION From the two values of K the constants ΔH_0 and C in Eq. (1-17) may be determined. First Δa, Δb, and Δc must be obtained from heat-capacity data. For the reaction

$$C_2H_4(g) + H_2O(g) \rightarrow C_2H_5OH(g)$$

these are

$$\Delta = C_2H_5OH - C_2H_4 - H_2O$$

$$\Delta a = 6.990 - 2.830 - 7.256 = -3.096$$

$$\Delta b = 0.039741 - 0.028601 - 0.002298 = 0.008842$$
$$\Delta c = (-11.926 + 8.726 - 0.283) \times 10^{-6} = -3.483 \times 10^{-6}$$

Substituting these values in Eq. (1-17), we have, at 145°C,

$$R_g \ln (6.8 \times 10^{-2}) = -\frac{\Delta H_0}{418} - 3.096 \ln 418 + \frac{0.00884}{2}(418)$$

$$-\frac{3.483 \times 10^{-6}}{6}(418)^2 + CR_g$$

or

$$\frac{\Delta H_0}{418} - CR_g = -R_g \ln (6.8 \times 10^{-2}) - 3.096 \ln 418$$
$$+ 0.00442(418) - (0.580 \times 10^{-6})(418)^2 = -11.59 \quad \text{(A)}$$

and at 320°C

$$\frac{\Delta H_0}{593} - CR_g = -R_g \ln (1.9 \times 10^{-3}) - 3.096 \ln 593$$
$$+ 0.00442(593) - (0.580 \times 10^{-6})(593)^2 = -4.91 \quad \text{(B)}$$

Equations (A) and (B) may be solved simultaneously for ΔH_0 and C. The results are

$$\Delta H_0 = -9460 \text{ cal}$$

$$C = -5.56$$

---

† H. M. Stanley et. al., *J. Soc. Chem. Ind.*, **53**, 205 (1934); R. H. Bliss and B. F. Dodge, *Ind. Eng. Chem.*, **29**, 19 (1937).

Then the expression for K as a function of temperature is

$$\ln K = \frac{9,460}{R_g T} - \frac{3.096}{R_g} \ln T + \frac{0.00442}{R_g} T - \frac{0.580 \times 10^{-6}}{R_g} T^2 - 5.56$$

or

$$\ln K = \frac{4,760}{T} - 1.558 \ln T + 0.00222T - 0.29 \times 10^{-6}T^2 - 5.56$$

Example 1-3 Estimate the maximum conversion of ethylene to alcohol by vapor-phase hydration at 250°C and 500 psia. Use the equilibrium data of Example 1-2 and choose an initial steam-ethylene ratio of 5.

SOLUTION The equilibrium constant at 250°C can be evaluated from the equation for K developed in Example 1-2:

$$\ln K = \frac{4,760}{523} - 1.558 \ln 523 + 0.00222(523) - 0.29 \times 10^{-6}(523^2) - 5.56$$

$$= -5.13$$

$$K = 5.9 \times 10^{-3}$$

It is necessary to assume that the gas mixture is an ideal solution. Then Eq. (1-12) is applicable, and

$$5.9 \times 10^{-3} = K_y \frac{f'_A}{f'_E f'_W} \tag{A}$$

The fugacities of the pure components can be determined from generalized correlations† and are evaluated at the temperature and pressure of the equilibrium mixture:

$$\frac{f'_A}{p_t} = 0.84 \qquad \text{for ethanol}$$

$$\frac{f'_E}{p_t} = 0.98 \qquad \text{for ethylene}$$

$$\frac{f'_W}{p_t} = 0.91 \qquad \text{for water}$$

Substituting these data in Eq. (A), we have

$$K_y = \frac{y_A}{y_E y_W} = 5.9 \times 10^{-3} \frac{0.98(0.91)}{0.84} \frac{500}{14.7} = 0.21 \tag{B}$$

† For example from J. M. Smith and H. C. Van Ness, "Introduction to Chemical Engineering Thermodynamics," 1975, McGraw-Hill Book Company, New York, 3d ed., Figs. 7-5 and 7-6.

If the initial steam-ethylene ratio is 5 and a basis of 1 mole of ethylene is chosen, a material balance gives the following moles for equilibrium conditions:

$$Ethanol = z$$
$$Ethylene = 1 - z$$
$$Water = 5 - z$$
$$Total\ moles = \overline{6 - z}$$

Then

$$y_A = \frac{z}{6 - z}$$

$$y_E = \frac{1 - z}{6 - z}$$

$$y_W = \frac{5 - z}{6 - z}$$

Substituting in Eq. (B), we have

$$0.21 = \frac{z(6 - z)}{(1 - z)(5 - z)}$$

$$z^2 - 6.0z + 0.868 = 0$$

$$z = 3.0 \pm 2.85 = 5.85 \qquad or\ 0.15$$

The first solution is greater than unity and is impossible. Therefore $z = 0.15$, indicating that 15% of the ethylene could be converted to ethanol, provided that equilibrium were achieved.

In this reaction increasing the temperature decreases K and the conversion. Increasing the pressure increases the conversion. From an equilibrium standpoint the operating pressure should be as high as possible (limited by condensation) and the temperature as low as possible. A catalyst is required to obtain an appreciable rate, but all the catalysts that are presently available require a temperature of at least 150°C for a reasonably fast rate. Even at this temperature the catalysts which have been developed as yet do not give the equilibrium conversion. In this instance both equilibrium and reaction rate affect the reaction process.

Chemical equilibrium—multiple reactions In many chemical reactors undesirable reactions occur as well as the required reaction. Then the production of desired product relative to the production of side products (i.e., the selectivity—see Chap. 2) is a critical factor. In such cases the equilibrium composition becomes important since it represents the ultimate composition that could be obtained regardless of the rates of the reactions. For each reaction Eqs. (1-12) and (1-13) can be combined to give a relationship between the mole fractions of each chemical species involved in the reaction. Then, the conservation of mass principle (stoichiometry) allows this relationship to be expressed in terms of one unknown, the extent of the reaction. Simultaneous solution of the resulting set of equations, one for each independent reaction, gives the extent of each reaction and the desired equilibrium composition. However, the solution can be difficult when the equations are nonlinear. In such cases it is more convenient to use another method based upon the thermodynamic requirement that the total free energy of the system be a minimum at the equilibrium composition. The procedure is to formulate an expression for the total free energy and then minimize this quantity with respect to composition changes at a constant temperature and pressure. This minimization must be carried out in such a way that the mass of each element (i.e., the number of atoms of each element) is conserved. The usual approach applies the Lagrange method of undetermined multipliers and is briefly summarized in the following paragraphs.†

The first step is to formulate the expressions for the conservation of mass. Let the subscript k identify a particular element. Then define A_k to be the total number of atomic weights of the kth element in the system. This number is known from the initial or feed composition. Further, let a_{ik} be the number of atoms of the kth element in each molecule of chemical species i. Conservation of mass of each element k requires that the following equation be satisfied

$$\sum_i n_i a_{ik} = A_k$$

or

$$\sum_i (n_i a_{ik}) - A_k = 0 \qquad (k = 1, 2, \ldots, w) \qquad (1\text{-}20)$$

where n_i is the number of moles of species i in the mixture at equilibrium.

Next, the previous equation for each element k is multiplied by an undetermined constant λ_k (the Lagrange multipler) giving

$$\lambda_k \left[\sum_i (n_i a_{ik}) - A_k \right] = 0 \qquad (k = 1, 2, \ldots, w)$$

Then, these w equations, one for each element, are added. Since their sum must remain equal to zero, we may write

$$\sum_k \lambda_k \left[\sum_i n_i a_{ik} - A_k \right] = 0 \qquad (1\text{-}21)$$

Addition of the total free energy F_t to Eq. (1-21) will give a new quantity M, and M will still be equal to F_t since the right-hand side of Eq. (1-21) is zero. Therefore, minimizing M will also give a minimum value for F_t. The minimization is accomplished by differentiating M with respect to the moles of each chemical species i and setting this derivative equal to zero. Thus

$$M = F_t + \sum_k \lambda_k \left[\sum_i (n_i a_{ik}) - A_k \right]$$

---

† The summary here follows closely the more detailed discussion given by J. M. Smith and H. C. Van Ness, "Introduction to Chemical Engineering Thermodynamics," 3d ed., p. 419, 1975.

and

$$\left(\frac{\partial M}{\partial n_i}\right)_{T,\,P,\,n_j} = \left(\frac{\partial F_t}{\partial n_i}\right)_{T,\,P,\,n_j} + \sum_k \lambda_k a_{ik} = 0 \qquad (1\text{-}22)$$

The subscript n_j means that the differentiation is carried out at constant values of the number of moles of all the components except n_i. The first term on the right-hand side of Eq. (1-22) is the chemical potential μ_i. This is related, for gas-phase reactions (where the standard state is the ideal gas at 1 atm pressure), to the fugacity as follows:

$$\left(\frac{\partial F_t}{\partial n_i}\right)_{T,\,P,\,n_j} = \mu_i = RT \ln f_i + F_i^\circ \qquad (1\text{-}23)$$

Here F_i° is the standard-state free energy of pure species i. Let F_i° be chosen equal to zero for the elements. Then for compounds F_i° is equal to the standard-free energy of formation for species i, that is, equal to $\Delta F_{f_i}^\circ$. Using this equality for F_i° in Eq. (1-23), and substituting the resultant expression for $(\partial F_t/\partial n_i)_{T,\,P,\,n_j}$ in Eq. (1-22), gives

$$\Delta F_{f_i}^\circ + RT \ln f_i + \sum_k \lambda_k a_{ik} = 0 \qquad (i = 1, 2, \ldots, N) \qquad (1\text{-}24)$$

The fugacity, f_i, of each chemical species at equilibrium conditions in the mixture is related to the composition of the mixture. Thus, the ϵ are N equilibrium equations of the form of Eq. (1-24), one for each chemical species. Also, there are w conservation equations of the form of Eq. (1-20), one for each element. These $N + w$ equations can be solved for the $N + w$ unknowns, which are the n_i, of which there are N, and the λ_k of which there are w. The n_i results determine the equilibrium composition since the mole fraction is $y_i = n_i \Big/ \left(\sum_i n_i \right)$.

If the reaction mixture is an ideal gas, f_i is simply related to n_i by the expression

$$f_i = p_t y_i = p_t n_i \Big/ \left(\sum_i n_i \right)$$

Then Eq. (1-24) can be expressed directly in terms of the equilibrium composition; that is

$$\Delta F_{f_i}^\circ + RT \ln \left(p_t \frac{n_i}{\sum_i n_i} \right) + \sum_k \lambda_k a_{ik} = 0 \qquad (i = 1, 2, \ldots, N) \qquad (1\text{-}25)$$

The equations are linear and the calculations for n_i can easily be carried out by computer. Example 1-4 illustrates the method.

Example 1-4 Hydrogen and carbon monoxide are to be produced by the methane-stream reaction:

$$CH_4 + H_2O \rightarrow 3H_2 + CO$$

However, undesirable carbon dioxide is also produced by the watergas reaction

$$CO + H_2O \rightarrow CO_2 + H_2$$

The feed consists of 2 moles of CH_4 and 3 moles of H_2O. Calculate the fraction of the methane that is converted to CO and to CO_2, if equilibrium could be attained in a reactor operating at 1000 K and 1 atm. The values of ΔF_f° at these conditions are:

$$(\Delta F_f^\circ)_{CH_4} = 4{,}610 \text{ cal/(g mol)}$$

$$(\Delta F_f^\circ)_{H_2O} = -46{,}030 \text{ cal/(g mol)}$$

$$(\Delta F_f^\circ)_{CO} = -47{,}940 \text{ cal/(g mol)}$$

$$(\Delta F_f^\circ)_{CO_2} = -94{,}610 \text{ cal/(g mol)}$$

SOLUTION The required values of A_k are determined from the initial numbers of moles, and the values of a_{ik} come directly from the chemical formulas of the species. These are shown in the accompanying table.

	Element k		
	Carbon	Oxygen	Hydrogen
	A_k = no. of atomic wts of k in system		
	$A_C = 2$	$A_O = 3$	$A_H = 14$
Species i	a_{ik} = no. of atoms of k per molecule of i		
CH_4	$a_{CH_4, C} = 1$	$a_{CH_4, O} = 0$	$a_{CH_4, H} = 4$
H_2O	$a_{H_2O, C} = 0$	$a_{H_2O, O} = 1$	$a_{H_2O, H} = 2$
CO	$a_{CO, C} = 1$	$a_{CO, O} = 1$	$a_{CO, H} = 0$
CO_2	$a_{CO_2, C} = 1$	$a_{CO_2, O} = 2$	$a_{CO_2, H} = 0$
H_2	$a_{H_2, C} = 0$	$a_{H_2, O} = 0$	$a_{H_2, H} = 2$

At 1(atm) and 1000(K) the assumption of ideal gases is justified, and the fugacity, $f_i = p_t y_i$. Since $p_t = 1$(atm), Eq. (1-25) can be written:

$$\frac{\Delta F_{f_i}^\circ}{RT} + \ln \frac{n_i}{\sum_i n_i} + \sum_k \frac{\lambda_k}{RT} a_{ik} = 0$$

The five equations for the five species then become

$$CH_4: \qquad \frac{4{,}610}{RT} + \ln \frac{n_{CH_4}}{\sum n_i} + \frac{\lambda_C}{RT} + \frac{4\lambda_H}{RT} = 0$$

$$H_2O: \qquad \frac{-46{,}030}{RT} + \ln \frac{n_{H_2O}}{\sum n_i} + \frac{2\lambda_H}{RT} + \frac{\lambda_O}{RT} = 0$$

$$CO: \qquad \frac{-47{,}940}{RT} + \ln \frac{n_{CO}}{\sum n_i} + \frac{\lambda_C}{RT} + \frac{\lambda_O}{RT} = 0$$

$$\text{CO}_2: \qquad \frac{-94{,}610}{RT} + \ln\frac{n_{\text{CO}_2}}{\sum n_i} + \frac{\lambda_C}{RT} + \frac{2\lambda_O}{RT} = 0$$

$$\text{H}_2: \qquad\qquad\qquad \ln\frac{n_{\text{H}_2}}{\sum n_i} + \frac{2\lambda_H}{RT} = 0$$

The three conservation equations [Eq. (1-20)] are:

For carbon: $\qquad n_{\text{CH}_4} + n_{\text{CO}} + n_{\text{CO}_2} = 2$

For hydrogen: $\qquad 4n_{\text{CH}_4} + 2n_{\text{H}_2\text{O}} + 2n_{\text{H}_2} = 14$

For oxygen: $\qquad n_{\text{H}_2\text{O}} + n_{\text{CO}} + 2n_{\text{CO}_2} = 3$

Simultaneous, computer solution of these eight equations, with $R = 1.987$ cal/(g mol)(K) gives

$$\begin{aligned}
n_{\text{CH}_4} &= 0.1722\\
n_{\text{H}_2\text{O}} &= 0.8611\\
n_{\text{H}_2} &= 5.7934\\
n_{\text{CO}} &= 1.5172\\
n_{\text{CO}_2} &= \underline{0.3107}\\
\sum n_i = n &= 8.6546 \text{ moles}
\end{aligned}$$

The λ_k values (which are not pertinent) are: $\lambda_C = 1584$, $\lambda_O = 49{,}870$ and $\lambda_H = 3994$.

From these results, the total fraction of methane reacted is $(2 - 0.1722)/2$, or 0.914. The fraction of the CH_4 that reacts to CO is $1.5172/2 = 0.759$ and the fraction that goes to CO_2 is $0.3107/2$, or 0.155. Hence, of the methane that reacts, $0.155/0.914 = 0.17$ or 17% goes to undesirable CO_2 and 83% to CO.

1-5 Classification of Reactors

Chemical reactors vary widely in size and shape and in method of operation. Small flasks and beakers are common forms used in chemical laboratories for liquid-phase reactions. At the other extreme in size are the large cylindrical tubes used in the petroleum industry, for example, for cracking of hydrocarbons. Even larger are the "retorts" employed for in-situ production of oil from the kerogen in oil shale. Here the reaction between the carbon residue in the rock and oxygen provides the heat necessary to liquefy and decompose the kerogen into oil.

In the laboratory beaker a charge of reactants is added, brought to reaction temperature, held at this condition for a predetermined time, and then the product is removed. This *batch reactor* is characterized by the variation in extent of reaction and properties of the reaction mixture with *time*. The hydrocarbon-cracking reactor operates in a different way. There is a continuous, steady flow of reactants in and products out. This is the *continuous-flow type*, in which the extent of reaction and properties such as temperature and composition may vary with *position* in the reactor but not with time. Hence, one classification of reactors is according to method of operation.

Another classification is according to shape. If the laboratory vessel is equipped with an efficient stirrer, the composition and temperature of the reaction mass will tend to be the same in all parts of the reactor. A vessel in which the properties are uniform is called a *stirred-tank* (or well-mixed) *reactor* (STR). If there is no mixing in the direction of flow, and complete mixing in the transverse direction, in the cylindrical vessel for hydrocarbon processing, another ideal type is realized: the *ideal tubular-flow*, or *plug-flow*, *reactor*. Here the reaction mass consists of elements of fluid in the axial direction that are independent of each other, each one having a different composition, temperature, etc. This classification is of basic significance in design, because simple forms of the conservation equations for mass and energy are applicable for both ideal reactors. All possible mixing conditions between the extremes of stirred tank and plug flow may occur in actual reactors. Note that operating conditions and reactor accessories as well as shape can influence mixing conditions. Thus, stirrer design and speed (rpm) can affect the approach to well-mixed conditions in a tank-type vessel. Similarly, flow rate and method of feed injection can influence the approach to plug flow in a continuous, tubular-flow reactor. While well-mixed conditions would be difficult to achieve in a tubular reactor, a nozzle-type feed line and a high flow rate would introduce some mixing in the axial direction.

The two classifications, batch or continuous, and tank or tube, are not necessarily related. Thus, the laboratory-beaker reactor has been described as being operated batchwise, but it could be transformed into a continuous-flow type. This would be done by adding tubes for continuous addition of reactants and withdrawal of products. Also, a tubular reactor can be operated batchwise. This is not done by adding a stirrer since it is difficult to achieve a uniform concentration in a long tube by stirring. However, the same result is possible by recycling the reaction mixture at a high rate through the tube in a closed system. This arrangement is called a *batch-recycle* reactor.

A third classification refers to the number of phases in the reaction system. This classification is significant because it influences the number and importance of mass and energy transfer steps that must be accounted for in the design problem. As an illustration consider the two forms of batch-operated, stirred-tank reactors. In Fig. 1-3a the reaction mixture of liquid reactants A and B and product C is homogeneous. With an efficient stirrer the composition will be nearly uniform

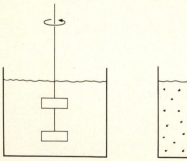

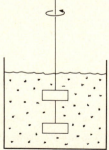

(a) Homogeneous
  stirred-tank reactor

(b) Two-phase (slurry)
  stirred-tank reactor

Figure 1-3 Homogeneous and heterogeneous stirred-tank reactors.

throughout the tank, and no mass transfer resistances due to concentration gradients will exist. In Fig. 1-3b a catalyst consisting of small particles of a solid are added to increase the rate of reaction. Even with efficient stirring, there can be a mass transfer resistance between the bulk fluid and the surface of the catalyst particles. This is because the particles, being small, tend to move with the liquid. There will be a layer of nearly stagnant fluid (with respect to the particle) surrounding each particle. Reactants A and B must be transferred through this layer by diffusion in order to reach the catalyst surface. This diffusion resistance will result in a difference in concentration, for both A and B, between the bulk fluid and the catalyst surface. Hence, the global rate concept noted earlier must be used to account for the coupling of intrinsic kinetics (at the catalyst surface) and mass transfer. It is important to note that the necessity of considering mass transfer as well as intrinsic kinetics is due to the heterogeneous nature of the system. The catalytic nature of the reaction is not responsible for the diffusion effect. The same interaction of intrinsic kinetics and diffusion can occur in *noncatalytic* heterogeneous reactions such as smelting of ores (for example, $ZnS(s) + \frac{3}{2}O_2(g) \rightarrow ZnO(s) + SO_2(g)$).

A common heterogeneous reactor of the tubular-flow type is the gas–solid catalytic form shown in Fig. 1-4a. Here the reacting fluid flows through a bed of relatively large (of the order of a centimeter) catalyst particles held in a stationary position—hence the name *fixed-bed* reactor. The oxidation of sulfur dioxide with V_2O_5 particles, mentioned at the beginning of this chapter, is an example.

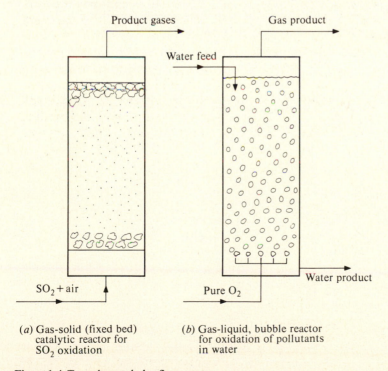

(a) Gas-solid (fixed bed) catalytic reactor for SO_2 oxidation

(b) Gas-liquid, bubble reactor for oxidation of pollutants in water

Figure 1-4 Two-phase tubular-flow reactors.

Again, due to the heterogeneous nature of the system there may exist a concentration difference between bulk gas and catalyst because of diffusion resistance in the gas near the surface of the particle.

The two previous illustrations involved solid catalyst phases. Similar effects can exist in reactions between two fluid phases. For example, reactors involving two liquid phases are common. Figure 1-5 shows an alkylation reactor in which an olefin such as butylene reacts with isobutane to form C_8-isomers (alkylate). The two liquid hydrocarbon streams enter the bottom of the tubular-flow vessel where they are dispersed as bubbles into a cocurrent, continuous, stream of liquid HF which acts as a catalyst. Phase separation occurs at the top of the reactor where the lighter alkylate product is removed, and the heavier HF stream is recycled to the bottom of the reactor. Since the olefin and isobutane must reach the acid to react, interphase mass transfer is involved as well as the intrinsic kinetics of the reaction in the acid phase.

Reactions between a gas and a liquid can be carried out in either tank or tubular-flow vessels. For example, hexamethylenetetramine is made by bubbling ammonia gas through an aqueous solution of formaldehyde contained in a stirred tank such as shown in Fig. 1-3a. Alternately, bubble-type reactors may be used. In this arrangement, the gas phase is dispersed as bubbles at the bottom of a tubular vessel. The bubbles rise through the continuous liquid phase which is flowing downward, as shown in Fig. 1-4b. The removal of organic pollutants from water by noncatalytic oxidation with pure O_2 is an example.

Another type of heterogeneous reactor is the three-phase version in which gas and liquid reactants are brought into contact with solid catalyst particles. The effect of physical processes on reactor performance now becomes more complex than for two-phase systems, because both gas-liquid and liquid-solid interphase

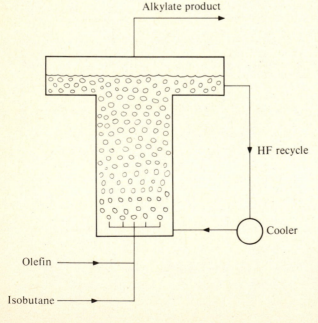

Alkylate product

HF recycle

Cooler

Olefin

Isobutane

Figure 1-5 Liquid-liquid heterogeneous tubular-flow reactor (for alkylation of olefins and isobutane).

transport effects may be coupled with the intrinsic rate. The most common arrangements are the *slurry* and *trickle-bed* forms shown in Fig. 1-6. The three-phase slurry reactor is similar to the two-phase *tank* type shown in Fig. 1-3*b*, except that provision is made for dispersion of reactant gas as bubbles into the bottom of the reactor. The reactor may operate continuously as a steady-flow system with respect to both gas and liquid phases, as indicated in Fig. 1-6*a*. An example is the polymerization of ethylene in a slurry of solid catalyst particles in a solvent such as cyclohexane. Alternately, a fixed charge of liquid may be added to the stirred tank initially and gas added continuously; that is, the reactor is batch with respect to the liquid phase. For example, some hydrogenation reactions, such as hydrogenation of oils in a slurry of nickel catalyst particles, are carried out in this way.

In the trickle-bed form, gas and liquid flow concurrently downward over a fixed-bed of catalyst particles contained in a *tubular* reactor. This three-phase arrangement is used widely for hydrodesulfurization of liquid petroleum fractions, as illustrated in Fig. 1-6*b*.

There are numerous modifications of the classifications that have been discussed. For example, a variety of tank and tubular-shaped reactors are employed either on a commercial scale or in the laboratory. Some of these are illustrated in Fig. 1-7. As mentioned, the well-mixed condition can be closely approached in a tubular reactor by recycling, as indicated in Fig. 1-7*a*. Operation can be continuous as shown in the figure, or batch, by closing feed and product lines. The fixed-bed catalytic type can be modified from the tubular form of Fig. 1-4*a* by arranging for radial flow through the catalyst bed, as illustrated in Fig. 1-7*c*. In

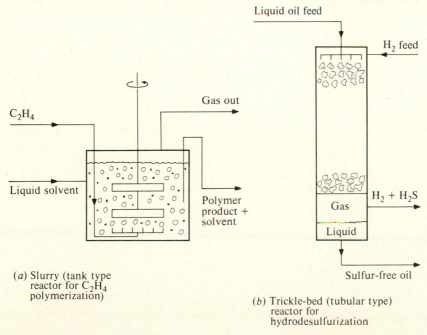

(*a*) Slurry (tank type reactor for C_2H_4 polymerization)

(*b*) Trickle-bed (tubular type) reactor for hydrodesulfurization

Figure 1-6 Three-phase reactors.

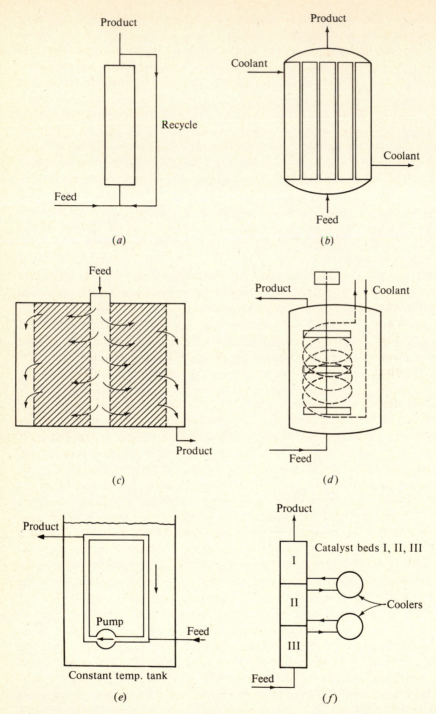

Figure 1-7 Typical reactors. (*a*) Tubular-flow recycle reactor. (*b*) Multitube-flow reactor. (*c*) Radial-flow catalytic reactor. (*d*) Stirred-tank reactor with internal cooling. (*e*) Loop reactor. (*f*) Reactor with intercoolers.

comparison with the tubular form, this arrangement provides for a lower pressure drop for the same catalyst volume.

In heterogeneous catalytic reactions, the catalyst often loses activity with operating time. If this decrease is rapid and severe, it is desirable to regenerate the catalyst continuously without shutting down the reactor. The fluidized-bed system shown in Fig. 1-8 provides an effective way to achieve this objective. The reactants enter and leave the reactor, which contains a fluidized bed of small (100 to 200 mesh) catalyst particles. Part of the catalyst is continually withdrawn to another tubular vessel, the regenerator. Here the catalyst particles are regenerated and returned to the reactor. Catalytic cracking of petroleum fractions is carried out in such an arrangement. The catalyst loses activity because of carbon deposition from complete cracking. The carbon is burned off the catalyst with air in the regenerator. The activity of the newer catalysts is so high that much of the cracking can occur in the line returning the regenerated catalyst to the reactor. This is termed *transfer-line cracking*.

The lime kiln is another form of gas-solid heterogeneous reactor. In this noncatalytic system, the gas phase flows continuously through a large-diameter tube, in one direction, while the reacting solid particles of $CaCO_3$ are moved in the other direction by slow rotation of the inclined tube.

Sometimes a tank-type, homogeneous reactor is indicated, but the operating pressure is so high that large-diameter vessels of the form shown in Fig. 1-4 are not feasible. The same mixing conditions can be achieved by using a recycle reactor, but with small-diameter tubes. A "loop" reactor of this form is shown in Fig. 1-7e.

Heat-transfer requirements can affect both the shape of a reactor and method of operation. For example, removal of large amounts of energy in a tank-type reactor can be achieved by introducing cooling coils to supply the needed heat-transfer area (Fig. 1-7d). Similarly, in a tubular-flow reactor the heat-transfer rate can be enhanced by increasing the number and decreasing the diameter of the tubes arranged in parallel (Fig. 1-7b). This arrangement has been used, for example, in the oxidation of naphthalene or xylenes to pththalic anhydride which are highly exothermic reactions. A fixed-bed reactor can be arranged with divided beds and intercoolers to remove or supply heat. The sketch in Fig. 1-7f shows a reactor of three sections of catalyst beds with intercoolers between the sections. An advantage of the fluidized-bed reactor (Fig. 1-8) is that, even for highly exothermic or endothermic reactions, mixing is such that the temperature is nearly the same in all parts of the reactor; that is, isothermal operation is approached.

There are some general relations between the physical nature of the reaction mixture and the type of reactor used in practice. Homogeneous gas-phase reactions are normally carried out in continuous-tubular-flow reactors rather than the tank type, either batch or flow. For liquid-phase and liquid-solid phase, heterogeneous reactions, both tank and tubular-flow reactors are employed. Batch-operated tank reactors are often used for small-scale production and when flexible operating conditions (temperature and pressure) are required. Such systems frequently involve costly reactants and products, as in the pharmaceutical industry.

In summary, the three classifications of reactors that have been discussed are:

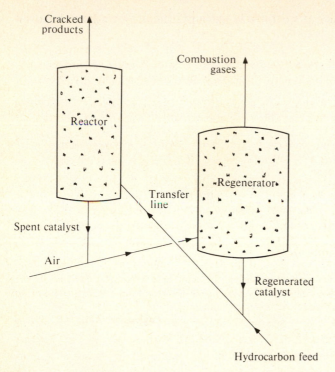

Figure 1-8 Fluidized-bed reactor-regenerator system for catalytic cracking of petroleum fractions.

(1) batch or continuous, (2) tank or tubular, and (3) homogeneous or heterogeneous. We shall consider the design of the various types starting with Chap. 4. Homogeneous reactors are taken up in Chaps. 4 to 6 and heterogeneous reactors in Chaps. 10 to 14. Since the complete development of a commercial reactor requires obtaining intrinsic rates, interpretation of laboratory reactor data will be considered in parallel with the design of the large-scale equipment. The design procedure for homogeneous and heterogeneous reactors is essentially the same and consists of the following steps: (1) formulate the conservation equations for the specific type of reactor involved, (2) introduce appropriate expressions for the pertinent mass and energy transfer rates, (3) introduce the appropriate rate-of-reaction equation, and (4) solve the resulting conservation equations.

Prior to Chap. 4 we must consider the subject of chemical kinetics in order to know how to represent the intrinsic rate of a chemical reaction. This is done in Chap. 2. In Chap. 3, general forms of the conservation equations are developed and applied to the two extreme classifications considered in this section; that is, to ideal plug-flow and ideal-stirred tank reactors.

BIBLIOGRAPHY

There are several text books which have as their objective the application of kinetics to chemical reactor design. The subject has been treated differently in each of these texts, and each is worthwhile for supplementary reading. A partial list is as follows:

Aris, Rutherford, "Elementary Chemical Reactor Analysis," Prentice-Hall, Inc., Englewood Cliffs, N.J., 1969. A rather broad, analytical book in which theory of the general case is presented first and then followed with applications.

Carberry, J. J., "Chemical and Catalytic Reaction Engineering," McGraw-Hill Book Company, New York, 1976. A general text on chemical reaction engineering with emphasis on heterogeneous reactors.

Denbigh, K. G., and J. C. R. Turner, "Chemical Reactor Theory," 2d ed., Cambridge University Press, Cambridge, 1971. This clearly written book provides an excellent treatment of important problems in chemical reactor design.

Hill, Charles G., Jr., "An Introduction to Chemical Engineering Kinetics and Reactor Design," John Wiley and Sons, New York, 1977. A broad text covering both homogeneous and heterogeneous reactions with emphasis on mathematical modeling of different types of reactors.

Kramers, H., and K. R. Westerterp, "Chemical Reactor Design and Operation," Academic Press, Inc., New York, 1963. A concise treatment of a few selected subjects of reactor design, including optimization and residence-time distributions.

Levenspiel, Octave, "Chemical Reaction Engineering," 2d ed., John Wiley & Sons, Inc., New York, 1972. A complete text for undergraduate students. The emphasis is rather more on homogeneous and noncatalytic reactors than on catalytic ones. Effects of residence-time distribution on reactor performance are treated in detail.

Petersen, E. E., "Chemical Reaction Analysis," Prentice-Hall, Inc., Englewood Cliffs, N.J., 1965. A more advanced text emphasizing fluid-solid catalytic reactions.

PROBLEMS

1-1 Methanol can be manufactured by the gas-phase reaction

$$CO + 2H_2 \rightarrow CH_3OH(g)$$

For one feed composition, the equilibrium gas at 400 K and 1 atm contains 40 mol % H_2, and the only other species present are CO and CH_3OH. At 400 K, the equilibrium constant is 1.52 and $\Delta H° = -22{,}580$ cal/(mol CO).

a. Assuming ideal gas behavior, what is the complete composition of the equilibrium gas?

b. For the same feed composition would the equilibrium gas at 500 K and 1 atm contain more or less than 40% H_2?

1-2 For the ammonia synthesis reaction

$$\tfrac{3}{2}H_2 + \tfrac{1}{2}N_2 \rightarrow NH_3$$

what is the standard-state free-energy change and the equilibrium constant at 773 K? Heat capacity data are as follows (T = degrees Kelvin (K), c_p = cal/(g mol)

$$N_2: \qquad c_p = 6.83 + 0.90 \times 10^{-3}\,T$$

$$H_2: \qquad c_p = 6.52 + 0.78 \times 10^{-3}\,T$$

$$NH_3: \qquad c_p = 7.11 + 6.0 \times 10^{-3}\,T$$

1-3 Using K from Prob. 1-2, what is the maximum conversion of nitrogen to ammonia at 773 K and 250 atm with a feed gas containing only H_2 and N_2 in the molar ratio of 3? What would be the conversion for a feed of 4.5 mol of H_2 per mol of N_2? Estimate the fugacities at 250 atm and 500 K using a suitable correlation of pure-component fugacities (see a standard thermodynamics text such as that by Smith and Van Ness). Assume the gas mixture behaves as an ideal solution.

1.4 Water vapor is 1.85 mol % dissociated at 2000°C and 1 atm total pressure, at equilibrium. Calculate the equilibrium dissociation at 25°C and 1 atm.

1-5 Consider a reactor in which two gas-phase reactions occur:
1. $A + B \rightarrow C + D$
2. $A + C \rightarrow 2E$

At the reaction temperature, $K_{p1} = 2.667$ and $K_{p2} = 3.200$. The total pressure is 10 atm, and the feed to the reactor has a composition of 2 moles of A and 1 mole of B. Calculate the composition of the reactor effluent if equilibrium is attained with respect to both reactions.

1-6 Calculate the dissociation pressure of $Ag_2O(s)$ at 200°C. Use the following data:

$$\Delta H° = 6950 \text{ cal/g mol}$$
$$\Delta F° = 2230 \text{ cal/g mol} \quad \text{at } 25°C$$

$$c_p = \begin{array}{ll} 5.60 + 1.5 \times 10^{-3}T & \text{for } Ag(s) \\ 6.50 + 1.0 \times 10^{-3}T & \text{for } O_2(g) \end{array}$$

where T is in degrees Kelvin.

1-7 Assuming that the value of K_p for the methanol-synthesis reaction is 9.28×10^{-3} at 1 atm pressure and 300°C, what are the numerical values of the following quantities[†] at this temperature?
  (a) K at $p_t = 1$ atm
  (b) K_p at $p_t = 10$ atm
  (c) K_p at $p_t = 50$ atm
  (d) K at 10 and 50 atm total pressure
  (e) K_y at 1, 10, and 50 atm total pressure

1-8 Newton and Dodge[‡] and von Wettberg and Dodge[§] measured the composition of equilibrium mixtures of CO, H_2, and CH_3OH in the methanol synthesis. Compute the values of K and $\Delta F°$ at 309°C from the following data taken from their work:

$$t = 309°C$$
$$p_t = 170 \text{ atm}$$

The equilibrium gas analysis, in mol%, is

Hydrogen	= 60.9
Carbon monoxide	= 13.5
Methanol	= 21.3
Inerts	= 4.3
Total	= 100.0

1-9 The complete results referred to in Prob. 1-8 are as follows:

$1000/T$, K^{-1}	1.66	1.73	1.72	1.75	1.82	1.81	1.82	1.82
log K	−4.15	−3.75	−3.65	−3.30	−3.10	−3.20	−3.00	−2.90
$1000/T^{-1}$	1.83	1.88	1.91	1.91	1.92	2.05	2.05	2.05
log K	−2.95	−2.60	−2.70	−3.00	−2.30	−2.30	−2.15	−2.35

From this information determine the best relationship between K and T in the form

$$\ln K = A \frac{1}{T} + B$$

[†] Use correlations in standard thermodynamics tests to estimate the required pure-component fugacities.
[‡] *J. Am. Chem. Soc.*, **56**, 1287 (1934).
[§] *Ind. Eng. Chem.*, **22**, 1040 (1930).

1-10 The determination of K in Prob. 1-8 was based on direct measurement of equilibrium compositions. Use the calorimetric data below and the third law to prepare a plot of log K vs. $1/T$ for the methanol synthesis. Include a temperature range of 298 to 800 K. Compare the graph with the result obtained in Prob. 1-9.

The entropy of CO gas at 298.16 K in the ideal-gas state at 1 atm is 47.30 cal/(g mole)(K). A similar value for hydrogen is 31.21. The heat of vaporization for methanol at 298.16 K is 8943.7 cal/g mole, and the vapor pressure at 298.16 K is 0.1632 atm. Heat of formation of CH_3OH in the ideal-gas state at 1 atm is $-48,490$ cal/g mol. Low-temperature specific-heat and heat-of-transition data for methanol are as follows:

T, K	18.80	21.55	24.43	27.25	30.72	34.33	37.64	40.87	43.93
c_p, cal (g mol)(°C)	1.109	1.512	1.959	2.292	2.829	3.437	3.962	4.427	4.840
	48.07	56.03	59.53	63.29	69.95	73.95	77.61	81.48	85.52
	5.404	6.425	6.845	7.252	8.001	8.392	8.735	9.001	9.295
	89.29	93.18	97.22	111.14	111.82	117.97	118.79	121.44	125.07
	9.693	9.939	10.23	11.23	11.48	11.64	11.64	11.74	12.18
	129.38	133.71	147.86	152.29	153.98	164.14	166.23	167.75	181.09
	12.28	12.64	12.97	13.69	14.12	11.29	11.63	11.68	16.60
T, K	185.10	189.06	196.77	210.34	235.84	256.34	273.58	285.15	292.01
c_p	16.67	16.77	16.78	16.97	17.41	17.70	18.30	18.70	19.11

Methanol crystals undergo a phase transition at 157.4 K for which $\Delta H = 154.3$ cal/g mole. The melting point is 175.22 K, and the heat of fusion is 757.0 cal/g mole. Specific-heat data at temperatures above 298.16 K are as follows:

T, K	298.16	300	400	500	600	700	800
c_p, CH_3OH, cal/(g mol)(°C)	10.8	10.8	12.7	14.5	16.3	17.8	19.2

T, °C	25	100	200	300	400	500	600
c_p, CO, cal/(g)(°C)	0.249	0.250	0.253	0.258	0.264	0.271	0.276
c_p, H_2, cal/(g)(°C)	3.42	3.45	3.47	3.47	3.48	3.50	3.53

1-11 Limestone ($CaCO_3$) decomposes upon heating to yield lime (CaO) and carbon dioxide. Determine the temperature at which limestone exerts a decomposition pressure of 1 atm.

1-12 The free energy for gas-phase mixtures of n-pentane and neopentane at 1 atm and 400 K is given by the expression

$$F = 9,600 \, y_1 + 8,990 \, y_2 + 800(y_1 \ln y_1 + y_2 \ln y_2)$$

where subscripts 1 and 2 refer to n-pentane and neopentane, respectively, and F is the free energy in cal/(g mol of mixture).

(a) For the isomerization reaction between n-C_5 and neo-C_5 determine the equilibrium composition at 1 atm and 400 K.

(b) Plot a curve of F vs. y_1 for the mixture at 1 atm and 400 K and mark the location of the *reaction* equilibrium point found in (a).

1-13 Butadiene is produced by the gas-phase, catalytic dehydrogenation of 1-butene

$$C_4H_8 \rightarrow C_4H_6 + H_2$$

In order to suppress side reactions and maintain the temperature level, the butene feed is mixed with high-temperature steam before entering the reactor. Estimate the reactor effluent temperature, if reaction equilibrium is attained, required to convert 30% of the butene to butadiene. The reactor pressure is 2 atm and the feed consists of 12 mol of steam per mol of butene. Free-energy data $\Delta F°$ (cal/(g mol) are:

	600 K	700 K	800 K	900 K
1,3-butadiene	46,780	50,600	54,480	58,400
1-butene	36,070	42,730	49,450	56,250

1-14 Hydrogen cyanide can be formed in small amounts by the acetylene-nitrogen reaction:

$$N_2(g) + C_2H_2(g) \rightarrow 2HCN(g)$$

For a stoichiometric feed to a reactor, what is the composition of the reactor effluent if equilibrium is reached? Consider two sets of operating conditions:

(a) 1 atm and 573 K

(b) 200 atm and 573 K

At 573 K, $\Delta F° = 7190$ cal/(mol of C_2H_2).

CHEMICAL KINETICS

In this chapter we want to learn how to formulate expressions for the intrinsic rate for chemical reactions. This means studying what variables are important and how they affect the rate. At present, rates cannot be predicted, so they must be measured. To do this, it is necessary to use a reactor, preferably a small-scale, laboratory unit. Rates cannot be measured directly but must be obtained by interpreting data measured in a reactor. Such data normally consist of concentrations of reactants and products, and the specific results depend upon the type of reactor employed. This is because the concentrations usually are influenced by such physical processes as convection as well as by the reaction itself. The analysis of reactor data to obtain the intrinsic rate is, in essence, the reverse of the design procedure for a commercial-scale reactor. However, the problem is simplified in that the small-scale reactor can be built so as to extract the intrinsic rate accurately and easily from the measured concentrations. That is, we can eliminate *most* of the complexities, introduced by the coupling of physical processes with intrinsic kinetics, by proper design of laboratory reactors. *All* of these complexities can be eliminated if (1) the concentration of a reactant or product and the temperature are the same in all points of the reaction mixture, and (2) if the concentrations can change only due to reaction. The first requirement is satisfied for a homogeneous batch, well-mixed reactor (the stirred-tank form shown in Fig. 1-3a). The second requirement arises since the concentration of a reactant can change not only due to it reacting, but because the volume changes. The volume will be constant if the temperature, pressure, and, for a gaseous reaction, the moles of reaction mixture are constant. We are concerned in this chapter primarily with the variables that affect the intrinsic rate. At this point, it is not desirable to complicate the procedure for obtaining an intrinsic rate equation by including effects of physical processes. Hence, it will be assumed that these two requirements are satisfied; that is, only well-mixed batch reactors of constant volume are considered. Then, in Chaps. 3, 4, and 6, for homogeneous reactions, and in Chaps. 10 and 11, for heterogeneous reactions, we will consider how to extract intrinsic rates from reactor data where these two restrictions are not satisfied.

2-1 Rate of Homogeneous Reactions

The rate of a homogeneous reaction is defined as the change in moles (due to reaction) of a reactant, or product, per unit time per unit volume of the reaction mixture. With the *restrictions* listed in the previous paragraph, the rate of *production* of species i may be expressed as

$$\mathbf{r}_i = \frac{1}{V}\frac{dn_i}{dt} = \frac{dC_i}{dt} \tag{2-1}$$

where n_i and C_i are the number of moles and concentration of chemical species $\mathbf{i}$. If $\mathbf{i}$ is a product, $\mathbf{r}_i$ will be positive; for a reactant $\mathbf{r}_i$ will be negative.[†] Also, for the same reaction, the numerical value of the rate will vary depending upon which product or reactant is used (unless all the stoichiometric coefficients are equal). This dependency can be eliminated, and the progress of a reaction expressed by one variable, ξ, *the extent of reaction.* Consider the general homogeneous reaction

$$aA + bB \rightarrow cC + dD \tag{2-2}$$

and suppose that initially there are $(n_A)_0$ moles of A, $(n_B)_0$ moles of B, etc. The rate of change of moles of one chemical species is related to that of any other by the stoichiometry of the reaction. Thus

$$-\frac{1}{a}\frac{dn_A}{dt} = -\frac{1}{b}\frac{dn_B}{dt} = \frac{1}{c}\frac{dn_C}{dt} = \frac{1}{d}\frac{dn_D}{dt} \tag{2-3}$$

Define the extent of reaction as

$$d\xi = \frac{-dn_A}{a} = \frac{-dn_B}{b} = \frac{dn_C}{c} = \frac{dn_D}{d} = \frac{dn_i}{v_i} \tag{2-4}$$

where v_i is the stoichiometric coefficient of species $\mathbf{i}$. If $\mathbf{i}$ refers to a reactant, v_i is negative, and for a product v_i is positive. Substituting dn_i from Eq. (2-4) in Eq. (2-1) gives an expression for the rate in terms of ξ:

$$\mathbf{r}_i = \frac{v_i}{V}\frac{d\xi}{dt}$$

or

$$\frac{\mathbf{r}_i}{v_i} = \frac{1}{V}\frac{d\xi}{dt} \tag{2-5}$$

[†] In contrast, the rate of *disappearance* of a reactant will be positive. We will usually define the rate so that it is positive [see Eq. (2-9)].

This result shows that the rate divided by the stoichiometric number is independent of the choice of reactant or product for **i**. From Eq. (2-1) this ratio r_i/v_i can
be expressed in terms of concentration as

$$\frac{r_i}{v_i} = \frac{1}{v_i}\frac{dC_i}{dt} \tag{2-5a}$$

Equation (2-5a) is a useful way of maintaining consistency in reporting rate
values, since r_i/v_i must be the same for all chemical species participating in the
reaction.

The extent of reaction is closely related to the *conversion x*. This latter quantity is defined as the fraction of a reactant that reacts. Thus, for reaction (2-2) the
conversion of species A is

$$x_A = \frac{(n_A)_0 - (n_A)}{n_{A0}} \tag{2-6}$$

If Eq. (2-4) is integrated from $\xi = 0$, $n_i = (n_i)_0$

$$n_i - (n_i)_0 = v_i(\xi - 0) = v_i\xi \tag{2-7}$$

Applying Eq. (2-7) to reactant A

$$n_A - (n_A)_0 = -a\xi$$

and substituting $n_A - (n_A)_0$ in Eq. (2-6) gives

$$x_A = \frac{a\xi}{(n_A)_0} \tag{2-8}$$

Note that ξ has the units of moles while the conversion x is dimensionless.

We consider r_i to be a property of a reaction system. Not a property in the
sense that it is a function of the equilibrium state of the system, but rather that, for
a specific homogeneous reaction, it is a function of temperature and concentrations of the reactants and products. In Secs. 2-2 to 2-8 the nature of the function
$r_i = f(C_i, T)$ is considered. Then these results are used in the remainder of the
chapter to formulate equations for the intrinsic rate from experimental data. This
is done for various types of single reactions and for complex (multiple reaction)
systems. In all cases the data are for conditions where Eq. (2-1) is justified.

2-2 Fundamentals of Rate Equations—Effect of Concentration

The early workers in kinetics found that simple relations existed between rates of
reaction and concentrations of reactants. Berthelot and St. Gilles† discovered that
the rate of esterification was proportional to the first power of the concentration
of ethanol and to the first power of the concentration of acetic acid. The rate is
said to be first order with respect to each reactant. In general terms, suppose that
the rate of disappearance of A by the *irreversible* reaction

$$aA + bB \rightarrow cC + dD$$

† M. Berthelot and L. P. St. Gilles, *Ann. Phys.*, **63**, 385 (1862).

may be written

$$\mathbf{r}_A = -\frac{dC_A}{dt} = kC_A^\alpha C_B^\beta \qquad (2\text{-}9)$$

Then α is the order of the reaction with respect to A, and β is the order with respect to B. The proportionality constant k, called the reaction rate constant, is independent of concentrations. Its dependence on temperature is discussed in Sec. 2-5.

It is not necessary that order and stoichiometry agree; that is, α is not necessarily equal to a and β to b. For example the rate of the reaction

$$2NO + 2H_2 \rightarrow N_2 + 2H_2O \qquad (2\text{-}10)$$

has been found† to be first order in hydrogen ($\beta = 1$) and second order in NO ($\alpha = 2$), while stoichiometry would require that the rate be second order with respect to both reactants. As kinetic data for many reactions have accumulated over the years, it has become increasingly clear that formation of the final products from the original reactants usually occurs in a series of simple steps. Herein lies the explanation for the difference between order and stoichiometric coefficients. Each step consists of a single reaction in which only one or two atomic bonds are broken or formed. The rates of the individual steps will normally differ from each other, and the rate of the overall reaction will be determined primarily by the slowest of these steps. The *mechanism* of a reaction is the sequence of steps that describes how the final products are formed from the original reactants. If the mechanism is known, it is usually possible to evaluate a rate equation such as (2-9) and, hence, the order of reaction. For example, suppose that Eq. (2-10) occurs in two steps

$$2NO + H_2 \rightarrow N_2 + H_2O_2 \qquad (2\text{-}11)$$

$$H_2O_2 + H_2 \rightarrow 2H_2O \qquad (2\text{-}12)$$

If reaction (2-12) is fast with respect to reaction (2-11), the rate will be determined by the first step. This would explain the observed, first-order dependency on the hydrogen concentration. The individual steps which constitute the reaction mechanism are termed *elementary* reactions. In an elementary step the molecules react exactly as the equation is written. For example, reaction (2-12) occurs by one molecule of hydrogen peroxide reacting with one molecule of hydrogen. The rate of each of these steps normally follows the stoichiometry of the reaction of that step. That is, order and stoichiometry agree for most elementary reactions. Thus, we may refine our concept of mechanism by stating that the mechanism is the sequence of *elementary* steps that determine the overall reaction.

The term molecularity is used by kineticists to designate the number of molecules involved in an elementary reaction. The most common elementary step is a bimolecular reaction, for example Eq. (2-12). A few elementary steps are unimolecular (step 1 of the N_2O_5 decomposition described in Sec. 2-3) or ter-

---

† C. N. Hinshelwood and T. E. Green, *J. Chem. Soc.*, **129**, 730 (1926).

molecular (Eq. 2-11). If a reaction involves four or more species, it most always represents a combination of two or more elementary steps. Note that a unimolecular reaction is normally a first-order process and a bimolecular reaction a second-order process.

RATE EQUATIONS FROM PROPOSED MECHANISMS

Accepting the concept that most all practical reactions occur by a sequence of elementary steps, how is a rate equation developed from this sequence? There are two parts to this problem: (1) How is the sequence of elementary steps chosen, i.e., how is the mechanism chosen? (2) How is a rate equation for the overall reaction developed from the mechanism? The first part must account for the fact that available experimental data may contain little or no information about the identification or concentration of intermediate chemical species. If information only about reactants and products for the overall reaction is available, there is no procedure for arriving at a unique mechanism. However, accumulated data over the years provide the kineticist with some working rules to follow in proposing a reaction mechanism. Edwards, Greene, and Ross[†] have proposed a set of such guidelines. Some are straightforward, such as the necessity for the elementary steps to be sufficient to explain the formation of all the observed products, and others are less evident. Choosing a mechanism is the activity of the kineticist, rather than the engineer, so it will not be considered here. However, from time to time we will use a mechanism in order to develop a rate equation, as in Sec. 2-3. Two points should be emphasized: (1) in experimental studies of kinetics as much information as possible should be obtained about reaction intermediates and products, including how their concentration changes with extent of the overall reaction, and (2) where several steps are involved, more than one mechanism can lead to the same overall rate equation, that is, an equation that agrees with experimental results based upon concentrations of overall reactants and products. From this it is clear that information about intermediate species is critical if the correct mechanism is to be identified.

After a mechanism is chosen, the second part of the problem is to formulate a rate equation in terms of concentrations of the reactants and products for the overall reaction. In principle, this can be accomplished by eliminating the concentrations of the intermediate species by combining rate equations (of the form of Eq. 2-9) written for each elementary step. However, except for the simplest mechanisms, this leads to equations so complex as to be useless for reactor design. Hence, the analysis is usually based upon some assumptions. First, considerable simplification is realized if the reactions are taken to be irreversible. This is always true during the initial stages of a reaction, because there are few product molecules. Also, it is true at any stage, if the equilibrium constant is large. Further, one of two additional hypotheses are usually made and these are discussed in the next two sections.

---

† J. O. Edwards, E. F. Greene, and J. Ross, *Chem. Education,* **45,** 381 (1968).

2-3 Rate-Controlling Step

If one of the elementary steps in a mechanism occurs at a much slower rate than the others, this step will determine the rate of the overall reaction. This assumption was used to explain the kinetics of the nitric oxide–hydrogen reaction (Eq. 2-10).

Consider a second example, the decomposition of N_2O_5:

$$2N_2O_5 \rightarrow 4NO_2 + O_2, \tag{2-13}$$

Ogg[†] found the rate to be first-order with respect to N_2O_5. The reaction is not likely to be an elementary one and probably occurs by the following three-step mechanism, with the first step occurring twice:

1. $N_2O_5 \underset{k_1'}{\overset{k_1}{\rightleftharpoons}} NO_2 + NO_3$

2. $NO_2 + NO_3 \overset{k_2}{\rightarrow} NO + O_2 + NO_2$ (slow)

3. $NO + NO_3 \overset{k_3}{\rightarrow} 2NO_2$

where the first step represents a reversible decomposition with forward rate constant k_1 and reverse rate constant k_1'. In the second step NO_2 does not react but affects the decomposition of NO_3. If the second step is inherently much slower than the others, the rate is given by

$$-\frac{1}{2}\frac{dC_{N_2O_5}}{dt} = \frac{dC_{O_2}}{dt} = r = k_2(C_{NO_2})(C_{NO_3}) \tag{2-14}$$

The concentration of NO does not appear in this rate equation, because the third step is fast with respect to the second. Since the first step also is fast, and reversible, we may relate the concentrations of the three species by an equilibrium equation:

$$K_1 = \frac{C_{NO_2}C_{NO_3}}{C_{N_2O_5}} \tag{A}$$

where K_1 is the equilibrium constant for this first step. Utilizing Eq. (A) to eliminate $(C_{NO_2})(C_{NO_3})$, the rate expression may be written

$$r = (k_2 K_1)C_{N_2O_5} \tag{B}$$

By using the assumption that the kinetics is controlled by one step in the mechanism, the rate is shown to be first-order in N_2O_5, in agreement with the observed results. Note that it would be erroneous to assume that Eq. (2-13) is an elementary step, even though the stoichiometric numbers could be divided by 2, implying that one molecule of N_2O_5 reacts. This result agrees with the observed first-order dependency on $C_{N_2O_5}$, but it would be incorrect. The N_2O_5 molecule is too complex, contains too many bonds, to be expected to decompose all the way to the simple molecules NO and O_2 in one step.

The assumption of a rate-controlling step in a mechanism will be used again in Chap. 9 for heterogeneous catalytic reactions.

[†] R. A. Ogg, Jr., *J. Chem. Phys.*, **15**, 337, 613 (1947).

2-4 Stationary-State Approximation

In the three-step mechanism proposed in the previous section for the decomposition of N_2O_5, both NO_3 and NO are intermediates. It was supposed that these species rapidly disappear according to the third step. Under these conditions the concentration of either of these species cannot build up to a significant level as the reaction proceeds. Since the values initially are zero, a logical approximation is that

$$\frac{dC_{NO_3}}{dt} = 0 \tag{A}$$

and

$$\frac{dC_{NO}}{dt} = 0 \tag{B}$$

These expressions can be used as an alternate procedure (to that of Sec. 2-3) to derive a rate equation for N_2O_5 decomposition. Noting that NO_3 is produced and consumed by reversible step 1, and consumed in steps 2 and 3, Eq. (A) may be expressed as follows:

$$\frac{dC_{NO_3}}{dt} = 0 = k_1 C_{N_2O_5} - k_1' C_{NO_2} C_{NO_3} - k_2 C_{NO_2} C_{NO_3} - k_3 C_{NO} C_{NO_3} \tag{C}$$

This equation can be solved for the concentration of intermediate NO_3

$$C_{NO_3} = \frac{k_1 C_{N_2O_5}}{k_2 C_{NO_2} + k_3 C_{NO} + k_1' C_{NO_2}} \tag{D}$$

Applying Eq. (B) in the same way,

$$\frac{dC_{NO}}{dt} = 0 = k_2 C_{NO_2} C_{NO_3} - k_3 C_{NO} C_{NO_3}$$

or

$$C_{NO} = \frac{k_2}{k_3} C_{NO_2} \tag{E}$$

This result gives C_{NO} in terms of the concentration of a final product NO_2. Equation (E) can be used for C_{NO} in Eq. (D) to give an expression for C_{NO_3} in terms of concentrations of reactants and final products:

$$C_{NO_3} = \frac{k_1 C_{N_2O_5}}{(2k_2 + k_1')C_{NO_2}} \tag{F}$$

Finally Eq. (F) can be substituted in Eq. (2-14) to obtain the rate equation

$$-\frac{1}{2}\frac{dC_{N_2O_5}}{dt} = r = \frac{k_1 k_2}{2k_2 + k_1'} C_{N_2O_5} \tag{G}$$

This result also is consistent with the first-order kinetics that are observed experimentally. The rate constants in Eq. (G) and in Eq. (B) of Sec. 2-3 are different.

However, $k_2 \ll k'_1$ since step 2 in the mechanism is assumed to be slow. Also, if step 1 is fast and reversible, its net rate, while not zero, will be much less than the rate in either the forward or reverse direction. Hence, with little error, the forward and reverse rates for step 1 may be equated to each other:

$$k_1 C_{N_2O_5} = k'_1 C_{NO_2} C_{NO_3} \qquad \text{(H)}$$

Comparison with Eq. (A) of Sec. 2-3 gives

$$K_1 = \frac{k_1}{k'_1} \qquad (2\text{-}15)$$

for this *elementary* step. With these two modifications, Eq. (G) becomes identical to Eq. (B) of Sec. 2-3:

$$\mathbf{r} = \frac{k_1 k_2}{k'_1} C_{N_2O_5} = k_2 K_1 C_{N_2O_5}$$

Both the *controlling step* and *stationary-state approximation methods* lead to the same expression for the overall rate.

As expected, the stationary-state approximation becomes more accurate when the intermediate species become more reactive; that is, as the rate constant for the destruction of the intermediate species increases. The validity of the approximation as a function of the value of the rate constant is examined in Prob. 2-26. Such large rate constants are characteristic of most free-radical reaction systems and, indeed, it was for such systems that the stationary-state approximation was first developed.[†] Free radical reactions are considered in Sec. 2-8.

Equation (2-15), which relates rate constants and the thermodynamic equilibrium constant, is an important conclusion for reversible reactions. Its application and limitations are discussed in Sec. 2-7.

2-5 Effect of Temperature—Arrhenius Equation

Up to this point we have considered the influence only of concentration on the rate. The *specific rate constant k* in Eq. (2-9) includes the effects of all other variables. The most important of these is temperature, but others may be significant. For example, a reaction may be primarily homogeneous but have appreciable wall or other surface effects. In such cases k will vary with the nature and extent of the surface. A reaction may be homogeneous but also require a miscible catalyst. An example is the reaction for the inversion of sugar, where the acid acts as a catalyst. In these instances k may depend on the concentration and nature of the catalytic substance. When the concentration effect of the catalyst is

---

† M. Bodenstein, *Z. Physik Chem.* **85**, 329 (1913); J. A. Christiansen, *Kgl. Danske Videnskab. Selskab., Mat.-Fys. Medd,* **1**, 14 (1919); K. F. Herzfeld, *Ann. Physik.* **59**, 635 (1919); M. Polanyi, *Z. Elektrochem.* **26**, 50 (1920).

known, it is better to include the catalyst concentration in Eq. (2-9), so that k is independent of all concentrations.

The dependency of k on temperature for an elementary process follows the Arrhenius equation

$$k = Ae^{-E/R_gT} \tag{2-16}$$

where A is the frequency (or preexponential) factor and E is the activation energy. Combining Eqs. (2-16) and (2-9) yields

$$r = -\frac{dC_A}{dt} = \mathbf{A}e^{-E/R_gT}C^{\alpha}_A C^{\beta}_B \tag{2-17}$$

This provides a description of the rate in terms of the measurable variables, concentration and temperature. Rigorously it is limited to an elementary process, because the Arrhenius equation is so restricted. However, the exponential effect of temperature often accurately represents experimental rate data for an overall reaction, even though the activation energy is not clearly defined and may be a combination of E values for several of the elementary steps.

The Arrhenius expression was originally derived[†] from thermodynamic considerations. For an elementary reaction whose rates are rapid enough to achieve a dynamic equilibrium the *van't Hoff equation* states that

$$\frac{d \ln K}{dT} = \frac{\Delta H^{\circ}}{R_g T^2} \tag{2-18}$$

Suppose the reaction is

$$A + B \underset{k_1'}{\overset{k_2}{\rightleftarrows}} C \tag{2-19}$$

with forward and reverse rate constants of k_2 and k_1'. Then, from Eq. (2-15) the equilibrium and rate constants are related by

$$K = \frac{k_2}{k_1'}$$

Using this result, Eq. (2-18) may be written

$$\frac{d \ln k_2}{dT} - \frac{d \ln k_1'}{dT} = \frac{\Delta H}{R_g T^2} \tag{2-20}$$

The right-hand side of Eq. (2-20) can be divided into two enthalpy changes, ΔH_1 and ΔH_2, such that

$$\Delta H = \Delta H_2 - \Delta H_1 \tag{2-21}$$

[†] S. Arrhenius, *Z. Physik Chem.* **4**, 226 (1889).

Then Eq. (2-20) may be split into two equations, one for the forward reaction and another for the reverse reaction, which will have a difference in agreement with (2-21):

$$\frac{d(\ln k_2)}{dT} = \frac{\Delta H_2}{R_g T^2} \tag{2-22}$$

$$\frac{d(\ln k_1')}{dT} = \frac{\Delta H_1}{R_g T^2} \tag{2-23}$$

Integrating either equation, and setting the integration constant equal to $\ln A$, gives a result of the form of the Arrhenius equation, Eq. (2-16):

$$k = Ae^{-\Delta H/R_g T} \tag{2-24}$$

An alternate derivation is based on the concept of an intermediate state, often called a *transition* or *activated state*, which is a postulate of the transition-state theory (Sec. 2-6). Suppose that product C of the reaction

$$A + B \rightarrow C \tag{2-25}$$

is formed by decomposition of an activated form of reactants A and B, which will be designated $(AB)^*$. Then the reaction occurs by two elementary steps,

$$1. \quad A + B \rightleftarrows (AB)^* \tag{2-26}$$

$$2. \quad (AB)^* \rightarrow C \tag{2-27}$$

If the first step is comparatively rapid in both forward and reverse directions, $(AB)^*$ will be in equilibrium with A and B so that its concentration is given by

$$C_{AB^*} = K^* C_A C_B \tag{2-28}$$

where K^* is the equilibrium constant for the formation of $(AB)^*$. The rate of reaction (rate of formation of C) is then given by the rate of the first-order decomposition step. With Eq. (2-28), this may be expressed as

$$\mathbf{r} = k^* C_{AB^*} = k^* K^* C_A C_B \tag{2-29}$$

If we integrate the van't Hoff equation, Eq. (2-18), replacing K with K^* and $\Delta H°$ with ΔH^*, the result is

$$K^* = Ie^{-\Delta H^*/R_g T} \tag{2-30}$$

where I is the constant of integration. Combining Eqs. (2-29) and (2-30) gives

$$\mathbf{r} = k^* Ie^{-\Delta H^*/R_g T} C_A C_B \tag{2-31}$$

Comparison with Eq. (2-16) shows that

$$k = \mathbf{A}e^{-\Delta H^*/R_g T} \tag{2-32}$$

where $\mathbf{A} = k^* I$. Equation (2-32) is also of the form of the Arrhenius equation.

Since ΔH^* is the energy required to form the activated state $(AB)^*$ from A and B, $e^{-\Delta H^*/R_g T}$ is the *Boltzmann expression* for the fraction of molecules having an energy ΔH^* in excess of the average energy. This gives some meaning to the activation energy E in the Arrhenius equation. The diagram in Fig. 2-1 shows that this value is the energy barrier that must be overcome to form $(AB)^*$, and ultimately, product C.

The value of Eq. (2-16) rests substantially on the accuracy with which it represents experimental rate-temperature data (see Example 2-1). When measured rates do not agree with the theory it is usually found that the reaction is not an elementary step and the mechanism changes with temperature, or that physical resistances are affecting the measurements. In other words, Eq. (2-16) correlates remarkably well the rate measurements for single reactions free of diffusion and thermal resistances.

The Arrhenius equation provides no basis for discerning the value of E. However, Fig. 2-1 indicates that the activation energy must be greater than the heat of the overall reaction, ΔH, for an endothermic case. Also the energy of activation for the reverse reaction is less than that for the forward reaction. The reverse is true for an exothermic reaction. Since rates normally increase with temperature, the activation energy is positive. Equation (2-16) shows that a numerical value for E can be obtained by plotting experimental data for rate constants at different temperatures. The logarithmic form of Eq. (2-16) is

$$\ln k = -\frac{E}{R_g}\left(\frac{1}{T}\right) + \ln A \tag{2-33}$$

Hence, a plot of $\ln k$ vs. $1/T$, called an Arrhenius plot, yields a slope equal to $-E/R_g$. Calculation of E is illustrated in Example 2-1. If the concentration dependence of the rate is not known, but rate vs. temperature data are available for *constant* concentrations, E can be obtained from a plot of $\ln r$ vs. $1/T$. As an illustration consider Eq. (2-17). The concentration function $C_A^\alpha C_B^\beta$ would have a constant, unknown value. Hence it could be combined with **A** to give a new constant, A'. Then Eq. (2-33) is applicable if $\ln \mathbf{r}$ replaces $\ln k$ and A' replaces A.

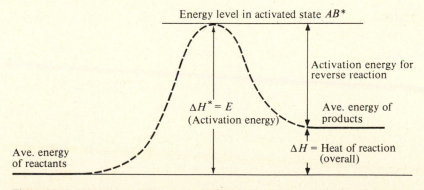

Figure 2-1 Energy levels of initial, activated, and final states (endothermic reaction).

Example 2-1 Wynkoop and Wilhelm† studied the rate of hydrogenation of ethylene, using a copper–magnesium oxide catalyst, over restricted pressure and composition ranges. Their data may be interpreted with a first-order rate expression of the form

$$\mathbf{r} = (k_1)_p \, p_{H_2} \tag{A}$$

† Raymond Wynkoop and R. H. Wilhelm, *Chem. Eng. Prog.*, **46**, 300 (1950).

Table 2-1 Data for hydrogenation of ethylene

Run	$(k_1)_p \times 10^5$, g mol/(s)(atm)(cm³)	T, °C	$1/T \times 10^3$, K^{-1}
1	2.70	77	2.86
2	2.87	77	2.86
3	1.48	63.5	2.97
4	0.71	53.3	3.06
5	0.66	53.3	3.06
6	2.44	77.6	2.85
7	2.40	77.6	2.85
8	1.26	77.6	2.85
9	0.72	52.9	3.07
10	0.70	52.9	3.07
11	2.40	77.6	2.85
12	1.42	62.7	2.98
13	0.69	53.7	3.06
14	0.68	53.7	3.06
15	3.03	79.5	2.83
16	3.06	79.5	2.83
17	1.31	64.0	2.97
18	1.37	64.0	2.97
19	0.70	54.5	3.05
20	0.146	39.2	3.20
21	0.159	38.3	3.21
22	0.260	49.4	3.10
23	0.322	40.2	3.19
24	0.323	40.2	3.19
25	0.283	40.2	3.19
26	0.284	40.2	3.19
27	0.277	39.7	3.20
28	0.318	40.2	3.19
29	0.323	40.2	3.19
30	0.326	40.2	3.19
31	0.312	39.9	3.19
32	0.314	39.9	3.19
33	0.307	39.8	3.19

Source: Raymond Wynkoop and R. H. Wilhelm, Chem. Eng. Progr., 46, 300 (1950).

where **r** is the rate of reaction, in g mol/(cm³)(s), and p_{H_2} is the partial pressure of hydrogen, in atmospheres. With this rate equation $(k_1)_p$ will be reported in g mol/(cm³)(s)(atm). The results for $(k_1)_p$ at various temperatures are given in Table 2-1.

(a) What is the activation energy from rate equation (A)?

(b) What would it be if the rate equation were expressed in terms of the concentration of hydrogen rather than the partial pressure?

SOLUTION (a) In the last column of Table 2-1 the reciprocal of the absolute temperature is shown for each run. Figure 2-2 is a plot of $(k_1)_p$ vs. $1/T$ on semilogarithmic coordinates. It is apparent that the data describe a straight line, except for runs 8, 20, 21, and 22. It has been suggested that water vapor may have caused the low rates in these cases.† The line shown in the figure may be located by fitting the data points by the least squares technique. Equation (2-33) takes the form

$$\ln (k_1)_p = \ln A - \frac{E_p}{R_g} \frac{1}{T} \qquad (B)$$

† Raymond Wynkoop and R. H. Wilhelm. *Chem. Eng. Prog.*, **46**, 300 (1950).

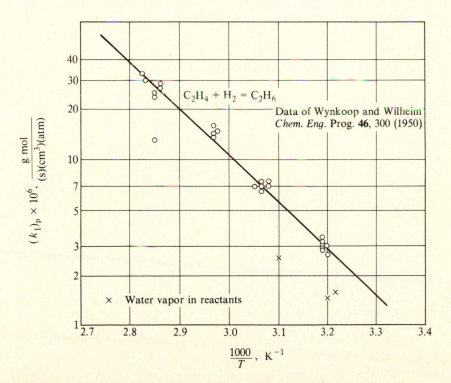

Figure 2-2 Plot of Arrhenius equation for hydrogenation of ethylene.

If (T_i, k_i) represents one of n data points, the values of A and E_p/R_g which describe the least square fit are

$$-\frac{E_p}{R_g} = \frac{n \sum_{i=1}^{n} (\ln k_i)(1/T_i) - \left(\sum_{i=1}^{n} 1/T_i\right) \sum_{i=1}^{n} \ln k_i}{n \sum_{i=1}^{n} (1/T_i)^2 - \left(\sum_{i=1}^{n} 1/T_i\right)^2} \tag{C}$$

$$\ln A = \frac{\sum \ln k_i \sum (1/T_i)^2 - \sum 1/T_i \sum (1/T_i \ln k_i)}{n \sum (1/T_i)^2 - (\sum 1/T_i)^2} \tag{D}$$

Carrying out the summations indicated for all the data points, with the data in Table 2-1 for k_i and T_i, we find

$$-\frac{E_p}{R_g} = -6{,}460$$

$$E_p = 6{,}460 R_g = 12{,}800 \text{ cal/g mol}$$

A quicker but less accurate method is to draw, visually, a straight line through the data plotted as $\ln (k_1)_p$ vs. $1/T$, measure its slope, and multiply by R_g to obtain the activation energy. Note that an approximate value of E can be obtained from data at but two temperatures.

(b) For gaseous reactions the rate equation can be expressed in terms of concentrations or pressures. Equation (A) is the pressure form for this example. In terms of concentrations, the rate is

$$\mathbf{r} = (k_1)_C C_{H_2}$$

Expressing $(k_1)_C$ in the Arrhenius form and then differentiating gives

$$(k_1)_C = A_C e^{-E_C/R_g T}$$

and

$$\frac{d[\ln (k_1)_C]}{d(1/T)} = -\frac{E_C}{R_g} \tag{E}$$

This may be related to Eq. (A) by noting that for an ideal-gas mixture the concentration of H_2 is

$$C_{H_2} = \frac{p_{H_2}}{R_g T}$$

Then

$$\mathbf{r} = \frac{(k_1)_C}{R_g T} p_{H_2} \tag{F}$$

Comparison of Eqs. (A) and (F) gives the relationship between the two rate constants,

$$(k_1)_C = (k_1)_p R_g T \tag{G}$$

Differentiating the logarithmic form of Eq. (G) and using Eqs. (B) and (E), we obtain

$$\frac{d[\ln (k_1)_C]}{d(1/T)} = \frac{d[\ln (k_1)_p]}{d(1/T)} + \frac{d(\ln T)}{d(1/T)}$$

$$-\frac{E_C}{R_g} = -\frac{E_p}{R_g} - T$$

or

$$E_C = E_p + R_g T \tag{H}$$

Thus the activation energy, in principle, depends on whether the rate equation is expressed in terms of concentrations or partial pressures. Also, the difference between E_C and E_p depends on the temperature. In practice this difference is seldom significant. In this example, at a temperature of 77°C,

$$E_C - E_p = 2(350 \text{ K}) = 700 \text{ cal/g mol}$$

This difference of 6% is too small to be discerned from rate measurements of the usual precision.

In this example metric units (cgs system) have been employed. These units and commonly used English units will be used in subsequent chapters. However, illustrations are also given in SI units. It may be helpful here to note the conversion of common kinetic quantities. In the SI system energy is expressed in joules (J) or kilojoules (kJ). Hence, the activation energy of Ex. 2-1 may be expressed as

$$(E_p)_{SI} = 12,800(4.186 \text{ J/cal}) = 53,600 \text{ J/mol} \qquad \text{or } 53,600 \text{ kJ/kg mol}$$

Note that the gas constant R_g in SI units is

$$(R_g)_{SI} = 1.985 \frac{\text{cal}}{(\text{mol})(\text{K})} (4.186 \text{ J/cal})$$

$$= 8.314 \text{ J/(mol)(K)}$$

First-order rate constants in $(s)^{-1}$ are the same in either the cgs or SI system. Rate constants of order n have units of $(\text{time})^{-1}(\text{concn})^{1-n}$. Thus, for a second-order reaction, k_2 in $cm^3/(\text{mol})(s)$ should be multiplied by $(10^{-6} \text{ m}^3/\text{cm}^3)(10^3 \text{ mol/k mol})$ or 10^{-3} to convert to a k_2 reported as $m^3/(\text{k mol})(s)$.

2-6 Prediction of Reaction Rates—Theories of Kinetics

If the frequency factor A and energy E could be evaluated from molecular properties of the reacting species, one would have a means of *predicting* reaction rates for elementary steps. No kinetics experiments would be needed. The science of kinetics has not yet developed to this stage. Nevertheless, it is useful to summarize briefly the status of theories to predict rates.

An elementary form of the kinetic theory of gases inspired the *collision* theory. In this view reaction occurs when the collision of reacting molecules evolves enough energy to provide the necessary activation energy,† that is, to surmount the energy level shown in Fig. 2-1. This idea led to a rate expression based upon the frequency of molecular collisions and the fraction of the collisions that evolved the required minimum energy.

Example 2-2 Use the collision theory to estimate the specific reaction rate for the decomposition of hydrogen iodide, $2HI \rightarrow I_2 + H_2$. Assume that the collision diameter σ is 3.5 Å (3.5×10^{-8} cm), and employ the activation energy of 44,000 cal/g mol determined experimentally by Bodenstein.‡ Also evaluate the frequency factor.

SOLUTION According to the collision theory, the rate for the reaction, $A + B \rightarrow$ products, is given by

$$r = kC_A C_B = C_A C_B \sigma_{AB}^2 \left(8\pi R_g T \frac{M_A + M_B}{M_A M_B} \right)^{1/2} e^{-E/R_g T} \qquad (2\text{-}34)$$

where M_A and M_B are the molecular weights of A and B and σ_{AB} is the collision diameter. The exponential quantity represents the fraction of the collisions releasing an energy equal to E or greater. The preexponential quantity is the collision rate per cm³ of reaction mixture when the concentrations are in mol/cm³. The temperature is 321.4°C.

For the reaction $2HI \rightarrow I_2 + H_2$,

$$M_A = M_B = M_{HI} = 128$$

The other numerical quantities required are

$$R_g = k_B n = (1.38 \times 10^{-16})(6.02 \times 10^{23})$$

$$= 8.30 \times 10^7 \text{ erg/(K)(g mol)} \qquad \text{or } 1.98 \text{ cal/(g mol)(K)}$$

$$\sigma_{AB} = 3.5 \times 10^{-8} \text{ cm}$$

$$E = 44,000 \text{ cal/g mol}$$

$$T = 273 + 321.4 = 594.6 \text{ K}$$

Substituting these values in Eq. (2-34) yields

$$k = (3.5 \times 10^{-8})^2 \left[8\pi(8.30 \times 10^7)594.6 \left(\frac{2}{128} \right) \right]^{1/2} e^{-44,000/R_g T}$$

$$= 1.70 \times 10^{-10} e^{-37.4} \qquad \text{cm}^3/(\text{molecule})(s)$$

To convert this result to the usual units of cm³/(g mol)(s) it should be multiplied by Avogadro's number, 6.02×10^{23} molecules/mol

$$k = 6.02 \times 10^{23} \times 1.70 \times 10^{-10} e^{-37.4}$$

$$= 1.02 \times 10^{14} e^{-37.4} = 5.7 \times 10^{-3} \text{ cm}^3/(\text{g mol})(s) \qquad (A)$$

---

† M. Polanyi, *Z. Elektrochem* **26**, 48 (1920).
‡ M. Bodenstein, *Z. Physik Chem.*, **100**, 68 (1922).

As we shall see in Example 2-7, the rate constant from Kistiakowsky's experimental data is 2.0×10^{-3} cm³/(g mol)(s). For reactions involving more complex molecules the experimental rates are usually much less than the theory predicts.

Comparison of the form of Eq. (A) and the Arrhenius expression shows that the frequency factor is

$$A = 1.0 \times 10^{14} \text{ cm}^3/\text{(g mol)(s)}$$

The collision theory has been found to give results in good agreement with experimental data for a number of bimolecular gas reactions. The decomposition of hydrogen iodide considered in Example 2-2 is an illustration. The theory has also been satisfactory for several reactions in solution involving simple ions. However, for many other reactions the predicted rates are much too large. Predicted frequency factors lie in the rather narrow range of 10^{12} to 10^{14}, while measured values may be several orders of magnitude less. The deviation appears to increase with the complexity of the reactant molecules. (Moreover, unimolecular decompositions are difficult to rationalize by the collision theory.)

Beginning about 1930 the principles of quantum mechanics were applied to this problem by Eyring, Polanyi, and their coworkers, and the result is known as the *activated-complex*, or *transition-state*, *theory*.[†] In this theory reaction is still presumed to occur as a result of collisions between reacting molecules, but what happens after collision is examined in more detail. This examination is based on the concept that molecules possess vibrational and rotational, as well as translational, energy levels.

The essential postulate is that an activated complex (or transition state) is formed from the reactants, and that this subsequently decomposes to the products. The activated complex is assumed to be in thermodynamic equilibrium with the reactants. Then the rate-controlling step is the decomposition of the activated complex. The concept of an equilibrium activation step followed by slow decomposition is equivalent to assuming a time lag between activation and decomposition into the reaction products. It is the answer proposed by the theory to the question of why all collisions are not effective in producing a reaction.

These ideas may be illustrated by a simple reaction between A and B to form a product C. If the activated complex is designated by $(AB)^*$, the overall process can be written as

$$A + B \rightleftharpoons (AB)^* \rightarrow C$$

Since equilibrium is assumed for the first step, the concentration of $(AB)^*$ is that determined by the equilibrium constant. Then the rate of the overall reaction is equal to the product of the frequency of decomposition of the complex and its equilibrium concentration, or

$$\mathbf{r} = vC_{AB^*} \qquad \text{molecules/(s)(cm}^3) \qquad (2\text{-}35)$$

† Samuel Glasstone, K. J. Laidler, and Henry Eyring, "The Theory of Rate Processes," McGraw-Hill Book Company, New York, 1941.

where v has units of $(\text{seconds})^{-1}$ and concentration C_{AB^*} is in molecules per cubic centimeter.

Using thermodynamic principles to express C_{AB^*} in terms of C_A and C_B, and quantum mechanics to evaluate the decomposition frequency v, the rate is given by

$$\mathbf{r} = \frac{k_B T}{h} \frac{\gamma_A \gamma_B}{C^\circ \gamma_{AB}} (e^{-\Delta F^*/R_g T}) C_A C_B \tag{2-36}$$

where k_B = Boltzmann's constant, 1.380×10^{-16} erg/K$(1.38 \times 10^{-23}$ J/K)
h = Planck's constant, 6.024×10^{-27} erg(s)$[6.024 \times 10^{-34}$ J(s)]
C° = standard-state concentration used to define the activity coefficient γ; i.e.,

$$a_i = \gamma_i C_i / C_i^\circ$$

Hence, the reaction rate constant is

$$k = \frac{k_B T}{hC^\circ} \left(\frac{\gamma_A \gamma_B}{\gamma_{AB}} \right) e^{-\Delta F^*/R_g T} = \frac{k_B T}{hC^\circ} \left(\frac{\gamma_A \gamma_B}{\gamma_{AB}} \right) e^{\Delta S^*/R_g - \Delta H^*/R_g T} \tag{2-37}$$

The latter form follows from the thermodynamic relation $\Delta F = \Delta H - T \Delta S$.

Comparison of Eq. (2-37) with the Arrhenius equation shows that

$$A = \frac{k_B T}{hC^\circ} \left(\frac{\gamma_A \gamma_B}{\gamma_{AB}} \right) e^{\Delta S^*/R_g} \tag{2-38}$$

$$E = \Delta H^* \tag{2-39}$$

These two relations are the predictions of transition-state theory for the frequency factor and the energy of activation.

The collision theory [Eq. (2-34)] does not offer a method for calculating the activation energy. Transition-state theory suggests that E is the enthalpy change for formation of the activated complex from the reactants [Eq. (2-39)]. To predict this enthalpy we must know exactly what the activated complex is, i.e., we must know its structure. Even then the prediction of enthalpy from molecular-structure data by statistical mechanics is an uncertain operation for any but the simplest molecule. Eckert and Boudart† have illustrated the calculation procedures with the hydrogen iodide decomposition reaction. If an activation energy is available from experimental measurements, the theory need be used only for estimating the frequency factor from Eq. (2-38). Again the structure of the activated complex is necessary, this time to calculate the entropy of activation, ΔS^*.

Uncertainties about the structure of the activated complex and the assumptions involved in computing its thermodynamic properties seriously limit the practical value of the theory. However, it does provide qualitative interpretation of how molecules react and a reassuring foundation for the empirical rate expres-

---

† C. A. Eckert and M. Boudart, *Chem. Eng. Sci.* **18**, 144 (1963).

sions inferred from experimental data. The effect of temperature on the frequency factor is difficult to evaluate from rate measurements. This is because the strong exponential function in the Arrhenius equation effectively masks the temperature dependency of A. Equation (2-38) suggests that A is proportional to T, the collision theory indicates a $T^{1/2}$ dependency [Eq. (2-34)], and the Arrhenius relationship [Eq. (2-16)] implies that A is unaffected by temperature. It is normally impossible to measure rates of reaction with sufficient sensitivity to evaluate these differences.

Equation (2-35) states that the rate is proportional to the concentration of the activated complex. Similarly, in the collision theory [Eq. (2-34)] it is assumed that the concentration determines the collision frequency and the rate. However, if the concepts of thermodynamics were followed, the rate might be assumed proportional to activity. If the activity replaced concentration in Eq. (2-35), the activity coefficient of the activated complex would not be needed. The expression for the rate constant would then be

$$k = \frac{k_B T}{h} \gamma_A \gamma_B e^{\Delta S^*/R_g - \Delta H^*/R_g T} \tag{2-40}$$

instead of Eq. (2-37). Since the activity coefficient is a function of pressure, k values predicted from the two equations would vary differently with pressure.

Eckert and Boudart[†] analyzed rate data for the decomposition of HI in this way. Their results were more compatible with Eq. (2-37), suggesting that the rate is proportional to the *concentration* of the activated complex. In this text the rate equation will be written in terms of concentrations.

For a detailed discussion of the status of predictive theories and for correlation and methods of estimation of rate constants, the books by Benson and by Laidler are available.[‡§]

2-7 Rate and Equilibrium Constants

It has been mentioned that for an elementary reaction the equilibrium constant K is equal to the ratio of the forward and reverse rate constants [Eq. (2-15)]. This follows from the *principle of microscopic reversibility* which states[§] that at equilibrium the rates in the reverse and forward directions are the same; that is, the *net* rate is zero. The question arises as to whether the ratio of the forward and reverse rate constants is equal to the equilibrium constant for a nonelementary process. Consider the overall reaction

$$A \rightleftharpoons C \tag{A}$$

[†] Ibid.

[‡] S. W. Benson, "Thermochemical Kinetics," 2d ed., John Wiley, New York, 1976.

[§] K. J. Laidler, "Theories of Chemical Reaction Rates," McGraw-Hill Book Company, New York, 1969.

[*] For a proof of this principle see K. J. Laidler, "Chemical Kinetics," 2d ed., p. 110, McGraw-Hill Book Company, New York, 1965.

to occur by the following sequence of elementary steps:

$$1. \quad A \underset{k_1'}{\overset{k_1}{\rightleftharpoons}} B \tag{B}$$

$$2. \quad B \underset{k_2'}{\overset{k_2}{\rightleftharpoons}} C \tag{C}$$

If the system is at complete equilibrium, that is, equilibrium with respect to both reactions 1 and 2, we may write

$$K = \left(\frac{C_C}{C_A}\right)_{eq} \tag{D}$$

where K is the equilibrium constant for the overall reaction [Eq. (A)]. Applying the principle of microscopic reversibility to *elementary* reactions (B) and (C)

$$\left(\frac{C_B}{C_A}\right)_{eq} = \frac{k_1}{k_1'} \quad \text{and} \quad \left(\frac{C_C}{C_B}\right)_{eq} = \frac{k_2}{k_2'}$$

From these two expressions,

$$\left(\frac{C_C}{C_A}\right)_{eq} = \frac{k_2 \, k_1}{k_2' \, k_1'} \tag{E}$$

Comparing Eqs. (D) and (E) yields

$$K = \frac{k_2 \, k_1}{k_2' \, k_1'} \tag{F}$$

Equation (F) is the correct relation between the equilibrium and rate constants rather than Eq. (2-15). This can be demonstrated in the following way. At complete equilibrium, the concentration of intermediate B will be constant. Hence, we can write

$$\frac{dC_B}{dt} = 0 = k_1 C_A - k_1' C_B - k_2 C_B + k_2' C_C$$

or

$$C_B = \frac{k_1 C_A + k_2' C_C}{k_1' + k_2} \tag{G}$$

The proper expression for the rate of a *disappearance* of A is

$$\mathbf{r} = -\frac{dC_A}{dt} = k_1 C_A - k_1' C_B \tag{H}$$

Introducing Eq. (G) for C_B

$$\mathbf{r} = k_1 C_A - \frac{k_1'(k_1 C_A + k_2' C_C)}{k_1' + k_2}$$

$$\mathbf{r} = \frac{k_1 k_2}{k_1' + k_2} C_A - \frac{k_1' k_2'}{k_1' + k_2} C_C \tag{I}$$

This equation satisfies the requirement that at equilibrium ($\mathbf{r} = 0$), K is given by Eq. (F). However, suppose we made measurements for the rate of disappearance of A starting with only A, so that B and C were not present—a nonequilibrium condition. The rate then is given by

$$-\frac{dC_A}{dt} = k_1 C_A$$

These experiments would determine k_1. Similarly, if measurements were made with pure C only, k_2' would be obtained. It is *not* correct to use these nonequilibrium results to write an expression for the net rate of disappearance of A as

$$\mathbf{r} = -\frac{dC_A}{dt} \neq k_1 C_A - k_2' C_C \qquad \text{(J)}$$

Similarly, this expression could *not* be applied at equilibrium ($\mathbf{r} = 0$) to write

$$K = \left(\frac{C_C}{C_A}\right)_{eq} \neq \frac{k_1}{k_2'} \qquad \text{(K)}$$

It is concluded that the ratio of forward- and reverse-rate constants is equal to K for a nonelementary reaction only if the rate constants are evaluated at near equilibrium conditions. Further, such rate constants will not be equal to k values for the single steps but rather combinations of rate constants, as illustrated by Eq. (I). The failure of Eq. (K) is consistent with the fact that reaction orders [α, β in Eq. (2-9)] are not necessarily equal to the stoichiometric numbers, v_i.

2-8 Chain Reactions

Several important types of industrial reactions, such as hydrocarbon cracking and photochlorination, have been shown to consist of a series of elementary steps in which one or more active intermediates are continually regenerated. This regeneration makes such systems different in type from all the sequences of elementary steps considered earlier. The first step in a chain reaction is the production of the active intermediate, an atom or free radical. Since this process requires breaking a bond in a stable molecule, considerable energy is often required [for example, about 35 kcal/mol for dissociation of Cl_2]. As a result the activation energy is large and the rate of the initiation step is low. To enhance the initiation rate the temperature can be increased or an outside energy source, such as radiant energy (photochemical initiation), can be provided. Initiation is followed by the propagation steps in which product is produced and the atom or free radical regenerated. The rates of propagation steps are normally large since reactive atoms or radicals are involved. The activation energy for these processes is low. The rate of production of product can be very high even though the rate of initiation is slow. This is because once one atom or free radical is generated, it can, through the propagation steps, produce many molecules of stable product. This explains the high quantum yields of some photochemical processes. One type of evidence suggesting a chain mechanism is an induction period. This is caused by the need to remove small quantities of substances that act to remove free radicals or atoms. The rate will be zero during the time period that these substances, frequently impurities such as oxygen, are being used up to scavenge the free radicals. Another evidence of a chain mechanism is very high rates (e.g., explosions) when two or more free radicals are generated for every radical consumed. The rate of chain reactions is limited by the removal of the reactive intermediate by the third step in the mechanism—termination reactions. The three types of elementary steps (initiation, propagation, and termination) determine the net rate of production of products. Since the free radicals or atoms are very reactive, the stationary-state hypothesis (Sec. 2-4) is a good approximation. Applied to each active intermediate, this approximation permits expressing the rate in terms of concentrations of stable species. As an illustration, consider the chlorination of propane (PrH). The experimental evidence suggests the following sequence:

Initiation

$$Cl_2 \xrightarrow{k_1} 2Cl$$

Propagation

$$Cl + PrH \xrightarrow{k_2} Pr + HCl$$

$$Pr + Cl_2 \xrightarrow{k_3} PrCl + Cl$$

Termination

$$\left. \begin{array}{l} Cl + Cl \xrightarrow{k_4} Cl_2 \\[2mm] Cl + Pr \xrightarrow{k_5} PrCl \end{array} \right\} \text{homogeneous}$$

$$\left. \begin{array}{l} Cl + W \xrightarrow{k_6} \quad \text{end product} \\[2mm] Pr + W \xrightarrow{k_7} \quad \text{end product} \end{array} \right\} \text{heterogeneous}$$

In the initiation step the chlorine atoms can be obtained from intermolecular collisions of molecules heated to a high temperature (thermal activation). Alternately, the chlorine molecule can absorb radiant energy of the proper wavelength and dissociate (photochemical activation). The propagation steps produce products (propyl chloride, PrCl, and HCl), and also regenerate chlorine atoms. The radical Pr is C_3H_7. It and Cl are chain carriers. Only a small amount of initiation is required to start the chain and lead to large rates of product formation. As shown by the reactions, termination may be homogeneous (occurring by intermolecular collisions) or heterogeneous (occurring by collision with the wall, W, of the reactor).

When the propagation steps produce two chain carriers for every one consumed, the extra radical may cause further propagation or it may be destroyed by termination processes. If the extra radical is not destroyed, the growth in propagation (and product formation) approaches infinity, resulting in an explosion. The oxidation of hydrogen is an example. After initiation produces the hydrogen atom, H, this may react with O_2 producing *two* chain carriers, OH and O. Both chain carriers then propagate further to form product and regenerate themselves. Thus the sequence of reactions is

$$
\begin{array}{ll}
1. & H + O_2 \rightarrow OH + O \\
2. & OH + H_2 \rightarrow H_2O + H \\
3. & O + H_2 \rightarrow OH + H
\end{array} \Biggr\} \text{propagation}
$$

Reactions such as reaction 1 are called *chain-branching steps*.

Polymerization does not occur by a normal chain-reaction sequence, since the activated reactant is not regenerated. There is no chain carrier. Nevertheless, the overall process can be effectively analyzed in many instances as a combination of initiation, propagation, and termination steps. For example, suppose P_r represents a reactive polymer containing r molecules of monomer and M_{r+n} represents an inactive polymer of $r + n$ molecules of monomer. The polymerization process might then be described by the following reactions, in which M is the monomer feed and P_1 an activated form of the monomer:

Initiation

$$M \rightarrow P_1$$

Propagation

$$M + P_1 \rightarrow P_2$$

$$M + P_2 \rightarrow P_3$$

$$\dots\dots\dots\dots\dots$$

$$M + P_{r-1} \rightarrow P_r$$

Termination

$$P_r + P_n \rightarrow M_{n+r}$$

$$P_{r-1} + P_n \rightarrow M_{r-1+n}$$

$$\dots\dots\dots\dots\dots\dots$$

The kinetics problems of interest in chain reactions, as in all complex systems, are to predict the conversion and product distribution as a function of time from the rate equations for the individual reactions, or to decide on the reactions involved and evaluate their rate constants from experimental data on the conversion and product distribution. The methods are illustrated with a photochemical reaction in Example 2-3. Polymerization will be discussed in more detail in Chap. 4, after we have considered the nature of reactor performance.

Example 2-3 Experimental measurements for the photochlorination of propane at 25°C and 1 atm pressure showed that the rate of consumption of Cl_2 was independent of propane, second order in chlorine, and first order in light intensity.† If the controlling termination step is the heterogeneous termination of propyl radicals, show that the description of elementary steps given earlier for the chlorination of propane satisfactorily explains the experimental data.

SOLUTION If h is Planck's constant and v is the frequency of radiation, the initiation reaction for a photochemical activation may be written as

$$Cl_2 + hv \rightarrow 2Cl$$

The rate of production of Cl atoms depends on the volumetric rate of adsorption of radiant energy, Ia, according to the expression

$$r_i = 2\phi_i Ia \tag{A}$$

where ϕ_i is the quantum yield of the initiation step.‡ The rate of energy absorption is equal to the intensity of radiation, I, multiplied by the absorptivity α and the chlorine concentration, so that Eq. (A) becomes

$$r_i = 2\phi_i \alpha C_{Cl_2} I = 2k_1 C_{Cl_2} I$$

where the equivalent rate constant for the initiation step is $k_1 = \phi_i \alpha$. Suppose that the concentrations of intermediates rapidly attain constant, low values; that is, the stationary-state hypothesis is valid:

$$\frac{dC_{Pr}}{dt} = \frac{dC_{Cl}}{dt} = 0 \tag{B}$$

We can write two independent equations from Eq. (B),§ one for Pr and the other for Cl, using the elementary steps given on page 58.

$$\frac{dC_{Pr}}{dt} = 0 = k_2 C_{Cl} C_{PrH} - k_3 C_{Pr} C_{Cl_2} - k_7 C_{Pr}$$

$$\frac{dC_{Cl}}{dt} = 0 = 2\phi_i \alpha C_{Cl_2} I - k_2 C_{Cl} C_{PrH} + k_3 C_{Pr} C_{Cl_2}$$

Adding these two expressions gives C_{Pr} in terms of the concentration of stable species and the rate constants,

$$C_{Pr} = \frac{2\phi_i \alpha C_{Cl_2} I}{k_7} \tag{C}$$

The rate of consumption of Cl_2 or production of propyl chloride is obtained from the second of the propagation reactions.¶ Thus

$$-\frac{dC_{Cl_2}}{dt} = k_3 C_{Pr} C_{Cl_2} \tag{D}$$

---

† A. E. Cassano and J. M. Smith, *AIChE J.*, **12**, 1124 (1966).

‡ The quantum yield is the molecules of product produced per quantum of energy absorbed. For a primary step, ϕ approaches unity. Note that the coefficient 2 in Eq. (A) takes account of the production of two atoms of chlorine for each molecule of Cl_2.

§ In writing these expressions only the one termination step involving k_7 is needed, since it is assumed to control the termination process.

¶ Since the propagation steps are rapid with respect to initiation, the chlorine consumed by the initiation reaction can be neglected.

Using Eq. (C) in Eq. (D) gives

$$-\frac{dC_{Cl_2}}{dt} = \frac{2k_3\phi_i}{k_7}\alpha C_{Cl_2}^2 I = \frac{2k_3 k_1}{k_7}C_{Cl_2}^2 I \qquad (E)$$

This is the desired result, giving the rate as second order in chlorine and first order in light intensity. By choosing different termination steps to be controlling, we can obtain different expressions for dC_{Cl_2}/dt, some involving C_{Pr}, C_{Cl_2}, and I to other powers. Comparison of the various results with experimentally determined rates allows us to choose the best form of the rate equation and to evaluate the ratio of rate constants; for example, $k_3 k_1/k_7$ in Eq. (E). From measurements only on stable species, individual values of k cannot, in general, be established.

No mention has been made of the effects of wavelength, measurement of light intensity, and other factors involved in photochemical studies. These questions and others are discussed in the literature.†

EVALUATION OF RATE EQUATIONS FROM LABORATORY DATA

The chemical engineer needs a numerical equation for the intrinsic rate in order to design a commercial-scale reactor. Referring to Eq. (2-17) he will need to know the form of the concentration function, including numerical values for the constants that occur in the function, and a value for the activation energy.‡ In Sec. 2-5 the method of evaluating E was considered so we need only be concerned with isothermal systems here. The information available normally will be laboratory-scale data for concentration as a function of time. From this, an expression for the form of Eq. (2-9) which best fits the data needs to be determined. This often requires a trial-and-error procedure, comparing various proposed rate equations with the data. Since the reaction is most likely to be a combination of several elementary processes, auxiliary knowledge can be very useful in the first step of postulating rate expressions. Such information would include identification of intermediates and knowledge of rate equations found to fit data for other examples of the same type of reaction. The treatment in the preceding sections provides a basis for postulating the form of the rate equation from the mechanism. However, the problem of finding an appropriate concentration function is different, and much simpler, than establishing a mechanism for the reaction. All the chemical engineer requires is an equation for the rate that will be accurate over the range of conditions expected in the commercial-scale reactor. While it is useful to know the mechanism, it is not a requirement for reactor design. Conversely, a satisfactory rate equation does not usually provide enough information to establish the reaction mechanism.

---

† J. G. Calvert and J. N. Pitts, Jr., "Photochemistry," John Wiley & Sons, Inc., New York, 1966.

‡ Equation (2-17) applies to a single, irreversible reaction. Complex and reversible reactions require similar information. Some examples of these systems are included in Secs. 2-10 and 2-11.

In the following discussion, it is supposed that the kinetics data are reproducible and accurate. A variety of laboratory procedures for following the course of a reaction are available. Hill[†] has summarized many of these procedures and also listed requirements for obtaining suitable data. Remember that in all cases in this chapter the treatment is restricted to *homogeneous batch reactions at constant volume*.

The comparison of experimental kinetics with proposed rate equations can be done in two ways:

1. The *integration method*, which consists of comparing observed and predicted concentrations as a function of time. To use this method the rate equation must be integrated to provide a predicted C_i-vs.-t relation. Such integrations are summarized for irreversible, reversible, and complex reactions in the sections that follow.
2. The *differential method* requires differentiation of the C_i-vs.-t experimental data to obtain a rate derived from experiment. This rate is then compared with that obtained from the proposed rate equation.

Both methods are shown in some of the examples (Examples 2-4 to 2-6) that follow. Also, some of the experimental techniques (e.g., *initial-rate* and *half-life* methods) used in kinetics studies are explained. The integration of simple rate equations, in preparation for using the integration method, is considered next.

2-9 Concentration-vs.-Time Equations for Single, Irreversible Reactions

Zero Order Zero order means that the rate is independent of the concentration. This may occur in two situations: when the rate is intrinsically independent of concentration and when the species is in such abundant supply that its concentration is nearly constant during reaction. In the latter case the dependency of the rate on concentration cannot be detected, and apparent zero order prevails. Thus in the oxidation of NO to NO_2 in the presence of a large excess of O_2, the rate is zero order in O_2.

For a zero-order reaction at constant density Eq. (2-9) becomes

$$-\frac{dC_A}{dt} = k_0 \tag{2-41}$$

Integrating from an initial condition of $C_A = C_{A0}$ yields

$$C_A - C_{A0} = -k_0 t \tag{2-42}$$

The distinguishing feature of a zero-order reaction is that the concentration of reactant decreases linearly with time. It is difficult to cite a homogeneous reaction that is intrinsically zero order, although many reactions have apparent zero-order characteristics when the concentration of the species is large. However, in some

[†] C. G. Hill, Jr., "An Introduction to Chemical Engineering Kinetics and Reactor Design," John Wiley and Sons, New York, 1977.

heterogeneous reactions where the solid phase acts as a catalyst the rate is zero order. An example is the decomposition of NH_3 on platinum and tungsten surfaces.†

Equation (2-42) can be used with measurements of concentration vs. time to determine if a reaction is zero order and to evaluate k. If two reactants, A and B, are involved, experiments can be carried out with A in large excess, so that the rate equation is independent of C_A. Then the concentration of B can be varied and its order determined. This *isolation* technique is useful for evaluating the effect of one reactant in the presence of another.

It may be simpler to measure the time when a certain fraction of reactant has disappeared than to obtain concentration-vs.-time data. Common practice is to obtain the time required for one-half of the reactant to disappear. Defining this half-life as $t_{1/2}$, we have from Eq. (2-42)

$$\tfrac{1}{2}C_{A0} - C_{A0} = k_0 \, t_{1/2}$$

or

$$t_{1/2} = \frac{C_{A0}}{2k_0} \tag{2-43}$$

Half-life data can be used with Eq. (2-43) to evaluate k_0, as an alternate to using Eq. (2-42).

First Order Equation (2-9) for a first-order rate is

$$-\frac{dC_A}{dt} = k_1 C_A \tag{2-44}$$

If the initial condition is $C_A = C_{A0}$, integration yields

$$-\ln \frac{C_A}{(C_A)_0} = k_1 t \tag{2-45}$$

This result shows that a linear relation between $\ln C_A/(C_A)_0$ and t suggests a first-order reaction. The half-life is given by

$$k_1 t_{1/2} = -\ln \tfrac{1}{2} \tag{2-46}$$

or

$$t_{1/2} = \frac{1}{k_1} \ln 2 \tag{2-47}$$

Equations (2-45) and (2-47) show that the half-life and fraction of reactant remaining are independent of initial concentration for a first-order reaction.

Among the numerous examples of homogeneous, first-order reactions are the rearrangement of cyclopropane to propylene,‡ certain cis-trans isomerizations, and the inversion of sucrose.

---

† C. N. Hinshelwood and R. E. Burk, *J. Chem. Soc.*, **127**, 1051, 1114 (1925).
‡ T. S. Chambers and G. B. Kistiokowsky, *J. Am. Chem. Soc.*, **56**, 399 (1934).

Second Order Two types of second-order reactions are of interest:

Type I: $A + A \rightarrow P$

$$-\frac{dC_A}{dt} = k_2 C_A^2 \tag{2-48}$$

Type II: $A + B \rightarrow P$

$$-\frac{dC_A}{dt} = k_2 C_A C_B \tag{2-49}$$

For type I reactions, integration of Eq. (2-48) yields

$$\frac{1}{C_A} - \frac{1}{(C_A)_0} = k_2 t \tag{2-50}$$

In terms of the half-life, this becomes

$$t_{1/2} = \frac{1}{k_2 (C_A)_0} \tag{2-51}$$

Note that for a zero-order reaction $t_{1/2}$ is directly proportional to $(C_A)_0$; for a first-order reaction it is independent of $(C_A)_0$, and for a second-order reaction inversely proportional to $(C_A)_0$. Two well-known examples of type I reactions are the decomposition of HI in the gas phase and dimerization of cyclopentadiene in either gas or liquid phase.

In considering type II, let us examine first the case where order and stoichiometry are not the same. That is, the rate is second order but **a** moles of A and **b** moles of B (a and $b \neq 1$) react according to the equation

$$aA + bB \rightarrow \text{products}$$

Suppose that at zero time the number of moles of A and B are $(n_A)_0$ and $(n_B)_0$. Introducing the extent of reaction ξ from Eq. (2-7), the number of moles at any time is given by

$$n_A = (n_A)_0 - a\xi$$

and

$$n_B = (n_B)_0 - b\xi$$

Since in this chapter we consider the concentration to change only because the number of moles changes (that is, the volume is constant), the previous equations may be written in terms of concentrations as

$$C_A = (C_A)_0 - \frac{a}{V}\xi; \qquad C_B = (C_B)_0 - \frac{b}{V}\xi \tag{A}$$

$$\frac{dC_A}{dt} = -\frac{a}{V}\frac{d\xi}{dt} \tag{B}$$

Substituting these expressions in Eq. (2-49) gives

$$\frac{a}{V}\frac{d\xi}{dt} = k_2 \left[(C_A)_0 - \frac{a}{V}\xi \right]\left[(C_B)_0 - \frac{b}{V}\xi \right] \tag{C}$$

Integrating from $\xi = 0$ at $t = 0$, and replacing ξ with C_A and C_B, through Eqs. (A) and (B), gives

$$\left[\frac{1}{(C_B)_0 - \frac{b}{a}(C_A)_0}\right] \ln \frac{C_B}{C_A} - \ln \frac{(C_B)_0}{(C_A)_0} = k_2 t \tag{2-52}$$

Since C_B is related to C_A through Eqs. (A) and (B), Eq. (2-52) can be expressed in terms of C_A alone. The result is

$$\frac{1}{(C_B)_0 - \frac{b}{a}(C_A)_0} \left\{ \ln \left\{ \frac{1}{C_A}\left[(C_B)_0 - \frac{b}{a}\{(C_A)_0 - C_A\} \right] \right\} - \ln \frac{(C_B)_0}{(C_A)_0} \right\} = k_2 t \tag{2-53}$$

If the initial concentrations are in stoichiometric proportions, $(C_B)_0 = (b/a)(C_A)_0$, Eq. (2-49) becomes

$$-\frac{dC_A}{dt} = \frac{b}{a} k_2 \left[(C_A)_0 - \frac{a}{V}\xi\right]^2 = \frac{b}{a} k_2 C_A^2 \tag{2-54}$$

This rate expression is the same as Eq. (2-48) for type I, except that the rate constant is multiplied by b/a. By analogy to Eq. (2-50), the solution is

$$\frac{1}{C_A} - \frac{1}{(C_A)_0} = \frac{b}{a} k_2 t \tag{2-55}$$

If $(C_B)_0 = (C_A)_0$ and $a = b = 1$, a type II second-order reaction is identical to type I and the solution is given by Eq. (2-50).

Concentration-time data can easily be analyzed to test second-order kinetics. For type I (or type II with stoichiometric proportions of A and B initially), Eq. (2-50) indicates that the data should give a straight line if $1/C_A$ is plotted against t. For type II, Eq. (2-52) shows that a plot of $\log C_B/C_A$ vs. t should be linear. In this case the slope $[C_{B_0} - (b/a)C_{A_0}]k_2$ will be positive or negative, depending on the stoichiometric coefficients a and b and the initial concentrations.

In the following example first- and second-order rate equations are used to interpret data for an isothermal, constant-density, liquid system.

Equations (2-42), (2-45), (2-50), and (2-53) all are of the form $f(C_i) = kt$. Hence, as Hill[†] has pointed out, a plot of $f(C_i)$ vs. t should be a straight line with a slope equal to k. With experimental C_i-vs.-t data, the correct order would be determined by which function of C_i best fit the linear requirement.

Example 2-4 The liquid-phase reaction between trimethylamine and n-propyl bromide was studied by Winkler and Hinshelwood[‡] by immersing sealed glass tubes containing the reactants in a constant-temperature bath. The results at 139.4°C are shown in Table 2-2. Initial solutions of trimethylamine and n-propyl bromide in benzene, 0.2-molal, were mixed, sealed in glass tubes,

† Hill, op. cit.
‡ C. A. Winkler and C. N. Hinshelwood, *J. Chem. Soc.*, 1147 (1935).

Table 2-2

Run	t, min	Conversion, %
1	13	11.2
2	34	25.7
3	59	36.7
4	120	55.2

and placed in the constant-temperature bath. After various time intervals the tubes were removed and cooled to stop the reaction, and the contents were analyzed. The analysis depends on the fact that the product, a quaternary ammonium salt, is completely ionized. Hence the concentration of bromide ions can be estimated by titration.

From this information determine the first-order and second-order specific rates, k_1 and k_2, assuming that the reaction is irreversible over the conversion range covered by the data. Use both the integration and the differential method, and compare the results. Which rate equation best fits the experimental data?

SOLUTION The reaction may be written

$$N(CH_3)_3 + CH_3CH_2CH_2Br \rightarrow (CH_3)_3(CH_2CH_2CH_3)N^+ + Br^-$$

For this constant-volume system rate equations (2-44) and (2-49) are applicable for the first- and second-order possibilities. If T denotes trimethylamine and P n-propyl bromide, these rate expressions are

$$r = -\frac{dC_T}{dt} = k_1 C_T \tag{A}$$

$$r = -\frac{dC_T}{dt} = k_2 C_T C_p \tag{B}$$

Integration Method For the first-order case, the integrated form of Eq. (A) is Eq. (2-45); that is,

$$-\ln\frac{C_T}{(C_T)_0} = k_1 t \tag{C}$$

In the second-order case the stoichiometric coefficients are equal, and $(C_T)_0 = (C_p)_0 = 0.1$ molal. Hence Eq. (B) reduces to a type 1 second-order equation

$$-\frac{dC_T}{dt} = k_2 C_T^2 \tag{D}$$

The integrated solution is Eq. (2-50)

$$\left(\frac{1}{C_T} - \frac{1}{(C_T)_0}\right) = k_2 t \tag{E}$$

The conversion x is the fraction of the reactant that has been consumed. For constant volume,

$$x = \frac{(C_T)_0 - C_T}{C_{T_0}}$$

or

$$C_T = C_{T_0}(1 - x) \tag{F}$$

The calculation of k_1 and k_2 will be illustrated for the first run. From Eq. (F),

$$C_T = (C_T)_0(1 - 0.112) = 0.1(0.888)$$

Substituting in Eq. (C),

$$k_1 = \frac{1}{t} \ln \frac{(C_T)_0}{C_T} = \frac{1}{13(60)} \ln \frac{0.1}{0.0888} = 1.54 \times 10^{-4} \text{ s}^{-1}$$

For the second-order possibility, from Eq. (E)

$$k_2 = \frac{1}{t(C_T)_0}\left(\frac{1}{1-x} - 1\right) = \frac{0.112}{60(13)(0.1)(1 - 0.112)}$$

$$= 1.63 \times 10^{-3} \text{ liter/(g mol)(s) or } 1.63 \times 10^{-3} \text{ m}^3/\text{(kg mol)(s)}$$

Table 2-3 shows the results obtained in a similar way for the four runs. The k_1 values show a definite trend with time, and therefore the first-order mechanism does not satisfactorily explain the kinetic data. The k_2 values not only are more nearly identical, but the variations show no definite trend.

The graphical procedure for applying the integration method is to plot the function of C_T given by Eqs. (C) and (E), vs. t. The concentration C_T for each time value is obtained from Eq. (F). The resulting graphs, shown in Fig. 2-3, confirm that the second-order equation is preferred. Not only do the data conform more closely to the required linear relationship, but there is less scatter. The slope of the curve establishes the value of k_2. From the second-order line the slope is

$$\text{Slope} = k_2 = 0.102 \text{ liter/(g mol)(min)}$$

$$= 1.69 \times 10^{-3} \text{ liter/(g mol)(s) or m}^3/\text{(kg mol)(s)}$$

Table 2-3 Specific reaction rates for trimethylamine and n-propyl bromide reaction

Run	t, s	$k_1 \times 10^4$, s^{-1}	$k_2 \times 10^3$, liter/(g mol)(s) or m^3/(kg mol)(s)	C_B, g mol/liter or kg mol/m^3
1	780	1.54	1.63	0.0112
2	2,040	1.46	1.70	0.0257
3	3,540	1.30	1.64	0.0367
4	7,200	1.12	1.71	0.0552
			(1.67 av)	

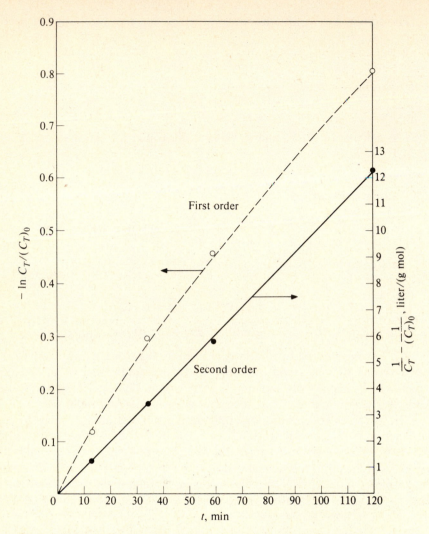

Figure 2-3 Graph of integration-method results for trimethylamine reaction.

Differential Method The moles of bromide ions B produced are equal to the moles of trimethylamine reacted. Hence, for this constant volume reaction,

$$C_B = (C_T)_0 - C_T$$

then C_B can be calculated from the conversion data using Eq. (F); that is,

$$C_B = x(C_T)_0$$

A plot of C_B vs. time of reaction is shown in Fig. 2-4. The slope of this curve at any point is equal to the rate of reaction, since

$$\mathbf{r} = -\frac{dC_T}{dt} = \frac{dC_B}{dt}$$

Slopes determined from the curve are given in Table 2-4.

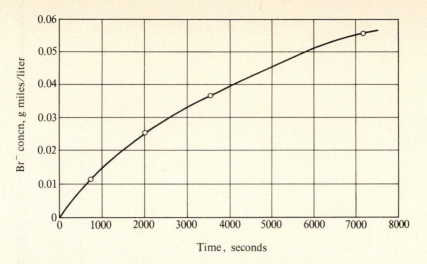

Figure 2-4 Concentration vs. time for reaction between $(CH_3)_3N$ and $CH_3CH_2CH_2Br$.

A convenient graphical procedure for determining the reaction order by the differential method is to plot the logarithmic form of the rate equation. For the first-order case the logarithmic form of Eq. (A) is

$$\log r = \log k_1 + \log C_T \qquad \text{(G)}$$

Similarly, if the reaction is second order, from Eq. (D) we have

$$\log r = \log k_2 + \log C_T^2 = \log k_2 + 2 \log C_T \qquad \text{(H)}$$

For the first-order case $\log r$ plotted against $\log C_T$ should yield a straight line with a slope of 1.0. For the second-order case the result should be a straight line of slope of 2.0, in accordance with Eq. (H). A plot of the data in Table 2-4 is shown in Fig. 2-5. While there is some scattering, the points do suggest a straight line of a slope approximately equal to 2.0. For comparison, lines with slopes of both 2.0 and 1.0 have been included on the plot. The equation of the solid line (slope 2.0) is

$$\log r = -2.76 + 2.0 \log C_T$$

Table 2-4

Concentration, g mol/liter		$r = dC_B/dt$
C_B	C_T	g mol/(liter)(s)
0.0	0.10	1.58×10^{-5}
0.01	0.09	1.38×10^{-5}
0.02	0.08	1.14×10^{-5}
0.03	0.07	0.79×10^{-5}
0.04	0.06	0.64×10^{-5}
0.05	0.05	0.45×10^{-5}

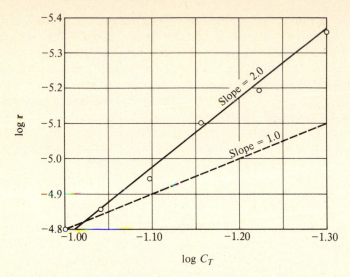

Figure 2-5 Rate vs. concentration of trimethylamine.

By comparison with Eq. (H),

$$\log k_2 = -2.76$$

$$k_2 = 1.73 \times 10^{-3} \text{ liter/(g mol)(s)}$$

This value agrees well with the result, 1.67×10^{-3}, obtained by the integration method.

Both methods show that the second-order mechanism is preferable. However, the failure of the first-order assumption is perhaps more clearly shown by the differential method than the integration approach. The data in Fig. 2-5 do not approach a slope of 1.0 at all closely, but the k_1 values in Table 2-3 are of the same magnitude, differing from an average value by not more than 17%. This is because the integration process tends to mask small variations.

Third-order reactions are uncommon. Fractional orders and rate equations with concentration-dependent terms in the denominator may occur when the reaction represents a sequence of several elementary steps. Procedures for establishing the order and rate constants for these cases are similar to those already considered. Experimental data that suggest fractional-order rate equations should be examined carefully for effects of physical resistances. Sometimes these effects, rather than a sequence of elementary processes, can be responsible for the fractional order. An example is the study of the hydrochlorination of lauryl alcohol with zinc chloride as a homogeneous catalyst:†

$$CH_3(CH_2)_{10}CH_2OH(l) + HCl(g) \rightarrow CH_3(CH_2)_{10}CH_2Cl(l) + H_2O(l)$$

† H. A. Kingsley and H. Bliss, *Ind. Eng. Chem.*, **44**, 2479 (1952).

This homogeneous, constant volume reaction was carried out by dissolving gaseous HCl in a stirred vessel containing the alcohol. The resulting concentration-time data could be correlated with a rate equation half-order in alcohol concentration. However, the rate constant was found to vary with the gas (HCl) flow rate into the reactor, suggesting that the observed rate was influenced by the resistance to diffusion of dissolved HCl in the liquid phase. A method of analysis which took into account the diffusion resistance indicated that the chemical step was probably first order in dissolved HCl and zero order with respect to lauryl alcohol. Such combinations of intrinsic kinetics and mass transfer are considered when heterogeneous reactions are discussed (Chap. 10).

2-10 Concentration-vs.-Time Equations for Reversible Reactions

For an *elementary* process the ratio† of the forward- and reverse-rate constants is equal to the equilibrium constant, for example, Eq. (2-15). Hence, the net rate of reaction can be expressed in terms of the forward rate constant k and the equilibrium constant. Then the integrated form of this rate expression can be used with concentration-time data to evaluate k, just as for irreversible reactions. The procedure is illustrated for first- and second-order rate equations.

First Order If k_1 and k_1' are the forward- and reverse-rate constants for the elementary process

$$A \rightleftarrows B$$

then

$$-\frac{dC_A}{dt} = k_1 C_A - k_1' C_B \qquad (2\text{-}56)$$

The concentration of B can be expressed in terms of C_A by a mass balance. At constant volume, and since the number of moles is constant, the concentration of B is its initial value $(C_B)_0$ plus the concentration of A that has reacted; that is,

$$C_B = (C_B)_0 + C_{A_0} - C_A \qquad (2\text{-}57)$$

Combining this result with Eq. (2-56) gives

$$-\frac{dC_A}{dt} = (k_1 + k_1')C_A - k_1'[(C_A)_0 + (C_B)_0] \qquad (2\text{-}58)$$

At equilibrium the forward- and reverse-reaction rates are equal. Eq. (2-56) becomes

$$k_1(C_A)_{eq} = k_1'(C_B)_{eq}$$

or

$$\frac{(C_B)_{eq}}{(C_A)_{eq}} = K = \frac{k_1}{k_1'} \qquad (2\text{-}59)$$

† See Sec. 2-7 for a discussion of this ratio for nonelementary reactions.

where K is the equilibrium constant. Eliminating k'_1 from Eq. (2-58) by using Eq. (2-59) yields

$$-\frac{dC_A}{dt} = k_1 \left\{ \frac{K+1}{K} C_A - \frac{1}{K} [(C_A)_0 + (C_B)_0] \right\} \qquad (2\text{-}60)$$

Now, applying Eq. (2-57) at equilibrium conditions to find $(C_B)_{eq}$ and substituting this result in Eq. (2-59), we have

$$K = \frac{(C_B)_0 + (C_A)_0 - (C_A)_{eq}}{(C_A)_{eq}}$$

or

$$(C_B)_0 + (C_A)_0 = (C_A)_{eq}(K+1) \qquad (2\text{-}61)$$

With this value for $(C_B)_0 + (C_A)_0$ we can express Eq. (2-60) in terms of $C_A - (C_A)_{eq}$

$$-\frac{dC_A}{dt} = k_1 \frac{K+1}{K} [C_A - (C_A)_{eq}]$$

or

$$-\frac{dC'_A}{dt} = k_R C'_A \qquad (2\text{-}62)$$

where

$$k_R = \frac{k_1(K+1)}{K} \qquad (2\text{-}63)$$

and

$$C'_A = C_A - (C_A)_{eq} \qquad (2\text{-}64)$$

Equation (2-62) is similar to Eq. (2-44) for an irreversible reaction, so the integrated form is analogous to Eq. (2-45); that is,

$$-\ln \frac{C_A - (C_A)_{eq}}{(C_A)_0 - (C_A)_{eq}} = k_R t \qquad (2\text{-}65)$$

If experimental concentration-time data are plotted as the left-hand side of Eq. (2-65) vs. t, the result is a straight line with a slope of k_R. If the equilibrium constant is known, k_1 can be evaluated from Eq. (2-63). Note that $(C_A)_{eq}$ is determined solely by K and the initial concentrations by means of Eq. (2-61).

Illustrations of first-order reversible reactions are gas-phase cis-trans isomerizations, isomerizations in various types of hydrocarbon systems, and the racemization of α and β glucoses. An example of a catalytic reaction is the ortho-para hydrogen conversion on a nickel catalyst. This reaction is used to illustrate other forms of Eq. (2-62) in the following example.

Example 2-5 The ortho-para hydrogen reaction has been studied at $-196°C$ and constant pressure in a flow reactor, with a nickel-on-Al_2O_3 catalyst.† The rate data can be explained with an expression of the form

$$r = k(y_{eq} - y)_p \tag{A}$$

where y_p is the mole fraction of para hydrogen. Show that this expression follows from Eq. (2-62).

SOLUTION The reaction is

$$O - H_2 \rightleftarrows p - H_2$$

so that C_A in Eq. (2-62) refers to the concentration of ortho hydrogen. Assuming hydrogen is an ideal gas at reaction conditions and that P is the total pressure, we have

$$C_A = \frac{P}{R_g T} y_0 = \frac{P}{R_g T}(1 - y_p) \tag{B}$$

also

$$(C_A)_{eq} = \frac{P}{R_g T}(1 - y_{eq})_p$$

Substituting these results in Eq. (2-62), using (2-63) and (2-64), yields

$$-\frac{dC_A}{dt} = r = \frac{k_1(K + 1)}{K} \frac{P}{R_g T}(y_{eq} - y)_p \tag{C}$$

Equation (C) is equivalent to Eq. (A), with

$$k = \frac{k_1(K + 1)}{K} \frac{P}{R_g T} \tag{D}$$

Second Order Consider an elementary, second-order reaction with $v_i = 1$, that is

$$A + B \rightleftarrows C + D$$

The rate equation is

$$-\frac{dC_A}{dt} = k_2 C_A C_B - k_2' C_c C_D = k_2 \left[C_A C_B - \frac{1}{K} C_c C_D \right] \tag{2-66}$$

where K is the equilibrium constant. To express all the concentrations in terms of one variable we could use the *extent of reaction* ξ, as was done in Sec. 2-9. It is simpler for a constant-volume system to use a concentration as the variable.

---

† N. Wakao, P. W. Selwood, and J. M. Smith, *AIChE J.*, **8**, 478 (1962).

Suppose initially only A and B are present at concentrations $(C_A)_0$ and $(C_B)_0$. Choose as the variable C_c. The other concentrations, in terms of C_c, are

$$C_A = (C_A)_0 - C_c$$

$$C_B = (C_B)_0 - C_c$$

$$C_D = C_c$$

Substituting these concentration equations in Eq. (2-66) gives

$$\frac{1}{k_2}\frac{dC_c}{dt} = [(C_A)_0 - C_c][(C_B)_0 - C_c] - \frac{1}{K}C_c^2 \qquad (2\text{-}67)$$

This may be written

$$\frac{1}{k_2}\frac{dC_c}{dt} = \alpha + \beta C_c + \gamma C_c^2 \qquad (2\text{-}68)$$

where

$$\left.\begin{aligned} \alpha &= (C_A)_0(C_B)_0 \\[4pt] \beta &= -[(C_A)_0 + (C_B)_0] \\[4pt] \gamma &= 1 - \frac{1}{K} \end{aligned}\right\} \qquad (2\text{-}69)$$

The solution of Eq. (2-68) with $C_c = 0$ at $t = 0$, is

$$\frac{1}{q^{1/2}}\ln\frac{2\gamma C_c/(\beta - q^{1/2}) + 1}{2\gamma C_c/(\beta + q^{1/2}) + 1} = k_2 t \qquad (2\text{-}70)$$

where,

$$q = \beta^2 - 4\alpha\gamma \qquad (2\text{-}71)$$

If the initial concentrations and the equilibrium constant are known, α, β, and γ can be evaluated. Then experimental concentration vs. time data, for example for reactant A [note $C_c = (C_A)_0 - C_A$], are sufficient to plot the left-hand side of Eq. (2-70) vs. t. If the second-order rate equation is satisfactory, a straight line will be obtained with a slope equal to k_2. For other forms of rate equations and different initial conditions the integration procedure can be more difficult. Alternately, if stoichiometric initial concentrations apply for second-order kinetics, Eqs. (2-70) and (2-71) are simplified. Example 2-6 illustrates this case. In this example both the integration method with Eqs. (2-70) and (2-71) and the differential method are used to analyze the experimental data.

The development for reversible reactions has been for elementary steps, since the relation $K = k/k'$ has been used to eliminate the rate constant in the reverse direction. For nonelementary reactions this relation is valid only if the experimental data are such that near equilibrium exists for all the individual steps (see Sec. 2-7).

Example 2-6† The reaction between methyl iodide and dimethyl-*p*-toluidine in nitrobenzene solution forms an ionized quaternary ammonium salt. It can be studied kinetically in the same manner as the trimethylamine reaction considered in Example 2-4. The data shown in Table 2-5 were obtained from an initial solution containing methyl iodide and dimethyl-*p*-toluidine in concentrations of 0.05 g mole/liter.

In view of the results for Example 2-4, and assuming that the equilibrium constant for this reaction is 1.43, what rate equation best fits the experimental data?

SOLUTION The reaction may be written as

$$CH_3I + N{-}R \rightleftharpoons CH_3{-}N^+{-}R + I^-$$

If it is supposed to be second order and reversible, Eq. (2-66) is applicable. Expressing the rate in terms of the iodide concentration, we have

$$\frac{dC_I}{dt} = k_2 C_M C_T - k_2' C_I C_N \tag{A}$$

The initial concentrations of reactants are equal and those of the products are zero. If C_I is taken as the iodide concentration and C_M the concentration of either reactant, Eq. (2-67) becomes

$$\frac{1}{k_2}\frac{dC_I}{dt} = [(C_M)_0 - C_I]^2 - \frac{1}{K}C_I^2 \tag{B}$$

Integration Method The solution of Eq. (B) is given by Eqs. (2-70) and (2-71), where $C_c = C_I$ and

$$\alpha = (C_M)_0^2$$

$$\beta = -2(C_M)_0$$

$$\gamma = \frac{K-1}{K}$$

$$q = 4(C_M)_0^2/K$$

† From K. J. Laidler, "Chemical Kinetics," McGraw-Hill Book Company, New York, 1950.

Table 2-5

Run	t, min	Fraction of toluidine reacted
1	10.2	0.175
2	26.5	0.343
3	36.0	0.402
4	78.0	0.523

Substituting these values in Eq. (2-70) and simplifying gives

$$\frac{2k_2(C_M)_0}{K^{1/2}} t = \ln \frac{(C_M)_0 + C_1(K^{-1/2} - 1)}{(C_M)_0 - C_1(K^{-1/2} + 1)} \tag{C}$$

The data are given as the fraction x of the toluidine (or, in this example, methyliodide) reacted. Hence $C_1 = x(C_M)_0$. Then Eq. (C) becomes

$$k_2 = \frac{K^{1/2}}{2(C_M)_0} \ln \frac{1 + x(K^{-1/2} - 1)}{1 - x(K^{-1/2} + 1)} \tag{D}$$

Substituting numerical values for K and $(C_M)_0$ gives the following expression relating x and t:

$$k_2 = \frac{1.43^{1/2}}{2(0.05)t} \ln \frac{1 - 0.165x}{1 - 1.835x}$$

This expression and the data in Table 2-5 can be used to calculate a value of k_2 for each run. The results are shown in Table 2-6.

As a matter of interest, the values of k_2, evaluated with the assumption that the reaction is irreversible, are shown in the last column of the table. They were computed from Eq. (2-50) written in terms of C_1

$$k_2 t = \frac{C_1}{(C_M)_0[(C_M)_0 - C_1]}$$

The steady trend in k_2 indicates that the irreversible assumption is a poor one.

In studies of rate data, trends in k are of more significance than random variations. The former suggest that the assumed order is open to question, while the latter suggest errors in the experimental data. Of course, if the precision of the data is poor, random variations can mask trends in computed k values.

Differential Method If the rate of reaction is taken as dC_1/dt, Eq. (B) can be written, in logarithmic form, as

$$\log \mathbf{r} = \log \frac{dC_1}{dt} = \log k_2 + \log \left\{ [(C_M)_0 - C_1]^2 - \frac{1}{K} C_1^2 \right\} \tag{E}$$

We can evaluate the rate by differentiating the C_1-vs.-t data, and then comparing the results with the second-order reversible form of Eq. (E). Introducing numerical values,

$$\log \mathbf{r} = \log k_2 + \log \{(0.05 - C_1)^2 - 0.70C_1^2\} \tag{F}$$

Table 2-6

Run	θ, s	k_2, liters/(g mol)(s)	k_2 (neglecting reverse reaction)
1	612	7.05×10^{-3}	6.93×10^{-3}
2	1,590	7.06×10^{-3}	6.57×10^{-3}
3	2,160	7.06×10^{-3}	6.23×10^{-3}
4	4,680	7.97×10^{-3}	4.68×10^{-3}

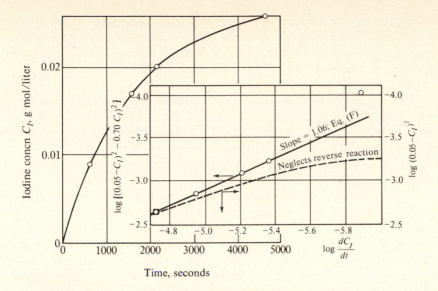

Figure 2-6

If the data follow second-order kinetics, Eq. (F) shows that a log plot of **r** vs. $[(0.05 - C_1)^2 - 0.70C_1^2]$ should give a straight line with a slope of unity.

Figure 2-6 shows a plot of C_1 vs. t. Slopes of this curve give the rate values shown in Table 2-7. Also shown is a plot of Eq. (F). Observe that the first four points establish a line with a slope close to unity, as required by the second-order mechanism. The last point deviates from the line, just as the k value for this point obtained by the integration method was not in agreement with the other values.

The complex equations for reversible reactions can be avoided if measurements are made before much reaction has occurred. Under these conditions the concentrations of the products will be small, making the reverse rate negligible. Then the data can be analyzed as though the system were irreversible to determine the forward-rate constant. With this result and the equilibrium constant (for an elementary reaction), the rate constant for the reverse reaction can be obtained. This *initial-rate* approach is frequently used to simplify kinetic studies. Besides the fact that the reverse reaction is eliminated,

Table 2-7

t, s	C_1 g mol/liter	$\dfrac{dC_1}{dt}$	$\log \dfrac{dC_1}{dt}$	$(0.05 - C_1)^2 - 0.70C_1^2$	$\log\{(0.05 - C_1)^2 - 0.70C_1^2\}$
0	0	1.93×10^{-5}	-4.71	25.0×10^{-4}	-2.60
612	0.00875	1.12×10^{-5}	-4.95	16.5×10^{-4}	-2.78
1,590	0.0171	0.62×10^{-5}	-5.20	8.75×10^{-4}	-3.05
2,160	0.0201	0.42×10^{-5}	-5.37	6.13×10^{-4}	-3.21
4,680	0.0261	0.13×10^{-5}	-5.89	0.96×10^{-4}	-4.01

the composition of the reaction system is usually known more precisely at the initial state than at subsequent times. This is because compositions at later times are generally evaluated from a limited experimental analysis plus assumptions that certain reactions have occurred.

For nonelementary reactions, rate constants from data at initial conditions may differ from k values obtained at later stages, as explained in Sec. 2-7. By comparing rate constants and reaction orders obtained at different extents of reaction, conclusions may be drawn about the reaction mechanism. When the rate equation determined from the concentration dependency at initial conditions is different from that determined at finite conversions, some change arising from the presence of intermediate, or stable products, is indicated. When the order of a reaction determined from data at significant conversions is less than the order at initial conditions, the rate is decreasing less rapidly with time than expected. This suggests that the products of the reaction somehow speed up the rate. Such systems are called *autocatalytic*. Conversely, when the order at significant conversions is greater than that established at initial conditions, the products of reaction must inhibit the reaction.†

Example 2-7 illustrates the initial-rate method of analysis.

Example 2-7 The interpretation of kinetic data for gaseous reactions is similar to that for liquid systems. The analysis for a reversible case is well illustrated by the vapor-phase decomposition of hydrogen iodide,

$$2HI \rightleftarrows H_2 + I_2$$

This reaction has been carefully studied by a number of investigators,‡ and it is generally considered one of the most certain examples of a second-order reaction, at least at low pressures. The equilibrium values of the fraction x of HI decomposed can be accurately represented by Bodenstein's equation,

$$x_{eq} = 0.1376 + 7.22 \times 10^{-5}t + 2.576 \times 10^{-7}t^2 \qquad t = °C$$

Kistiakowsky used a static experimental method to study the reaction. Pure hydrogen iodide was sealed in glass bulbs, immersed in a constant-temperature bath for various time intervals, and then removed and cooled, and the contents were analyzed for all three chemical species. The initial pressure (and hence the initial concentration) of HI and the size of the reaction bulb were varied over a wide range. The data obtained at an average temperature of 321.4°C are given in Table 2-8.

From this information estimate the specific reaction-rate constants [liter/(g mol)(s)] for the forward and reverse reactions, both of which may be taken as second order.

---

† M. Letort, *Bull. Soc. Chim. France*, **9**, 1 (1942).

‡ M. Bodenstein, *Z. Physik Chem.*, **13**, 56 (1894); *Z. Physik Chem.*, **22**, 1 (1897); *Z. Physik Chem.*, **29**, 295 (1898); H. A. Taylor, *J. Phys. Chem.*, **28**, 984 (1924); G. B. Kistiakowsky, *J. Am. Chem. Soc.*, **50**, 2315 (1928).

Table 2-8

Run	t, s	% of HI decomposed	Volume of reaction bulb, cm³	$(C_{HI})_0$, g mol/liter
1	82,800	0.826	51.38	0.02339
2	172,800	2.567	59.80	0.03838
3	180,000	3.286	51.38	0.04333
4	173,100	3.208	51.38	0.04474
5	81,000	2.942	7.899	0.1027
6	57,560	2.670	7.899	0.1126
7	61,320	4.499	7.899	0.1912
8	19,200	2.308	7.899	0.3115
9	18,000	2.202	7.899	0.3199
10	16,800	2.071	7.899	0.3279
11	17,400	2.342	7.899	0.3464
12	17,700	2.636	7.899	0.4075
13	18,000	2.587	7.899	0.4228
14	23,400	4.343	7.899	0.4736
15	6,000	2.224	3.28	0.9344
16	5,400	1.903	0.778	0.9381
17	8,160	3.326	0.781	1.138
19	5,400	2.741	0.713	1.231

SOLUTION The extent of reaction, noted by the percentage of HI decomposed, is always low. The series of runs constitute *initial rate* data, with each run corresponding to a different concentration of HI. Reasonably accurate results should be obtained by neglecting the reverse reaction. This will be checked by first including the small effect of the reverse reaction.

If the rate is followed by the concentration of iodine, Eq. (2-67) becomes

$$\frac{1}{k_2}\frac{dC_I}{dt} = [(C_{HI})_0 - 2C_{I_2}]^2 - \frac{1}{K}(C_{I_2})^2$$

If the rate equation is integrated, with the initial condition $C_{I_2} = 0$ at $t = 0$, the result is very similar to Eq. (C) of the previous example:

$$\frac{2k_2(C_{HI})_0}{K^{1/2}}t = \ln\frac{(C_{HI})_0 + C_{I_2}(K^{-1/2} - 2)}{(C_{HI})_0 - C_{I_2}(K^{-1/2} + 2)} \qquad (A)$$

The equilibrium constant is related to the concentrations at equilibrium by

$$K = \left(\frac{C_{I_2}C_{H_2}}{C_{HI}^2}\right)_{eq}$$

If the fraction of HI decomposed is x, then

$$C_{I_2} = C_{H_2} = 1/2(C_{HI})_0 x$$
$$C_{HI} = (C_{HI})_0(1 - x)$$

Hence

$$K = \frac{1}{4} \frac{x_{eq}^2}{(1 - x_{eq})^2}$$

From the expression given for x_{eq} at 321.4°C,

$$x_{eq} = 0.1376 + 7.221 \times 10^{-5}(321.4) + 2.576 \times 10^{-7}(321.4)^2 = 0.1873$$

Hence

$$K = \frac{1}{4} \frac{(0.1873)^2}{(1 - 0.1873)^2} = 0.0133$$

Substituting this value of K in Eq. (A) and introducing the fraction decomposed, we have

$$k_2 = \frac{1}{2(8.67)(C_{HI})_0 t} \ln \frac{(C_{HI})_0 + \frac{1}{2}(C_{HI})_0(8.67 - 2)x}{(C_{HI})_0 - \frac{1}{2}(C_{HI})_0(8.67 + 2)x}$$

or

$$k_2 = \frac{1}{2(8.67)(C_{HI})_0 t} \ln \frac{1 + 3.335x}{1 - 5.335x} \tag{B}$$

The experimental data for x can be used directly in Eq. (B) to compute values of the specific reaction rate k_2. However, another form of the expression is more useful when the x values are very low, as in this case (the maximum value of x is 0.04499 for run 7). Equation (B) may be written in the form

$$k_2 = \frac{1}{2(8.67)(C_{HI})_0 t} \ln \left(1 + \frac{8.67x}{1 - 5.335x}\right) \tag{C}$$

The use of Eq. (C) may be illustrated with run 1:

$$k_2 = \frac{1}{2(8.67)(0.02339)(82{,}800)} \ln \left(1 + \frac{8.67(0.00826)}{1 - 5.335(0.00826)}\right)$$

$$= 2.14 \times 10^{-6} \text{ liter/(g mol)(s)}$$

The results for the other runs are summarized in Table 2-9. The average value of k_2 is 1.99×10^{-6}. For the reverse reaction

$$k_2' = \frac{k_2}{K_1} = \frac{1.99 \times 10^{-6}}{0.0133} = 1.50 \times 10^{-4} \text{ liter/(g mol)(s) or m}^3/\text{(kg mol)(s)}$$

If the reverse reaction is neglected, the rate equation, in terms of C_{I_2} is

$$\frac{dC_{I_2}}{dt} = k_2 C_{HI}^2 = k_2[(C_{HI})_0 - 2C_{I_2}]^2$$

Table 2-9

Run	Conversion	$k_2 \times 10^6$, liter/(s)(g mol)
1	0.00826	2.14
2	0.02567	2.01
3	0.03286	2.20
4	0.03208	2.17
5	0.02942	1.92
6	0.02670	2.08
7	0.04499	2.04
8	0.02308	1.99
9	0.0202	1.80
10	0.02071	1.77
11	0.02342	2.00
12	0.02636	1.90
13	0.02587	1.75
14	0.04343	2.08
15	0.02224	2.05
16	0.01903	1.93
17	0.03326	1.87
19	0.02741	2.15

Integrating with $C_{I_2} = 0$ at $t = 0$ gives

$$k_2 = \frac{1}{2t}\left[\frac{1}{(C_{HI})_0 - 2C_{I_2}} - \frac{1}{(C_{HI})_0}\right]$$

or, in terms of conversion of HI,

$$k_2 = \frac{1}{2t(C_{HI})_0}\left(\frac{x}{1-x}\right) \tag{D}$$

Using the data for run 1:

$$k_2 = \frac{1}{2(82{,}800)(0.02339)}\left[\frac{0.00826}{1 - 0.00826}\right]$$

$$= 2.15 \times 10^{-6} \text{ liter/(g mol)(s)}$$

This is nearly the same k_2 as obtained from Eq. (C). For runs with higher conversions, the deviation would be somewhat larger, but in all cases the reverse reaction is not significant for this essentially initial-rate data.

The integrated forms of rate equations for simple reactions developed in Sections 2-9 and 2-10 are summarized in Table 2-10.

Table 2-10 Rate equations for simple reactions

Reaction	Order	Rate equation	Integrated forms
		Reversible reactions	
$A \rightarrow B$	Zero	$\dfrac{-dC_A}{dt} = k_0$	$C_A = (C_A)_0 - k_0 t$
			$t_{1/2} = \dfrac{(C_A)_0}{2k_0}$
		Irreversible reactions	
$A \rightarrow B$	First	$-\dfrac{dC_A}{dt} = k_1 C_A$	$\ln \dfrac{C_A}{(C_A)_0} = -k_1 t$
			$t_{1/2} = \dfrac{1}{k_1} \ln 2$
$A + A \rightarrow P$	Second, type I	$-\dfrac{dC_A}{dt} = k_2 C_A^2$	$\dfrac{1}{C_A} - \dfrac{1}{(C_A)_0} = k_2 t$
			$t_{1/2} = \dfrac{1}{k_2 (C_A)_0}$
$aA + bB \rightarrow P$	Second, type II	$-\dfrac{dC_A}{dt} = k_2 C_A C_B$	$\dfrac{1}{(C_B)_0 - (b/a)(C_A)_0}$ $\left\{ \ln\left[\dfrac{1}{C_A} \left\{ (C_B)_0 - \dfrac{b}{a}[(C_A)_0 - C_A] \right\} \right] \right.$ $\left. - \ln \dfrac{(C_B)_0}{(C_A)_0} \right\} = k_2 t$
		Reversible reactions	
$A \rightleftarrows B$	First $\rightleftarrows$ first	$-\dfrac{dC_A}{dt} = k_1 C_A$ $\qquad - k_1' C_B$	$\dfrac{C_A - (C_A)_{eq}}{(C_A)_0 - (C_A)_{eq}} = e^{-k_R t}$ $k_R = \dfrac{k_1(K + 1)}{K}$ $(C_A)_{eq} = \dfrac{(C_B)_0 + (C_A)_0}{K + 1}$
$A + B \rightleftarrows C + D$	Second $\rightleftarrows$ second	$-\dfrac{dC_A}{dt} = k_2 C_A C_B$ $\qquad - k_1 C_c C_D$	$k_2 q^{1/2} t = \ln \dfrac{\{2\gamma C_c/(\beta - q^{1/2})\} + 1}{\{2\gamma C_c/(\beta + q^{1/2})\} + 1}$ $C_c = (C_A)_0 - C_A](C_c)_0 = (C_D)_0 = 0$ $\alpha, \beta, \gamma,$ and q defined by Eq. (2-69)

ANALYSIS OF COMPLEX RATE EQUATIONS

In a *complex* reaction system, stable products (in contrast to unstable intermediates) are produced by more than one reaction. Some of the products may be more desirable than others. For example, in the air oxidation of ethylene the desired product is ethylene oxide, but complete oxidation to carbon dioxide and water also occurs. The more important performance factors may be the production rate of ethylene oxide and its purity in the reaction products, rather than the total amount of ethylene reacted. To characterize this performance two parameters are used: yield and selectivity. The *yield* of a specific product may be defined as the fraction of reactant converted to the specific product. The *point selectivity* is the ratio of the *rate* of production of one product to the rate for another product. With multiple products there is a separate selectivity based on each pair of products. The *overall*, or integrated, *selectivity* is the ratio of the *amount* of one product produced to the amount of another. Selectivity and yield are related to each other through the total conversion, i.e., the total fraction of reactant converted to all products.

As an illustration of these terms consider the parallel reaction system

$$A \overset{k_1}{\underset{k_2}{\rightarrow}} \begin{matrix} B \\ \searrow \\ C \end{matrix} \tag{2-72}$$

Suppose the total conversion of A is x_t, consisting of the conversion x_B of reactant A to B and the conversion x_C to C. The *yield* of B is simply x_B and that of C is x_C. The amount of a product produced is proportional to the yield. Hence the *overall selectivity* of B is the ratio of the yields of B and C,

$$S_o = \frac{x_B}{x_C} \tag{2-73}$$

If both reactions are first order and irreversible, the *point selectivity* is

$$S_p = \frac{dC_B/dt}{dC_C/dt} = \frac{k_1 C_A}{k_2 C_A} = \frac{k_1}{k_2} \tag{2-74}$$

Under the restriction of constant volume followed in this chapter, the amount of a product produced is proportional to its concentration. Thus the overall selectivity can also be written as

$$S_o = \frac{C_B}{C_C} \tag{2-75}$$

The simple form of Eq. (2-74) shows that selectivity and yield calculations can advantageously be carried out by dividing the rate for one reaction by that for another, eliminating time in the process. Since yield and selectivity are usually more important than total conversion for complex-reaction systems, this procedure will be emphasized in the following section. The possible combinations of simultaneous, parallel, and consecutive reactions are very large. A few irreversible first-order cases will be analyzed in Sec. 2-11. For more complex reaction orders the same procedure can be followed, but explicit solutions may not be possible for such cases.

2-11 First-order Complex Reactions

Consider first the parallel reactions described by Eq. (2-72). The rates of formation of the components are given by the following three equations:

$$-\frac{dC_A}{dt} = (k_1 + k_2)C_A \tag{2-76}$$

$$\frac{dC_B}{dt} = k_1 C_A \tag{2-77}$$

$$\frac{dC_c}{dt} = k_2 C_A \tag{2-78}$$

Our goal is to integrate these equations to establish C_B and C_c for any C_A. Then all yields and selectivities can be obtained from equations like (2-73). In this simple example Eq. (2-76) could be immediately integrated to give $C_A = f(t)$. This result used in Eqs. (2-77) and (2-78) allows these two equations to be integrated so that C_B and C_c are also known as a function of time. However, in more complicated cases (see Example 2-9) integrating with respect to time is not as easy. Therefore the solution for this simple case is illustrated by dividing Eqs. (2-77) and (2-78) by Eq. (2-76), to eliminate time:

$$\frac{dC_B}{dC_A} = -\frac{k_1}{k_1 + k_2} \tag{2-79}$$

$$\frac{dC_c}{dC_A} = -\frac{k_2}{k_1 + k_2} \tag{2-80}$$

If Eqs. (2-79) and (2-80) are integrated with the conditions that at $t = 0$, $C_A = (C_A)_0$ and $C_B = C_c = 0$, then the yields of B and C are

$$x_B = \frac{C_B}{(C_A)_0} = \frac{k_1}{k_1 + k_2}\left(1 - \frac{C_A}{(C_A)_0}\right) = \frac{k_1}{k_1 + k_2} x_t \tag{2-81}$$

$$x_c = \frac{C_c}{(C_A)_0} = \frac{k_2}{k_1 + k_2}\left(1 - \frac{C_A}{(C_A)_0}\right) = \frac{k_2}{k_1 + k_2} x_t \tag{2-82}$$

where x_t is the total conversion of A to B plus C.

From Eqs. (2-81) and (2-82), the overall selectivity of B is

$$S_o = \frac{x_B}{x_C} = \frac{k_1}{k_2} \tag{2-83}$$

The point selectivity, given by Eq. (2-74) for first-order parallel reactions, is also equal to k_1/k_2. Although point and overall selectivities are identical for this type of first-order system, the two selectivities differ for most complex reactions.

Next let us consider a set of *consecutive* reactions,

$$A \xrightarrow{k_1} B \xrightarrow{k_3} D$$

and take $C_B = C_D = 0$ and $C_A = (C_A)_0$ at $t = 0$. The rates are

$$\frac{dC_A}{dt} = -k_1 C_A \tag{2-84}$$

$$\frac{dC_B}{dt} = k_1 C_A - k_3 C_B \tag{2-85}$$

$$\frac{dC_D}{dt} = k_3 C_B \tag{2-86}$$

Dividing Eqs. (2-85) and (2-86) by Eq. (2-84) yields

$$\frac{dC_B}{dC_A} = -1 + \frac{k_3 C_B}{k_1 C_A} \tag{2-87}$$

$$\frac{dC_D}{dC_A} = -\frac{k_3 C_B}{k_1 C_A} \tag{2-88}$$

Since Eq. (2-87) is a linear first-order differential equation, it has an analytic solution. With the stated initial condition the result can be expressed in terms of the yield of B,

$$x_B = \frac{C_B}{(C_A)_0} = \frac{k_1}{k_1 - k_3} \left\{ \left(\frac{C_A}{(C_A)_0}\right)^{k_3/k_1} - \frac{C_A}{(C_A)_0} \right\} \tag{2-89}$$

This expression for C_B can be substituted in Eq. (2-88). Then integration of Eq. (2-88) gives

$$x_D = \frac{C_D}{(C_A)_0} = \frac{k_1}{k_1 - k_3} \left\{ 1 - \left(\frac{C_A}{(C_A)_0}\right)^{k_3/k_1} \right\} - \frac{k_3}{k_1 - k_3} \left(1 - \frac{C_A}{(C_A)_0} \right) \tag{2-90}$$

From these two expressions the overall selectivity x_B/x_D is seen to depend on the fraction unconverted, $C_A/(C_A)_0$, as well as on the rate constants. This means that the yield of B and the overall selectivity will vary with time. This is in contrast to the result for parallel reactions, Eq. (2-83).

Example 2-8 If $C_B = C_D = 0$, initially, for the consecutive reaction system described by Eqs. (2-84) to (2-86), what is the time at which the yield of B is a maximum? What is the maximum yield?

SOLUTION Equation (2-89) gives the yield of B in terms of $C_A/(C_A)_0$. This ratio can be expressed as a function of time by integrating Eq. (2-84). The result is

$$\frac{C_A}{(C_A)_0} = e^{-k_1 t} \tag{A}$$

Then Eq. (2-89) becomes

$$x_B = \frac{k_1}{k_1 - k_3} (e^{-k_3 t} - e^{-k_1 t}) \tag{B}$$

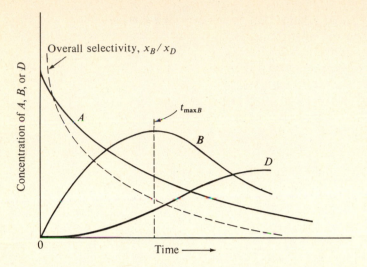

Figure 2-7 Concentration vs. time for consecutive-reaction system $A \rightarrow B \rightarrow D$ for $B = D = 0$ at $t = 0$.

To obtain the maximum value of x_B we differentiate Eq. (B) with respect to time and set the derivative equal to zero:

$$\frac{dx_B}{dt} = 0 = \frac{k_1}{k_1 - k_3}(-k_3 e^{-k_3 t} + k_1 e^{-k_1 t})$$

$$t_{\max B} = \frac{\ln (k_1/k_3)}{k_1 - k_3} \tag{C}$$

Substituting this value for the time in Eq. (B) gives the maximum yield,

$$(x_B)_{\max} = \left(\frac{k_1}{k_3}\right)^{k_3/(k_3 - k_1)} \tag{D}$$

The form of the concentration-vs.-time curves are shown in Fig. 2-7. The selectivity x_B/x_D, obtained by dividing Eq. (2-89) by (2-90) is very large at low time values (corresponding to $C_A/(C_A)_0 \rightarrow 1$) and decreases at larger times. The selectivity curve is shown dotted in Fig. 2-7.

Assemblies of first-order differential equations with time as the independent variable are ideally suited for solution by analog computation. Hence complex kinetics equations of the type considered in this section may conveniently be solved with an analog computer. This is illustrated in the problems at the end of the chapter.

Example 2-9 Benzene is chlorinated in the liquid phase in a kettle-type reactor operated on a semibatch basis; i.e., the reactor is initially charged with liquid benzene, and then chlorine gas is bubbled into the well-agitated solution. The reactor is equipped with a reflux condenser which will condense the benzene and chlorinated products but will not interfere with the removal of hydrogen chloride. Assume that the chlorine is added sufficiently slowly that (1) the chlorine and hydrogen chloride concentrations in the liquid phase are small and (2) all the chlorine reacts.

At the constant operating temperature of 55°C the significant reactions are the three substitution ones leading to mono-, di-, and trichlorobenzene.

$$1. \quad C_6H_6 + Cl_2 \xrightarrow{k_1} C_6H_5Cl + HCl$$

$$2. \quad C_6H_5Cl + Cl_2 \xrightarrow{k_2} C_6H_4Cl_2 + HCl$$

$$3. \quad C_6H_4Cl_2 + Cl_2 \xrightarrow{k_3} C_6H_3Cl_3 + HCl$$

MacMullin† found the ratios of the rate constants to have the following values at 55°C:

$$\frac{k_1}{k_2} = 8.0 \qquad \frac{k_2}{k_3} = 30$$

Find the yields of each product as a function of the moles of chlorine added per mole of benzene charged to the reactor.‡ The holdup in the reflux condenser is negligible.

SOLUTION§ The process described is neither flow nor batch, but semibatch in nature. However, with assumptions which are reasonably valid, the problem can be reduced to one of constant volume. If the density of the solution remains constant and the hydrogen chloride vaporizes and leaves the solution, the volume of the liquid-phase reaction will be constant. Then the restrictions of this chapter are met and Eq. (2-9) is the proper expression for the rate. Assume that the reactions are second-order. Then the rate of disappearance of benzene, determined entirely by the first reaction, is

$$-\frac{dC_B}{dt} = k_1 C_B C_{Cl_2} \tag{A}$$

Similarly the net rates of formation of the mono-, di-, and trichlorobenzenes (M, D, T) are

$$\frac{dC_M}{dt} = k_1 C_B C_{Cl_2} - k_2 C_M C_{Cl_2} \tag{B}$$

$$\frac{dC_D}{dt} = k_2 C_M C_{Cl_2} - k_3 C_D C_{Cl_2} \tag{C}$$

$$\frac{dC_T}{dt} = k_3 C_D C_{Cl_2} \tag{D}$$

---

† R. B. MacMullin, *Chem. Eng. Progr.*, **44**, 183 (1948).

‡ From the standpoint of determining rate equations from experimental data it would be more appropriate to reverse this example, i.e., to require the evaluation of the ratios of the rate constants from given composition curves. The calculations are essentially the same in both cases.

§ This problem was originally solved by MacMullin in a somewhat different manner.

These four rate equations, along with a mass balance, can be solved for the desired yields of the products (in terms of the amount of benzene reacted) by eliminating time as a variable. The expressions cannot be solved directly to give compositions as a function of time because the magnitudes of the individual rate constants are not known (only their ratios are known). Although the rate equations are second order, the chlorine concentration appears in all the expressions and cancels out when time is eliminated by dividing one rate equation by another. Hence, the reaction system is equivalent to the consecutive first-order system already considered.

If Eq. (B) is divided by Eq. (A) to eliminate time

$$\frac{dC_M}{dC_B} = -1 + \frac{k_2}{k_1}\frac{C_M}{C_B} \tag{E}$$

This is similar to Eq. (2-87), and so its solution can be written immediately by reference to Eq. (2-89). Thus, the yield of monochlorobenzene is

$$x_M = \frac{C_M}{(C_B)_0} = \frac{k_1}{k_1 - k_2}\left\{\left(\frac{C_B}{(C_B)_0}\right)^{k_2/k_1} - \frac{C_B}{(C_B)_0}\right\}$$

As the volume of the reaction liquid is constant, the concentration is proportional to the number of moles. Also, all yields are ratios of the concentration of the component to the initial concentration of benzene. Hence, $C_M/(C_B)_0 = n_M/(n_B)_0$, etc. Since $(n_B)_0 = 1.0$ mole, the previous equation for x_M may be written

$$x_M = n_M = \frac{n_B}{1 - \alpha}(n_B^{\alpha-1} - 1) \tag{F}$$

where $\alpha = k_2/k_1$.

Similarly, Eq. (C) can be divided by Eq. (A) to give

$$\frac{dn_D}{dn_B} = -\alpha\frac{n_M}{n_B} + \beta\frac{n_D}{n_B} \tag{G}$$

where $\beta = k_3/k_1$. Equation (F) can be used in Eq. (G) to replace n_M with a function of n_B. Then another linear first-order differential equation in n_D results. Integrating this, and noting that $(n_D)_0 = 0$ when $(n_B)_0 = 1$, gives the yield of D,

$$x_D = \frac{n_D}{(n_B)_0} = n_D = \frac{\alpha}{1 - \alpha}\left(\frac{n_B}{1 - \beta} - \frac{n_B^\alpha}{\alpha - \beta}\right) + \frac{\alpha n_B^\beta}{(\alpha - \beta)(1 - \beta)} \tag{H}$$

The concentration of trichlorobenzene can be determined by difference, with a mass balance of aromatic components. Since there is 1.0 mole of benzene, and 1 mole of each chlorobenzene is produced per mole of benzene, the total moles is constant and equal to unity. Hence

$$1.0 = n_B + n_M + n_D + n_T \tag{I}$$

Equations (F), (H), and (I) give the moles of n_M, n_D, and n_T in terms of n_B. The corresponding amount of chlorine added can be determined from a mass balance on chlorine. If n_{Cl_2} represents the total moles of chlorine added (or reacted) per mole of benzene, then

$$n_{Cl_2} = n_M + 2n_D + 3n_T \tag{J}$$

As an illustration of the numerical calculations choose the point at which one-half the benzene has reacted. Then $n_B = 0.5$. It is given that

$$\alpha = \frac{1}{8} = 0.125$$

$$\beta = \frac{k_3}{k_1} = \frac{k_3 k_2}{k_2 k_1} = \frac{1}{30}\left(\frac{1}{8}\right) = 0.00417$$

Then in Eq. (F)

$$x_M = n_M = \frac{1}{0 - 0.125}(0.5^{0.125} - 0.5) = 0.477$$

Equations (H) and (I) give

$$x_D = C_D = \frac{0.125}{1 - 0.125}\left(\frac{0.5}{1 - 0.00417} - \frac{0.50^{0.125}}{0.125 - 0.00417}\right)$$

$$+ \frac{0.125(0.50)^{0.00417}}{(0.125 - 0.00417)(1 - 0.00417)} = 0.022$$

$$n_T = 1 - n_B - n_M - n_D = 1 - 0.50 - 0.477 - 0.022 = 0.001$$

Finally, from Eq. (J),

$$n_{Cl_2} = 0.477 + 2(0.022) + 3(0.001) = 0.524$$

Hence, with 0.524 mole of chlorine reacted per mole of benzene, most of the product is monochlorobenzene, with a little dichlorobenzene, and a negligible quantity of trichlorobenzene.

To obtain the composition at a much later time choose $n_B = 0.001$. Then, proceeding in the same manner, we find most of the product to be mono- and dichlorobenzene, with a little trichlorobenzene substituted product. The results for a range of n_{Cl_2} up to 2.14 are summarized in Table 2-11. Note that the

Table 2-11 Composition of chlorinated benzenes

	Mole fractions, or yields per mole of initial benzene							
Benzene	1.0	0.50	0.10	0.01	0.001	10^{-4}	10^{-10}	10^{-20}
Monochlorobenzene	0	0.477	0.745	0.632	0.482	0.362	0.064	0.004
Dichlorobenzene	0	0.022	0.152	0.353	0.509	0.625	0.877	0.852
Trichlorobenzene	0	0.001	0.003	0.005	0.008	0.013	0.059	0.144
Total	1.000	1.000	1.000	1.000	1.000	1.000	1.000	1.000
Moles of chlorine used per mole of original benzene	0	0.524	1.06	1.35	1.52	1.65	1.99	2.14

maximum yield of monochlorinated product is obtained when approximately 1 mole of Cl_2 has been reacted, and the maximum yield of dichlorinated product results when about 2 moles of Cl_2 have been reacted. Selectivities for any two products can easily be found by taking ratios of the yields, Eq. (2-73). Since the problem has been solved on a basis of 1 mole of initial benzene, n_M, n_D, n_T also are equivalent to mole fractions.

2-12 Precision of Kinetics Measurements

Errors in experimental kinetics data may arise from random causes or inherent difficulties in the system. The latter type can be corrected once the performance of the system is fully understood. For example, in kinetics studies erroneous rates may be caused by some appreciable unknown reaction not accounted for in the treatment of the data. Random errors, such as temperature fluctuations in a thermostat and uncertainty in concentration measurements, can be reduced by improvements in apparatus and technique but usually cannot be eliminated. These residual random errors may be evaluated from the known precision of the experimental observations. It is important in kinetics to be able to calculate the precision of rates of reaction, rate constants, and activation energies from the errors in the measurements.

As an illustration consider the precision of rate constants. The fractional error in dependent variable Ω, which is a function of independent variables α_i, is given by

$$\left(\frac{\Delta\Omega}{\Omega}\right)^2 = \sum_{i=1}^{m} \left[\frac{\partial(\ln\Omega)}{\partial(\ln\alpha_i)}\right]^2 \left(\frac{\Delta\alpha_i}{\alpha_i}\right)^2 \tag{2-91}$$

The starting point in using this expression is the relation between the quantity we want to know the precision of and the experimental observations used. For our example this is the relation between the rate constant and the rate and concentrations. Suppose the reaction is second order and of the form

$$\mathbf{r} = k_2 C_A C_B$$

then the required relationship is

$$k_2 = \frac{\mathbf{r}}{C_A C_B} \tag{2-92}$$

where $\mathbf{r}$, C_A, and C_B are the three independent variables (α_i) and k is the dependent variable (Ω). Evaluating the partial derivatives from Eq. (2-92) and substituting them in Eq. (2-91) yields

$$\left(\frac{\Delta k}{k}\right)^2 = \left[\frac{\partial(\ln k)}{\partial(\ln \mathbf{r})}\right]^2 \left(\frac{\Delta\mathbf{r}}{\mathbf{r}}\right)^2 + \left|\frac{\partial(\ln [k])}{\partial(\ln [C_A])}\right|^2 \left(\frac{\Delta C_A}{C_A}\right)^2 + \left|\frac{\partial(\ln k)}{\partial(\ln C_B)}\right|^2 \left(\frac{\Delta C_B}{C_B}\right)^2$$

or

$$\left(\frac{\Delta k}{k}\right)^2 = \left(\frac{\Delta\mathbf{r}}{\mathbf{r}}\right)^2 + \left(\frac{\Delta C_A}{C_A}\right)^2 + \left(\frac{\Delta C_B}{C_B}\right)^2 \tag{2-93}$$

This result shows that the squares of the fractional errors in the individual measurements are additive. If the precision of the rate measurements is 8% and that of each concentration is 4%, the error in k will be

$$\left(\frac{\Delta k}{k}\right)^2 = 0.08^2 + 0.04^2 + 0.04^2 = 0.0096$$

$$\frac{\Delta k}{k} = 0.098 \qquad \text{or } 9.8\%$$

The rate is not a direct measurement but is itself calculated from observations of such variables as time and concentration. Its precision, arbitrarily chosen in the example as 8%, should be based on an evaluation similar to that illustrated for obtaining the error in k.

The evaluation of k from data for rates and concentrations should be carried out by statistically sound methods, provided sufficient data points are available. For example, Eq. (2-92) shows that a linear relationship exists between the product $C_A C_B$ and $\mathbf{r}$. Hence k_2 should be determined from the slope of the line of $\mathbf{r}$ vs. $C_A C_B$ determined by a least-mean-square fit of the data points. This technique was illustrated in Example 2-1, where the best fit of the Arrhenius equation to k-vs.-T data was used to evaluate the activation energy.

Errors in activation energy can be evaluated from the Arrhenius equation by the same procedure described for finding the errors in k. The precision of E will depend on the uncertainty in k and T. Since E is based on differences in k and T values, the same errors in k and T will result in smaller errors in E as temperature range covered by the data is increased.

BIBLIOGRAPHY

Benson, S. W., "The Foundations of Chemical Kinetics," McGraw-Hill Book Company, New York, 1960. The stationary-state approximation mentioned in Sec. 2-3 is defined and illustrated on pp. 50–53. In chap. XII the collision theory is considered in detail, complications related to the energy distribution of molecules and the steric factor are discussed, and the results are compared in depth with those from the transition-state theory.

——— "Thermochemical Kinetics," John Wiley & Sons, Inc., New York, 1968. Methods are proposed for estimating rate constants for a variety of homogeneous reactions.

Boudart, Michel, "Kinetics of Chemical Processes," Prentice-Hall, Inc., Englewood Cliffs, N.J., 1968. A concise presentation of the fundamental concepts of kinetics for homogeneous and heterogeneous reactions, including a chapter on the application and validity of the stationary-state approximation.

Frost, A. A., and R. G. Pearson, "Kinetics and Mechanism," John Wiley & Sons, Inc., New York, 1961. Provides an excellent discussion of reaction mechanisms with applications.

Laidler, K. J., "Chemical Kinetics," 2d ed., McGraw-Hill Book Company, New York, 1965. A general text on chemical kinetics.

Semenov, N. N., "Problems in Chemical Kinetics and Reactivity," vols. I and II translated by M. Boudart, Princeton University Press, Princeton, N.J., 1958. A large reservoir of information on theories of kinetics and experimental data.

PROBLEMS

2-1 A common rule of thumb is that the rate of reaction doubles for each 10°C rise in temperature. What activation energy would this suggest at a temperature of 25°C?

2-2 The rate of an overall reaction $A + 2B \rightarrow C$ has been found to be first-order with respect to both A and B. What possible mechanisms do these results suggest? $A + B \rightarrow AB$
$$AB + B \rightarrow C$$

2-3 The overall reaction for the thermal decomposition of acetaldehyde is

$$CH_3CHO \rightarrow CH_4 + CO$$

A chain-reaction sequence of elementary steps proposed to explain the decomposition is as follows:

Initiation

$$\overset{A}{CH_3CHO} \xrightarrow{k_1} \overset{B}{CH_3 \cdot} + \overset{C}{CHO \cdot}$$

Propagation

$$\overset{B}{CH_3 \cdot} + \overset{A}{CH_3CHO} \xrightarrow{k_2} \overset{C}{CH_3CO \cdot} + \overset{D}{CH_4}$$

$$\overset{D}{CH_3CO \cdot} \xrightarrow{k_3} \overset{B}{CH_3 \cdot} + \overset{E}{CO}$$

Termination

$$\overset{B}{CH_3 \cdot} + \overset{B}{CH_3 \cdot} \xrightarrow{k_4} \overset{F}{C_2H_6}$$

Use the stationary-state approximation to derive an expression for the overall rate of decomposition. Do order and stoichiometry agree in this case?

2-4 Using the collision theory, calculate the rate constant at 300 K for the decomposition of hydrogen iodide, assuming a collision diameter of $3.5A$ and an activation energy of 44 kg-cal (based on a rate constant in concentration units).

2-5 The homogeneous dimerization of butadiene has been studied by a number of investigators[†] and found to have an experimental activation energy of 23,960 cal/g mol, as indicated by the specific-reaction rate,

$$k = 9.2 \times 10^9 e^{-23,960/R_gT} \qquad cm^3/(g\ mol)(s)$$

(based on the disappearance of butadiene). (*a*) Use the transition-state theory, to predict a value of A at 600 K for comparison with the experimental result of 9.2×10^9. Assume that the structure of the activated complex is

$$\overset{|}{CH_2} - CH = CH - CH_2 - CH_2 - \overset{|}{CH} - CH = CH_2$$

and use the group-contribution method (see Sec. 1-4) to estimate the thermodynamic properties required. (*b*) Use the collision theory to predict a value of A at 600 K and compare it with the experimental result. Assume that the effective collision diameter is 5×10^{-8} cm.

2-6 From the transition-state theory and the following thermodynamic information, calculate the rate constant for the given unimolecular reactions. Assume ideal-gas behavior.

[†] W. E. Vaughan, *J. Am. Chem. Soc.*, **54**, 3863 (1932); G. B. Kistiakowsky and F. R. Lacher, *J. Am. Chem. Soc.*, **58**, 123 (1936); J. B. Harkness, G. B. Kistiakowsky, and W. H. Mears, *J. Chem. Phys.*, **5**, 682 (1937).

Reaction	T, K	ΔH^*, cal/g mol	ΔS^*, cal/(g mol)(K)
Decomposition of methyl azide, CH_3N_3	500	42,500	8.2
Decomposition of dimethyl ether, CH_3OCH_3	780	56,900	2.5

Source: O. A. Hougen and K. M. Watson, "Chemical Process Principles," vol. III, "Kinetics and Catalysis," John Wiley & Sons, Inc., New York, 1947.

2-7 A second-order, elementary reversible reaction of the form

$$A + B \rightleftarrows C + D$$

has forward-rate constants as follows:

$$k_2 = \begin{cases} 10.4 \text{ liter/(g mol)(s)} & \text{at } 230°C \\ 45.4 \text{ liter/(g mol)(s)} & \text{at } 260°C \end{cases}$$

The standard-state entropy and enthalpy changes for the overall reaction are approximately independent of temperature and are given by $\Delta H° = 8400$ cal/g mol and $\Delta S° = -2.31$ cal/(g mol)(K). Derive expressions for the forward- and reverse-rate constants as functions of temperature.

2-8 The reaction

$$CH_3COCH_3 + HCN \rightleftarrows (CH_3)_2C{\overset{\displaystyle CN}{\underset{\displaystyle OH}{\big<}}}$$

was studied in aqueous solution by Svirbely and Roth.[†] In one run with initial concentrations of 0.0758 normal for HCN and 0.1164 normal for acetone, the following data were obtained:

t, min	4.37	73.2	172.5	265.4	346.7	434.4
C_{HCN}, normal	0.0748	0.0710	0.0655	0.0610	0.0584	0.0557

Determine a reasonable rate equation from these data. ($K_e = 13.87$ liter/mol).

2-9 Kinetics data at 25°C for the conversion of tert-butyl bromide to tert-butyl alcohol in a solvent of 90% acetone and 10% water are given in the following table:

Time, h	Concn. of $(CH_3)_3CBr$ g mol/liter
0	0.1039
3.15	0.0896
4.10	0.0859
6.20	0.0776
8.20	0.0701
10.0	0.0639
13.5	0.0529
18.3	0.0353
26.0	0.0270
30.8	0.0207

† J. Am. Chem. Soc., **75**, 3109 (1953).

The reaction is

$$(CH_3)_3CBr + H_2O \rightarrow (CH_3)_3COH + HBr$$

What rate equation is consistent with the data?

2-10 With HCl as a homogeneous catalyst the rate of esterification of acetic acid and alcohol is increased. At 100°C the rate of the forward reaction is

$$r_2 = k_2 C_H C_{OH} \qquad \text{g mol/(liter)(min)}$$

$$k_2 = 4.76 \times 10^{-4} \text{ liter/(g mol)(min)}$$

and the rate of the reverse reaction is

$$r'_2 = k'_2 C_E C_W \qquad \text{g mol/(liter)(min)}$$

$$k'_2 = 1.63 \times 10^{-4} \text{ liter/(g mol)(min)}$$

where C_H = concentration of acetic acid
C_{OH} = concentration of alcohol
C_E = concentration of ester
C_W = concentration of water

An initial mixture consists of equal masses of 90 wt % aqueous solution of acid and 95 wt % solution of ethanol. For constant-volume conditions calculate the conversion of acid to ester for various times of reaction. Assuming complete miscibility, estimate the equilibrium conversion.

2-11 The following data have been reported[†] for the irreversible reaction (at 64°C) between sodium ethoxide (NaOC$_2$H$_5$) and ethyl dimethyl sulfonium iodide [C$_2$H$_5$(CH$_3$)$_2$SI] using ethanol as a solvent.

Time	Concentrations[‡]	
min	NaOC$_2$H$_5$	C$_2$H$_5$(CH$_3$)$_2$SI
0	22.55	11.53
12	20.10	9.08
20	18.85	7.83
30	17.54	6.52
42	16.37	5.35
51	15.72	4.10
63	14.96	3.94
∞	11.02	0

Two irreversible reactions are possible:

$$NaOC_2H_5 + C_2H_5(CH_3)_2SI \rightarrow NaI + C_2H_5OC_2H_5 + S(CH_3)_2$$

$$NaOC_2H_5 + C_2H_5(CH_3)_2SI \rightarrow NaI + C_2H_5OH + C_2H_4 + S(CH_3)_2$$

The rate of reaction appears to be the same for the two reactions.
What rate equation do these data suggest? Use the integration method?

2-12 Solve Problem 2-11 by the differential method.

2-13 For two consecutive reversible reactions (liquid-phase)

$$\text{1.} \quad A \leftrightarrows B$$

$$\text{2.} \quad B \rightleftarrows C$$

† E. E. Hughes, et al., *J. Chem. Soc.*, 2072 (1948).
‡ The values given are proportional to concentrations in mol/liter.

the forward-rate constants and equilibrium constants K are

$k_2 \approx 0.0125$

$$k_1 = 1 \times 10^{-3} \text{ min}^{-1} \qquad K_1 = 0.8$$

$K_2 \approx 0.017$

$$k_2 = 1 \times 10^{-2} \text{ min}^{-1} \qquad K_2 = 0.6$$

$t_1 t_2 = \dfrac{c_B}{c_A} \dfrac{c_c}{c_B} = \dfrac{c_c}{c_A}$

$c_c = k_1 k_2 c_A$

If the initial concentration of A is 1.0 molal, plot the concentration of A vs. time from 0 to 1000 min. Both reactions are first order.

2-14 The thermal decomposition of nitrous oxide (N_2O) in the gas phase at 1030 K is studied in a constant-volume vessel at various initial pressures of N_2O. The half-life data so obtained are as follows:

p_0, mmHg	52.5	139	290	360
$t_{1/2}$, s	860.0	470	255	212

$k_1 \theta_{1/2} = \dfrac{(2^{n-1} - 1)}{n-1} c_A^{0 \, 1-n}$

$c_A^0 = \dfrac{p_0}{RT}$

Determine a rate equation that fits these data.

2-15 It has been postulated that the thermal decomposition of diethyl ether occurs by the following chain mechanism:

Initiation

$$(C_2H_5)_2O \xrightarrow{k_1} CH_3 \cdot + \cdot CH_2OC_2H_5$$

Propagation

$$CH_3 \cdot + (C_2H_5)_2O \xrightarrow{k_2} C_2H_6 + \cdot CH_2OC_2H_5$$

$$\cdot CH_2OC_2H_5 \xrightarrow{k_3} CH_3 \cdot + CH_3CHO$$

Termination

$$CH_3 \cdot + \cdot CH_2OC_2H_5 \xrightarrow{k_4} \text{end products}$$

Show that the stationary-state approximation suggests that the rate of decomposition is first order in ether concentration.

2-16 Consider the first-order, irreversible reaction sequence

$$A \xrightarrow{k_1} B \xrightarrow{k_2} C \xrightarrow{k_3} D$$

Determine the profiles of concentration vs. time for A, B, C, and D on an analog computer. Solve the problem in the following steps:

Step 1 To limit the variables range from 0 to 1.0, first write the differential equations expressing the rates in dimensionless form by using the new variables

$$A^* = \frac{C_A}{(C_A)_0} \qquad B^* = \frac{C_B}{(C_A)_0} \qquad C^* = \frac{C_c}{(C_A)_0} \qquad D^* = \frac{C_D}{(C_A)_0} \qquad t^* = k_1 t$$

where $(C_A)_0$ is the initial concentration of A. Note that the initial conditions are $t^* = 0$, $A^* = 1$, and $B^* = C^* = D^* = 0$.

Step 2 Next prepare a block diagram showing the hookup of integrators, inverters, summers, and potentiometers needed to solve the differential equations for A^*, B^*, C^*, and D^*. Suppose $k_2/k_1 = k_3/k_1 = 1.0$.

Step 3 Hook up an analog computer according to the block diagram and display the outputs A^*, B^*, C^*, and D^* vs. t^* on an oscilloscope or xy plotter.

2-17 Solve Prob. 2-16 by analytical integration of the differential equations (dimensionless form).

2-18 Aqueous solutions of diazobenzene decompose irreversibly according to the reaction

$$C_6H_5N_2Cl(aq) \rightarrow C_6H_5Cl(aq) + N_2(g)$$

The kinetics are first order. In one experiment,[†] at 50°C the initial concentration of $C_6H_5N_2Cl$ was 10 g/liter and the following amounts of N_2 were liberated:

Reaction time, min	6	9	12	14	18	20	22	24	26	30
N_2 liberated, cm³ at 50°C, 1 atm		19.3	26.0	32.6	36.0	41.3	43.3	45.0	46.5	48.4	50.3

Complete decomposition of the diazo salt liberated 58.3 cm³ of N_2. Calculate an accurate value for the rate constant.

2-19[‡] Suppose that a gaseous reaction between A and B is studied kinetically by making isothermal measurements of the half-life period for several initial compositions of reactants. The results for each of four different initial conditions are as follows:

$(p_A)_0$, mmHg	500	125	250	250
$(p_B)_0$, mmHg	10	15	10	20
$(t_{1/2})_B$, min	80	213	160	80

If the rate is first order with respect to component A and second order with respect to B, what is the numerical value of the specific reaction rate?

2-20[§] The decomposition of nitrogen dioxide follows a second-order rate equation. Data at different temperatures are as follows:

T, K	592	603	627	651.5	656
k_2, cm³/(g mol)(s)	522	755	1700	4020	5030

Compute the energy of activation E from this information. The reaction is written

$$2NO_2 \rightarrow 2NO + O_2$$

2-21 Reactants A and B are placed in a reaction vessel at zero time, where $(C_A)_0 = (C_B)_0$. The following reactions occur at constant volume

$$1. \quad A + B \xrightarrow{k_1} C$$

$$2. \quad A + C \xrightarrow{k_2} D$$

where C is the desired product. If both reactions are second order, derive an expression for the selectivity of C with respect to D in terms of the total conversion of A. Also determine the total

---

[†] O. A. Hougen and K. M. Watson, "Chemical Process Principles," vol. III, "Kinetics and Catalysis," John Wiley & Sons, Inc., New York, 1947.

[‡] From A. A. Frost and R. G. Pearson, "Kinetics and Mechanism," John Wiley & Sons, Inc., New York, 1953 (by permission).

[§] Ibid.

conversion at which the selectivity will be a maximum if $k_2/k_1 = 1.0$. Will the maximum conversion of A to C occur at the same total conversion as that for which the selectivity of C with respect to D is a maximum?

2-22 The parallel, first-order, irreversible reaction system

$$1. \quad A \xrightarrow{k_1} B$$

$$2. \quad A \xrightarrow{k_2} C$$

consists of three components, so that the reaction path can be conveniently represented on a triangular diagram.† Suppose that initially only A is present at a concentration $(C_A)_0$. Assume constant density, so that the sum of the concentrations of all the components will be constant and equal to $(C_A)_0$. Let one apex of the equilateral triangle represent a reaction mixture containing 100% A, the second 100% B, and the third 100% C. Use equations in Sec. 2-11 to show how the composition of the reaction mixture can be represented on the diagram. Specifically, draw the reaction path from zero to complete conversion of A when $k_2/k_1 = 2$. What path would be followed if $k_2/k_1 = 0$?

2-23 The reaction

$$A_2 + 2B \rightarrow 2C$$

occurs by the mechanism:

$$1. \quad A_2 \underset{k_1'}{\overset{k_1}{\rightleftarrows}} 2A$$

$$2. \quad A + B \underset{k_2'}{\overset{k_2}{\rightleftarrows}} C$$

If the first step is reversible and slow with respect to the second reversible step, what is the rate equation for the net disappearance of A?

2-24 Suppose the consecutive first-order reactions described in Sec. 2-11 occur at constant density in a batch reactor, with an initial mixture containing only A at a concentration $(C_A)_0$. Show on a triangular diagram the reaction paths for three cases: $k_3/k_1 = 0.5$, 1.0, and 2.0.

2-25 The second-order reactions

$$1. \quad A + B \xrightarrow{k_1} C$$

$$2. \quad A + A \xrightarrow{k_2} A_2$$

occur at constant density, with an initial mixture containing only A and B, each at the same concentration. For a batch reactor, show the reaction path on a triangular diagram for $k_2/k_1 = 1.0$.

2-26 The mechanism of the decomposition of ozone may be written:

$$O_3 \xrightarrow{k_1} O + O_2$$

$$O + O_3 \xrightarrow{k_2} 2O_2$$

Assume that the second step is independent of ozone concentration (due to special experimental conditions in the laboratory system). Then a first-order rate constant, $k_3 = k_2 C_{O_3}$, is applicable.

(a) Solve the rate equations to give C_{O_3}, C_O and C_{O_2} as a function of time, under conditions such that $C_{O_2} = C_O = 0$ and $C_{O_3} = (C_{O_3})_0$ at $t = 0$.

(b) Plot curves of $C_{O_3}/(C_{O_3})_0$, $C_{O_2}/(C_{O_3})_0$ and $C_O/(C_{O_3})_0$ vs. time for $k_1 = 10^3$ (s)$^{-1}$ and $k_3 = 10^3$ (s)$^{-1}$.

---

† The representation of reaction paths on triangular diagrams is described in detail in Kramers and Westerterp, "Chemical Reactor Design and Operation," p. 47, Academic Press, Inc., New York, 1963.

(c) Next suppose that the oxygen atom is very reactive with respect to ozone so that $k_3 = 10^6$ $(s)^{-1}$. Again plot the concentration vs. time curves for the three species.

(d) What are the times, t_{max}, where the concentration of atomic oxygen reaches a maximum for the two values of k_3? What are the maximum concentrations of atomic oxygen?

(e) What values would t_{max} and $(C_O)_{max}$ approach as $k_3/k_1 \to \infty$? Suppose the stationary-state approximation is applied to this system, treating atomic oxygen as the reactive intermediate. If this approximation is to be valid, what can be said about the relative values of k_1 and k_3?

2-27 The gaseous reaction $H_2 + Br_2 \rightleftharpoons 2HBr$ has been extensively studied, starting with the work of Bodenstein and Lind.† They showed that the rate could be represented by the equation

$$r = \frac{dC_{HBr}}{dt} = \frac{k_2 C_{H_2} C_{Br_2}^{1/2}}{1 + k_i C_{HBr}/C_{Br_2}}$$

This equation with a denominator term involving concentrations suggests that a complex mechanism is involved. Investigators have proposed the following chain of atomic reactions

$$Br_2 \xrightarrow{k_1} 2Br\cdot \quad \text{(initiation)}$$

$$Br\cdot + H_2 \xrightarrow{k_2} H\cdot + HBr$$

$$H\cdot + Br_2 \xrightarrow{k_3} Br\cdot + HBr$$

$$H\cdot + HBr \xrightarrow{k_4} H_2 + Br\cdot$$

$$2Br\cdot \xrightarrow{k_5} Br_2 \quad \text{(termination)}$$

Alternately, the following set of reversible reactions could be suggested for the mechanism:

$$Br_2 + M \underset{k_6'}{\overset{k_6}{\rightleftharpoons}} 2Br + M \quad \text{(initiation)}$$

$$Br\cdot + H_2 \underset{k_7'}{\overset{k_7}{\rightleftharpoons}} HBr + H\cdot$$

$$H\cdot + Br_2 \underset{k_8'}{\overset{k_8}{\rightleftharpoons}} HBr + Br\cdot$$

where M is any gaseous molecule which has sufficient energy to cause dissociation of Br_2 to $Br\cdot$.

Show that both mechanisms lead to the rate equation proposed by Bodenstein and Lind.

2-28 Consider that the photochlorination of propane occurs according to the reactions in Sec. 2-8. If the controlling termination step is the heterogeneous termination of chlorine radicals,

$$Cl + wall \to end\ product$$

derive a rate equation for the overall reaction. What would be the form of the rate equation if the second-order homogeneous termination of Cl were controlling (i.e., according to the reaction $Cl + Cl \to Cl_2$)?

2-29 The kinetics of a second-order irreversible liquid-phase reaction of the form $A + B \to C$ are studied in a constant-volume apparatus. Starting with equal concentrations of 1.0 mol/liter for A and B, the reaction is stopped after 30 min, at which time about 20% of the reactants have disappeared. Random errors will amount to about 5 s in the time readings and 0.002 mol/liter in the concentration measurements. Estimate the fractional error in the rate constants computed from such data.

† M. Bodenstein and S. C. Lind, Z. *Physik Chem.* **57**, 168 (1906).

2-30 Consider the reaction sequence

$$A \xrightarrow{k_1} B \xrightarrow{k_2} C$$

where both reactions are first order and irreversible. The reactions occur isothermally under homogeneous, well-stirred conditions.

It is desired to establish criteria regarding permissible values of the rate constants for the use of the stationary state approximation (for component B) for calculating the rate of production of C. More specifically, suppose that the approximation is satisfactory if the ratio $(C/A_0)_e/(C/A_0)_s \geq \beta$, where β is close to but less than unity. Suppose also that the criteria should be valid for all reaction times corresponding to conversions of A greater than α (α is close to zero). Develop the criterion for a batch reactor with the initial condition $A = A_0$, $B = C = 0$, where A, B, and C represent concentrations of the components. The concentration ratio $(C/A_0)_e$ is the exact value and $(C/A_0)_s$ is the result from the stationary state approximation.

THREE

DESIGN FUNDAMENTALS AND MASS CONSERVATION EQUATIONS FOR IDEAL REACTORS

The first part of this chapter presents a qualitative discussion of reactor design—questions that the chemical engineer must answer before proposing equipment and operating conditions. Following this, the fundamental concept of reactor design is introduced. This is the use of conservation principles to predict reactor performance. In this chapter the quantitative treatment is limited to isothermal operation of the two ideal reactor types (stirred tank and plug flow), although illustrations of deviations from the ideal types are discussed in Sec. 3-5. The mass balance is the only conservation principle required. Multiple reactions, multiple reactors, and variations in operating conditions are presented in Chap. 4, still for the ideal reactor types and for isothermal conditions. Then in Chap. 5 quantitative treatment for nonisothermal reactors is considered. This requires application of the conservation-of-energy principle as well as that of mass. Since nonisothermal conditions are common in flow reactors, some general discussion of temperature effects is given in the last part of this chapter (Sec. 3-7) in order that we not give the impression that reactor performance can be predicted using only mass conservation equations.

While the illustrations in Chaps. 3 to 5 are for homogeneous reactions, the form of the conservation equations is generally applicable. Hence, in the later chapters where heterogeneous systems are considered, we will frequently refer back to the equations in Chaps. 3 and 5.

The mass conservation equation includes a term involving the rate of reaction. This rate is the same property of the system as that defined in Chap. 2. In fact, for a well-mixed batch reactor operating at constant volume, the conservation equation reduces to exactly the same expression, relating concentration to the rate, as that given in Chap. 2 [Eq. (2-1)]. For other reactor types and operating conditions, we will see that the rate is related in a different way to concentrations in the reactor.

3-1 Reactor Design

The first decisions that must be made in design are to choose the type of reactor and method of operation. Common types and operating methods for reactors were discussed in Sec. 1-5, Classification of Reactors. The next step is to evaluate the *reactor performance*, that is, the reactor size, composition of the product stream, and the required operating conditions. The latter include temperatures, pressures, and compositions within the reactor. The first decisions may not be possible without carrying out the performance calculations for several types of reactors. Ultimately, the choice of reactor would be made on the basis of profit, safety, and environmental factors. The kind of reaction system may be a primary factor in determining the type of reactor and method of operation. For example, batch-operated tank reactors would not be used normally for gas-phase reactions because the production rate of product would be uneconomically low. In contrast, such reactors may be well suited for liquid phase reactions involving expensive materials (e.g., pharmaceutical industry) and where careful control of conditions is more important than labor costs.†

The data that are available usually consist of the temperature, pressure, and composition and flow rates of the feed streams (or initial charge for a batch reactor). This information along with the required production rate of desired product constitutes the *design conditions*. With these data and the proposed type of reactor and method of operation, reactor performance can be evaluated. For the ideal reactors considered in this chapter the evaluation requires only solution of the mass conservation equations.

There can be many combinations of size and operating conditions for a given reactor type which satisfy the design conditions. The optimum design from a profit standpoint depends on raw materials, initial and operating costs, and the market value of the end products. First, the method of calculating reactor performance needs to be established and then an optimization technique used to find the most profitable design.

Generally some of the conditions, such as feed composition, will change with time. Such changes may be abrupt, as in switching to feed that contains material of a different composition, or they may be mild cyclical changes. The problem in *reactor control* is determining how to change operating conditions so that the reactor returns to optimum performance as quickly as possible. First, specific operating conditions must be chosen to use as control points. Then a policy or strategy is developed for responding to fluctuations in such a way as to maximize profit. The resulting control procedure may be manual. Alternately, it may be a semicomputerized procedure in which a computer is used to determine rapidly the

---

† A more complete discussion of advantages and disadvantages, both practical and technical, of tank-type reactors is given in "Chemical Reactor Design for Process Plants," chap. 8, by H. F. Rase, John Wiley & Sons, Inc., New York, 1977.

value of the profit function, but the indicated changes in operating conditions are made manually. Finally, it may be possible to employ a direct-digital or closed-loop procedure, in which adjustments to operating conditions are made automatically on signal from the computer. All control procedures, to be successful, require a knowledge of how to design the reactor for a set of constant design conditions.

The optimum design requires iterative numerical work; hence machine computation greatly simplifies the optimization task. In this book we shall limit the problem to answering design questions for a single set of design conditions. Even for such a limited scope we shall find that the numerical integrations often require repetitive calculations well suited for machine solution.

As mentioned in Sec. 1-1, the first step in a logical design procedure is to obtain a suitable expression for the rate of the chemical reaction process, and this requires experimental data. The data can be obtained in one of three ways:

1. *From a bench-scale laboratory reactor designed to operate at nearly constant temperature.* Operating conditions and reactor type are chosen to facilitate separating the effects of mass and heat transfer (the physical processes) from the observed measurements, so that the rate of the chemical step can be accurately evaluated. This is the most successful of the three methods. The examples in Chap. 2 were all illustrations of this procedure; an isothermal, well-mixed batch reactor was used to obtain concentration vs. time data. Isothermal, flow reactors also may be used in obtaining data from which a rate equation can be obtained. Analysis of data from such laboratory flow reactors is discussed in Chap. 4.
2. *From a small-scale reactor (pilot plane) in which the composition, temperature, and pressure may change.* Here calculations similar to, but the reverse of, the design steps are required to evaluate the rate of the chemical reaction. Accurate separation of mass and heat transfer effects from the intrinsic chemical step may be difficult because of changes in temperature and composition.
3. *From a commercial-scale reactor which happens to be already available.* The problems in obtaining an expression for the chemical rate are similar to those in method 2 but are usually even more severe, because there is less instrumentation, and hence fewer data.

As an illustration of the first two methods, consider the oxidation of sulfur dioxide. Suppose that an air-SO_2 mixture flows over particles of solid catalyst in a tubular-flow reactor. In the bench-scale study the reactants would be passed over a very small amount of catalyst, and the rate of production of sulfur trioxide would be determined by measuring the rates of flow and composition of the inlet and exit streams.† This production rate, divided by the mass of catalyst, would represent the global rate of reaction,‡ for example, in grams of sulfur trioxide per

---

† In flow systems such small-scale reactors are commonly called *differential reactors*, since the changes in temperature and composition in the reactor are small.

‡ For a reaction requiring a solid catalyst the rate is usually based on a unit mass of catalyst, rather than on a unit volume as used in Chap. 2. The two rates are directly related through the bulk density of the bed of catalyst (see Chap. 9).

hour per gram of catalyst. It would approach a local rate rather than an average value, because the amount of catalyst is small enough that the temperature, pressure, and composition changes in passing over the catalyst bed are similarly small.† In the second approach the amount of catalyst in the pilot-plant reactor is sufficient to cause considerable conversion, and the temperature and composition may change appreciably as the mixture flows through the reactor. Since the rate of reaction is a function of these variables, it varies from location to location, and the measured production of sulfur trioxide represents an integrated average of all the local rates. To reduce the measured results (called *integral-conversion*, or *integral-reactor*, data) to the rate of the chemical step requires a procedure that is the reverse of the design calculations. The integral method mentioned in Chap. 2 could be used. First, a promising rate equation is assumed. Then the local rates are integrated through the catalyst bed, with the effects of diffusion and heat transfer taken into account. Finally, the predicted conversions are compared with the experimental results. Repetition of this procedure will result in an equation for the rate of the chemical step. Because of the inaccuracy in accounting correctly for the composition and temperature changes, it is difficult to arrive at a rate equation that is wholly satisfactory.

Data obtained in both bench and pilot-plant equipment are valuable, and it is common practice to carry out investigations with both before building the commercial-scale reactor. The first yields a better rate equation and more knowledge about the kinetics of the reaction; i.e., it tells the engineer more accurately just what variables affect the rate of the chemical step and how they influence the course of the reaction. This information is particularly valuable in case it is necessary to predict how the commercial-scale plant will be affected by a change of operating conditions not specifically considered in the pilot-plant work.

A bench-scale study alone leaves the engineer largely dependent on correlations and estimates for the mass and energy transport terms in the conservation equations needed to calculate reactor performance. Experimental data obtained in a pilot plant can provide a check on the suitability of the design procedure.‡ As an illustration, consider the extreme case where the kinetic studies are carried out in a batch tank reactor with diffusion resistances, and the commercial unit is to be a tubular-flow reactor. The diffusion rates will not be the same under batch and flow conditions. Hence, the observed global rate in the batch reactor will not be directly applicable for design calculations in the commercial unit. While the importance of such effects can be estimated for different types of reactors, as explained in Chap. 10, the uncertainties in the estimates are sometimes so large that experimental verification from pilot-plant data is desirable. There are similar

---

† Note that the change in composition between the inlet and exit streams must be large enough for precise measurement; otherwise the rate of conversion in the reactor cannot be accurately established. This restriction imposes a limitation on the applicability of the method. If precise analytical methods of determining small composition changes are not available for the particular reaction, a close approach to a local value of the rate cannot be ascertained.

‡ From this standpoint the objective of the pilot plant is to obtain a model of the reactor, that is, an understanding of how the physical processes affect the performance of a reactor. In contrast, the objective of the bench-scale reactor is to obtain a model for the chemical kinetics, that is, a rate equation.

problems with temperature differences arising from heat-transfer considerations. Even where the kinetic studies are carried out in a flow system similar to that to be employed in the large plant, pilot-plant investigations provide invaluable information on such important factors as temperature distribution in the reactor, materials and methods of construction, and the effect of specialized design features.†

The procedures for extracting a rate equation from experimental data and the procedure for designing a commercial-scale reactor using an available rate equation are similar. In principle, one is the reverse of the other. However, if a bench-scale unit is used to obtain the rate equation, there is enough flexibility in design to eliminate many of the transport effects. From this standpoint the interpretation of bench-scale data is easier than the reactor design problem.

3-2 Conservation of Mass in Reactors

The course of a reaction can be followed by monitoring the mass of a particular molecular species (reactant or product). In a single-reaction system one reactant is usually limiting because of cost. For example, in the air oxidation of SO_2 the sulfur dioxide is the limiting reactant. Of course, the balance may be written for each component, and for the total mass. However, this is not necessary because the composition of the reaction mixture can be expressed in terms of one variable, the conversion or the extent ξ of reaction (Sec. 2-1). This is done by using the composition of the original reactants and the stoichiometry of the reaction (see Examples 3-1, 3-2). For multiple reactions one mass balance can be written for each reaction. Each balance can be expressed in terms of one dependent variable, the conversion for that reaction.

The conversion x is the fraction of a reactant that has been converted into products. When there is only one reaction, there is no uncertainty in this definition. When a reactant can undergo simultaneous or successive reactions to multiple products, both the total conversion of reactant and the conversions to specific products are important. The conversion to a given product was defined in Chap. 2 as the yield of that product. For example, in the air oxidation of ethylene both ethylene oxide and carbon dioxide are products. It is customary to speak of the yield of ethylene oxide or conversion to ethylene oxide.

> **Example 3-1** Liquid benzene is chlorinated by bubbling gaseous chlorine into a well-stirred tank reactor containing the benzene. Three reactions can occur:
>
> 1. $C_6H_6 + Cl_2 \rightarrow C_6H_5Cl + HCl$
>
> 2. $C_6H_5Cl + Cl_2 \rightarrow C_6H_4Cl_2 + HCl$
>
> 3. $C_6H_4Cl_2 + Cl_2 \rightarrow C_6H_3Cl_3 + HCl$
>
> Initially the reactor contains $(N_B)_0$ moles of benzene. Then a total of N_{Cl_2} $[N_{Cl_2} < 3(N_B)_0]$ moles of chlorine *per mole of benzene* is added to the reactor

---

† A more detailed discussion of the functions of bench-scale and pilot-plant reactors is given by J. M. Smith, *Chem. Eng. Prog.*, **64**, 78 (1968).

slowly, so that no unreacted chlorine leaves the reactor. Also, the concentrations of dissolved unreacted chlorine and dissolved HCl are small.† If the density of the reaction mixture is constant, express the concentrations of mono-, di-, and trichlorobenzene in terms of their corresponding conversions x_M, x_D, and x_T.

SOLUTION Chlorine is the limiting reactant, so the conversions will be based on this component. Let N_B, N_M, N_D, and N_T represent the moles of benzene and mono-, di-, and trichlorobenzene per mole of original benzene. Since 1 mole of chlorine is required for 1 mole of monochlorobenzene,

$$x_M = \frac{N_M}{N_{Cl_2}} \tag{A}$$

However, 2 moles of chlorine are needed to produce 1 mole of dichlorobenzene. Therefore

$$x_D = \frac{2N_D}{N_{Cl_2}} \tag{B}$$

and similarly,

$$x_T = \frac{3N_T}{N_{Cl_2}} \tag{C}$$

The number of moles of unreacted benzene will be

$$N_B = 1 - (N_M + N_D + N_T) = 1 - (x_M + \tfrac{1}{2}x_D + \tfrac{1}{3}x_T)N_{Cl_2} \tag{D}$$

The initial concentration of benzene in the reactor is $C_{B_0} = (N_B)_0/V$. Equations (A) to (D) give the number of moles of each component per mole of initial benzene. Therefore the concentrations at any conversion will be

$$C_M = \frac{(N_B)_0 N_M}{V} = C_{B_0} N_{Cl_2} x_M$$

$$C_D = \tfrac{1}{2}C_{B_0} N_{Cl_2} x_D$$

$$C_T = \tfrac{1}{3}C_{B_0} N_{Cl_2} x_T$$

$$C_B = \frac{(N_B)_0 N_B}{V} = [1 - (x_M + \tfrac{1}{2}x_D + \tfrac{1}{3}x_T)N_{Cl_2}]C_{B_0}$$

These four equations give the desired relationships for the concentrations in terms of the conversions for a constant-density system. All three conversions are not independent, because it is supposed that there is no unreacted chlorine; that is, their sum must be unity. This is made clear by writing a mass balance of chlorine,

Total chlorine fed = total chlorine in products

$$N_{Cl_2} = N_M + 2N_D + 3N_T \tag{E}$$

† The HCl produced leaves the reactor as a gas.

Introducing the conversions with Eqs. (A) to (C), we have

$$N_{Cl_2} = x_M N_{Cl_2} + x_D N_{Cl_2} + x_T N_{Cl_2}$$

or

$$1 = x_M + x_D + x_T$$

Example 3-2 When ethylene is oxidized with air by means of a silver catalyst at low temperature (200 to 250°C), two reactions occur:

1. $C_2H_4(g) + \frac{1}{2}O_2(g) \rightarrow C_2H_4O(g)$
2. $C_2H_4(g) + 3O_2(g) \rightarrow 2CO_2(g) + 2H_2O(g)$

If the conversion of ethylene is x_1 by reaction 1 and x_2 by reaction 2, express the molal composition of the reaction mixture in terms of the conversions and the ratio a, moles of air per mole of ethylene in the feed. What is the yield and selectivity of ethylene oxide?

SOLUTION If a basis of 1 mole of C_2H_4 is chosen, and $N_{C_2H_4O}$ and N_{CO_2} represent the moles of these components at any conversions, x_1 and x_2, then

$$x_1 = \frac{N_{C_2H_4O}}{1} \quad \text{or } N_{C_2H_4O} = x_1$$

and

$$x_2 = \frac{\frac{1}{2}N_{CO_2}}{1} \quad \text{or } N_{CO_2} = 2x_2$$

The moles of the other components will be

$$H_2O = 2x_2$$
$$C_2H_4 = 1 - x_1 - x_2$$
$$N_2 = 0.79a$$
$$O_2 = 0.21a - (\tfrac{1}{2}x_1 + 3x_2)$$
$$\text{Total moles} = 1 + a - \tfrac{1}{2}x_1$$

Then the mole fractions of each component are as follows:

$$C_2H_4O = \frac{x_1}{1 + a - \tfrac{1}{2}x_1}$$

$$CO_2 = \frac{2x_2}{1 + a - \tfrac{1}{2}x_1}$$

$$H_2O = \frac{2x_2}{1 + a - \tfrac{1}{2}x_1}$$

$$C_2H_4 = \frac{1 - x_1 - x_2}{1 + a - \frac{1}{2}x_1}$$

$$N_2 = \frac{0.79a}{1 + a - \frac{1}{2}x_1}$$

$$O_2 = \frac{0.21a - (\frac{1}{2}x_1 + 3x_2)}{1 + a - \frac{1}{2}x_1}$$

The total conversion of ethylene is $x_1 + x_2$. The yield of ethylene oxide is x_1, and its overall selectivity is x_1/x_2.

For any reactor the conservation principle requires that the *mass* of species i in an element of reactor volume ΔV obeys the following statement:

$$\left\{ \begin{array}{c} \text{Rate of } i \text{ into} \\ \text{volume element} \end{array} \right\} - \left\{ \begin{array}{c} \text{rate of } i \text{ out} \\ \text{of volume element} \end{array} \right\} + \left\{ \begin{array}{c} \text{rate of production of } i \\ \text{within the volume element} \end{array} \right\}$$

$$= \left\{ \begin{array}{c} \text{rate of accumulation of } i \\ \text{within the volume element} \end{array} \right\} \quad (3\text{-}1)$$

A key feature of this statement is the size of the volume element and its relation to the rate-of-production term. The element should be small enough that the concentration and temperature are uniform throughout its volume. Otherwise, the reaction rate **r** appearing in the rate of production term will not be constant throughout the volume element, and the utility of Eq. (3-1) is reduced. Of course, this restriction always may be satisfied by choosing a three-dimensional differential volume element. If this is done, the species continuity equation is obtained. This more general equation may then be integrated over the reactor volume for a *specific* type of reactor to yield a macroscopic mass balance for that *particular* reactor form. Alternately, Eq. (3-1) can be applied directly to a specific reactor type, taking advantage of being able to choose finite volume elements for some reactors. While both methods of obtaining the relationship between concentration and intrinsic rate have advantages, the second normally will be followed in this introductory text. However, the first method is illustrated in Example 3-4.

Mass conservation equations are developed in this chapter for the two extreme reactor types: ideal stirred tank, and ideal plug flow (refer to Sec. 1-5). Then deviations from these ideal equations are discussed in Chap. 6 (for homogeneous reactions) and in Chap. 13 (for heterogeneous reactions).

3-3 The Ideal Stirred-Tank Reactor

The stirred-tank reactor (STR) may be operated as a steady-state flow type†
(Fig. 3-1a) or a batch type (Fig. 3-1b). The key feature of this reactor is that the mixing is complete so that the properties (e.g., concentration, temperature) of the

---

† The flow form of an STR does not have to operate at steady state. For example, the mass feed rate may be different from the exit rate of product. Such transient or semibatch operation is considered in Chap. 4. The mass balance equations in this chapter are not applicable for such semibatch conditions.

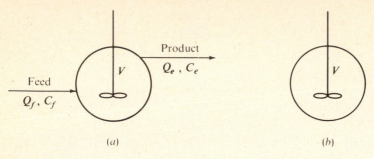

Figure 3-1 Ideal stirred-tank reactors classified according to method of operation. (a) Flow (steady-state). (b) Batch.

reaction mixture are uniform in all parts of the vessel. For a flow reactor the properties of the reaction mixture are the same as those in the exit (or product) stream. This means that the volume element used in Eq. (3-1) can be taken as the volume V of the *entire* contents. Also, the composition and temperature at which reaction takes place are the same as the composition and temperature of any exit stream. Thus, the reaction occurs at a rate corresponding to the *product* temperature and composition, and not the values for the feed.

Batch STR (Fig. 3-1b) Since there are no feed or exit streams, only the last two terms in Eq. (3-1) exist. For a volume element V, the mass balance becomes

$$\mathbf{r}_i V = \frac{d(C_i V)}{dt} \tag{3-2}$$

where C_i is the concentration of species i at any time. If the volume of the reaction mixture is constant, Eq. (3-2) becomes

$$\mathbf{r}_i = \frac{dC_i}{dt} \quad \text{(constant volume)} \tag{2-1} \text{ or } \text{(3-3)}$$

where $\mathbf{r}_i$ is the intrinsic rate of production, defined in Sec. 2-1. Note that Eq. (3-3) is identical to (2-1). This is because the restrictions of uniform concentration and temperature are satisfied in an ideal STR and the volume has been assumed to be constant. STR are normally used only for liquid-phase reactions. Since density changes with temperature and composition for liquids are small, the assumption of constant volume is usually reasonable.

Steady-State Flow STR (Fig. 3-1a) For steady state, the last term in Eq. (3-1) is zero. If Q is the total volumetric flow rate and subscripts f and e refer to feed and exit streams, the mass balance is

$$Q_f(C_i)_f - Q_e(C_i)_e + \mathbf{r}_i V = 0$$

or

$$Q_f(C_i)_f - Q_e(C_i)_e = -\mathbf{r}_i V \tag{3-4}$$

For equal flow rates $Q_e = Q_f = Q$, Eq. (3-4) reduces to

$$(C_i)_f - (C_i)_e = -\mathbf{r}_i\left(\frac{V}{Q}\right) = -\mathbf{r}_i\bar{\theta} \qquad \text{(constant flow rates)} \qquad (3\text{-}5)$$

where $\bar{\theta}\ (=V/Q)$ is the *average* residence time of the fluid in the reactor volume.

Equations (3-3) and (3-5) are applicable for calculating the rate of reaction from concentration vs. time data for a batch, laboratory reactor (as illustrated in Chap. 2) or from concentration difference vs. flow rate (Q) data obtained in a continuous flow, laboratory reactor. Alternately, such equations are useful for predicting concentrations vs. time or $(C_i)_f - (C_i)_e$ vs. Q in large-scale reactors. If i is a product, $\mathbf{r}_i$ is positive, while for a reactant $\mathbf{r}_i$ is negative.

Example 3-3 Express Eqs. (3-2) to (3-5) in terms of conversion of *reactant* (A).

SOLUTION In Sec. 2-1 [Eq. (2-6)] the conversion x_A is defined as the fraction of A that has been converted to products by the reaction: that is

$$x_A = \frac{(n_A)_0 - n_A}{(n_A)_0} \qquad (2\text{-}6)$$

A. For the *batch* STR, $(n_A)_0$ signifies the moles of A at $t = 0$. If C is a molal concentration

$$n_A = VC_A$$

$$(n_A)_0 = V_0(C_A)_0$$

and Eq. (2-6) becomes

$$x_A = \frac{V_0 C_{A0} - VC_A}{V_0 C_{A0}} \qquad (3\text{-}6)$$

Differentiating with respect to time

$$\frac{d(VC_A)}{dt} = -V_0 C_{A0} \frac{dx_A}{dt}$$

Comparison with Eq. (3-2), with species i taken as reactant A, gives

$$\frac{dx_A}{dt} = -\frac{V}{V_0 C_{A0}} \mathbf{r}_A \qquad (3\text{-}7)$$

where $\mathbf{r}_A$ is the rate of production of A, a negative quantity for a reactant. If the reactor volume does not change as the reaction proceeds, $V = V_0$, and

$$\frac{dx_A}{dt} = -\frac{1}{C_{A0}} \mathbf{r}_A \qquad \text{(constant volume)} \qquad (3\text{-}8)$$

For application to designing a reactor when the rate is known, Eq. (3-7) can be written in an integrated form as follows:

$$t - 0 = -V_0 C_{Ao} \int_{x_A=0}^{x_A} \frac{dx_A}{V r_A} \tag{3-9}$$

where the initial condition is taken as $x_A = 0$ at $t = 0$. Note that the product $V_0 C_{Ao}$ is the number of moles of reactant A in the reactor initially. The importance of arranging the expression in this way is that time, the independent design variable, is separated from the dependent variables characteristic of the chemical reaction. If the rate r and the density of the reaction mixture are known as a function of conversion, the value of the integral for any desired conversion can be evaluated without reference to reaction equipment. Then the various combinations of time and mass of charge that will give the required production rate of product can be examined separately. Expressed differently, Eq. (3-9) relates the time to an integral dependent on the series of intensive states experienced by the reaction mixture.

For constant volume, $V_0 = V$ and Eq. (3-9) becomes

$$t = -C_{Ao} \int_0^{x_A} \frac{dx_A}{r_A} \tag{3-10}$$

Equation (3-10) is the integrated form of Eq. (3-3) written in terms of conversion.

B. For the *flow* STR, the conversion x_A is the fraction of reactant A in the feed that has been converted to products. Hence, Eq. (2-6) takes the form

$$x_A = \frac{Q_f (C_A)_f - Q_e (C_A)_e}{Q_f (C_A)_f} \tag{3-11}$$

where x_A is the conversion in the exit (product) stream. Equation (3-11) can be combined with Eq. (3-4) to express the performance of the flow STR in terms of x_A

$$\frac{V}{(C_A)_f Q_f} = -\frac{x_A}{r_A} \tag{3-12}$$

The product $(C_A)_f Q_f$ is the molal flow rate of reactant A, which we will designate as F_A. Hence, Eq. (3-12) may be written:

$$\frac{V}{F_A} = -\frac{x_A}{r_A} \tag{3-13}$$

Equation (3-13) is valid both for $Q_e \neq Q_f$ and $Q_e = Q_f$, while Eq. (3-5) is limited to constant flow rates. However, this advantage is seldom important because STR are normally used only for liquid reaction systems for which density changes are small. For this reason equations for STR are usually written in terms of concentrations and will be used in this form, primarily, in later chapters. For gas phase reactions, changes in total number of moles due to the reaction itself can cause large changes in both molal concentration and flow rates (at constant pressure). Hence, in tubular-flow reactors, commonly used for gas-phase reactions, mass balance equations in terms of conversion are preferable. This is discussed in the next section.

3-4 The Ideal Tubular-Flow (Plug-Flow) Reactor

As mentioned in Chap. 1 (Sec. 1-5) the *ideal* tubular-flow reactor is one in which there is no mixing in the direction of flow and complete mixing perpendicular to the direction of flow (i.e., in the radial direction). Figure 3-2 represents such a reactor. Concentrations will vary along the length coordinate, z, but not the radial coordinate, r. Except when isothermal operation is possible, temperature will also vary with z. We conclude that the rate of reaction will vary with reaction length. Therefore, the volume element in the mass balance [Eq. (3-1)] must be differential in length, but can extend across the entire diameter of the reactor. Tubular-flow reactors are normally operated at steady state so that properties at any position are constant with respect to time.† For such steady-state operation, Eq. (3-1) applied to the volume element ΔV, becomes

$$QC_i|_V - QC_i|_{V+\Delta V} + \mathbf{r}_i \Delta V = 0$$

Taking the limit as $\Delta V \to 0$, gives

$$\frac{d(QC_i)}{dV} = \mathbf{r}_i \tag{3-14}$$

The volumetric flow rate Q, as well as C_i, can vary significantly with reactor length due to changes in temperature and number of total moles for a gas-phase reaction. For these reasons it is usually preferable to express Eq. (3-14) in terms of conversion rather than concentration. This takes advantage of the constancy of the feed rate of reactant F_A. The relation between concentration and conversion is

† At startup or shutdown such reactors are not at steady state. Mass balance equations for such transient conditions in a tubular-flow reactor are considered in Chaps. 4 and 5.

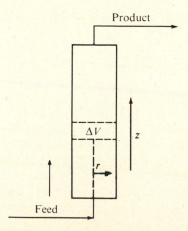

Product

Feed

ΔV

z

r

Figure 3-2 Ideal tubular-flow reactor.

given by Eq. (2-6). At any point z along the reactor where the conversion is x_A and flow rate is Q, this relation may be written

$$x_A = \frac{Q_f C_{Af} - Q C_A}{Q_f C_{Af}} \tag{3-15}$$

Since $Q_f C_{Af}$ is the constant molal feed rate F_A of reactant A, Eq. (3-15) can be differentiated with respect to reactor volume to give

$$\frac{d(Q C_A)}{dV} = -Q_f C_{Af} \frac{dx_A}{dV} \tag{3-16}$$

Now, if species i is reactant A, Eq. (3-14) may be combined with Eq. (3-16) to give the mass balance in terms of conversion

$$\frac{dx_A}{dV} = -\frac{\mathbf{r}_A}{Q_f C_{Af}} = -\frac{\mathbf{r}_A}{F_A} \tag{3-17}$$

or, integrating formally,

$$\frac{V}{F_A} = -\int_{x_A=0}^{x_A} \frac{dx_A}{\mathbf{r}_A} = \int_{x_A=0}^{x_A} \frac{dx_A}{\mathbf{r}_p} \tag{3-18}$$

The last equality shows that the minus sign can be deleted if the rate $\mathbf{r}_p$ is for the production of product rather than reactant. In Eqs. (3-17) and (3-18) there are only two dependent variables x_A and $\mathbf{r}_A$ rather than the three (Q, C_i, $\mathbf{r}_i$) in Eq. (3-14). Equations (3-17) and (3-18) are the common forms of mass balance for a plug-flow reactor (PFR). They are applicable for variable temperature and total moles for a gas-phase reaction, and, of course, are applicable for liquid-phase systems. As written, Eq. (3-18) is restricted to the case of zero conversion in the reactor feed. On occasion,† a partially converted feed enters the reactor. For this condition, F_A should be the molal feed rate of A corresponding to zero conversion. If this is $(F_A)_0$, the moles reacted in the volume element is $(F_A)_0\, dx$ and according to Eq. (3-17) this is equal to $-\mathbf{r}_A\, dV$. In integrated form

$$\frac{V}{(F_A)_0} = -\int_{x_{Af}}^{x_A} \frac{dx_A}{\mathbf{r}_A} \tag{3-18a}$$

where x_{Af} represents the conversion in the reactor feed. Conversion in this case is given by the following form of Eq. (3-15)

$$x_A = \frac{F_{A0} - Q C_A}{F_{A0}} = \frac{Q_0 C_{A0} - Q C_A}{Q_0 C_{A0}} \tag{3-15a}$$

where Q_0 and C_{A0} are for unconverted ($x_A = 0$) conditions. The conversion of the feed stream is

$$x_{Af} = \frac{Q_0 C_{A0} - Q_f C_{Af}}{Q_0 C_{A0}} \tag{3-15b}$$

† For example, in recycle reactors (Sec. 4-10).

For the special case of constant volume flow rate Q, Eq. (3-14) may be written

$$\mathbf{r}_i = Q\frac{dC_i}{dV} = \frac{dC_i}{d(V/Q)} = \frac{dC_i}{d\theta} \qquad \text{(constant } Q\text{)} \qquad (3\text{-}19)$$

Here θ is the residence time of an element of fluid in the reactor (θ is the same for all elements of fluid in a PFR—see Chap. 6). For these restricted conditions Eq. (3-19) for a PFR is the same as the mass balance for a batch STR of constant volume [Eq. (3-3)], except that the real time variable t is replaced by the residence time variable θ. Note that the mass balance expressions for a batch STR [Eq. (3-3)] and for a PFR [Eq. (3-17)] are differential equations. The rate varies with time in the batch reactor and with axial position in the tabular-flow reactor. In contrast, the mass balance for a continuous-flow STR, Eq. (3-13) is an algebraic equation.

Example 3-4 Derive the species continuity equation and then apply this result to a plug-flow reactor to obtain Eqs. (3-14) and (3-17).

SOLUTION We first derive the species continuity equation for reactant A.[†] Consider a differential volume element ($\Delta x \, \Delta y \, \Delta z$) in a reaction mixture. The homogeneous reacting fluid flows through the volume element with a vector velocity $\mathbf{v}$, of which the components in the x, y, and z directions are v_x, v_y, v_z. The flux (mol/(time)(area)) of A, in general, will consist of a convective term due to the velocity and a diffusion contribution. Suppose this total flux is the vector $\mathbf{N}_A$ with components $(N_A)_x$, $(N_A)_y$, $(N_A)_z$ in the three directions. The molal flow rates of A in and out of the element are:

$$\text{Rate of } A \text{ in} = (N_A)_x\Big|_x (\Delta y \, \Delta z) + (N_A)_y\Big|_y (\Delta x \, \Delta z) + (N_A)_z\Big|_z (\Delta x \, \Delta y)$$

$$\text{Rate of } A \text{ out} = (N_A)_x\Big|_{x+\Delta x} (\Delta y \, \Delta z) + (N_A)_y\Big|_{y+\Delta y} (\Delta x \, \Delta z) + (N_A)_z\Big|_{z+\Delta z} (\Delta x \, \Delta y)$$

The rate of *production* of A in the volume element is $\mathbf{r}_A(\Delta x \, \Delta y \, \Delta z)$ and the rate of accumulation of A within the element is $(\partial C_A/\partial t)(\Delta x \, \Delta y \, \Delta z)$. Substituting these expressions in Eq. (3-1) and dividing by $(\Delta x \, \Delta y \, \Delta z)$ yields[‡]

$$\frac{(N_A)_x\big|_{x+\Delta x} - (N_A)_x\big|_x}{\Delta x} + \frac{(N_A)_y\big|_{y+\Delta y} - (N_A)_y\big|_y}{\Delta y}$$

$$+ \frac{(N_A)_z\big|_{z+\Delta z} - (N_A)_z\big|_z}{\Delta z} - \mathbf{r}_A = -\frac{\partial C_A}{\partial t}$$

[†] A more careful and complete derivation is given in texts on transport phenomena; e.g., "Transport Phenomena," R. B. Bird, W. E. Stewart, and E. N. Lightfoot, Chap. 18, John Wiley & Sons, Inc., New York, 1960.

[‡] After changing signs of all terms.

If now the limit is taken as Δx, Δy, and Δz approach zero, the species continuity equation is obtained:

$$\frac{\partial (N_A)_x}{\partial x} + \frac{\partial (N_A)_y}{\partial y} + \frac{\partial (N_A)_z}{\partial z} - \mathbf{r}_A = -\frac{\partial C_A}{\partial t} \tag{3-20}$$

In vector notation Eq. (3-20) may be written:

$$-\mathbf{V} \cdot \mathbf{N}_A + \mathbf{r}_A = \frac{\partial C_A}{\partial t} \tag{3-21}$$

where the scalar (dot) product $\mathbf{V} \cdot \mathbf{N}_A$ is the divergence of A.

Next Eq. (3-21) is integrated over the reactor volume V. In integrating the divergence term, we can replace the volume integral with an area integral using the Gauss divergence theorem.[†] That is,

$$\int (\mathbf{V} \cdot \mathbf{N}_A) \, dV = \int \mathbf{N}_A \cdot \mathbf{n} \, dA$$

where $\mathbf{n}$ is a unit vector directed perpendicularly outward from the surface enclosing the volume V. With this substitution, Eq. (3-21) may be written

$$-\int \mathbf{N}_A \cdot \mathbf{n} \, dA + \int \mathbf{r}_A \, dV = \frac{\partial}{\partial t} \int C_A \, dV \tag{3-22}$$

Equation (3-22) is a general form of the macroscopic mass balance for component A. It is applicable for any method of operating any type of reactor. When this equation is applied to a plug-flow reactor (Fig. 3-2), the flux vector $\mathbf{N}_A$ will have a finite value only for the cross-sectional areas of the feed and exit streams (A can't pass through the walls of the reactor). Further, in a plug-flow reactor, there will be no contribution to the flux due to diffusion, but only the convective term due to the velocity so that $\mathbf{N}_A = v_z C_A$. With these simplifications, and noting that the reactor operates at steady state, Eq. (3-22) reduces to the form

$$(C_A v_z A)_f - (C_A v_z A)_e + \int \mathbf{r}_A \, dV = 0 \tag{A}$$

Here subscripts f and e refer to feed and exit streams. It also has been assumed that the concentrations and velocities are uniform over the areas A_f and A_e; otherwise area-average values must be employed. Since $v_z A = Q$ and for a plug-flow reactor, $A_f = A_e = A$, Eq. (A) may be written:

$$(QC_A)_f - (QC_A)_e + \int \mathbf{r}_A \, dV = 0 \tag{B}$$

[†] "Advanced Engineering Mathematics," C. R. Wylie, Jr., 3d ed., p. 572, McGraw-Hill Book Co., New York (1960). Also see "Transport Phenomena," R. B. Bird, W. E. Stewart, and E. D. Lightfoot, p. 213, John Wiley & Sons, Inc., New York, 1960.

Differentiating with respect to volume and noting that the volume increases in the direction from feed to exit, Eq. (B) becomes

$$\frac{d(QC_A)}{dV} = \mathbf{r}_A \tag{C}$$

which is identical with Eq. (3-14). Next we follow the same procedure as employed in the text to obtain Eq. (3-17) from (3-14). That is, Eq. (3-16) is used to express Eq. (C) in terms of conversion to give Eq. (3-17).

Example 3-5 The vapor phase decomposition of acetaldehyde

$$CH_3CHO(g) \rightarrow CH_4(g) + CO(g)$$

results in a decrease in total moles. Suppose this thermal decomposition reaction is studied in an ideal-tubular flow reactor which operates isothermally and at essentially constant pressure. The feed rate of pure acetaldehyde is F_A mol/s, corresponding to a volumetric rate of Q m³/s. The reaction is second order so the rate of production of reactant A is:

$$\mathbf{r}_A = -kC_A^2 \quad [\text{mol}/(\text{s})(\text{m}^3)]$$

$$C_A = \text{moles/m}^3$$

For this reaction with a change in moles, demonstrate the advantage of Eq. (3-17) vs. (3-14) for calculating the amount of reaction in a reactor volume V.

SOLUTION To integrate Eq. (3-17) all that is necessary is to express the rate in terms of the conversion x_A. We assume that the mixture follows ideal gas behavior. Then the concentration of acetaldehyde at any point in the reactor is

$$C_A = \frac{N_A}{Q} \tag{A}$$

where N_A and Q are the molal flow rate of CH_3CHO and the total volumetric flow rate. The two quantities can be expressed in terms of the conversion by the stoichiometry of the reaction. If we take as a basis a feed flow of 1.0 mol/m³ of CH_3CHO, and x_A is the conversion, the molal flow rate of each component will be:

$$CH_3CHO = 1 - x_A$$

$$CO = x_A$$

$$CH_4 = x_A$$

$$\overline{\text{Total moles} = 1 + x_A}$$

Then Eq. (A) in terms of x_A is

$$C_A = \frac{1 - x_A}{(R_g T/p_t)(1 + x_A)} \tag{B}$$

where $R_g T/P$ is the volume per mole at the operating temperature and pressure of the reactor. Equation (B) can be used to give the desired expression for r_A in terms of x:

$$r_A = \frac{-k}{(R_g T/p_t)^2} \left(\frac{1 - x_A}{1 + x_A}\right)^2 \tag{C}$$

Equation (C) is next substituted into Eq. (3-17) and integrated formally, in accordance with Eq. (3-18).

$$\frac{V}{F_A} = \left(\frac{R_g T}{p_t}\right)^2 \frac{1}{k} \int_0^{x_A} \left(\frac{1 + x_A}{1 - x_A}\right)^2 dx_A \tag{D}$$

This result is an easily integrable equation giving the conversion x_A corresponding to a reactor volume V. Note that by using the conversion form of the mass balance the changing volumetric flow rate, due to a change in total moles, does not make the problem complex. Similarly, no difficulties are introduced if Q varies because the temperature is not uniform in the reactor. However, in this case, the rate constant k in Eq. (D) becomes a variable and would be placed within the integral. Then the temperature as a function of position would need to be known before the integration could be carried out. This type of problem is considered in Chap. 5.

If Eq. (3-14) were to be used, it could be integrated formally and written

$$V = \frac{d(QC_A)}{r_A} = \int_{C_{Af}}^{C_{Ae}} \frac{Q}{r_A} dC_A + \int_{Q_f}^{Q_e} \frac{C_A}{r_A} dQ \tag{E}$$

To integrate Eq. (E) we need the relationship between Q and C_A. This is most easily obtained by expressing each variable in terms of x_A. Eq. (B) gives C_A in terms of x_A, and the volumetric flow rate is

$$Q = \frac{R_g T}{p_t} (1 + x_A) \tag{F}$$

The conversion can be eliminated between Eqs. (B) and (F) to obtain $Q = f(C_A)$. Since r_A is given in terms of C_A, Eq. (E) could then be integrated to yield the exit conversion C_{Ae} for any reactor volume V. However, the solution of Eq. (E) is much more cumbersome than the solution of Eq. (D).

We conclude that only when the volumetric flow rate is constant is it convenient to use Eq. (3-14).

3-5 Deviations from Ideal Reactors

We have been able to develop rather simple mass-conservation equations for the ideal versions of the stirred-tank and tubular-flow reactors. When the mixing criteria of these ideal forms are not satisfied, mathematical expressions for the conservation equations become more difficult. In this section types of deviations are discussed qualitatively so that we recognize that the idealized equations presented in Secs. 3-3 and 3-4 are not always applicable. Then in Chap. 6 mathematical treatment is presented for nonideal reactors in which isothermal, homogeneous reactions occur. In Chaps. 12 and 13 nonideal behavior in heterogeneous reactors is considered.

As an example of nonideal operation, imagine a continuous-flow, *tank* reactor of poor design (inadequate mixing) such that pockets of nearly stagnant fluid exist, as shown by regions marked *S* in Fig. 3-3*a*. The conversion will become very high in the nearly stagnant fluid, but this fluid will remain in the reactor for a long time. Because the volume available for flow has been reduced by the stagnant regions, the bulk of the fluid will spend less time in the reactor and, hence, have less time to react. The result will be an average conversion in the exit stream which is less than that for the ideal type. Figure 3-3*b* shows another type of deviation, caused by bypassing or short-circuiting of the fluid. Here a part of the fluid entering the tank follows a shortened path to the exit and maintains its identity (does not mix) while doing so. Again, the conversion in the exit stream is reduced below that of the ideal stirred-tank reactor. While these are extreme cases attributable to poor design, it is clear that actual reactors may deviate to some degree from ideal behavior, and the behavior of a specific reactor will depend on the extent of mixing.

Deviations from ideal *tubular-flow* behavior are also possible. Two kinds of deviations are: (1) some mixing in the longitudinal direction and (2) incomplete mixing in the radial direction. Figure 3-4*a* and *b* illustrates these two effects. In Fig. 3-4*a* the inlet and exit nozzles are such that vortices and eddies produce mixing in the longitudinal direction. Figure 3-4*b* represents the situation where the fluid is in laminar flow, forming a parabolic velocity profile across the tube. Since the molecular-diffusion process is relatively slow, the annular elements of fluid flow through the reactor only slightly mixed in the radial direction. Also, the fluid near the wall will have a longer residence time in the reactor than for ideal tubular-flow performance, while the fluid near the center will have a shorter residence time. The result again is a decrease in conversion. Bypassing (channeling) can also occur in a fixed-bed, fluid-solid catalytic reactor, as shown in Fig. 3-4*c*. The nonuniform packing arrangement results in a higher velocity near the tube wall where the porosity is greatest. Hence, the time for reaction (residence time) near the wall will be less than that at the center of the tube. This means that the composition of the fluid will vary radially, so that the requirement of complete radial mixing is not satisfied.

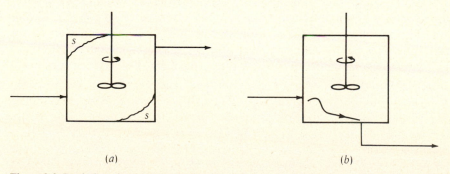

(a)   (b)

Figure 3-3 Deviations from ideal stirred-tank performance: (*a*) stagnant regions; (*b*) bypassing.

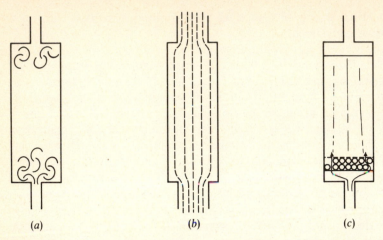

Figure 3-4 Deviations from ideal tubular-flow performance: (*a*) longitudinal mixing due to vortices and turbulence; (*b*) laminar-flow (poor radial mixing); (*c*) bypassing in fixed-bed catalytic reactor.

Deviations from ideal performance divide into two classifications. The first is a flow arrangement in which elements of fluid do not mix, but follow separate paths through the reactor (segregated flow). These elements are retained in the reactor for different times; that is, they have different residence times. The second is a flow arrangement whereby adjacent elements of fluid partially mix (micromixing). The effects of these deviations on the conversion can be evaluated, provided we know the distribution of residence times in the fluid leaving the reactor *and* the extent of micromixing. Such complete information is seldom available. However, for well-defined cases of micromixing the effect of a distribution of residence times on the conversion can be evaluated. Procedures for treating these factors quantitatively are considered in Chap. 6. Usually the effects are relatively small, although for special reactors, such as a stirred-tank type with internal coils and a viscous reaction mixture, they may be large. Because of the great influence of temperature on the rate of reaction, unaccounted-for temperature profiles in a reactor are often more significant.†

3-6 Space Velocity

The term V/F_A in Eqs. (3-13) and (3-18) for flow reactors determines the size of the reactor necessary to obtain the conversion x_A for a reactant feed rate F_A. Except for the inconsistency of comparing volume and mass, V/F_A is a measure of the reaction time. This can be indicated more directly by expressing the reactant feed rate in terms of the total volumetric flow rate of feed, Q_f. If C_{A_f} is the concentration of reactant in the feed, $F_A = C_{A_f} Q_f$, so that

$$\frac{V}{F_A} = \frac{V}{C_{A_f} Q_f}$$

† See K. G. Denbigh, "Chemical Reactor Theory," pp. 44, 63, Cambridge University Press, New York, 1965.

or

$$\frac{V}{Q_f} = C_{Af}\left(\frac{V}{F_A}\right)$$ (3-23)

The ratio, V/Q_f, has the units of time. Its reciprocal, Q_f/V, is called the *space velocity*. This term is often used in practice, along with conversion, to describe overall reactor performance. For example, it is common practice to plot conversion vs. space velocity to describe the effect of feed rate on performance of a flow reactor.†

The quantity V/Q_f is equal to the residence time in a plug-flow reactor only if the volumetric flow rate is constant throughout the reactor. The residence time depends upon the change in flow rate *through* the reactor as well as V/Q_f. The change in Q in turn depends upon the variation in temperature, pressure, and number of moles (for a gaseous reaction). The effect of a change in number of moles is illustrated in Example 3-6.

Example 3-6 Acetaldehyde vapor is decomposed in an *ideal* tubular-flow reactor according to the reaction

$$CH_3CHO \rightarrow CH_4 + CO$$

The reactor is 3.3 cm (0.033 m) ID and 80 cm (0.80 m) long and is maintained at a constant temperature of 518°C. The acetaldehyde vapor is measured at room temperature and slightly above atmospheric pressure. For consistency, the measured flow rate is corrected to standard conditions (0°C and 1 atm, or 273 K and 101 kPa) before the space velocity is reported. In one run, at a reported space velocity of 8.0 h^{-1} (2.22×10^{-3} s^{-1}), 35% of the acetaldehyde is decomposed in the reactor. The second-order specific rate constant is 0.33 liter/(s)(g mol) [or 0.33 m^3/(s)(kg mol)] at 518°C, and the reaction is irreversible. The pressure is essentially atmospheric. Calculate the actual residence time and compare it with V/Q_f.

SOLUTION The time required for an element of reaction mixture to flow through an element of reactor volume dV is

$$d\theta = \frac{dV}{Q} = \frac{dV}{N_t v}$$ (3-24)

where v is the volume per mol of total molal flow rate N_t. Both N_t and v can vary position within the reactor. Since these quantities are directly related to

† In many cases (particularly in the petroleum and petrochemical industries) the feed is a liquid at room temperature (and is so metered) while the reaction mixture is a gas at the higher temperatures in the reactor. For these cases it is common to report performance as conversion (and selectivity) vs. LHSV, the *liquid hourly space velocity*. This is $(Q_f)_L/V$ where $(Q_f)_L$ is the metered liquid flow rate of the feed. Even for gaseous feeds the space velocity is normally reported in terms of volumes measured at metering conditions or standard conditions (see Example 3-6) rather than at reactor operating temperatures and pressures.

conversion, and not to V, the integration of Eq. (3-24) is more easily carried out by replacing dV with dx, using Eq. (3-17). This gives

$$d\theta = -F_A \frac{dx_A}{N_t v \mathbf{r}_A} \tag{3-25}$$

In this expression F_A (the feed rate of reactant) is a constant, while N_t will vary along the reactor length, if there is a change in number of moles.

In our example v is constant and equal to $R_g T/p$, since p and T are constant. N_t varies, but in Example 3-5, we found the total molal flow rate to be $(1 + x_A)$ for a feed rate of pure acetaldehyde equal to unity. Hence, for an acetaldehyde feed rate of F_A,

$$N_t = F_A(1 + x_A) \tag{A}$$

The rate in terms of conversion is given by Eq. (C) of Example 3-5. Substituting these expressions for v, N_T, and $\mathbf{r}_A$ in Eq. (3-25) yields

$$d\theta = \frac{R_g T}{p_t k} \frac{(1 + x_A)}{(1 - x_A)^2} dx_A \tag{B}$$

Integration from $x_A = 0$ to $x_A = 0.35$ gives

$$\theta = \frac{R_g T}{p_t k} \left[\frac{2}{1 - x_A} + \ln (1 - x_A) \right]_0^{0.35}$$

$$= \frac{0.082(273 + 518)}{1(0.33)} \left[\frac{2}{1 - 0.35} + \ln (1 - 0.35) - 2 \right] = 127 \text{ s}$$

The *space velocity* of 8.0 h^{-1} is the value of V/Q_f where Q_f is measured at 0°C and 1 atm. Hence,

$$\frac{V}{Q_f} = \frac{1}{8} = 0.125 \text{ h} \quad \text{or} \quad 450 \text{ s}$$

The major difference between θ and V/Q_f arises because the space velocity was based on a flow rate at a standard temperature of 273°C. Thus, if the space velocity were corrected to the reactor temperature, it would be

$$Q_f/V \text{ (at 791 K)} = 8.0 \frac{791}{273} = 23.2 \text{ h}^{-1}$$

Then

$$V/Q_f = \frac{1}{23.2} (3,600) = 155 \text{ s}$$

The difference between this value and the actual θ (127 s) is due to the increase in the number of moles occurring as a result of reaction.

3-7 Temperature Effects†

In a few cases the heat of reaction is so low that heat exchange with the surroundings is sufficient to eliminate temperature changes. The design problem for such isothermal reactors is greatly simplified because the variation in rate of reaction with temperature need not be considered. The isomerization of n-butane (ΔH at $25°C = -1600$ cal/g mol) is an example of this class of reactions. Even when the heat of reaction is moderate, it may be possible to approach isothermal operation by adding or removing heat from the reactor. In alkylation processes for producing isooctane from isobutane and butenes, the heat of reaction is about $-17,000$ cal/g mol at $25°C$. However, by cooling the mixture in the reactor with external cooling jackets it is possible to reduce the temperature variation to 20 to $40°F$.

When the heat of reaction is large, sizable temperature variations may be present even though heat transfer between the reactor and surroundings is facilitated. In such cases it is necessary to consider the effect of temperature on the rate of reaction. Reactors operating in this fashion are termed *nonisothermal* or *nonadiabatic*.

Reactors (both flow and batch) may also be insulated from the surroundings so that their operation approaches adiabatic conditions. If the heat of reaction is significant, there will be a change in temperature with time (batch reactor) or position (flow reactor). In the PFR reactor this temperature variation will be limited to the direction of flow; i.e., there will be no radial variation in a tubular-flow reactor. We shall see in Chap. 13 that the design procedures are considerably simpler for adiabatic operation.

In addition to heat exchange with the surroundings, there are other methods of approaching isothermal operating conditions. For example, in the dehydrogenation of butylenes to butadiene the temperature must be maintained at a rather high level (1200 to 1400°F) for a favorable equilibrium conversion. However, the endothermic nature of the reaction means that the reaction mixture will cool as it flows through the reactor bed. It is both difficult and expensive to transfer heat to the reaction mixture at this high temperature level by external heating. Instead, high-temperature steam is added directly to the butenes entering the reactor. The large quantity of steam serves to maintain the reaction mixture at a high temperature level.‡ Another device frequently employed for tank reactors is internal cooling or heating coils. A modification of the same principle is illustrated in tubular-flow reactors for the oxidation of naphthalene to phthalic anhydride. The flow reactor is divided into a large number of small tubes rather than a single large-diameter tube. Each small tube is surrounded with cooling fluid (often a boiling liquid) which absorbs the heat of reaction. Under actual operating conditions some naphthalene is oxidized completely to carbon dioxide and water vapor,

---

† The influence of heat effects on reactor type, including a description of various methods of cooling and heating reactors and of energy recovery, is given in " Chemical Reactor Design for Process Plants," chap. 6, by H. F. Rase, John Wiley & Sons, Inc., New York, 1977.

‡ Steam has other advantages; in particular, it reduces polymerization.

so that the total heat of reaction per gram mole of naphthalene is as high as $-570,000$ cal. By careful control of temperature to prevent overheating, it is possible to reduce CO_2 production and so increase the selectivity to phthalic anhydride.

In summary, the operation of commercial reactors falls into three categories: isothermal, adiabatic, and the broad division of nonisothermal. In the third category attempts are made to approach isothermal conditions, but the magnitude of the heat of reaction or the temperature level prevents attaining this objective. Quantitative calculations for isothermal and nonisothermal homogeneous reactors are given in Chaps. 4 and 5, respectively.

3-8 Mechanical Features

Tank Reactors† The batch reactor is a kettle or tank, or it may be a closed loop of tubing provided with a circulating pump. It should have a number of accessories to be operated satisfactorily. It generally must be closed, except for a vent, to prevent loss of material, pollution, and danger to the operating personnel. For reactions carried out under pressure the vent is replaced by a safety valve.

High-pressure conditions frequently introduce complications in the design and greatly increase the initial cost. For tanks the top closure must be able to withstand the same maximum pressure as the rest of the autoclave. At medium pressures a satisfactory closure can be assembled by using bolts or studs and suitable flanges and gaskets. The seal is obtained by tightening the six or more bolts holding the flange to the head. Such a closure is illustrated in the batch reactor shown in Fig. 3-5. For higher pressure (above approximately 5000 lb/in.² abs) this type of construction is not desirable because of the very high stresses that the bolts must withstand. The preferred design is one in which the

---

† An extensive discussion of mechanical features of tank reactors is given in chap. 8, Chemical Reactor Design for Process Plants, H. F. Rase, John Wiley & Sons, Inc., New York, 1977.

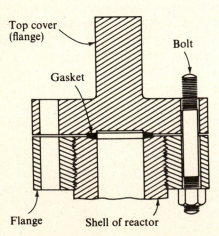

Top cover (flange)

Bolt

Gasket

Flange

Shell of reactor

Figure 3-5 Conventional flange-and-bolt closure for batch reactor.

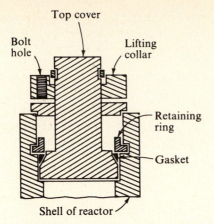

Top cover

Bolt hole

Lifting collar

Retaining ring

Gasket

Shell of reactor

Figure 3-6 Closure for high-pressure batch reactor.

pressure itself seals the vessel, and increases in pressure do not cause a corresponding increase in stress in the bolts. An example of this self-sealing closure is shown in Fig. 3-6. The pressure acting on the head is transmitted to the gasket, which is confined by the reactor wall, the head, and a retaining ring. The internal pressure pushes the head against the gasket, augmenting the force exerted by the bolts through the lifting collar. The problems encountered in designing batch reactors for medium- and high-pressure operation have been studied in some detail.† Loop reactors are well suited for high pressure, because the tubing has a smaller diameter than a tank.

It is necessary to agitate the reaction mixture in tank reactors. This can be done mechanically with stirrers operated by a shaft extending through the reactor wall. In high-pressure reactors rather complicated packing glands are needed around the shaft to prevent leakage. A typical design is illustrated in Fig. 3-7, where the mechanical details of a reactor, which is also jacketed, are shown.

Batch reactors vary in construction from ordinary steel tanks to glass-lined equipment, depending on the properties of the reaction mixture for which they are to be used. In pilot-plant operations either stainless-steel or glass-lined reactors are ordinarily used because of their corrosion resistance, and hence their general applicability to a variety of systems. In commercial-scale equipment it may be more economical to use ordinary steel even though corrosion is significant. In the food and pharmaceutical industries it is frequently necessary to use glass-lined or stainless-steel equipment to ensure purity of the product.‡

Tubular Reactors§ Flow reactors may be constructed in one of a number of ways. The conventional thermal cracking units in the petroleum industry are examples of a noncatalytic type. The petroleum fraction is passed through a

---

† D. B. Gooch, *Ind. Eng. Chem.*, **35**, 927 (1943); E. L. Clark, P. L. Golber, A. M. Whitehouse, and H. H. Storch, *Ind. Eng. Chem.*, **39**, 1955 (1947); F. D. Moss, *Ind. Eng. Chem*, **45**, 2135 (1945).

‡ See J. H. Perry, "Chemical Engineers' Handbook," 5th ed., McGraw-Hill Book Company, New York, 1973.

§ See chap. 9, J. H. Rase, op. cit.

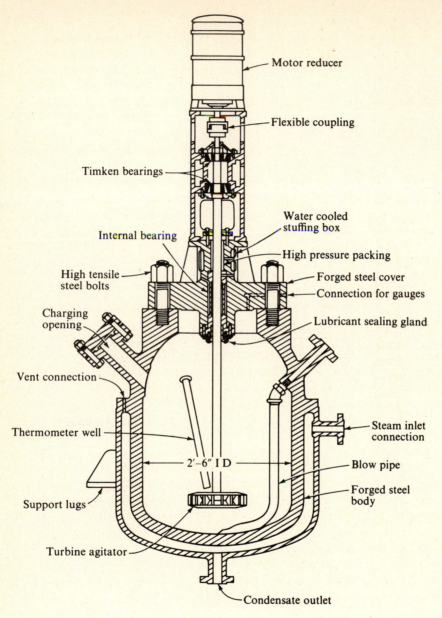

Figure 3-7 Jacketed batch reactor. [*By permission D. B. Gooch, Ind. Eng. Chem., 35, 927 (1943).*]

number of lengths of alloy-steel tubes placed in series on the inside of the walls and roof of a furnace. Heat is transferred by convection and radiation to the tube surface in order to raise the temperature of the fluid to the reaction level (600 to 1300°F) and to supply the endothermic heat of reaction. Flow reactors may consist of a tank or kettle, much like a batch reactor, with provision for continuously adding reactants and withdrawing product. The tank type is not suitable for reactions such as thermal cracking, where large quantities of thermal energy must be supplied, because of the low-heat-transfer surface per unit volume of reactor. Tank-type flow reactors are advantageous for conversions that require a long reaction time. In such reactors it is possible to obtain essentially complete mixing by mechanical agitation. Under these conditions the composition, temperature, and pressure are uniform through the vessel. In the tubular type, where the length is generally large with respect to the tube diameter, the forced velocity in the direction of flow is sufficient to retard mixing in the axial direction; that is, it is possible to approach plug-flow performance.

A large number of commercially important reactions are of the fluid-solid catalytic class. Examples are the catalytic cracking of petroleum, oxidation of sulfur dioxide, ammonia synthesis, dehydrogenation of butane to butene, oxidative dehydrogenation of butene to butadiene, and oxidation of naphthalene or xylenes to phthalic anhydride. In this group the solid catalyst may be held in a fixed position while the fluid moves through it (fixed-bed reactors), or, much smaller catalyst particles may be suspended in the fluid phase by the motion of the fluid (fluidized-bed reactor), or the solid particles may be in point-to-point contact and fall slowly by gravity through the fluid (moving-bed reactor).

A fixed-bed reactor built as a single large-diameter tube is less costly than a multitubular design. However, the latter arrangement may be required when large quantities of heat must be transferred to the surroundings, as in the case of a highly exothermic reaction, e.g., oxidation reactions such as conversion of naphthalene or xylenes to phthalic anhydride. The smaller the tube diameter, the larger the ratio of heat-transfer surface to mass of reaction mixture in the tube and the easier it is to limit temperature changes between inlet and exit. Of course, the low capacity of small tubes means that a larger total number of tubes must be built into the reactor in parallel to achieve a given production rate. Another means of reducing temperature variations in fixed-bed reactors is to divide the catalyst bed into sections, with heating or cooling coils placed between the sections (see Chap. 1, Fig. 1-7).

All these devices to reduce temperature gradients in the fixed-bed reactor are corrective rather than preventive. In solid catalytic reactors the potentially large temperature variations in the direction of flow are due to the fact that the solid catalyst is unable to mix and reach a more uniform temperature. Near the entrance to the bed the rate of reaction is high and large quantities of heat are evolved (for an exothermic reaction), while near the exit, where the rate is low, there is a relatively small evolution of heat. Because the heat-transfer rate from pellet to pellet and between pellet and gas is small, each layer of catalyst in the bed is, in effect, partially insulated from adjacent layers. This effectively prevents the flow of heat from the entrance to the exit of the catalyst bed, which in turn causes significant temperature gradients. The fluidized-bed reactor does away with the source of this difficulty, the stationary condition of the bed. The rapid movement of the small catalyst particles goes a long way toward eliminating temperature variations within the solid phase. Any one particle may be near the entrance of the reactor at one instant and near the exit at the next. This rapid mixing tends to equalize both fluid- and solid-phase temperatures, so that the entire reactor system approaches a uniform temperature. The small size (about 20 to 100 microns) of the fluidized particles provides a large heat-transfer area per unit mass and in this way increases the heat-transfer rates between solid and gas phases. Although the resultant motion of the gas is upward through the reactor, usually neither plug-flow nor complete mixing describes the flow pattern (see Chap. 13).

An important advantage of the fluidized-bed reactor over the fixed-bed type is that the catalyst can be regenerated without disturbing the operation of the reactor. In fluidized-bed catalytic cracking units a portion of the solid particles is continuously removed from the reactor and regenerated in a separate unit. The regeneration is accomplished by burning off the carbon with air, and the reactivated catalyst is continuously returned to the reactor proper. In the fixed-bed reactor the closest approach to continuous operation obtainable with a catalyst of limited life is to construct two or more identical reactors and to switch streams from one to the other when the catalyst needs to be regenerated.

A disadvantage of fluidized-bed reactors is that the equipment is large. In order that the solid particles will not be blown out the top of the reactor, the gas velocity must be low. This, in turn, necessitates large-diameter vessels and increases the initial cost. There are also losses of catalyst fines from the reactor, necessitating expensive dust-collection equipment in the exit streams.

Moving-bed systems do not permit the uniformity of temperature achieved in fluidized-bed reactors, but they do allow continuous handling of the solid phase. This is advantageous for catalyst regeneration. For example, the regeneration of the charcoal adsorbent for the hypersorption process† of separating hydrocarbons has been accomplished in a moving-bed reactor. The deactivated charcoal is added to the top of the regenerator, and steam is passed upward through the slowly moving bed of solid to strip out and react with the adsorbed hydrocarbons. Moving-bed systems are also employed for fluid-solid noncatalytic reactions; e.g., blast furnaces, lime kilns, and smelters (see Chap. 14).

PROBLEMS

3-1 A pellet of uranium dioxide actually consists of a mixture of UO_2 and UO_3 with a composite formula $UO_{2.14}$. The pellet is reacted with HF gas, and the extent of reaction is followed by weighing the pellet at successive time increments. The reactions, which have equal rates, are

$$1. \quad UO_2(s) + 4HF(g) \rightarrow UF_4(s) + 2H_2O(g)$$

$$2. \quad UO_3(s) + 2HF(g) \rightarrow UO_2F_2(s) + H_2O(g)$$

Derive a relationship between the initial weight of the pellet, m_0, its weight at any time, m, and a composite conversion expressed as the fraction of $UO_{2.14}$ reacted.

3.2‡ Desired product C is produced in an ideal tubular-flow reactor by the second-order reaction

$$A + B \rightarrow C \qquad \mathbf{r}_C = k_1 C_A C_B$$

Unwanted product A_2 is also produced by a second-order reaction

$$A + A \rightarrow A_2 \qquad \mathbf{r}_{A_2} = k_2 C_A^2$$

---

† Clyde Berg, *Trans. AIChE*, **42**, 685 (1946).
‡ This problem was supplied by Professor J. B. Butt.

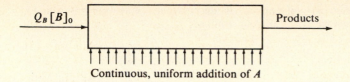

$Q_B [B]_0$ Products

Continuous, uniform addition of A

For the maximum selectivity of C to A_2 it is necessary to keep the concentration of A as low as possible. To approach this ideal, only B is added to the inlet to the reactor, and A is added uniformly along the length, as indicated in the figure. The molal flow rate of B is $Q_B C_{B_0}$, where Q_B is the volumetric flow rate entering the reactor and C_{B_0} is the concentration of B in the feed. Reactant A is added through the reactor wall in a stream whose concentration is C_{A_0}. The uniform rate of addition of this stream is dQ_A/dV, as volumetric flow rate per unit volume of reactor. Assuming that the density is constant, write mass balances for reactants A and B for this reator operating at steady state.

3-3 Acids such as HF catalyze the isomerization of normal hydrocarbons to branched chain compounds. Such compounds have a higher octane number in internal combustion engines. Hence, the development of isomerization processes with high selectivity and conversion is important for producing high-octane motor fuel without environmentally undesirable additives such as tetraethyl lead. In the laboratory, at 200 lb/in^2 gauge and 60°F, an n-hexane stream is isomerized in a continuous, stirred-tank reactor with a liquid, acid stream. This reaction mixture consists of two, intimately mixed liquid phases, the acid and hydrocarbon. The product stream from the reactor is separated into the acid and hydrocarbon phases, and the acid phase recycled back to the reactor feed. Hence, the hydrocarbons make one pass through the reactor, but the acid phase remains in the system. The pressure of 200 lb/in^2 gauge is maintained with hydrogen to reduce cracking reactions. While many products are formed, the major components, in one series of runs, can be represented by the following analyses:

Analysis of reaction product

Hydrocarbon residence time, min	On-stream time, h	Cracked products†	n-hexane	Neo-hexane	Cyclo-hexane	2-methyl-pentane
			Weight %			
33	26	10	16	11	48	15
33	53	11	18	10	48	13
33	70	14	21	8	48	9
33	125	17	27	4	48	4

† Includes gas as well as liquid cracked products.

The feed composition is:

	wt %
$n - C_6H_{14}$	52
methylcyclopentane	40
cyclohexane	8
	100

Numerous reactions occur, but the chief products can be explained by three isomerization and one cracking reaction:

1. n-C$_6$H$_{14}$ $\rightleftarrows$ 2.2-dimethyl butane (neohexane)
2. n-C$_6$H$_{14}$ $\rightleftarrows$ 2-methylpentane
3. methylcyclopentane $\rightleftarrows$ cyclohexane
4. n-C$_6$H$_{14}$ $\rightleftarrows$ cracked products

(a) Calculate the total conversion of n-hexane and the conversion to neohexane (the product with the highest octane number) for each of the four samples.

(b) From the results in (a) what can be said about (1) the effect of time on the activity of the acid catalyst for the isomerization of n-hexane, and (2) the effect of acid age on the extent of the undesirable cracking reactions? How does catalyst age affect the total conversion of n-hexane?

3-4 Derive the mass conservation expression [Eq. (3-13)] for a continuous flow STR from the species continuity equation [Eq. (3-21)]. In the feed and exit streams the mass flux due to diffusion is negligible with respect to the flux due to the velocity of the streams. Also derive Eq. (3-9) for a batch STR from Eq. (3-21). The volume of the reaction mixture is constant.

3-5 In an otherwise ideal stirred-tank reactor (flow) it is estimated that 10% of the reactor is occupied by stagnant fluid (Fig. 3-3a). If there were no stagnant region, the conversion in the effluent would be 60% for a first-order reaction. What conversion is expected in the actual reactor?

3-6 In studying the kinetics of the homogeneous gas-phase reaction between sulfur vapor and methane. Fisher[†] reported conversions for various space velocities. These space velocities were defined as the volumetric flow rate in milliliters per hour divided by the total volume of empty reactor in cubic centimeters. The flow rate is based on all the sulfur considered to be S_2 and is referred to 0°C and 1 atm pressure.

From the fact that the operating pressure was 1.0 atm and the temperature 600°C, compute the values of V/F corresponding to the space velocities given in the reference for runs 55, 58, 57, 78, and 79. V/F is the ratio of the volume of reactor to the molal feed rate, in gram moles per hour. Also determine the true contact time for a slug of reaction mixture for each of the runs; i.e., the time it takes for a slug of gas to pass through the reactor.[‡]

3-7 The production§ of toluene from benzene and xylenes was studied by Johanson and Watson[§] in a standard 1-in.-pipe reactor, with a silica-alumina catalyst. At the reactor temperature of 932°F the reaction mixture is the vapor phase. However, the benzene and xylenes were measured and pumped separately into the system as liquids by means of a proportioning pump. Hence the space velocity was reported on a liquid-hourly basis; that is, as the ratio of the feed rate, in cubic centimeters of liquid per hour, to the total volume of the reactor, in cubic centimeters. The feed consisted of an equimolal mixture of reactants, and the liquid rates were corrected to 60°F before reporting the following information:

Liquid-hourly space velocity, h^{-1}	0.5	0.25	1.0	2.0	2.0	4.0
Reactor pressure, lb/in.² abs	20	20	65	65	115	115

The reactor contained 85 g of catalyst packed in a volume of 135 cm³, and the densities of benzene and xylenes at 60°F may be taken as 0.879 and 0.870 g/cm³, respectively.

From the data determine corresponding ratios of mass of catalyst to feed rate, expressed in units of g catalyst/[(g mol)/(h)].

3-8 Convert the liquid-hourly space velocities in Prob. 3-7 to a gas basis, that is, to space velocities defined as the ratio of the gas-flow rate at reaction conditions to the total reactor volume. Then calculate the actual contact time for each run. The gases may be assumed to obey the perfect-gas law. The reaction does not result in a change in number of total moles,

$$C_6H_6 + C_6H_4(CH_3)_2 = 2C_6H_5CH_3$$

† R. A. Fisher, *Ind. Eng. Chem.*, **42**, 704 (1950).
‡ See Fisher's work (or Table 4-5) for the conversion values for the runs.
§ L. N. Johanson and K. M. Watson, *Natl. Petrol. News*, Aug. 7, 1946.

3-9 The gas-phase decomposition of formic acid

$$HCOOH(g) \rightarrow H_2O(g) + CO(g)$$

is studied in an ideal, plug-flow reactor. In one run a space velocity of 1.29 min^{-1} yields a conversion of 60%. The reactor operates isothermally and the feed is pure HCOOH. The space velocity is based upon a volumetric feed rate measured at standard conditions ($0°C$, 1 atm) while the reactor operates at $150°C$ and 1 atm. What is the actual residence time θ? How does θ compare with the value of V/Q_f, where Q_f is based upon the reactor temperature of $150°C$? The decomposition rate at $150°C$ is irreversible and first order with $k = 2.46 \text{ min}^{-1}$.

FOUR

ISOTHERMAL REACTORS
FOR HOMOGENEOUS REACTIONS

The objective in this chapter is to use the rate equations developed in Chap. 2, with the mass conservation expressions presented in Chap. 3, to design isothermal reactors for homogeneous reactions. After discussing single, stirred-tank and once-through, plug-flow reactors in Secs. 4-1 to 4-5, combinations and modifications of these ideal forms are considered in Secs. 4-6 and 4-7. Then special operating arrangements are analyzed, semibatch operation in Sec. 4-8 and plug-flow, recycle reactors in Secs. 4-9 and 4-10. Reactors for heterogeneous reactions, especially those involving solid catalysts, are more widely used than those for homogeneous reactions. However, the mass balances studied in this chapter are of the same form as those for catalytic reactions. Hence, the methods and form of equations developed here will provide a foundation for the material in the second half of the book. In Secs. 4-1 to 4-5, the procedure is, first, to examine each reactor type and operation for single reactions, and then to consider multiple reaction systems with emphasis on selectivity. In all cases the flow within any one reactor is of the ideal type, either stirred-tank or plug-flow, as defined in Chap. 3.

We have mentioned that reactor analysis has two parts: interpretation of small-scale reactor data to obtain a rate equation, and the design problem of predicting the performance of a reactor once the rate expression is known. In Chap. 2 the first part was illustrated in detail, for constant density, batch, stirred-tank reactors. In this chapter only the special case of variable density (gaseous reactions with a change in total moles) is examined with the objective of obtaining a rate expression. The remainder of the material on batch reactors is directed toward the design problem. For flow reactors both types of analysis are treated. When but one reaction occurs, the design objective is to determine the conversion as a function of reactor size. This requires a numerical expression for the reaction rate. For multiple reactions, both reactor size and product distribution (selectivity) are design objectives. Usually, the latter is more important, because the expense of separating undesirable products from the effluent stream is significant. Product distribution can be determined without numerical values for each reaction rate. It is necessary to know only ratios of rates or rate constants, as illustrated in Example 4-9. If numerical values of rates are available, both conversion and product distribution can be determined. A problem of this type is considered in Example 4-7.

Homogeneous reactions may be either noncatalytic (thermal) or catalytic. For example, one process for the production of ethylene dichloride consists of the reaction between ethylene and chlorine in the presence of bromine, with all three materials in the vapor phase. The bromine reacts with chlorine to form the unstable bromine chloride, which behaves as a catalyst in accordance with the following reactions:

1. $\quad C_2H_4 + BrCl \rightarrow C_2H_4BrCl$

2. $\quad C_2H_4BrCl + Cl_2 \rightarrow C_2H_4Cl_2 + BrCl$

Illustrations of noncatalytic homogeneous reactions are numerous and include the thermal cracking of hydrocarbons, polymerization, condensation, the combustion of gaseous fuels such as natural gas, and various inorganic reactions in aqueous solution.

IDEAL BATCH REACTORS

Batch reactors are seldom employed on a commercial scale for gas-phase reactions because the quantity of product that can be produced in a reasonably sized reactor is small. Their chief use for gaseous reactions is in kinetic studies. Batch reactors are often used, however, for liquid-phase reactions, particularly when the required production is small. Batch reactors are generally more expensive to operate than continuous units but can provide more flexibility and control. However, the initial cost of a continuous system may be higher owing to the instrumentation required. Therefore, for relatively high-priced products (such as pharmaceuticals) where operating expense is not a predominant factor in the total cost, batch reactors are commonly employed.

4-1 Design Procedure—Batch Reactors

For liquid-phase reactions the reactor volume usually does not change significantly with the extent of reaction. Hence, Eq. (3-10) is applicable for evaluating the time necessary to obtain the desired conversion. The difficulty in integrating Eqs. (3-10) depends on the number of variables influencing the rate of reaction. For example, if the rate of formation of the desired product depends on only one irreversible reaction, the expression for **r** will be simpler than if reversible or multiple reactions are involved.

The following example concerning the rate of esterification of butanol and acetic acid in the liquid phase illustrates the design problem of predicting the time-conversion relationship for an isothermal, single-reaction, batch reactor.

Example 4-1 Leyes and Othmer[†] studied the formation of butyl acetate in a batch-operated reactor at 100°C, with sulfuric acid as a homogeneous catalyst. The original feed contained 4.97 moles butanol/mole acetic acid, and the catalyst concentration was 0.032% by weight H_2SO_4. The following rate equation was found to correlate the data when an excess of butanol was used:

$$r_A = -kC_A^2$$

where C_A is the concentration of acetic acid in gram moles per millimeter, and r is the rate of reaction, in gram moles of acetic acid produced per milliliter per minute. For a ratio of butanol to acid of 4.97 and a sulfuric acid concentration of 0.032% by weight, the reaction-velocity constant was

$$k = 17.4 \text{ cm}^3/(\text{g mol})(\text{min})$$

Densities of mixtures of acetic acid, butanol, and butyl acetate are not known. Reported densities for each of the three compounds at 100°C were

$$\text{Acetic acid} = 0.958 \text{ g/cm}^3$$
$$\text{Butanol} = 0.742$$
$$\text{Butyl acetate} = 0.796$$

Although the density of the reaction mixture varies with conversion, the excess of butanol reduces the magnitude of the change. Therefore, as an approximation, the density of the mixture was assumed constant and equal to 0.75 g/cm³.

(a) Calculate the time required to obtain a conversion of 50%. (b) Determine the size of reactor and the original mass of reactants that must be charged to the reactor in order to produce ester at the average rate of 100 lb/h. Only one reactor will be used, and this unit will be shut down 30 min between batches for removal of product, cleaning, and startup. Assume that the reaction mixture is well mixed.

SOLUTION The molecular weights are

$$\text{Ester} = 116$$
$$\text{Butanol} = 74$$
$$\text{Acetic acid} = 60$$

(a) The concentration of acetic acid, C_A, is related to the conversion by $C_A = (C_A)_0(1 - x)$ where $(C_A)_0$ is the initial acid concentration. Substituting this expression in the rate equation gives

$$r_A = -k(C_A)_0^2(1 - x)^2$$

[†] C. E. Leyes and D. F. Othmer, *Ind. Eng. Chem.*, **36**, 968 (1945).

Since the volume is constant, Eq. (3-10) is applicable. Replacing r_A with the rate equation in terms of x gives

$$t = (C_A)_0 \int_0^{x_1} \frac{dx}{k(C_A)_0^2(1-x)^2} = \frac{1}{k(C_A)_0} \int_0^{x_1} \frac{dx}{(1-x)^2} \tag{A}$$

Integrating, and substituting a final value of $x_1 = 0.50$, leads to the result

$$t = \frac{1}{(C_A)_0}\left(\frac{1}{1-x_1} - \frac{1}{1-0}\right) = \frac{1}{k(C_A)_0}(2-1) = \frac{1}{k(C_A)_0} \tag{B}$$

The initial concentration of acetic acid is

$$(C_A)_0 = \frac{1(0.75)}{4.97(74) + 1(60)} = 0.0018 \text{ g mole/cm}^3$$

From Eq. (B), the time required for a conversion of 50% is

$$t = \frac{1}{17.4(0.0018)} = 32 \text{ min} \qquad \text{or } 0.53 \text{ h}$$

(b) The production rate of the reactor, in pounds of ester per hour, in terms of the pounds of acid charged, m_A, will be

$$100 = \frac{(m_A/60)(116)(0.5)}{0.53 + 0.50}$$

This expression allows 30 min for shutdown time per charge and takes into account that the conversion is 50%. Thus

$$m_A = 106 \text{ lb acetic acid/charge}$$

$$\text{Total charge} = 106 + 4.97(\tfrac{74}{60})(106) = 756 \text{ lb (343 kg)}$$

The volume occupied by the charge will be

$$V = \frac{756}{0.75(62.4)(0.1337)} = 121 \text{ gal (0.458 m}^3)$$

The reactor must be large enough to handle 121 gal of reaction mixture. The charge would consist of 106 lb of acid and 650 lb of butanol.

When the rate of the reverse reaction is significant or when more than one reaction is involved, the mechanics of solving the design equation may become more complex, but the principle is the same. Equation (3-10) is applicable, but the more complicated rate function, $r = f(x)$, may lead to complex analytical solutions or require numerical integration.

The calculations for a reversible reaction are illustrated in Example 4-2 for the esterification of ethyl alcohol.

Example 4-2 In the presence of water and hydrochloric acid (as a catalyst) the rate of esterification, in g mol/(liter)(min), of acetic acid and ethyl alcohol at 100°C is given by

$$r_2 = kC_H C_{OH} \qquad k = 4.76 \times 10^{-4} \text{ liter/(min)(g mol)}$$

The rate of the reverse reaction, the hydrolysis of the ester in the same concentration of catalyst, is

$$r'_2 = k'C_E C_W \qquad k' = 1.63 \times 10^{-4} \text{ liter/(min)(g mol)}$$

$$CH_3COOH + C_2H_5OH \underset{k'}{\overset{k}{\rightleftharpoons}} CH_3COOC_2H_5 + H_2O$$

(a) A reactor is charged with 100 gal of an aqueous solution containing 200 lb of acetic acid, 400 lb of ethyl alcohol, and the same concentration of HCl as used to obtain the reaction-velocity constants. What will be the conversion of acetic acid to ester after 120 min of reaction time? The density may be assumed constant and equal to 8.7 lb/gal; neglect the water vaporized in the reactor. (b) What is the equilibrium conversion?

SOLUTION (a) The net rate of formation of ester is obtained by combining the rate expressions for the forward and reverse reactions,

$$r = kC_H C_{OH} - kC_E C_W$$

The initial concentrations of acid H, alcohol OH, and water W are as follows:

$$(C_H)_0 = \frac{200}{100(60)} \frac{454(1,000)}{0.1337(30.5)^3} = 4.00 \text{ g mol/liter}$$

$$(C_{OH})_0 = \frac{400}{100(46)} \frac{454(1,000)}{0.1337(30.5)^3} = 10.8 \text{ g mol/liter}$$

$$(C_W)_0 = \frac{8.7(100) - (200 + 400)}{100(18)} \frac{454(1,000)}{0.1337(30.5)^3}$$

$$= 18.0 \text{ g mol/liter}$$

With the conversion based on the acid, as required by the statement of the problem, the concentrations at any time are

$$C_H = 4.0(1 - x)$$

$$C_{OH} = 10.8 - 4x$$

$$C_E = 4.0x$$

$$C_W = 18 + 4x$$

These relationships between concentrations and conversion rest on the assumption of a constant density during the reaction. The problem could still be solved without this assumption, provided data on the variation in density with conversion were available. Substituting the expressions for concentration in the rate equation for production of ester, we have

$$r = k[4(1 - x)(10.8 - 4x)] - k'[4x(18 + 4x)]$$

The numerical values for k and k' result in the following equation for $\mathbf{r}$, in g mol/(liter)(min):

$$\mathbf{r} = (0.257 - 0.499x + 0.062x^2)(8 \times 10^{-2})$$

Substituting this in the design equation for constant volume [Eq. (3-10)] yields†

$$t = \frac{(C_H)_0}{8 \times 10^{-2}} \int_0^{x_1} \frac{dx}{0.257 - 0.499x + 0.062x^2}$$

$$= 50 \int_0^{x_1} \frac{dx}{0.257 - 0.499x + 0.062x^2}$$

This expression can be integrated to give

$$t = \frac{50}{0.430} \left[\ln \frac{0.125x - 0.499 - 0.430}{0.125x - 0.499 + 0.430} \right]_0^{x_1}$$

$$= \frac{50}{0.430} \ln \frac{(0.125x_1 - 0.929)(0.069)}{(0.125x_1 - 0.069)(0.929)}$$

The conversion is desired for a time of 120 min. Hence

$$\frac{120(0.430)}{50} = 1.03 = \ln \frac{(0.125x_1 - 0.929)0.069}{(0.125x_1 - 0.069)0.929}$$

Solving for the conversion gives $x_1 = 0.365$, or 36.5% of the acid is converted to ester. It is interesting to compare this result with that based on neglecting the reverse reaction. Under this condition the rate equation is

$$\mathbf{r} = kC_H C_{OH} = k(C_H)_0(1 - x)((C_{OH})_0 - (C_H)_0 x)$$

The design equation becomes

$$t = (C_H)_0 \int_0^{x_1} \frac{dx}{k(C_H)_0(1 - x)[(C_{OH})_0 - (C_H)_0 x]}$$

$$= \frac{1}{k(C_{OH})_0 - (C_H)_0} \int_0^{x_1} \left(\frac{1}{1 - x} - \frac{(C_H)_0}{(C_{OH})_0 - (C_H)_0 x} \right) dx$$

$$= \frac{1}{k((C_{OH})_0 - (C_H)_0)} \left[-\ln (1 - x_1) + \ln \frac{(C_{OH})_0 - (C_H)_0 x_1}{(C_{OH})_0} \right]$$

Simplifying, we have

$$\frac{(C_{OH})_0 - (C_H)_0 x_1}{(C_{OH})_0(1 - x_1)} = e^{tk((C_{OH})_0 - (C_H)_0)} = e^{120(4.76 \times 10^{-4})(10.8 - 4.0)} = 1.474$$

---

† Equation (3-10) is written in terms of the rate of production of a reactant A, a negative quantity. In terms of the rate of production of a product, the ester, the minus sign is not needed.

Solving for x_1, we find

$$x_1 = \frac{0.474(C_{OH})_0}{1.474(C_{OH})_0 - (C_H)_0} = \frac{0.474(10.8)}{1.474(10.8) - 4.0} = 0.43$$

If the reverse reaction is neglected, the conversion is in error by

$$[(43 - 36.5)/36.5]100 \qquad \text{or } 18\%$$

This deviation would increase as equilibrium is approached.

(b) The expression already available for the net rate of reaction is

$$\mathbf{r} = (0.257 - 0.499x + 0.0626x^2)(8 \times 10^{-2})$$

At equilibrium the net rate must be zero. Hence the equilibrium conversion is determined by the expression

$$0.257 - 0.499x_e + 0.0626x_e^2 = 0$$

Solving,

$$x_e = 0.55 \qquad \text{or } 55\%$$

In these examples it has been assumed that a negligible amount of reaction occurs during heating and cooling periods. If only one reaction is involved, the time computed on the assumption that reaction takes place only at the operating temperature will be a conservative value. However, in reactions of industrial interest it is likely that by-products will be created. Then the yield of desired product will be reduced during heating and cooling cycles in a batch operation, provided an undesired reaction is favored at temperatures below the operating temperature. The magnitude of this effect will depend on the time spent for heating and cooling in relation to reaction time. In serious cases the problem may be solved by preheating the reactants separately prior to mixing, or by feeding one reactant continuously to a batch charge of the coreactant (semibatch operation, Sec. 4-8). Rapid quenching after reaction may be realized by rapid discharge of the mass (blowdown) or by rapid coolant injection (directly or indirectly contacted with the product mass).

4-2 Rate Equations from Batch-Reactor Measurements; Total Pressure Method for Gaseous Reactions

For gaseous reactions with a change in total number of moles, the total pressure will also change.† This pressure change is uniquely related to the extent of reaction. Hence, measuring total pressure vs. time in a laboratory, batch reactor is a suitable method for studying kinetics.

As an illustration of the method, consider the gaseous reaction

$$2NO_2 \rightarrow N_2O_4$$

---

† For reactions with no change in total moles, variation in total pressure normally would be insufficient for use in studying kinetics. Such a change would be due solely to changes in the deviation of the reaction mixture from ideal gas behavior.

in a constant-volume, batch reactor. A mass balance relates the number of moles of NO_2 and N_2O_4. If, initially, the number of moles of NO_2 is N_o and no N_2O_4 is present, the total moles at any time is

$$N_t = N_{N_2O_4} + N_{NO_2} = N_{N_2O_4} + (N_o - 2N_{N_2O_4}) = N_o - N_{N_2O_4}$$

Then the total pressure will be

$$p_t = \frac{N_t R_g T}{V} = \frac{N_o}{V} R_g T - \frac{N_{N_2O_4}}{V} R_g T$$

If the initial total pressure is p_o, the initial number of moles is

$$N_o = p_o \frac{V}{R_g T}$$

Hence, N_o can be eliminated from the equation for p_t to give

$$p_t = p_o - \left(\frac{N_{N_2O_4}}{V}\right) R_g T = p_o - C_{N_2O_4} R_g T$$

or

$$C_{N_2O_4} = \frac{1}{R_g T}(p_o - p_t) \tag{4-1}$$

Similarly, we can express C_{NO_2} in terms of p_t and p_o by noting that

$$N_{NO_2} = N_o - 2N_2O_4$$

or

$$C_{NO_2} = \frac{N_o}{V} - 2C_{N_2O_4} = \frac{p_o}{R_g T} - 2C_{N_2O_4}$$

Using Eq. (4-1) for $C_{N_2O_4}$,

$$C_{NO_2} = \frac{p_o}{R_g T} - 2(p_o - p_t)\frac{1}{R_g T} = \frac{1}{R_g T}(2p_t - p_o) \tag{4-1a}$$

Now we can use these relations between concentrations and total pressure to evaluate rate equations. For example, suppose that at the temperature involved the rate is second order and irreversible. Equation (2-1) for the rate becomes

$$\mathbf{r} = \frac{dC_{N_2O_4}}{dt} = k_2(C_{NO_2}^2) \tag{4-2}$$

This rate equation can be expressed in terms of total pressure by differentiating Eq. (4-1) to obtain $dC_{N_2O_4}/dt$ and substituting Eq. (4-1a) for C_{NO_2}. The result is

$$-\frac{1}{R_g T}\frac{dp_t}{dt} = k_2\left(\frac{2p_t - p_o}{R_g T}\right)^2$$

or

$$-\frac{dp_t}{dt} = \frac{k_2}{R_g T}(2p_t - p_o)^2 \tag{4-2a}$$

This rate equation can be tested with p_t vs. t data by either the integral or differential method. If the integral method (Sec. 2-9) is employed, Equation (4-2a) is integrated, using the initial condition $p_t = p_o$ at $t = 0$, to give

$$\frac{1}{2}\left(\frac{1}{2p_t - p_o} - \frac{1}{p_o}\right) = \frac{k_2}{R_g T}(t - 0)$$

or

$$k_2 = \frac{R_g T}{2t}\left(\frac{1}{2p_t - p_o} - \frac{1}{p_o}\right) \tag{4-3}$$

Total pressure-vs.-time data can be used directly in Eq. (4-3) to evaluate k_2 and test the suitability of second-order kinetics.

Other illustrations of the total pressure method of studying kinetics are Probs. 4-6, 4-7, and 4-11.

IDEAL TUBULAR-FLOW (PLUG-FLOW) REACTORS

Tubular-flow reactors are used both as laboratory units, where the purpose is to obtain a rate equation, and for commercial-scale production. In fact, it is a desirable and common practice to develop a rate equation for data obtained in a laboratory-sized reactor of the same form as the proposed large-scale unit. The interpretation of laboratory-reactor data is discussed in Sec. 4-3 and then the design problem is considered in Sec. 4-4.

4-3 Interpretation of Data from Laboratory Tubular-Flow Reactors

Interpretation of data to obtain a rate equation requires calculations which are the reverse of those for design. The two procedures differ because in the laboratory it may be feasible to operate at nearly constant temperature, and possibly nearly constant composition; in the commercial unit it may be possible to approach isothermal conditions, but constant composition in a tubular-flow reactor is impossible. These differences simplify the analysis of laboratory results, as illustrated in Example 4-3.

The starting point for evaluating the rate of reaction in a *tubular-flow reactor* is Eq. (3-14) or (3-17). If the reactor is small enough, the change in composition of the fluid as it flows through the volume will be slight. In addition, suppose that heat-transfer conditions are such that the temperature change is also slight. Since composition and temperature determine the rate r will also be nearly constant throughout the reactor. In other words, a local rate is measured, corresponding to the average composition and temperature in the reactor. An apparatus of this type is called a *differential reactor*. Since r_i is a constant, integration of Eq. (3-14) gives

$$\frac{\Delta(QC_i)}{\Delta V} = r_i \tag{4-4}$$

where ΔV is the volume of the differential reactor. The quantity $\Delta(QC_i)$ is the change in molal rate of component i due to flow through the reactor. If Eq. (3-17) is used, the differential-reactor equation in terms of *reactant A* is:

$$\frac{F_A(\Delta x_A)}{\Delta V} = -r_A \tag{4-5}$$

The following example illustrates how the rate is calculated and the error that is involved in assuming that it is constant.

Example 4-3 The homogeneous reaction between sulfur vapor and methane has been studied in a small silica-tube reactor of 35.2 cm^3 volume.† In a particular run at 600°C and 1 atm pressure the measured quantity of carbon disulfide produced in a 10-min run was 0.10 g. Assume that all the sulfur present is the molecular species S_2. The sulfur-vapor (considered as S_2) flow rate was 0.238 g mol/h in this steady-state run.

(a) What is the rate of reaction, expressed in g mol of *carbon disulfide produced*/(h)(cm^3 of reactor volume)? (b) The rate at 600°C may be expressed by the second-order equation

$$r = k p_{CH_4} p_{S_2}$$

where p is partial pressure, in atmospheres.‡ Use the rate determined in (a) and this form of the rate equation to calculate the specific reaction rate in units of g mol/(cm^3)(atm^2)(h). The methane flow rate was 0.119 g mol/h, and the H_2S and CS_2 concentrations in the reactants were zero. (c) Also compute the value of k without making the assumption that the rate is constant and that average values of the partial pressures may be used; that is, consider the equipment to operate as an integral, rather than a differential, reactor. Compare the results and comment on the suitability of the apparatus as a differential reactor.

SOLUTION Consider the reaction

$$CH_4 + 2S_2 \rightarrow CS_2 + 2H_2S$$

(a) The rate of formation of carbon disulfide in gram moles per hour is (mol wt of $CS_2 = 76$)

$$\Delta(QC_{CS_2}) = \frac{(0.10)}{(76)}\frac{60}{10} = 0.0079$$

---

† R. A. Fisher and J. M. Smith, *Ind. Eng. Chem.*, **42**, 704 (1950).

‡ Heretofore, we have expressed rate equations in terms of concentrations. An alternate procedure is to use partial pressures. For ideal gases $(pV = NRT)$, the relation between C_A and p_A is $C_A = N_A/V = p_A/R_g T$. At constant temperature p_A is proportional to C_A so that the two methods are equivalent, as far as kinetics studies are concerned. The numerical values of the two rate constants, k_p and k_c will differ by the appropriate powers of $R_g T$, for example $(R_g T)^2$ for a second-order reaction and $(R_g T)$ for the first-order case. When the temperature changes in the reactor, the two methods suggest a different relation between k and T. This is because T appears in the concentration-temperature relation. However, the importance of temperature in this relation is dwarfed by the exponential effect of temperature on k according to the Arrhenius equation, as illustrated in Chap. 2, Example 2-1. Hence, for practical purposes rate equations for gaseous reactions may be expressed in terms of either concentrations or partial pressures.

Then the rate of reaction, according to Eq. (4-4) will be

$$r_{CS_2} = 0.0079 \frac{1}{35.2} = 2.2 \times 10^{-4} \text{ g mol/(h)(cm}^3)$$

(b) With the assumption that at 600°C and 1 atm pressure the components behave as perfect gases, the partial pressure is related to the mole fraction by the expression

$$p_{CH_4} = p_t y_{CH_4} = 1 y_{CH_4}$$

where y_{CH_4} represents the mole fraction of methane. The average composition in the reactor will be that corresponding to a carbon disulfide rate of $(0 + 0.0079)/2 = 0.0040$ g mol/h. At this point the molal rates of the other components will be

$$CS_2 = 0.0040 \text{ g mol/h}$$

$$S_2 = 0.238 - 2(0.0040) = 0.230$$

$$CH_4 = 0.119 - 0.0040 = 0.1150$$

$$H_2S = 2(0.0040) = 0.0079$$

$$\text{Total} = 0.357 \text{ g mol/h}$$

The partial pressures will be

$$p_{CH_4} = \frac{0.1150}{0.357} = 0.322 \text{ atm}$$

$$p_{S_2} = \frac{0.230}{0.357} = 0.645 \text{ atm}$$

From the rate equation, using the value of $\mathbf{r}$ obtained in (a), we have

$$k = \frac{r_{CS_2}}{p_{CH_4} p_{S_2}} = \frac{2.2 \times 10^{-4}}{0.322(0.645)}$$

$$= 1.08 \times 10^{-3} \text{ g mol/(cm}^3)(\text{atm}^2)(\text{h})$$

(c) If the variations in rate through the reactor are taken into account, the mass balance developed in Chap. 3 for tubular-flow reactor should be used; this is Eq. (3-18),

$$\frac{V}{F_{CH_4}} = \int_0^{x_1} \frac{dx}{r_{CS_2}} \qquad \text{(A) \quad or \quad (3-18)†}$$

where the conversion x is based upon methane and F_{CH_4} is the methane

---

† By expressing the rate in terms of the product, the minus sign in Eq. (3-18) is eliminated; i.e., the rate of formation of methane is equal to $-r_{CS_2}$.

feed rate. At a point in the reactor where the conversion of methane is x, the molal flow rate of each component will be

$$CS_2 = 0.119x$$

$$S_2 = 0.238(1 - x)$$

$$CH_4 = 0.119(1 - x)$$

$$H_2S = 0.238x$$

$$\text{Total} = 0.357 \text{ g mol/h}$$

With this information, the rate equation may be written in terms of x as

$$r_{CS_2} = k y_{CH_4} y_{S_2} = k \frac{0.119(0.238)(1 - x)^2}{0.357^2} \tag{B}$$

Equation (A) may be integrated from the entrance to the exit of the reactor by means of Eq. (B). Thus

$$\frac{V}{F_{CH_4}} = \frac{4.5}{k} \int_0^{x_1} \frac{dx}{(1 - x)^2} = \frac{4.5}{k} \frac{x_1}{1 - x_1}$$

Solving for k, we have

$$k = \frac{4.5}{V/F_{CH_4}} \frac{x_1}{1 - x_1}$$

The conversion of methane at the exit of the reactor, in moles of methane reacted per mole of methane in the feed, is

$$x_1 = \frac{0.0079}{0.119} = 0.0664$$

Hence the specific reaction rate is

$$k = \frac{4.5}{35.2/0.119} \left(\frac{0.0664}{1 - 0.0664} \right) = 1.08 \times 10^{-3} \text{ g mol/(cm}^3)(\text{atm}^2)(\text{h})$$

in good agreement with the result in (b).

    In the foregoing example the form of the rate equation (second order) was given. Normally the objective is to determine the kinetics of the reaction from experimental data. To accomplish this objective in a tubular-flow reactor many runs similar to the one discussed in the example would be made, each with a different feed composition but at the same temperature. Then various forms of the rate equation would be compared with the experimentally determined rates to establish the best rate expression. Subsequent runs at different temperatures could be made to establish the activation energy. The procedure is analogous to the initial-rate method of studying kinetics in batch reactors, as described in Sec. 2-10. In Example 4-3 the change in composition in the reactor is sufficiently small that a rate corresponding to the average composition may be used to evaluate the rate.

In other words, the concept of a differential reactor is satisfactory. If the conversion had been considerably larger than $x = 0.066$, this would not have been true. By the type of calculation illustrated in part (c) the error in the use of the differential-reactor assumption can be evaluated for any conversion level and for any chosen reaction order. The magnitude of the error depends upon the form of the rate equation. It has been shown† that the differential-reactor assumption can be used up to 25% conversion with no more than 5% error for most any type of kinetics, at isothermal conditions.

While the differential-reactor system was satisfactory in Example 4-3, in many cases there are difficulties with this approach. If the small differences in concentration between entering and exit streams cannot be determined accurately, the method is not satisfactory. Also, when the heat of reaction or the rate, or both, is particularly high, it may not be possible to operate even a small reactor at conditions approaching constant temperature and constant composition. These difficulties in achieving *differential* reactor operation can sometimes be circumvented by operating the tubular reactor with a recycle stream. With this arrangement, discussed in Secs. 4-9 and 4-10, the conversion and heat effect *per pass* can be very low and yet the concentration differences between feed and discharge streams be large.

The limitation to plug flow is not important in a differential reactor. For low conversions deviations from plug flow, regardless of cause, do not result in significant concentration changes.

When the conditions of approximately constant rate cannot be met in a once-through reactor, the measured conversion data will represent the integrated value of the rates in all parts of the reactor. A laboratory unit operated in this way is termed an *integral reactor*. The problem of obtaining a rate equation is essentially one of either *differentiating* the measured conversion data to give point values of the rate, or *integrating* an assumed form for the rate equation and comparing the results with the observed conversion data. To give either process a reasonable chance of success, as many variables as possible should remain constant. The ideal situation is one in which the only variable is composition, i.e., where the temperature is the same in all parts of the reactor. This type of integral operation is just one step removed from that of a differential reactor. If both the temperature and the composition change significantly, the process of differentiating the experimental data to obtain a rate equation is of doubtful accuracy and, indeed, is seldom successful.

These conclusions concerning the ease of interpretation of laboratory data in terms of a rate equation are summarized in Table 4-1.

A complete set of integral-reactor data consists of measurements of the conversion for different flow rates through the reactor, with each run made at constant reactants ratio, pressure, and temperature (if possible). Then an additional set of conversion-vs.-flow runs is made at a different reactants ratio‡ but the same pressure and temperature. This procedure is continued until data are obtained

---

† H. A. Massaldi and J. A. Maymo, *J. Catal.* **14**, 611 (1969).

‡ The effect of intermediate or final products on the rate can be ascertained by adding these components to the feed. This is a particularly valuable procedure when a differential reactor is used.

Table 4-1 Experimental methods of obtaining rate data for once-through, tubular-flow reactors

Type of reactor	Characteristics	Interpretation of data
Differential	Nearly constant temperature and concentrations (i.e., constant rate)	Rate data obtained directly; interpretation simple.
Integral (A)	Nearly constant temperature (rate depends upon concentrations only)	Interpretation of integral data usually satisfactory by graphical differentiation or fitting of integral-conversion curves.
Integral (B)	Temperature and concentrations both change in reactor	Interpretation complicated by temperature variation; if effect of temperature on rate is known from independent measurements, interpretation possible, in principle.

over the entire range of reactants ratio and temperature that may be employed in the commercial reactor. A useful means of summarizing the results is to prepare graphs of V/F vs. conversion at constant values of reactants ratio and p_t. Such graphs for representing the experimental data are suggested by the form of the design equation for flow reactors [Eq. (3-18)]. The shape of these curves is determined by the nature of the rate, and they provide some qualitative information about the rate equation.

The two procedures for treating integral-reactor data are equivalent to the integration and differential methods described in Sec. 2-9 for batch systems. In the integration approach, a rate equation is assumed and then the design expression is integrated. This integration gives a relationship between V/F and x which may be compared with the experimental data. The final step is the choice of the rate equation that gives the best agreement with the experimental V/F-vs.-x curves for all conditions of reactants ratio, temperature, and pressure. The differential method entails differentiating the V/F-vs.-x curves graphically to obtain the rate of reaction as a function of composition. Various assumed rate equations can then be tested for agreement with the rate-vs.-composition data.

Perhaps the clearest way of explaining these methods of interpreting laboratory data is to carry out specific examples in some detail. In Example 4-4 the integration and differential methods are applied to a simple, single-reaction system. Example 4-5 involves a more complicated, single-reaction system and in Example 4-6 there are several reactions.

Example 4-4 A kinetic study is made of the decomposition of acetaldehyde at 518°C and 1 atm pressure in a flow apparatus. The reaction is

$$CH_3CHO \rightarrow CH_4 + CO$$

Acetaldehyde is boiled in a flask and passed through a reaction tube maintained by a surrounding furnace at 518°C. The reaction tube is 3.3 cm ID and

Table 4-2

Rate of flow, g/h	130	50	21	10.8
Fraction of acetaldehyde decomposed	0.05	0.13	0.24	0.35

80 cm long. The flow rate through the tube is varied by changing the boiling rate. Analysis of the products from the end of the tube gives the results in Table 4-2.

What is a satisfactory rate equation for these data?

SOLUTION A second-order rate equation for production of acetaldehyde, $r_A = -k_2(C_A)^2$, will be tested by both the integral and the differential methods. To utilize either, it is necessary to express the rate in terms of the conversion of acetaldehyde. This may be accomplished by applying material balances and the ideal-gas law.

The molal flow rate of acetaldehyde entering the reaction tube is F_A. At a point where the conversion is x it will be $N_A = F_A(1 - x)$. The molal rates of the other components will be

$$\text{Methane} = xF_A$$
$$\text{Carbon monoxide} = xF_A$$
$$\text{Total flow rate} = F_A(1 + x)$$

From the ideal-gas law, the concentration of acetaldehyde will be:

$$C_A = \frac{N_A}{Q} = \frac{1 - x}{(1 + x)(R_g T/p_t)}$$

Then, the second-order rate expression in terms of conversion is

$$r_A = -k_2(C_A)^2 = -k_2\left(\frac{p_t}{R_g T}\right)^2\left(\frac{1 - x}{1 + x}\right)^2 \tag{A}$$

Integration method Equation (A) may be substituted in the design expression, Eq. (3-18), to evaluate the second-order assumption:

$$\frac{V}{F_A} = \frac{1}{k_2(p_t/R_g T)^2}\int_0^{x_1} \frac{dx}{[(1 - x)/(1 + x)]^2}$$

Integration yields

$$k_2\left(\frac{p_t}{R_g T}\right)^2\frac{V}{F_A} = \frac{4}{1 - x} + 4\ln(1 - x) + x - 4 \tag{B}$$

which provides the relationship between V/F_A and x to be tested with the experimental data. Since

$$V = \pi\frac{3.3^2}{4}(80) = 684 \text{ cm}^3$$

the experimental data may be expressed in terms of V/F_A, as shown in Table 4-3. Equation (B) may be used to compute a value of k for each of the sets of x

and V/F_A values given in the table. For example, at $x = 0.13$, substituting in Eq. (B), we obtain

$$k_2 \left[\frac{1}{0.082(518 + 273)} \right]^2 (2,160) = \frac{4}{1 - 0.13} + 4 \ln (1 - 0.13) + 0.13 - 4$$

$$k_2 = 0.33 \text{ liter/(g mol)(s)}$$

Values of k_2 for the other three sets, shown in the last column of the table, are in good agreement with each other.

Before proceeding with the differential method, some discussion is necessary about the term V/F_A in Eq. (3-18). The experimental data were obtained in Example 4-4 in the normal way by varying the feed rate through the same reactor. Referring to Eq. (3-18), V was constant but F_A varied from run to run. In contrast, Eq. (3-18) was obtained from (3-17) by integrating *through* a reactor (variable volume) operated at constant feed rate F_A. The question arises: Can Eq. (3-18) be used to relate data obtained at various F_A but constant V? As noted in Sec. 3-6, V/F_A is related to the residence time of an element of fluid in the reactor. Hence, the question is whether the same effect of residence time is obtained by changing the feed rate as by changing the reactor volume. For example, would the same conversion change be observed by doubling the feed rate as by reducing the reactor to one-half its original volume. The key point is that the velocity in the two cases would be different. If a change in velocity has no effect on the conversion (for the same residence time), then changing F_A is equivalent to changing V. This will be true for a plug-flow reactor because it is an ideal type where there is assumed to be no axial dispersion, and no radial gradients. It is the magnitude of these deviations from ideal flow that are influenced by velocity. Hence, for *ideal* reactors we can write Eqs. (3-17) and (3-18) in the following way:

$$d \left(\frac{V}{F_A} \right) = - \frac{dx_A}{\mathbf{r}_A} \tag{3-17a}$$

or

$$\left(\frac{V}{F_A} \right) = - \int_0^{x_A} \frac{dx_A}{\mathbf{r}_A} \tag{3-18a}$$

where the parentheses indicate that the separate effects of V and F_A can be combined in ratio form.

Table 4-3

Conversion	Feed rate		V/F_A, (liter)(s)/g mol	k_2, liter/(g mol)(s)
	g/h	g mol/s		
0.05	130	0.000825	828	0.32
0.13	50	0.000316	2,160	0.33
0.24	21	0.000131	5,210	0.32
0.35	10.8	0.0000680	10,000	0.33

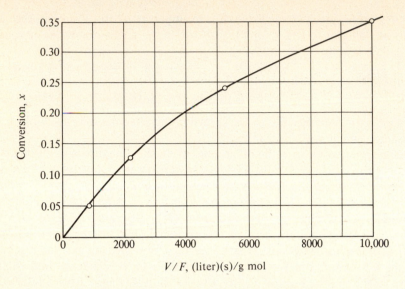

Figure 4-1 Conversion data for decomposition of acetaldehyde.

Differential method The experimental V/F_A data are plotted against conversion in Fig. 4-1. According to Eq. (3-17a), the slope, $dx/d(V/F_A)$, of this curve at any conversion gives the rate of reaction $(-\mathbf{r}_A)$ at that point. The slopes of the curve at the experimental conversions are given in the second column of Table 4-4. These slopes may be compared with the assumed rate equation either by plotting $\ln \mathbf{r}$ vs. $\ln [(1 - x)/(1 + x)]$ according to Eq. (A) and noting whether or not a straight line of slope 2 is obtained, or by computing values of k_2 directly from Eq. (A). Following this second procedure, at $x = 0.13$ we have

$$\mathbf{r} = -4.9 \times 10^{-5} = -k_2 \left[\frac{1}{0.082(791)} \right]^2 \left(\frac{1 - 0.13}{1 + 0.13} \right)^2$$

$$k_2 = 0.35 \text{ liter/(g mol)(s)}$$

The results at other conversions are shown in the third column of Table 4-4. Although there is some variation from point to point, there is no significant

Table 4-4

Conversion	Slope from Fig. 4-1, g mole/(liter)(s)	k_2, liter/(s)(g mol)	k_1, s^{-1}
0.05	6.2×10^{-5}	0.32	0.0045
0.13	4.9×10^{-5}	0.35	0.0041
0.24	2.8×10^{-5}	0.33	0.0030
0.35	2.0×10^{-5}	0.35	0.0027

trend. Hence the differential method also confirms the validity of a second-order rate equation. The variation is due to errors associated with the measurement of slopes of the curve in Fig. 4-1.

First-order rate equation For comparison, the differential method will be applied to a first-order mechanism. The rate expression replacing Eq. (A) is

$$\mathbf{r} = -k_1 C_A = -k_1 \frac{p_t}{R_g T} \frac{1-x}{1+x} \tag{C}$$

This mechanism may be tested by computing k_1 values from the experimental rates of reaction according to Eq. (C). For example, at $x = 0.13$

$$-\mathbf{r} = \text{slope} = 4.9 \times 10^{-5} = k_1 \frac{1}{0.082(791)} \frac{1-0.13}{1+0.13}$$

$$k_1 = 0.0041 \text{ s}^{-1}$$

The k_1 values for other conversions (fourth column of Table 4-4) show a distinct trend toward lower values at higher conversions. This indicates that the first-order rate is not a likely one.

Example 4-5 A flow reactor consisting of a 1-in. stainless-steel pipe 6 in. long and packed with inert rock salt is used to study the noncatalytic homogeneous reaction

$$\frac{4}{b} S_b + CH_4 \rightarrow 2H_2S + CS_2$$

The measurements are carried out at atmospheric pressure in the vapor phase at 600°C. From available data on the rate of dissociation of the sulfur species, it is reasonable to assume that the reactions

$$S_8 \rightleftharpoons 4S_2$$

$$S_6 \rightleftharpoons 3S_2$$

are very fast with respect to the combination of sulfur vapor with methane. Accordingly, the relative amounts of S_8, S_6, and S_2 are determined by thermodynamic equilibrium calculations. The void volume, measured by benzene displacement, was 35.2 cm³. The conversion x *of methane to carbon disulfide* was measured for various flow rates and initial reactants ratios, and the results are shown in Table 4-5. (S_2) represents the total amount of sulfur vapor present, expressed as S_2. Hence

$$(S_2) = N_{S_2} + 3N_{S_6} + 4N_{S_8}$$

where the N values are the number of moles of each sulfur species.

For this example assume, that the only species of sulfur that exists is S_2. Then test the assumptions of first- and second-order rate equations with the experimental data.

Table 4-5

Run	Flow rate, g mol/h			Reactants ratio, moles (S_2)/moles CH_4	Conversion of CH_4, x
	$(CH_4)_f$	$(S_2)_f$	$(CS_2)_p$		
55	0.02975	0.0595	0.0079	2	0.268
58	0.0595	0.119	0.0086	2	0.144
57	0.119	0.238	0.0078	2	0.066
59	0.119	0.238	0.0072	2	0.060
56	0.238	0.476	0.0059	2	0.025
75	0.0595	0.119	0.0079	2	0.133
76	0.02975	0.0595	0.0080	2	0.269
77	0.119	0.238	0.0069	2	0.058
78	0.0893	0.0893	0.0087	1	0.0975
79	0.119	0.0595	0.0096	0.5	0.0807

SOLUTION It is convenient to convert the data to a conversion x_t *based on the total feed.* For example, for run 55

$$F_t = 0.02975 + 0.0595 = 0.0892 \text{ g mol/h}$$

$$\frac{V}{F_t} = \frac{35.2}{0.0892} = 395$$

$$x_t = \frac{0.0079}{0.0892} = 0.0890$$

where F_t is the molal feed rate. Table 4-6 shows the corresponding values of x_t and V/F_t. These results are also plotted in Fig. 4-2. Note that

$$x_t = \frac{\text{moles product}}{\text{moles total feed}} = x \frac{\text{moles } CH_4 \text{ in feed}}{\text{moles total feed}} = x \frac{1}{1+a}$$

Table 4-6

Run	Conversion		V/F_t, $(cm^3)(h)/g \text{ mol}$	$-\log[1 - (a+1)(x_t)_1]$ or $-\log(1 - x_1)$	$k_1 \times 10^4$	$(k_1)_s \times 10^4$
	x_1	$(x_t)_1$				
55	0.268	0.0890	395	0.135	7.87	3.93
58	0.144	0.0480	197	0.0675	7.90	3.95
57	0.066	0.022	98.6	0.0297	6.93	3.46
59	0.060	0.020	98.6	0.0269	6.29	3.15
56	0.025	0.00833	49.3	0.0110	5.14	2.57
75	0.133	0.0443	197	0.0620	7.25	3.62
76	0.269	0.0895	395	0.136	7.95	3.97
77	0.058	0.0193	98.6	0.0269	6.05	3.02
78	0.0975	0.0487	197	0.045	5.30	5.45
79	0.0807	0.0538	197	0.0368	4.31	9.90

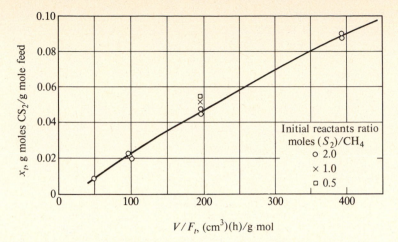

Figure 4-2 Experimental conversion data for the sulfurization of methane.

or

$$x = (a + 1)x_t$$

where a is the moles of sulfur in the feed per mole of methane.

Two first-order mechanisms may be assumed, one with respect to methane and one with respect to sulfur vapor (S_2). Starting with the fist-order assumption for methane, and applying Eq. (3-18), we have

$$\frac{V}{F_t} = \int_0^{(x_t)_1} \frac{dx_t}{r} = \int_0^{(x_t)_1} \frac{dx_t}{k_1 p_{CH4}} = \int_0^{(x_t)_1} \frac{dx_t}{k_1 y_{CH4}} \tag{A}$$

where r is the positive rate of production of carbon disulfide. For ideal gases

$$p_{CH4} = y_{CH4} P_t = y_{CH4}$$

The mole fraction of methane varies with the conversion. Hence to relate y_{CH4} to x_t we must know the number of moles of each component as a function of conversion. Since we are assuming here that all the sulfur present is S_2, the number of moles of S_2 and CH_4 are simple to ascertain. For a mole ratio of S_2 to CH_4 of 2, which is the case for all runs except 78 and 79, the mole fraction entering the reactor is

$$y_{CH4} = \frac{1}{2 + 1} = 0.333$$

If a basis of 1 mole of CH_4 and 2 moles of S_2 is chosen, at a point in the reactor where the conversion is x_t, the number of moles of each component will be

$$CS_2 = x = 3x_t$$
$$H_2S = 2x = 6x_t$$
$$CH_4 = 1 - x = 1 - 3x_t$$
$$S_2 = 2 - 6x_t$$

Total moles $= 3$

Thus the mole fractions are

$$y_{CH_4} = \frac{1 - 3x_t}{3} = 0.333 - x_t$$

$$y_{S_2} = \frac{2 - 6x_t}{3} = 0.667 - 2x_t$$

More generally, if the initial mole ratio of sulfur to CH_4 is a, then

$$\left. \begin{aligned} y_{CH_4} &= \frac{1 - (1 + a)x_t}{1 + a} = \frac{1}{1 + a} - x_t \\ y_{S_2} &= \frac{a - 2(1 + a)x_t}{1 + a} = \frac{a}{1 + a} - 2x_t \end{aligned} \right\} \tag{B}$$

By the integral method, Eq. (A) can now be integrated with Eq. (B) for methane to yield

$$\frac{V}{F_t} = \frac{1}{k_1} \int_0^{(x_t)_1} \frac{dx_t}{1/(1 + a) - x_t} = \frac{1}{k_1} \ln \frac{1/(1 + a) - (x_t)_1}{1/(1 + a)}$$

$$= -\frac{1}{k_1} \ln \left[1 - (a + 1)(x_t)_1 \right] = -\frac{1}{k_1} \ln (1 - x_1) \tag{C}$$

where $(x_t)_1$ is the conversion leaving the reactor. The V/F_t-vs.-$(x_t)_1$ data shown in Table 4-6 can now be used to test Eq. (C). Instead of plotting curves obtained from Eq. (C) with different k values for comparison with the experimental curve (Fig. 4-2), it is more effective to use one of two other approaches:

1. Plot the experimental V/F_t-vs.-x data given in Table 4-6 as V/F_t vs. $-\ln (1 - x_1)$. This method of plotting should result in a straight line if the assumed rate expression is correct. Figure 4-3 shows that this is not true if the rate is first order with respect to methane. At a constant reactants ratio of $a = 2.0$ the points fall on a straight line, but the data for other ratios deviate widely from the line.
2. From the data in Table 4-6, compute values of k_1 required by Eq. (C) for each of the runs. If these values are essentially constant, the assumed mechanism fits the data. For this case the resulting k_1 values are given in the next to last column of Table 4-6. Again, the runs for different reactants ratios show that the assumed rate equation does not fit the data.

As we shall see later, the assumed rate equation may be complex enough to prevent analytical integration of the design equation (A). However, in such cases it is still possible to evaluate the integral $\int dx_t/r$ graphically and then apply either method 1 or method 2 to determine the suitability of the rate assumption. Both methods are general and hence applicable to any reaction.

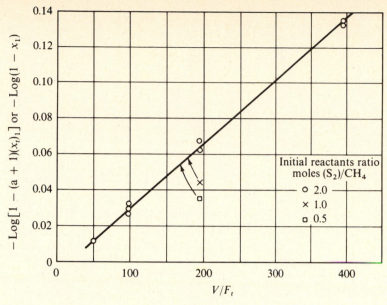

Figure 4-3 Test of first-order mechanism with respect to CH_4.

Equation (3-18), for an assumed mechanism which is first order with respect to sulfur vapor S_2, is obtained by substituting y_{S_2} from Eq. (B) for y_{CH_4} in Eq. (A). Thus,

$$\frac{V}{F_t} = \int_0^{(x_t)_1} \frac{dx_t}{(k_1)_s y_{S_2}} = \frac{1}{(k_1)_s} \int_0^{(x_t)_1} \frac{dx_t}{a/(1+a) - 2x_t}$$

$$= -\frac{1}{2(k_1)_s} \ln \frac{a - 2(a+1)(x_t)_1}{a} = \frac{-1}{2(k_1)_s} \ln \left(1 - \frac{2x_1}{a}\right) \qquad (D)$$

For $a = 2$, Eq. (D) becomes the same as Eq. (C) if $(k_1)_s = \frac{1}{2}k_1$. Hence the test of this mechanism is whether or not the two runs for $a = 1.0$ and $a = 0.5$ are either brought in line with the other data on a plot similar to Fig. 4-3 or else give the same value of $(k_1)_s$. The values of $(k_1)_s$ are given in the last column of Table 4-6. It is clear that the results for the runs with $a = 1.0$ and 0.5 are far removed from the rest of the data. Hence *neither* first-order mechanism is satisfactory.

If a second-order mechanism is tried, again with the assumption that all the sulfur vapor is present as S_2, Eq. (3-18) becomes

$$\frac{V}{F_t} = \int \frac{dx_t}{k_2 p_{CH_4} p_{S_2}} = \frac{1}{k_2} \int \frac{dx_t}{y_{CH_4} y_{S_2}}$$

$$= \frac{1}{k_2} \int \frac{dx_t}{\{[1/(1+a)] - x_t\}\{[a/(1+a)] - 2x_t\}}$$

This expression can be simplified and integrated to†

$$\frac{V}{F_t} = \frac{a+1}{k_2} \int_0^{x_1} \frac{dx}{(1-x)(a-2x)}$$

$$= \frac{a+1}{a-2} \frac{1}{k_2} \left[\ln\left(1 - \frac{2}{a}x_1\right) - \ln\left(1 - x_1\right) \right] \qquad (E)$$

$$= \frac{3}{2k_2} \frac{x_1}{1 - x_1} \qquad \text{for } a = 2$$

Figure 4-4 shows a plot of V/F_t vs. $[(a+1)/(a-2)] \times \{\ln [1 - (2/a)x_1] - \ln (1 - x_1)\}$, or $\frac{3}{2}x_1/(1 - x_1)$ for $a = 2$. It is apparent that this correlation is an improvement over the previous ones, since the point for $a = 1.0$ is close to the straight line. However, the data point for $a = 0.5$ is still not satisfactorily correlated. From this preliminary investigation, neglecting the S_6 and S_8 species, we conclude that a second-order rate is more appropriate than a first-order one.

The next example concerns multiple reactions where all the reactions except one are very fast. Under these conditions only one rate constant exists; the extent of the other reactions are determined by the thermodynamic equilibrium constants (See Sec. 1-4).

† Equation (E) becomes indeterminate when $a = 2$. Hence it is necessary to substitute $a = 2$ prior to integration and obtain the particular solution.

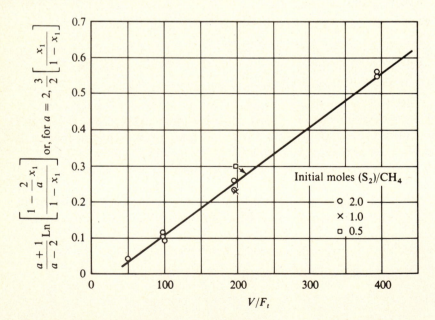

Figure 4-4 Test of second-order reaction mechanism.

Example 4-6 Reanalyze the data given in Example 4-5 by taking into account the existence of S_8, S_6 and S_2.

(a) Assume that S_2 is the sulfur vapor species that reacts with methane according to a second-order rate equation.

(b) Test whether a second-order rate equation with S_6 or S_8 and CH_4 would be preferable to that with S_2.

SOLUTION (a) To take into account the presence of the S_6 and S_8 species we shall assume equilibrium among S_2, S_6, and S_8, utilizing the following equilibrium constants at 600°C:

$$\tfrac{1}{4}S_8 \rightarrow S_2 \qquad K_p = 0.930 \ (\text{atm})^{3/4}$$

$$\tfrac{1}{3}S_6 \rightarrow S_2 \qquad K_p = 0.669 \ (\text{atm})^{2/3}$$

This information, along with the stoichiometry of the reaction, can be used to develop expressions analogous to Eqs. (B) of example 4-5 for the relation between y_{S_2}, or y_{CH_4}, and the conversion. Suppose that there are a moles of sulfur vapor per mole of CH_4 entering the reactor. However, let a refer to the total moles of sulfur considered as S_2, not to the sum of the moles of S_2, S_6, and S_8. If α_0, β_0, and γ_0 represent the moles of S_2, S_6, and S_8 entering the reactor, the total moles of sulfur, considered as S_2, will be

$$a = a_0 + 3\beta_0 + 4\gamma_0$$

At a point in the reactor where the conversion is x, the moles of each component will be

$$CS_2 = x$$

$$H_2S = 2x$$

$$CH_4 = 1 - x$$

$$S_2 = \alpha$$

$$S_6 = \beta$$

$$S_8 = \gamma$$

$$\text{Total moles} = \alpha + \beta + \gamma + 2x + 1$$

The conversion has been defined in terms of the methane reacted. It can also be written in terms of the sulfur used, providing a relationship between x, α, β, and γ. If (S_2) represents the total moles of sulfur (considered as S_2) present at any point in the reactor, then

$$x = \tfrac{1}{2}[a - (S_2)]$$

but

$$S_2 = \alpha + 3\beta + 4\gamma$$

so that

$$x = \frac{a - \alpha - 3\beta - 4\gamma}{2} \tag{A}$$

Two additional relationships among x, α, β, and γ must exist because of the dissociation equilibria:

$$0.930 = \frac{p_{S_2}}{p_{S_8}^{1/4}} = \frac{y_{S_2}}{y_{S_8}^{1/4}} = \frac{\alpha}{\gamma^{1/4}(\alpha + \beta + \gamma + 2x + 1)^{3/4}} \tag{B}$$

$$0.669 = \frac{p_{S_2}}{p_{S_6}^{1/3}} = \frac{y_{S_2}}{y_{S_6}^{1/3}} = \frac{\alpha}{\beta^{1/3}(\alpha + \beta + \gamma + 2x + 1)^{2/3}} \tag{C}$$

Equations (A) to (C) permit the evaluation of α, β, and γ at any conversion and reactants ratio a. Then the mole fractions are immediately obtainable from the expressions

$$y_{S_2} = \frac{\alpha}{\alpha + \beta + \gamma + 2x + 1}$$

$$y_{CH_4} = \frac{1 - x}{\alpha + \beta + \gamma + 2x + 1}$$

To illustrate the method of calculation let us take $a = 2$. Entering the reactor, $x = 0.0$, and Eqs. (A) to (C) become

$$2 = \alpha + 3\beta + 4\gamma$$

$$\gamma^{1/4}(\alpha + \beta + \gamma + 1)^{3/4} = \frac{\alpha}{0.93}$$

$$\beta^{1/3}(\alpha + \beta + \gamma + 1)^{2/3} = \frac{\alpha}{0.669}$$

Solving these by trial gives

$$\alpha = 0.782 \qquad y_{S_2} = \frac{0.782}{2.171} = 0.360$$

$$\beta = 0.340 \qquad y_{S_6} = \frac{0.340}{2.171} = 0.156$$

$$\gamma = 0.049 \qquad y_{S_8} = \frac{0.049}{2.171} = 0.023$$

$$CH_4 = 1.000 \qquad y_{CH_4} = \frac{1.000}{2.171} = 0.461$$

$$\text{Total} = 2.171 \qquad \sum y = 1.000$$

The mole fractions corresponding to any conversion can be evaluated in a similar manner. The results of such calculations are shown in Fig. 4-5,

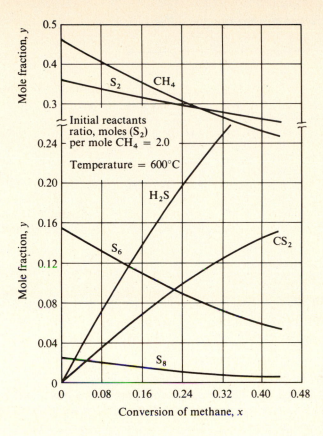

Figure 4-5 Composition of reaction mixture vs. conversion for the sulfurization of methane.

where composition of the reaction mixture is plotted against x. Information for other values of a are obtained in an analogous fashion.

The necessary data are now at hand for testing the second-order rate expression. However, the calculational procedure used in Ex. 4–5 is not applicable because the mass balance equation

$$\frac{V}{F_t} = \frac{1}{k_2'} \int_0^{(x_t)_1} \frac{dx_t}{y_{CH_4} y_{S_2}} \qquad (D)$$

cannot be integrated analytically. Two alternatives are possible. The integral method entails graphical evaluation of the integral in Eq. (D) for each experimental point, with the measured value of the conversion as the upper limit in each case. This procedure leads to values of $\int_0^{(x_t)_1} dx_t / y_{CH_4} y_{S_2}$ for each value of V/F_t given in Table 4-6. The results are shown in the fourth column of Table 4-7 and are plotted in Fig. 4-6, with V/F_t as the abscissa and $\int_0^{(x_t)_1} dx_t / y_{CH_4} y_{S_2}$ as the ordinate. If the assumed rate equation is a satisfactory interpretation of the data, Fig. 4-6 should be a

Table 4-7 Test of reaction mechanisms for sulfurization of methane

Run	Reactants ratio	V/F_t	$\int_0^{(x_t)_1} dx_t/y_{CH_4} y_{S_b}$		
			$b = 2$	$b = 6$	$b = 8$
55	2	395	0.762	2.11	17.0
58	2	197	0.347	0.890	6.64
57	2	98.6	0.144	0.350	2.48
59	2	98.6	0.130	0.315	2.23
56	2	49.3	0.052	0.123	0.855
75	2	197	0.316	0.800	5.88
76	2	395	0.765	2.17	17.1
77	2	98.6	0.126	0.303	2.15
78	1	197	0.315	1.27	12.1
79	0.5	197	0.388	2.97	36.8

straight line with a slope equal to k_2'. This is apparent if Eq. (D) is rearranged as

$$\int_0^{(x_t)_1} \frac{dx_t}{y_{CH_4} y_{S_2}} = k_2' \frac{V}{F_t}$$

It is evident from the plot that this condition is satisfied.

The other approach, the differential method, entails graphical differentiation of the data for comparison with the differential form of Eq. (D). If the data are plotted as x_t vs. V/F_t, as in Fig. 4-2, the slope of the curve is equal to the rate of reaction. Thus, from Eq. (D),

$$d\left(\frac{V}{F_t}\right) = \frac{dx_t}{k_2' y_{CH_4} y_{S_2}}$$

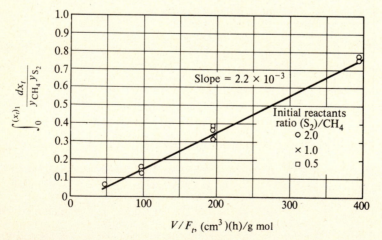

Figure 4-6 Test of rate equation $r = k_2' p_{CH_4} p_{S_2}$.

or

$$\frac{dx_t}{d(V/F_t)} = \text{slope} = k_2' y_{CH_4} y_{S_2}$$

If the slope is taken at various values of the conversion from Fig. 4-2 and the corresponding values of y_{CH_4} and y_{S_2} are read from Fig. 4-5, and this is done for each reactants ratio, complete information on the effect of composition on the rate is available. The equation $\mathbf{r} = k_2' y_{CH_4} y_{S_2}$ can then be tested by observing the constancy in computed values of k_2'.

(b) The results of part (a) indicate that the assumption of S_2 as the active sulfur-vapor species in the second-order mechanism gives a satisfactory interpretation of the conversion data. However, it is important to know whether equally good correlations would be obtained by assuming that S_6 or S_8 is the reactive form. To do this Eq. (3-18) may be written for the other species

$$\frac{V}{F_t} = \frac{1}{k_6'} \int_0^{(x_t)_1} \frac{dx_t}{y_{CH_4} y_{S_6}} \tag{E}$$

$$\frac{V}{F_t} = \frac{1}{k_8'} \int_0^{(x_t)_1} \frac{dx_t}{y_{CH_4} y_{S_8}} \tag{F}$$

Figures 4-7 and 4-8, which show V/F_t plotted vs. the integrals in (E) and (F), were prepared by graphical integration, with Fig. 4-5 used for the relations of x_t to y_{S_6} and y_{S_8}. The values of the integrals are given in the last two columns of Table 4-7. We see that in contrast to Fig. 4-6, based on S_2, the data do not fall on a straight line as required by Eqs. (E) and (F). Also, the points for various reactants ratios are not in agreement. We

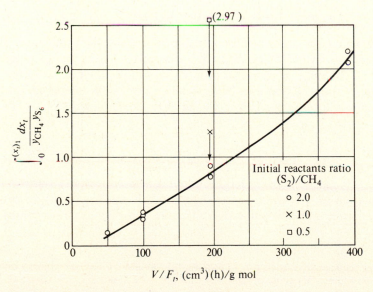

Figure 4-7 Test of rate equation $\mathbf{r} = k_6' p_{CH_4} p_{S_6}$.

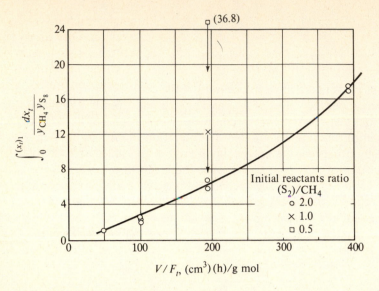

Figure 4-8 Test of rate equation $\mathbf{r} = k'_8 \, p_{CH_4} \, p_{S_8}$.

conclude that a second-order mechanism based on either S_6 or S_8, as the reactive sulfur species, does not agree with the experimental data.

The preferred equation for the rate of reaction is

$$\mathbf{r} = k'_2 \, p_{CH_4} p_{S_2}$$

From Fig. 4-6, the slope of the straight line is 2.2×10^{-3} g mol/(h)(cm^3) (atm^2). Since this number is equal to k'_2, the final expression for the rate of reaction at 600°C is†

$$\mathbf{r} = 2.2 \times 10^{-3} p_{CH_4} p_{S_2}$$

It is not advisable to make a firm statement regarding the mechanism of a reaction from data as limited as those given in Example 4-6. It is best to consider that the final rate equation is a satisfactory interpretation of the kinetic data. The problem of interpreting laboratory kinetic data in terms of a rate equation for *catalytic* reactions will be considered in Chaps. 9 to 12.

4-4 Design Procedure—Tubular-Flow Reactors

When a satisfactory rate expression has been found, as described in Examples 4-4 to 4-6, this equation for $\mathbf{r}$ is employed in the mass balance [Eq. (3-18)] to determine the size of reactor as a function of conversion and operating conditions.

---

† By evaluating the specific-reaction rate k'_2 at other temperatures an activation energy can be determined. For this aspect of the problem see R. A. Fisher and J. M. Smith, *Ind. Eng. Chem.*, **42**, 704 (1950).

For multiple reactions the same procedure as for single reactions is applicable. However, selectivity is important as well as conversion. One method of solution is to formulate one mass balance around the reactor for each reaction. The choice of species upon which to base the mass balance is arbitrary, but usually a limiting, reactant or product is chosen. Once the extent of reaction is established for one species, the amounts of the other species are fixed by the stoichiometry of the reaction (e.g., see Sec. 2-1 where the single variable is the extent of reaction or the conversion). Alternately, mass balances may be written around the reactor for each species. For some problems (see Example 4-9), this procedure is simpler than writing a balance for each reaction.

The first part of Example 4-7 illustrates the procedure for a single reaction. In the second part a two-reaction system is considered.

Example 4-7 Hougen and Watson,[†] in an analysis of Kassell's data for the homogeneous, vapor-phase dehydrogenation of benzene in a tubular-flow reactor considered two reactions

$$1. \quad 2C_6H_6(g) \rightarrow C_{12}H_{10}(g) + H_2(g)$$
$$2. \quad C_6H_6(g) + C_{12}H_{10}(g) \rightarrow C_{18}H_{14}(g) + H_2(g)$$

The rate equations are:

$$\mathbf{r}_1 = 14.96 \times 10^6 e^{-15,200/T}\left(p_B^2 - \frac{p_D p_H}{K_1}\right) \text{lb moles benzene reacted/(h)(ft}^3)$$

$$\mathbf{r}_2 = 8.67 \times 10^6 e^{-15,200/T}\left(p_B p_D - \frac{p_T p_H}{K_2}\right)$$

lb moles triphenyl produced or diphenyl reacted/(h)(ft^3)

where p_B = partial pressure of benzene, atm
p_D = partial pressure of diphenyl, atm
p_T = partial pressure of triphenyl, atm
p_H = partial pressure of hydrogen, atm
T = temperature, K
K_1, K_2 = equilibrium constants for the two reactions in terms of partial pressures.

The data on which the rate equations are based were obtained at a total pressure of 1 atm and temperatures of 1265° and 1400°F in a 0.5-in. tube 3 ft long.

It is now proposed to design a tubular reactor which will operate at 1 atm pressure and 1400°F. (*a*) Determine the total conversion of benzene to di- and triphenyl as a function of space velocity. (*b*) Determine the reactor volume required to process 10,000 lb/h of benzene (the feed is pure benzene) as a

[†] O. A. Hougen and K. M. Watson, "Chemical Process Principles," vol. 3, p. 846, John Wiley & Sons, Inc., New York, 1947.

function of the total conversion. First carry out the solution with the assumption that only reaction 1 occurs, and then proceed to the solution for the two consecutive reactions. (c) What is the space velocity for which the concentration of diphenyl is a maximum? Assume that the reactor will be operated isothermally and that no other reactions are significant.

SOLUTION (a) Since the reactor is isothermal, the equilibrium constants K_1 and K_2 will have fixed values. They may be estimated at 1400°F from equations developed by Hougen and Watson, by methods described in Sec. 1-4. The results are

$$K_1 = 0.312$$

$$K_2 = 0.480$$

As these values are not large, the reverse reactions may be important. At 1400°F (1033 K), in terms of the rate of disappearance of benzene, the two rates are

$$r_1 = (14.96 \times 10^6)e^{-14.7}\left(p_B^2 - \frac{p_D p_H}{0.312}\right) = 6.23\left(p_B^2 - \frac{p_D p_H}{0.312}\right) \quad \text{(A)}$$

$$r_2 = (8.67 \times 10^6)e^{-14.7}\left(p_B p_D - \frac{p_T p_H}{0.480}\right) = 3.61\left(p_B p_D - \frac{p_T p_H}{0.480}\right) \quad \text{(B)}$$

Mass balance Eq. (3-18) may be written for each reaction. If the mass balances are based upon benzene, the two equations are:†

$$\frac{V}{F} = \int \frac{dx_1}{r_1} \quad \text{(C)}$$

$$\frac{V}{F} = \int \frac{dx_2}{r_2} \quad \text{(D)}$$

where the conversion x_1 is the pound moles of benzene disappearing by reaction 1 per pound mole of feed, and the conversion x_2 is the pound moles of benzene disappearing by reaction 2 per pound mole of feed.

If we take as a basis 1.0 mole of entering benzene, the moles of each component at conversions x_1 and x_2 are

$$H_2 = \tfrac{1}{2}x_1 + x_2$$

$$C_{12}H_{10} = \tfrac{1}{2}x_1 - x_2$$

$$C_6H_6 = 1 - x_1 - x_2$$

$$C_{18}H_{14} = x_2$$

$$\text{Total moles} = 1.0$$

† Since only benzene is in the feed, we need no subscript on F; $F_{C_6H_6} = F = $ total molal feed rate.

Since the total moles equals 1.0, regardless of x_1 and x_2, the mole fractions of each component are also given by these quantities. If the components are assumed to behave as ideal gases, the partial pressures are

$$p_H = 1.0(\tfrac{1}{2}x_1 + x_2) = \tfrac{1}{2}x_1 + x_2$$

$$p_D = \tfrac{1}{2}x_1 - x_2$$

$$p_B = 1 - x_1 - x_2$$

$$p_T = x_2$$

With these relationships, rate equations (A) and (B) can be expressed in terms of x_1 and x_2 as

$$\mathbf{r}_1 = 6.23\left[(1 - x_1 - x_2)^2 - \frac{(\tfrac{1}{2}x_1 - x_2)(\tfrac{1}{2}x_1 + x_2)}{0.312}\right] \tag{E}$$

$$\mathbf{r}_2 = 3.61\left[(1 - x_1 - x_2)(\tfrac{1}{2}x_1 - x_2) - \frac{x_2(\tfrac{1}{2}x_1 + x_2)}{0.480}\right] \tag{F}$$

Equations (E) and (F) can be substituted in the design equations (C) and (D) and values of exit conversions x_1 and x_2 computed for various values of V/F.

Simplified solution (reaction 2 neglected) Since $x_2 = 0$, Eq. (E) simplifies to

$$\mathbf{r} = 6.23[(1 - x_1)^2 - 0.801(x_1)^2] \tag{G}$$

This may be substituted in design Eq. (C) and integrated directly:

$$\frac{V}{F} = \frac{1}{6.23}\int_0^{x_1} \frac{dx_1}{(1 - x_1)^2 - 0.801(x_1)^2}$$

$$= \frac{1}{6.23}\int_0^{x_1} \frac{dx_1}{(1 - 0.105x_1)(1 - 1.895x_1)}$$

$$= \frac{0.559}{6.23}\ln\frac{1 - 0.105x_1}{1 - 1.895x_1} \tag{H}$$

Table 4-8 gives the values of V/F corresponding to successive values of x_1 obtained by substituting in Eq. (H). A particular V/F value represents the ratio of volume of reactor to feed rate necessary to give the corresponding conversion. Figure 4-9 is a plot of the data given in the table.

The equilibrium conversion is 52.8%, as that is the value corresponding to infinite time. The same result could have been obtained by using the equilibrium constant $K_1 = 0.312$ to compute the equilibrium yield by the method described in Sec. 1-4.

In many cases analytical integration of the design equation [Eq. (C)] is not possible. However, numerical integration can be carried out by

Table 4-8

x_1, lb moles C_6H_6 reacted/ lb mole feed	V/F, ft³ (h)/ (lb mole feed)	Space velocity, h⁻¹
0.0	0	∞
0.1	0.0179	21,200
0.2	0.0409	9,280
0.3	0.0725	5,230
0.4	0.123	3,080
0.5	0.260	1,460
0.52	0.382	990
0.528†	∞	0

† Equilibrium conditions.

numerous methods.† The Euler method is simple but can lead to large truncation errors. The fourth-order Runge-Kutta procedure reduces this error but requires more extensive and more complicated calculations. We will illustrate both methods in this elementary case where comparison with an analytical solution is possible. In the examples in the remainder of

† For a description of numerical methods for solving differential equations see "Applied Numerical Methods," p. 363, Brice Carnahan, H. A. Luther, and James O. Wilkes, John Wiley & Sons, Inc., New York (1969) and "Digital Computations for Chemical Engineers," p. 86, Leon Lapidus, McGraw-Hill Book Company, New York (1962).

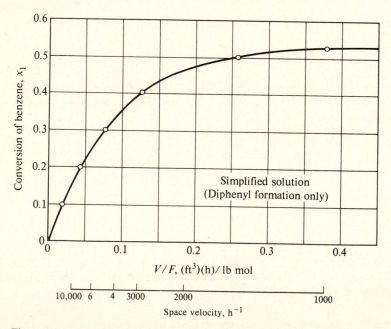

Figure 4-9 Dehydrogenation of benzene (simplified solution).

the book we will normally use the more accurate Runge-Kutta procedure, referring to this example for the working equations. Both Euler and Runge-Kutta methods are simple to program for machine computation. When a program is already available for the Runge-Kutta procedure, the more complicated working equations require no additional personal effort.

In the Euler method an average rate, $\bar{r}$, is assumed for an increment of the independent variable; in our example this is $\Delta(V/F)$. Then Eq. (C) may be written in difference form as

$$\Delta x_1 = \bar{r}_1 \, \Delta(V/F) \tag{I}$$

The calculational procedure is illustrated by calculating x_1 for the first increment of V/F, starting at the reactor entrance. The rate at the entrance is given by Eq. (G) with $x_1 = 0$:

$$(r_1)_0 = 6.23(1 - 0) = 6.23 \text{ lb mol } C_6H_6$$

$$\text{converted}/(h)(ft^3)$$

Choose an increment $\Delta(V/F) = 0.0179$ ft^3/(lb mol feed)(h) in order to correspond to the first tabulated result in Table 4-8. In the *simple* Euler method the rate at the entrance to the V/F increment is chosen as $\bar{r}_1$. Then the conversion at the exit of the increment is, according to Eq. (I):

$$(x_1)_1 = (x_1)_0 + \Delta x_1 = 0 + 6.23(0.0179) = 0.111$$

The rate at the exit of this first increment is, from Eq. (G),

$$(r_1)_1 = 6.23[(1 - 0.111)^2 - 0.801(0.111)^2] = 4.85$$

A more accurate result can be obtained by the *modified* Euler method. In this modification the value of $\bar{r}_1$ is taken as the arithmetic average of the rates entering and leaving the increment; that is,

$$\bar{r}_1 = \frac{(r_1)_0 + (r_1)_1}{2} = \frac{6.23 + 4.85}{2} = 5.54$$

With this revised value, the conversion at the exit of the first increment is

$$(x_1)_1 = 0 + 5.54(0.0179) = 0.099$$

In order to illustrate the Runge-Kutta method, we carry out the calculations for the same first increment. Equation (C) in differential form may be written

$$\frac{dx_1}{d(V/F)} = r_1(x_1) \tag{J}$$

For one dependent variable (x_1), the working equations are:†

$$k_0 = \Delta(V/F)\,\mathbf{r}_1(x_{1_n}) \tag{4-6}$$

$$k_1 = \Delta(V/F)\,\mathbf{r}_1(x_{1_n} + k_0/2) \tag{4-7}$$

$$k_2 = \Delta(V/F)\,\mathbf{r}_1(x_{1_n} + k_1/2) \tag{4-8}$$

$$k_3 = \Delta(V/F)\,\mathbf{r}_1(x_{1_n} + k_2) \tag{4-9}$$

where n designates the nth increment of V/F. The term such as $\mathbf{r}_1(x_{1_n} + k_0/2)$ indicates that the reaction rate is evaluated for a conversion $x_1 = x_{1_n} + k_0/2$. Using these equations the conversion leaving increment $n + 1$ is obtained from the conversion leaving increment n by the expression:

$$(x_1)_{n+1} = (x_1)_n + \tfrac{1}{6}(k_0 + 2k_1 + 2k_2 + k_3) \tag{4-10}$$

We start the calculations at the reactor entrance $(n = 0)$ and choose $\Delta(V/F) = 0.0179$. The rate at the reactor entrance has already been found to be 6.23. Then from Eq. (4-6)

$$k_0 = 0.0179(6.23) = 0.1115$$

From Eq. (G)

$$\mathbf{r}_1(x_{1_0} + k_0/2) = r_1(0 + 0.1115/2)$$

$$= 6.23[(1 - 0.0557)^2 - 0.801(0.0557)^2]$$

$$= 5.54$$

and from Eq. (4-7)

$$k_1 = 0.0179(5.54) = 0.099$$

In a similar way we find for k_2 and k_3

$$k_2 = 0.101$$

$$k_3 = 0.100$$

Finally, Eq. (4-10) is used to determine the conversion leaving the first increment. Thus,

$$(x_1)_1 = (x_1)_0 + \tfrac{1}{6}(k_0 + 2k_1 + 2k_2 + k_3)$$

$$= 0 + \tfrac{1}{6}(0.111 + 0.198 + 0.202 + 0.100)$$

$$= 0.101$$

We can now summarize the results obtained by the various methods. The analytical integration (the correct result) gave $x_1 = 0.100$ at $V/F = 0.0179$. With the simple Euler method, $x_1 = 0.11$, a 11% error. The modified Euler procedure and the Runge-Kutta method gave

† Ibid.

$x_1 = 0.099$ and 0.101, respectively. Both of these latter values are within 1% of the correct result.

With any of the numerical methods the next step is to repeat the calculations for successive increments of V/F until the required conversion is attained. The errors accumulate as the number of increments increase, particularly for the simpler Euler method. Also, the numerical integrations become more accurate as the increment size is decreased.

Correct solution (both reactions considered) If the rate equations (E) and (F) are substituted into the differential forms of the corresponding mass balances, Eqs. (C) and (D), one obtains

$$\frac{dx_1}{d(V/F)} = r_1(x_1, x_2) = 6.23\left[(1 - x_1 - x_2)^2 - \frac{(\frac{1}{2}x_1 - x_2)(\frac{1}{2}x_1 + x_2)}{0.312}\right] \tag{K}$$

$$\frac{dx_2}{d(V/F)} = r_2(x_1, x_2) = 3.61\left[(1 - x_1 - x_2)(\frac{1}{2}x_1 - x_2) - \frac{x_2(\frac{1}{2}x_1 + x_2)}{0.480}\right] \tag{L}$$

These two equations cannot be integrated independently since r_1 and r_2 are functions of *both* x_1 and x_2. Therefore, a numerical solution is indicated. Since two dependent variables are involved (x_1 and x_2) the working equations are different from those used in the simplified solution. We will illustrate both the modified Euler and Runge-Kutta methods.

For the modified Euler method, Eqs. (K) and (L) are written in difference form as

$$\Delta x_1 = \bar{r}_1 \, \Delta\left(\frac{V}{F}\right) \tag{M}$$

$$\Delta x_2 = \bar{r}_2 \, \Delta\left(\frac{V}{F}\right) \tag{N}$$

The procedure is illustrated by calculating x_1 and x_2 for the first increment of $\Delta(V/F)$, starting at the entrance to the reactor. The rates at the entrance are given by Eqs. (E) and (F) with $x_1 = x_2 = 0$

$$r_{1_0} = 6.23(1 - 0) = 6.23 \text{ lb mol } C_6H_6 \text{ converted/(h)(ft}^3)$$

$$r_{2_0} = 3.61(0) = 0$$

Choose an increment of $\Delta(V/F) = 0.005$ ft^3/(lb mol feed)(h). To initiate the calculations, initial rates are taken as average values over this interval. Then conversions at the end of the interval are, according to Eqs. (M) and (N),

$$x_{1_1} = 0 + 0.005(6.23) = 0.03115$$

$$x_{2_1} = 0 + 0.005(0) = 0$$

The rates at the end of the interval are, from Eqs. (E) and (F):

$$r_{1_1} = 6.23 \left[(1 - 0.03115)^2 - \frac{1}{4} \frac{0.03115^2}{0.312} \right] = 6.23(0.938) = 5.84$$

$$r_{2_1} = 3.61 \left[(1 - 0.03115) \left(\frac{0.03115}{2} - 0 \right) \right] = 0.0544$$

According to the modified Euler method, better values of the rates in the first increment of $\Delta(V/F)$ are arithmetic averages of the rates entering and leaving the increment. Thus,

$$\bar{r}_1 = \frac{(r_1)_0 + (r_1)_1}{2} = \frac{6.23 + 5.84}{2} = 6.04$$

$$\bar{r}_2 = \frac{(r_2)_0 + (r_2)_1}{2} = \frac{0 + 0.0544}{2} = 0.0272$$

With these revised values the conversions at the end of the first increment are, again according to Eqs. (M) and (N),

$$x_{1_1} = 0 + 0.005(6.04) = 0.0302$$

$$x_{2_1} = 0 + 0.005(0.0272) = 0.00014$$

The working equations for the Runge-Kutta procedure for two dependent variables [x_1 and x_2 in Eqs. (K) and (L)] are:[†]

$$\left. \begin{aligned} k_0 &= \Delta\left(\frac{V}{F}\right) r_1(x_{1_n}, x_{2_n}) \\[2ex] m_0 &= \Delta\left(\frac{V}{F}\right) r_2(x_{1_n}, x_{2_n}) \end{aligned} \right\} \tag{4-11}$$

$$\left. \begin{aligned} k_1 &= \Delta\left(\frac{V}{F}\right) r_1\left(x_{1_n} + \frac{k_0}{2}, x_{2_n} + \frac{m_0}{2}\right) \\[2ex] m_1 &= \Delta\left(\frac{V}{F}\right) r_2\left(x_{1_n} + \frac{k_0}{2}, x_{2_n} + \frac{m_0}{2}\right) \end{aligned} \right\} \tag{4-12}$$

$$\left. \begin{aligned} k_2 &= \Delta\left(\frac{V}{F}\right) r_1\left(x_{1_n} + \frac{k_1}{2}, x_{2_n} + \frac{m_1}{2}\right) \\[2ex] m_2 &= \Delta\left(\frac{V}{F}\right) r_2\left(x_{1_n} + \frac{k_1}{2}, x_{2_n} + \frac{m_1}{2}\right) \end{aligned} \right\} \tag{4-13}$$

$$\left. \begin{aligned} k_3 &= \Delta\left(\frac{V}{F}\right) r_1(x_{1_n} + k_2, x_{2_n} + m_2) \\[2ex] m_3 &= \Delta\left(\frac{V}{F}\right) r_2(x_{1_n} + k_2, x_{2_n} + m_2) \end{aligned} \right\} \tag{4-14}$$

† Ibid.

A term such as $r_1(x_{1_n} + k_1/2, x_{2_n} + m_1/2)$ indicates that the rate of reaction is evaluated at conversions $x_{1_n} + k_1/2$ and $x_{2_n} + m_1/2$ from Eq. (E). Using these equations for the k and m values, x_1 and x_2 at increment $n + 1$ are obtained from the conversions at n by the expressions:

$$\left. \begin{aligned} x_{1_{n+1}} &= x_{1_n} + \tfrac{1}{6}(k_0 + 2k_1 + 2k_2 + k_3) \\ x_{2_{n+1}} &= x_{2_n} + \tfrac{1}{6}(m_0 + 2m_1 + 2m_2 + m_3) \end{aligned} \right\} \tag{4-15}$$

The rates at the reactor entrance are those already calculated: $r_1(x_{1_0}, x_{2_0}) = 6.23$ and $r_2(x_{1_0}, x_{2_0}) = 0$. For the increment, $\Delta(V/F) = 0.005$ ft^3/lb mol feed)(h), Eqs. (4-11) yield

$$k_0 = 0.005(6.23) = 0.0311$$

$$m_0 = 0.005(0) = 0$$

These values of k_0 and m_0 are next used to evaluate r_1 at $x_{1_0} + k_0/2$ and $x_{2_0} + m_0/2$. Then Eqs. (4-12) give

$$k_1 = 0.005(6.03) = 0.0302$$

$$m_1 = 0.005(0.0277) = 0.00014$$

This procedure is continued using k_1 and m_1 to evaluate k_2 and m_2, etc. from Eqs. (4-13) and (4-14). The results are

$$k_2 = 0.005(6.04) = 0.0302; \, k_3 = 0.0302\dagger$$

$$m_2 = 0.005(0.02684) = 0.00013; \, m_3 = 0.00013\dagger$$

Finally, the conversions at the end of the first increment, at $V/F = 0.005$ are obtained from Eqs. (4-15).

$$x_{1_1} = 0 + \tfrac{1}{6}[0.0311 + 2(0.0302) + 2(0.0302) + 0.0302] = 0.0302$$

$$x_{2_1} = 0 + \tfrac{1}{6}[0 + 2(0.00014) + 2(0.00013) + 0.00013] = 0.00011$$

These conversions for the first increment are essentially the same as those obtained by the simpler, modified Euler method. However, as the number of increments increases, the Runge-Kutta procedure usually is more accurate.

By repeating such calculations for successive increments, the results from zero to the equilibrium conversion shown in Table 4-9 and in Fig. 4-10 are obtained. Space velocities, (V/Q_f) are calculated from Eq. (3-23) and based upon a flow rate evaluated at 60°F and 1 atm.

(b) The reactor volume required to process 10,000 lb/h of benzene may be computed from the V/F data in Table 4-9. For a total conversion of 41.7%, for example, $V/F = 0.12$ and the reactor volume is

$$V = 0.12F = 0.12\frac{10,000}{78} = 15.4 \text{ ft}^3$$

---

† k_3 and m_3 are equal to k_2 and m_2, respectively, in this case because the conversions are so low. In general this would not be true.

Table 4-9 Conversion vs. V/F for the dehydrogenation of benzene

V/F	Space velocity, h^{-1}	Conversion			Composition of mixture (molal)			
		x_1	x_2	x_t	C_6H_6	$C_{12}H_{10}$	$C_{18}H_{14}$	H_2
0	∞	0	0	0	1.000	0	0	0
0.005	75,800	0.0302	0.00011	0.030	0.970	0.0150	0.00014	0.0152
0.010	37,900	0.0586	0.00051	0.059	0.941	0.0288	0.00051	0.0298
0.020	18,950	0.110	0.0018	0.112	0.888	0.0534	0.0018	0.0571
0.040	9,500	0.197	0.0062	0.203	0.797	0.092	0.0062	0.104
0.060	6,320	0.264	0.0119	0.276	0.724	0.120	0.0119	0.143
0.080	4,740	0.316	0.0180	0.334	0.666	0.140	0.0180	0.176
0.100	3,790	0.356	0.0242	0.380	0.620	0.154	0.0242	0.202
0.120	3,160	0.387	0.0302	0.417	0.583	0.163	0.0302	0.224
0.140	2,710	0.411	0.0359	0.447	0.533	0.170	0.0359	0.241
0.180	2,110	0.446	0.0459	0.491	0.509	0.177	0.0459	0.268
0.220	1,720	0.466	0.0545	0.521	0.479	0.179	0.0545	0.289
0.260	1,460	0.478	0.0614	0.540	0.460	0.177	0.0614	0.301
0.300	1,260	0.487	0.0671	0.554	0.446	0.175	0.0671	0.310
0.400	950	0.494	0.0770	0.571	0.429	0.170	0.0770	0.324
∞†	0	0.496	0.091	0.587	0.413	0.157	0.091	0.339

† Equilibrium conditions.

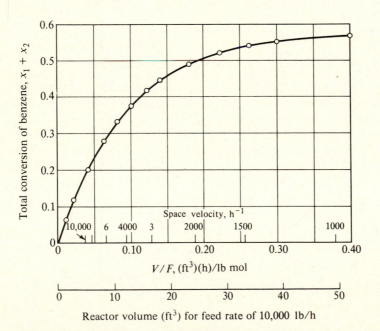

Figure 4-10 Dehydrogenation of benzene.

If a 4-in.-ID tubular reactor were used, a length of about 175 ft would be required. The complete curve of volume vs. total conversion is also shown in Fig. 4-10.

The equilibrium total conversion of $49.6 + 9.1 = 58.7\%$ and the equilibrium composition are shown as the values for V/F at infinity in Table 4-9. These results could also have been obtained from Eqs. (E) and (F). Thus at equilibrium, $r_1 = r_2 = 0$, and

$$(1 - x_1 - x_2)^2 = \frac{(\frac{1}{2}x_1 - x_2)(\frac{1}{2}x_1 + x_2)}{0.312}$$

$$(1 - x_1 - x_2)(\frac{1}{2}x_1 - x_2) = \frac{x_2(\frac{1}{2}x_1 + x_2)}{0.480}$$

Simultaneous solution of these two equations gives the same values, $x_1 = 0.496$ and $x_2 = 0.091$.

The rates of each reaction as a function of V/F are shown in Fig. 4-11. In both cases the rates fall off toward zero as V/F approaches high values. The rate of reaction 2 also approaches zero at very low values of V/F because diphenyl is not present in the benzene entering the reactor.

(c) The last four columns in Table 4-9 give the composition of the reaction mixture. With a fixed feed rate, increasing V/F values correspond to moving through the reactor, i.e., increasing reactor volume. Hence these columns show how the composition varies with position in the reactor. Note that the mole fraction of benzene decreases continuously as the

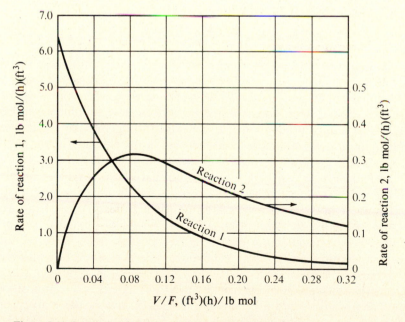

Figure 4-11 Rates of reaction for dehydrogenation of benzene.

reaction mixture proceeds through the reactor, while the diphenyl content reaches a maximum at a space velocity of about 1720 h^{-1}, typical of the concentration of an intermediate product in consecutive reactions.

Example 4-7 illustrates the design procedure for a homogeneous flow reactor operating at constant temperature and pressure. When the pressure drop in the reactor is significant and the rate of reaction varies with pressure, modification is necessary. This involves no new concepts. In comparison with Example 4-7, the differences are as follows:

1. The effect of pressure on the rate of reaction must be accounted for; i.e., rate equations (A) and (B) would be used instead of (E) and (F), which are based upon 1 atm pressure.
2. The same stepwise numerical procedure is applicable, except that the pressure at any point in the reactor must be computed and the appropriate value used in Eqs. (A) and (B).

The pressure drop is due to friction, potential-energy changes, and kinetic-energy changes in the flow process in the reactor. Hence the pressure drop may be computed from the Bernoulli equation, with suitable auxiliary relationships, such as the Fanning equation for the friction. Since the pressure drop is small for flow in empty pipes, which is the usual case for homogeneous reactors, the effect of pressure may often be neglected. This is not invariably the case. For example, thermal cracking of hydrocarbons is accomplished by passing the fluid through pipes which may be several hundred feet long. In such instances the pressure drop may be of the order of 100 lb/in.^{2} at an absolute pressure of 500 lb/in.^{2} abs.

IDEAL STIRRED-TANK, FLOW REACTORS

For steady-state operation of single, ideal stirred-tank reactors, the species mass balances developed in Chap. 3 were Eq. (3-4) for unequal flow rates of feed and effluent and Eq. (3-5) for equal rates. These *algebraic* equations can be used to calculate the effluent concentration obtained for a given reactor volume and feed flow rate Q_f. The rate of production r_i is constant throughout the reactor and corresponds to the concentrations and temperature in the effluent. Similarly, the concentrations in the reactor are equal to those of the effluent.

Since the temperature will be the same in all parts of this type of reactor, isothermal operation is always achieved as long as steady-state conditions prevail. However, the reactor temperature may be different from that of the feed stream, because of either the heat of reaction or the energy exchange with the surroundings. The treatment in this chapter is restricted to cases where the feed and reactor temperatures are the same. (The more general case will be considered in Chap. 5, along with nonisothermal behavior.) Under these conditions Q_e will be equal to Q_f, unless the density of the reaction mixture changes with composition. If density vs. composition data are available so that Q_e may be calculated from Q_f, the more general equation (3-4) can be used. For equal flow rates, Eq. (3-5) shows that the key factor determining reactor performance is the *average* residence time V/Q, rather than V and Q separately. In contrast to a plug-flow reactor, there is a distribution of residence times of the fluid in a stirred-tank reactor. This distribution function is given in Chap. 6.

Single-stirred tank reactors are considered first (Sec. 4-5). Next, multiple stirred-tanks in series (Sec. 4-6) are discussed, and this is followed by a comparison of plug-flow and stirred-tank reactors (Sec. 4-7). Finally, a brief treatment is given for non-steady-state (semibatch) operation of tank reactors (Sec. 4-8).

4-5 Single Stirred-Tank Reactors

The performance of single reactors is illustrated with three examples consisting of a single reaction, two sequential reactions, and a multireaction, polymerization system. Calculating the conversions and selectivity for multiple reactions is simpler for a stirred-tank than for a plug-flow reactor. This is because it is generally easier to solve an array of algebraic equations of the form of Eq. (3-5) than it is to integrate an array of differential equations of the form of Eq. (3-17).

Example 4-8 Eldridge and Piret† have investigated the hydrolysis of acetic anhydride in stirred-tank flow reactors of about 1800 cm³ volume at temperatures from 15 to 40°C. Their experimental results for several volumetric feed rates are shown in Table 4-10. Independent determination by these investigators from other experiments in batch-operated reactors resulted in the following first-order equations for the rate of *disappearance* of anhydride, in g mol/(cm³)(min):

$$r = \begin{cases} 0.0567C & \text{at } 10°C \\ 0.0806C & \text{at } 15°C \\ 0.1580C & \text{at } 25°C \\ 0.380C & \text{at } 40°C \end{cases}$$

where C is the concentration of acetic anhydride, in gram moles per cubic centimeter.

For each run compute the percent of hydrolysis and compare it with the observed value in the table. In all cases the feed temperature is the same as the temperature of the reaction mixture. Since the concentrations are low, the densities of feed and product stream are the same.

SOLUTION The hydrolysis reaction,

$$(CH_3CO)_2O + H_2O \rightarrow 2CH_3COOH$$

would be expected to be second order. In this case, since dilute aqueous solutions were employed, the water concentration is essentially constant, so

† J. W. Eldridge and E. L. Piret, *Chem. Eng. Progr.*, **46**, 290 (1950).

Table 4-10

Entering anhydride concentration, mol/cm³	Volumetric feed rate, cm³/min	% hydrolysis of anhydride	T, °C
2.1×10^{-4}	378	25.8	15
1.4×10^{-4}	582	33.1	25
1.37×10^{-4}	395	40.8	25
1.08×10^{-4}	555	15.3	10
0.52×10^{-4}	490	16.4	10
0.95×10^{-4}	575	55.0	40
0.925×10^{-4}	540	55.7	40
1.87×10^{-4}	500	58.3	40
2.02×10^{-4}	88.5	88.2	40

that a first-order equation is satisfactory. Equation (3-5) is applicable because the densities of feed and product stream are the same.

Substituting the rate equation $\mathbf{r} = kC$ into Eq. (3-5) written for acetic anhydride, and solving the resulting expression for the exit concentration C_1, gives

$$C_1 = \frac{C_0}{1 + k(V/Q)} = \frac{C_0}{1 + k\bar{\theta}}$$

The fraction of the anhydride hydrolyzed, y, is equal to $(C_0 - C_1)/C_0$. Hence,

$$y = 1 - \frac{C_1}{C_0} = 1 - \frac{1}{1 + k\bar{\theta}} \tag{A}$$

Equation (A) gives the fraction hydrolyzed in terms of $\bar{\theta}$ and the reaction-rate constant. For the first set of data in the table, at 15°C,

$$k = 0.0806 \text{ min}^{-1} \qquad \bar{\theta} = \frac{1,800}{378} \text{ min}$$

$$k\bar{\theta} = 0.0806 \frac{1,800}{378} = 0.384 \text{ (dimensionless)}$$

Substitution in Eq. (A) gives

$$y = 1 - \frac{1}{1 + 0.384} = 0.277 \quad \text{or} \quad 27.7\%$$

This agrees reasonably well with the experimental value of 25.8%.

Table 4-11 shows the results of similar calculations for the other sets of data. The computed values are based on the concept of complete mixing and hence on the assumption of uniform composition and temperature through-out the mass. The agreement between the calculated results and the experimental ones is a measure of the validity of the complete mixing assumption.

Tests with successively less stirring can be used to study the level of agitation required for complete mixing. MacDonald and Piret† have made such studies for a first-order reaction system.

Example 4-9 At high pressures acrilonitrile and isobutylene react homogeneously to amines, from which polymers with unusual properties can be prepared. The first reaction produces a monoamine that subsequently reacts with additional acrilonitrile to produce a diamine. The two reactions are:

$$1. \quad \overset{B}{H_2C}\!=\!C(CH_3)_2 + \overset{A}{H_2C}\!=\!CHCN \overset{k_1}{\longrightarrow} \overset{M}{H_2C}\!=\!C(CH_3)CH_2CH_2CH_2CN$$

$$2. \quad \overset{M}{H_2C}\!=\!C(CH_3)CH_2CH_2CH_2CN + \overset{A}{H_2C}\!=\!CHCN$$

$$\overset{k_2}{\longrightarrow} \overset{D}{H_2C}\!=\!C\!\!\begin{array}{c} \diagup CH_2CH_2CH_2CN \\ \diagdown CH_2CH_2CH_2CN \end{array}$$

The desired product is the diamine. High molecular weight compounds (as undesirable by-products) are also produced, perhaps by polymerization of the diamine. Neglect these degradation reactions.

Determine the conversion of acrilonitrile (A) to monoamine (M) and to diamine (D) as a function of the total conversion of acrilonitrile (A). Also, calculate the conversion of isobutylene (B) and the overall selectivity of diamine to monoamine, both in terms of the total conversion of acrilonitrile.

A single, stirred-tank reactor operating isothermally is to be used with a feed composition of 3 moles of isobutylene per mole of acrilonitrile. Suppose that the ratio of the rate constants for the two reactions is $k_1/k_2 = 2$. Also assume that both reactions are second order and irreversible.

† R. W. MacDonald and E. L. Piret, *Chem. Eng. Progr.*, **47**, 363 (1951).

Table 4-11

Volumetric feed rate, cm³/min	% hydrolysis	
	Experimental	Calculated from Eq. (A)
378	25.8	27.7
582	33.1	32.8
395	40.8	41.8
555	15.3	15.5
490	16.4	17.2
575	55.0	54.4
540	55.7	55.9
500	58.3	57.8
88.5	88.2	88.5

SOLUTION Since only *ratios* of rate constants are given, reactor volumes required for specific feed rates and conversions cannot be calculated. However, as illustrated for batch reactors in Example 2-9, selectivities can be determined.

Second-order rate equations written for the *disappearance* of the indicated species are

$$r_A = k_1 C_A C_B + k_2 C_A C_M = \text{rate of } \textit{disappearance} \text{ of acrilonitrile} \quad \text{(A)}$$

$$r_B = k_1 C_A C_B \quad \text{(B)}$$

$$r_M = -k_1 C_A C_B + k_2 C_A C_M \quad \text{(C)}$$

$$r_D = -k_2 C_A C_M \quad \text{(D)}$$

For simplicity the subscript e has been omitted, but all concentrations are those for the effluent stream.

Mass-balance equations of the form of Eq. (3-5) may be written for each of the two reactions. By using as variables the extent of reaction ξ (see Sec. 2-1) these two equations can be expressed in terms of the two variables. However, it is simpler to solve this problem by writing mass balances for species B, M, and D and then employing the two ratios of these three equations. Using rate equations (B) to (D), Eq. (3-5) applied to species B, M, and D gives

$$\frac{C_{B_f} - C_B}{\theta} = r_B = k_1 C_A C_B \quad \text{(E)}$$

$$\frac{0 - C_M}{\theta} = r_M = -k_1 C_A C_B + k_2 C_A C_M \quad \text{(F)}$$

$$\frac{0 - C_D}{\theta} = r_D = -k_2 C_A C_M \quad \text{(G)}$$

The ratio of Eqs. (F) to (E) eliminates θ, and provides a relation between C_M and C_B:

$$C_M = \frac{k_1(C_{B_f} - C_B)}{(k_1 - k_2)C_B + k_2 C_{B_f}} C_B \quad \text{(H)}$$

Dividing this equation by C_{B_f} gives

$$\frac{C_M}{C_{B_f}} = \frac{k_1(C_B/C_{B_f})(1 - C_B/C_{B_f})}{(k_1 - k_2)(C_B/C_{B_f}) + k_2} \quad \text{(I)}$$

The conversion of isobutylene is

$$x_B = 1 - C_B/C_{B_f}$$

Hence, in terms of x_B, Eq. (I) becomes

$$\frac{C_M}{C_{B_f}} = \frac{\alpha(1 - x_B)x_B}{x_B + \alpha(1 - x_B)}$$

where $\alpha = k_1/k_2$. We wish to know the conversion x_M of acrilonitrile to mono-amine. This is

$$x_M = \frac{C_M}{C_{A_f}} = \frac{C_M}{C_{B_f}}\frac{C_{B_f}}{C_{A_f}}$$

So, multiplying (I) by (C_{B_f}/C_{A_f}) will yield x_M:

$$x_M = \frac{C_M}{C_{A_f}} = \frac{\alpha(1 - x_B)x_B}{x_B + \alpha(1 - x_B)}\left(\frac{C_{B_f}}{C_{A_f}}\right) \tag{K}$$

By a similar procedure, the ratio of Eqs. (G) to (E) gives a relationship between C_D and C_B in terms of C_M. Using the expression [Eq. (H)] already obtained for C_M in terms of C_B, we can evaluate C_D/C_{B_f}. This leads to the following result for x_D:

$$x_D = \frac{C_D}{C_{A_f}} = \frac{x_B^2}{x_B + \alpha(1 - x_B)}\left(\frac{C_{B_f}}{C_{A_f}}\right) \tag{L}$$

The total conversion of acrilonitrile is found by writing a total mass balance of nitrogen, which is present in compounds A, M, and D. This balance is:

$$C_{A_f} = C_A + C_M + 2C_D \tag{M}$$

Since $x_A = 1 - C_A/C_{A_f}$, dividing Eq. (M) by C_{A_f} yields

$$x_A = x_M + 2x_D \tag{N}$$

Substituting Eqs. (K) and (L) for x_M and x_D gives x_A in terms of x_B:

$$x_A = \left(\frac{C_{B_f}}{C_{A_f}}\right)x_B\left(1 + \frac{x_B}{x_B + \alpha(1 - x_B)}\right) \tag{P}$$

Equations (K), (L), and (P) provide the several conversions of acrilonitrile in terms of the conversion of isobutylene, x_B. Hence, we can choose various values of x_B and calculate x_M, x_D, and x_A. These individual conversions can be plotted versus the total conversion of acrilonitrile to give the required solution. The results for $\alpha = 2$ and $C_{B_f}/C_{A_f} = 3$ are shown in Fig. 4-12.

The overall selectivity of diamine to monoamine is x_D/x_M. This ratio is also plotted in Fig. 4-12.

The low conversion of isobutylene and the low selectivity of diamine to monoamine suggest that for efficient performance the product stream would have to be separated, and the isobutylene, acrilonitrile, and monoamine recycled back to the reactor feed.

Example 4-10 An ideal continuous stirred-tank reactor is used for the homogeneous polymerization of monomer M. The volumetric flow rate is Q, the volume of the reactor is V, and the density of the reaction solution is invariant with composition. The concentration of monomer in the feed is M_0. The polymer product is produced by an initiation step and a consecutive series of

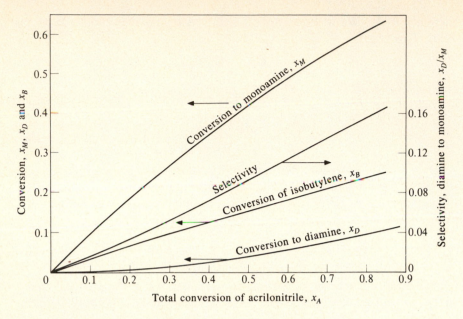

Figure 4-12 Conversion and selectivity in a stirred-tank reactor.

propagation reactions. The reaction mechanism and rate equations may be described as follows, where P_1 is the concentration of the activated monomer and $P_2, \ldots, P_n$ are concentrations of polymer molecules containing n monomer units:

Initiation

$$M \to P_1 \qquad \mathbf{r} = k_i M \qquad \text{mol/(time)(vol.)}$$

Propagation

$$P_1 + M \to P_2$$
$$P_2 + M \to P_3$$
$$\mathbf{r} = k_p P_n M \qquad \text{mol/(time)(vol.)}$$
$$\cdots\cdots\cdots\cdots$$
$$P_n + M \to P_{n+1}$$

All the propagation steps have the same second-order rate constant k_p.

If k_i, k_p, M_0, Q, and V are known, derive expressions for (a) the conversion of monomer, (b) the concentration of polymer P_n in the product stream, and (c) the weight fraction w_n of the polymer product (this is a form of the molecular weight distribution of the product).

SOLUTION (a) Equation (3-5) is valid and may be applied to each polymer product and to the monomer. Solving Eq. (3-5) for the rate ($\mathbf{r}$ = rate of disappearance of reactant, a *positive quantity*) and applying it to the monomer, we obtain

$$\mathbf{r}_m = \frac{M_0 - M}{V/Q} = \frac{M_0 - M}{\bar{\theta}} \tag{A}$$

The monomer is consumed by the initiation step and all the propagation reactions, and so its rate of disappearance is

$$\mathbf{r}_m = k_i M + k_p M \sum_{n=1}^{\infty} P_n \tag{B}$$

Combining Eqs. (A) and (B) gives

$$\frac{M_0 - M}{\bar{\theta}} = k_i M + k_p M \sum_{n=1}^{\infty} P_n \tag{C}$$

Writing Eq. (A) for P_1, noting that there is no P_1 in the feed, and inserting the appropriate expressions for the rate, we obtain

$$\mathbf{r}_1 = -k_i M + k_p P_1 M = \frac{0 - P_1}{\bar{\theta}} = -\frac{P_1}{\bar{\theta}}$$

This may be written

$$k_i M = \frac{P_1}{\bar{\theta}} + k_p P_1 M \tag{D}$$

or

$$P_1 = \frac{k_i M}{(1/\bar{\theta}) + k_p M} \tag{E}$$

Similarly, for P_2 and P_n the mass balances are

$$k_p M P_1 = \frac{P_2}{\bar{\theta}} + k_p P_2 M \tag{F}$$

$$k_p M P_{n-1} = \frac{P_n}{\bar{\theta}} + k_p P_n M \tag{G}$$

Solving for P_2 from Eq. (F) and P_n from Eq. (G) gives

$$P_2 = \frac{k_p M P_1}{(1/\bar{\theta}) + k_p M} \tag{H}$$

$$P_n = \frac{k_p M P_{n-1}}{(1/\bar{\theta}) + k_p M} \tag{I}$$

Inserting Eq. (E) in Eq. (H) to eliminate P_1 and proceeding in a similar way for $P_2, P_3, \ldots, P_{n-1}$, we can eliminate all intermediate concentrations to obtain

$$P_n = \frac{k_i M^n k_p^{n-1}}{(1/\bar{\theta} + k_p M)^n} = \frac{k_i}{k_p} \frac{1}{\{1 + 1/\bar{\theta} k_p M\}^n} \tag{J}$$

Summing up all the mass balances for the individual species written in the form of Eqs. (D), (F), and (G) gives a simple relationship between the concentration of unreacted monomer and the concentrations of all the polymer products,

$$k_i M = \frac{P_1}{\bar{\theta}} + \frac{P_2}{\bar{\theta}} + \cdots = \frac{1}{\bar{\theta}} \sum_{n=1}^{\infty} P_n \tag{K}$$

Combining this result with Eq. (C) provides a relationship between M and M_0,

$$\frac{M_0 - M}{\bar{\theta}} = k_1 M + k_i k_p M^2 \bar{\theta} \tag{L}$$

Since the conversion of monomer is

$$x = \frac{M_0 - M}{M_0} \tag{M}$$

Eq. (L) may be written

$$x = k_i \bar{\theta}(1 - x)\{1 + k_p \bar{\theta} M_0(1 - x)\} \tag{N}$$

This is the answer to the first part of the problem. Note that the conversion depends on the initial monomer concentration as well as on the rate constants and $\bar{\theta}$.

(b) If we know x, we can find M from Eq. (M) and then calculate P_n from Eq. (J) for any value of n.

(c) The weight fraction of P_n in the polymer product is the weight of polymer P_n divided by the weight of total polymer product. The latter is equal to the weight of monomer that has reacted. If W_0 is the molecular weight of monomer, the weight fraction of P_n is

$$w_n = \frac{P_n n W_0}{(M_0 - M)W_0} = \frac{n P_n}{M_0 - M} \tag{P}$$

Substituting Eq. (J) for P_n in Eq. (P) gives

$$w_n = \frac{k_i n}{k_p(M_0 - M)} \frac{1}{\{1 + 1/\bar{\theta}k_p M)\}^n} \tag{Q}$$

As an illustration of these results Eqs. (N) and (Q) have been used to calculate the conversion and molecular weight distribution for the case

$$k_1 = 10^{-2} \text{ s}^{-1}$$

$$k_p = 10^2 \text{ cm}^3/(\text{g mol})(\text{s})$$

$$M_0 = 1.0 \text{ g mol}/(\text{liter})$$

Calculations were carried out for average residence times ($\bar{\theta} = V/Q$) from 3 to 1000 s. Figure 4-13 shows the conversion and w_n for several products.

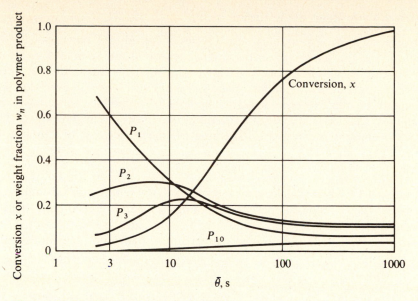

Figure 4-13 Distribution of products in stirred-tank reactor for polymerization.

In this range of $\bar{\theta}$ the conversion increases from 3.7 to 97.3%. The weight fraction of P_1 decreases continuously with residence time. This is expected, since w refers to the weight fraction of product (not of total effluent from the reactor), and at low $\bar{\theta}$ the product will be nearly all P_1. At the other extreme, w_{10} is low but increases continuously, since higher products $(m > 10)$ are not formed in significant amounts. Curves are also shown for two products of intermediate molecular weight, P_2 and P_3. These curves show maxima at intermediate $\bar{\theta}$.

The curves in Fig. 4-13 are helpful in choosing a reactor size appropriate for producing a specified product distribution. For example, if the objective is to make the maximum amount of trimer (P_3), the reactor volume should be such that $\bar{\theta}$ is about 13 s. This problem does not allow for destruction of active polymer P_n by termination reactions. This is often the case and leads to a more symmetrical distribution of molecular weights. Denbigh[†] has considered this and similar polymerization problems.

It was possible to solve this polymerization problem in closed form because we were dealing with algebraic equations, that is, with an ideal stirred-tank reactor. The same problem in a tubular-flow reactor would require the solution of a series of integral equations, which is possible only by numerical methods.

† K. G. Denbigh, *Trans. Faraday Soc.*, **43**, 648 (1947).

4-6 Stirred-Tank Reactors in Series

In some cases it may be desirable to use a series of stirred-tank reactors, with the exit stream from the first serving as the feed to the second, and so on. For constant density the exit concentration or conversion can be solved by consecutive application of Eq. (3-5) to each reactor. Eldridge and Piret[†] have derived solutions for a number of rate expressions and for systems of reversible, consecutive, and simultaneous reactions. Graphical procedures have also been developed. The kinds of calculations involved are illustrated for the simple case of a first-order reaction in Example 4-11.

Example 4-11 Acetic anhydride is to be hydrolyzed in three stirred-tank reactors operated in series. Suppose that each has a volume of 1800 cm³, that the temperature is constant and equal to 25°C, and that the feed rate to the first reactor is 582 cm³/min. Compute the percent of hydrolysis accomplished in the three reactors.

SOLUTION From Table 4-11 (25°C, 582 cm³/min), the fraction hydrolyzed in the stream leaving the first reactor is 0.328. If the anhydride concentration leaving the first reactor is designated as C_1, and that leaving the second is C_2, Eq. (3-5) applied to the second reactor is

$$C_2 = \frac{C_1}{1 + k\bar{\theta}_2}$$

or

$$\frac{C_1 - C_2}{C_1} = (\text{fraction hydrolyzed in reactor 2}) = 1 - \frac{1}{1 + k\bar{\theta}_2}$$

Since $\bar{\theta}_2 = \bar{\theta}_1$, the fraction hydrolyzed in the second reactor will be the same as in the first. Therefore the following equations may be written for the three reactors:

$$\frac{C_0 - C_1}{C_0} = 0.328 \qquad \text{or } C_1 = (1 - 0.328)C_0 = 0.672C_0$$

$$\frac{C_1 - C_2}{C_1} = 0.328 \qquad \text{or } C_2 = 0.672C_1 = (0.672)^2 C_0$$

$$\frac{C_2 - C_3}{C_2} = 0.328 \qquad \text{or } C_3 = 0.672C_2 = (0.672)^3 C_0$$

The equation for C_3 may be generalized to the form

$$C_n = (0.672)^n C_0 = \left(\frac{1}{1 + k\bar{\theta}_i}\right)^n C_0$$

---

† J. W. Eldridge and E. L. Piret, *Chem. Eng. Progr.*, **46**, 290 (1950).

or

$$1 - \frac{C_n}{C_0} = 1 - \frac{1}{(1 + k\bar{\theta}_i)^n} = \text{fraction hydrolyzed in } n \text{ reactors} \qquad \text{(A)}$$

where n is the number of reactors in series, and $\bar{\theta}_i$ is the residence time in each stage. For $n = 3$, the fraction hydrolyzed is $1 - 0.672^3 = 0.697$, or 69.7%.

If the restriction of equal residence times in each reactor is removed, Eq. (A) of Example 4-11 becomes

$$x_n = 1 - \frac{1}{(1 + k\bar{\theta}_1) \cdots (1 + k\bar{\theta}_i) \cdots (1 + k\bar{\theta}_n)} \qquad \text{(4-16)}$$

where x_n is the conversion in the effluent from the nth reactor and $\bar{\theta}_i = V_i/Q$ is the mean residence time in the ith reactor.

Graphical methods can be used to obtain the conversion from a series of reactors and have the advantage of displaying the concentration in each reactor. Moreover, no additional complications are introduced when the rate equation is not first order. As an illustration of the procedure consider three stirred-tank reactors in series, each with a different volume, operating as shown in Fig. 4-14a. The density is constant, so that at steady state the volumetric flow rate to each reactor is the same. The flow rate and reactant concentration of the feed (Q and C_0) are known, as are the volumes of each reactor. We construct a graph of the rate of reaction $\mathbf{r}$ vs. reactant composition. The curved line in Fig. 4-14b shows how the rate varies with C according to the rate equation, which may be of any order.

Equation (3-5), applied to the first stage and written for acetic anhydride, is

$$\frac{V_1}{Q} = \bar{\theta}_1 = \frac{C_0 - C_1}{\mathbf{r}_1}$$

where $\mathbf{r}_1$ is the rate of *disappearance* of anhydride. This expression may be rearranged in the form of a linear relationship between $\mathbf{r}$ and C:

$$\mathbf{r}_1 = \left(\frac{Q}{V_1}\right)C_0 - \left(\frac{Q}{V_1}\right)C_1$$

For the well-mixed, stirred-tank reactor $\mathbf{r}_1$ and C_1 will represent a point on the rate curve. The location of this point and, hence, the concentration C_1 leaving the first reactor, is found by the following procedure. First, locate C_0 at the proper point on the abscissa (point A in Figure 4-14b). Then draw a straight line with a slope of $-Q/V_1$ through C_0. The previous equation shows that the intersection of the straight line with the rate curve gives C_1. The second stage is constructed by locating C_1 on the abscissa (point C) and then drawing a straight line from C with a slope of $-Q/V_2$. The intersection of this line with the reaction curve (point D) establishes the effluent concentration C_2 from the second reactor. Similar construction for the third reactor is shown in Fig. 4-14b, and leads to the desired concentration C_3 (point F) leaving the third reactor.

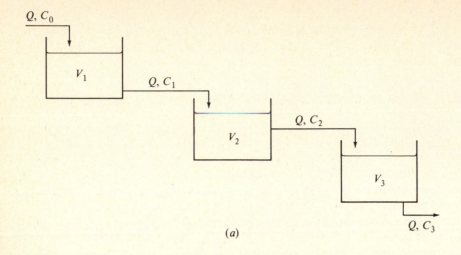

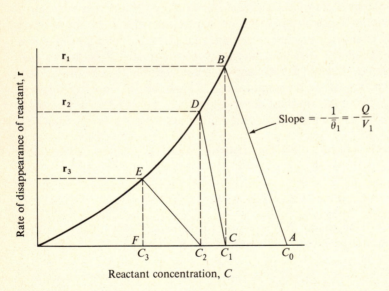

Figure 4-14 (*a*) Steady operation of three stirred-tank reactors in series. (*b*) Graphical solution for stirred-tank reactors in series.

4-7 Comparison of Stirred-Tank and Tubular-Flow Reactors

The stirred-tank reactor has certain advantages because of the uniform temperature, pressure, and composition attained as a result of mixing. As mentioned, it is possible to operate such a reactor under isothermal conditions even when the heat of reaction is high—an impossibility in the usual tubular type. When a small temperature variation is desired, for example, to minimize side reactions or avoid unfavorable rates, the opportunity for isothermal operation at the optimum temperature is a distinct advantage. Stirred-tank reactors, by virtue of their large volumes (and hence their large V/F values) provide a long residence time. This, combined with the isothermal nature of the reactor, permits operation at the optimum temperature for a long reaction time. For rate equations of certain types the selectivity in multiple reactions may be greater in tank reactors than in tubular-flow reactors for the same residence time. For other forms of the rate equations the reverse is true. Examples later in this section illustrate this point.

For high-pressure reactions it is usually necessary, because of cost considerations, to use small-diameter tubular reactors rather than tank types. Tank reactors that are to be operated at high pressures require a large wall thickness and complex sealing arrangements for the mixer shaft, factors which increase both initial and maintenance costs. Stirred-tank performance may be achieved in a recycle form of tubular-flow reactor, as illustrated in Fig. 1-7a. The diameter may be minimized for high-pressure operation by constructing the reactor in the form of a closed loop of tubing with entrance and exit connections and with a recycle pump in the loop.†

The rate of heat transfer per unit mass of reaction mixture is generally lower in the conventional tank type than in a small-diameter tubular reactor, chiefly because of the lower ratio of surface area (available for heat transfer) to volume in the tank reactors and their lower heat-transfer coefficients. So, in instances where the heat of reaction is high it may be desirable to use a tubular reactor. For example, various thermal reactions of hydrocarbons require significant amounts of thermal energy at an elevated temperature level. This would be difficult to accomplish with a large-diameter reactor because of the limited external heat-transfer surface (per unit mass of reaction mixture) and the low coefficient of heat transfer from the oil in the tank to the tank wall. In the tubular reactors (pipe stills) used in industry the coefficient of heat transfer can be increased by forcing the oil through the tubes at a high speed. It is also apparent that severe difficulties would arise in attempting to provide for efficient stirring under reaction conditions (800 to 1200°F, 300 to 800 lb/in.² abs). The tubular-loop reactor operated at high circulation rates can give complete mixing in a small-diameter tube and high rates of heat transfer. By introducing small solid particles that are free to move, a type is obtained in which there is considerable mixing. Such fluidized-bed reactors also give improved heat-transfer coefficients between the fluid and the wall.

In summary, stirred-tank reactors have been employed on a commercial scale mainly for liquid-phase reaction systems at low or medium pressures. Stirred-tank reactors can be used when the heat of reaction is high, but only if the temperature level obtained in their isothermal operation is satisfactory from other standpoints. If the reactions are endothermic and a high temperature is required, tubular reactors are usually indicated. However, a tank type may be employed as a semibatch unit for a highly exothermic reaction. For example, the production of hexamethylenetetramine by reacting ammonia and formaldehyde (in aqueous solution) is highly exothermic, but the rate of reaction is rapid and 100% conversion is possible over a range of temperature of at least 80 to 100°C. By adjusting

---

† Practical advantages of loop reactors are discussed in *Process Engineering*, p. 62, December 1973 and *Hydrocarbon Processing*, **55**, 99 (June 1976).

the rate of feed and reactor volume, it is possible to add the feed at 20°C and remove enough heat to keep the reaction mixture below 100°C.

Denbigh and coworkers† have discussed the technical advantages and disadvantages of continuous-stirred-tank reactors, especially in comparison with batch-operated tank reactors. Stead, Page, and Denbigh‡ describe experimental techniques for evaluating rate equations from stirred-tank data. Rase§ presents practical and technical aspects of both plug-flow and stirred-tank reactors.

In stirred-tank equipment the reaction occurs at a rate determined by the composition of the exit stream from the reactor. Since the rate generally decreases with the extent of conversion, the tank reactor operates at the lowest point in the range between the high rate corresponding to the composition in the reactor feed and the low rate corresponding to the exit composition. In the tubular type maximum advantage is taken of the high rates corresponding to low conversions in the first part of the reactor. This means that the tank reactor must have a larger volume for a given feed rate (larger V/F value). Of course, this reasoning does not take into account the effects of side reactions or temperature variations; these may offset this disadvantage of the tank reactor, as illustrated in Example 5-3. Also, the total volume required in a tank-flow reactor can be reduced by using several small units in series. The relation between volumes required in stirred-tank and tubular-flow reactors can be illustrated by reference to a constant volume first-order reaction. Equation (3-5) is applicable for the stirred-tank reactor and gives

$$\frac{V_s}{Q} = \frac{C_0 - C}{\mathbf{r}} = \frac{C_0 - C}{kC}$$

where C refers to the concentration of reactant and $\mathbf{r}$ to its rate of *disappearance*. In terms of conversion of reactants

$$x = \frac{C_0 - C}{C_0} = \frac{k(V_s/Q)}{1 + k(V_s/Q)} \tag{4-17}$$

For the tubular-flow case Eq. (3-18) can be used:

$$\frac{V_p}{Q} = C_0 \int \frac{dx}{\mathbf{r}} = C_0 \int_0^x \frac{dx}{kC_0(1 - x)} = \frac{-1}{k} \ln(1 - x)$$

or

$$x = 1 - e^{-k(V_p/Q)} \tag{4-18}$$

Equations (4-17) and (4-18) are plotted in Fig. 4-15 as conversion vs. $k(V/Q)$. For equal flow rates $k(V/Q)$ is proportional to the volume. It is clear that for any conversion the volume required is largest for the tank reactor and that the difference increases with residence time. We can obtain a direct measure of the ratio of

† K. G. Denbigh, *Trans. Faraday Soc.*, **40**, 352 (1944), **43**, 648 (1947); K. G. Denbigh, M. Hicks, and F. M. Page, *Trans. Faraday Soc.*, **44**, 479 (1948).

‡ B. Stead, F. M. Page, and K. G. Denbigh, *Disc. Faraday Soc.*, **2**, 263 (1947).

§ H. F. Rase, "Chemical Reactor Design for Process Plants," chap. 6, John Wiley & Sons, New York, 1977.

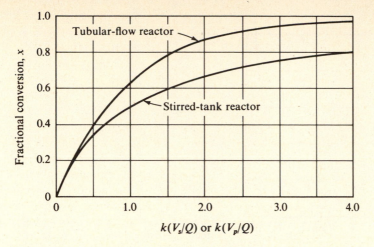

Figure 4-15 Conversion in stirred-tank and plug-flow reactors.

volume V_s of the stirred-tank reactor to volume V_p of the tubular-flow (plug-flow) reactor at the same conversion by equating Eqs. (4-17) and (4-18).

$$\frac{k(V_s/Q)}{1 + k(V_s/Q)} = 1 - e^{-k(V_p/Q)}$$

If α is the ratio of volumes, $\alpha = V_s/V_p$, then the previous equation can be written in terms of V_p and α.

$$\frac{\alpha k(V_p/Q)}{1 + \alpha k(V_p/Q)} = 1 - e^{-k(V_p/Q)} \tag{4-19}$$

If we now replace $k(V_p/Q)$ in Eq. (4-19) with the function of x from Eq. (4-18) and solve for α, we have the ratio of volumes as a function of conversion,

$$\alpha = \frac{x}{(x - 1)\ln(1 - x)} \tag{4-20}$$

This result is plotted in Fig. 4-16 and shows that at low conversions there is little to be gained in using a tubular-flow reactor, but at conversions of 70% or larger more than twice as much volume is required for a stirred-tank unit.[†]

Selectivity may also be different in stirred-tank and tubular-flow reactors. It has been shown[‡] that, depending on the kinetics and nature of the multiple reactions, selectivity obtained in a stirred-tank reactor may be less, the same as, or greater than that for a tubular-flow reactor. Examples of reaction systems for each result are given in Table 4-12. The order of the rate equation is assumed to follow the stoichiometry for each reaction. Since the stirred-tank reactor corresponds to

---

[†] Reactor volume vs. conversion for other forms of rate equations and for different reactor types is given by Wirges and Shah, *Hydrocarbon Processing*, **55**, (April 1976).

[‡] T. E. Corrigan, G. A. Lessells, and M. J. Dean, *Ind. Eng. Chem.*, **60**, 62 (1968).

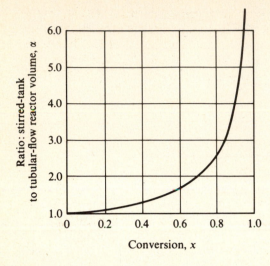

Figure 4-16 Ratio of volumes required for stirred-tank and tubular-flow (plug-flow) reactors.

complete mixing and the tubular-flow unit to no axial mixing, the table shows the effect of mixing upon selectivity. The following example illustrates the method of establishing the conclusions given in Table 4-12.

Example 4-12 Develop equations for the selectivity of product B with respect to D for reaction system 1 of Table 4-12 for stirred-tank and tubular-flow reactors. Assume isothermal conditions and constant density. Let capital letters designate concentrations. In the feed $A = A_0$ and $B = D = 0$.

SOLUTION The reaction sequence is

$$A \xrightarrow{k_1} |B$$

$$B \xrightarrow{k_3} D$$

Table 4-12 Effect of mixing on selectivity for various types of reaction systems

Reaction system	Reaction type	Overall selectivity†
1. $A \to B$ $\quad B \to D$	Consecutive (first order)	$S_S < S_p$ (selectivity of B with respect to D)
2. $A + B \to R$ $\quad A + R \to S$	Consecutive (second order)	$S_S < S_p$ (selectivity of R with respect to S)
3. $A \to B$ $\quad A \to C$	Parallel (equal, first order)	$S_p = S_S$ (selectivity of B with respect to C)
4. $A + B \to C$ $\quad A + B \to D$	Parallel (equal, second order)	$S_p = S_S$ (selectivity of C with respect to D)
5. $A + B \to C$ $\quad 2A \to D$	Parallel (unequal order with respect to A)	$S_S > S_p$ (selectivity of C with respect to D)

† Selectivity, as defined in Chap. 2, is the ratio of the yield of one product to that of another.

In Sec. 2-11 this system was analyzed for a constant-volume batch reactor. Since the tubular-flow reactor will have the same form of mass balance (see Sec. 3-4) as that for the batch reactor the results in Sec. 2-11 can be applied here if t is replaced with the residence time V/Q. Hence the overall selectivity for a *tubular-flow* reactor is given by the ratio of Eqs. (2-89) and (2-90). Since $A/A_0 = 1 - x_t$, this ratio is

$$S_p = \frac{x_B}{x_D} = \frac{[k_1/(k_1 - k_3)][(1 - x_t)^{k_3/k_1} - (1 - x_t)]}{[k_1/(k_1 - k_3)][1 - (1 - x_t)^{k_3/k_1}] - [k_3/(k_1 - k_3)]x_t} \tag{A}$$

For the *stirred-tank* case Eq. (3-5) may be written for components A, B, and D† as follows:

$$\bar{\theta} = \frac{V}{Q} = \frac{A_0 - A}{k_1 A} \quad \text{or } A = \frac{A_0}{1 + k_1 \bar{\theta}} \tag{B}$$

$$\bar{\theta} = \frac{V}{Q} = \frac{0 - B}{k_3 B - k_1 A} \quad \text{or } B = \frac{k_1 \bar{\theta} A}{1 + k_3 \bar{\theta}} \tag{C}$$

$$\bar{\theta} = \frac{V}{Q} = \frac{0 - D}{-k_3 B} \quad \text{or } D = k_3 \bar{\theta} B \tag{D}$$

From Eqs. (B) and (C),

$$\frac{B}{A_0} = x_B = \frac{k_1 \bar{\theta}}{(1 + k_1 \bar{\theta})(1 + k_3 \bar{\theta})} \tag{E}$$

Using this result in Eq. (D) gives

$$\frac{D}{A_0} = x_D = \frac{k_1 \bar{\theta} k_3 \bar{\theta}}{(1 + k_1 \bar{\theta})(1 + k_3 \bar{\theta})} \tag{F}$$

Then the selectivity S_s in the stirred-tank reactor will be

$$S_s = \frac{x_B}{x_D} = \frac{1}{k_3 \bar{\theta}} \tag{G}$$

This result may be expressed in terms of the total conversion of A by noting, from Eq. (B), that

$$k_1 \bar{\theta} = \frac{A_0}{A} - 1 = \frac{x_t}{1 - x_t}$$

Using this result in Eq. (G) to eliminate $\bar{\theta}$ gives

$$S_s = \frac{k_1}{k_3} \frac{1 - x_t}{x_t} \tag{H}$$

Equations (A) and (H) can be employed to calculate S_p and S_s for any conversion. The results are shown in Fig. 4-17 for $k_1/k_3 = 2$. The *relative* position of

---

† Alternately, mass balance equations could be written for each reaction and a specific component, and then a total mass balance used to obtain the three required equations (See Examples 4-7, 4-9).

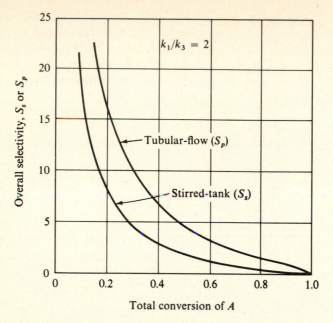

Figure 4-17 Selectivity for consecutive reactions in stirred-tank and tubular-flow reactors.

the two curves would be the same for other values of k_1/k_3. Note that the selectivity of B with respect to D is greater in the tubular-flow reactor for all conversions, although the difference approaches zero as the conversion approaches zero.

Table 4-12 compares selectivities for *single* reactors. For some reaction systems a combination of stirred-tank and tubular-flow units may give higher selectivities than a single reactor of the same total volume. The possible combinations of number and arrangement of reactors and reaction systems are huge. However, the approach to selectivity evaluation is the same and follows the methods described in Example 4-12. A simple illustration of the conversion obtained in a combination of reactors is given in the following example.

Example 4-13 A dilute aqueous solution of acetic anhydride is to be hydrolyzed continuously at 25°C. At this temperature the rate equation for the disappearance of anhydride is

$$\mathbf{r} = 0.158C, \text{ g mol/(cm}^3)\text{(min)}$$

where C is the concentration of anhydride, in gram moles per cubic centimeter. The feed rate to be treated is 500 cm³/min of solution, with an anhydride concentration of 1.5×10^{-4} g mol/cm³.

There are two 2.5-liter and one 5-liter reaction vessels available, with excellent agitation devices.

(a) Would the conversion be greater if the one 5-liter vessel were used as a steady-flow tank reactor or if the two 2.5-liter vessels were used as reactors in series? In the latter case all the feed would be sent to the first reactor and the product from that would be the feed to the second reactor.

(b) Would a higher conversion be obtained if the two 2.5-liter vessels were operated in parallel; that is, if 250 cm^3/min of feed were fed to each reactor and then the effluent streams from each reactor joined to form the final product?

(c) Compare the conversions calculated in parts (a) and (b) with that obtainable in a tubular-flow reactor of 5-liter volume.

(d) Would the conversion be increased if a tank-flow reactor of 2.5 liters were followed with a 2.5-liter tubular-flow reactor?

Assume that the density of the solutions is independent of concentrations and that operation is steady state.

SOLUTION Since this is a first-order constant-density reaction, Eqs. (4-17) and (4-18) give the conversions for single-stirred-tank and ideal tubular-flow reactors in terms of residence time V/Q. For multiple-stirred-tank reactors Eq. (A) of Example (4-11) is applicable.

(a) For a single 5-liter vessel, $\bar{\theta} = 5000/500 = 10$ min. From Eq. (4-17),

$$x = \frac{0.158(10)}{1 + 0.158(10)} = 0.612$$

For two 2.5-liter reactors in series,

$$\bar{\theta}_1 = \bar{\theta}_2 = \frac{2,500}{500} = 5 \text{ min}$$

Substituting in Eq. (A) of Example 4-11 gives

$$x = 1 - \frac{1}{(1 + k\bar{\theta})^2} = 1 - \frac{1}{[1 + 0.158(5)]^2} = 0.688$$

(b) For a 2.5-liter reactor with a feed rate of 250 cm^3/min, $\bar{\theta} = 10$ min. Hence the conversion will be the same as for the 5-liter reactor with $Q = 500$ cm^3/min, that is, 0.612.

(c) For a single tubular-flow reactor, from Eq. (4-18),

$$x = 1 - e^{-0.158(5000/500)} = 1 - 0.206 = 0.794$$

(d) In the first reactor, $\bar{\theta}_1 = 2500/500 = 5$ min. Hence the conversion in the product stream from the first reactor will be

$$x_1 = \frac{k\bar{\theta}_1}{1 + k\bar{\theta}_1} = \frac{0.158(5)}{1 + 0.158(5)} = 0.442$$

When the conversion in the feed stream to a tubular-flow reactor is x_1 rather than zero, integration of Eq. (3-18) gives

$$\theta = C_0 \int_{x_1}^{x_2} \frac{dx}{\mathbf{r}} = C_0 \int \frac{dx}{kC_0(1 - x)} = -\frac{1}{k}[\ln(1 - x_2) - \ln(1 - x_1)]$$

or

$$x_2 = 1 - (1 - x_1)e^{-k\theta}$$

Table 4-13

Type	Conversion, %
Single-stirred tank (5 liters)	61.2
Two stirred tanks in parallel (each 2.5 liters)	61.2
Two stirred tanks in series (each 2.5 liters)	68.8
Stirred tank followed by tubular-flow reactor (each 2.5 liters)	74.6
Single tubular-flow reactor (5 liters)	79.4

The residence time in the second tubular-flow reactor is also 5 min. With a feed of conversion $x_1 = 0.442$, the final conversion would be

$$x_2 = 1 - (1 - 0.442)e^{-0.158(5)} = 1 - 0.254 = 0.746$$

The various results, arranged in order or increasing conversion, are shown in Table 4-13.

In the previous example increasing the number of stirred-tank reactors from one to two (with the same total residence time) caused an increase in conversion from 61.2% to 68.8%. Further increase in number of tank reactors in series would lead to a maximum conversion of 79.4%, the value for a tubular-flow reactor with the same residence time. An infinite number of stirred-tank reactors in series is equivalent to a tubular-flow reactor, provided the total residence time is the same. This may be demonstrated by deriving Eq. (4-18) from Eq. (A) of Example 4-11, which is applicable for equal residence time in each stirred tank. For a total residence time of $\bar{\theta}_t$, Eq. (A) becomes

$$x = 1 - \frac{1}{(1 + k\bar{\theta}_t/n)^n} \tag{4-21}$$

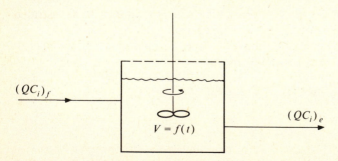

Figure 4-18 Semibatch reactor (stirred-tank type).

It is known that

$$\lim_{n \to \infty} \left(1 + \frac{\alpha}{n}\right)^n = e^{\alpha}$$

If we use this result with $\alpha = k\bar{\theta}_t$, Eq. (4-21) becomes

$$x = 1 - e^{-k\bar{\theta}_t}$$

which is the same as Eq. (4-18).

4-8 Non-Steady Flow (Semibatch) Reactors

A tank-type reactor which does not operate at steady-state can be described as a semibatch reactor. Semibatch behavior occurs when a tank-flow reactor is started up, when its operating conditions are changed from one steady state to another, or when it is shut down. Purging processes in which an inert material is added to the reactor can also be classified as semibatch operation.

In addition to applications arising from short-period deviations from steady-state, the semibatch reactor often is used for its own particular characteristics. For example, it is sometimes advantageous to add all of one reactant initially and then add the other reactant continuously. When the heat of reaction is large, the energy evolution can be controlled by regulating the rate of addition of one of the reactants. In this way limited heat-transfer characteristics of tank reactors can be partially eliminated. This form of operation also allows for a degree of control of concentration of the reaction mixture, and hence rate of reaction, that is not possible in batch or continuous-flow reactors. Another example is the case in which the reactants are all added initially to the vessel but one of the products is removed continuously, as in the removal of water by boiling in esterification reactions. The advantage here is an increase in rate, owing to the removal of one of the products of a reversible reaction and to increased concentrations of reactants.

The mass-balance equations for semibatch operation (illustrated in Fig. 4-18) may include all four of the terms in the general balance, Eq. (3-1). The feed and withdrawal streams can cause changes in composition and volume of the mixture in the reactor, in addition to such changes due to the reaction itself. Many operating alternatives exist. One reactant may be present in the initial charge to the reactor and the second reactant added continuously, periodically, or at a continuously varying rate. Similarly, product can be removed in a variety of ways. After presenting general equations, semibatch reactor problems will be illustrated with two special cases (Examples 4-14 and 4-15).

In terms of the molal concentration C_i of species i, volumetric flow rate Q, reactor volume V, Eq. (3-1) takes the form:

$$(QC_i)_f - (QC_i)_e + \mathbf{r}_i V = \frac{d[V(C_i)_e]}{dt} \qquad (4\text{-}22)\dagger$$

---

† Note that C_i and $\mathbf{r}_i$ in the reactor are equal to their values in the effluent stream for an ideal stirred tank. Also note that Eq. (4-22) reduces to the usual form for a batch reactor [Eq. (3-2)], if there are no feed or effluent streams, and to Eq. (3-4) for a continuous-flow steady-state reactor.

where r_i is the molal rate of production of i. In addition to the species mass balance, a *total* mass balance may be written. Let the mass densities of the feed and effluent streams be ρ_f and ρ_e. Then Eq. (3-1), written for the total flow is

$$Q_f\rho_f - Q_e\rho_e + 0 = \frac{d(V\rho_e)}{dt} \tag{4-23}$$

The two expressions are sufficient to determine the effluent concentration $(C_i)_e$ and reactor volume as a function of time, provided that the flow rates Q_f and Q_e are known as a function of time, densities are known, r_i is known as a function of concentrations, and the initial concentrations are known. Frequently, numerical integration is required, even for isothermal situations.

An important case in which analytical integration is possible is when the feed and exit flow rates, feed composition, and density are all constant, and the reaction is first order. Mason and Piret† have analyzed single and cascades (reactors in series) of stirred-tank reactors operating under these restrictions. The results are a reasonable representation of the behavior for many systems under startup and shutdown periods. With constant density, the concentration accounts fully for changes in amount of reactant. Also, for constant density and constant flow rates, Eq. (4-23) shows that the reactor volume V will remain constant. Under these restrictions Eq. (4-22) may be written

$$Q(C_i)_f - Q(C_i)_e + r_i V = V\frac{d(C_i)_e}{dt}$$

or

$$(C_i)_f - (C_i)_e + r_i\bar{\theta} = \bar{\theta}\frac{d(C_i)_e}{dt} \tag{4-24}$$

If i is a reactant, its rate of production is $r_i = -kC_i$. Then if temperature is constant, Eq. (4-24) is a linear differential equation which can be integrated analytically. In terms of $\bar{\theta}$, it may be written

$$\frac{d(C_i)_e}{dt} + \left(\frac{1}{\bar{\theta}} + k\right)(C_i)_e = \frac{(C_i)_f}{\bar{\theta}} \tag{4-25}$$

Example 4-14 Acetic anhydride is hydrolyzed at 40°C in a semibatch system operated by initially charging the stirred-tank reactor with 10 liters of an aqueous solution containing 0.50×10^{-4} g mol anhydride/cm³. The vessel is heated to 40°C, and at that time a feed solution containing 3.0×10^{-4} g mol anhydride/cm³ is added at the rate of 2 liters/min. Product is withdrawn at the same rate. The density may be assumed constant, and the rate of disappearance of anhydride is

$$r = kC \quad \text{g mol/(cm}^3)(\text{min})$$

$$k = 0.380 \text{ min}^{-1}$$

† D. R. Mason and E. L. Piret, *Ind. Eng. Chem.*, **42**, 817 (1950); *Ind. Eng. Chem.*, **43**, 1210 (1951).

Determine the concentration of the solution leaving the reactor as a function of time.

SOLUTION Equation (4-25) is applicable, and will be applied to anhydride. Then $(C_i)_f/\bar{\theta}$ is a constant. Thus

$$\bar{\theta} = \frac{V}{Q} = \frac{10,000}{2,000} = 5 \text{ min}$$

$$\frac{(C_i)_f}{\bar{\theta}} = \frac{3.00 \times 10^{-4}}{5} = 6 \times 10^{-5} \text{ g mol/(cm}^3)(\text{min})$$

$$\frac{1}{\bar{\theta}} + k = \frac{1}{5} + 0.380 = 0.580 \text{ min}^{-1}$$

The integrated solution of Eq. (4-25) is

$$(C_i)_e = \frac{(C_i)_f}{\bar{\theta}} \frac{1}{(1/\bar{\theta}) + k} + Ie^{-(1/\bar{\theta}+k)t}$$

The constant of integration I may be obtained by noting that at $t = 0$ and $(C_i)_e = 0.50 \times 10^{-4}$

$$I = 0.50 \times 10^{-4} - \frac{(C_i)_f}{\bar{\theta}} \frac{1}{(1/\bar{\theta}) + k}$$

$$= 0.50 \times 10^{-4} - \frac{6 \times 10^{-5}}{0.580} = -5.34 \times 10^{-5}$$

Then the final expression for the acetic anhydride concentration leaving the reactor, as a function of time, is

$$(C_i)_e = \frac{6 \times 10^{-5}}{0.580} - 5.32 \times 10^{-5}e^{-0.58t}$$

$$= 10.3 \times 10^{-5} - 5.34 \times 10^{-5}e^{-0.58t}$$

Table 4-14 shows values of $(C_i)_e$ for increasing time, measured from the instant of addition of the feed stream. These results indicate that after 5 min the reactor is operating at very nearly steady-state conditions. Hence, the concentration at this time could also be obtained directly from Eq. (3-5) for steady-state operation.

Table 4-14

t, min	C_1, g mol/cm^3
0	5.00×10^{-5}
1	7.35×10^{-5}
2	8.67×10^{-5}
3	9.40×10^{-5}
5	10.0×10^{-5}
∞	10.3×10^{-5}

As a second illustration of semibatch behavior, consider the reactor shown in Fig. 4-18 to be operated with no withdrawal stream but with continuous addition of feed. For cases like this where the volume in the reactor varies with time, it is more convenient to use as a variable in the mass balance the moles of species i, (rather than its concentration). For our case of no withdrawal Eq. (4-22) becomes

$$F_i + r_i V = \frac{d(n_i)_e}{dt} \tag{4-26}$$

where F_i is the molal feed rate of i and n_i is the number of moles of i in the reactor at any time t. Since density changes due to reaction are usually small for liquid mixtures (it is for liquids that semibatch reactors are normally used), we assume constant density. Then Eq. (4-23) reduces to

$$\frac{dV}{dt} = Q_f \tag{4-27}$$

Equations (4-26) and (4-27) are a coupled set of differential equations which may be solved simultaneously for n_i and the reactor volume as a function of time. If the feed rate is constant, Eq. (4-27) can be immediately integrated to give

$$V = V_0 + Q_f t \tag{4-28}$$

where V_0 is the initial volume in the reactor. For this case Eq. (4-28) can be substituted for V in Eq. (4-26) so that there is but one differential equation to solve. Example 4-15 is an illustration of this case for a second-order reaction.

Example 4-15 The esterification of acetic acid and ethyl alcohol is to be carried out in a semibatch stirred-tank reactor at a constant temperature of 100°C. The alcohol is added initially as 400 lb of pure C_2H_5OH. An aqueous solution of acetic acid is then added at a rate of 3.92 lb/min for 120 min. The solution contains 42.6% by weight acid. Assume that the density is constant and equal to that of water.

The reaction is reversible, and the specific reaction rates may be taken as the same as those used in Example 4-2:

$$CH_3COOH + C_2H_5OH \rightleftharpoons CH_3COOC_2H_5 + H_2O$$

$$k = 4.76 \times 10^{-4} \text{ liter/(g mol)(min)}$$

$$k' = 1.63 \times 10^{-4} \text{ liter/(g mol)(min)}$$

Compute the conversion of acetic acid to ester as a function of time from $t = 0$ until the last amount of acid is added (120 min).

SOLUTION Equations (4-26) and (4-28) are applicable. Take species i to be acetic acid and define the conversion x as the ratio of the moles of acid reacted to the total moles added up to that time. Then, if subscript H designates acetic acid,

$$n_H = F_A t(1 - x) \tag{A}$$

From Example 4-2, the rate of production of acid is

$$\mathbf{r}_H = -(kC_H C_{OH} - k'C_E C_W) \tag{B}$$

The rate can be expressed in terms of conversion by first formulating equations for the concentrations. All the concentrations can be written as functions of x_A and t through the stoichiometry of the reaction and with Eq. (4-28). Thus the concentrations (in the reactor) omitting the subscript e, are

$$C_H = \frac{n_H}{V} = \frac{F_H t(1-x)}{V_0 + Q_f t} \tag{C}$$

$$C_{OH} = \frac{n_{OH}}{V} = \frac{(n_{OH})_0 - xF_H t}{V_0 + Q_f t} \tag{D}$$

$$C_W = \frac{n_W}{V} = \frac{F_W t + xF_H t}{V_0 + Q_f t} \tag{E}\dagger$$

$$C_E = \frac{n_E}{V} = \frac{xF_H t}{V_0 + Q_f t} \tag{F}$$

where $(n_{OH})_0$ represents the moles of alcohol in the reactor initially and F_W the molal feed rate of water. Substituting these equations for the concentration in Equation (B) gives the rate as a function of x:

$$-\mathbf{r}_H(x, t) = \frac{kF_H t(1-x)[(n_{OH})_0 - xF_H t]}{(V_0 + Q_f t)^2} - \frac{k'(F_w t + xF_H t)xF_H t}{(V_0 + Q_f t)^2} \tag{G}$$

All the quantities in Eq. (4-26) can be expressed in terms of x and t. Substituting Eqs. (A) and (4-28) for n_H and V, the mass balance (Eq. 4-26) becomes:

$$F_H + [\mathbf{r}_H(x, t)](V_0 + Q_f t) = F_H(1-x) - F_H t \frac{dx}{dt} \tag{H}$$

This may be written in the form:

$$\frac{dx}{dt} = R(x, t) = \left[\frac{-\mathbf{r}_H(x, t)(V_0 + Q_f t)}{F_H} - x\right]\frac{1}{t} \tag{I}$$

The rate $\mathbf{r}_H$ in terms of x and t, is given by Eq. (G). Equation (I) cannot be integrated analytically. However, it is of the same form as Eq. (J) of Example 4-7, except that there is but one dependent variable, x. Hence, such methods as the Runge-Kutta may be used to obtain a numerical solution. The procedure is to start at $t = 0$ and calculate the conversion at successive increments of time, Δt. If subscript n designates the number of time increments, x_{n+1} is obtained from x_n by the formula

$$x_{n+1} = x_n + \tfrac{1}{6}(k_0 + 2k_1 + 2k_2 + k_3) \tag{4-29}$$

† The first term in the numerator is the moles of water introduced in the feed and the second represents the moles of water produced by the reaction.

where the k's are found by evaluating $R(x, t)$ for values of x and t given by the expressions:

$$k_0 = (\Delta t)R(x_n, t_n)$$
$$k_1 = (\Delta t)R(x_n + k_0/2, t_n + \Delta t/2)$$
$$k_2 = (\Delta t)R(x_n + k_1/2, t_n + \Delta t/2) \qquad \text{(4-30)}†$$
$$k_3 = (\Delta t)R(x_n + k_2, t_n + \Delta t)$$

The known quantities are:

$$F_H = \frac{3.92(0.426)}{60} = 0.0278 \text{ lb mol of acetic acid/min}$$

$$F_w = \frac{3.92(1 - 0.426)}{18} = 0.125 \text{ lb mol of water/min}$$

$$(n_{OH})_0 = \frac{400}{46} = 8.70 \text{ mol of alcohol}$$

$$V_0 = 400/\rho_w = 400/59.8 = 6.69 \text{ ft}^3$$

$$Q_f = 3.92/59.8 = 0.0656 \text{ ft}^3/\text{min}‡$$

From Example 4-2, the rate constants in $\text{ft}^3/(\text{lb mol})(\text{min})$ are:

$$k = 4.76 \times 10^{-4} \frac{1}{28.32}(454) = 7.63 \times 10^{-3}$$

$$k' = 1.63 \times 10^{-4} \frac{1}{28.32}(454) = 2.62 \times 10^{-3}$$

With these values Eq. (I) gives for $R(x, t)$

$$R(x, t) = \left[\frac{(6.69 + 0.0656t)[(0.0278)kt(1 - x)(8.70 - 0.0278xt - 0.0278k'xt(0.125t + 0.0278xt)]}{0.0278(6.69 + 0.0656t)^2} - x \right]\frac{1}{t} \quad \text{(J)}$$

Equation (I) can now be solved numerically for x at any t using Eq. (J) for $R(x, t)$.§| The volume of the reaction mixture at any time is obtained from Eq. (4-28) which becomes, in this problem,

$$V = 6.69 + 0.0656t \qquad \text{(K)}$$

---

† Equations (4-29) and (4-30) are the working expressions for the single-dependent variable form of the fourth-order Runge-Kutta method, analogous to Eqs. (4-6) to (4-15) for two dependent variables.
‡ 59.8 lb/ft^3 is the density of water at 100°C.
§ $R(x, t)$ is indeterminate at $x = 0$, $t = 0$ since no acid reactant is present. To initiate the calculation, $R(0, 0)$, and, hence, k_0, may be *estimated* and then corrected after x for the first increment is obtained.

Table 4-15 Semibatch-reactor design for esterification of acetic acid with ethyl alcohol

t, min	V, ft^3	Conversion of acid	r, lb mol acid reacted/(min)(ft^3)
0	6.69	0	0
5	7.02	0.024	1.82×10^{-4}
10	7.34	0.045	3.2×10^{-4}
20	8.00	0.083	5.2×10^{-4}
40	9.31	0.141	7.0×10^{-4}
60	10.6	0.184	7.3×10^{-4}
80	11.9	0.215	6.9×10^{-4}
100	13.2	0.240	6.3×10^{-4}
120	14.5	0.259	5.7×10^{-4}

The results for the entire time interval 0 to 120 min, determined by continuing these stepwise calculations,† with $\Delta t = 1$ min, are summarized in Table 4-15. The conversion of acid added is greater at low times than at later values, indicating that the acetic acid is not reacting as fast as it is added. This is due to the relatively high feed rate of acid and the increasing importance of the reverse reaction as the products are produced.

As has been mentioned, when reverse reactions are important, continuous removal of one or more of the reaction products will increase the conversion obtainable in a given time. Thus one reactant could be charged to the reactor, a second reactant could be added continuously, and one of the products could be withdrawn continuously.‡ This form of semibatch reactor can be treated by modifying the design equation (4-22), the volume equation (4-23), and the expressions for the concentration, to take into account the withdrawal of material.

RECYCLE REACTORS

In a *continuous-flow recycle reactor* part of the effluent is returned to the feed as indicated in Fig. 4-19. If the fraction of the effluent that is recycled is reduced to zero, the ideal plug-flow reactor (no mixing axially) is achieved. At the other extreme of a very large fraction recycled, ideal stirred-tank performance (complete mixing) is approached. Suppose that there is no product or fresh feed stream, but instead a well-mixed reservoir is introduced, as shown in Fig. 4-20. With a high recycle rate the system now behaves essentially as a tank-type *batch* reactor. This arrangement is called a *batch-recycle reactor*.

---

† This solution of the problem is approximate because of the assumption of constant density of the reaction mixture. However, if density-vs.-composition data were available, the same stepwise method of calculation could be employed to obtain a more precise solution. Equation (K) would have to be modified to take into account density differences.

‡ For example by distilling the reaction mixture.

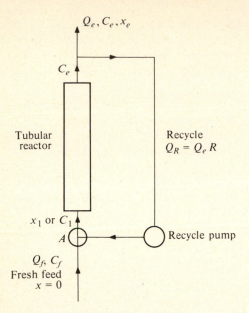

Figure 4-19 Continuous-flow recycle reactor.

By changing the recycle rate, the continuous-flow reactor can be operated with any desired degree of mixing. Such flexibility can have several advantages. For example, for some complex reaction schemes an intermediate degree of mixing (corresponding to an intermediate recycle rate) gives the maximum selectivity for an intermediate product.† For high-pressure reactions where complete mixing is desired, the expense of building a large-diameter tank could be prohibi-

† B. G. Gillespie and J. J. Carberry, *Chem. Eng. Sci.*, **21**, 472 (1966).

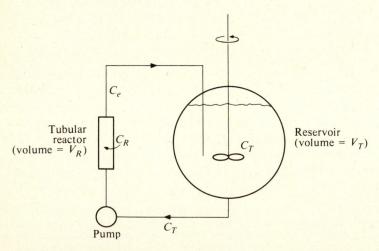

Figure 4-20 Laboratory, batch-recycle reactor.

tive. The same results may be obtained with a small-diameter tube operated as a recycle reactor (or the loop reactor of Fig. 1-7e) operating at a high recycle rate.

Both batch and flow recycle arrangements are often useful for laboratory studies of kinetics when operated at high recycle rates. One reason is that with a high recycle rate, the *conversion per pass* through the reactor can be made very low. This provides *differential-reactor* operation, yet the difference in concentrations between feed and discharge can be relatively large. Recycle operation can eliminate the problem of accurately measuring small concentrations or concentration differences. For example, in the continuous-flow recycle reactor the difference between concentrations in the fresh feed and effluent stream can be relatively large and yet the change in concentration from entrance to exit of the reactor (or conversion per pass) can be very small. A similar result is obtained in the batch-recycle unit. The reaction time (run time of the experiment) can be lengthened so that large changes in concentration in the reservoir (Fig. 4-20) will be observed, yet the concentration change per pass through the reactor will be low.

Batch-recycle operation is also useful for studying reactions which are slow. By operating a long time, enough reaction occurs to provide large changes in concentration in the reservoir. In a once-through tubular reactor the changes in concentration may be too low to measure accurately. The same advantage exists in comparing batch and tubular flow reactors without recycle.

The equations for describing the behavior of isothermal recycle reactors can be derived from the species mass balance, Eq. (3-1). This is done in the following paragraphs for the batch and continuous-flow types.

4-9 Batch-Recycle Reactors

First write a mass balance around the entire system of reactor and reservoir in Fig. 4-20. The volume of the connecting lines and the pump are assumed to be negligible. Also, it is supposed that reaction occurs only in the reactor.† Then the rate of production of species i is $\int \mathbf{r}_i \, dV_R$. There will be a contribution to the accumulation term from both the reactor of volume V_R and reservoir of volume V_T. These volumes will be assumed to be constant and also the contents of the tank are supposed to be well mixed. With these restrictions, Eq. (3-1) takes the form (eliminating subscript i for convenience):

$$\int_0^{V_R} \mathbf{r} \, dV_R = \frac{d}{dt} \int C_R dV_R + V_T \frac{dC_T}{dt} \qquad (4\text{-}31)$$

The concentration $(C)_R$ varies with position in the tubular reactor. It can be expressed in terms of the recycle flow rate Q_R and the rate $\mathbf{r}_i$ by using the mass balance (Eq. 3-14) for a tubular-flow reactor. However, the major application of batch recycle reactors is under conditions of *differential*-reactor operation. Then the change in concentration from entrance to exit of the reactor is very small so

---

† This limitation of reaction to the reactor is achieved for a reaction catalyzed by a solid by placing catalyst only in the reactor. Hence, the batch recycle arrangement is often used for catalytic reactions (see Chap. 12). The reaction can be restricted to the reactor in other ways (see Example 4-16).

that $C_R \approx C_T$. Also, for differential operation **r** is nearly constant throughout the reactor. Under these additional restrictions Eq. (4-31) reduces to

$$\mathbf{r} = \left(\frac{V_R + V_T}{V_R}\right)\frac{dC_T}{dt} \tag{4-32}$$

This is the equation for the rate of reaction in a batch-recycle reactor which operates in the *differential* mode. Such reactors have been used to establish rate equations from experimental measurements.† Example 4-16 is an illustration.

It is important in the design of experiments to determine the conditions such that differential operation is a reasonably valid assumption. To do this a species mass balance needs to be written around only the reactor in Fig. 4-20. Suppose that the volumetric flow rate Q is constant and that the concentration of i in the reactor effluent is C_e. Then Eq. (3-1) becomes

$$QC_T - QC_e + \int_0^{V_R} \mathbf{r}\ dV_R = \frac{d}{dt}\int C_R dV_R \tag{4-33}$$

Eliminating $d/dt \int C_R dV_R$ between Eqs. (4-31) and (4-33) and solving for $C_T - C_e$ gives

$$C_T - C_e = -\frac{V_T}{Q}\left(\frac{dC_T}{dt}\right) \tag{4-34}$$

The concentration C_R in the reactor will vary from C_T to C_e. Hence, $C_R \approx C_T$ if $(C_T - C_e) \to 0$. If this requirement is met, Eq. (4-32) is valid. Substituting dC_T/dt from this equation into Eq. (4-34) yields the desired criterion in terms of operating conditions and the rate of reaction,

$$C_T - C_e = -\frac{V_R}{Q}\left(\frac{V_T}{V_R + V_T}\right)\mathbf{r} \qquad \text{for}\quad (C_T - C_e) \to 0 \tag{4-35}$$

where **r** is the rate of production of species i. We conclude that for any reaction rate, $C_T - C_e \to 0$ and differential reactor operation is obtained, *provided* that V_R/Q is small; that is, a high recycle flow rate and small reactor volume are employed.

Example 4-16 During the induction period in the homogeneous photopolymerization of acrilamide, dissolved oxygen is consumed by reacting with the initiator added to the acrilamide solution. The kinetics of the disappearance of oxygen has been studied in a batch-recycle reactor.‡ The course of the reaction was followed by measuring the oxygen concentration in the reservoir

---

† For example, see Bruno Boval and J. M. Smith, *Chem. Eng. Sci.* **28**, 1961(1973) or A. E. Cassano, T. Matsurra, and J. M. Smith, *Ind. Eng. Chem. Fundam.* **7**, 655 (1968) for applications of a batch, recycle reactor for homogeneous, isothermal reactions.

‡ Miguel Ibârra and J. M. Smith, *AIChE J.*, **20**, 404 (1974).

at various times. For a particular run the data were

$$Q = 46.7 \times 10^{-6} \text{ m}^3/\text{s}$$

$$V_T = 6.6 \times 10^{-3} \text{ m}^3$$

$$V_R = 0.232 \times 10^{-3} \text{ m}^3$$

$$C_T = 2.40 \times 10^{-4}, \text{ k mol/m}^3, \qquad \text{at } t = 0$$

t, s	100	300	500	700	1000
$C_T \times 10^4$, k mol/m^3	2.31	2.20	2.09	1.90	1.73

From this information calculate: (*a*) the rate of reaction as a function of oxygen concentration and (*b*) the concentration difference, $C_T - C_e$, through the reactor (in order to check the suitability of differential-reactor operation).

SOLUTION (*a*) We assume that Eq. (4-32) is valid and then check this assumption (part *b*). The C_{O_2} data are first plotted vs. *t*. The slope of the resulting curve for any C_T gives dC_T/dt at that concentration. Normally, the slope would change, indicating from Eq. (4-32) that the rate changed with concentration. In this particular case the data establishes a straight line, as seen in Fig. 4-21. The constant rate suggests a zero-order dependency on oxygen concentration over the range of C_{O_2} that is involved.†

† Such zero-order kinetics for the induction period for removal of the initiator poison (oxygen) in photopolymerizations is well established. (P. J. Flory, "Principles of Polymer Chemistry," pp. 132–135, Cornell University Press, Ithaca, New York, 1969).

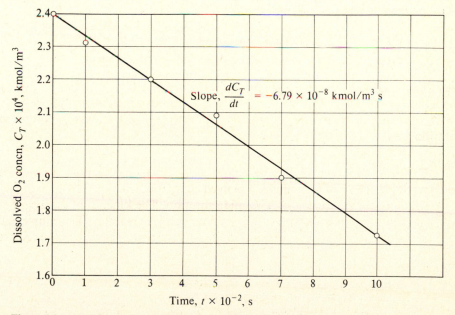

Figure 4-21 Batch-recycle reactor data for induction-period kinetics.

From the slope of the line in Fig. 4-21, Eq. (4-32) gives the production rate of oxygen as

$$r = \frac{(0.232 + 6.60) \times 10^{-3}}{0.232 \times 10^{-3}} (-6.79 \times 10^{-8})$$

$$= -2.0 \times 10^{-6} \text{ k mol/(m}^3\text{)(s)}$$

(*b*) The concentration difference, $C_T - C_e$, is calculable immediately from Eq. (4-35).

$$C_T - C_e = -\frac{V_R}{Q} \frac{V_T}{V_R + V_T} r$$

$$= -\frac{0.232 \times 10^{-3}}{46.7 \times 10^{-6}} \frac{6.6 \times 10^{-3}}{(0.232 + 6.60) \times 10^{-3}} (-2.0 \times 10^{-6})$$

$$= 9.6 \times 10^{-6} \text{ k mol/m}^3$$

This is a small value compared to C_T; differential-reactor operation is approached, and Eq. (4-32) is valid. The highest conversion per pass is obtained when C_T is the lowest, that is, at 1000 s. At this point

$$(\Delta x)_{\text{pass}} = \frac{C_T - C_e}{C_T} = \frac{9.6 \times 10^{-6}}{1.73 \times 10^{-4}} = 0.056 \qquad \text{or} \quad 5.6\%$$

4-10 Flow Recycle Reactors

The continuous-flow reactor operates at steady state. Such reactors may be used for laboratory studies of kinetics or on a large scale. In the *laboratory*, differential operation (low conversion per pass) is desired so as to simplify evaluating the rate of reaction (see Sec. 4-3). We first discuss this aspect and then consider commercial-scale recycle reactors where the conversion is large.

A mass balance around the *entire* system in Fig. 4-19 gives, following Eq. (3-1), and omitting the subscript *i*,

$$Q_f C_f - Q_e C_e + \int_0^{V_R} r \, dV_r = 0 \qquad (4\text{-}36)$$

where subscripts *f* and *e* refer to the *fresh* feed and the reactor effluent streams. Suppose that the reactor has a small volume and operates with a large recycle rate so that the change in concentration between entrance and exit of the reactor itself is small. For such differential-reactor operation the rate will be nearly the same in all parts of the reactor. Then the integral in Eq. (4-36) for the rate of production of *i* becomes $r V_R$, and Eq. (4-36) may be written

$$r = -\frac{Q_f C_f - Q_e C_e}{V_R} \qquad (4\text{-}37)$$

For a liquid-phase reaction, or an isothermal, gaseous reaction with no change in moles, the flow rate Q_e will be nearly the same as that of the *fresh* feed, Q_f. Then Eq. (4-37) becomes:

$$\mathbf{r} = -\frac{Q_f}{V_R}(C_f - C_e) \qquad (4\text{-}38)$$

Equations (4-37) or (4-38) can be used to calculate the rate of reaction from measurements of the concentrations, flow rate, and reactor volume. Such recycle reactors are common, particularly for catalytic reactions. Hence, these equations will be used again in Chaps. 12 and 13.

We now must examine the operating conditions required for Eq. (4-37) to be valid. To do this a species mass balance is written around *only* the reactor in Fig. 4-19. The concentration C_1 entering the reactor is the value resulting from the combination of the recycle stream and the fresh feed. For this situation Eq. (3-1) gives

$$(Q_f + Q_R)C_1 - (Q_R + Q_e)C_e + \int_0^{V_R} \mathbf{r}\, dV_R = 0 \qquad (4\text{-}39)$$

where Q_R is the flow rate of the recycle stream. We wish to find the conditions for which $(C_1 - C_e)$ approaches zero, for that will result in differential-reactor operation. By taking the integral equal to $\mathbf{r}V_R$, Eq. (4-39) gives the required conditions. To simplify the system, we will assume equal flow rates for the fresh feed and product streams,† that is, $Q_f = Q_e$. Then Eq. (4-39) becomes

$$C_1 - C_e = -\left(\frac{V_R}{Q_R + Q_e}\right)\mathbf{r} = -\left[\frac{V_R}{Q_e(R+1)}\right]\mathbf{r} \qquad (4\text{-}40)$$

The second equality is in terms of a recycle ratio defined as Q_R/Q_e, the ratio of volumetric flow rates of recycle and product streams.

Equation (4-40) shows again that differential reactor performance will be approached for a high recycle rate and a low reactor volume. Equation (4-40) can be used to estimate the recycle flow rate $Q_R = RQ_e$ required to reduce the change in concentration in the reactor to any assigned value.

The *commercial-scale* recycle unit may operate either with a significant change in concentration through the reactor (low or intermediate recycle ratio) or as a stirred-tank reactor (large recycle ratio). In the latter case Eq. (4-37) gives the desired production rate $Q_e C_e$. Note that the rate would be evaluated at the concentration C_e in the product stream.

When there is a significant change in concentration, Eq. (4-36) is valid but it is simpler to use the tubular-flow reactor equation based upon the flow rate of unconverted reactant A. This is Eq. (3-18a), derived for a tubular-flow reactor with partially converted feed. The quantity $(F_A)_0$ represents the hypothetical feed rate of A, corresponding to A in the fresh feed plus A in the recycle, if there were no conversion in the recycle stream. If the recycle stream were completely unconverted, the concentration of A would be that of the fresh feed. For simplicity, consider the

---

† For an isothermal, *differential* reactor operating at nearly constant pressure, Q_e will be nearly equal to Q_f, even for a gaseous reaction with a change in number of moles.

special case where the volumetric flow rate is unaffected by the reaction so that $Q_e = Q_f$. Then

$$F_{A_0} = Q_R C_{A_f} + Q_f C_{A_f} = C_{A_f}(Q_R + Q_f)$$

or, in terms of the recycle ratio, $R = (Q_R/Q_f)$,

$$F_{A_0} = C_{A_f} Q_f(R + 1) \tag{4-41}$$

Equation (3-18a) now becomes:

$$\frac{V}{Q_f} = -C_f(R + 1) \int_{x_1}^{x_e} \frac{dx}{r} \tag{4-42}$$

where x_1 represents the conversion entering the reactor and x_e is the conversion in the effluent. The value of x_1 (or C_1) is a function of the recycle ratio. In a design problem R would normally be specified while x_1 is unknown. For our case of $Q_f = Q_e$, it is easier to write Eq. (4-42) in terms of concentrations. Based upon unconverted feed, and deleting subscript A,

$$x = \frac{C_f - C}{C_f}$$

or

$$dx = -\frac{1}{C_f} dC$$

With this substitution Eq. (4-42) becomes

$$\frac{V}{Q_f} = (R + 1) \int_{C_1}^{C_e} \frac{dC}{r} \tag{4-43}$$

The mixed-feed concentration, C_1, is obtained by a mass balance around the mixing point P in Fig. 4-19. Thus

$$(Q_R + Q_f)C_1 = Q_f C_f + Q_R C_e = Q_f(C_f + RC_e)$$

or

$$C_1 = \frac{Q_f(C_f + RC_e)}{RQ_f + Q_f} = \frac{C_f + RC_e}{R + 1} \tag{4-44}$$

Equations (4-43) and (4-44) can be used to calculate the volume required to change the concentration of A in a feed stream of rate Q_f from C_{A_f} to C_{A_e}.[†]

Note that for no recycle, $C_1 = C_f$ from Eq. (4-44). Then Eq. (4-43) reduces to the mass balance for a once-through, plug-flow reactor. Thus, for constant Q the integrated form of Eq. (3-14) is

$$\frac{V}{Q} = \int_{C_f}^{C_e} \frac{dC}{r}$$

[†] A more general development of Eq. (4-43) applicable to gaseous reactions involving a change in flow rates is given by Octave Levenspiel, "Chemical Reaction Engineering," 2d ed., John Wiley & Sons, Inc., New York, 1972.

This is identical to Eq. (4-43), with $R = 0$. When R is large, the rate becomes a constant and Eq. (4-43) reduces to

$$\frac{V}{Q_f} = (R + 1)\frac{C_e - C_1}{\mathbf{r}}$$

Introducing Eq. (4-44) for C_1 gives

$$\frac{V}{Q_f} = \left[\frac{R + 1}{\mathbf{r}}\right]\frac{C_e(R + 1) - (C_f + RC_e)}{R + 1} = \frac{C_e - C_f}{\mathbf{r}}$$

The result is identical with Eq. (3-5) for a stirred-tank reactor. We conclude that Eq. (4-43), with (4-44), is consistent with the physical behavior expected of a flow recycle reactor at the two extremes of recycle rate.

PROBLEMS

4-1 Nitrous oxide decomposes approximately according to a second-order rate equation. The specific reaction rate in the forward direction,

$$2N_2O \rightarrow 2N_2 + O_2$$

is $k = 977$ cm^3/(g mol)(s) at 895°C. Calculate the fraction decomposed at 1.0 and 10 s and at 10 min in a constant-volume batch reactor. The rate of the reverse reaction is negligible and the initial pressure (all N_2O) is 1 atm.

4-2 The saponification reaction

$$NaOH\ (aq)\ +\ C_2H_5(CH_3COO) \rightarrow Na(CH_3COO)\ (aq)\ +\ C_2H_5OH\ (aq)$$

is second order and irreversible at low conversions. A laboratory, well-stirred tank reactor is charged with an aqueous solution containing NaOH and ethyl acetate, both of initial concentrations equal to 0.1 normal. After 15 minutes the conversion of the ethyl acetate is 18%.

For an initial charge containing NaOH and ethyl acetate in equal concentrations of 0.2 normal, what time would be required to obtain a conversion of 30% in a stirred-tank, commercial-scale, batch reactor? What volume is necessary to produce 50 kg of sodium acetate?

4-3 An aqueous solution of ethyl acetate is to be saponified with sodium hydroxide. The initial concentration of ethyl acetate is 5.0 g/liter and that of caustic soda is 0.10 normal. Values of the second-order rate constant, in liter/(g mol)(min), are

$$k = \begin{cases} 23.5 & \text{at } 0°C \\ 92.4 & \text{at } 20°C \end{cases}$$

The reaction is essentially irreversible. Estimate the time required to saponify 95% of the ester at 40°C.

4-4 It has been mentioned that gaseous reactions are more suitably carried out on a commercial scale in flow equipment than in batch reactors. Consider the following example. Watson† has studied the thermal (noncatalytic) cracking of butenes at 1 atm pressure in a flow reactor. The rate equation determined from his experimental data is

$$\log k_1 = -\frac{60,000}{4.575T} + 15.27$$

† K. M. Watson, *Chem. Eng. Progr.*, **44**, 229 (1948).

where k_1 = butenes cracked, g mol/(h)(liter)(atm), and T is in degrees Kelvin. Although the feed consists of a number of different butenes and the products vary from coke to butadiene, the irreversible reaction may be considered first order,

$$C_4H_8 \rightarrow C_4H_6 + H_2$$

It is desired to crack butenes in a batch-type reactor which will operate at 1200°F and will be equipped for efficient agitation. The initial charge to the reactor will consist of 1 lb mole of butenes and 10 lb moles of steam. Under these conditions the change in total moles during the course of the reaction can be neglected. (a) Determine the time required for a conversion of 30% of the butenes. (b) Determine the reactor volume required. (c) Suppose that the initial charge consists of 10 moles of steam per mole of hydrocarbon, as before, but this time the hydrocarbon fraction contains 60 mol % butenes and 40% butadiene. The butadiene may undergo two reactions: cracking and polymerization to the dimer. Assuming that the rates of these reactions are known, outline a method of determining the conversion of butenes and of butadiene for a given reaction time.

4-5 The decomposition of phosphine is irreversible and first order at 650°C,

$$4PH_3(g) \rightarrow P_4(g) + 6H_2(g)$$

The rate constant (s^{-1}) is reported as

$$\log k = -\frac{18,963}{T} + 2 \log T + 12.130$$

where T is in degrees Kelvin. In a closed vessel (constant volume) initially containing phosphine at 1 atm pressure, what will be the pressure after 50, 100, and 500 s. The temperature is maintained at 650°C.

4-6 Smith[†] has studied the gas-phase dissociation of sulfuryl chloride, SO_2Cl_2, into chlorine and sulfur dioxide at 279.2°C. The total-pressure method was employed to follow the course of the reaction. Under constant-volume conditions the results were as follows:

t, min	3.4	15.7	28.1	41.1	54.5	68.3	82.4	96.3
p_t, mmHg	325	335	345	355	365	375	385	395

What reaction order do these data suggest? The conversion is 100% at infinite time.

4-7 The thermal decomposition of dimethyl ether in the gas phase was studied by Hinshelwood and Askey[‡] by measuring the increase in pressure in a constant-volume reaction vessel. At 504°C and an initial pressure of 312 mmHg the following data were obtained:

t, s	390	777	1,195	3,155	∞
p_t, mmHg	408	488	562	779	931

Assuming that only ether was present initially and that the reaction is

$$(CH_3)_2O \rightarrow CH_4 + H_2 + CO$$

determine a rate equation for the decomposition. What is the numerical value of the specific-reaction rate at 504°C?

† D. F. Smith, *J. Am. Chem. Soc.* **47**, 1862 (1925).
‡ C. N. Hinshelwood and P. J. Askey, *Proc. Roy. Soc. (London)* **A115**, 215 (1977).

4-8 A first-order homogeneous gas-phase reaction, $A \rightarrow 3R$, is first studied in a constant-pressure batch reactor. At a pressure of 2 atm and starting with pure A, the volume increases by 75% in 15 min. If the same reaction is carried out in a constant-volume reactor, and the initial pressure is 2 atm, how long is required for the pressure to reach 3 atm?

4-9 Suppose the initial mixture in Prob. 4-8 consisted of 70 mol % A and 30 mol % helium (inert) at a total pressure of 2 atm. In a constant-pressure reactor, what would be the increase in volume in 15 min?

4-10† A stage in the production of propionic acid, C_2H_5COOH, is the acidification of a water solution of the sodium salt according to the reaction

$$C_2H_5COONa + HCl \rightarrow C_2H_5COOH + NaCl$$

The reaction rate may be represented by a second-order reversible equation. Laboratory data on the rate of reaction are obtained by taking 10 cm^3 samples of the reaction solution at varying times and neutralizing the unreacted HCl with 0.515-normal NaOH. The original amount of acid is determined from a sample taken at zero time. The temperature is 50°C, and the initial moles of HCl and C_2H_5COONa are the same. The data are as follows:

t, min	0	10	20	30	50	∞
NaOH required, cm^3	52.5	32.1	23.5	18.9	14.4	10.5

Determine the size of an agitated batch reactor to produce propionic acid at an average rate of 1000 lb/h. Twenty minutes is required for charging the reactor and heating to 50°C, and 10 min is necessary for cooling and removing the products. The final conversion is to be 75% of the sodium propionate. The initial charge to the reactor will contain 256 lb of C_2H_5COONa and 97.5 lb of HCl per 100 gal. Assume that the density of the reaction mixture is 9.9 lb/gal and remains constant.

4-11 The reaction mechanism for the decomposition of nitrogen pentoxide is complex, as described in Sec. 2-2. However, a satisfactory rate equation can be developed by considering the two reactions

$$1. \quad 2N_2O_5 \rightarrow 2N_2O_4 + O_2$$

$$2. \quad N_2O_4 \rightleftarrows 2NO_2$$

Reaction 2 is rapid with respect to reaction 1, so that nitrogen dioxide and nitrogen tetroxide may be assumed to be in equilibrium. Hence only reaction 1 need be considered from a kinetic standpoint. Calculate the specific reaction rate for this reaction (which is essentially irreversible) from the following total-pressure data‡ obtained at 25°C:

t, min	0	20	40	60	80	100	120	140	160	∞
p_t, mmHg	268.7	293.0	302.2	311.0	318.9	325.9	332.3	338.8	344.4	473.0

It may be assumed that only nitrogen pentoxide is present initially. The equilibrium constant K_p for the dissociation of nitrogen tetroxide into nitrogen dioxide at 25°C is 97.5 mmHg.

4-12 The production of carbon disulfide from methane and sulfur vapor can be carried out homogeneously or with a solid catalyst. Also, some solid materials act as a poison, retarding the reaction. The following data were obtained on a flow basis at a constant temperature of 625°C and with an initial reactants ratio of 1 mole of CH_4 to 2 moles of sulfur vapor (considered as S_2). The first set of data was

---

† From W. F. Stevens, "An Undergraduate Course in Homogeneous Reaction Kinetics," presented at Fourth Summer School for Chemical Engineering Teachers, Pennsylvania State University, June 27, 1955.

‡ F. Daniels and E. H. Johnston, *J. Am. Chem. Soc.*, **43**, 53 (1921).

obtained with the reactor empty (effective volume 67.0 cm^3), and the second set was obtained after packing the reactor with a granular material (7 mesh) which reduced the void volume to 35.2 cm^3.

Set	Run	Feed rate, g mol/h CH$_4$	S$_2$	Production of CS$_2$, g mol/h	Conversion of methane
1	1	0.417	0.834	0.0531	0.127
	2	0.238	0.476	0.0391	0.164
	3	0.119	0.238	0.0312	0.262
2	1	0.119	0.238	0.0204	0.171
	2	0.178	0.357	0.0220	0.123

Was the granular material acting as a catalyst or as a poison in this case?

4-13 Butadiene and steam (0.5 mole steam/mole butadiene) are fed to a tubular-flow reactor which operates at 1180°F and a constant pressure of 1 atm. The reactor is noncatalytic. Considering only the reversible polymerization reaction to the dimer, determine (a) the length of 4-in-ID reactor required to obtain a conversion of 40% of the butadiene with a total feed rate of 20 lb mol/h and (b) the space velocity, measured as (liters feed gas)/(h)(liters reactor volume at 1180°F and 1 atm), required to obtain a conversion of 40%.

The polymerization reaction is second order and has a specific reaction-rate constant given by

$$\log k = -\frac{5,470}{T} + 8.063$$

where k is C$_4$H$_6$ polymerized, in g mol/(liter)(h)(atm^2), and T is in degrees Kelvin. The reverse (depolymerization) reaction is first order. At 1180°F (911 K) the equilibrium constant for the reaction is 1.27.

4-14 A special characteristic of bacterial oxidation is that the bacteria not only catalyze the oxidation of carbonaceous compounds, but also the carbon compound provides "fuel" for growth of the bacteria. As an illustration, consider a stirred-tank reactor operating continuously at steady state, and at constant temperature. The feed consists of an aqueous solution of glucose (the carbonaceous material) which does not contain bacteria. The concentration of glucose in the feed is C_{S_0}, and its volumetric flow rate into the reactor is Q. The mixture in the tank contains bacteria in a concentration C_B, and the product stream has a flow rate equal to Q. This product stream contains bacteria and glucose in concentrations C_B and C_S (both in mg/liter).

The two major reactions may be written (in a highly simplified manner) as follows:

(1) $\quad$ C$_6$H$_{12}$O$_6$ + B → nB + inert products $\qquad$ ($n>1$)

(2) $\quad$ C$_6$H$_{12}$O$_6$ + O$_2$ $\overset{B}{\rightarrow}$ H$_2$O + CO$_2$ + inert products

The Monod rate equations have been found to fit some types of bacterial oxidations. For conditions where the oxygen supply is adequate, these equations can be written for reactions 1 and 2 as

$$R_x = \frac{\mu_m C_S}{K + C_S} C_B$$

$$R_S = \frac{1}{Y}\frac{\mu_m C_S}{K + C_S} C_B$$

R_x = rate of formation of bacteria (cell mass), mg/(liter)(h)
R_S = total rate of disappearance of glucose (substrate), mg/(liter)(h)
R_x represents the growth of bacteria according to reaction 1, while
R_S is the *total* rate of disappearance of glucose by both reactions.

Notice that reaction 2 is catalyzed by the bacteria. The symbol Y represents the yield, or ratio of rates of production of bacteria to the total rate of consumption of glucose. Other symbols are defined in part (d).

(a) For steady-state operation, derive a relationship between Q/V (called the dilution rate, D) and the concentration C_S of glucose in the product stream.

(b) If C_S is very large with respect to K, corresponding to a very large excess of glucose in the feed stream, how is the relationship, of part (a) simplified?

(c) The result obtained in part (b) gives a specific relation between the dilution rate D and the rate constant, or constants, in the Monod equations. If the actual dilution rate is increased above the value given by this specific relation, for example by increasing the flow rate, what would happen to the bacteria in the reactor?

(d) For a particular case, the following data apply:

$$Q = 1.0 \text{ liter/h}$$
$$V = \text{volume of stirred tank reactor} = 4 \text{ liters}$$
$$\mu_m = \text{rate constant in Monod equation} = 0.5 \text{ (h)}^{-1}$$
$$K_S = \text{constant in Monod rate equation} = 15 \text{ mg/liter}$$
$$C_{S_0} = 80 \text{ mg/liter} = \text{concn of glucose in feed}$$
$$Y = \text{yield (assume constant)} = 0.5$$

Calculate (1) the concentration of glucose in the product stream leaving the reactor; (2) the concentration of bacteria in the product stream leaving the reactor.

(e) For the conditions of part (d), what would be the maximum dilution rate that could be used and yet avoid the problem referred to in part (c)?

4-15 A mixture of butenes and steam is to be thermally (noncatalytically) cracked in a tubular-flow reactor at a constant temperature of 1200°F and a constant pressure of 1.0 atm. Although the feed consists of a number of different butenes† and the products vary from coke to butadiene, the rate of reaction may be adequately represented by the first-order mechanism

$$\mathbf{r}_1 = \frac{\varepsilon}{\rho} k_1 p_4$$

$$\log k_1 = -\frac{60,000}{4.575T} + 15.27$$

The rate was determined experimentally in a reactor packed with inert quartz chips, and the reactor to be designed in this problem will also be so packed. The data and notation are as follows:

$$\mathbf{r}_1 = \text{g moles butenes cracked/(g quartz chips)(h)}$$
$$\varepsilon = \text{void fraction} = 0.40$$
$$\rho = \text{bulk density of bed packed with quartz chips} = 1,100 \text{ g/liter}$$
$$p_4 = \text{partial pressure of butenes, atm}$$
$$T = \text{temperature, K}$$

The ratio of steam to butenes entering the reactor will be $10:1.0$ on a molal basis. Under these conditions the change in total number of moles during the course of the reaction can be neglected.

(a) Determine the conversion as a function of size of reactor. Also prepare a plot of conversion of butenes versus two abscissas: (1) lb quartz chips(W)/lb mole butene feed per hour, covering a range of values from 0 to 3000, and (2) space velocity, defined as (ft^3 feed)/(h)(ft^3 void volume) at 1200°F. What total volume of reactor would be required for a 20% conversion with a butenes feed rate of 5 lb mol/h?

(b) Suppose that the feed consists of 10 moles of steam per mole of total hydrocarbons. The hydrocarbon fraction is 60 mole % butenes and 40 mole % butadiene. Consider that the butenes react

---

† See *Chem. Eng. Progr.*, **44**, 229 (1948).

as in (a) and that the butadiene may undergo two reactions, cracking and polymerization to the dimer. The rate for cracking is

$$r_2 = \frac{\varepsilon}{\rho} \bar{k}_2 p_4''$$

$$\log k_2 = -\frac{30,000}{4.575T} + 7.241$$

where r_2 is butadiene cracked, in g mol/(g quartz chips)(h), and p_4'' is partial pressure of butadiene, in atmospheres; the rate for polymerization to the dimer is

$$r_3 = \frac{\varepsilon}{\rho} k_3 (p_4'')^2$$

$$\log k_3 = -\frac{25,000}{4.575T} + 8.063$$

where r_3 is butadiene polymerized, in g mol/(g quartz chips)(h).

Determine the conversion of butenes and of butadiene as a function of W/F from 0 to 3000 lb chips/(lb mole feed per hour). Assume that the total number of moles is constant and neglect all reactions except those mentioned.

4-16 The following conversion data were obtained in a tubular-flow reactor for the gaseous pyrolysis of acetone at 520°C and 1 atmosphere. The reaction is

$$CH_3COCH_3 \rightarrow CH_2{=}C{=}O + CH_4$$

Flow rate, g/h	130.0	50.0	21.0	10.8
Conversion of acetone	0.05	0.13	0.24	0.35

The reactor was 80 cm long and had an inside diameter of 3.3 cm. What rate equation is suggested by these data?

4-17 A small pilot plant for the photochlorination of hydrocarbons consists of an ideal tubular-flow reactor which is irradiated, and a recycle system, as shown in the sketch. The HCl produced is separated at the top of the reactor, and the liquid stream is recycled. The Cl_2 is dissolved in the hydrocarbon (designated as RH_3) before it enters the reactor. It is desired to predict what effect the type of reactor operation will have on the concentration ratio $[RH_2Cl]/[RHCl_2]$ in the product stream.

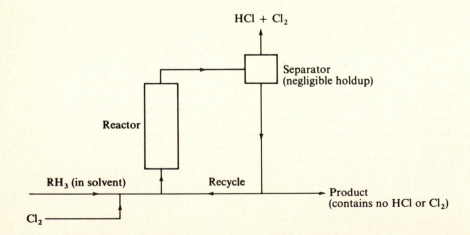

Determine this ratio, as a function of total conversion of RH_3, for two extremes: zero reflux ratio (ratio of recycle flow rate to product-stream flow rate) and infinite reflux ratio. The concentration of hydrocarbon in the feed entering the reactor is $[RH_3]_0$. The hydrocarbon is dissolved in an inert solvent in which Cl_2 is also soluble. There is a stoichiometric excess of Cl_2 fed to the reactor. The reactor operates isothermally, and the HCl product remains in solution until it reaches the separator at the top of the reactor.

Plot curves of $[RH_2Cl]/[RHCl_2]$ vs. conversion for each of the conditions and for three types of kinetics. The reactions are

$$1. \quad RH_3 + Cl_2 \rightarrow RH_2Cl + HCl$$

$$2. \quad RH_2Cl + Cl_2 \rightarrow RHCl_2 + HCl$$

The three types of kinetics are as follows:

(a) First-order (with equal rate constants), so that the rates of production, in g mol/(vol.)(time), are

$$r_{RH_3} = -k_1 C_{CR_3}$$

$$r_{RH_2Cl} = k_1 C_{CR_3} - k_1 C_{RH_2Cl}$$

(b) Second-order (with equal rate constants),

$$r_{RH_3} = -k_2 C_{RH_3} C_{Cl_2}$$

$$r_{RH_2Cl} = k_2 C_{RH_3} C_{Cl_2} - k_2 C_{RH_2Cl} C_{Cl_2}$$

(c) Chain kinetics, for which the *elementary* steps are

Initiation

$$Cl_2 + hv \rightarrow 2Cl^* \qquad r = \phi\alpha l C_{Cl_2} = k_1 C_{Cl_2}$$

Propagation

$$Cl^* + RH_3 \xrightarrow{k_2} RH_2^* + HCl$$

$$RH_2^* + Cl_2 \xrightarrow{k_3} RH_2Cl + Cl^*$$

$$Cl^* + RH_2Cl \xrightarrow{k_2} RHCl^* + HCl$$

$$RHCl^* + Cl_2 \xrightarrow{k_3} RHCl_2 + Cl^*$$

Termination

$$RH_2^* \xrightarrow{k_6} \text{end product}$$

$$RHCl^* \xrightarrow{k_6} \text{end product}$$

In solving the problem for this case, use the stationary-state hypothesis for the intermediates (free radicals Cl^*, RH_2^*, $RHCl^*$) to obtain rate equations for r_{RH_3} and r_{RH_2Cl} analogous to the rate equations for the first-order and second-order cases. Note that the rate constant for the termination steps is usually much less than that for the propagation steps.

4-18 Reconsider Example 4-9 by taking into account high molecular weight products. Assume that the rate of production of these products can be represented by the dimerization of the diamine:

$$2H_2C = C{\overset{\displaystyle (CH_2)_3CN}{\underset{\displaystyle (CH_2)_3CN}{\Big\langle}}} \quad \xrightarrow{k_3} \quad \text{polymer products}$$

where k_3 is a second-order rate constant. For the same conditions as in Example 4-9, and for the ratio $k_1/k_3 = 2$, calculate the conversion of acrilonitrile to monoamine, to diamine, and to polymer products as a function of the total conversion of acrilonitrile.

4-19 Two irreversible reactions have the rate constants shown:

$$A \xrightarrow{k_1} B \qquad k_1 = 0.25 \text{ h}^{-1}$$

$$B \xrightarrow{k_2} C \qquad k_2 = 0.13 \text{ h}^{-1}$$

The reactions are both first-order and occur in a constant density liquid phase. Consider a case where a feed stream contains A in a concentration $C_A = 4.0$ lb mol/ft^3 and $C_B = C_C = 0$.

(a) Calculate the average residence time (V/Q) required for one half of the A to react for the following two cases:

    (i) Ideal plug-flow reactor

    (ii) Ideal stirred-tank reactor

    (b) Calculate the concentration of B leaving the reactor for the two cases in part A.

4-20 A homogeneous liquid-phase polymerization is carried out in a completely mixed stirred-tank reactor which has an average residence time of 33.6 s. The concentration of monomer in the feed stream is 1.0×10^{-3} g mol/cm^3. The polymerization reactions follow a two-step process:

1. An initiation reaction producing an active form of the monomer, P_1. This reaction has a rate constant $k_i = 0.10 \text{ s}^{-1}$.
2. Propagation reactions, where the monomer reacts with successive polymers of the form P_n. These are all second order with the same rate constant, $k_p = 500 \text{ cm}^3/(\text{g mol})(\text{s})$.

(a) Of the total polymer content of the exit stream, what is the weight-fraction distribution of polymer molecules from P_1 to P_{10}? (b) For the same reactor and flow rates but another reaction, the initiation rate constant is the same and monomer feed concentration is again 1.0×10^{-3} g mol/cm^3. The weight-fraction distribution of the total polymer in the exit stream in this case is

Polymer	P_1	P_2	P_3	P_4	P_7	P_{10}	P_{20}
Weight fraction	0.0180	0.0314	0.0409	0.0470	0.0546	0.0503	0.0250

What rate constant for the propagation reactions is indicated by these data?

4-21 The following irreversible first-order reactions occur at constant density:

$$A \xrightarrow{k_1} B \xrightarrow{k_2} C$$

$$k_1 = 0.15 \text{ min}^{-1} \qquad k_2 = 0.05 \text{ min}^{-1}$$

This reaction system is to be analyzed in continuous-flow reactors with a volumetric feed rate of 5 ft^3/min and feed composition $C_A = C_{A_0}$ and $C_B = C_C = 0$. For the highest production rate of B, which of the following reactors is preferable?

    (a) A single-stirred tank of volume $V = 10$ ft^3

    (b) Two stirred tanks in series, each with a volume of 5 ft^3

    (c) Two stirred tanks in parallel, each of 5 ft^3 volume and with the feed stream split equally between them

    (d) A plug-flow (ideal tubular-flow) reactor with a volume of 10 ft^3

4-22 Acetic anhydride is hydrolyzed in three stirred-tank reactors operated in series. The feed flows to the first reactor $(V = 1$ liter) at a rate of 400 cm^3/min. The second and third reactors have volumes of 2 and 1.5 liters, respectively. The temperature is 25°C, and at this condition the first-order irreversible rate constant is 0.158 min^{-1}. Use a graphical method to calculate the fraction hydrolyzed in the effluent from the third reactor.

4-23 Suppose in Prob. 4-22 that the first reactor is operated at 10°C, the second at 40°C, and the third at 25°C. The additional rate constants are 0.0567 min^{-1} (at 10°C) and 0.380 min^{-1} (at 40°C). Determine the fraction hydrolyzed in the effluent from the third reactor.

4-24 Benzene is to be chlorinated in the liquid phase in a kettle-type reactor operated on a steady-state basis. Liquid benzene is added continuously, and the liquid product and gaseous hydrogen chloride are removed continuously. The chlorine gas is bubbled continuously into the liquid reaction mixture in the kettle. The rate of reaction may be assumed large enough that there is no unreacted chlorine in the reaction products. Also, the concentrations of chlorine and HCl in the reaction mixture will be small. The density of the liquid mixture may be assumed constant.

At the constant operating temperature of 55°C the significant reactions are the three substitution ones leading to mono-, di-, and trichlorobenzene. Each reaction is second-order and irreversible. The three reactions are

$$1. \quad C_6H_6 + Cl_2 \xrightarrow{k_1} C_6H_5Cl + HCl$$

$$2. \quad C_6H_5Cl + Cl_2 \xrightarrow{k_2} C_6H_4Cl_2 + HCl$$

$$3. \quad C_6H_4Cl_2 + Cl_2 \xrightarrow{k_3} C_6H_3Cl_3 + HCl$$

It was noted in Chap. 2 that at 55°C the ratios of the rate constants are

$$\frac{k_1}{k_2} = 8.0$$

$$\frac{k_2}{k_3} = 30$$

Under the proposed operating conditions the composition of the liquid product will be constant for any one run. Different products will be obtained for different ratios of benzene and chlorine fed to the reactor. Compute the composition of the liquid product for the case where 1.4 moles chlorine/mole benzene are fed to the reactor.

4-25 Reconsider Prob. 4-24 for the case where, instead of a single reactor, two reactors ($V_1 = V_2$) are used. The liquid stream enters the first reactor (as pure benzene) and flows from the first reactor to the second, and finally the product is withdrawn from the second reactor. Gaseous hydrogen chloride is withdrawn from each reactor. Plot the composition of the product vs. moles of total chlorine added per mole of benzene. Cover a range of the latter variable from 0 to 2.5. One-half the total chlorine is added to each reactor.

4-26 The successive irreversible reactions

$$A \xrightarrow{k_1} B \xrightarrow{k_2} C$$

are first order. They are carried out in a series of identical stirred-tank reactors operating at the same temperature and at constant density. Derive an expression for the concentration of B in the effluent in terms of the number of reactors, the total average residence time, and the rate constants. The feed stream contains no B or C. What is the ratio C_B/C_{A_0} for the specific case where $k_2 = 0.1$ h^{-1}, $k_1 = 0.05$ h^{-1}, and total residence time is 1.5 h for 6 reactors? What is the maximum value of C_B/C_{A_0} and the corresponding number of reactors?

4-27 A constant-density reaction is carried out in a stirred-tank reactor of volume V. The reaction is first order and irreversible, with a rate constant k_1. The volumetric feed rate is Q. Under these conditions the conversion in the product stream is x [given by Eq. (4-17)].

(a) If one-half of the product stream is recycled and the makeup feed rate is reduced to $Q/2$, what will be the change in conversion in the product stream? What will be the change in the production rate of the product? (b) If one-half of the product stream is recycled, but the fresh-feed rate is maintained equal to Q, what will be the effects on the conversion and production rate?

4-28 Two stirred tanks of volume V_1 and $2V_1$ are available for carrying out a first-order irreversible reaction at constant density and temperature. If the flow rate of the feed stream is Q, which of the following arrangements would give the highest production rate of product?

(a) Parallel operation of the two reactors, with equal average residence times in each
(b) Parallel operation, with different average residence times
(c) Series operation, with the feed stream entering the larger reactor
(d) Series operation, with the feed entering the smaller reactor

4-29 Repeat Example 4-14 with the modification that the effluent from the first reactor is fed to a second reactor. The second reactor originally contains 10 liters of an anhydride solution of concentration 0.50×10^{-4} g mol/cm^3. Product is withdrawn from reactor 2 at a constant rate of 2 liter/min. Temperatures in both are 40°C, and all other conditions are the same as in Example 4-14.

(a) Determine the concentration of anhydride in the solution leaving the second reactor from zero time until steady-state conditions are reached. (b) Suppose that reactor 2 was originally empty and that its capacity is 10 liters. After it is filled, product is withdrawn at the rate of 2 liters/min. What would be the concentration of the first anhydride solution leaving the second reactor?

4-30 Ethyl acetate is to be produced in a 100 gal reactor. The reactor originally holds 100 gal of solution containing 20% by weight ethanol and 35% by weight acetic acid. Its density is 8.7 lb/gal; assume that this value remains constant for all compositions. The reactor is heated to 100°C and this temperature is maintained. Pure ethanol is added at a rate of 2 gal/min (17.4 lb/min). Solution is withdrawn at the same volumetric rate. For these conditions the rate data are

$$\mathbf{r} = kC_H C_{OH} \qquad k = 4.76 \times 10^{-4} \text{ liter/(g mol)(min)}$$

$$\mathbf{r}' = k'C_E C_W \qquad k' = 1.63 \times 10^{-4} \text{ liter/(g mol)(min)}$$

where H, OH, W, and E refer to acid, alcohol, water, and ester. The concentrations are expressed in gram moles per liter.

Determine the concentration of ester in the product stream for time values from 0 to 1 hour. What is the percent conversion of the total amount of ethanol added? Assume that no water is vaporized in the reactor.

4-31 Ethyl acetate is to be saponified by adding a 0.05 normal solution of sodium hydroxide continuously to a kettle containing the ethyl acetate. The reactor is initially charged with 100 gal of an aqueous solution containing 10 g ethyl acetate/liter. The sodium hydroxide solution is added at a rate of 1.0 gal/min until stoichiometric amounts are present. The reaction is relatively fast and irreversible, with a specific reaction rate of 92 liter/(g mol)(min) at 20°C. Assuming that the contents of the kettle are well mixed, determine the concentration of unreacted ethyl acetate as a function of time. At what time will this concentration be a maximum? Assume constant density.

4-32 Isoprene [CH_2=$C(CH_3)CH$=CH_2] can be produced from isobutylene and formaldehyde. The principle reactions are:

1. $HCHO + CH_2$=$C(CH_3)CH_3 \rightarrow CH_2$=$C(CH_3)CH_2CH_2OH$
   3 methyl, 3-butene, 1-ol

2. H_2C=$C(CH_3)CH_2CH_2OH \rightarrow CH_2$=$C(CH_3)CH$=$CH_2 + H_2O$

3. $HCHO + HCHO \rightarrow CH_3 \overset{\displaystyle O}{\overset{\displaystyle \|}{-O-CH}}$

4. $2CH_2$=$C(CH_3)CH_2CH_2OH + HCHO \rightarrow [CH_2$=$C(CH_3)CH_2CH_2O]_2CH_2 + H_2O$

The third and fourth reactions form undesirable by-products, methyl formate, and heavies (high molecular weight products). The first reaction does not require a catalyst but must be carried out at high pressure (2000 lb/in.^{2} abs) and at about 200°C. The second reaction, forming isoprene, can be accomplished at 1 atm pressure (at about 300°C) in the vapor phase using a solid catalyst. This second step is relatively simple and would follow the first step in separate equipment in the plant. This second reaction does not occur under any conditions without a catalyst. All reactions are irreversible.

Consider the design of a reactor for the first, homogeneous reaction. A tank-type reactor is not desirable because of the high pressure. It is proposed to use a recycle, "loop" reactor (see Fig. 1-7e). Commercially, the reactor would operate at steady state with continuous feed and discharge streams and an adjustable recycle rate.

(a) To obtain kinetic data for design purposes, a laboratory-scale model has been built and operated. The ID of the tube is $\frac{1}{2}$ in. and the total volume of the loop is 200 cm³. Most of the data are obtained with this reactor operated batchwise (no feed or discharge). Also, the pump is run at a high speed, circulation rate = 5 liter/min, so that the reaction mixture is completely mixed at any instant. Initially, the system volume is completely filled with reactants at room temperature (at which condition the rate of reaction is negligible). The run is initiated by starting the pump and by supplying condensible Dowtherm to the chamber in which the reactor and pump (canned-rotor type) are located. This raises the temperature quickly to 200°C. Material is withdrawn to maintain the pressure at 2000 lb/in.² abs, but the total volume removed is small with respect to 200 cm³. Then micro samples of reaction mixture are withdrawn periodically for analysis by chromatography. The reaction mixture is completely miscible regardless of composition:

1. From the given batch data develop a suitable rate equation for the first step. Neglect side reactions 3 and 4. This is a realistic assumption because the data are obtained at high initial ratios of isobutylene to formaldehyde. Also, the volume change due to reaction is negligible.
2. After batch-data were obtained, the reactor is operated as a continuous flow, recycle unit. Feed and product rates are constant at 1.0 cm³/min, the circulation rate is maintained at 5 liter/min, and the feed stream contains only isobutylene and formaldehyde in a mole ratio (isobutylene/HCHO) of 10. What conversion of HCHO would be expected in the product stream from the reactor operated in this way?
3. The *average residence* time of the fluid in the reactor in (2) is 200/1.0 = 200 min. Would the conversion in (2) be expected to be the same as that corresponding to a *reaction* time of 200 min in the batch operation in part 1?

Conversion (of HCHO) vs. time data in batch experiments:

t, min	60	90	120	180	300	400
conversion[†]	0.100	0.147	0.188	0.270	0.409	0.506

† Determined from the analysis of the samples for HCHO content.

4-33 The photochemical decomposition of acetone (in the gas phase) has been studied in a batch-recycle reactor.[†] While the reaction sequence is complex, the overall reaction at 97°C and 870 mmHg may be expressed, approximately, as

$$CH_3COCH_3 \rightarrow CO + C_2H_6$$

The operating conditions were as follows (refer to nomenclature of Sec. 4-9 and to Fig. 4-20):

$$Q = 3.33 \times 10^{-4} \text{ m}^3/\text{s}$$

$$V_T = 5.50 \times 10^{-3} \text{ m}^3$$

$$V_R = 62.6 \times 10^{-6} \text{ m}^3$$

† A. E. Cassano, T. Matsurra, and J. M. Smith, *Ind. Eng. Chem. Fundam.*, 7, 655 (1968).

The course of the reaction was followed by analyzing very small gas samples for CO and C_2H_6. In one run, for which the initial concentration of acetone, C_{A_0}, was 5.7×10^{-3} kmol/m^3, the results were:

t, h†	1.25	2.25	3.25	4.25
$C_{C_2H_6}$, kmol/m^3	2.0×10^{-5}	6.5×10^{-5}	8.5×10^{-5}	11.8×10^{-5}

† These times were corrected for an induction period, of about 1.75 h.

Data for other runs at different initial concentrations gave the following results, reported as conversion of acetone x_A:

	Initial acetone concn, $C_{A_0} \times 10^3$ kmol/m^3			
	4.62	1.99	1.33	1.27
Time t, h	Conversion of acetone, $100 x_A$, %			
1.0	0.60	1.00	1.10	1.15
2.0	1.10	1.75	2.00	2.05
3.0	1.70	2.55	3.00	3.10
4.0	2.20	3.35	3.90	4.00
5.0	2.60	4.30	5.00	5.20

(a) Calculate values of the *conversion per pass* through the reactor. Does the system operate as a *differential* reactor?

(b) Using the stationary-state hypothesis, and with some assumptions about the reaction mechanism, the following rate expression for the photodecomposition of acetone can be derived:

$$r = \frac{k_1 C_A}{1 + k_2 C_A}$$

How do the data compare with a linearized form of the derived rate equation?

4-34 A first-order, irreversible reaction, $A \to B$, is studied in a flow, recycle reactor (Fig. 4-19). Derive an equation for the reactor volume in terms of the conversion in the effluent stream, the rate constant, k, and the recycle ratio, R. The volumetric flow rates of the feed and effluent are equal ($Q_f = Q_e$). The recycle volumetric flow rate is $Q_R = RQ_e$. The fresh feed contains no product (B).

FIVE

NONISOTHERMAL REACTORS

Nonuniform temperatures are common in tubular-flow or batch reactors. Also, continuous-flow tank reactors often operate at temperature levels different from the temperature of the feed or surroundings. The temperature may be varied deliberately (by heat exchange with the surroundings) in order to achieve maximum rates or selectivity. More frequently temperature changes arise because the inherent heat of reaction is significant and heat exchange with the surroundings is limited. Adiabatic operation results when the heat exchange is negligible. Large commerical reactors are more likely to approach adiabatic operation than isothermal conditions, except in those few cases, such as isomerization processes, where the heat of reaction itself is negligible. The design of nonisothermal reactors normally requires the simultaneous solution of an energy balance with the mass conservation equations developed in Chap. 3. The general form of the energy equation is discussed in Sec. 5-1. This is followed with illustrations of the design procedure for tank and for tubular-flow reactors in Secs. 5-2 to 5-4. In all cases ideal behavior is assumed: either well-stirred-tank or plug-flow, tubular reactors. When but one reaction is involved, the desired results of the calculations are the conversion and temperature vs. time or vs. reactor volume. For multiple reactions, selectivity as well as conversion is important. Both objectives are illustrated in Examples 5-1 to 5-3.

Multiple steady states in continuous-flow, stirred-tank reactors are discussed in Sec. 5-5. The particular steady state that is attained depends upon the initial state. Hence, Sec. 5-5 is closely related to the dynamic behavior during startup, a subject treated in the study of semibatch reactors in Sec. 5-6. Finally, optimum temperatures in reactors are discussed in the last part of the chapter, Sec. 5-7. The optimum-temperature question in tubular-flow reactors for reversible exothermic reactions is particularly interesting because the profile is not uniform.

As an introduction to the effect of temperature changes, it is instructive to consider the qualitative behavior of adiabatic flow reactors. Such reactors are simple to analyze since the energy balance can be solved, without reference to the mass balance, to obtain a relation between temperature and conversion. This eliminates the stepwise numerical procedure necessary for nonadiabatic and non-isothermal operation. Normally, heat exchange with the surroundings modifies rather than completely changes the temperature effects in adiabatic reactors. Therefore, the results for adiabatic conditions provide a helpful insight into temperature effects. Figure 5-1 illustrates the solution of the energy equation for endothermic and exothermic reactions. These nearly straight lines† express the fact that, for adiabatic operation, all the heat of reaction is reflected in a temperature change of the reaction mixture.

Consider now a single, irreversible reaction, first in a tubular-flow reactor. As the conversion increases with reactor length the temperature would also increase (for an exothermic reaction), as determined by the line in Fig. 5-1. Since the rate is a function of temperature as well as conversion, the nature of the temperature-length profile will depend upon the activation energy as well as the order of the reaction. Typical profiles of rate, conversion, and temperature for an exothermic reaction are shown in Fig. 5-2a. Note that the rate first increases due to the temperature rise, and then goes through a maximum. The rate approaches zero as the conversion approaches 100% because the reactant concentration is decreasing to zero. However, the volume required for a given conversion will be less for

---

† They would be exactly straight if the specific heat of the reaction mixture and heat of reaction are independent of temperature and composition [as shown in Sec. 5-3, Eq. (5-17)].

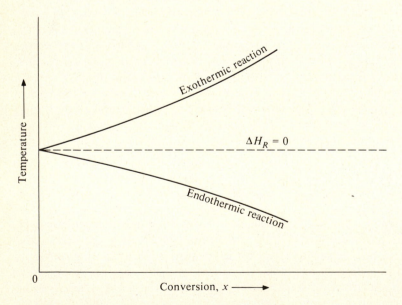

Figure 5-1 Temperature vs. conversion in adiabatic reactors.

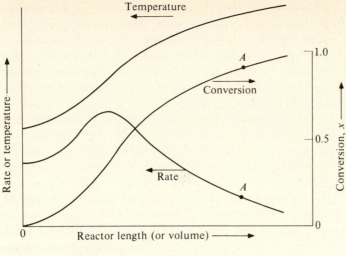

(*a*) Irreversible reaction

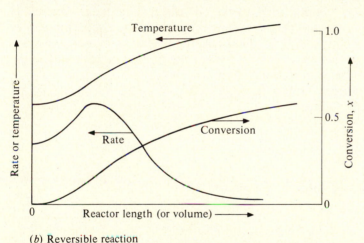

(*b*) Reversible reaction

Figure 5-2 Rate, temperature, and conversion profiles for exothermic reactions in adiabatic flow reactors.

nonisothermal operation than in an isothermal reactor. Next, consider a stirred-tank reactor. The temperature and composition will be uniform throughout the vessel and this means that reactor behavior can be represented by a point on each of the profiles in Fig. 5-2. Suppose the desired conversion is high, for example corresponding to point *A*. The figure shows that the stirred-tank reactor would operate at a constant, lower rate than at any point (except at the outlet) in the tubular-flow reactor. The required volume would be larger for a stirred-tank unit. For either type, permissible temperatures may be limited by undesirable side reactions, or other factors. In these cases, effective design depends upon efficient removal of energy.

For a reversible, exothermic reaction in an adiabatic reactor, the profiles would be similar to those in Fig. 5-2a but with some differences. The *maximum* (equilibrium) conversion is limited to less than 100% and would decrease at higher temperatures. The reverse rate becomes more important as the temperature rises so that the net forward rate decreases sharply as the temperature and reactor length increase. The curves in Fig. 5-2b illustrate the profiles for the case where equilibrium is approached at the reactor outlet.

For endothermic reactions in adiabatic tubular reactors, the rate decreases continuously with reactor length, both due to disappearance of reactant and the drop in temperature. This results in a sharply falling rate profile and a convex-rising conversion profile, as shown in Fig. 5-3. The conversion in a nonisothermal reactor will be less than the conversion for isothermal operation. Adding energy to reduce the temperature drop along the reactor length will increase the conversion. If the reaction is reversible, adding energy provides the further advantage of increasing the maximum (equilibrium) conversion. A practical example is the dehydrogenation of butenes to butadiene. Only at an elevated temperature is the equilibrium conversion high enough for the process to be economically feasible. Also, only at high temperatures will the reaction rate be sufficient to approach equilibrium conversion in a reactor of reasonable volume. Energy can be added in many ways: adding a high temperature diluent (e.g., steam) to the feed to absorb the heat of reaction, circulating a hot fluid through a jacket around the reactor, by a simultaneous exothermic reaction involving the reactants or other components, or from heaters between sections of the reactor.

Effects of nonisothermal operation on selectivity in multireaction systems are summarized at the end of Sec. 5-4.

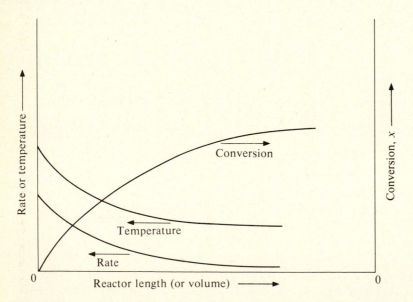

Figure 5-3 Rate, temperature, and conversion profiles for endothermic reactions in adiabatic flow reactors.

5-1 Energy Conservation Equations

In Chap. 3, the species mass balance was introduced in general form as Eq. (3-1). A similar word statement for the conservation of energy in reactors may be written:

$$\begin{Bmatrix} \text{Energy in streams} \\ \text{entering volume} \\ \text{element} \end{Bmatrix} - \begin{Bmatrix} \text{energy in} \\ \text{streams leaving} \\ \text{volume element} \end{Bmatrix} + \begin{Bmatrix} \text{energy transferred} \\ \text{from surroundings} \\ \text{into volume element} \end{Bmatrix}$$

$$= \begin{Bmatrix} \text{accumulation of} \\ \text{energy within} \\ \text{volume element} \end{Bmatrix} \quad (5\text{-}1)$$

This is a statement of the first law of thermodynamics written so as to apply for either flow or batch operating conditions. As in Chap. 3 the volume element is the largest element within which the variation in any property is negligible.

In a *batch* process, Eq. (5-1) reduces to

$$\begin{Bmatrix} \text{Energy transferred} \\ \text{from surroundings} \\ \text{into volume element} \end{Bmatrix} = \begin{Bmatrix} \text{accumulation of} \\ \text{energy within} \\ \text{volume element} \end{Bmatrix} \quad (5\text{-}2)$$

In chemical reactors the energy exchange with the surroundings is normally due only to heat, Q. Then for a time interval dt, where the heat transferred is dQ, Eq. (5-2) becomes[†]

$$dQ = dU$$

where U is the internal energy of the reaction mixture. The difference between the change in energy and change in enthalpy of a system is generally unimportant in reactors. This means that, for batch reactors, the foregoing equation may be written

$$dQ = dH \quad (5\text{-}3)$$

where dQ and dH refer to quantities in a time increment dt.

In *flow* reactors the energy in the flowing streams includes kinetic and potential energy contributions in addition to the enthalpy. However, when reactions occur these two contributions are normally unimportant. Then for *steady-state* operation of a flow reactor, Eq. (5-1) becomes

$$H'_f - H'_e + Q' = 0 \quad (5\text{-}4)[‡]$$

where H' and Q' refer to energy *rates* and subscripts f and e designate feed and effluent streams for the appropriate volume element of the reactor. For a tank

---

[†] In thermodynamics the first law for a batch process is expressed as $\Delta U = Q - W$. If $W \to 0$, $Q = \Delta U$. (For example see "Introduction to Chemical Engineering Thermodynamics" by J. M. Smith and H. G. Van Ness, 3d ed., p. 25, McGraw-Hill Book Company, New York, 1975.

[‡] The first law of thermodynamics for a steady-state process is written $(H + \text{P.E.} + \text{K.E.})_e - (H + \text{P.E.} + \text{K.E.})_f = Q' - W'$. The terms P.E. and K.E. refer to potential and kinetic energy. If $W' \to 0$, $\Delta(\text{P.E.}) \to 0$ and $\Delta(\text{K.E.}) \to 0$, this expression is the same as Eq. (5-4).

reactor the volume element is the entire volume of the reactor while for a tubular flow form a differential volume must be used (see Secs. 3-3 and 3-4).

For semibatch reactors, all four terms of Eq. (5-1) may be important. If kinetic and potential energy contributions are neglected as well as any energy transferred as work, the energy balance for a time dt is

$$(H'_f - H'_e) \, dt + dQ = dH \qquad (5-5)$$

Here, as in Eq. (5-3) dQ and dH refer to *quantities* of energy rather than rates. In writing Eq. (5-5) the accumulation of energy in the reaction has been assumed to be equal to the accumulation of enthalpy, ΔH, as was done in writing Eq. (5-3) for batch reactors.

Equations (5-3), (5-4), and (5-5) are the starting expressions for treating heat effects in nonisothermal reactors operating as batch, continuous flow, or semi-batch processes. In the following sections these expressions are employed, along with mass balances, for calculating reactor performance (temperature, conversions, selectivity vs. reactor volume, or time).

5-2 Batch, Stirred-Tank Reactors

Equations (3-7) [or Eq. (3-8) for constant volume] and (5-3) are the mass-and-energy-conservation equations applicable for batch operation of stirred-tank reactors. For the energy equation to be useful, the enthalpy change dH needs to be expressed in terms of temperature and reaction rate (or conversion). This is primarily a thermodynamic procedure. Suppose the total mass and constant-pressure specific heat of the reaction mixture are m_t and c_p. In an increment of time dt, the mixture undergoes a temperature change dT and also a composition change determined by the amount of reaction. We wish to evaluate the enthalpy change in this time increment. Since enthalpy is a state property, any sequence of states may be used to determine dH, without regard to the actual process (path). Heats of reaction are more likely to be available, and accurately known, at a reference temperature, $T_r = 298$ K. Therefore, the preferable path will involve the reaction at T_r and consist of the following three steps:

1. Cool the reaction mixture at constant pressure from T to T_r. The enthalpy change is

$$\Delta H_1 = m_t \int c_p \, dT \approx m_t c_p (T_r - T)$$

The second equality requires that c_p be independent of temperature. Also, c_p is the specific heat for the composition of the mixture at time t.

2. Carry out the reaction at T_r. The amount of reaction will be $(\mathbf{r}V)dt$ moles in time dt, where $\mathbf{r}$ is the molal rate of disappearance of reactant (the same reactant as used to formulate the mass balance). If ΔH_R is the molal heat of reaction at T_r, the enthalpy change for this constant temperature and constant pressure process is

$$dH_2 = \Delta H_R(\mathbf{r}V) \, dt$$

3. Heat the reaction mixture from T_r to $T + dT$, the temperature at the end of the time increment. For this third step:

$$\Delta H_3 \approx m_t c_p (T + dT - T_r)$$

The required enthalpy change is the sum of ΔH_1, dH_2 and ΔH_3. If we neglect changes in c_p with composition as well as with temperature, the sum is

$$dH = m_t c_p \, dT + \Delta H_R (\mathbf{r}V) \, dt \qquad (5\text{-}6)$$

The heat exchange with the surroundings may be expressed in terms of an overall heat transfer coefficient h_o, surroundings temperature† T_s, and heat transfer area A_h. Thus, the amount of energy transferred to the reactor in time dt is

$$dQ = h_o A_h (T_s - T) \, dt \qquad (5\text{-}7)$$

Finally, we substitute Eqs. (5-6) and (5-7) in (5-3) to obtain an energy equation in terms of temperature and rate of disappearance of reactant:

$$h_o A_h (T_s - T) = \Delta H_R (\mathbf{r}V) + m_t c_p \frac{dT}{dt} \qquad (5\text{-}8)$$

Many assumptions regarding c_p and the relationship between dH and dU have been made in obtaining Eq. (5-8). However, the errors introduced are nearly always negligible with respect to uncertainties in the reaction rate and the magnitude of the remaining terms.

Equation (5-8) and the mass balance, along with an expression for the reaction rate, are sufficient to calculate the composition and temperature of the reaction mixture as a function of time. A stepwise numerical solution of the two differential equations is usually required because of the exponential effect of temperature in the rate equation.

The energy balance can be expressed in terms of conversion by combining the mass balance, Eq. (3-7),‡ with Eq. (5-8) to eliminate $\mathbf{r}$:

$$h_o A_h (T_s - T) = \Delta H_R (V_0 C_{A0}) \frac{dx_A}{dt} + m_t c_p \frac{dT}{dt} \qquad (5\text{-}9)$$

where x_A is the conversion of reactant A. This form is advantageous for adiabatic operation. Thus, when $dQ = 0$, Eq. (5-9) reduces to a form not involving time:

$$m_t c_p \, dT = -\Delta H_R (V_0 C_{A0}) \, dx_A \qquad (5\text{-}10)$$

If again, c_p is assumed to be independent of temperature and composition, Eq. (5-10) can be integrated immediately to give a relationship between temperature and conversion:

$$T - T_0 = -\frac{\Delta H_R (V_0 C_{A0})}{m_t c_p} (x - 0) \qquad (5\text{-}11)$$

---

† For example, the temperature in the internal cooling (or heating) coils or jacket surrounding the reactor.

‡ The negative sign in Eq. (3-7) disappears if $\mathbf{r}_A$ is the rate of *disappearance* of reactant A.

where T_0 and $x = 0$ are the initial temperature and conversion. Note that $(V_0 C_{A_0})$ is initial moles of reactant A. For adiabatic operation Eq. (5-11) eliminates the need for simultaneous stepwise solution of the mass and energy balances. Both adiabatic and nonadiabatic design calculations are illustrated in the following example.

Example 5-1 In a study of the production of drying oils by the decomposition of acetylated castor oil, Grummitt and Fleming† were able to correlate decomposition data on the basis of a first-order reaction written as

$$\text{Acetylated castor oil } (l) \rightarrow CH_3COOH(g) + \text{drying oil } (l)$$

$$\mathbf{r} = (k/60)C$$

where $\mathbf{r}$ is rate of decomposition, in kg of acetic acid produced per second per cubic meter, and C is concentration of acetic acid, in kg per m^3, equivalent to acetylated castor oil. Data obtained over the temperature range 295 to 340°C indicated an activation energy of 44,500 cal/g mol in accordance with the following expression for the specific-reaction-rate constant k:

$$\ln k = \frac{-44{,}500}{R_g T} + 35.2$$

where T is in degrees Kelvin.

If a batch reactor initially contains 227 kg of acetylated castor oil at 340°C (density 0.90) and the operation is adiabatic, plot curves of conversion (fraction of the acetylated oil that is decomposed) and temperature vs. time. It is estimated that the endothermic heat effect for this reaction is 62,760 J/mol of acetic acid vapor. The acetylated oil charged to the reactor contains 0.156 kg of equivalent acetic acid per kg of oil; i.e., complete decomposition of 1 kg of the oil would yield 0.156 kg of acetic acid. Assume that the specific heat of the liquid reaction mixture is constant and equal to 2.51×10^3 J/(kg)(K). Also assume that the acetic acid vapor produced leaves the reactor at the temperature of the reaction mixture.

SOLUTION The applicable mass conservation expression, assuming no change in volume, is Eq. (3-8):

$$\frac{dx}{dt} = \frac{1}{C_0}\mathbf{r}$$

or

$$\frac{dt}{dx} = \frac{C_0}{\mathbf{r}} \tag{A}$$

---

† *Ind. Eng. Chem.*, **37**, 485 (1945).

where C_0 and **r** are measured in terms of equivalent acetic acid.† The concentration-conversion relation is

$$C = C_0(1 - x)$$

If we replace **r** with the rate equation expressed in terms of conversion and temperature Eq. (A) becomes

$$\frac{dt}{dx} = \frac{C_0}{(k/60)C_0(1 - x)}$$

$$\frac{dt}{dx} = f(T, x) = \frac{60}{[\exp(35.2 - 44500/R_g T)](1 - x)} \tag{B}$$

Since the reactor operates adiabatically, Eq. (5-11) is applicable. If we take 1 kg of oil as a basis, the initial *moles* of equivalent acetic acid is $V_0 C_{A_0} = 0.156/60$. Substituting numbers in Eq. (5-11) yields

$$T - T_0 = -\frac{62{,}760}{2.51 \times 10^3}\left(\frac{0.156/60}{10^{-3}}\right)(x - 0) \tag{C}$$

or

$$T = T_0 - 65x = (340 + 273) - 65x \tag{D}$$

The right-hand side of Eq. (B) can now be expressed in terms of x only by substituting Eq. (D) for the temperature. Hence, Eq. (B) can be solved numerically, for example, by the Runge-Kutta procedure described in Examples 4-7 and 4-15. The working equations for this, one dependent-variable problem are:

$$k_0 = (\Delta x)f(x_n, T_{x_n})$$

$$k_1 = (\Delta x)f\left(x_n + \frac{\Delta x}{2}, T_{(x_n + \Delta x/2)}\right)$$

$$k_2 = (\Delta x)f\left(x_n + \frac{\Delta x}{2}, T_{(x_n + \Delta x/2)}\right)$$

$$k_3 = (\Delta x)f(x_n + \Delta x, T_{(x_n + \Delta x)})$$

$$t_{n+1} = t_n + \tfrac{1}{6}(k_0 + 2k_1 + 2k_2 + k_3)$$

where Eqs. (B) and (D) give $f(T, x)$ and n represents the number of increments of Δx.

---

† The reaction rate used in mass or energy conservation equations can be expressed either in mass or moles. This is the first example in which we have used mass. Of course, the conservation equations must be consistent. If r in Eq. (3-8) is expressed in terms of mass, the concentration must also be so expressed.

For the first increment choose $\Delta x = 0.1$. Since $x = 0$ at $t = 0$, Eq. (D) gives $T = T_0 = 613$ K. Then from Eq. (B),

$$f(x_0, T_{x_0}) = f(0,613°) = \frac{60}{\left\{\exp\left[35.2 - \dfrac{44500}{1.98(613)}\right]\right\}(1 - 0)} = 259 \text{ s}$$

The k values are

$$(k_0) = 0.1\, f(0,613°) = 0.1\,(259) = 25.9 \text{ s}$$

$$k_1 = 0.1\, f(0 + 0.1/2, T_{0+0.1/2})$$

$$= 0.1\, f(0.05, 610°)† = 0.1(327) = 32.7 \text{ s}$$

$$k_2 = 0.1\, f(0 + 0.1/2, T_{0+0.1/2}) = 32.7 \text{ s}$$

$$k_3 = 0.1\, f(0 + 0.1, T_{0.1}) = f(0.1, 606.5°) = 42.7 \text{ s}$$

and

$$t_1 = t_0 + \tfrac{1}{6}(25.9 + 2(32.7 + 2(32.7) + (42.7)$$

$$= 0 + 33 = 33 \text{ s}$$

Results for subsequent increments are given in Table 5-1 and in Fig. 5-4.

The curves of temperature and conversion vs. time show the necessity of supplying energy (as heat) to a highly endothermic reaction if large conversions are desired. In the present case, where no energy was supplied, the temperature decreased so rapidly that the reaction essentially stopped after a conversion of 50% was reached. If, instead of operating adiabatically, a constant rate of energy $Q' = 52,700$ J/s had been added to the reactor, the energy balance, according to Eqs. (5-9) and (5-7), would have been

$$m_t c_p \frac{dT}{dt} = -\Delta H_R(V_0 C_{A_0})\frac{dx_A}{dt} + \frac{dQ}{dt}$$

or, numerically

$$227(2.51 \times 10^3)\frac{dT}{dt} = -62760\left(\frac{0.156/60}{10^{-3}}\right)227\frac{dx}{dt} + Q'$$

† The temperature of 610 K is obtained from Eq. (D) with $x = 0.10/2$.

Table 5-1

Conversion	T, K	t, s
0	613	0
0.10	606	33
0.20	600	78
0.30	594	162
0.40	587	307

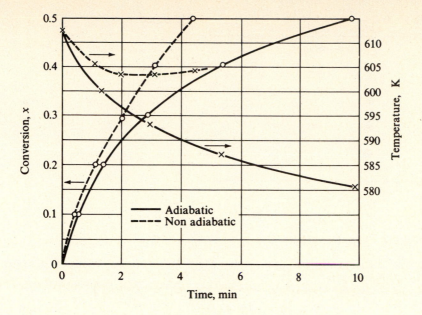

Figure 5-4 Temperature and conversion vs. time.

Integrating from $x = 0$, $T = T_0$ at $t = 0$, this expression becomes

$$T = T_0 + 1.75 \times 10^{-6}Q'(t - 0) - 65(x - 0) \qquad (D')$$

Equation (B) is still applicable for this nonadiabatic case and can be integrated numerically. However, Eq. (D') must be used to determine the temperature. The working equations for the Runge-Kutta method are:

$$k_0 = (\Delta x)f(x_n, T_{x_n, t_n})$$

$$k_1 = (\Delta x)f\left(x_n + \frac{\Delta x}{2}, T_{x_n + \Delta x/2, t_n + k_0/2}\right)$$

$$k_2 = (\Delta x)f\left(x_n + \frac{\Delta x}{2}, T_{x_n + \Delta x/2, t_n + k_1/2}\right)$$

$$k_3 = (\Delta x)f(x_n + \Delta x, T_{x_n + \Delta x, t_n + k_2})$$

The subscripts on T indicate the values of x and t to use in Eq. (D') to evaluate the temperature.

The temperatures and times obtained by these numerical calculations for various conversions are given in Table 5-2 and are also shown in Fig. 5-4. Comparison of the results for adiabatic and nonadiabatic operation demonstrates the advantage of adding heat. For example, a conversion of 30% is obtained in about one-third less time when 52,700 J/s, (3000 Btu/min) are added continuously to the reactor.

Table 5-2

	Nonadiabatic		Adiabatic	
Conversion	T, K	t, s	T, K	t, s
0.0	613	0	613	0
0.10	609	27	606	33
0.20	606	65	600	78
0.30	604	115	594	162
0.40	603	177	587	307
0.50	604	251	580	570

5-3 Tubular-Flow Reactors

Since a differential volume element is necessary for a tubular reactor, it is appropriate to write Eq. (5-4) as

$$dQ' = dH' \tag{5-12}$$

where the differentials refer to a change over a volume element rather than an element of time as in Eq. (5-3). That is, dH' is the change in energy *rate* between the streams entering and leaving the volume element and dQ' is the *rate* of energy transfer into the differential volume element.† The enthalpy difference will again be due to the temperature change dT and a composition change associated with the reaction. Hence the procedure used in Sec. 5-2 for deriving an expression for dH is applicable here. If F_t is the *total* molal feed flow rate through the reactor, and c_p is the molal heat capacity (assumed constant) of the reaction mixture,

$$dH' = F_t c_p \, dT + \Delta H_R(\mathbf{r} \, dV) \tag{5-13}$$

where $\mathbf{r} \, dV$ is the moles of reactant *disappearing* per unit time in the reactor volume dV, and ΔH_R is the molal heat of reaction.

If the heat transfer rate can be expressed in terms of an overall coefficient h_o and the surrounding temperature, T_s,

$$dQ' = h_o(T_s - T) \, dA_h \tag{5-14}$$

Here dA_h is the effective heat transfer area in the volume element.

Substituting Eqs. (5-13) and (5-14) into (5-12) yields

$$h_o(T_s - T) \, dA_h = \Delta H_R(\mathbf{r} \, dV) + F_t c_p \, dT \tag{5-15}$$

---

† For nonideal reactors (where the plug flow assumption is not justified) axial dispersion contributions to the energy entering and leaving the reactor may be important. This is most likely to be significant in heterogeneous catalytic reactions and is considered in Chap. 13. In the present chapter the assumptions of a plug-flow reactor, no axial mixing, and no radial gradients, are retained.

This energy equation can be expressed in terms of conversion by using the mass balance, Eq. (3-17), to eliminate the rate. The result is:

$$h_o(T_s - T)\, dA_h = \Delta H_R F\, dx + F_t c_p\, dT \qquad (5\text{-}16)$$

Consistent mass or molal units may be employed in either of the two terms on the right in Eqs. (5-15) and (5-16). We have chosen to express F (molal feed rate of reactant) and F_t in molal units.

For adiabatic operation and constant c_p, Eq. (5-16) reduces to an easily integrable form not involving the volume of the reactor. If at $V = 0$, $T = T_f$ and $x = 0$ the result is

$$T - T_f = -\frac{\Delta H_R}{c_p}\left(\frac{F}{F_t}\right)(x - 0) \qquad (5\text{-}17)$$

Note that the ratio F (in moles)/F_t (in moles) is equal to the mole fraction of reactant in the feed (if $x = 0$ in the feed).

Simultaneous solution of Eq. (5-15), the mass balance [Eq. (3-17)], and the rate equation establishes the temperature and composition as a function of reactor volume. For adiabatic operation, the differential mass and energy balances need not be solved simultaneously. Equation (5-17) can be used in the rate equation to eliminate temperature as a variable. Then, this modified rate equation can be substituted in the mass balance to give a single differential equation (see Prob. 5-3). For the nonadiabatic case the calculations are the same as those for batch reactors (Example 5-1), except that the reactor volume replaces time as the independent variable. The procedure is not repeated here for single reactions; instead the treatment for multiple-reaction systems is illustrated. The most significant point is the effect of temperature on the product distribution (selectivity) in the effluent stream. Example 5-2 concerns a nonisothermal reactor in which two *parallel* reactions occur.[†] The desired reaction has the higher activation energy. After the example, a case of *consecutive* reactions with different activation energies is considered.

Example 5-2 It is proposed to design a pilot plant for the production of allyl chloride. The reactants consist of 4 moles propylene/mole chlorine and enter the reactor at 392°F. The reactor will be a vertical tube of 2 in. ID. If the combined feed rate is 0.85 lb mol/h, determine the conversion to allyl chloride as a function of tube length. The pressure may be assumed constant and equal to 29.4 lb/in.² abs.

The reactants will be preheated separately to 392°F and mixed at the entrance to the reactor. At this low temperature explosion difficulties on mixing are not serious. The reactor will be jacketed with boiling ethylene glycol, so that the inside-wall temperature will be constant and equal to 392°F. The inside-heat-transfer coefficient may be taken as 5.0 Btu/(h)(ft²)(°F).

---

† Kinetics data and selectivity for an important set of homogeneous reactions, the thermal cracking of light hydrocarbons, have been analyzed by C. F. Froment et al. in *AIChE J.* **23**, 93 (1977).

Additional data and notes The basic development of the allyl chloride process has been reported by Groll and Hearne[†] and Fairbairn, Cheney, and Cherniavsky.[‡] It was found that the three chief reactions were allyl chloride formation,

1. $Cl_2 + C_3H_6 \rightarrow CH_2=CH-CH_2Cl + HCl$

    the parallel addition reaction giving 1,2-dichloropropane,

2. $Cl_2 + C_3H_6 \rightarrow CH_2Cl-CHCl-CH_3$

    and consecutive chlorination of allyl chloride to give 1,3-dichloro-l-propene

3. $Cl_2 + CH_2=CH-CH_2Cl \rightarrow CHCl=CH-CH_2Cl + HCl$

To simplify the kinetic treatment of the problem we shall consider only the first two reactions. The heats of reaction are shown in Table 5-3. The molal heat capacities c_p will be assumed constant and equal to the values given in Table 5-4.

No information is published for the rate-of-reaction equations, although it is known that reaction 2 takes place at temperatures as low as 212°F, while reaction 1 has an insignificant rate below 392°F. As the temperature increases above 392°F, the rate of reaction 1 increases rapidly, until at 932°F it is several times as fast as that of reaction 2. From this general information and some published data on the effect of residence time and temperature on the conversion, the proposed rate equations are

$$r_1 = 206{,}000e^{-27{,}200/R_gT}p_{C_3H_6}p_{Cl_2}; \ (T, °R)$$

$$r_2 = 11.7e^{-6{,}860/R_gT}p_{C_3H_6}p_{Cl_2}; \ (T, °R)$$

where r_1 and r_2 are in lb moles of Cl_2 disappearing per hour per cubic foot, T is in degrees Rankine, and the partial pressure p is in atmospheres.

SOLUTION Since there will be heat transfer from the reaction gases to the glycol jacket, the temperature in the reactor will depend on the length z, and the design calculations have to be carried out by the stepwise integration of the mass and energy conservation equations. One mass balance may be written for each reaction. It is convenient in this problem to define the conversion

[†] H. P. A. Groll and G. Hearne, *Ind. Eng. Chem.*, **31**, 1530 (1939).
[‡] A. W. Fairbairn, H. A. Cheney, and A. J. Cherniavsky, *Chem. Eng. Progr.*, **43**, 280 (1947).

Table 5-3

ΔH Btu/lb mole	298 K
Reaction 1	$-48{,}000$
Reaction 2	$-79{,}200$

Table 5-4

Component	c_p, Btu/(lb mol)(°R)
Propylene (g)	25.3
Chlorine (g)	8.6
Hydrogen chloride (g)	7.2
Allyl chloride (g)	28.0
1,2--Dichloropropane (g)	30.7

as moles of chlorine reacted per mole of *total* feed rather than per mole of chlorine in the feed. If this conversion is $(x_t)_1$ for reaction 1 and $(x_t)_2$ for 2, the mass-balance equations, Eq. (3-17), are

$$\mathbf{r}_1 \, dV = F_t \, d(x_t)_1 \tag{A}$$

$$\mathbf{r}_2 \, dV = F_t \, d(x_t)_2 \tag{B}$$

where F_t is the total *molal* feed rate.

The appropriate energy balance is Eq. (5-15).

$$h_o(T_s - T) \, dA_h = (\mathbf{r}_1 \, \Delta H_{R_1} + \mathbf{r}_2 \, \Delta H_{R_2}) \, dV + F_t c_p \, dT$$

If z is the reactor length and d is its diameter,

$$dV = \frac{\pi d^2}{4} \, dz$$

$$dA_h = \pi d \, dz$$

Using these relations in the previous expression gives

$$(F_t \, c_p) \, dT = \pi d \, h_0(T_s - T) \, dz - (\mathbf{r}_1 \, \Delta H_{R_1} + \mathbf{r}_2 \, \Delta H_{R_2})\left(\frac{\pi d^2}{4}\right) dz$$

or

$$\frac{dT}{dz} = \frac{\pi d \, h_0(T_s - T) - (\mathbf{r}_1 \Delta H_{R_1} + \mathbf{r}_2 \, \Delta H_{R_2})\dfrac{\pi d^2}{4}}{(F_t c_p)} \tag{C}$$

Equations (A) to (C) and the expressions for the rates $\mathbf{r}_1$ and $\mathbf{r}_2$ can be solved numerically for $(x_t)_1$ and $(x_t)_2$ and for T as a function of reactor length z. To accomplish this we first express Eqs. (A) and (B) in terms of z:

$$\frac{d(x_t)_1}{dz} = \frac{\pi d^2}{4F_t} \mathbf{r}_1 \tag{D}$$

$$\frac{d(x_t)_2}{dz} = \frac{\pi d^2}{4F_t} \mathbf{r}_2 \tag{E}$$

Also the equations for $\mathbf{r}_1$ and $\mathbf{r}_2$ need to be written in terms of $(x_t)_1$, $(x_t)_2$ and T. If 4 moles of propylene per mole of chlorine enter the reactor, at an axial position where the conversions are $(x_t)_1$ and $(x_t)_2$, the moles of each component are:

$$\text{Chlorine}\dagger = 1 - 5(x_t)_1 - 5(x_t)_2$$

$$\text{Propylene} = 4 - 5(x_t)_1 - 5(x_t)_2$$

$$\text{Allyl chloride} = 5(x_t)_1$$

$$\text{Dichloropropane} = 5(x_t)_2$$

$$\text{Hydrogen chloride} = 5(x_t)_1$$

$$\text{Total moles} = 5[1 - (x_t)_2]$$

If we $\quad$ ve that all the components behave as ideal gases at $p_t = 29.4$ lb/in.^{2} abs, $\quad$ partial pressures of chlorine and propylene, in atmospheres, are given by

$$p_{C_3H_6} = \frac{29.4}{14.7}\left[\frac{4 - 5(x_t)_1 - 5(x_t)_2}{5[1 - (x_t)_2]}\right] = 2\frac{0.8 - (x_t)_1 - (x_t)_2}{1 - (x_t)_2}$$

$$p_{Cl} = \frac{29.4}{14.7}\left[\frac{1 - 5(x_t)_1 - 5(x_t)_2}{5[1 - (x_t)_2]}\right] = 2\frac{0.2 - (x_t)_1 - (x_t)_2}{1 - (x_t)_2}$$

These expressions can be substituted in the proposed rate equations to relate the rates to the conversions and the temperature. The results are

$$r_1 = 824{,}000e^{-13{,}700/T}\frac{[0.8 - (x_t)_1 - (x_t)_2][0.2 - (x_t)_1 - (x_t)_2]}{[1 - (x_t)_2]^2} \qquad (F)$$

$$r_2 = 46.8e^{-3{,}460/T}\frac{[0.8 - (x_t)_1 - (x_t)_2][0.2 - (x_t)_1 - (x_t)_2]}{[1 - (x_t)_2]^2} \qquad (G)$$

Substituting Eqs. (F) and (G) for r_1 and r_2 allows Eqs. (C), (D), and (E) to be expressed in terms of the three dependent variables T, $(x_1)_t$, and $(x_2)_t$ and the independent variable z.

Values of d, h_o, T_s, ΔH_{R_1}, and ΔH_{R_2} are given. The heats of reaction do not change significantly with temperature. Assuming ideal gases, the term $(F_t c_p)$ in Eq. (C) is given by

$$(F_t c_p) = \sum_i F_i c_{p_i}$$

At a location where the conversions are $(x_t)_1$ and $(x_t)_2$

$$(F_t c_p) = F_t\underbrace{[0.2 - (x_t)_1 - (x_t)_2]c_{p_{Cl_2}}}_{\text{chlorine}} + F_t\underbrace{[0.8 - (x_t)_1 - (x_t)_2]c_{p_{C_3H_6}}}_{\text{propylene}}$$

---

† The coefficient 5 appears in this equation because x_t is the conversion based on 1 mole of *total* feed. Note that the maximum value for the sum of $(x_t)_1$ and $(x_t)_2$ is 0.2, which corresponds to complete conversion of the chlorine to either allyl chloride or dichloropropane.

For example, entering the reactor $(x_t)_1 = (x_t)_2 = 0$ so that

$$(F_t c_p) = 0.85(0.2)(8.6) + 0.85(0.8)(25.3)$$

$$= 18.5 \; \text{Btu/(h)(°F)}$$

These results establish numerical values for all the coefficients in Eqs. (C), (D), and (E).

The numerical procedure is somewhat different from that used in Examples 4-7, 4-15, and 5-1 since we need to solve three coupled differential equations with three dependent variables. The modified Euler method will be used first to obtain approximate results for the first increment. By doing this the physical aspects of the problem and method of solution are clearly displayed. Then, more accurate results from the fourth-order Runge-Kutta procedure are presented.

Modified Euler method Equations (C), (D), and (E) are first written in difference form:

$$F_t \, c_p \, \Delta T = \left[\pi d \, h_o(T_s - T)_{\text{av}} - (\mathbf{r}_1 \, \Delta H_{R_1} + \mathbf{r}_2 \, \Delta H_{R_2}) \frac{\pi d^2}{4} \right] (\Delta z) \quad \text{(C')†}$$

$$\Delta(x_t)_1 = \bar{\mathbf{r}}_1 \frac{\pi d^2}{4F_t} (\Delta z) \tag{D'}$$

$$\Delta(x_t)_2 = \bar{\mathbf{r}}_2 \frac{\pi d^2}{4F_t} (\Delta z) \tag{E'}$$

A solution procedure is as follows:

1. From the known initial conversion $[(x_t)_1 = 0$ and $(x_t)_2 = 0]$ and temperature (200°C), $\mathbf{r}_1$ and $\mathbf{r}_2$ entering the reactor are computed.
2. An arbitrary increment of reactor length Δz is chosen. The smaller this increment, the more accurate the solution (and the more time consuming the calculations).
3. For the chosen Δz, first estimates of the conversion occurring within the increment are obtained from Eqs. (D') and (E'). It is convenient to assume that the average values of $\mathbf{r}_1$ and $\mathbf{r}_2$ are equal to the initial values evaluated in step 1.
4. The change in temperature within the increment ΔT is determined from Eq. (C'). An estimate of the average temperature difference $(T_s - T)_{\text{av}}$ in the increment is required in order to evaluate the heat-loss term. Once ΔT has been computed, the estimate of $(T_s - T)_{\text{av}}$ can be checked. Hence a trial-and-error calculation is necessary to evaluate ΔT.
5. From the conversion and temperature at the end of the first increment, as determined in steps 3 and 4, the rate of reaction is computed at this

† Since the inside-wall temperature is known, the heat transfer with the surroundings can be evaluated from $h_o(T_s - T)_{\text{av}} \pi d \, \Delta z$, where h_o is the inside-film coefficient and T_s is the inside-wall-surface temperature. The overall coefficient U is not necessary in this case.

position in the reactor. Then steps 3 and 4 are repeated, using for $\bar{r}_1$ and $\bar{r}_2$ the arithmetic average of the values at the beginning and end of the increment. This, in turn, will give more precise values of the conversion and temperature at the end of the first increment and permit a third estimate of the average values of the rates. If this third estimate agrees with the second, the next increment of reactor length is chosen and the procedure is repeated.

The calculations according to these steps are as follows:

Step 1 The rates of reaction at the reactor entrance are given by Eqs. (F) and (G), with $T = (200 + 273)(1.8) = 852°R$ and $(x_t)_1 = (x_t)_2 = 0$:

$$r_1 = 824{,}000e^{-16.1}(0.16) = 0.0135 \text{ lb mol/(h)(ft}^3)$$

$$r_2 = 46.8e^{-4.06}(0.16) = 0.129 \text{ lb mol/(h)(ft}^3)$$

Step 2 An increment of reactor length is chosen as $\Delta z = 4.0$ ft.

Step 3 Assuming that the rates computed in step 1 are average values for the increment, the first estimates of the conversion in the increment are given by Eqs. (D') and (E'):

$$\frac{\pi d^2}{4} = \frac{\pi}{4}\left(\frac{2}{12}\right)^2 = 0.0218 \text{ ft}^2$$

$$\frac{\pi d^2}{4F_t} = \frac{0.0218}{0.85} = 0.0257$$

hence

$$\Delta(x_t)_1 = 0.0135(0.0257)(4.0) = 0.0014$$

$$(x_t)_1 = 0 + 0.0014 = 0.0014$$

Similarly, for the second reaction we obtain

$$\Delta(x_t)_2 = 0.129(0.0257)(4.0) = 0.0133$$

$$(x_t)_2 = 0 + 0.0133 = 0.0133$$

Step 4 If the average temperature difference $(T_s - T)_{av}$ for the first increment is *estimated* to be $-20°F$, substitution in Eq. (C') gives

$$18.5 \, \Delta T = \pi \tfrac{2}{12}(5)(-20)(4) - [0.0135(-48{,}000)$$

$$+ 0.129(-79{,}200)](0.0218)(4)$$

$$\Delta T = \frac{946 - 210}{18.5} = 40°F$$

$$T_1 = 852 + 40 = 892°R \qquad \text{temperature at end of first increment}$$

$T_s - T$ at entrance $= 0$

$T_s - T$ at end of increment $= 852 - 892 = -40°F$

$$(T_s - T)_{av} = \frac{0 + (-40)}{2} = -20° \qquad \text{vs.} \; -20° \text{ assumed}$$

Step 5 At the end of the first increment the first estimate of the conversions and temperature is

$$(x_t)_1 = 0.0014$$
$$(x_t)_2 = 0.0133$$
$$T_1 = 852 + 40 = 892°R$$

Substitution of these conditions in rate equations (F) and (G) yields

$$r_1 = 0.0256 \qquad r_2 = 0.143$$

A second, and more accurate, estimate of the average values of the rates for the first increment can now be made:

$$\bar{r}_1 = \frac{0.0135 + 0.0256}{2} = 0.0195$$

$$\bar{r}_2 = \frac{0.129 + 0.143}{2} = 0.136$$

The second estimate of the conversion in the increment, obtained from these rates and Eqs. (C') and (D'), is

$$\Delta(x_t)_1 = 0.0195(0.0257)(4.0) = 0.0020$$
$$(x_t)_1 = 0 + 0.0020 = 0.0020$$
$$\Delta(x_t)_2 = 0.136(0.0257)(4.0) = 0.0140$$
$$(x_t)_2 = 0 + 0.0140 = 0.0140$$

The second estimate of ΔT, determined from Eq. (C'), is

$$\Delta T = 41°F$$
$$T_1 = 852 + 41 = 893°R$$

At the revised values of $(x_t)_1$, $(x_t)_2$, and T (at the end of the first increment) the rates are

$$r_1 = 0.0258$$
$$r_2 = 0.144$$

Since these values are essentially unchanged from the previous estimate, the average rates will also be the same, and no further calculations for this increment are necessary.

The same calculations can be repeated for successive increments until the required conversion is attained.

Runge-Kutta Method The procedure for extending the Runge-Kutta method to any number of dependent variables is available.† The working equations for our case of three dependent variables are an extension of those for two dependent variables given in Example 4-7 [Eqs. (4-11) to (4-15)]. With the working equations and a choice of Δz we can solve Eqs. (C), (D), and (E) to obtain $(x_t)_1$, $(x_t)_2$, and T for any reactor length. Results up to $z = 20$ ft are given by the solid curves in Fig. 5-5 and 5-6 and in Table 5-5 (nonadiabatic conditions). They are based upon $\Delta z = 1$ ft.

A comparison of the results of the modified Euler and Runge-Kutta methods at $z = 4$ ft is as follows:

	Modified Euler	Runge-Kutta
Reactor length, z, ft	4	4
Comparison per mole feed		
$(x_t)_1$	0.0020	0.0021
$(x_t)_2$	0.0140	0.0142
Temperature, R	893	895

The agreement is good, particularly in view of the larger increment size (4 ft) used for the modified Euler method.

These calculations indicate two trends. First, the rate of reaction 1 is relatively low with respect to reaction 2 at low temperatures but increases

† "Applied Numerical Analysis" by Bruce Carnahan, H. A. Luther, and James O. Wilkes, pp. 376–379; John Wiley & Sons Inc., New York, 1969. This reference includes computer programs for carrying out the calculations for solving a multiple-variable problem.

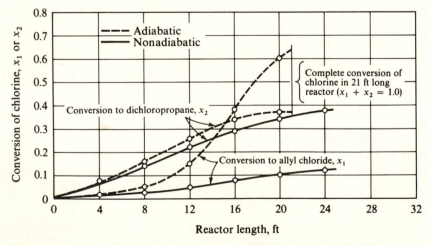

Figure 5-5 Conversion curves for allyl chloride production in a tubular reactor.

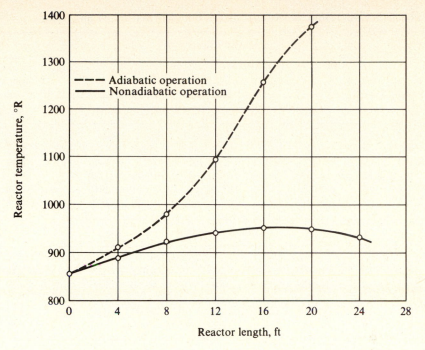

Figure 5-6 Temperature profiles for allyl chloride production in a tubular reactor.

rapidly with temperature, while the rate of reaction 2 is not very sensitive to temperature changes. This is because the activation energy for reaction 1 is greater than that for reaction 2. The second trend is the increase in rate of heat transfer to the surroundings with an increase in reactor length. This effect, which is a consequence of the increase in $(T - T_s)$, offsets the rise in temperature due to the exothermic nature of the reactions, and results ultimately in a decrease in temperature with reactor length. This length is reached when the heat transferred from the reactor tube to the surroundings is greater than the heat evolved from the reactions. Figure 5-6 indicates that this occurs about 18 ft from the entrance to the reactor and is at too low a temperature for the rate of reaction 1 to become large with respect to that of reaction 2. Therefore the conversion to allyl chloride never reaches a high value. Most of the product is dichloropropane, and the selectivity for allyl chloride is always less than unity, as noted in the last column of Table 5-5. A reactor of this type is not well suited for the production of allyl chloride. Adiabatic operation would eliminate the maximum in the temperature and result in higher temperatures, which in turn would favor the production of allyl chloride. The only change in the calculations for such an adiabatic case is that the term $\pi d h_o(T_s - T)$ in Eq. (C) would be zero. For comparison, the temperature and conversion results for such an adiabatic reactor are also shown in Figs. 5-5 and 5-6 and Table 5-5. The selectivity now is greater than unity at high conversions.

Even adiabatic operation results in the formation of considerable amounts of the undesirable dichloropropane. This occurs in the first part of

Table 5-5 Conversion vs. reactor length for chlorination of propylene

Nonadiabatic operation

Reactor volume, ft^3	Reactor length ft^3	Conversion per mole feed				Conversion of chlorine in feed			Selectivity $S_o = x_1/x_2$
		$(x_t)_1$	$(x_t)_2$	Total	T, °R	x_1	x_2	Total	
0	0	0.0	0.0	0.0	852	0.0	0.0	0.0	
0.087	4	0.0021	0.0142	0.0163	895	0.0105	0.0710	0.081	0.15
0.174	8	0.0056	0.0294	0.0350	925	0.0280	0.1470	0.175	0.19
0.261	12	0.0102	0.0444	0.0546	942	0.0510	0.2220	0.273	0.23
0.349	16	0.0151	0.0582	0.0733	949	0.0755	0.2910	0.366	0.26
0.436	20	0.0196	0.0704	0.0900	948	0.0980	0.3520	0.450	0.28

Adiabatic operation

Reactor volume, ft^3	Reactor length ft^3	$(x_t)_1$	$(x_t)_2$	Total	T, °R	x_1	x_2	Total	S_o
0	0	0.0	0.0	0.0	852	0.0	0.0	0.0	
0.087	4	0.0023	0.0145	0.0168	910	0.0115	0.0725	0.0840	0.16
0.174	8	0.0086	0.0318	0.0404	986	0.0430	0.159	0.202	0.27
0.261	12	0.0274	0.0514	0.0788	1100	0.137	0.257	0.394	0.53
0.349	16	0.0814	0.0688	0.1502	1281	0.407	0.344	0.751	1.18
0.436	20	0.1228	0.0739	0.1967	1390	0.614	0.369	0.983	1.66

the reactor, where the temperature of the flowing mixture is low. A tubular-flow reactor is less desirable for this reaction system than a stirred-tank unit. The same reaction system is illustrated in Example 5-3 for a stirred-tank reactor.

Consider next a pair of consecutive reactions where the intermediate component is the desired product. Suppose also that the desired product is produced by the reaction with the lower activation energy. To illustrate the effect of type of reaction on selectivity, assume that the tubular reactor operates isothermally. The two reactions of Example 4-9 starting with isobutylene (B) and acrilonitrile (A) provide an illustration. Suppose now the desired product is the monoamine (M) rather than the diamine (D). The two homogeneous irreversible second-order reactions are

$$1. \quad B + A \rightarrow M$$

$$2. \quad M + A \rightarrow D$$

Assume that the density of the system is constant and that the specific reaction rates for the true reactions are

$$k_1 = A_1 e^{-E_1/R_g T} \tag{5-18}$$

$$k_2 = A_2 e^{-E_2/R_g T} \tag{5-19}$$

The problem is to find the constant operating temperature in a tubular-flow reactor for which the yield of M will be a maximum. We would also like to know the value of the yield and the overall selectivity of M with respect to D at this condition. For these objectives it is not necessary to know the individual values of A_1, A_2, E_1, and E_2 in Eqs. (5-18) and (5-19), but only the ratio k_2/k_1. Suppose this is defined by

$$\frac{A_2}{A_1} = 7.95 \times 10^{11} \tag{5-20}$$

$$E_2 - E_1 = 16{,}700 \text{ cal/g mol} \tag{5-21}$$

and that we take the permissible range of operating temperatures as 5 to 45°C.† Since $E_1 < E_2$, it is anticipated that the lowest temperature, 5°C, would give the best yield of M. We wish to verify this and calculate the maximum yield and the selectivity.

The reactions represent a common type of substitution reaction; for example, the successive chlorination of benzene considered in Example 2-9 is of this form. In that example yields of primary and secondary products were obtained for an isothermal batch reactor. As pointed out in Sec. 3-4, the results for batch reactors may be used for tubular-flow reactors if the time is replaced by the *residence* time. In Example 2-9 the yield equations were expressed in terms of fraction of benzene unreacted rather than time. Identical results apply for the tubular-flow reactor. The yield of M is given by Eq. (F) of Example 2-9,‡ the monoadduct replacing the monochlorobenzene; that is,

$$x_M = \frac{C_M}{C_{B_0}} = \frac{1}{1 - \beta}\left[\left(\frac{C_B}{C_{B_0}}\right)^\beta - \frac{C_B}{C_{B_0}}\right] \tag{5-22}$$

where

$$\beta = \frac{k_2}{k_1} = \frac{A^2 e^{-E_2/R_g T}}{A_1 e^{-E_1/R_g T}} = 7.95 \times 10^{11} e^{-16{,}700/R_g T} \tag{5-23}$$

The yield of monoamine, M, described by Eq. (5-22), has a maximum value at an intermediate value of C_B/C_{B_0}, or at an intermediate value of the conversion $(1 - C_B/C_{B_0})$. We first find this local maximum, and then express it as a function of β. Once this is done, we can find the value of β, and hence the temperature, that gives the overall maximum within the allowable temperature range.

To determine the local maximum, let us set the derivative of x_M with respect to C_B/C_{B_0} (call this z) equal to zero; that is,

$$\frac{dx_M}{dz} = 0 = \frac{1}{1 - \beta}[\beta(z)^{\beta-1} - 1]$$

or

$$z = C_B/C_{B_0} = \left(\frac{1}{\beta}\right)^{1/(\beta-1)} \tag{5-24}$$

---

† Probably the rates are unrealistically high for this temperature range.

‡ Actually, by the equation immediately preceding Eq. (F). The same result is also given by Eq. (2-89), which was developed for consecutive first-order reactions. As shown in Example 2-8, in this case the results are the same for first- or second-order kinetics.

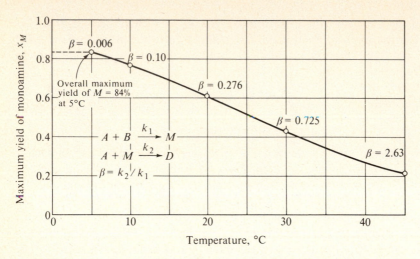

Figure 5-7 Maximum yield of intermediate product (M) for consecutive reactions.

This is the value of C_B/C_{Bo} for the local maximum of x_β. Substituting this value in Eq. (5-22) gives the local maximum in terms of β,[†]

$$(x_M)_{\max} = \frac{1}{1-\beta}\left[\left(\frac{1}{\beta}\right)^{1/(\beta-1)}\left(\frac{1}{\beta}-1\right)\right] = \left(\frac{1}{\beta}\right)^{\beta/(\beta-1)} \qquad (5\text{-}25)$$

From Eq. (5-23) we can find the value of β for any temperature over the range 5 to 45°C. Substituting these results for β in Eq. (5-25) gives the corresponding local-maximum yields of M. The results are shown in Fig. 5-7, plotted as $(x_M)_{\max}$ vs. temperature. As expected, the overall maximum occurs at 5°C and is about 0.84. The corresponding value of β is 0.06.

The total conversion at maximum yield conditions is, from Eq. (5-24),

$$x_t = 1 - C_B/C_{Bo} = 1 - \left(\frac{1}{\beta}\right)^{1/(\beta-1)}$$

$$= 1 - \left(\frac{1}{0.06}\right)^{1/(0.06-1)} = 1 - 0.05 = 0.95 \qquad \text{or } 95\%$$

From the stoichiometry of the reactions, the mass balance for isobutylene is

$$C_B = C_{Bo} - C_M - C_D$$

or

$$\frac{C_B}{C_{Bo}} = 1 - \frac{C_M}{C_{Bo}} - \frac{C_D}{C_{Bo}}$$

$$x_D = \frac{C_D}{C_{Bo}} = 1 - \frac{C_B}{C_{Bo}} - \frac{C_M}{C_{Bo}} = x_t - x_M \qquad (5\text{-}26)$$

[†] This same result was obtained in Example for 2-8, Eq. (D), for a batch reactor.

Hence

$$x_D = 0.95 - 0.84 = 0.11$$

Of the total conversion of 95%, 84% is to monoamine and 11% to diamine. The overall selectivity of monoamine to diamine is $0.84/0.11 = 7.6$ at the maximum yield of monoamine. The selectivity is not necessarily a maximum at the conditions of maximum yield. If the separation of M from D in the product stream requires an expensive process, it might be more profitable to operate the reactor at a conversion level at which M is less than 84% but the selectivity is higher than 7.6. The selectivity may be evaluated for any total conversion and β from Eqs. (5-22) and (5-26); thus the selectivity is

$$S_o = \frac{x_M}{x_D} = \frac{1/(1-\beta)[(1-x_t)^\beta - (1-x_t)]}{x_t - 1/(1-\beta)[(1-x_t)^\beta - (1-x_t)]} \tag{5-27}$$

Equation (5-27) indicates that S_o decreases sharply with x_t. The relationship is shown for 5°C by the dotted curve in Fig. 5-8. This curve suggests that if a selectivity of 20 is required to reduce separation costs, the reactor should be designed to give a conversion of about 73%, at which point the yield of monoamine is only 70%. In this case the maximum yield occurs at a conversion of 95%, while the maximum selectivity is at a conversion approaching zero. Figure 5-8 also includes the curves for yield vs. total conversion for several temperatures.

Note that as conditions have been attained for the maximum yield of M, the reactor has become relatively large. At 5°C the reaction rate will be low and the total conversion is approaching 100%. Whether it is advisable to operate at these

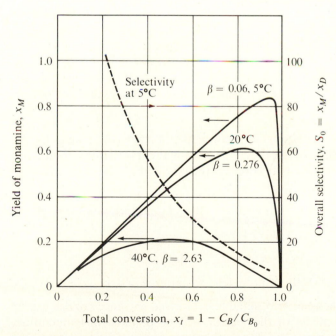

Figure 5-8 Yield of intermediate product (M) vs. total conversion for consecutive reactions.

conditions for maximum conversion depends on the economics of reactor costs, separation costs, and price of the monoamine product.

We can predict qualitatively what effect nonisothermal operations would have on the selectivity. For example, if the reactions were exothermic, it would be preferable to have as much heat transfer as possible in order to operate at low temperatures. Furthermore, a tubular-flow reactor would be preferable to a stirred tank (opposite conclusion from that of Example 5-2) since the latter would favor the production of diamine. This is because the concentration of monoamine would be higher and the temperature is also likely to be higher.

5-4 Continuous Stirred-Tank Reactors

The ideal stirred-tank reactor operates isothermally at a constant rate. However, an energy balance is needed to predict the constant temperature when the heat of reaction is sufficient (or the heat exchange between the surroundings and reactor is insufficient) to cause a difference between feed and reactor temperatures. The proper energy balance is the algebraic relation presented as Eq. (5-4) and the mass balance is Eq. (3-13). Sometimes simultaneous solution of the mass and energy equations is required. For example, if a reactor of known volume is to be used, the calculation of the conversion and the temperature requires a simultaneous trial-and-error solution of Eqs. (3-13) and (5-4), with the rate equation expressing the temperature dependency of r. In contrast, if the objective is the reactor volume needed to obtain a specified conversion, Eq. (5-4) can be solved independently for the reactor-exit temperature T_e. The rate at T_e can be found from the rate equation and used in Eq. (3-13) to obtain the reactor volume. When more than one reaction occurs, a separate mass balance is necessary for each reaction.

A stirred tank may give better or worse selectivities than a tubular-flow unit in multiple-reaction systems. As usual, the key point is the relative values of the activation energies. In particular, for a set of exothermic reactions, where the desired product is formed by the reaction with the higher activation energy, the stirred tank is advantageous. The production of allyl chloride considered in Example 5-2 is an example. The performance of a stirred-tank reactor for this system is discussed next.

Example 5-3 Consider the design of a continuous stirred-tank reactor for the production of allyl chloride from propylene, using the reaction-rate data given in Example 5-2. So that we may compare the two types of reactors, the same feed condition will be employed:

$$F_t = 0.85 \text{ mol/h } (2.36 \times 10^{-7} \text{ kg mol/h})$$

$$T = 200°C \text{ (473 K)}$$

$$p = 29.4 \text{ lb/in.}^2 \text{ abs (202 kPa)}$$

Molal ratio of propylene to chlorine = 4.0

The operation is adiabatic. Also, suitable baffles and entrance nozzles will be used so that the contents of the reactor will be of uniform temperature, pressure, and composition, even for this gaseous system.

Using the heats of reaction and heat capacities given in Example 5-2, determine the conversion of chlorine to allyl chloride expected for a range of sizes of reactors (i.e., reactor volumes).

SOLUTION The rate of each reaction (allyl chloride and dichloropropane formation) will be a constant and should be evaluated at the temperature and composition of the steam leaving the reactor. The temperature is determined by Eq. (5-4). The enthalpy rates are constant (steady-state operation). They can be written immediately for the feed and effluent streams by applying Equation (5-13) to a stirred-tank reactor. The rate is the same throughout the reactor and F_t and c_p are also constant. Hence, Eq. (5-13) can be integrated between feed and effluent streams to give

$$H'_e - H'_f = F_t c_p(T_e - T_f) + \Delta H_R(\mathbf{r} V) \tag{5-28}$$

The mass balance, Eq. (3-13) can be used to eliminate the rate from Eq. (5-28) and express the enthalpy change in terms of conversion

$$H'_e - H'_f = F_t c_p(T_e - T_f) + F\,\Delta H_R(x_e - x_f) \tag{5-29}$$

Substituting this expression in Eq. (5-4) yields

$$F_t c_p(T_e - T_f) + F\,\Delta H_R(x_e - x_f) = h_o A_h(T_s - T_e) \tag{5-30}$$

where the heat transfer rate Q' has been expressed in terms of the temperature difference $(T_s - T_e)$ as in Eq. (5-14). Equation (5-30) is a general expression for the energy balance in a continuous-flow stirred-tank reactor. If the feed stream is unreacted, $x_f = 0$. Also, note that F and F_t are the feed rate of reactant and the total feed rate. In this example there are two reactions and $Q' = 0$. Hence, Eq. (5-30) becomes

$$F_t c_p(T_e - T_f) + F(\Delta H_{R_1} x_1 + \Delta H_{R_2} x_2) = 0$$

If, as in Example 5-2, the conversions are based upon the total feed rate, the energy balance becomes

$$F_t c_p(T_e - T_f) + F_t(\Delta H_{R_1} x_{t_1} + \Delta H_{R_2} x_{t_2}) = 0 \tag{A}$$

If only one reaction were involved, Eq. (A) could be used to calculate T_e for a series of arbitrarily chosen conversions $(x_t)_1$. Then each value of $(x_t)_1$ and T_e would fix a rate $\mathbf{r}_1$. Substitution of these rates in the mass balance, Eq. (3-13), would give the reactor volumes corresponding to the conversions. However, in this case two equations of the form of (3-13), one for each reaction, must be satisfied,

$$(x_t)_1 - 0 = \mathbf{r}_1 \frac{V}{F_t} \tag{B}$$

$$(x_t)_2 - 0 = \mathbf{r}_2 \frac{V}{F_t} \tag{C}$$

The rate expressions are obtained by substituting T_e for the temperature in Eqs. (F) and (G) of Example 5-2. Thus

$$r_1 = 824{,}000e^{-13{,}700/T_e} \frac{[0.8 - (x_t)_1 - (x_t)_2][0.2 - (x_t)_1 - (x_t)_2]}{[1 - (x_t)_2]^2} \quad \text{(D)}$$

$$r_2 = 46.8e^{-3{,}460/T_e} \frac{[0.8 - (x_t)_1 - (x_t)_2][0.2 - (x_t)_1 - (x_t)_2]}{[1 - (x_t)_2]^2} \quad \text{(E)}$$

Algebraic Eqs. (A) to (E) provide the relations needed to calculate the five unknowns $(x_t)_1$, $(x_t)_2$, T_e, r_1, and r_2 at different values of the reactor volume V. One procedure which is not tedious is first to choose a value of T_e. Then, from the ratio of Eqs. (B) and (C), using (D) and (E) for r_1 and r_2, we obtain the ratio $(x_t)_1/(x_t)_2$. Employing this ratio in Eq. (A) will give separate values for each conversion. Finally, the corresponding reactor volume can be obtained from either Eq. (B) or Eq. (C). This approach will be illustrated by including the numerical calculations for an exit temperature of 1302°R (450°C),

$$\frac{(x_t)_1}{(x_t)_2} = \frac{r_1}{r_2} = \frac{824{,}000e^{-13{,}700/1{,}302}}{46.8e^{-3{,}460/1.302}} = 6.77$$

Using this ratio in Eq. (A) and noting that the heat capacity of the feed $F_t c_p$ was determined in Example 5-2 as 18.5 Btu/(h)(°F), we have

$$18.5(T_f - T_e) + 0.85[6.77(x_t)_2(48{,}000) + (x_t)_2(79{,}200)]$$

$$= 18.5(852 - 1{,}302) + 343{,}000(x_t)_2 = 0$$

$$(x_t)_2 = 0.0243$$

$$(x_t)_1 = 6.77(0.0243) = 0.164$$

The reactor volume required for these conversions is, from Eq. (B),

$$0.164 = \frac{V}{0.85}(824{,}000e^{-13{,}700/1.302})$$

$$\times \frac{(0.8 - 0.164 - 0.024)(0.2 - 0.164 - 0.024)}{(1 - 0.0243)^2}$$

$$V = \frac{0.164(0.85)}{0.167} = 0.83 \text{ ft}^3 \ (0.023 \text{ m}^3)$$

The corresponding values of the conversions and volume for other temperatures are summarized in Table 5-6. Comparing the results with those of Example 5-2, we see that the adiabatic stirred-tank reactor gives much higher yields and selectivities for allyl chloride than the tubular-flow type for the same reactor volume. In the tubular-flow equipment considerable dichloropropane is formed in the initial sections of the reactor, where the temperature is relatively low. This is avoided in the adiabatic tank reactor by operation at a constant temperature high enough to favor allyl chloride formation. For example, if the adiabatic tank reactor is operated at 450°C, 82% of the chlorine is converted to allyl chloride and 12% is converted to dichloropropane;

Table 5-6 Conversion vs. reactor volume for adiabatic tank reactor: allyl chloride production

Reactor (or exit) temperature		Conversion per mole feed		Conversion of chlorine in feed		Reactor volume, ft^3	Selectivity $S_o = x_1/x_2$
°R	°C	$(x_t)_1$	$(x_t)_2$	x_1	x_2		
960	260	0.0098	0.0237	0.049	0.119	0.12	0.41
1032	300	0.0282	0.0324	0.14	0.162	0.15	0.86
1122	350	0.0660	0.0341	0.33	0.171	0.18	1.93
1212	400	0.114	0.0298	0.57	0.149	0.24	3.82
1257	425	0.138	0.0273	0.69	0.136	0.34	5.07
1302	450	0.164	0.0243	0.82	0.121	0.83	6.78

the total conversion is 94%. In the adiabatic tubular reactor of Example 5-2 the products contained much greater amounts of dichloropropane for all reactor volumes. These conclusions are summarized in Fig. 5-9, where the fraction of chlorine converted to each product is shown plotted against reactor volume for the tubular and the tank reactors.

In the tubular-flow reactor in Example 5-2 it was necessary to solve simultaneously three differential equations, while for the tank reactor only simultaneous solution of algebraic equations was required.

Examples 5-2 and 5-3 are for a pair of exothermic, *parallel* reactions where the reaction producing the desired product has the higher activation energy. The results show that for both selectivity and conversion a stirred-tank reactor is preferred over the tubular-flow type.

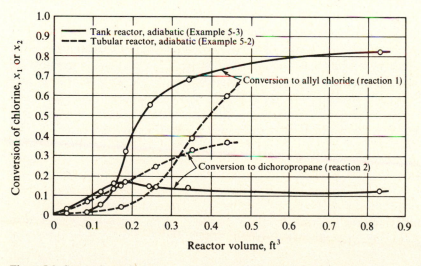

Figure 5-9 Comparison of tubular and tank reactors for allyl chloride production.

The most significant factors in analyzing the selectivity in nonisothermal reactors are the activation energies. Suppose that the activation energy for the reaction producing the desired product is greater than the E values for the other reactions. Then higher temperatures increase the selectivity. For example, for exothermic reactors adiabatic operations either stirred tank or tubular flow gives a higher selectivity than isothermal operation. If energy is added, the selectivity will be increased above that in the adiabatic reactor for either exothermic or endothermic reactions. Example 5-2 is an illustration for exothermic reactions. The key point is that the temperature should be as high as possible. The same conclusion applies for either parallel or consecutive reactions. The selectivity comparison between stirred tank and tabular flow depends upon activation energies, and, in general, on the kinetics of the reactions and whether they are exothermic or endothermic.

When the desired product is produced by the reaction with the lowest activation energy, selectivity (but not conversion) is favored by lower temperatures. Hence, for endothermic reactions adiabatic operation gives a higher selectivity than an isothermal reactor. For exothermic reactions transfer of energy to the surroundings improves selectivity. Note that if the temperatures become too low, the very low rates may result in unrealistically large reactor volumes. The procedures illustrated in Examples 5-2 and 5-3 are applicable for evaluating selectivity and conversion for any type of reaction system† for either stirred-tank or tubular-flow reactors.

5-5 Stable Operating Conditions in Stirred-Tank Reactors‡

In Example 5-3 the temperature and conversion leaving the reactor were obtained by simultaneous solution of the mass and energy balances. The results for each temperature in Table 5-6 represented such a solution and corresponded to a different reactor, i.e., a different reactor volume. However, the numerical trial-and-error solution required for this multiple-reaction system hid important features of reactor behavior. We now reconsider the *steady-state* performance of a stirred-tank reactor for a simple single-reaction system.

Suppose an exothermic irreversible reaction with first-order kinetics is carried out in an adiabatic stirred-tank reactor, as shown in Fig. 5-10. The permissible, or stable, operating temperatures and conversions can be shown analytically and graphically for this system by combining the mass and energy balances. If the density is constant, Eq. (4-17) represents the steady-state mass balance for this case. Since V/Q is the average residence time, Eq. (4-17) may be written

$$x = \frac{k\bar{\theta}}{1 + k\bar{\theta}} \tag{5-31}$$

---

† See Octave Levenspiel, "Chemical Reaction Engineering," 2d ed., chap. 8, John Wiley & Sons, New York, 1972. Also K. G. Denbeigh, "Chemical Reactor Theory," chaps. 5 and 6, Cambridge University Press, 1956.

‡ For more complete discussions of this subject see C. van Heerden, *Ind. Eng. Chem.*, **45**, 1242 (1953); K. G. Denbigh, *Chem. Eng. Sci.*, **8**, 125 (1958).

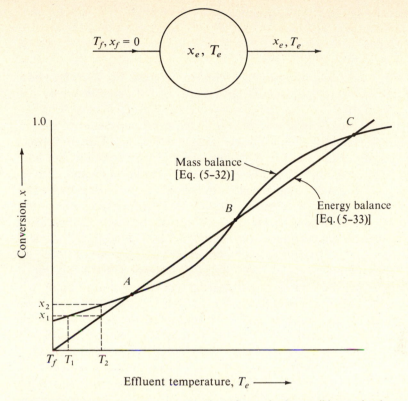

Figure 5-10 Temperature vs. conversion for a first-order irreversible reaction in an adiabatic stirred-tank reactor.

In terms of temperature, this takes the form

$$x = \frac{\bar{\theta}\mathbf{A}e^{-E/R_g T_e}}{1 + \bar{\theta}\mathbf{A}e^{-E/R_g T_e}} \tag{5-32}$$

where E is the activation energy and $\mathbf{A}$ is the frequency factor. At a fixed $\bar{\theta}$ (that is, for a given reactor), Eq. (5-32) expresses the result of the combined effects of temperature and reactant concentration on the rate as referred to in Fig. 5-2. At low conversion levels the conversion increases approximately exponentially with temperature, since the exponential term in the denominator is small with respect to unity. At high temperatures the reactant composition, and hence the rate, approaches zero; then the exponential term dominates the denominator and the conversion approaches a constant value. For an irreversible reaction Eq. (5-32) shows that this value is unity (100% conversion). The S-shaped curve representing this T-vs.-x relationship is sketched in Fig. 5-10.

The steady-state energy balance [Eq. (5-30)] for adiabatic operation and zero conversion in the feed is

$$x_e - 0 = -\frac{F_t c_p(T_e - T_f)}{F(\Delta H_R)}$$

For constant-density systems $F_t = \rho Q$ and $F = C_f Q$, so that

$$x_e = \frac{-\rho c_p}{C_f(\Delta H_R)}(T_e - T_f) \tag{5-33}$$

Since the heat of reaction usually varies little with temperature, Eq. (5-33) shows a linear relationship between $T_e - T_f$ and conversion and is represented by the straight line in Fig. 5-10.

For a given reactor and kinetics the operating temperature and conversion in the effluent stream are established by the simultaneous solution of Eqs. (5-32) and (5-33). The intersections shown in Fig. 5-10 indicate that such solutions are possible at three points, A, B, and C. Stable steady-state operation cannot exist at other temperatures. For example, suppose that the initial temperature is below the temperature at point A, say, at T_1. Figure 5-10 shows that the conversion required by the mass balance is greater than that corresponding to the energy balance. Thus the energy evolved at the conversion x_1, corresponding to T_1 from Eq. (5-32), will raise the temperature to T_2. This is the temperature corresponding to x_1 in Eq. (5-33). The conversion at T_2 is x_2, from Eq. (5-32). Hence a further heating of the reaction mixture occurs. This transient heating will continue until point A is reached. At initial temperatures between points A and B the reaction rate is too small to justify steady-state operation, and the reaction mixture cools to the point A. Initial temperatures between B and C would be similar to temperatures below point A, so that transient heating will occur until point C is reached. At initial temperatures above C transient cooling will take place until the temperature drops to C. The temperature-vs.-time relationships may be evaluated by solving the mass and energy balances for non-steady-state operation. This is done in Sec. 5-6.

Point B is different from A and C. After small initial displacements from B the system does not return to B, whereas disturbances from points A and C are followed by a return to these stable points.

The relative position of the mass-balance curve and the energy-balance line in Fig. 5-10 depends on the chemical properties (A, E and ΔH_R) and physical properties (ρ and c_p) of the system and the operating conditions ($\bar{\theta}$ and C_f), according to Eqs. (5-32) and (5-33). These properties and conditions determine whether or not stable operating conditions are possible and how many stable operating points exist. For example, consider a series of experiments in which the only property that changes is C_f. Since θ, A, and E are constant, the relation between x and T determined by Eq. (5-32) is fixed, as shown by the S-shaped curve in Fig. 5-11. For experiment 1, which has the lowest C_f, the energy-balance line will be steep and will intersect the mass-balance curve at a small value of $T_e - T_f$. This point is the only stable operating condition. The reactor will operate at a low conversion and at a temperature only slightly above the feed temperature. For experiment 2, with an intermediate value of C_f, there will be two stable operating conditions, points A and C in Fig. 5-11, and a metastable point, B. This is the situation described in Fig. 5-10. Experiment 3 is with a large feed concentration; there is just one intersection, and this occurs at nearly complete conversion. As shown in Fig. 5-11, point D, the reactor temperature is far above T_f.

Stability behavior in heterogeneous reactions is similar. For example, the

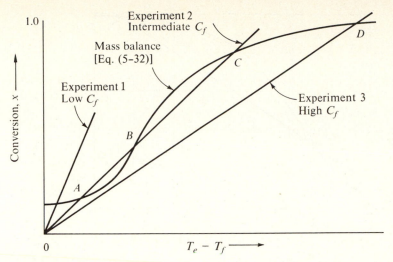

Figure 5-11 Temperature rise vs. conversion as a function of heat of reaction in an adiabatic stirred-tank reactor.

interaction of chemical and physical processes (rates of chemical reaction and mass and heat transfer) in reactions between gases and solid particles causes results analogous to those discussed here. We shall return to this subject in Chap. 10.

Calculations of stable operating conditions are illustrated in the following example.

Example 5-4 A first-order homogeneous (liquid-phase) reaction is carried out in an ideal stirred-tank reactor. The concentration of reactant in the feed is 3.0 kmol/m³ and the volumetric flow rate is 60×10^{-6} m³/s. The density and specific heat of the reaction mixture are constant at 10^3 kg/m³ and 4.19 kJ/kg K (or 1.0 cal/g K) respectively. The reactor volume is 18×10^{-3} m³. There is no product in the feed stream and the reactor operates adiabatically. The heat and rate of reaction are

$$\Delta H_R = -2.09 \times 10^8 \text{ J/k mol (or } -50{,}000 \text{ cal/g mol)}$$

$$r = 4.48 \times 10^6 \text{ C exp}\left(-\frac{62{,}800}{R_g T}\right), \text{ k mol/(m)}^3\text{(s)}$$

where C is the reactant concentration in k mol/m³; T is in degrees Kelvin, and $R_g = 8.314$ J/(g mol)(K). The activation energy in these SI units is 62,800 J/g mol. If the feed stream is at 298 K, what are the steady-state conversions and temperatures in the product stream?

SOLUTION Stable operating conditions are given by the simultaneous solution of Eqs. (5-32) and (5-33). Since $A = 4.48 \times 10^6$ s⁻¹, the dimensionless quantity $\bar{\theta} A$ is

$$\bar{\theta} A = \frac{18 \times 10^{-3}}{60 \times 10^{-6}} (4.48 \times 10^6) = 1.34 \times 10^9$$

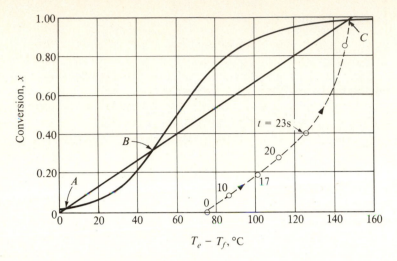

Figure 5-12 Temperature rise vs. conversion.

Then Eq. (5-32) becomes

$$x_e = \frac{1.34 \times 10^9 \exp\left(-62{,}800/R_g T\right)}{1 + 1.34 \times 10^9 \exp\left(-62{,}800/R_g T\right)} \tag{A}$$

This mass-balance relation between x_e and T_e is shown as the S-shaped curve in Fig. 5-12. With the given numerical values, Eq. (5-33) is

$$x_e = \frac{-(10^3)(4.19) \times 10^3}{3.0(-2.09 \times 10^8)} (T_e - T_f) = \tfrac{1}{150}(T_e - T_f) \tag{B}$$

This energy-balance relation is represented by the straight line in Fig. 5-12.

The intersections of the two curves in Fig. 5-12 give the possible operating temperatures and conversions. The intersections at both ends of the curves are stable operating points and give the results shown in Table 5-7. Point B at $T = 298 + 48 = 346$ K$(73°C)$ represents a metastable point, as described in connection with Fig. 5-10. If the reaction had been started in the usual way by adding feed at 25°C, point A would be the steady-state operating condition at $25 + 3 = 28°C$ and the conversion would be only 1.5%. To obtain a high conversion the initial temperature would have to be above 73°C (point B). Then the reaction temperature would increase to $25 + 147 = 172°C$ and the conversion would be 98%. The dynamic behavior of the reactor (dotted curve), showing the temperature-time relation as this steady state is approached, is evaluated in Example 5-5.

Table 5-7

Intersection	$T_e - T_f$, °C	T, K	Conversion, %
A	3	301	1.5
C	147	445	98.0

5-6 Semibatch Reactors

There are many variations of semibatch operating conditions. Mass flow rates or feed temperatures can change with time. Also, heat exchange rates between reactor and surroundings can be a function of time. These dynamic conditions arise either involuntarily, as in startup or shutdown periods for continuous-flow reactors, or they may be voluntarily imposed to achieve desired reactor behavior. In Sec. 4-8, semibatch operation of *stirred-tank* reactors was considered for isothermal conditions. Here we will again limit the treatment to stirred-tank reactors but investigate different types of nonisothermal operation.

First, consider a simple example of the startup period for a continuous-flow reactor (Fig. 5-13). The mass flow rates and temperature and composition of the feed stream do not vary with time, but the initial temperature, T_0, of the reactor contents is different from the feed temperature T_f. Under these conditions the reactor temperature and composition of the product stream will change with time until a steady-state condition is reached. Suppose also that the reactor operates adiabatically. Then Eq. (5-5) becomes

$$(H'_f - H'_e)\, dt = dH \tag{5-34}$$

We have already developed expressions for the enthalpy changes. Equation (5-6) for a *batch* reactor gives the change in enthalpy of the contents of the reactor in time dt. Equation (5-29) is applicable for the difference in enthalpy rates of the feed and effluent streams in a stirred-tank *flow* reactor. Hence, Equation (5-34) may be written

$$F_t c_p(T_f - T_e) + F\, \Delta H_R(x_f - x_e) = m_t c_p \frac{dT_e}{dt} + \Delta H_R(\mathbf{r}V) \tag{5-35}$$

The term $(\mathbf{r}V)$ on the right-hand side is the rate of *disappearance* of reactant in the vessel. It may be expressed in terms of conversion by using the mass balance for a batch reactor. Suppose the volume of the reaction mixture is constant. Then Eq. (3-8) is applicable so that Eq. (5-35) becomes

$$F_t c_p(T_f - T_e) + F\, \Delta H_R(x_f - x_e) = m_t c_p \frac{dT_e}{dt} + \Delta H_R V C_0 \frac{dx}{dt} \tag{5-36}$$

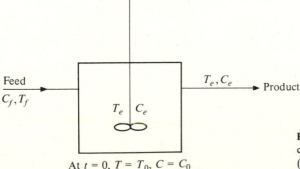

Feed
C_f, T_f

$T_e \quad C_e$

T_e, C_e → Product

At $t = 0$, $T = T_0$, $C = C_0$

Figure 5-13 Startup period for a continuous-flow, stirred-tank reactor (C = reactant concentration).

where C_0 is the *initial* reactant concentration in the reactor. Equations (5-35) or (5-36) are suitable forms of the energy balance for this startup problem. One of them, along with the appropriate mass balance developed in Chap. 4 [for example, Eq. (4-22)],† and the initial conditions of reactant concentration and temperature, are sufficient to determine how the effluent conversion x_e and temperature T_e vary with time.

Ultimately, a steady state is reached at which time the dynamic equations reduce to Eqs. (5-32) and (5-33) of Sec. 5-5. As noted in Sec. 5-5, more than one steady state may be possible.

Example 5-5 shows how Eqs. (5-36) and (4-22) can be used to evaluate the dynamic behavior ultimately leading to the steady-state solutions discussed in Example 5-4.

Example 5-5 Consider the dynamic behavior of the reactor and reaction system of Example 5-4. Initially, the reactor contains 18 liters (18×10^{-3} m³) of solution with a reactant concentration of 3.0 mol/liter (3.0 kmol/m³). The initial temperature is 373°C. Determine the concentration and temperature of the reactor effluent as a function of time. What is the temperature and conversion in the effluent at steady state? The feed temperature, concentration, and flow rate are the same as in Example 5-4.

SOLUTION Since the density is constant, Q and V are constant. Then the mass balance of reactant [Eq. (4-22)] simplifies to:

$$C_f - C_e - r(V/Q) = (V/Q)\frac{dC_e}{dt}$$

where r is the rate of *disappearance* of reactant. In terms of the average residence time, $\bar{\theta} = V/Q$, and the first-order rate equation, this expression becomes

$$\frac{dC_e}{dt} = \frac{C_f - C_e}{\bar{\theta}} - kC_e \qquad \text{(A)}$$

where k is a function of temperature. Since the conversion in the feed is zero, the relationship between C_e and x_e is

$$x_e = \frac{C_f - C_e}{C_f} \qquad \text{(B)}$$

and

$$\frac{dC_e}{dt} = -C_f\frac{dx_e}{dt} \qquad \text{(C)}$$

Substituting Eqs. (B) and (C) in Eq. (A) yields

$$\frac{dx_e}{dt} = f_1(x_e, T_e) = -\frac{x_e}{\bar{\theta}} + k(1 - \dot{x}_e) \qquad \text{(D)}$$

---

† For the case of constant mass flow rates considered in this section only Eq. (4-22), the species mass balance, and not the total balance, Eq. (4-23), is required.

In this example $C_0 = C_f$ so that $C_0\, dx$ in Eq. (5-36) is equal to $C_f\, dx_e$. With this equality, Eq. (5-36), becomes

$$F_t c_p (T_f - T_e) - F(\Delta H_R)x_e = m_t c_p \frac{dT_e}{dt} + \Delta H_R V C_f \frac{dx_e}{dt} \qquad \text{(E)}$$

For constant density, $F_t = Q\rho$, $F = QC_f$, and $m_t = V\rho$. Therefore, Eq. (E) may be written

$$\frac{dT_e}{dt} = f_2(x_e, T_e) = \frac{T_f - T_e}{\bar\theta} - C_f \frac{(\Delta H_R)}{\rho c_p}\left[\frac{x_e}{\bar\theta} + \frac{dx_e}{dt}\right] \qquad \text{(F)}$$

By substituting Eq. (D) for dx_e/dt, Eq. (F) takes the form

$$\frac{dT_e}{dt} = \frac{T_f - T_e}{\bar\theta} - \frac{\Delta H_R}{\rho c_p} kC_f(1 - x_e) \qquad \text{(G)}$$

with initial condition

$$T_e = T_0 \quad \text{and} \quad C_e = C_0 = C_f \quad \text{at} \quad t = 0 \qquad \text{(H)}$$

Equations (D) and (F) can be solved numerically with (H). If we use the Runge-Kutta method for two dependent variables, x_e and T_e, the working equations for the nth element of time Δt are (following Example 4-7):

$$k_0 = (\Delta t)f_1(x_{e_n}, T_{e_n})$$

$$m_0 = (\Delta t)f_2(x_{e_n}, T_{e_n})$$

etc., for k_1, k_2, k_3, and m_1, m_2, m_3, as in Example 4-7. Then

$$x_{e_{n+1}} = x_{e_n} + \tfrac{1}{6}(k_0 + 2k_1 + 2k_2 + k_3)$$

$$T_{e_{n+1}} = T_{e_n} + \tfrac{1}{6}(m_0 + 2m_1 + 2m_2 + m_3)$$

The numerical values of the constants are:

$$C_f = 3.0 \times 10^{-3} \text{ g mol/cm}^3 (3.0 \text{ k mol/m}^3)$$

$$T_0 = 373 \text{ K}$$

$$T_f = 298 \text{ K}$$

$$\frac{(C_f)\,\Delta H_R}{\rho C_p} = \frac{(3 \times 10^{-3})(-5 \times 10^4)}{1.0(1.0)} = -150 \text{ K}$$

$$\bar\theta = V/Q = \frac{18000}{60} = 300 \text{ s}$$

Also,

$$k = 4.48 \times 10^6 \exp\left(-\frac{62{,}800}{R_g\, T_e}\right) \text{ s}^{-1}$$

where $R_g = 8.314$ J/(g mol)(K).

Using these values and choosing $\Delta t = 10$ s, we have for the first increment, from Eq. (D),

$$k_0 = 10 f_1(x_{e_0}, T_{e_0}) = 10 \left[0 + 4.48 \times 10^6 \exp\left(\frac{-62,800}{R_g(373)}\right)(1-0)\right] = 0.071$$

and from Eq. (F)

$$m_0 = 10 f_2(x_{e_0}, T_{e_0}) = 10 \left[\frac{T_f - T_e}{\theta} - \frac{C_f \Delta H}{\rho c_p}\left(\frac{x_e}{\theta} + \frac{dx_e}{dt}\right)\right]$$

$$= 10 \left[\frac{298 - 373}{300} + 150(0 + 0.0071)\right]$$

$$m_0 = 10 \left|\frac{-75}{300} + 1.07\right| = 8.2 \text{ K}$$

Continuing the calculations to obtain k_1, k_2, and k_3 and m_1, m_2, and m_3 [using equations analogous to Eqs. (4-12) to (4-15)] yields the following results for the first time increment:

$$x_{e_1} = 0 + \tfrac{1}{6}[0.071 + 2(0.084) + 2(0.088) + 0.112] = 0.088$$

Hence, the conversion at the end of 10 seconds will be 8.8%, and the temperature is

$$T_{e_1} = 373 + \tfrac{1}{6}[8.2 + 2(10.3) + 2(10.8) + 14.4] = 373 + 11 = 386 \text{ K}$$

More accurate calculations, taking $\Delta t = 0.1$ s give $x_e = 0.091$ and $T_e = 384$ K after 10 seconds. Similar calculations for further time increments give the results shown in Fig. 5-12 by the dotted curve. The curve suggests a steady-state conversion corresponding to point C, as found in Example 5-4. These values can be checked by considering the steady-state forms of the mass and energy balances, Eqs. (D) and (F). The time derivatives will be zero so that these two equations reduce to

$$x_e = \frac{k\bar\theta}{1 + k\bar\theta} \tag{I}$$

and

$$x_e = \frac{\rho c_p}{C_f \Delta H_R}(T_f - T_e) \tag{J}$$

These expressions are the same as Eqs. (5-31) and (5-33) developed in Sec. 5-5 for the steady-state solution. Hence, simultaneous solution for T_e and x_e will give 445 K (172°C) and 98% conversion, as found in Example 5-4. The steady-state corresponds to the case where the energy evolved due to the reaction just balances the energy required to heat the feed to the reaction temperature T_e. In this example we started with $T_e = T_0 = 373$ K so that the steady-state temperature 445 K was higher than T_0. If T_0 had been below, $73 + 273 = 344$ K, the steady-state value would have been lower (301 K), as shown in Example 5-4.

As a second illustration of semibatch operation, consider a stirred tank in which one reactant is in the initial charge and the second is continuously fed to the reactor. This arrangement permits flexibility in temperature control. It is useful when the heat of reaction is high and temperatures must be controlled within narrow limits. By reducing the feed rate the temperature change in the reactor can be maintained within any required value. Suppose there is no product stream. Then Fig. 4-18 describes the reactor and Example 4-15 treated the design calculations for an isothermal case. Here we consider a special case of nonisothermal operation. Equation (5-5) is applicable with $H'_e = 0$. Equation (5-6) gives dH. To evaluate H'_f it is convenient to take, as a reference for enthalpy, zero conversion and the reactor temperature T_e (a variable). Suppose that the conversion in the feed stream is zero. Then H'_f above the reference state is

$$H'_f = F_t c_p (T_f - T_e) \tag{5-37}$$

Substituting these results in Equation (5-5) gives

$$F_t c_p (T_f - T_e)\, dt + U A_h (T_s - T_e)\, dt = m_t c_p\, dT + \Delta H_R (\mathbf{r} V)\, dt$$

or

$$F_t c_p (T_f - T_e) + U A_h (T_s - T_e) = m_t c_p \frac{dT}{dt} + \Delta H_R (\mathbf{r} V) \tag{5-38}$$

where the heat transferred from the surroundings to the reactor is written in terms of an overall heat transfer coefficient U and the heat transfer area A_h.

When temperature control is critical, the reaction rate is often very fast with respect to the rate of heat transfer. Then the factors that determine the design are the rate of energy exchange with the surroundings, the feed temperature, and the feed rate. Under these circumstances only the energy balance is required. The reaction may be assumed to be at thermodynamic equilibrium, so that the rate equation and mass balance are not required. The rate of reaction in the entire reactor is equal to the rate of addition of reactant in the feed stream. The quantity $\mathbf{r} V$ is equal to $C_f Q_f$, and Eq. (5-38) becomes

$$m_t c_p \frac{dT_e}{dt} = F_t c_p (T_f - T_e) + U A_h (T_s - T_e) - \Delta H_R (Q_f C_f) \tag{5-39}$$

Solution of this expression shows how the temperature varies with time for various combinations of feed rate, feed temperature, and heat-exchange rate. The conversion obtained under such conditions is always the equilibrium value corresponding to the temperature at the end of the process. Its application to a practical problem is illustrated in Example 5-6.

Example 5-6 Hexamethylenetetramine (HMT) is to be produced in a semibatch reactor by adding an aqueous ammonia solution (25 wt $\%$ NH_3) at the rate of 2 gal/min (1.26×10^{-4} m^3/s) to an initial charge of 238 gal (0.901 m^3) (at 25°C) of formalin solution containing 42$\%$ by weight formaldehyde. The original temperature of the formalin solution is raised to 50°C in order to start the reaction. The temperature of the NH_4OH solution is 25°C. The heat of reaction in the liquid phase may be assumed independent of temperature and concentration and taken as -960 Btu/lb (-2.23×10^6 J/kg) of HMT. If the reactor can be operated at a temperature of 100°C, the rate of reaction is very fast in comparison with the rate of heat transfer with the surroundings. Temperatures higher than 100°C are not desirable because of vaporization and increase in pressure.

It is proposed to cool the reactor by internal coils through which water is passed. The overall heat-transfer coefficient between the stirred reaction mixture and the cooling water will be 85 Btu/(h)(ft^2)(°F) [483 J/(s)(m)^2(K) or 0.483 kJ/(s)(m)^2(K)]. The water rate through the coils is such that its temperature varies little, and an average value of 25°C may be used. Given the following data, calculate the length of 1-in.-OD tubing required for the cooling coils.

$$\text{Density of ammonia solution} = 0.91 \text{ g/cm}^3 \ (0.91 \times 10^{-3} \text{ kg/m}^3)$$

$$\text{Density of formalin (42\%) at 25°C} = 1.10 \text{ g/cm}^3 \ (1.10 \times 10^{-3} \text{ kg/m}^3)$$

Specific heat of reaction mixture (assume constant),
$$c_p = 1.0 \text{ Btu/(lb)(°F)} \ [4.19 \text{ kJ/(kg)(K)}]$$

Specific heat of 25 wt% NH_3 solution,
$$c_p = 1.0 \text{ Btu/(lb)(°F)} \ [4.19 \text{ kJ/(kg)(K)}]$$

The rate of the reverse reaction is negligible.

SOLUTION Since the rate is very rapid and the reaction is irreversible, the ammonia in the inlet stream will be completely converted to HMT just as soon as it is added to the reactor according to the reaction

$$4NH_3 + 6HCHO \rightarrow N_4(CH_2)_6 + 6H_2O$$

Since 4 moles of ammonia are required for 6 moles of formaldehyde, the total amount of ammonia required to react with all the charge of formalin solution will be

$$(NH_3)_t = \frac{238(8.33)(1.10)(0.42)}{30} \frac{4}{6}(17) = 346 \text{ lb} \ (157 \text{ kg})$$

From the ammonia feed rate of 2 gal/min, the total time of reaction will be

$$t_t = \frac{346}{2(8.33)(0.91)(0.25)} = 91.3 \text{ min}$$

The heat-transfer surface is to be sufficient to prevent the temperature from exceeding 100°C. Hence at 100°C, dT/dt in Eq. (5-39) will be zero. At temperatures below 100°C the driving force $T_e - T_s$, will be insufficient to transfer enough energy to the cooling coils to maintain a constant temperature. At the start of the addition of ammonia the last term in Eq. (5-39), which is positive for an exothermic reaction, will be greater than the sum of the first and second terms. Hence, the temperature of the reaction mixture will increase. From a practical standpoint this heating period would be reduced to a minimum by shutting off the flow of cooling water until the temperature reaches 100°C.

To determine the required heat-transfer area, Eq. (5-39) may be used when the temperature is 100°C and $dT_e/dt = 0$. Thus

$$UA_h(T_e - T_s) = -\Delta H_R(QC_f)_{NH_3} + F_t c_p(T_f - T_e)$$

or

$$85A_h(100 - 25)1.8 = -\Delta H_R(Q_f C_f)_{NH_3} + F_t(1.0)(25 - 100)1.8 \qquad (A)$$

The heat of reaction is -960 Btu/(lb of HMT). On the basis of NH_3, noting that the molecular weight of HMT is 140,

$$\Delta H = -\frac{960(140)}{4} = -33,600 \text{ Btu/(lb mol of } NH_3) \text{ or } -78,000 \text{ kJ/k mol}$$

The total, and NH_3, feed rates are

$$F_t = 2(60)(8.33)(0.91) = 910 \text{ lb/h } (0.114 \text{ kg/s})$$

$$Q_f C_f = F_t w_t/17 = 910(0.25)/17$$

$$= 13.3 \text{ lb mol of } NH_3/h \ (0.00168 \text{ k mol/s})$$

where w_t is the weight fraction NH_3 in the feed.

Substituting these values in Eq. (A) and solving for the heat transfer area, we obtain

$$A_h = \frac{-(-33,600)13.3 + 910(1.0)(-75)1.8}{85(75)1.8} = 28.3 \text{ ft}^2$$

In SI units,

$$A_h = \frac{-(-78000)(0.00168) + 0.114(4.19)(-75)}{0.483(75)} = 2.62 \text{ m}^2$$

If the heat-transfer coefficient of 85 is based on the outside area of the tubes, the length L of 1-in.-OD coil is

$$L = \frac{28.3}{\pi D} = \frac{28.3(12)}{\pi} = 108 \text{ ft}$$

An appropriate size of the reactor can be obtained by noting that the total mass of mixture at the end of the process will be

$$910\left(\frac{91.3}{60}\right) + 238(8.33)(1.10) = 3,560 \text{ lb}$$

If the density of the HMT solution is 72 lb/ft³, the minimum reactor volume is 50 ft³. A cylindrical vessel 4 ft in diameter and 6 ft in height would provide 33% excess capacity. If the 1-in. tubing were wound into a 3-ft diameter coil, approximately 12 loops would be needed.

The length of time necessary to raise the reaction temperature from its initial value, 50°C, to 100°C can be obtained by integrating Eq. (5-39). With the water rate shut off $UA_h(T_e - T_s)$ is negligible, and the expression becomes

$$\int_{T_0}^{T_e} \frac{dT_e}{F_t c_p(T_f - T_e) - \Delta H_R(Q_f C_f)} = \int_0^t \frac{dt}{(m_0 + F_t t)c_p}$$

where m_t has been replaced by $m_0 + F_t t$. The initial condition is $m_t = m_0$ and $T_e = T$. If ΔH_R and c_p are constant, this equation may be integrated to yield

$$-\frac{1}{F_t c_p} \ln \frac{-\Delta H_R(Q_f C_f) + F_t c_p(T_f - T_e)}{-\Delta H_R(Q_f C_f) + F_t c_p(T_f - T_0)} = \frac{1}{F_t c_p} \ln \frac{m_0 + F_t t}{m_0} \qquad \text{(B)}$$

Equation (B) relates the temperature and time during the heating period. We are given that $T_0 = 50° + 253°$ and $T_f = 25° + 273°$. Also

$$\Delta H_R(Q_f C_f) = -33{,}600(13.3) = -449{,}000 \text{ Btu/h}$$

$$m_0 = 238(8.33)1.10 = 2180 \text{ lb}$$

Using these values in Eq. (B), the time required for T_e to reach 100°C is

$$-\frac{1}{1(910)} \ln \frac{449{,}000 + (910)(1)(25 - 100)(1.8)}{449{,}000 + (910)(1)(25 - 50)(1.8)} = \frac{1}{910(1)} \ln \frac{2{,}180 + 910t}{2{,}180}$$

$$\ln (1 + 0.418t) = -\ln \frac{326{,}000}{408{,}000}$$

$$1 + 0.418t = 1.25$$

or

$$t = 0.60 \text{ h} \qquad \text{or } 36 \text{ min}$$

In summary, the reaction temperature would rise from 50 to 100°C in 36 min after the ammonia feed is started, provided water is not run through the cooling coil. After 36 min the water flow would be started in order to maintain the reactor temperature at 100°C. After a total time of 91 min sufficient ammonia would have been added to convert all the formaldehyde to HMT.

5-7 Optimum Temperature Profiles

In the previous sections we calculated conversion, selectivity, and temperatures for various reactors under conditions where the temperature sequence was fixed. That is, these quantities were determined by the energy and mass balances applicable for *specified* operating conditions. A different problem in nonisothermal reactors is the calculation of the temperature sequence required for optimum reactor performance. For multiple-reaction systems this normally means the temperature sequence needed to give maximum selectivity for the desired product. Some conclusions about such optimum temperatures were given at the end of Sec. 5-4. For single reactions (practically, this corresponds to the situation where side reactions are not significant) the problem is to determine the temperature sequence which minimizes the reactor volume (in a flow reactor) for a given conversion. While we will discuss flow reactors here, the analogous problem in a batch reactor is to minimize the time for a given conversion.

The minimum volume will be achieved if the reaction rate is a maximum at all positions in the reactor [for example, see Eq. (3-18)]. The optimum temperature at any position will be that for which the rate is a maximum at any conversion level. Therefore, the relation $\mathbf{r}(T, x)$ is the basic information necessary to determine optimum temperatures. This relationship is determined by the reaction order (kinetics) and the activation energy.

For an *irreversible* reaction Eq. (2-17) gives for the rate of disappearance of reactant

$$\mathbf{r}(T, x) = A(e^{-E/R_g T})f(x) \tag{5-40}$$

where $f(x)$ is usually a decreasing function of x.† Since E is positive, the rate increases with temperature at any composition. Hence the optimum temperature sequence will be the highest temperature that is practical. This conclusion applies for both exothermic and endothermic reactions. Properties of construction materials and the ultimate importance of side reactions can limit the temperature.

For a *reversible* reaction, Eq. (5-40) becomes

$$\mathbf{r}(T, x) = A(e^{-E/R_g T})f(x) - A'(e^{-E'/R_g T})g(x) \tag{5-41}$$

where E' and $g(x)$ apply for the reverse reaction. The function $g(x)$ is an increasing function of x since the concentrations of products increases with conversion. For an endothermic reaction, $E > E'$ (see Fig. 2-1). Again the rate increases with temperature for any conversion. We conclude that for *reversible, endothermic* reactions, the optimum temperature sequence is also the maximum permissible temperature.

For an *exothermic, reversible* reaction‡ E' must be greater than E. Hence, the rate of the reverse reaction increases more rapidly with temperature than that of the forward reaction. Equation (5-41) also shows that the reverse reaction will be slow at low conversions, while the forward reaction rate will be fast. For the rate to be a maximum at any conversion the temperature should be high at low conversions to take advantage of the predominant rate of the forward reaction, and lower at high conversions where the rate of the reverse reaction is high. Hence, for reversible, exothermic reactions the optimum temperature sequence will be continually decreasing temperature. This is illustrated in Fig. 5-14 where the *net* rate from Eq. (5-41) is plotted versus temperature, each curve corresponding to a constant conversion. An optimum temperature program is determined by connecting the maxima in the rate vs. temperature curves in Fig. 5-14, and is shown by the dashed curve.

Suppose the feed stream to a tubular-flow reactor is unconverted ($x = 0$) and its maximum temperature is T_A. The rate at the reactor entrance corresponds to point A in Fig. 5-14. If a conversion x_5 is required, the optimum rate at the outlet would be at point B. The requirement of a maximum in the rate vs. T curves cannot be achieved in the first part of the reactor because the maximum temperature is T_A. Hence, the best that can be done is to operate isothermally at T_A until the conversion at point C is reached, and then follow the dashed line to the required conversion. The optimum temperature path is ACB in Fig. 5-14. If instead of the optimum temperature sequence, the reactor was operated isothermally at the feed temperature, the reaction rate sequence would be that

---

† For example for a first-order reaction, with a feed concentration of reactant C_f, $f(x) = C_f(1 - x)$.

‡ The discussion about exothermic, reversible reactions follows that given by K. G. Denbigh in "Chemical Reactor Theory," Cambridge University Press, 1965.

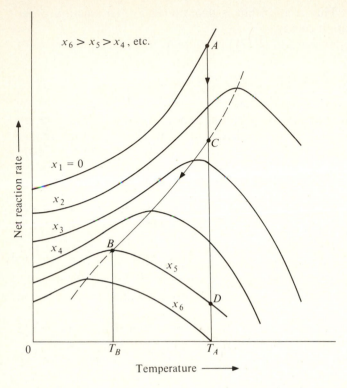

Figure 5-14 Optimum temperature sequence for an exothermic reversible reaction.

corresponding to the vertical line ACD. Such rates would be lower than those for the optimum sequence at every conversion, except from $x = 0$ to x at point C. Hence, the reactor volume would be greater.†

A stirred-tank reactor operates at constant temperature and conversion. If the required conversion is x_5, the optimum condition, that is, the maximum rate, would require operation at a temperature T_B. Since the rate is less at point B than at point A, the volume of the stirred-tank reactor would be greater than that for the plug-flow type. These conclusions are demonstrated in Example 5-7.

Example 5-7 There are many industrially important reactions that are of the exothermic, reversible type: the oxidation of SO_2, hydration of ethylene to produce ethyl alcohol, and ammonia and methanol syntheses. These reactions all require a catalyst, and the kinetics of catalytic reactions is not discussed until Chap. 9. However, we can illustrate the optimum temperature problem by considering a first-order, gaseous, reversible reaction of the form

$$A \rightleftharpoons B$$

$$\mathbf{r} = kC_A - k'C_B, \text{ k mol/(m)}^3\text{(s)}$$

---

† The question of optimum temperatures in various types of reactors has been studied extensively; for example R. Aris, "The Optimal Design of Chemical Reactors," Academic Press, New York, 1961.

where

$$k = A \exp\left(-\frac{E}{R_g T}\right) = 0.02 \exp\left(-\frac{29,300}{R_g T}\right), (s)^{-1}$$

$$k' = A' \exp\left(-\frac{E'}{R_g T}\right) = 0.33 \exp\left(-\frac{46,000}{R_g T}\right), (s)^{-1}$$

If the equilibrium constant K is equal to k/k' (see Sec. 2-7)

$$K = \frac{k}{k'} = 0.061 \exp\left(-\frac{E - E'}{R_g T}\right) = 0.061\left(\exp\frac{16,700}{R_g T}\right)$$

At 873 K, $K = 0.61$, while at 473 K, $K = 4.3$. For a maximum permissible feed temperature of 873 K and a feed rate $F_A(\text{k mol/s})$ of reactant at a feed concentration C_{A_f}, what is the optimum temperature profile in a plug-flow reactor? A conversion of 60% is required and the feed concentration of A is C_{A_f} and $C_{B_f} = 0$. What would be the required reactor volume? If the reactor operated isothermally at 873 K, what is the V/F_A required for 60% conversion? To simplify the calculations, neglect the effect of temperature on the concentration (i.e., assume constant density).

SOLUTION We first express the rate in terms of the conversion (of A) by noting that

$$x = \frac{C_{A_f} - C_A}{C_{A_f}} = \frac{C_B}{C_{A_f}}$$

Hence,

$$\mathbf{r} = C_{A_f}\left[A(1 - x) \exp\left(-\frac{E}{R_g T}\right) - A'x \exp\left(-\frac{E'}{R_g T}\right)\right] \tag{A}$$

If Eq. (A) is differentiated with respect to T (at constant conversion) and the derivative set equal to zero, the following requirement for maximum rate is obtained:

$$\frac{x}{(1 - x)} = \frac{EA}{E'A'} \exp\left[-(E - E')/R_g T\right] \tag{B}$$

or

$$T, \text{K} = \frac{(E' - E)/R_g}{\ln\left(\dfrac{A'E}{AE}\dfrac{x}{1 - x}\right)} = \frac{2010}{\ln\left(20.2\dfrac{x}{1 - x}\right)} \tag{C}$$

Solving this equation for T for various values of x gives the results shown in the first two columns of Table 5-8. The optimum temperature exceeds the limit of 873 K until the conversions reaches 0.33. Hence, the temperature sequence will be an isothermal path at 873 K up to 33% conversion, followed by the decreasing temperatures in Table 5-8 until $x = 0.60$.

Table 5-8 Optimum temperature sequence for Example 5-7

Conversion x	Temperature °C	Max rate, r/C_{A_f}, (s)$^{-1}$	Rate at 873 K r/C_{A_f}, (s)$^{-1}$
0		3.52×0^{-4}	3.52×10^{-4}
0.1	2250	2.59×10^{-4}	2.59×10^{-4}
0.2	979	1.82×10^{-4}	1.82×10^{-4}
0.3	664	0.73×10^{-4}	0.73×10^{-4}
0.33	600	0.45×10^{-4}	0.45×10^{-4}
0.38	.....	....	0
0.4	505	0.23×10^{-4}	
0.5	399	0.18×10^{-4}	
0.6	319	0.037×10^{-4}	

The reaction volume is given by Eq. (3-18). For $x = 0$ to 0.33, the rate is obtained from Eq. (A) by substituting $T = 873$ K:

$$r = C_{A_f}\left[0.02 \exp\left(-\frac{29,300}{R_g(873)}\right)(1-x) - 0.33 \exp\left(-\frac{46,000}{R_g T}\right)(x)\right] \quad \text{(D)}$$

$$r = C_{A_f}[3.52 \times 10^{-4}(1-x) - 5.78 \times 10^{-4}x] \quad \text{(E)}$$

For $x \geq 0.33$ the corresponding x and T given in the first two columns of Table 5-8 are used in Eq. (D) to calculate r. All of these rates are given in the third column of the table. Then V/F_A may be obtained by integrating Eq. (3-18) graphically, by plotting $1/r$ vs. x_A according to the equation

$$\frac{V}{F_A} = \int_0^{0.6} \frac{dx_A}{r} \quad \text{(F)}$$

The results give V/F_A equal to $0.23 \times 10^4/C_{A_f}$ for $x = 0$ to 0.33, and $2.35 \times 10^4/C_{A_f}$ for $x = 0.33$ to 0.6. The total value is

$$\frac{V}{F_A} = 2.58 \times 10^4/C_{A_f}, \text{ m}^3/(\text{k mol/s})$$

For an isothermal reactor operating at 873 K, Eq. (F) would be applicable with rate values determined at 873 K from Eq. (E). These rates are tabulated in the fourth column of Table 5-8 and indicate a zero value by the time the conversion has reached 0.38. In other words, at 873 K the equilibrium conversion is 38%. Regardless of the reactor volume, it is not possible to achieve the required 60% conversion. This situation would correspond to a conversion x_6 in Fig. 5-14. A higher conversion could be reached by operating isothermally at a lower temperature than 873 K. However, the required V/F_A for 60% conversion would always be greater than $2.58 \times 10^5/C_{A_f}$.

A continuous change in temperature as indicated in Example 5-7 is not practical. However, the optimum can be approached by operating different parts of the reactor at different temperature levels (i.e., discontinuous temperature changes).

One way of accomplishing this is to divide the reactor into sections with inter-coolers between sections (see Fig. 1-7f). An important aspect of the optimum calculations is that one then knows the *minimum* reactor volume for a given set of conditions. Such a result can be useful for comparison with actual behavior.

PROBLEMS

5-1 The liquid-phase hydrolysis of dilute aqueous acetic anhydride solutions is second order and irreversible, as indicated by the reaction

$$(CH_3CO)_2O + H_2O \rightarrow 2CH_3COOH$$

A batch reactor for carrying out the hydrolysis is charged with 200 liters of anhydride solution at 15°C and a concentration of 2.16×10^{-4} g mol/cm^3. The specific heat and density of the reaction mixture are essentially constant and equal to 0.9 cal/(g)(°C) and 1.05 g/cm^3, respectively. The heat of reaction may be assumed constant at $-50,000$ cal/g mol. The rate has been investigated over a range of temperatures, of which the following results are typical:

t, °C	10	15	25	40
r, g mol/(cm^3)(min)	0.0567C	0.0806C	0.1580C	0.380C

where C is acetic anhydride concentration, in gram moles per cubic centimeter. (*a*) Explain why the rate expression can be written as shown in the table even though the reaction is second order. (*b*) If the reactor is cooled so that operation is isothermal at 15°C, what time would be required to obtain a conversion of 70% of the anhydride? (*c*) Determine an analytical expression for the rate of reaction in terms of temperature and concentration. (*d*) What time is required for a conversion of 70% if the reactor is operated adiabatically?

5-2 A reactor for the production of drying oils by the decomposition of acetylated castor oil is to be designed for a conversion of 70%. The initial charge will be 500 lb and the initial temperature 340°C, as in Example 5-1. In fact, all the conditions of Example 5-1 apply, except instead of adiabatic operation, heat will be supplied electrically with a cal-rod unit in the form of a 1-in.-OD coil immersed in the reaction mixture. The power input and the stirring in the reactor will be such that the surface tempera-ture of the heater is maintained constant at 700 K. The heat-transfer coefficient may be taken equal to 60 Btu/(h)(ft^2)(°F). What length of heater will be required if the conversion of 70% is to be obtained in 20 min?

5-3 The thermal (noncatalytic) decomposition of acetaldehyde,

$$CH_3CHO(g) \rightarrow CH_4(g) + CO(g)$$

is studied in an ideal tubular-flow reactor at a constant total pressure of p_t atm. Suppose pure ace-taldehyde vapor enters the reactor at T_0 K and a steady rate of F g/s. The heat of reaction and specific heat of the reaction mixture can be assumed constant and equal to ΔH cal/g mol and c_p cal/(g mol)(°C). The rate of reaction is given by the second-order equation

$$r = A(e^{-E/R_g T})p_A{}^2 \qquad \text{g mol/(s)(cm}^3) \tag{A}$$

where E is the activation energy, in calories per gram mole, and p_A is the partial pressure of acetal-dehyde, in atmospheres. If the reactor operates adiabatically, express the rate of reaction in terms of the conversion as the sole variable.

5-4 A tubular-flow reactor is to be designed for the production of butadiene from butene by the gas phase reaction:

$$C_4H_8 \rightarrow C_4H_6 + H_2$$

The composition of the feed is 10 moles of steam per mole of butene and no butadiene or hydrogen. The reactor operates at 2 atmospheres pressure with an inlet (feed) temperature of 1200°F. The reaction rate follows a first-order, irreversible equation for which the rate constant k as a function of temperature is:

T, K	922 (1200°F)	900	877	855	832
k	11.0		4.90	2.04	0.85	0.32

(k = g mol butene reacted/(h)(liter)(atm).

The heat of reaction may be taken as constant and equal to $\Delta H_R = 26,360$ cal/g mol. Similarly the specific heat of the feed stream may be regarded as constant and equal to 0.5 Btu/lb °R.

A. What would be the volume required for a conversion (of butenes) of 20% if the reactor were operated isothermally at 1200°F with a butene-plus-steam feed rate of 22 lb mol/h?
B. It is desired to determine the conversion as a function of reactor volume (ft³) for adiabatic operation if the feed rate is 2.0 lb mol of butene per hour and 20 lb mol of steam per hour. Demonstrate your ability to solve this type of problem by calculating the reactor volume for conversions (of butenes) of 10 and 20%. What changes in operating conditions could be employed to reduce the volume required for a given conversion?

5-5 Reconsider the pilot plant discussed in Example 5-2 for the production of allyl chloride. It has been proposed to reduce the extent of the side reaction to dichloropropane by preheating the feed to 300°C. To obtain a more uniform axial temperature profile, inert nitrogen will be added to the feed to give a composition corresponding to 5 moles of N_2, 4 moles of C_3H_6, and 1 mole of Cl_2. The total feed rate will be 0.85 lb mol/h. If all other conditions are also the same, calculate the temperature and conversion profiles for a tube length of 20 ft. Comment on this method of operation in comparison with that in Example 5-2 with respect to selectivity and production rate of allyl chloride.

5-6 Consider the irreversible constant-density reaction sequence

$$A + B \xrightarrow{k_2} C \qquad r_2 = k_2 C_A C_B$$

$$C \xrightarrow{k_1} D \qquad r_1 = k_1 C_c$$

The rate constants are

$$k_2 = A_2 e^{-E_2/R_g T}$$

$$k_1 = A_1 e^{-E_1/R_g T}$$

Component C is the desired product. The feed contains no C or D, and $C_A = C_{A_0}$, $C_B = C_{A_0}$.
(a) If an isothermal tubular-flow reactor is employed, develop equations for calculating the conversion of A for which the yield of C is a maximum and for the constant temperature at which the reactor should operate to give the highest value of the maximum yield of C. (b) Develop equations for predicting the conversion of A for which the selectivity of C with respect to D is a maximum.

5-7 Repeat Prob. 5-6 for a stirred-tank reactor.

5-8 The vapor-phase decomposition of phosphine, which is irreversible and first order, follows the reaction

$$4PH_3(g) \rightarrow P_4(g) + 6H_2(g)$$

Pure phosphine is fed to a tubular-flow reactor, operating at 1 atm and adiabatically with a feed temperature of 953 K. The reaction is endothermic, $\Delta H_R = 23,900$ J/mol of phosphine at 25°C. The molal heat capacities $(J/(mol)(K))$ are

$$P_4(g) \qquad C_p = 25.1 + 0.0040T$$

$$PH_3(g) \qquad C_p = 28.0 + 0.027T$$

$$H_2(g) \qquad C_p = 30.1$$

The rate constant k, $(s)^{-1}$, is the following function of temperature:

$$\ln k = +27.94 + 2 \ln T - 43,672/T \; j \; (K)$$

What volume to molal feed rate ratio (V/F) would be required to obtain a conversion of 10% in one pass through the reactor? What would be the conversion for the same V/F if the reactor operated isothermally at 953 K?

5-9 The gas-phase cracking of light hydrocarbons is a thermal (noncatalytic) process that requires a high temperature. Since the reactions are endothermic, energy must be supplied to maintain a high temperature. This can be accomplished, in part, by adding steam to the feed, but in addition it is desirable to add heat from the surroundings. This is done commercially by constructing the reactor of rows of tubes installed in a furnace (pipe still).

A laboratory study of such a cracking process indicates that the required conversion is obtained with a tube of length L_L, diameter d_L, for a hydrocarbon feed rate F_L and an average temperature difference $(T_w - T)_L$. T_w is the tube-wall temperature and T is the bulk temperature of the reaction mixture.

It is necessary to scale-up the laboratory reactor to pilot-plant size. An exact scale-up would require that the dimensionless axial and radial temperature profiles in the laboratory and pilot-plant tubes be the same. As an approximation, assume that it is sufficient for the dimensionless energy balances as written in Chap. 5 to be the same.

If the same conversion is to be obtained in the pilot-plant as in the laboratory reactor, what relations between reactor diameter and length and feed rate are required? Base the scale-up on the mass and energy balances. Do not consider pressure drop in the scale-up and neglect the effect of pressure on the rate of reaction.

5-10 Reconsider Example 5-3 for the case where the feed composition is $N_2 = 50$ mole %, $C_3H_6 = 40$ mole %, $Cl_2 = 10$ mole %. The feed temperature will be 300°C, with all other conditions the same as in Example 5-3. Determine conversions and temperatures for reactor volumes up to 0.5 ft^3.

5-11 A first-order irreversible (liquid-phase) reaction is carried out in a stirred-tank flow reactor. The density is 1.2 g/cm^3 and the specific heat is 0.9 cal/(g)(°C). The volumetric flow rate is 200 cm^3/s and the reactor volume is 10 liters. The rate constant is

$$k = 1.8 \times 10^5 e^{-12,000/R_g T} \qquad s^{-1}$$

where T is in degrees Kelvin. If the heat of reaction is $\Delta H_R = -46,000$ cal/g mole and the feed temperature is 20°C, what are the possible temperatures and conversions for stable, adiabatic operation at a feed concentration of 4.0 g mol/liter?

5-12 The reaction between sodium thiosulfate and hydrogen peroxide in dilute aqueous solution is irreversible and second order in thiosulfate.† The rate constant is the following function of temperature for the rate of disappearance of thiosulfate:

$$k = 6.85 \times 10^{14} \exp\left(-\frac{18300}{R_g T}\right), \; cm^3/(mole)(s)$$

Reaction stoichiometry indicates that 2 moles of H_2O_2 react with one mole of $Na_2S_2O_3$. The heat of reaction at 25°C is $\Delta H_R = -131,000$ cal/g mole.

† Kearns, D. L. and Manning, F. S., *AIChE J.* **15**, 660 (1969).

Kearns and Manning experimental studies in a stirred-tank reactor included the following conditions:

$$\text{Reactor volume} = 2790 \text{ cm}^3$$

$$\text{Feed temperature} = 25°C$$

$$\text{Feed rate} = 14.2 \text{ cm}^3/s$$

Consider adiabatic operation and feed concentrations of 2.04×10^{-4} g mol/cm³ and 4.08×10^{-4} g mol/cm³ of thiosulfate and hydrogen peroxide, respectively. What would be the conversion and temperature in the reactor effluent?

5-13 The exhaust gas from an internal combustion engine contains some unburned fuel and carbon monoxide. Combustion can be continued if an afterburner is placed in the exhaust line. Suppose such an afterburner operates as an adiabatic, stirred-tank reactor with an average residence time of 9 seconds. Consider only the further oxidation of CO and assume that, in excess air, the oxidation is first order (for CO of the order of 1%) and irreversible with the following rate constant:

$$k(s^{-1}) = 1.5 \times 10^{10} \exp\left(-\frac{272}{R_g T}\right)$$

$$E = 272 \text{ kJ/mol}$$

If the combustion gases (consider the properties to be those of air) enter at 1073 K, what are the steady-state values of the temperature and CO conversion in the effluent from the afterburner?

5-14 It is desired to evaluate the dynamic behavior of the reaction and reactor system of Example 5-4 at a different initial condition than that for Example 5-5. Suppose initially that the reactor contains 18×10^{-3} m³ of solution with a reactant concentration of 3.0 k mol/m³ at a temperature of 308 K. Determine curves of effluent temperature and conversion vs. time. Also calculate the steady-state temperature and conversion.

5-15 As noted in Sec. 5-7, for exothermic, reversible reactions increasing the temperature reduces the maximum (equilibrium) conversion but increases the forward rate. To obtain the maximum conversion a high temperature is needed at low conversions (where the reverse reaction is unimportant) and a lower temperature at higher conversions.

Consider a reversible first-order reaction $A \rightleftharpoons B$ for which, at 298 K,

$$\Delta F_{298}° = -2500 \text{ cal/g mol}$$

$$\Delta H_{298}° = -20,000 \text{ cal/g mol}$$

The reaction mixture is an ideal liquid solution (constant density) at all temperatures. (a) Assuming that $\Delta H°$ is constant, plot a curve of the equilibrium conversion vs. temperature from 0 to 100°C. (b) If the forward-rate constant is

$$k = 5 \times 10^8 e^{-12,500/R_g T} \qquad \text{min}^{-1}$$

determine the conversion in the effluent from an isothermal tubular-flow reactor for which the volumetric feed rate is 100 liters/min and the volume is 1500 liters. Calculate the conversion for a series of temperatures from 0 to 100°C and plot the results on the figure prepared for part (a). (c) Suppose that the maximum permissible temperature is 100°C and the concentration of A in the feed stream is 2 g mol/liter (the feed contains no B). Determine the maximum conversion obtainable in the reactor of part (b) if the temperature can be varied along the length of the reactor. First prepare curves (similar to Fig. 5-14) of net rate vs. temperature at constant conversion for several conversion levels. Then plot the optimum temperature profile, first as temperature vs. conversion and then as temperature vs. V/Q.

5-16 Consider the same reaction system, rate equation, feed rate, and composition as in Prob. 5-15. The reaction is to be carried out in two stirred-tank reactors, each with a volume of 750 liters. What should be the temperature (within the range 0 to 100°C) in each reactor in order to obtain the maximum conversion in the effluent from the second reactor? The graphical method for multiple-stirred-tank reactors (described in Sec. 4-6) may be helpful.

5-17 The HMT reactor described in Example 5-6 is to be redesigned so that the reaction temperature rises from 50 to 100°C as uniformly as possible as the ammonia solution is added. To accomplish this the rate of addition of ammonia must be varied. The cooling water will flow through the coils throughout the run. Determine the feed rate as a function of time, and the time required to add all the ammonia, to meet these operating requirements. All other conditions are the same as in Example 5-6.

SIX

DEVIATION FROM IDEAL-REACTOR PERFORMANCE

In Chap. 3 equations were developed for calculating the conversion for two extreme mixing states: the ideal stirred-tank reactor, corresponding to complete mixing, and the ideal tubular-flow reactor, corresponding to no axial mixing and uniform velocity in the direction of flow. The reasons for deviations from ideal forms in actual reactors were also pointed out (see Fig. 3-3 and 3-4). The objective in this chapter is to evaluate quantitatively the effect of deviations on the conversion in flow reactors. If the velocity and local rate of mixing (micromixing) of every element of fluid in the reactor were known and the differential mass balance could be integrated, an exact solution for the conversion could be obtained. Since such complete information is unavailable for actual reactors, approximate methods, using readily obtainable data and models of mixing, are necessary. It is difficult to measure velocities and concentrations within a reactor, but data can be obtained for the feed and effluent streams. Such end-effect, or response, data consist of measuring the effect observed in the effluent stream when a disturbance occurs in the feed, for example, when the concentration of an inert component in the feed is changed. In this chapter the treatment is restricted to a single reaction in a homogeneous reactor operated isothermally. Nonideal behavior in heterogeneous reactors will be discussed in Chap. 13.

6-1 Mixing Concepts and Models

Deviations from ideal flow can be classified in two types. In one type elements of fluid may move through the reactor at different velocities, causing channeling and "dead spots." For such behavior to occur, the elements of fluid must not completely mix locally, but remain at least partially *segregated* as they move through the reactor. The other deviation refers to the extent of the local or micromixing.

For example, there may be some local mixing or diffusion in the direction of flow in a tubular reactor.

We shall see in Sec. 6-3 that response measurements can be used to determine the residence-time distribution (RTD) of the elements of fluid in a reactor. However, the RTD is not enough to determine separately the extent of segregated flow in the reactor and the extent of micromixing. Hence residence-time information is not sufficient, in general, to evaluate the conversion in a nonideal reactor. Put another way, the RTD is affected by both types of deviations. There can be a large number of different mixing states, i.e., different extents of segregated flow and micromixing, which give the same RTD.†‡ However, for first-order kinetics the distinction between mixing states has no effect on the conversion (Sec. 6-7). In this case the RTD provides all the information necessary to calculate a correct conversion. Any mixing model which gives the actual RTD may be used to calculate the conversion for first-order kinetics.

We shall consider four methods of estimating deviations from ideal reactor performance. The first is to determine the actual RTD from experimental response data and then calculate the conversion by assuming the flow to be wholly segregated (Sec. 6-8). This model should be a good approximation, for example, for a tubular-flow reactor in which the flow is streamline. It would not describe a nearly ideal stirred-tank reactor, for here the fluid is nearly completely mixed immediately after it enters the reactor. In this method no error is introduced by an approximation of the RTD, since the actual RTD is used. An error does arise from the assumption of segregated flow when there may be some micromixing; as noted, this error disappears for first-order kinetics.

The other three methods are subject to both these errors, since both the form of the RTD and the extent of micromixing are assumed. In the *axial-dispersion model* the reactor is represented by allowing for axial diffusion in an otherwise ideal tubular-flow reactor. In this case the RTD for the actual reactor is used to calculate the best axial diffusivity for the model (Sec. 6-5), and this diffusivity is then employed to predict the conversion (Sec. 6-9). This is a good approximation for most tubular reactors with turbulent flow, since the deviations from plug-flow performance are small. In the third model the reactor is represented by a series of ideal stirred tanks of equal volume. Response data from the actual reactor are used to determine the number of tanks in series (Sec. 6-6). Then the conversion can be evaluated by the method for multiple stirred tanks in series (Sec. 6-10).

The final model is the plug-flow reactor with recycle shown in Fig. 4-19, and analyzed in Chap. 4 (Sec. 4-10). The reactor itself behaves as an ideal tubular type, but mixing is introduced by the recycle stream. When the recycle rate becomes

---

† Much of the development of mixing and residence-time-distribution concepts in reactors is due to P. V. Danckwerts, *Chem. Eng. Sci.*, **8**, 93 (1958), and T. N. Zwietering, *Chem. Eng. Sci.*, **11**, 1 (1959).

‡ Even though the flow is wholly segregated within the reactor, complete micromixing will occur at the exit. The conversion and other properties of the effluent stream are average values, based on a completely mixed stream. Zwietering has used the concept of the location in a reactor where complete mixing occurs as a parameter for describing the mixing state. For example, in an ideal stirred tank complete micromixing occurs as the feed enters. At the other extreme, in completely segregated flow micromixing occurs at the exit of the reactor.

very large, ideal stirred-tank performance is obtained, and when the recycle is zero, plug-flow operation results. The response data on the actual reactor are used to evaluate the recycle rate, and then the conversion is estimated for a plug-flow reactor with this recycle rate. The calculation of conversion in a nonideal reactor using this model is illustrated in Sec. 6-11.

In all these models except the segregated-flow approach, which employs the actual experimental response data, a single parameter is used to describe the data. One parameter does not always provide an adequate description. If the experimental response data cannot be described, for example, by a single axial diffusivity, the dispersion model may not give an accurate conversion. Many other, usually more elaborate models have been proposed.† Some employ more than one parameter in order to describe more accurately both the RTD and the extent of micromixing in the reactor.

In the next section RTD functions are discussed, and in Secs. 6-3 and 6-4 the RTD is evaluated from response data, with illustrations for ideal reactors.

6-2 A Residence-Time Distribution Function

The time it takes a molecule to pass through a reactor is called its *residence time* θ. Two properties of θ are important: the time elapsed since the molecule entered the reactor (its age) and the remaining time it will spend in the reactor (its residual lifetime). We are concerned mainly with the sum of these times, which is θ, but it is important to note that micromixing can occur only between molecules that have the same residual lifetime; molecules cannot mix at some point in the reactor and then unmix at a later point in order to have different residual lifetimes. A convenient definition of *residence-time distribution function* is the fraction $J(\theta)$ of the effluent stream that has a residence time less than θ. None of the fluid can have passed through the reactor in zero time, so $J = 0$ at $\theta = 0$. Similarly, none of the fluid can remain in the reactor indefinitely, so that J approaches 1 as θ approaches infinity. A plot of $J(\theta)$ vs. θ has the characteristics shown in Fig. 6-1a.

Variations in density, such as those due to temperature and pressure gradients, can affect the residence time and are superimposed on effects due to velocity variations and micromixing. We are concerned with micromixing in this chapter, and we shall therefore suppose that the density of each element of fluid remains constant as it passes through the reactor. Under these conditions the *mean residence time*, averaged for all the elements of fluid, is given by

$$\bar{\theta} = \frac{V}{Q}$$

where Q is the volumetric flow rate. For constant density, Q is the same for the feed as for the effluent stream. From the definition of $J(\theta)$ we can also say that $dJ(\theta)$ is the volume fraction of the effluent stream that has a residence time

† L. A. Spillman and O. Levenspiel, *Chem. Eng. Sci.*, **20**, 247 (1965); R. L. Curl, *AIChE J.*, **9**, 175 (1963); S. A. Shain, *AIChE J.*, **12**, 806 (1966); H. Weinstein and R. J. Adler, *Chem. Eng. Sci.*, **22**, 65 (1967); D. Y. Ng and D. W. T. Rippin, *Chem. Eng. Sci.*, **22**, 3, 247 (1967).

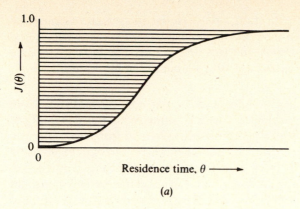

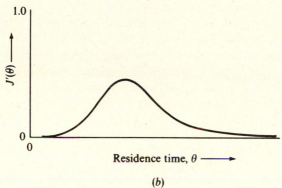

Figure 6-1 Residence-time distribution function.

between θ and $\theta + d\theta$. Hence the mean residence time is also given by

$$\bar{\theta} = \frac{\int_0^1 \theta \, dJ(\theta)}{\int_0^1 dJ(\theta)} = \int_0^1 \theta \, dJ(\theta) = \frac{V}{Q} \tag{6-1}$$

The shaded area in Fig. 6-1a represents $\bar{\theta}$.

The RTD can also be described in terms of the slope of the curve in Fig. 6-1a. This function, $J'(\theta) = dJ(\theta)/d\theta$, will have the shape usually associated with distribution curves, as noted in Fig. 6-1b. The quantity $J'(\theta) \, d\theta$ represents the fraction of the effluent stream with a residence time between θ and $\theta + d\theta$. Substituting $J'(\theta) \, d\theta$ for $dJ(\theta)$ in Eq. (6-1) gives an expression for $\bar{\theta}$ in terms of $J'(\theta)$,

$$\bar{\theta} = \int_0^\infty \theta J'(\theta) \, d\theta \tag{6-2}$$

6-3 Residence-Time Distributions from Response Measurements

The RTD, for a given reactor and flow rate, can be established from response-type experiments. In these experiments the concentration of an inert tracer is perturbed in the feed stream and its effect on the effluent stream is measured. The three most common perturbations are a step function, a pulse (square wave), and a sinusoidal wave. The relationships between the observed concentration-vs.-time curves and the RTD are examined here for step functions and pulses. The analysis of sinusoidal perturbations is more complex but is available in the literature.†

Step-function input Suppose a stream with a molecular concentration C_0 flows through a reactor at a constant volumetric rate Q. Imagine that at $\theta = 0$ all the molecules entering the feed are marked to distinguish them from the molecules which have entered prior to $\theta = 0$. Since the total concentration is not changed by the marking process, its value will be C_0 at any θ. However, the concentration C of marked molecules in the effluent will change with θ, since some marked molecules would spend longer than others in the reactor. The input and output (response) concentration ratios C/C_0 are shown generally for this situation in Fig. 6-2b and c. The exact shape of the response curve depends on the mixing state of the system.

At a time θ when the concentration of marked molecules in the effluent is C the rate of flow of these molecules will be CQ. All the marked molecules have entered the reactor in a time less than θ. By definition, $J(\theta)$ is the fraction of the total molecules that have this residence time range. Since the total flow rate of molecules is $C_0 Q$, the product $C_0 Q J(\theta)$ will also describe the flow rate of marked

† H. Kramers and G. Alberda, *Chem. Eng. Sci.*, **2**, 173 (1953).

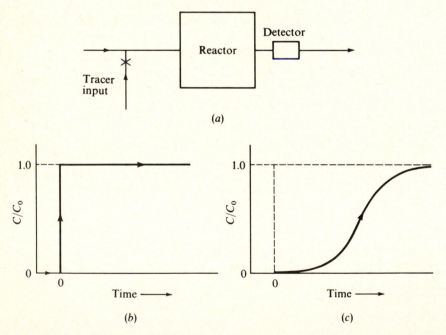

(a)

(b)   *(c)*

Figure 6-2 Response to a step-function input: (*a*) apparatus for input-response studies; (*b*) step-function input; (*c*) response in effluent.

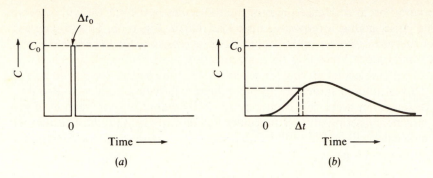

Figure 6-3 Response to a pulse input: (a) pulse input to feed; (b) response in effluent.

molecules. Equating these two expressions gives

$$CQ = C_0 Q J(\theta)$$

or

$$J(\theta) = \left(\frac{C}{C_0}\right)_{step} \tag{6-3}$$

In practice it is not possible to start marking the instant molecules start entering the reactor. Instead, a quantity of inert miscible substance is added to the feed stream in the apparatus shown in Fig. 6-2a. Provided that the tracer can be added quickly, that the tracer molecules move through the reactor in the same way as the feed stream, and that the detection of tracer in the effluent is rapid, the experimental results are satisfactory. In such an experiment the concentration of tracer in the feed takes the place of C_0 in Eq. (6-3), and C is the concentration of tracer molecules in the effluent. The mathematical description of this step-function input is

$$C = \begin{cases} 0 & \text{for } t < 0 \\ C_0 & \text{for } t > 0 \end{cases} \quad \text{at } z = 0 \tag{6-4}$$

Equation (6-3) shows that the C/C_0 response to a step-function input gives the RTD function $J(\theta)$ directly. This approach provides a simple experimental procedure for measuring the RTD for an actual reactor.

Pulse input Suppose the marking experiment is conducted in a different way. This time let all the molecules be marked only for a very short time interval at $\theta = 0$. The total molecules marked would be

$$M = C_0 Q \, \Delta t_0 \tag{6-5}$$

where Δt_0 is the marking interval and C_0 is again the total concentration of molecules. This input is shown graphically in Fig. 6-3a and is defined mathematically by

$$C = \begin{cases} 0 & \text{for } t < 0 \\ C_0 & \text{for } 0 < t < \Delta t_0 \quad \text{at } z = 0 \\ 0 & \text{for } t > \Delta t_0 \end{cases} \tag{6-6}$$

The variation in residence times of the molecules in the reactor will disperse the input pulse, giving a response curve like that in Fig. 6-3b. To evaluate the RTD we proceed as before and formulate two expressions for the marked molecules at some time θ. Since C is the concentration of marked molecules at θ, the number of such molecules leaving the reactor in the time period θ to $\theta + d\theta$ will be $CQ \, d\theta$. All the marked molecules in the effluent will have a residence time θ to $\theta + d\theta$ because they were added only at $\theta = 0$. By definition, the fraction of the effluent stream consisting of such molecules will be $dJ(\theta)$ or $J'(\theta) \, d\theta$. The number of such molecules will be $MJ'(\theta) \, d\theta$. Equating the two expressions for the number of marked molecules gives

$$CQ \, d\theta = MJ'(\theta) \, d\theta$$

or

$$J'(\theta) = \frac{(C)_{\text{pulse}} Q}{M} \tag{6-7}$$

Equation (6-7) shows that $J'(\theta)$ can be obtained from the measured response curve for a pulse input. In practice, instead of marking molecules we obtain this curve by introducing, as quickly as possible, a pulse of miscible tracer in the feed and detecting its concentration in the effluent (see Fig. 6-2a). The volumetric flow rate would be known, and M can be calculated. Since it is difficult to measure Δt_0 and C_0 accurately, M is better determined from the area of the response curve; that is,

$$M = Q \int_0^\infty C_{\text{pulse}} \, d\theta \tag{6-8}$$

This follows from the fact that all tracer molecules will ultimately appear in the effluent stream. With this formulation for M, Eq. (6-7) becomes

$$J'(\theta) = \frac{C_{\text{pulse}}}{\int_0^\infty C_{\text{pulse}} \, d\theta} \tag{6-9}$$

With this result we can evaluate $J'(\theta)$ for any reactor, using only the measured response curve to a pulse input. Since $dJ(\theta) = J'(\theta) \, d\theta$, Eq. (9) can be integrated from 0 to θ to give $J(\theta)$ in terms of C_{pulse}. Thus

$$J(\theta) = \int_0^\theta dJ'(\theta) \, d\theta = \frac{\int_0^\theta C_{\text{pulse}} \, d\theta}{\int_0^\infty C_{\text{pulse}} \, d\theta} \tag{6-10}$$

The relation between C_{step}-vs.-θ and C_{pulse}-vs.-θ curves can be obtained from Eq. (6-3) and (6-7). Differentiation of Eq. (6-3) yields

$$\frac{dJ(\theta)}{d\theta} = J'(\theta) = d(C/C_0)_{\text{step}}/d\theta$$

Substituting $J'(\theta)$ from Eq. (6-7) gives

$$C_{\text{pulse}} = \frac{M}{Q} [d(C/C_0)/d\theta]_{\text{step}}$$

Replacing M in terms of Q and C_0 from Eq. (6-5),

$$(C_{\text{pulse}}/C_0) = \Delta t_0 [d(C/C_0)_{\text{step}}/d\theta] \qquad (6\text{-}11)$$

Thus, the response curve to a pulse input is proportional to the derivative of the response curve for a step input.

6-4 Residence-Time Distributions for Reactors with Known Mixing Conditions

For reactors with known mixing characteristics the response curve and the RTD can be predicted; no experiments are necessary. As an illustration let us develop the RTD for the plug-flow reactor, a single ideal stirred-tank reactor, and a tubular reactor with laminar flow.

The characteristics of uniform velocity profile and no axial mixing in a *plug-flow reactor* require that the residence time be a constant, $\theta = V/Q$. The curve for response to a step-function input is shown in Fig. 6-4 as solid lines. From Eq. (6-3), the response curve is equal to $J(\theta)$. Then $J(\theta) = 0$ for $\theta < V/Q$ and $J(\theta) = 1$ for $\theta \geq V/Q$. The input and response curve for a pulse input would correspond to narrow peaks at $\theta = 0$ and $\theta = V/Q$, as shown in Fig. 6-5 (solid lines). The response curve, according to Eq. (6-7), is proportional to $J'(\theta)$.

For an ideal *stirred-tank reactor* $(C/C_0)_{\text{step}}$ can be calculated by writing a mass balance, Eq. (3-1), for a step-function input of tracer. The third term is zero, since there is no reaction. At a time θ after the tracer concentration in the feed is increased to C_0 the other terms in Eq. (3-1) give

$$C_0 Q \, \Delta\theta - CQ \, \Delta\theta = V \, \Delta C$$

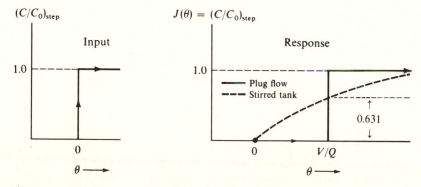

Figure 6-4 Response curves to a step input for ideal reactors.

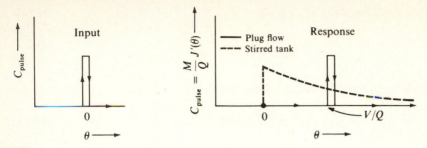

Figure 6-5 Response curves to a pulse input for ideal reactors.

where C is the effluent concentration at θ and ΔC is the change in concentration of tracer in the reactor during $\Delta\theta$. Dividing by $\Delta\theta$ and taking the limit as $\Delta\theta \to 0$ gives

$$\frac{dC}{d\theta} = \frac{Q}{V}(C_0 - C) = \frac{1}{\bar{\theta}}(C_0 - C) \tag{6-12}$$

With the initial condition $C = 0$ at $\theta \le 0$, the solution of Eq. (6-12) is

$$\left(\frac{C}{C_0}\right)_{\text{step}} = J(\theta) = 1 - e^{-(\theta/\bar{\theta})} \tag{6-13}$$

The curve described by Eq. (6-13), the dotted line in Fig. 6-4, indicates a large spread in residence times. The asymptotic tail shows that a few molecules remain in the reactor for a very long time. At the mean residence time $\bar{\theta} = V/Q$, $J(\theta) = 0.631$; or 63.1% of the effluent stream has a residence time less than the mean value.

A similar analysis for a pulse input would give a response curve of C_{pulse} vs. θ, but this can be obtained more easily by differentiating the $J(\theta)$ curve in Fig. 6-4, or its equivalent, Eq. (6-13). Thus, from Eq. (6-13)

$$J'(\theta) = \frac{1}{\bar{\theta}} e^{-(\theta/\bar{\theta})} \tag{6-14}$$

Substituting this expression in Eq. (6-7) gives the response for a pulse input

$$C_{\text{pulse}} = \frac{M}{Q} \frac{1}{\bar{\theta}} e^{-(\theta/\bar{\theta})} = C_0(\Delta t_0) \frac{1}{\bar{\theta}} e^{-(\theta/\bar{\theta})}$$

This equation shows that C_{pulse} will be greatest at $\theta = 0$ and will continually decrease toward zero as θ increases. Such a distribution curve, given as the dashed line in Fig. 6-5, shows that the most probable [largest $J'(\theta)\,d\theta$] residence time is at $\theta = 0$ for a stirred-tank reactor.

Figure 6-4 is useful for determining how closely an actual reactor fits one of the two ideal forms. The measured response curve may be superimposed on Fig. 6-4 to show the extent of the deviations from either plug-flow or stirred-tank behavior.

A tubular reactor with *laminar flow* has been mentioned as a good approximation to segregated flow. If the dispersion due to molecular diffusion is neglected, the approximation is exact. Since the flow is segregated and the velocity profile is known, the RTD can be calculated.† It is instructive to compare the calculated results with those for the ideal forms given in Fig. 6-4. The velocity in the axial direction for laminar flow is parabolic,

$$u(r) = 2\frac{Q}{\pi r_0^2}\left[1 - \left(\frac{r}{r_0}\right)^2\right] \tag{6-15}$$

where r is the radial position and r_0 is the tube radius. Since u is a function of r, the residence time also varies with r. If the reactor length is L, θ at any r is

$$\theta = \frac{L}{u} = \frac{\pi r_0^2}{2Q}\frac{L}{[1 - (r/r_0)^2]} \tag{6-16}$$

Since $L\pi r_0^2 = V$, this expression becomes

$$\theta = \frac{V/Q}{2[1 - (r/r_0)^2]} = \frac{\bar{\theta}}{2[1 - (r/r_0)^2]} \tag{6-17}$$

We first find how the volume fraction of the effluent varies with radius r. Then Eq. (6-17) is used to replace r with θ. The fraction of the effluent with a radius between r and $r + dr$ is

$$dJ(r) = dJ(\theta) = \frac{u(2\pi r\, dr)}{Q} \tag{6-18}$$

Note that the fraction $dJ(r)$ will have residence times between θ and $\theta + d\theta$. Hence, $dJ(r) = dJ(\theta)$.

Substituting Eq. (6-15) for u and simplifying gives

$$dJ(\theta) = \frac{4}{r_0^2}\left[1 - \left(\frac{r}{r_0}\right)^2\right]r\, dr \tag{6-19}$$

To replace r with θ in this expression we first differentiate Eq. (6-17) and then solve for $r\, dr$ to obtain

$$r\, dr = \frac{\bar{\theta}}{4}\frac{r_0^2}{\theta^2}\, d\theta \tag{6-20}$$

Substituting $1 - (r/r_0)^2$ from Eq. (6-17) and $r\, dr$ from Eq. (6-20) in Eq. (6-19) yields

$$dJ(\theta) = \frac{1}{2}\left(\frac{V}{Q}\right)^2\frac{d\theta}{\theta^3} = \frac{1}{2}\bar{\theta}^2\frac{d\theta}{\theta^3} \tag{6-21}$$

---

† The behavior of this type of reactor was first considered by K. G. Denbigh, *J. Appl. Chem.*, **1**, 227 (1951); and H. Kramers and K. R. Westerterp "Elements of Chemical Reactor Design and Operation," p. 85, Academic Press, Inc., New York, 1963.

or

$$\frac{dJ(\theta)}{d\theta} = J'(\theta) = \frac{1}{2}\frac{\bar{\theta}^2}{\theta^3} \tag{6-22}$$

Integration of Eq. (6-21) will give $J(\theta)$. The minimum residence time is not zero, but corresponds to the maximum velocity at the center of the tube. From Eq. (6-17), this is

$$\theta_{min} = \frac{1}{2}\frac{V}{Q} = \frac{1}{2}\bar{\theta} \tag{6-23}$$

Integration of Eq. (6-21) from θ_{min} to θ gives

$$J(\theta) = \frac{1}{2}\bar{\theta}^2 \int_{\frac{1}{2}\bar{\theta}}^{\theta} \frac{d\theta}{\theta^3} = 1 - \frac{1}{4}\left(\frac{\theta}{\bar{\theta}}\right)^{-2} \tag{6-24}$$

Equation (6-22) or Eq. (6-24) gives the desired RTD function for a tubular reactor with laminar flow.

$J(\theta)$ is plotted against $\theta/\bar{\theta}$ in Fig. 6-6; also shown are the curves for the two ideal reactors, taken from Fig. 6-4. The comparison brings out pertinent points about reactor behavior. Although the plug-flow reactor might be expected to be a better representation of the laminar case than the stirred-tank reactor, the RTD for the latter more closely follows the laminar-reactor curve for $\theta/\bar{\theta}$ from about 0.6 to 1.5. However, there is no possibility for θ to be less than 0.5 in the laminar-flow case. Hence the stirred-tank form is not applicable at all in the low θ region. At high θ the three curves approach coincidence. Conversions for these reactors and the various one-parameter models are compared in Secs. 6-8 to 6-11.

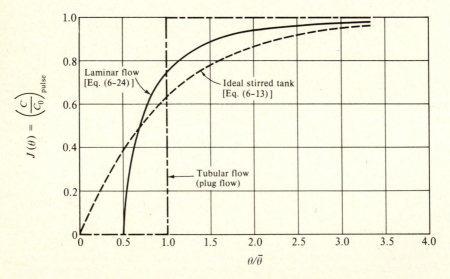

Figure 6-6 Residence-time distribution in a laminar-flow tubular reactor (segregated flow).

6-5 Interpretation of Response Data by the Dispersion Model

Here we wish to determine the axial diffusivity from response measurements, in preparation for using the dispersion model for conversion calculations. According to this model, the actual reactor can be represented by a tubular-flow reactor in which axial dispersion takes place according to the effective diffusivity D_L. It is supposed that the axial velocity u and the concentration are uniform across the diameter, as in a plug-flow reactor.

Imagine that a step function of inert tracer concentration C_0 is introduced into the feed at $\theta = 0$. By solving the transient mass balance, we can determine the response C/C_0 vs. θ as a function of D_L. In writing Eq. (3-1) for the mass balance we omit the term involving reaction but include axial dispersion in the input and output terms. The result is

$$\left[\left(-D_L\frac{\partial C}{\partial z} + uC\right)\pi r_0^2\right]_z \Delta\theta - \left[\left(-D_L\frac{\partial C}{\partial z} + uC\right)\pi r_0^2\right]_{z+\Delta z} \Delta\theta = \pi r_0^2\,\Delta z\,\Delta C$$

Canceling πr_0^2, dividing by $\Delta z\,\Delta\theta$, and taking the limit as $\Delta z \to 0$ gives

$$D_L\frac{\partial^2 C}{\partial z^2} - u\frac{\partial C}{\partial z} = \frac{\partial C}{\partial\theta} \tag{6-25}$$

The initial condition is

$$C = \begin{cases} 0 & \text{at } z > 0 \text{ for } \theta = 0 \\ C_0 & \text{at } z < 0 \text{ for } \theta = 0 \end{cases} \tag{6-26}$$

If no axial dispersion occurs in the feed line but does occur in the bed, the boundary condition at $z = 0$ is

$$-D_L\left(\frac{\partial C}{\partial z}\right)_{>0} + u(C)_{>0} = uC_0 \qquad \text{at } z = 0 \text{ for } \theta \geq 0 \tag{6-27}$$

where >0 designates the position just inside the reactor at $z = 0$. It has been shown† that at the exit the correct condition is

$$\frac{dC}{dz} = 0 \qquad \text{at } z = L \text{ for } \theta \geq 0 \tag{6-28}$$

The solution with these boundary conditions is difficult, but a good approximation, particularly when D_L is small, may be obtained by substituting for Eqs. (6-27) and (6-28) the boundary conditions

$$C = \begin{cases} C_0 & \text{at } z = -\infty \text{ for } \theta \geq 0 \\ 0 & \text{at } z = \infty \text{ for } \theta \geq 0 \end{cases} \tag{6-29}$$

---

† P. V. Danckwerts, *Chem. Eng. Sci.*, **2**, 1 (1953); J. F. Wehner and R. H. Wilhelm, *Chem. Eng. Sci.*, **6**, 89 (1959). The boundary conditions for this type of problem have been the subject of numerous reports. For a more complete analysis of the general problem see G. Standart, *Chem. Eng. Sci.*, **23**, 645 (1968); C. Y. Choi and D. D. Perlmutter, *Chem. Eng. Sci.* **31**, 250 (1976); W. D. Deckwer and E. A. Mahlmann, *Chem. Eng. Sci.* **31**, 1221 (1976); E. Wicke, Chemical Reaction Engineering Reviews, *Adv. Chem. Ser.* (148), 75, American Chemical Soc., Washington, D.C., (1975); K. B. Bischoff, *Chem. Eng. Sci.* **16**, 131 (1961).

The solution is most easily obtained by making the substitution

$$\alpha = \frac{z - u\theta}{\sqrt{4D_L\theta}} \tag{6-30}$$

Then the partial differential equation (6-25) becomes the ordinary differential equation

$$\frac{d^2C^*}{d\alpha^2} + 2\alpha\frac{dC^*}{d\alpha} = 0 \tag{6-31}$$

where C^* is the dimensionless concentration C/C_0. The boundary conditions now are

$$C^* = \begin{cases} 1 & \text{for } \alpha = -\infty \\ 0 & \text{for } \alpha = \infty \end{cases} \tag{6-32}$$

Equations (6-31) and (6-32) are readily solvable to give C^* as a function of α, or z and θ, through Eq. (6-30). Substituting $z = L$ gives the response at the end of the reactor as a function of θ; this is

$$C^*_{z=L} = \left(\frac{C}{C_0}\right)_{\text{step}} = \frac{1}{2}\left[1 - \text{erf}\left(\frac{1}{2}\sqrt{\frac{uL}{D_L}}\frac{1 - \theta/(L/u)}{\sqrt{\theta/(L/u)}}\right)\right] \tag{6-33}†$$

The mean residence time $\bar{\theta} = V/Q$, or L/u. Hence, Eq. (6-33) can be written so that C/C_0 is a function of D_L/uL (the reciprocal of the Peclet number) and $\theta/\bar{\theta}$.

$$\left(\frac{C}{C_0}\right)_{\text{step}} = \frac{1}{2}\left[1 - \text{erf}\left(\frac{1}{2}\sqrt{\frac{uL}{D_L}}\frac{1 - \theta/\bar{\theta}}{\sqrt{\theta/\bar{\theta}}}\right)\right] \tag{6-34}$$

Equation (6-34) is plotted in Fig. 6-7 with curves for various values of D_L/uL. When $D_L/uL = 0$ there is no axial diffusion and the reactor fulfills the requirements of plug-flow behavior. The curve for this case in Fig. 6-7 is the expected step-function response for a plug-flow reactor (see Fig. 6-4 or Fig. 6-6). The other extreme, $D_L/uL = \infty$, corresponds to an infinite diffusivity (complete mixing), and stirred-tank performance is obtained. The curve for $D_L/uL = \infty$ in Fig. 6-7 is identical to that described by Eq. (6-13) and shown in Figs. 6-4 and 6-6. The curves between the two extremes allow for intermediate degrees of axial mixing. The use of the dispersion model presupposes that in an actual reactor the mixing state, perhaps consisting of some segregated flow and some axial mixing, can be represented by a particular value of D_L/uL.

In Example 6-1 D_L/uL is evaluated with the response curve for the laminar-flow reactor described in Sec. 6-4 (Fig. 6-6). Example 6-2 treats the same problem, but with other response data.

---

† The *error function* erf is defined as

$$\text{erf}(y) = \frac{2}{\sqrt{\pi}}\int_0^y e^{-x^2}\,dx; \quad \text{erf}(\pm\infty) = \pm 1; \quad \text{erf}(0) = 0; \quad \text{erf}(-y) = -\text{erf}(y)$$

values of erf (y) are given in standard mathematical tables; see, for example, "Handbook of Chemistry and Physics," 46th ed., p. A-113, Chemical Rubber Publishing Company, Cleveland, Ohio, 1964.

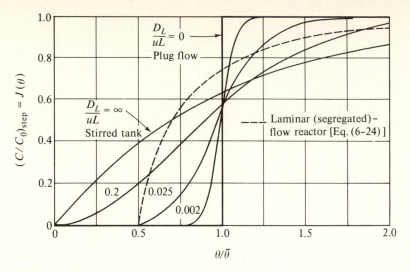

Figure 6-7 Response curves for dispersion model.

Example 6-1 Determine how well the dispersion model fits the segregated-flow (laminar-flow) reactor of Sec. 6-4.

SOLUTION Equation (6-24) gives the RTD, or $(C/C_0)_{step}$, for the laminar-flow tubular reactor. This equation, shown as a dashed line in Fig. 6-7, is in marked contrast to the dispersion curves. It is evident that introducing axial diffusion cannot account for the RTD given by segregated flow. The shape of the dashed curve is so different from those for the dispersion model that a value of D_L/uL cannot be chosen that will be even approximately correct. The effect on conversion of using a model which does not predict the correct RTD is illustrated in Sec. 6-8.

Example 6-2 Experimental response measurements on a continuous-flow tubular reactor give the following RTD:

$\theta/\bar{\theta}$	0	0.5	0.70	0.875	1.0	1.5	2.0	2 5	3.0
$J(\theta)$	0	0.10	0.22	0.40	0.57	0.84	0.94	0.98	0.99

Table 6-1 Conversion vs. RTD for first-order kinetics
$(k = 0.1 \text{ s}^{-1}, \bar{\theta} = 10 \text{ s})$

Type of reactor	Conversion	RTD	Example
Plug-flow reactor	0.63	Fig. 6-6, dashed line	
Actual tubular reactor	0.61	Fig. 6-9b	6-5
Stirred-tank reactor	0.50	Fig. 6-6, dashed curve	
Dispersion model	0.60	$D_L/uL = 0.117$	6-6
Series STR model	0.60	$n = 5$	6-7

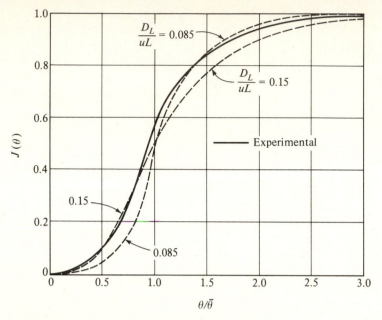

Figure 6-8 Fit of dispersion model to RTD data (Example 6-2).

Determine how well the dispersion model can be fitted to the RTD for this reactor and evaluate an appropriate D_L/uL.

SOLUTION Figure 6-8 shows the given RTD data and curves for two values of D_L/uL as computed from Eq. (6-34). In contrast to Example 6-1, the dispersion model fits the given RTD reasonably well. For $D_L/uL = 0.085$ the agreement is good for large $\theta/\bar{\theta}$, while the curve for $D_L/uL = 0.15$ fits well for small $\theta/\bar{\theta}$. An average value of 0.117 will be used in Sec. 6-9 for conversion calculations.

6-6 Interpretation of Response Data by the Series-of-Stirred-Tanks Model

In the series-of-stirred-tanks model the actual reactor is simulated by n ideal stirred tanks in series. The total volume of the tanks is the same as the volume of the actual reactor. Thus for a given flow rate the total mean residence time is also the same. The mean residence time per tank is $\bar{\theta}_t/n$. Figure 6-9a describes the situation. The objective is to find the value of n for which the response curve of the model would best fit the response curve for the actual reactor. To do this the relation between $(C/C_0)_{step}$ and n should be developed.

In Sec. 6-4 we found the desired result for $n = 1$; it was Eq. (6-13). The same method of writing a differential mass balance can be employed to find $(C/C_0)_{step}$ for any value of n. The result (see Example 6-3) is the series

$$(C_n/C_0)_{step} = J_n(\theta) = 1 - e^{-n\theta/\bar{\theta}_t}\left[1 + \frac{n\theta}{\bar{\theta}_t} + \frac{1}{2!}\left(\frac{n\theta}{\bar{\theta}_t}\right)^2 + \cdots + \frac{1}{(n-1)!}\left(\frac{n\theta}{\bar{\theta}_t}\right)^{n-1}\right]$$

$$(6-35)$$

The number of terms in brackets depends on n, the last term being the one with a power of $n\theta/\bar{\theta}_t$ equal to $n - 1$. For example, if $n = 1$, the last term is unity, or the first term in brackets. This result is the same as Eq. (6-13).

Figure 6-9b is a plot of Eq. (6-35) for various values of n. The similarity between Figs. 6-7 and 6-9b indicates that the axial-dispersion and series-of-stirred-tanks models give the same general shape of response curve. The analogy is exact for $n = 1$, for this curve in Fig. 6-9b agrees exactly with that in Fig. 6-7 for infinite dispersion, $D_L/uL = \infty$; both represent the behavior of an ideal stirred-tank reactor. Agreement is exact also at the other extreme, the plug-flow reactor ($n = \infty$ in Fig. 6-9b and $D_L/uL = 0$ in Fig. 6-7). The shapes of the curves for the two models are more nearly the same the larger the value of n.

The series-of-stirred-tanks model could not represent the RTD for the laminar-flow reactor shown in Fig. 6-6. However, the RTD data given in Example 6-2 can be simulated approximately. The dashed curve in Fig. 6-9b is a plot of this RTD. While no integer value† of n coincides with this curve for all $\theta/\bar{\theta}$,

---

† B. A. Buffham and L. G. Gibilaro [*AIChE J.*, **14**, 805 (1968)] have shown how the stirred-tanks-in-series model, with noninteger values of n, can be used to fit RTD data.

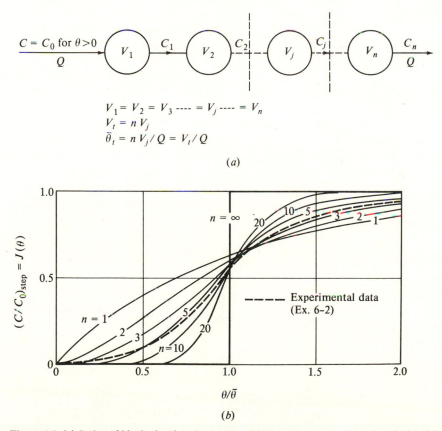

(a)

(b)

Figure 6-9 (*a*) Series of ideal stirred-tank reactors. (*b*) Response curves for series of stirred tanks.

the curve for $n = 5$ gives approximately the correct shape. Comparison of the fit in Figs. 6-8 and 6-9b indicates that about the same deviation exists between experimental and predicted RTD curves for either model.

Example 6-3 Develop Eq. (6-35) by deriving $(C_n/C_0)_{\text{step}}$ for $n = 2$ and 3 and then generalizing the results.

SOLUTION Figure 6-9a describes a step-function input of tracer of concentration C_0 for a series of ideal stirred-tank reactors. A mass balance on the j reactor in a series of n is, according to Eq. (3-1),

$$C_{j-1}Q - C_j Q = V_j \frac{dC_j}{d\theta}$$

or

$$\frac{dC_j}{d\theta} + \frac{n}{\bar{\theta}_t} C_j = \frac{n}{\bar{\theta}_t} C_{j-1} \tag{A}$$

since $\bar{\theta}_t = nV_j/Q$. The initial condition is $C_j = 0$ at $\theta = 0$. Integrating Eq. (A) formally with this initial condition gives

$$C_j = \frac{n}{\bar{\theta}_t} e^{-n\theta/\bar{\theta}_t} \int_0^\theta C_{j-1} e^{n\theta/\bar{\theta}_t} \, d\theta \tag{B}$$

Equation (B) may be integrated for successive stages in an n-stage system. For the first stage $C_{j-1} = C_0$, so that

$$C_1 = C_0 \frac{n}{\bar{\theta}_t} e^{-n\theta/\bar{\theta}_t} \int_0^\theta e^{n\theta/\bar{\theta}_t} \, d\theta$$

or

$$\frac{C_1}{C_0} = 1 - e^{-n\theta/\bar{\theta}_t} \tag{C}$$

When there is just one stage ($n = 1$) this becomes the familiar result for a single tank. In general Eq. (C) gives the response in the effluent from the first stage of an n-stage system.

Continuing to the second stage, we note that $C_{j-1} = C_1$ and that this is given by Eq. (C). Hence Eq. (B) becomes

$$C_2 = \frac{n}{\bar{\theta}_t} e^{-n\theta/\bar{\theta}_t} \int_0^\infty C_0(1 - e^{-n\theta/\bar{\theta}_t}) e^{n\theta/\bar{\theta}_t} \, d\theta \tag{D}$$

Integrating gives the response for the effluent from the second stage of an n-stage unit,

$$\frac{C_2}{C_0} = 1 - e^{-n\theta/\bar{\theta}_t} \left(1 + \frac{n\theta}{\bar{\theta}_t} \right) \tag{E}$$

For the third stage ($j = 3$) C_{j-1} in Eq. (B) becomes C_2, as given by Eq. (E). Proceeding as before, we find that the response in the effluent from the third stage is

$$\frac{C_3}{C_0} = 1 - e^{-n\theta/\bar{\theta}_t}\left[1 + \frac{n\theta}{\bar{\theta}_t} + \frac{1}{2}\left(\frac{n\theta}{\bar{\theta}_t}\right)^2\right] \tag{F}$$

By analogy with Eqs. (C), (E), and (F), the response from the fourth stage will be

$$\frac{C_4}{C_0} = 1 - e^{-n\theta/\bar{\theta}_t}\left[1 + \frac{n\theta}{\bar{\theta}_t} + \frac{1}{2}\left(\frac{n\theta}{\bar{\theta}_t}\right)^2 + \frac{1}{3!}\left(\frac{n\theta}{\bar{\theta}_t}\right)^3\right] \tag{G}$$

Hence the response from the nth stage of an n-stage system is described by Eq. (6-35).

6-7 Conversion in Nonideal Reactors

The maximum effect of the RTD on conversion is evident from the comparison of tubular-flow and stirred-tank reactors. Figure 4-15 shows such a comparison for first-order kinetics, and it is apparent that the differences are sizable at high conversion levels. For example, when $k_1\bar{\theta} = 4.0$ the conversion in the stirred-tank reactor is 80%, while in the tubular-flow unit it is 98%. The differences would be larger for second-order kinetics and smaller for half-order kinetics. These differences are for the extremes of RTD described by the two ideal reactors. For a reactor which followed neither ideal form, but showed an intermediate RTD, the conversion would be between the two extremes.

The other effect associated with nonideal flow depends on the extent of micromixing. As mentioned in Sec. 6-1, this error does not exist for first-order kinetics. This can be shown in a simple way. Consider two elements of fluid, of equal volume, moving through an increment ΔV of reactor volume. A single reaction occurs for which the rate is given by kC^a. Suppose the reactant concentration is C_1 in one element and C_2 in the other. Two extremes of micromixing may be imagined. In one, the elements are adjacent to each other and mix completely at the entrance to ΔV, giving a concentration $(C_1 + C_2)/2$. Afterwards, these elements flow on through the reactor increment ΔV in a completely mixed state. In the other case the two elements do not mix at all, but flow through the increment with the same residence time, retaining their individual concentrations. The rate of reaction for complete micromixing is, per unit volume

$$\mathbf{r}_m = k\left(\frac{C_1 + C_2}{2}\right)^a$$

For no micromixing (segregated flow) it is

$$\mathbf{r}_s = \tfrac{1}{2}(kC_1^a + kC_2^a) \tag{6-36}$$

The factor $\frac{1}{2}$ is needed because each element occupies one half of the volume increment. From Eq. (3-18) for the same residence times, V/Q, the ratio of the rates is equal to the ratio of the conversions. Thus

$$\frac{\Delta x_m}{\Delta x_s} = \frac{\mathbf{r}_m}{\mathbf{r}_s} = \frac{[(C_1 + C_2)/2]^\alpha}{(C_1^\alpha + C_2^\alpha)/2} \tag{6-37}$$

Evaluation of Eq. (6-37) for different values of α shows that

$$\frac{\Delta x_m}{\Delta x_s} = 1 \qquad \text{for } \alpha = 1 \text{ (first order)} \tag{6-38}$$

$$\frac{\Delta x_m}{\Delta x_s} > 1 \qquad \text{for } \alpha < 1 \tag{6-39}$$

$$\frac{\Delta x_m}{\Delta x_s} < 1 \qquad \text{for } \alpha > 1 \tag{6-40}$$

Thus for first-order kinetics the conversion will be the same regardless of the extent of micromixing. If the kinetics are second order or larger, then $\Delta x_m < \Delta x_s$, so that micromixing reduces the conversion. To obtain the maximum conversion for a given residence time for second-order kinetics a reactor with segregated flow is desirable. In contrast, for half-order reactions, a reactor with micromixing would give the higher conversion. The effect of micromixing on the conversion for non-first-order kinetics is generally less than the effect of the RTD, although exceptions can occur. Also, it should be remembered that unaccounted-for temperature differences in reactors can affect the isothermal conversion much more than deviations due either to the RTD or to micromixing. The magnitudes of these latter effects for representative cases are illustrated in later examples.

6-8 Conversion According to the Segregated-Flow Model

In segregated flow the conversion in each element is determined by the conversion-vs.-time relationship $x(\theta)$ as the element moves through the reactor. This relationship is given by the batch-reactor equations of Chap. 2 (see, for example, Table 2-10) according to the appropriate kinetics. Since the conversion in each element depends on its residence time, to obtain the average conversion we must divide the effluent into elements according to their residence time, that is, according to the RTD. The conversion in element is $x(\theta_i)\, dJ(\theta_i)$ or $x(\theta_i)J'(\theta_i)\, d\theta$. Therefore the average conversion in the effluent from the reactor will be

$$\bar{x} = \int_0^1 x(\theta)\, dJ(\theta) = \int_0^\infty x(\theta)J'(\theta)\, d\theta \tag{6-41}$$

To illustrate the use of Eq. (6-41) suppose we use the RTD for a stirred-tank reactor [Eq. (6-13)] but assume segregated flow. We wish to calculate the conversion for an irreversible first-order reaction. The function $x(\theta)$ is, from the second entry of Table 2-10,

$$x(\theta) = 1 - \frac{C}{C_0} = 1 - e^{-k_1\theta}$$

Using this function in Eq. (6-41) and $J'(\theta)$ given by Eq. (6-14), the mean conversion is

$$\bar{x} = \int_0^\infty (1 - e^{-k_1\theta}) \frac{1}{\bar{\theta}} e^{-\theta/\bar{\theta}}\, d\theta = \frac{k_1\bar{\theta}}{1 + k_1\bar{\theta}} \tag{6-42}$$

In contrast, suppose complete micromixing is assumed. Then, for the same RTD, an ideal stirred-tank reactor results. The conversion for a first-order reaction in this case is given by Eq. (4-17), which is identical to the above expression for segregated flow. This verifies the conclusion of Eq. (6-38) that the extent of micromixing does not affect conversion for first-order kinetics (as long as the correct RTD is used). The same development has been carried out for half-order and second-order kinetics,† and the results verify Eqs. (6-39) and (6-40). For example, the conversion was about 78% for a second-order reaction in segregated flow and 72% for complete micromixing, both based on the RTD of an ideal stirred-tank reactor. These figures are for $k_2 C_0 \bar{\theta} = 10$, a value that gives about the maximum effect of micromixing. For half-order reactions the maximum effect is about the same, but reversed. These results give an idea of the magnitude of deviations due to micromixing.

Example 6-4 Consider the laminar-flow reactor described in Sec. 6-4 and calculate the conversion for a first-order reaction for which $k_1 = 0.1 \text{ s}^{-1}$ and $\bar{\theta} = 10$ s.

SOLUTION The RTD for this reactor is given by Eq. (6-22) or Eq. (6-24). The flow is segregated, so that Eq. (6-41) is valid for calculating the conversion. Since $x(\theta)$ for a first-order reaction is

$$x(\theta) = 1 - e^{-k_1 \theta}$$

Eq. (6-41) becomes

$$\bar{x} = \int_{\frac{1}{2}\bar{\theta}}^{\infty} (1 - e^{-k_1 \theta}) \frac{1}{2} \frac{\bar{\theta}^2}{\theta^3} \, d\theta$$

Note that the lower limit for residence time is not zero for this reactor, but is instead $\frac{1}{2}\bar{\theta} = 5$ s. Inserting numerical values and integrating numerically, we obtain

$$\bar{x} = 50 \int_{5\text{ s}}^{\infty} (1 - e^{-0.1\theta}) \frac{d\theta}{\theta^3}$$

$$= 0.52$$

For the same $\bar{\theta}$ and k_1 the plug-flow result [Eq. (4-18)] is

$$\bar{x} = 1 - e^{k_1 \bar{\theta}} = 0.63$$

and the stirred-tank result [Eq. (4-17)] is

$$\bar{x} = \frac{k_1 \bar{\theta}}{1 + k_1 \bar{\theta}} = 0.50$$

For calculating conversion in this laminar-flow reactor the RTD for a stirred-tank reactor is more appropriate than that for a plug-flow reactor. This is not obvious from a comparison of the three RTDs shown in Fig. 6-6.

---

† H. Kramers and K. R. Westerterp, "Elements of Chemical Reactor Design and Operation," p. 87, Academic Press, Inc., New York, 1963; K. G. Denbigh, *J. Appl. Chem.*, **1**, 227 (1951).

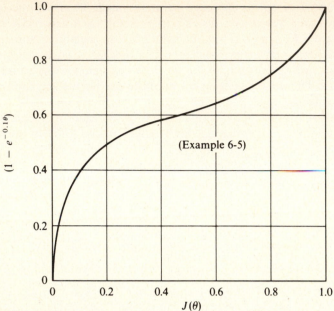

Figure 6-10 Conversion results according to segregated-flow model.

Example 6-5 Calculate the conversion for a first-order reaction ($k = 0.1$ s^{-1}, $\bar{\theta} = 10$ s), using the RTD given in Example 6-2 and assuming the flow to be segregated.

SOLUTION Equation (6-41) for this application is

$$\bar{x} = \int_0^1 (1 - e^{-0.1\theta})\, dJ(\theta)$$

where $J(\theta)$ is given by the data in Example 6-2. A plot of $(1 - e^{-0.1\theta})$ vs. $J(\theta)$ is shown in Fig. 6-10. According to the above equation for $\bar{x}$, they are under the curve from $J(\theta) = 0$ to $J(\theta) = 1$, is the conversion. Evaluation of this area gives $\bar{x} = 0.61$, which is the correct conversion. For the same values of $k_1 \bar{\theta}$, the results of Example 6-4 give conversions of 0.63 for a plug-flow reactor and 0.50 for a single, stirred tank.

The RTD of Example 6-2 is plotted as a dashed line in Figure 6-9b. This figure shows that not one, but five stirred tanks in series would best fit the RTD data. Thus the RTD comparison confirms that the single stirred tank would not provide a close simulation of the actual reactor. From this and the preceding example we have a good idea of how the RTD effects the conversion for first-order kinetics. The results are summarized in Table 6-1.

6-9 Conversion According to the Dispersion Model

In Sec. 6-5, a non-steady-state mass balance for a tubular-flow reactor (plug flow except for axial dispersion) was used to evaluate an effective diffusivity. Next we consider the problem of calculating the conversion when a reaction occurs in a dispersion-model reactor operated at steady-state conditions. Again a mass balance is written, this time for steady state and including reaction and axial-dispersion terms. As usual in this chapter the treatment is for isothermal conditions. It is considered now that the axial diffusivity is known.

Figure 6-11 shows a section Δz of the reactor with upward flow at a uniform velocity u. From Eq. (3-1), a mass balance around this section will be

$$\pi r_0^2 \left(-D_L \frac{dC}{dz} + uC \right)_z - \pi r_0^2 \left(-D_L \frac{dC}{dz} + uC \right)_{z+\Delta z} - \mathbf{r}\, \pi r_0^2\, \Delta z = 0$$

where $\mathbf{r}$ is the rate of reaction. Dividing by Δz and taking the limit as $\Delta z \to 0$ gives

$$D_L \frac{d^2C}{dz^2} - u \frac{dC}{dz} - \mathbf{r} = 0 \tag{6-43}$$

Note the similarities and differences between this steady-state equation and the transient one, Eq. (6-25). The boundary conditions for the two expressions are the same, that is, Eqs. (6-27) and (6-28). For first-order kinetics the solution is straightforward. Introducing $\mathbf{r} = k_1 C$ and using a dimensionless concentration $C^* = C/C_0$ and reactor length $z^* = z/L$, we obtain for the differential equation and boundary conditions

$$\frac{D_L}{uL} \frac{d^2C^*}{dz^{*2}} - \frac{dC^*}{dz^*} - k_1 \bar{\theta} C^* = 0 \tag{6-44}$$

$$C^* - \frac{D_L}{uL} \frac{dC^*}{dz^*} = 1 \qquad \text{at } z^* = 0 \tag{6-45}$$

$$\frac{dC^*}{dz^*} = 0 \qquad \text{at } z^* = 1 \tag{6-46}$$

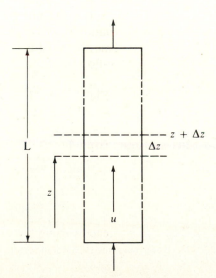

Figure 6-11 Axial section of tubular-flow reactor.

where $\bar{\theta} = L/u$. The solution of Eqs. (6-44) to (6-46) gives C as a function of z. Evaluating this function at the end of the reactor, $z = L$, we find

$$\frac{C}{C_0} = 1 - x = \frac{4\beta}{(1 + \beta)^2 e^{-(1/2)(uL/D_L)(1-\beta)} - (1 - \beta)^2 e^{-(1/2)(uL/D_L)(1+\beta)}} \quad (6\text{-}47)$$

where

$$\beta = \left(1 + 4k_1\bar{\theta}\frac{D_L}{uL}\right)^{1/2} \quad (6\text{-}48)$$

If D_L/uL, and of course $k\bar{\theta}$, are known, Eq. (6-47) gives the conversion predicted by the dispersion model, provided the reaction is first order. For most other kinetics numerical solution of the differential equation is necessary.†

As mentioned earlier, the dispersion model, like other models, is subject to two errors: inadequate representation of the RTD and improper allowance for the extent of micromixing. We can evaluate the first error for a specific case by using the dispersion model to obtain an alternate solution for Example 6-5. The second error does not exist for first-order kinetics. However, the maximum value of this error for second- and half-order kinetics was indicated in Sec. 6-8.

> **Example 6-6** In Example 6-2, $D_L/uL = 0.117$ was found to provide the best fit of the dispersion model to the reported RTD. Use this result to evaluate the conversion for a first-order reaction, with $k_1 = 0.1$ s^{-1} and $\bar{\theta} = 10$ s.
>
> SOLUTION From Eq. (6-48),
>
> $$\beta = [1 + 4(0.1)(10)(0.117)]^{1/2} = 1.21$$

Then, from Eq. (6-47),

$$1 - x = \frac{4(1.21)}{4.88e^{0.90} - 0.0445e^{-9.45}} = 0.40$$

or

$$x = 0.60$$

This result differs little from the value $x = 0.61$ obtained with the actual RTD (Example 6-5). We may conclude that the fit of the dispersion model to the actual RTD is adequate for conversion calculations.

6-10 Conversion According to the Series-of-Stirred-Tanks Model

The mass balance of reacting component for the first of a series of ideal stirred-tank reactors (see Eq. 3-5) is

$$C_0 - k\bar{\theta}_i C_1^\alpha = C_1 \quad (6\text{-}49)$$

---

† A. Burghardt and T. Zaleski [*Chem. Eng. Sci.*, **23**, 575 (1968)] have reviewed solutions of Eq. (6-43) for various forms of the rate equation.

where $\bar{\theta}_i$ = mean residence time (V/Q) in the first reactor
$\quad C_1$ = concentration of reactant in stream leaving the first reactor
$\quad \alpha$ = order of rate equation

If additional reactors with the same mean residence time are in series with the first one, the balance for the second reactor would be

$$C_1 - k\bar{\theta}_i C_2^\alpha = C_2 \tag{6-50}$$

and for the last reactor

$$C_{n-1} - k\bar{\theta}_i C_n^\alpha = C_n \tag{6-51}$$

Regardless of the value of α, the series of mass balances can be solved numerically for C_n and the conversion, provided $k\bar{\theta}_i$ and n are known. To use the stirred-tank model, n is found from response data, as illustrated in Sec. 6-6.

For first-order kinetics an analytical solution of Eqs. (6-49) to (6-51) is easily obtained; in fact, Eq. (4-16) gives the solution. For equal residence times in each reactor, $\bar{\theta}_i = \theta_t/n$, the result is

$$x = 1 - \frac{C_n}{C_0} = 1 - \frac{1}{(1 + k\bar{\theta}_t/n)^n} \tag{6-52}$$

This model of reactor behavior is illustrated in Example 6-7 by predicting the conversion from Eq. (6-52) for the same conditions as used in Examples 6-5 and 6-6.

Example 6-7 Using the series-of-stirred-tanks model to simulate the RTD of the reactor described in Example 6-2, predict the conversion for a first-order reaction for which $k = 0.1 \text{ s}^{-1}$ and $\bar{\theta} = 10 \text{ s}$.

SOLUTION In Sec. 6-6 we found that five stirred tanks in series fit the RTD of the reactor. With $n = 5$, the conversion from Eq. (6-52) is

$$x = 1 - \frac{1}{[1 + 0.1(10)/5]^5} = 1 - \frac{1}{(1.2)^5} = 1 - 0.40 = 0.60$$

This result is the same as that obtained with the dispersion model. Also, it deviates little from the result $x = 0.61$ obtained with the actual RTD (Example 6-5). In this case the estimated RTD obtained with either the dispersion or tanks-in-series model is a satisfactory representation for conversion calculations (see Table 6-1).

6-11 Conversion According to Recycle-Reactor Model

In this model the extent of mixing in the actual reactor is simulated by a flow-recycle reactor (see Fig. 4-19) with recycle ratio $R = Q_R/Q_e$. The evaluation of R from response data for a nonreactive system requires deriving an equation relating the effluent concentration (C_e) to the time, with R as a parameter. The derivation requires solution of a linear partial differential equation and is considered in Prob. 6-18. The resulting expression is analogous to Eq. (6-34) for the dispersion model and to Eq. (6-35) for the stirred-tanks-in-series model. In applying the model for conversion calculations, we can use Eq. (4-43) of Sec. 4-10 which gives the effluent concentration C_e in terms of the recycle ratio. Hence, if a given value of R is known to represent the RTD in the actual reactor, the conversion can be calculated. We illustrate the application of these equations in Example 6-8.

Example 6-8 Calculate the recycle ratio, in a flow recycle reactor, necessary to give the conversion of 61% for the actual reactor of Example 6-5.

SOLUTION For the first-order reaction of Example 6-5, Eq. (4-43) can be integrated to yield

$$\frac{V}{Q_f} = -(R+1) \int_{C_1}^{C_e} \frac{dC}{k_1 C} = -\frac{R+1}{k} \ln \left(\frac{C_e}{C_1}\right) \tag{A}$$

Substituting Eq. (4-44) for C_1 and noting that $V/Q_f = \bar{\theta}$ gives

$$\frac{k_1 \bar{\theta}}{R+1} = \ln \frac{1 + R(C_e/C_f)}{(C_e/C_f)(R+1)} \tag{B}$$

The ratio C_e/C_f is equal to C/C_0 in the nomenclature of this chapter. If we solve Eq. (B) for C_e/C_f the result is

$$\frac{C}{C_0} = \frac{C_e}{C_f} = \frac{1}{(R+1) \exp\left[k_1 \bar{\theta}/(R+1)\right] - R} \tag{6-53}$$

Equation (6-53) is the result [analogous to Eqs. (6-47) and (6-52)] that expresses C/C_0 in terms of the model parameter R.

Since $x_e = 1 - (C_e/C_f)$, Eq. (6-53) in terms of conversion becomes

$$x_e = 1 - \frac{1}{(R+1) \exp\left[k_1 \bar{\theta}/(R+1)\right] - R} \tag{C}$$

For $x_e = 0.61$ and $k_1 \bar{\theta} = 0.1(10) = 1.0$, Eq. (C) gives:

$$R = 0.20$$

We conclude that a model recycle reactor with a reflux ratio of 0.2 would correspond to a dispersion model with $D_L/uL = 0.117$ and a tanks-in-series model with 5 tanks. All models would give a conversion close to the actual value of 61% for this case of nearly plug-flow behavior.

The results of Examples 6-6 to 6-8 are for one case, but they are representative of the situation for many reactors. We saw in Secs. 6-7 and 6-8 that extremes of RTD can have large effects on the conversion at high conversion levels. However, with relatively simple models to estimate the RTD, little error need be involved. Put differently, if an engineer were to use an ideal stirred-tank reactor to simulate a nearly ideal tubular-flow unit at a high conversion level, the predicted conversion would be seriously in error. However, if the measured RTD or a reasonable model were employed, the result would be approximately correct. The residual error will be due to uncertainty in the extent of micromixing.

Residence time distributions greatly different from the ideals of plug flow or stirred tank can arise in several ways as shown in Figs. 3-3 and 3-4. One example is a stirred-tank reactor for polymerizing a viscous fluid, with two concentric cooling coils, as shown in Figure 6-12a. In such a situation the fluid between the coils, or between the coils and the reactor wall, may be nearly stagnant, while the fluid in the central section is well stirred. Here neither stirred-tanks-in-series nor dispersion models can fit the actual RTD. However, if response data are obtained, it is usually possible to develop a special model to fit the RTD. Thus a combination plug-flow and ideal stirred-tank reactor, as in Fig. 6-12b, might simulate the actual reactor. The flow rates to each reactor and their volumes provide adjustable parameters which can be chosen to fit the response data. The conversion in the exit stream from the simulation unit may be calculated by combining the results for each reactor as obtained from Eqs. (4-17) and (4-18). If this model is inadequate, additional plug-flow and stirred-tank units, of adjustable volumes,

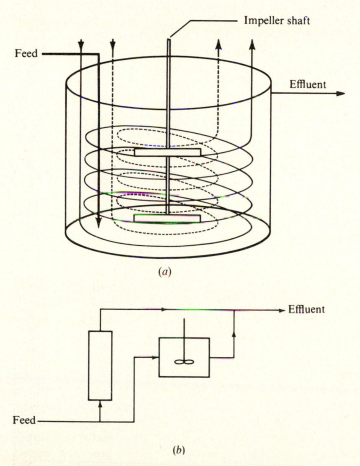

(a)

(b)

Figure 6-12 Polymerization tank reactor and simulation model.
(a) Polymerization reactor with two concentric cooling coils.
(b) Reactor model of parallel plug-flow and stirred-tank units.

may be added. While we have discussed only homogeneous reactions in this chapter, the presence of a second phase can cause an unusual RTD. A striking example is in the fluidized-bed reactor (Sec. 10-6) where the solid particles introduce nonidealities in the flow pattern of the gas stream.

Although the influence of nonideal flow on conversion may be only a small percentage, if the conversion level is high, the effect on subsequent separation units may be significant. Suppose that the conversion by an approximate model is 99% but proper accounting for nonideal flow would give 95%. Suppose also that the product must have a purity of 99.5%. The separation systems following the reactor would be considerably different in the two cases. One system would be designed to reduce the impurity from 1.0 to 0.5%, while the other would be for a reduction from 5.0 to 0.5%. Of course, errors in conversion estimates from any source, such as temperature variations, would have the same effect.

It is well to keep in mind that isothermal operation has been assumed throughout Chap. 6. It is difficult to treat analytically the nonisothermal problem. Yet, the effects of nonisothermal operation can be much greater than the mixing effects treated in this chapter.

Denbigh[†] and Levenspiel[‡] have provided useful guidelines for deciding when deviations (in conversion) from ideal tubular-flow performance are significant. In laminar flow, molecular diffusion in the axial direction causes little deviation if the reactor is reasonably long with respect to its diameter. Molecular diffusion in the radial direction may be important, particularly for gases, but it serves to offset the deviation from ideal performance caused by the velocity distribution. That is, radial diffusion tends to make the reactor behave more like an ideal tubular-flow unit. For turbulent flow, axial diffusion (which now includes eddy diffusion) can be significant. Its importance may be neglected if the Reynolds number is greater than 10^4 and the length is at least 50 times the diameter of the reactor. In turbulent flow, as in laminar flow, radial diffusion tends to improve the accuracy of the plug-flow model.

PROBLEMS

6-1 The RTD for a reactor is to be determined by tracer-concentration measurements. With a constant-density isothermal system, the following effluent concentration data are obtained in response to a pulse of tracer added to the feed:

t, min	0	5	10	15	20	25	30	35
Tracer concentration, g/cm^3	0	3.0	5.0	5.0	4.0	2.0	1.0	0

(a) Plot both $J(\theta)$ and $J'(\theta)$ vs. time on the same figure. (b) What is the mean residence time for the flow rate used?

† K. G. Denbigh, "Chemical Reactor Theory," pp. 50–51, Cambridge University Press, New York, 1965.

‡ Octave Levenspiel, Chemical Reaction Engineering, 2d ed. chap. 9, John Wiley & Sons, New York, 1972.

6-2 Response measurements to a step-function input are made for a reaction vessel. The following data are obtained for a given volumetric flow rate:

t, s	0	15	25	35	45	55	65	75	95
Tracer concentration, g/cm^3	0	0.5	1.0	2.0	4.0	5.5	6.5	7.0	7.7

(a) Plot the RTD expressed as $J(\theta)$ vs. time. (b) On the same graph plot $J'(\theta)$ vs. time. (c) What is $\bar{\theta}$ for this flow rate?

6-3 For the RTD of Prob. 6-1 determine (a) the value of D_L/uL which gives the most accurate fit of the dispersion model to the data, and (b), the most accurate integer value of n if the stirred-tanks-in-series model is used.

6-4 A liquid-phase reaction is currently carried out commercially in a series of three, equal volume, stirred-tank reactors operating isothermally. It is planned to replace these reactors with a single, tubular-flow reactor. To obtain the same RTD as in the stirred-tank equipment, and for the same V_t/Q, what degree of dispersion is necessary in the tubular reactor. That is, determine the value of D_L/uL if the flow in the tubular reactor could be represented by the dispersion model.

6-5 Repeat Prob. 6-3 for the RTD determined for Prob. 6-2.

6-6 Determine the RTD for an isothermal tubular-flow reactor in which the liquid is in laminar flow in an annulus of inner radius r_1 and outer radius $r_2(r_1/r_2 = \alpha)$. Neglect molecular diffusion. The velocity in the axial direction at any radius r (between r_1 and r_2) is given by

$$u(r) = A\left[1 - \left(\frac{r}{r_2}\right)^2 + \frac{1-\alpha^2}{\ln(1/\alpha)}\ln\frac{r}{r_2}\right]$$

Express the RTD, as $J(\theta)$, in terms of the mean residence time $\bar{\theta}$.

6-7 An ideal stirred-tank reactor followed by a plug-flow reactor is proposed as a model for the RTD of the reactor system in Prob. 6-1. The volumetric flow rate and combined volume of the two reactors will be the same as in Prob. 6-1. What ratio of the volume of the plug-flow and stirred-tank vessels would best represent the RTD? Comment on the suitability of the model.

6-8 Axial dispersion in a tubular-flow reactor becomes less important as the tube length increases (see Fig. 6-7). A homogeneous gas-phase reaction is carried out in a tube packed with inert particles to increase turbulence. The dispersion model (Sec. 6-5) fits flow in packed beds as well as flow in empty tubes, but the dimensionless model parameter is D_L/ud_p rather than D_L/uL (d_p is the diameter of the particles). At the flow rate used the particle Reynolds number is 50. At this Re for gases, D_L/ud_p is about 1.0. The particle diameter is 1.0 cm.

Suppose we want to obtain nearly plug flow and that this is achieved when the RTD corresponds to that for 10 stirred tanks in series. What length of tube would be required?

If we were satisfied with more axial dispersion, corresponding to an RTD of that for five stirred tanks in series, how much shorter would the tube be?

6-9 A series arrangement of one stirred-tank reactor and one plug-flow reactor is to be employed for an isothermal, homogeneous polymerization. The two reactors are of equal volume. A decision needs to be made: should the plug-flow reactor follow or precede the stirred tank?

(a) Sketch the RTD [as $(C/C_0)_{step}$ vs. θ] for both arrangements.
(b) Would the conversion of monomer be the same for the two arrangements if the reactions were first order?
(c) Would the conversions be the same if the reactions were second order?

6-10 Calculate the conversion in a laminar-flow tubular reactor for a second-order reaction $A + B \rightarrow C$ for which $k_2 = 100$ cm^3/(g mol)(s) and $\bar{\theta} = 10$ s. The feed concentration of both reactants is the same, 10^{-3} g mol/cm^3. Neglect molecular diffusion, so that the flow is segregated.

6-11 Calculate the conversion for the laminar-flow reactor of Prob. 6-10, using the dispersion model to represent the actual RTD and to calculate the conversion. The required solution of the nonlinear differential equation is given by Burghardt and Zaleski, *Chem. Eng. Sci.*, **23**, 575 (1968).

6-12 Calculate the conversion for the laminar-flow reactor of Prob. 6-10, using the stirred-tanks-in-series model to represent the RTD.

6-13 Reconsider the system in Prob. 6-10, with all conditions the same, except that volumes of different reactors required to obtain the same conversions (for the same flow rate) will be compared. Calculate the ratio of the volumes required for a laminar-flow reactor and a plug-flow reactor for several conversion levels between 0 and 100%. Do the results depend on the feed concentrations and the rate constant?

6-14 Repeat Prob. 6-13 for a first-order reaction for which $k_1 = 0.1 \text{ s}^{-1}$.

6-15 The flow characteristics of a continuous reactor are studied by suddenly introducing a quantity of miscible tracer into the feed stream. The concentrations of tracer in the effluent at various times after the instant of addition are:

t, min	0.1	0.2	1.0	2.0	5.0	10.	30
Tracer concentration, mg/liter	0.20	0.17	0.15	0.125	0.07	0.02	0.001

(*a*) What type of ideal reactor does the actual vessel most closely approach? (*b*) If an isothermal first-order reaction ($k_1 = 0.15 \text{ min}^{-1}$) occurs in the vessel, what conversion may be expected?

6-16 Another first-order reaction is to be studied at the same flow rate in the vessel characterized in Prob. 6-15. Measurements for this reaction in a plug-flow reactor, operating at the same mean residence time, give a conversion of 75.2%. (*a*) Calculate the value of the rate constant for this reaction. (*b*) Calculate the conversion expected for this reaction in the reactor of Prob. 6-15. (*c*) Are any assumptions made in obtaining the answer to part (*b*)? If the reaction were second order, would additional assumptions be necessary to use the same procedure for part (*b*)?

6-17 As a source of ethylene for petrochemicals, propane is thermally cracked to ethylene and methane. In one plant propane is passed through 16 ft. (long) sections of tubes. Each section is connected to the next with U-bends and the whole array is placed in a furnace.

The possible influence of nonideal flow is to be estimated. For this approximate study assume that the temperature in the tubes is constant and that the reaction gases are completely mixed in the U-bend at the end of each tube section and neglect the change in moles.

The cracking reactions are approximately first order and kinetics data indicate that three tube sections would be required to obtain 96% conversion of the propane for plug flow in each section. The flow rate is such that the Reynolds number in the tubes is 10,000. For these conditions the axial dispersion parameter, D_L/ud_t, is about 100 ($d_t = 4''$-tube diameter).

If the effect of nonideal flow is accounted for with the dispersion model, calculate

(*a*) the conversion at the end of 3 sections

(*b*) the number of sections required to obtain 96% conversion

6-18 A. Determine the RTD for a flow recycle reactor as a function of the recycle ratio R. This means deriving an equation for C_e/C_f as a function of time for a step-input of tracer to the apparatus shown in Fig. 4-19. Assume that $Q_e = Q_f$ and that at time zero a step function of tracer of concentration C_f is introduced. Prior to $t = 0$, an inert fluid with zero concentration has been flowing through the apparatus.

It is suggested that the derivation proceed as follows:

(1) Write a mass balance for tracer for a volume element ΔV within the plug-flow reactor.

(2) Write a boundary condition at the reactor entrance for $t \geq 0$. To find the concentration entering the reactor write a mass balance at the point of mixing the recycle stream and the fresh feed (point A in Fig. 4-19).

(3) Write an initial condition for the tracer concentration at any location within the reactor.
(4) Transform the partial differential equation obtained in step 1 to the Laplace domain. Then solve the resulting ordinary differential equation.
(5) Invert the transform solution in step 4 to give $C(t, V)$, where C is the concentration of tracer at time t and V is the volume at any length in the reactor.
(6) Evaluate $C(t, V)$ to apply at the exit of the reactor where $V = V_t$. This gives the desired result, $C_e(t)$. Express the result in terms of the residence time $\theta = V_t/Q_f$

    B. Sketch a plot of C_e/C_f vs. t/θ as prescribed by the solution obtained in A.
    C. Discuss how a value for the reflux ratio could be obtained from response data; i.e., from measurements of C_e vs. t for a step-function input of tracer.

SEVEN

HETEROGENEOUS PROCESSES, CATALYSIS, AND ADSORPTION

HETEROGENEOUS PROCESSES

The reaction rates and reactor designs considered thus far have been for homogeneous reactions. The remainder of the book is devoted to heterogeneous systems, particularly those involving fluid and solid phases. Such heterogeneous systems are very important because most chemical reaction processes require a solid catalyst. Accordingly, the main objective of the next three chapters is an understanding of the kinetics of reactions of fluids on solid catalysts. In this first of the three chapters, the essential differences between homogeneous and heterogeneous rates are presented. This is followed by a discussion of the general aspects of catalysis and adsorption. Chapter 8 is concerned with solid catalysts, particularly their physical properties, because these properties affect mass and energy transport within catalyst particles. Also included in Chap. 8 is a discussion of theories of behavior of solid catalysts and practical aspects such as preparation, life, etc. Then in Chap. 9 we discuss kinetics and rate equations for fluid-phase reactions occurring at the surface of solid catalysts. With this preparation we are in position to consider transport effects in heterogeneous processes (Chaps. 10 and 11) and reactor design (Chap. 13).

There are heterogeneous reactions other than those involving solid catalysts, for example, reactions between gases and solids such as in the reduction of metallic ores. Such gas-solid, noncatalytic reactions are taken up in Chap. 14. Also, there are important heterogeneous processes not involving solid phases. Examples are alkylation (liquid-liquid systems) and absorption and reaction processes such as the separation of CO_2 and H_2S from gases with amine solutions. These kinds of heterogeneous processes will be considered, but the major emphasis will be on fluid-solid catalytic reactions.

The phase boundaries inherent in heterogeneous systems require that transport processes (mass and energy transfer) as well as the intrinsic reaction rate be accounted for in reactor design. Consider the case of a gaseous reaction mixture and a solid catalyst. It is advantageous to use the same conservation equations for reactor design that were given in Chap. 3. Then much of our previous work on homogeneous systems (Chaps. 3, 4, and 5) will be applicable. This can be done if the reaction rate is expressed in terms of the bulk concentration and temperature of the gas phase. Such rates will include the effects of transport processes. In Equation (3-1) the rate term is based upon a volume element of reactor. If the volume element includes a heterogeneous region of reaction fluid and solid catalyst particles, the proper rate will incorporate mass and energy transfer from fluid to solid surface and within the particles (catalyst particles are normally porous). Such rates are called *global* or *overall* rates and their use allows us to apply many of the design equations given in Chaps. 3 to 5. In order to express the global rate in terms of bulk properties, expressions must be formulated for each of the steps in the overall process. The sequence of steps for converting reactants to products is:

1. Transport of reactants from the bulk fluid to the fluid-solid interface (external surface of catalyst particle)
2. Intraparticle transport of reactants into the catalyst particle (if it is porous)
3. Adsorption of reactants at interior sites of the catalyst particle
4. Chemical reaction of adsorbed reactants to adsorbed products (surface reaction—the intrinsic chemical step)
5. Desorption of adsorbed products
6. Transport of products from the interior sites to the outer surface of the catalyst particle
7. Transport of products from the fluid-solid interface into the bulk-fluid stream

At steady state the rates of all the individual steps will be the same. This equality can be used to develop a global rate equation in terms of the concentrations and temperatures of the bulk fluid. The derivation of such equations will be considered in detail in Chaps. 10 to 12, but a very simple treatment is given next to illustrate the nature of the problem.

7-1 Global Rates of Reaction

Consider an irreversible gas-phase reaction

$$A(g) \rightarrow B(g)$$

which requires a solid catalyst C. Suppose that the temperature is constant and that the reaction is carried out by passing the gas over a bed of *nonporous* particles of C. Since the catalyst is nonporous, steps 2 and 6 are not involved. The problem is to formulate the rate of reaction per unit volume† of the bed—that is, the *global rate for a catalyst particle*, r_v—in terms of the temperature and concentration of A in the bulk-gas stream. Note that these are the quantities that are measurable or can be specified (as design requirements), rather than the temperature and concentration at the gas-particle interface. Global rates of catalytic reactions are usually expressed per unit mass of catalyst, i.e., r_p. These are easily converted to rates r_v per unit volume by multiplying them by the bulk density ρ_B of the *bed* of catalyst particles.

The overall conversion of A to B in the bulk gas for this case of nonporous catalyst occurs according to steps 1, 3 to 5, and 7 in series. Let us further simplify the problem by supposing that steps 3 to 5 may be represented by a single first-order rate equation. Then the overall reaction process may be described in three steps: gas A is transported from the bulk gas to the solid surface, the reaction occurs at the interface, and finally, product B is transported from the catalyst surface to the bulk gas. Since the reaction is irreversible, the concentration of B at the catalyst surface does not influence the rate. This means that r_p can be formulated by considering only the first two steps involving A. Since the rates of these two steps will be the same at steady state, the disappearance of A can be expressed in two ways: either as the rate of transport of A to the catalyst surface,

$$r_p = k_m a_m (C_b - C_s) \tag{7-1}$$

or as the rate of reaction at the catalyst surface,

$$r_p = k C_s \tag{7-2}$$

In Eq. (7-1) k_m is the usual mass-transfer coefficient based on a unit of transfer surface, i.e., a unit of external area of the catalyst particle. In order to express the rate per unit mass‡ of catalyst, we multiply k_m by the external area per unit mass, a_m. In Eq. (7-2) k is the *reaction-rate constant* per unit mass of catalyst. Since a positive concentration difference between bulk gas and solid surface is necessary to transport A to the catalyst, the surface concentration C_s will be less than the bulk-gas concentration C_b. Hence Eq. (7-2) shows that the rate is less than it would be for $C_s = C_b$. Here the effect of the mass-transfer resistance is to reduce the rate. Figure 7-1 shows schematically how the concentration varies between bulk gas and catalyst surface.

---

† As defined, the volume element on which r_v is based must include at least one catalyst particle; otherwise the interphase physical effects cannot be considered. Yet this rate is used in such equations as (3-18) and (3-18a) as a rate applicable to a differential volume, i.e., as a point rate. This is an approximation which arises from treating the discrete nature of a bed of catalyst particles as a continuum. It is a necessary approximation because it is not yet possible to take into account variations in heat- and mass-transfer coefficients with position on the surface of a single catalyst particle. Thus average values of these coefficients over the particle surface are employed in formulating global rates. This is the value denoted by k_m in Eq. (7-1).

‡ Instead of rate per unit mass of catalyst, Eqs. (7-1) and (7-2) could be expressed as rates per unit external surface, in which case a_m would not be needed in Eq. (7-1) and k in Eq. (7-2) would be per unit of catalyst surface. Alternately, a rate per particle could be used. We shall generally use the rate per unit mass.

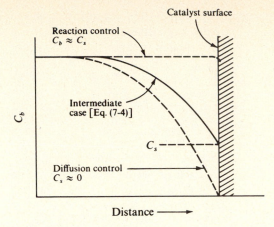

Figure 7-1 Mass transfer between fluid and catalyst surface.

The global rate can be expressed in terms of C_b by first solving for C_s from Eqs. (7-1) and (7-2); thus

$$C_s = \frac{k_m a_m}{k_m a_m + k} C_b \tag{7-3}$$

Then this result is substituted in either Eq. (7-1) or Eq. (7-2) to give

$$\mathbf{r}_p = \frac{k k_m a_m}{k_m a_m + k} C_b = \frac{1}{1/k + 1/k_m a_m} C_b \tag{7-4}$$

This is the expression for the global rate in terms of the bulk-reactant concentration. The second equality shows that the effects of reaction and mass transfer are additive. Since the terms are in the denominator, they are *resistances* to the rate. Thus we speak of a reaction resistance ($1/k$ term) and a mass-transfer resistance ($1/k_m a_m$ term). The concentration profile in this case is shown by the solid line in Fig. 7-1. It is a very restricted illustration of a global rate, since heat-transfer effects were not considered and only external mass transfer is involved (the catalyst particle is nonporous). These restrictions are removed in the detailed treatment in Chaps. 10 and 11, but this simple example illustrates the meaning of global rates of reaction for heterogeneous systems.

A hypothetical reaction has been used to develop the previous results. It is of interest to know the magnitude of the global rate for *real* situations. Fortunately, considerable experimental data are available. Measurements[†] for the oxidation of SO_2 with air on a platinum catalyst, deposited on $\frac{1}{8} \times \frac{1}{8}$-in. cylindrical pellets,[‡] gave a global rate of 0.0956 g mol/(h)(g catalyst). The bulk temperature of the gas was 465°C and the gas velocity in the catalyst bed was 350 lb/(h)(ft²). The bulk composition corresponded to an SO_2 partial pressure of 0.06 atm. At these conditions both mass- and heat-transfer resistances between the bulk gas and the surface of the catalyst pellets were important. Calculations with Eq. (7-1) indicated

---

[†] R. W. Olson, R. W. Schuler, and J. M. Smith, *Chem. Eng. Progr.*, **42**, 614 (1950).

[‡] The particles were of porous alumina, but the platinum was deposited on the outer surface, so that, in effect, a nonporous catalyst was used.

that the partial pressure of SO_2 dropped to 0.040 atm at the solid surface from the value of 0.06 atm in the bulk gas. This relatively large difference means that the diffusion resistance was large. If the rate were evaluated without consideration of this resistance—that is, if it were evaluated for $p_{SO_2} = 0.06$ atm at 465°C—it would be 0.333 g moles SO_2 reacted/(h)(g catalyst). At $p_{SO_2} = 0.04$ atm it would be 0.0730. Hence the error in neglecting the mass-transfer resistance would be large; that is, the rate so computed would be $(0.333/0.0730)(100)$, or 350% higher than the global rate.

Actually, the temperature at the catalyst surface is about 15°C above the bulk-gas temperature (see Example 10-2) because of heat-transfer resistance. Hence the reaction at the catalyst surface occurred at 480°C. According to the activation energy for this reaction, a 15°C temperature rise would increase the rate about 31%. Hence neglecting the heat-transfer resistance leads to a rate 31% less than the global value. If both thermal and diffusion resistances were neglected, the rate would be $(0.333/0.0956)(100)$, or 250% higher than the global rate. For this exothermic reaction the diffusion and thermal resistances have opposite effects on the rate. The example is extreme in that mass-transfer resistance was relatively large in comparison with reaction resistance. At lower temperatures and/or higher gas velocities past the catalyst pellet (higher k_m) the mass-transfer effect would be less.

Frequently external heat-transfer resistances are much larger than mass-transfer resistances, in contrast to the preceding example. As an illustration, consider the data of Maymo and Smith† for the oxidation of hydrogen with oxygen on a platinum-on-alumina catalyst. Rates and temperatures were measured for a single porous catalyst pellet (1.86 cm diameter) suspended in a well-mixed gas containing primarily hydrogen with a small percentage of oxygen and water vapor. Because of the turbulence in the gas, the concentration difference between bulk gas and pellet surface was negligible. However, there was a significant thermal resistance, so that the pellet-surface temperature was greater than the bulk-gas temperature. The pellet was porous with uniform distribution of platinum throughout. Hence there were internal resistances, both to mass and to heat transfer. In one particular run the measured rate of water production was 49.8×10^{-6} mol/(g catalyst)(s). This is the global value and includes the effects of both internal and external transport resistances. The bulk-gas, pellet-surface, and pellet-center temperatures were 89.9, 101, and 147.7°C, respectively. The rate of the chemical step on the surface was also measured by experiments with fine particles of catalyst for which there were no temperature or concentration differences between bulk gas and surface or interior of particle. This intrinsic rate was correlated by the equation

$$\mathbf{r}_p = 0.655 \, p_{O_2}^{0.804} e^{-5230/R_g T} \tag{7-5}$$

where p is in atmospheres and T is in degrees Kelvin. We can evaluate the effect of the external resistance in the following way. First calculate the rate at the surface temperature $(101 + 273 \text{ K})$ and $p_{O_2} = 0.0527$ atm, which is the oxygen partial

---

† J. A. Maymo and J. M. Smith, *AIChE J.*, **12**, 845 (1966).

pressure, either in the bulk gas or at the pellet surface. Substituting these quantities in Eq. (7-5) gives $r_p = 54.3 \times 10^{-6}$ mol/(g catalyst)(s). If the rate were evaluated at bulk-gas conditions (89.9°C and $p_{O_2} = 0.0527$ atm), the result from Eq. (7-5) would be 43.6×10^{-6} mol/(g catalyst)(s). Hence neglecting the external-temperature difference would give a rate [(54.3 − 43.6)/54.3](100), or 20%, less than the correct value.

This example also shows the effects of mass- and energy-transfer resistances within the catalyst pellet. The temperature increases toward the center of the pellet and increases the rate, but the oxygen concentration goes down, tending to reduce the rate. The global value of 49.8×10^{-6} is the resultant balance of both factors. Hence the net error in using the bulk conditions to evaluate the rate would be [(49.8 − 43.6)/49.8](100), 12.5%. In this case the rate increase due to external and internal thermal effects more than balances the adverse effect of internal mass-transfer resistance. The procedure for calculating the effects of internal gradients on the rate is presented in Chap. 11.

One additional illustration is of interest. This refers to the hydrofluorination of UO_2 pellets with $HF(g)$ according to the reaction

$$UO_2(s) + 4HF(g) \rightarrow UF_4(s) + 2H_2O(g)$$

This is a different kind of heterogeneous reaction—a gas-solid noncatalytic one. Let us examine the process at initial conditions ($t \rightarrow 0$), so that there has been no opportunity for a layer of $UF_4(s)$ to be formed around the UO_2 pellet. The process is much like that for gas-solid catalytic reactions. Hydrogen fluoride gas is transferred from the bulk gas to the surface of the UO_2 pellets and reacts at the pellet-gas interface, and H_2O diffuses out into the bulk gas. If the pellet is nonporous, all the reaction occurs at the outer surface of the UO_2 pellet, and only an external transport process is possible. Costa[†] studied this system by suspending spherical pellets 2 cm in diameter in a stirred-tank reactor. In one run, at a bulk-gas temperature of 377°C, the surface temperature was 462°C and the observed rate was $r = 6.9 \times 10^{-6}$ mol UO_2/(s)(cm² reaction surface). At these conditions the concentrations of HF gas were 1.12×10^{-5} g mol/cm³ at the surface of the pellet and 1.38×10^{-5} g mol/cm³ in the bulk gas. The intrinsic rate of reaction was found to be given by

$$r = 40 C_{HF} e^{-6070/R_g T} \quad \text{g mol/(cm}^2)(s) \tag{7-6}$$

where T is in degrees Kelvin and C_{HF} is in gram moles per cubic centimeter. If the rate is evaluated at bulk-gas conditions, Eq. (7-6) gives 5.0×10^{-6} g mol/(s)(cm²). Then the combined effects of an external temperature and concentration difference serve to increase the rate from 5.0×10^{-6} to 6.9×10^{-6} g mol/(cm²)(s), or 38%. In this case the temperature at the catalyst surface is 95°C higher than the bulk-gas temperature, and this has a dominating effect on the rate of reaction. The effect of external mass-transfer resistance is to reduce the rate of this first-order reaction by (1.38 − 1.12)/1.38, or 19%.

† E. C. Costa and J. M. Smith, *AIChEJ.*, **17**, 947 (1971).

7-2 Types of Heterogeneous Reactions

All the examples in Sec. 7-1 were of the gas-solid form. As mentioned earlier this is a most important type of heterogeneous system. Most salable chemicals are prepared by converting raw materials via chemical reactions. Such reactions usually require catalysts, and these are normally solids. Since the temperatures must be high for rapid rates, the reacting fluid commonly is in the gas phase. Examples of large-scale gas-solid catalytic reactions are the major hydrocarbon transformations: cracking, reforming, dehydrogenation (for example, butadiene and butenes from butane), isomerization, desulfurization, etc. Kinetics and reactor design for such systems have naturally received major attention, and most of our emphasis on design is directed to this type of flow reactor (Chap. 13).

Gas-solid heterogeneous reactions may be noncatalytic.[†] An example is the hydrofluorination of uranium dioxide pellets referred to in Sec. 7-1. Since one reactant is in the solid phase and is consumed, the rate of reaction varies with time. Hence such processes are intrinsically dynamic, in comparison with the stead-state operation of gas-solid catalytic reactors. The process for smelting ores such as zinc sulfide,

$$ZnS(s) + \tfrac{3}{2}O_2(g) \rightarrow ZnO(s) + SO_2(g)$$

is of this type. The conversion of $CaCO_3$ to CaO in the line kiln is another example. Yet another is the process for making HCl in a transport reactor[‡] from salt particles; the reaction is

$$2NaCl(s) + SO_3(g) + H_2O(g) \rightarrow Na_2SO_4(s) + 2HCl(g)$$

In many noncatalytic types a solid product builds up around the reacting core [for example, $Na_2SO_4(s)$ is deposited around the NaCl particles in the last illustration]. This introduces the additional physical processes of heat and mass transfer through a product layer around the solid reactant. A somewhat different form of noncatalytic gas-solid reaction is the regeneration of catalysts which have been deactivated by the deposition of a substance on the interior surface. The most common is the burning (with air) of carbon which has been gradually deposited on catalyst particles used in hydrocarbon reactions. Many of the physical and chemical steps involved here are the same as those for gas-solid catalytic reactions. The chief difference is the transient nature of the noncatalytic reaction. This type of heterogeneous reaction will be considered in Chap. 14. Another dynamic non-catalytic process is the separation or purification of gases or liquids by adsorption. Such processes are normally carried out in tubes packed with the adsorbent particles. Adsorption of the higher molecular weight hydrocarbons from gas streams in beds of activated carbon and removal of organic pollutants from water with activated carbon are examples.

[†] Julian Szekely, James W. Evans and Hong Yong Sohn, "Gas-Solid Reactions," Academic Press, New York, 1976.

[‡] A transport reactor is like a fluidized-bed reactor (Fig. 1-8), except that the fluidized particles move through and out of the reactor with the gas phase.

A steady-state example of noncatalytic reactions is the gas-liquid form in which a gaseous reactant is absorbed in the liquid and subsequently reacts in the liquid phase. The reactor in such cases is usually a countercurrent or cocurrent absorption column of the bubble type† (gas bubbling through a continuous liquid phase, spray column or a tube packed with inert particles). Examples are the separation of CO_2 from a gas by adsorption and reaction in an aqueous amine solution.‡

Liquid-solid reactions, where the solid phase is a catalyst, is a type of heterogeneous system that is common in the chemical and petroleum industries. Alkylation with $AlCl_3(s)$ catalyst is an example. In these systems the catalyst frequently forms complexes with reactants and/or products and becomes a poorly defined solid-liquid mixture, best described as a sludge. Analytical treatment in these cases is difficult. Three-phase systems involving a gas, liquid, and solid are also encountered. Hydrogenation of liquids normally is of this type. For example, a slurry may be formed of the catalyst particles in the oil to be hydrogenated. Then hydrogen is bubbled into the slurry. These reactions are carried out in stirred-tank reactors, where the high heat capacity of the tank contents simplifies temperature control of the exothermic hydrogenation reaction. Such gas-liquid-solid systems involve several physical steps, and indeed, resistance to diffusion of dissolved hydrogen to the catalyst particle is frequently significant. The quantitative treatment of combined physical and chemical steps in such slurry reactors is considered in Chaps. 10 and 13. Polymerization systems are often of this type. For example, ethylene is polymerized by dissolving it in a solvent containing suspended catalyst particles. The trickle-bed reactor is a somewhat different form of gas-liquid-solid system. Here gas and liquid, usually in cocurrent flow, pass over a bed of catalyst particles in a fixed bed. This form is used when the volatility of the liquid is so low that complete gasification is not practical. Sulfur compounds are removed from hydrocarbon liquids in such reactors. The gas phase is primarily hydrogen. Oxidation of liquids also is carried out in trickle-bed reactors (see Chaps. 10 and 13).

Liquid-liquid reactions are sometimes encountered. Alkylation of hydrocarbons with sulfuric acid or HF as a catalyst is an example.

Solid-solid noncatalytic reactions are important in ceramics manufacture.§ It appears that diffusion resistances may be important to some extent in all such systems. The diffusion process itself is difficult to define in solid-solid systems. At least two possibilities exist: volume diffusion in the solid and surface diffusion along interfaces and crystal boundaries.¶ Little is known about the kinetics of solid-solid reactions at the reacting interface because most measurements include diffusion effects.

---

† C. G. Hagberg and F. X. Krupa, *Proc. 4th Int. Symp. Chem. React. Eng.*, Sec. IX, p. 408, April, 1976, Heidelberg, DECHEMA, Frankfurt.

‡ P. V. Danckwerts, "Gas-Liquid Reactions," McGraw-Hill Book Company, New York, 1960.

§ W. Kingery, "Kinetics of High Temperature Processes," John Wiley & Sons, Inc., New York, 1959; G. Cohn, *Chem. Rev.*, **42**, 527 (1948).

¶ R. J. Arrowsmith and J. M. Smith, *Ind. Eng. Chem. Fund. Quart.*, **5**, 327 (1966).

CATALYSIS

The first step in determining global rates is to examine the processes associated with the solid surface (the adsorption, desorption, and intrinsic reaction steps in the list at the beginning of the chapter). We start out with a brief qualitative discussion of the nature of catalysis and then complete this chapter with an examination of adsorption. As kinetic information began to accumulate during the last century, it appeared that the rates of a number of reactions were influenced by the presence of a material which itself was unchanged during the process. In 1836 Berzelius† reviewed the evidence and concluded that a "catalytic" force was in operation. Among the cases he studied were the conversion of starch into sugar in the presence of acids, the decomposition of hydrogen peroxide in alkaline solutions, and the combination of hydrogen and oxygen on the surface of spongy platinum. In these three examples the acids, the alkaline ions, and the spongy platinum were the materials which increased the rate and yet were virtually unchanged by the reaction. Although the concept of a catalytic force proposed by Berzelius has now been discarded, the term "catalysis" is retained to describe all processes in which the rate of a reaction is influenced by a substance that remains chemically unaffected.

7-3 The Nature of Catalytic Reactions

Although the catalyst remains unchanged at the end of the process, there is no requirement that the material not take part in the reaction. In fact, present theories of catalyst activity postulate that the material does actively participate in the reaction. From the concept of the energy of activation developed in Chap. 2, the mechanism of catalysis would have to be such that the free energy of activation is lowered by the presence of the catalytic material. A catalyst is effective in increasing the rate of a reaction because it makes possible an alternative mechanism, each step of which has a lower free energy of activation than that for the uncatalyzed process. Consider the reaction between hydrogen and oxygen in the presence of platinum. According to the proposed concept, hydrogen combines with the platinum to form an intermediate substance, which then reacts with oxygen to provide the final product and reproduce the catalyst. It is postulated that the steps involving the platinum surface occur at a faster rate than the homogeneous reactions between hydrogen and oxygen.

The combination or complexing of reactant and catalyst is a widely accepted basis for explaining catalysis. For example, suppose the overall reaction

$$A + B \rightleftharpoons C$$

is catalyzed via two active centers, or catalytic sites, X_1 and X_2, which form complexes with A and B. The reaction is truly catalytic if the sequence of steps is

---

† J. J. Berzelius, *Jahresher. Chem.*, **15**, 237 (1836).

such that the centers X_1 and X_2 are regenerated after they have caused the formation of C. In a general way the process may be written

1. $A + X_1 \rightleftharpoons AX_1$

2. $B + X_2 \rightleftharpoons BX_2$

3. $AX_1 + BX_2 \rightleftharpoons C + X_1 + X_2$

Note that whereas X_1 and X_2 are occupied and regenerated a number of times, it does not necessarily follow that their catalyzing ability and/or number remain constant forever. For example, poisons can intervene to slowly remove X_1 and/or X_2 from the system, arresting the catalytic rate. What distinguishes this decline in catalytic activity from that of a noncatalytic reaction in which X_1 and X_2 are not regenerated is that the complexing-regenerating sequence occurs a great many times before X_1 and X_2 become inactive. In the noncatalytic sequence no regeneration of X occurs. Hence, while catalysts can deteriorate, their active lifetime is far greater than the time required for reaction.

A relatively small amount of catalyst can cause conversion of a large amount of reactant. For example, Glasstone† points out that cupric ions in the concentration of 10^{-9} mol/liter appreciably increase the rate of the oxidation of sodium sulfide by oxygen. However, the idea that a small amount of the catalyst can cause a large amount of reaction does not mean that the catalyst concentration is unimportant. In fact, when the reaction does not entail a chain mechanism, the rate of the reaction is usually proportional to the concentration of the catalyst. This is perhaps most readily understood by considering the case of surface catalytic reactions. In the reaction of hydrogen and oxygen with platinum catalyst the rate is frequently directly proportional to the platinum surface. Here a simple proportionality exists between platinum surface area and the number of centers X which catalyze the oxidation of hydrogen. While such a simple relationship may not often exist in solid-catalyzed reactions, in homogeneous catalysis there is frequently a direct proportionality between rate and catalyst concentration. For example, the hydrolysis of esters in an acid solution will depend on the concentration of hydrogen ion acting as a catalyst.

The position of equilibrium in a reversible reaction is not changed by the presence of the catalyst. This conclusion has been verified experimentally in several instances. For example, the oxidation of sulfur dioxide by oxygen has been studied with three catalysts: platinum, ferric oxide, and vanadium pentoxide. In all three cases the *equilibrium compositions* were the same.

An important characteristic of a catalyst is its effect on selectivity. An illustration is the decomposition of ethanol. Thermal decomposition gives water, acetaldehyde, ethylene, and hydrogen. If, however, ethanol vapor is suitably contacted with alumina particles, ethylene and water are the only products. In contrast, dehydrogenation to acetaldehyde is virtually the sole reaction when ethanol is reacted over a copper catalyst.

---

† S. Glasstone, "Textbook of Physical Chemistry," p. 1104, D. Van Nostrand Company, Inc., New York, 1940.

The general characteristics of catalysis may be summarized as follows:

1. A catalyst accelerates reaction by providing alternate paths to products, the activation energy of each catalytic step being less than that for the homogeneous (noncatalytic) reaction.
2. In the reaction cycle, active centers of catalysis first combine with at least one reactant and then are reproduced with the appearance of product. The freed center then recombines with reactant to produce another cycle, and so on.
3. Comparatively small quantities of catalytic centers are required to produce large amounts of product.
4. Equilibrium conversion is not altered by catalysis. A catalyst which accelerates the forward reaction in an equilibrium system is a catalyst for the reverse reaction.
5. The catalyst can radically alter selectivity.

Examples have been observed of negative catalysis, where the rate is decreased by the catalyst. Perhaps the most reasonable theory is that developed for chain reactions. In these cases it is postulated that the catalyst breaks the reaction chains, or sequence of steps, involved in the mechanism. For example, nitric oxide reduces the rate of decomposition of acetaldehyde or ethyl ether. Apparently nitric oxide has the characteristic of combining with the free radicals involved in the reaction mechanism. The halogens, particularly iodine, also act as negative catalysts in certain gaseous reactions. In the combination of hydrogen and oxygen, where a chain mechanism is probably involved, iodine presumably destroys the radicals necessary for the propagation of the chains.

Reactions in which the rate increases with *product* concentration also occur. Such reactions are termed autocatalytic. The most common examples are fermentation processes "catalyzed" by microorganisms such as yeasts, bacteria, and algae. The processes are complex but may be described approximately as a reaction between an organic feed and microorganism to produce products and more microorganisms. For example, the fermentation of sugar may be expressed:

$$\text{Organic feed} + \text{microorganism} \rightarrow \text{inert products} + \text{more yeast} \qquad (7\text{-}7)$$
$$\text{(e.g., sugars)} \qquad\qquad \text{(yeast)} \qquad\qquad \text{(alcohol)}$$

If a batch reactor initially contains a sugar solution and a small quantity of yeast, the rate will increase as the reaction proceeds, because the yeast produced is also a reactant. Ultimately, the depletion of sugar will cause the rate to reach a maximum and decrease toward zero. The microorganism does not act as a true catalyst since it is a reactant, but its presence does increase the rate. Such abnormal kinetics has been extensively treated in the literature, partly because the unusual rate vs. conversion relationship leads to interesting problems of reactor choice and optimization.[†][‡] The kinetic equations are illustrated in Prob. 7-5 for bacterial oxidation.

---

† K. B. Bischoff, *Can. J. Chem. Eng.*, **44**, 281 (1966).
‡ K. J. Laidler, "Chemical Kinetics," McGraw-Hill Book Company, New York, 1965.

7-4 The Mechanism of Catalytic Reactions

The concept that a catalyst provides an alternate mechanism for accomplishing a reaction, and that this alternate path is a more rapid one, has been developed in many individual cases. The basis for the theory is that first the catalyst and one or more of the reactants form an intermediate complex, a loosely bound compound which is unstable. This complex then takes part in subsequent reactions which result in the final products and the regenerated catalyst. Homogeneous catalysis can frequently be explained in terms of this concept. For example, consider catalysis by acids and bases. In aqueous solutions acids and bases can increase the rate of hydrolysis of sugars, starches, and esters. The kinetics of the hydrolysis of ethyl acetate catalyzed by hydrochloric acid can be explained by the following mechanism:

1. $CH_3COOC_2H_5 + H^+ \rightleftharpoons CH_3COOC_2H_5[H^+]$

2. $CH_3COOC_2H_5[H^+] + H_2O \rightleftharpoons C_2H_5OH + H^+ + CH_3COOH$

For this catalytic sequence to be rapid with respect to noncatalytic hydrolysis, the free energy of activation of reaction steps 1 and 2 must each be less than the free energy of activation for the noncatalytic reaction,

$$CH_3COOC_2H_5 + H_2O \rightleftharpoons CH_3COOH + C_2H_5OH$$

Similarly, the *heterogeneous* catalytic hydrogenation of ethylene on a solid catalyst might be represented by the steps

1. $\overset{\Delta F_1}{C_2H_4 + X_1 \rightleftharpoons C_2H_4X_1}$

2. $\overset{\Delta F_2}{H_2 + X_1C_2H_4 \rightleftharpoons C_2H_4[X_1]H_2}$

3. $\overset{\Delta F_3}{C_2H_4[X_1]H_2 \rightleftharpoons C_2H_6 + X_1}$

where X_1 is an active site on the solid catalyst and $C_2H_4[X_1]H_2$ represents the complex formed between the reactants and the catalyst. The *homogeneous* reaction, according to the absolute theory of reaction rates discussed in Chap. 2, would be written

$$C_2H_4 + H_2 \overset{\Delta F^*}{\rightleftharpoons} \overset{\text{Activated}}{\underset{}{\text{complex}}} C_2H_4 \cdot H_2 \to C_2H_6$$

where the free-energy change for the formation of the activated complex ΔF^* is the free energy of activation for the homogeneous reaction. The effectiveness of the catalyst is explained on the basis that the free energy of activation of each of the steps in the catalytic mechanism is less than ΔF^*.

These illustrations, particularly that for ethylene hydrogenation, are grossly oversimplified. They must be considered phenomenological models, not mechanisms. The actual mechanism of ethylene hydrogenation is quite complex. In spite of the considerable effort focused on this reaction, a mechanism satisfactory to all investigators has yet to be offered. The system does, however, provide an opportunity to compare homogeneous and heterogeneous rates. Using published data, Boudart† found that the homogeneous and catalytic rates can be expressed as

$$r_{hom} = 10^{27}e^{-43,000/R_g T} \tag{7-8}$$

$$r_{cat} = 2 \times 10^{27}e^{-13,000/R_g T} \quad \text{(CuO-MgO catalyst)} \tag{7-9}$$

At 600 K the relative rates are

$$\frac{r_{cat}}{r_{hom}} = e^{(43,000-13,000)/600R_g} \simeq 10^{11}$$

In this case the catalyst has caused a radical reduction in overall activation energy, presumably by replacing a difficult homogeneous step by a more easily executed surface reaction involving adsorbed ethylene. The results lead to the kinetics observed by Wynkoop and Wilhelm,‡ a reaction first order in H_2 and zero order in strongly adsorbed ethylene.

The three steps postulated for the catalytic hydrogenation of ethylene indicate that the rate may be influenced by both adsorption and desorption (steps 1 and 3) and the surface reaction (step 2). Extreme cases can be imagined: that step 1 or step 3 is slow with respect to step 2 or that step 2 is relatively slow. In the first two cases rates of adsorption or desorption are of interest since they determine the rate while in the third case the surface concentration of the adsorbed species, corresponding to equilibrium with respect to steps 1 and 3, is needed. In any event we should like to know the number of sites on the catalyst surface, or at least the surface area of the catalyst. These questions require a study of adsorption equilibrium and adsorption rates. This is the objective in Secs. 7-5 to 7-7.

The detailed mechanism of how solid catalysts operate requires knowledge of the structure of the complex between reactant and catalyst site. Understanding of this aspect to catalysis has been slow in developing. A brief review of suggested theories and experimental evidence is given in Sec. 8-4.

ADSORPTION

7-5 Surface Chemistry and Adsorption§

Even the most carefully polished surfaces are not smooth in a microscopic sense, but are irregular, with valleys and peaks alternating over the area. The regions of irregularity are particularly susceptible to residual force fields. At these locations the surface atoms of the solid may attract other atoms or molecules in the surrounding gas or liquid phase. Similarly, the surfaces of pure crystals have nonuni-

---

† M. Boudart, *Ind. Chim. Belg.*, **23**, 383 (1958).

‡ R. Wynkoop and R. H. Wilhelm, *Chem. Eng. Progr.*, **46**, 300 (1950).

§ A discussion of the thermodynamics and kinetics of adsorption is given by Alfred Clark, "The Theory of Adsorption and Catalysis," Academic Press, New York, 1970.

form force fields because of the atomic structure in the crystal. Such surfaces also have sites or active centers where adsorption is enhanced. Two types of adsorption may occur.

Physical adsorption† Physical adsorption is nonspecific and somewhat similar to the process of condensation. The forces attracting the fluid molecules to the solid surface are relatively weak, and the heat evolved during the exothermic adsorption process is of the same order of magnitude as the heat of condensation, 0.5 to 5 kcal/g mol. Equilibrium between the solid surface and the gas molecules is usually rapidly attained and easily reversible, because the energy requirements are small. The energy of activation for physical adsorption is usually no more than 1 kcal/g mol, since the forces involved in physical adsorption are weak. Physical adsorption cannot explain the catalytic activity of solids for reactions between relatively stable molecules, because there is no possibility of large reductions in activation energy. Reactions of atoms and free radicals at surfaces sometimes involve small activation energies, and in these cases physical adsorption may play a role. Also, physical adsorption serves to concentrate the molecules of a substance at a surface. This can be of importance in cases involving reaction between a chemisorbed reactant and a coreactant which can be physically adsorbed. In such a system the catalytic reaction would occur between chemisorbed and physically adsorbed reactants.

The amount of physical adsorption decreases rapidly as the temperature is raised and is generally very small above the critical temperatures of the adsorbed component. This is further evidence that physical adsorption is not responsible for catalysis. For example, the rate of oxidation of sulfur dioxide on a platinum catalyst becomes appreciable only above 300°C; yet this is considerably above the critical temperature of sulfur dioxide (157°C) or of oxygen (-119°C). Physical adsorption is not highly dependent on the irregularities in the nature of the surface, but is usually directly proportional to the amount of surface. However, the extent of adsorption is not limited to a monomolecular layer on the solid surface, especially near the condensation temperature. As the layers of molecules build up on the solid surface, the process becomes progressively more like one of condensation.

Physical-adsorption studies are valuable in determining the physical properties of solid catalysts. Thus the questions of surface area and pore-size distribution in porous catalysts can be answered from physical-adsorption measurements. These aspects of physical adsorption are considered in Chap. 8.

Chemisorption‡ The second type of adsorption is specific and involves forces much stronger than in physical adsorption. According to Langmuir's pioneer work,§ the adsorbed molecules are held to the surface by valence forces of the

---

† For a detailed treatment of physical adsorption see D. M. Young and A. D. Crowell, "Physical Adsorption of Gases," Butterworth & Co. (Publishers), London, 1962.

‡ For a detailed treatment of chemisorption see D. O. Hayward and B. M. W. Trapnell, "Chemisorption," 2d ed., Butterworth & Co. (Publishers), London, 1964.

§ I. Langmuir, *J. Am. Chem. Soc.*, **38**, 221 (1916).

same type as those occurring between atoms in molecules. He observed that a stable oxide film was formed on the surface of tungsten wires in the presence of oxygen. This material was not the normal oxide WO_3, because it exhibited different chemical properties. However, analysis of the walls of the vessel holding the wire indicated that WO_3 was given off from the surface upon desorption. This suggested a process of the type

$$3O_2 + 2W \rightarrow 2[W \cdot O_3]$$

$$2[W \cdot O_3] \rightarrow 2WO_3$$

where $[W \cdot O_3]$ represents the adsorbed compound. Further evidence for the theory that such adsorption involves valence bonds is found in the large heats of adsorption. Observed values are of the same magnitude as the heat of chemical reactions, 5 to 100 kcal/g mol.

Taylor† suggested the name *chemisorption* for describing this second type of combination of gas molecules with solid surfaces. Because of the high heat of adsorption, the energy possessed by chemisorbed molecules can be substantially different from that of the molecules alone. Hence the energy of activation for reactions involving chemisorbed molecules can be considerably less than that for reactions involving gas-phase molecules. It is on this basis that chemisorption offers an explanation for the catalytic effect of solid surfaces.

Two kinds of chemisorption are encountered: activated and, less frequently, nonactivated. *Activated chemisorption* means that the rate varies with temperature according to a finite activation energy in the Arrhenius equation. However, in some systems chemisorption occurs very rapidly, suggesting an activation energy near zero. This is termed *nonactivated chemisorption.*‡ It is often found that for a given gas and solid the initial chemisorption is nonactivated, while later stages of the process are slow and temperature dependent (activated adsorption).

An approximate, qualitative relation between temperature and quantity adsorbed (both physically and chemically) is shown in Fig. 7-2. Chemisorption is assumed to be activated in this case. When the critical temperature of the component is exceeded, physical adsorption approaches a very low equilibrium value. As the temperature is raised, the amount of activated adsorption becomes important because the rate is high enough for significant quantities to be adsorbed in a reasonable amount of time. In an ordinary adsorption experiment involving the usual time periods the adsorption curve actually rises with increasing temperatures from the minimum value, as shown by the solid line in Fig. 7-2. When the temperature is increased still further, the decreasing equilibrium value for activated adsorption retards the process, and the quantity adsorbed passes through a maximum. At these high temperatures even the rate of the relatively slow activated process may be sufficient to give results closely approaching equilibrium. Hence the solid curve representing the amount adsorbed approaches the dashed equilibrium value for the activated adsorption process.

---

† H. S. Taylor, *J. Am. Chem. Soc.*, **53**, 578 (1931).

‡ As an illustration, chemisorption of hydrogen on nickel at low temperatures is nonactivated; see G. Padberg and J. M. Smith, *J. Catalysis*, **12**, 172 (1968).

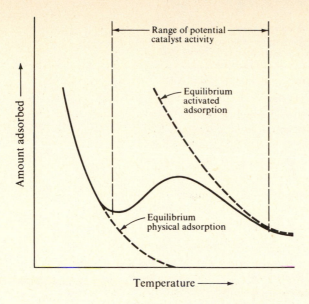

Figure 7-2 Effect of temperature on physical and activated adsorption.

It has been explained that the effectiveness of solid catalysts for reactions of stable molecules is dependent upon chemisorption. Granting this, the temperature range over which a given catalyst is effective must coincide with the range where chemisorption of one or more of the reactants is appreciable. This is indicated on Fig. 7-2 by the dashed vertical lines. There is a relationship between the extent of chemisorption of a gas on a solid and the effectiveness of the solid as a catalyst. For example, many metallic and metal-oxide surfaces adsorb oxygen easily, and these materials are also found to be good catalysts for oxidation reactions. When reactions proceed catalytically at low temperatures, Fig. 7-2 does not apply. In these cases catalysis is due to nonactivated chemisorption. Thus ethylene is hydrogenated on nickel at $-78°C$, at which temperature there would surely exist physical adsorption of the ethylene.

An important feature of chemisorption is that its magnitude will not exceed that corresponding to a monomolecular layer. This limitation is due to the fact that the valence forces holding the molecules on the surface diminish rapidly with distance. These forces become too small to form the adsorption compound when the distance from the surface is much greater than usual bond distances.

The differences between chemisorption and physical adsorption are summarized in Table 7-1.

In order to develop rate equations for catalytic reactions quantitative expressions for adsorption are necessary. Langmuir† proposed simple formulations for rates of adsorption and desorption of gases (applicable also to liquids) on solid surfaces. While his concern was with chemisorption, the concepts have been used for deriving a valuable relationship between the volume of a gas physically adsorbed and the surface area of the solid (see Sec. 8-1). Applied to chemisorption,

† I. Langmuir, *J. Am. Chem. Soc.*, **40**, 1361 (1918).

Table 7-1 Physical vs. chemical adsorption

Parameter	Physical adsorption	Chemisorption
Adsorbent	All solids	Some solids
Adsorbate	All gases below critical temperature	Some chemically reactive gases
Temperature range	Low temperature	Generally high temperature
Heat of adsorption	Low ($\approx \Delta H_{cond}$)	High; order of heat of reaction
Rate, activation energy	Very rapid, low E	Nonactivated, low E; activated, high E
Coverage	Multilayer possible	Monolayer
Reversibility	Highly reversible	Often irreversible
Importance	For determination of surface area and pore size	For determination of active-center area and elucidation of surface-reaction kinetics

the Langmuir treatment provides a systematic basis for developing rates of heterogeneous catalytic reactions and we use this approach in Chap. 9. Adsorption and desorption steps in the overall conversion of reactants to products (see list at beginning of this chapter) may be very fast and hence, occur at near equilibrium. Therefore, we need also to know the equilibrium relation between the concentration of adsorbent in the gas (or liquid) and the amount adsorbed. Such equilibrium isotherms, Langmuir and other forms, are considered in Sec. 7-6 and then rates of adsorption are discussed in Sec. 7-7.

7-6 Adsorption Isotherms

A. The Langmuir treatment of adsorption The derivations may be carried out by using as a measure of the amount adsorbed either the fraction of the surface covered or the concentration of the gas adsorbed on the surface. Both procedures will be illustrated, although the second is the more useful for kinetic developments (Chap. 9). The important assumptions are as follows:†

1. All the surface of the catalyst has the same activity for adsorption; i.e., it is energetically uniform. The concept of nonuniform surface with active centers can be employed if it is assumed that all the active centers have the same activity for adsorption and that the rest of the surface has none, or that an average activity can be used.
2. There is no interaction between adsorbed molecules. This means that the amount adsorbed has no effect on the rate of adsorption per site.

---

† It is also supposed that each site can accommodate only one adsorbed particle.

3. All the adsorption occurs by the same mechanism, and each adsorbed complex has the same structure.
4. The extent of adsorption is less than one complete monomolecular layer on the surface.

In the system of solid surface and gas, the molecules of gas will be continually striking the surface and a fraction of these will adhere. However, because of their kinetic, rotational, and vibrational energy, the more energetic molecules will be continually leaving the surface. An equilibrium will be established such that the rate at which molecules strike the surface, and remain for an appreciable length of time, will be exactly balanced by the rate at which molecules leave the surface.

The rate of adsorption $\mathbf{r}_a$ will be equal to the rate of collision $\mathbf{r}_c$ of molecules with the surface multiplied by a factor F representing the fraction of the colliding molecules that adhere. At a fixed temperature the number of collisions will be proportional to the pressure p of the gas (or its concentration), and the fraction F will be constant. Hence the rate of adsorption per unit of bare surface will be $\mathbf{r}_c F$. This is equal to kp, where k is a constant involving the fraction F and the proportionality between $\mathbf{r}_c$ and p.

Since the adsorption is limited to complete coverage by a monomolecular layer, the surface may be divided into two parts: the fraction θ covered by the adsorbed molecules and the fraction $1 - \theta$, which is bare. Since only those molecules striking the uncovered part of the surface can be adsorbed, the rate of adsorption per unit of total surface will be proportional to $1 - \theta$; that is,

$$\mathbf{r}_a = kp(1 - \theta) \tag{7-10}$$

The rate of desorption will be proportional to the fraction of covered surface

$$\mathbf{r}_d = k'\theta \tag{7-11}$$

The amount adsorbed at equilibrium is obtained by equating $\mathbf{r}_a$ and $\mathbf{r}_d$ and solving for θ. The result, called the *Langmuir isotherm*, is

$$\theta = \frac{kp}{k' + kp} = \frac{Kp}{1 + Kp} = \frac{v}{v_m} \tag{7-12}$$

where $K = k/k'$ is the adsorption equilibrium constant, expressed in units of (pressure)$^{-1}$. The fraction θ is proportional to volume of gas adsorbed, v, since the adsorption is less than a monomolecular layer. Hence Eq. (7-12) may be regarded as a relationship between the pressure of the gas and the volume adsorbed. This is indicated by writing $\theta = v/v_m$, where v_m is the volume adsorbed when all the active sites are covered, i.e., when there is a complete monomolecular layer.

The concentration form of Eq. (7-12) can be obtained by introducing the concept of an adsorbed concentration $\bar{C}$, expressed in moles per gram of catalyst. If $\bar{C}_m$ represents the concentration corresponding to a complete monomolecular layer on the catalyst, then the rate of adsorption, mol/(s)(g catalyst) is, by analogy with Eq. (7-10),

$$\mathbf{r}_a = k_c C_g(\bar{C}_m - \bar{C}) \tag{7-13}$$

where k_c is the rate constant for the catalyst and C_g is the concentration of adsorbable component in the gas. Similarly, Eq. (7-11) becomes

$$\mathbf{r}_d = k'_c \bar{C} \tag{7-14}$$

At equilibrium the rates given by Eqs. (7-13) and (7-14) are equal, so that

$$\bar{C} = \frac{K_c \bar{C}_m C_g}{1 + K_c C_g} \tag{7-15}$$

where now the equilibrium constant K_c is equal to k_c/k'_c and is expressed in units of $(\text{concn})^{-1}$, centimeters per gram mole. Since $\bar{C}/\bar{C}_m = \theta$, Eq. (7-15) may also be written

$$\theta = \frac{K_c C_g}{1 + K_c C_g} \tag{7-16}$$

which is a form analogous to Eq. (7-12), since C_g is proportional to p.

Equation (7-15) predicts that adsorption data should have the general form shown in Fig. 7-3. Note that at low values of C_g (or low surface coverages θ) the expression becomes a straight line with a slope equal to $K_c \bar{C}_m$. The data points in Fig. 7-3 are for the physical adsorption of n-butane on silica gel ($S_g = 832$ m^2/g) at 50°C.† The solid line represents Eq. (7-15), where

$$\bar{C}_m = 0.85 \times 10^{-3} \text{ g mol/g silica gel}$$

and the equilibrium constant of adsorption is

$$K_c = 4.1 \times 10^5 \text{ cm}^3/\text{g mol}$$

For this instance of physical adsorption Eq. (7-15) fits the data rather well. The measurements were made on mixtures of n-butane in helium (at 1 atm total pressure) to vary C_g. The percentage of n-butane in the gas corresponding to the concentration C_g is shown as a second abscissa in the figure. Also, θ is shown as a second ordinate. This was calculated from $\theta = \bar{C}/\bar{C}_m$. It is interesting to note that the isotherm is linear up to about $\theta = 0.10$, or a gas concentration of 1% n-butane in He. This means that the denominator in Eq. (7-15) is nearly unity at low concentrations. The region of linearity depends upon the adsorbent-adsorbate system. For example, for the same silica gel at 50°C, the linear isotherm was applicable up to 2% for adsorption of propane and to more than 4% for ethane.

Chemisorption data often do not fit Eq. (7-15). However, the basic concepts on which the Langmuir isotherm is based (the ideas of a dynamic equilibrium between rates of adsorption and desorption and a finite adsorption time) are sound and of great value in developing the kinetics of fluid-solid catalytic reactions. Equations (7-13) to (7-15) form the basis for the rate equations presented in Chap. 9.

---

† P. Schneider and J. M. Smith, *AIChE J.*, **14**, 769 (1968).

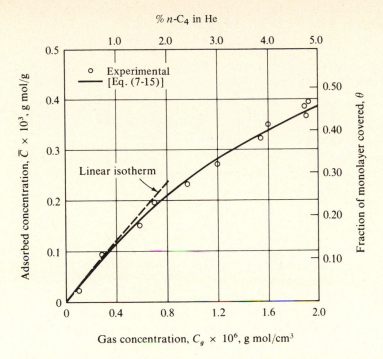

Figure 7-3 Adsorption-equilibrium data for n-butane on silica gel (surface area 832 m^2/g) at 50°C.

B. Other isotherms and heats of adsorption Experimentally, the five types of isotherms shown in Fig. 7-4 have been observed.[†] Note that the linear form of Eq. (7-15)

$$\bar{C} = (K_c \bar{C}_m)C_g \tag{7-17}$$

would fit all of the five isotherms at very low fluid concentrations. However, only curve I agrees with the Langmuir expression over the whole concentration range. The deviations from the Langmuir form are probably largely due to the postulates of sites of equal activity and no interaction between occupied and bare sites. It might be surmised that these assumptions correspond to a constant heat of adsorption. Indeed, it is possible[‡] to derive the Langmuir isotherm by assuming that ΔH_a is independent of θ. The heat of adsorption can be evaluated from adsorption-equilibrium data. First the Clausius-Clapeyron equation is applied to the two-phase system of gas and adsorbed component on the surface:

$$\left(\frac{dp}{dT}\right)_\theta = \frac{\Delta H_a}{T(V - V_a)}$$

[†] D. O. Hayward and B. M. W. Trapnell, "Chemisorption," 2d ed., Butterworth & Co. (Publishers), London, 1964.
[‡] K. J. Laidler, *J. Phys. Chem.*, **53**, 712 (1949).

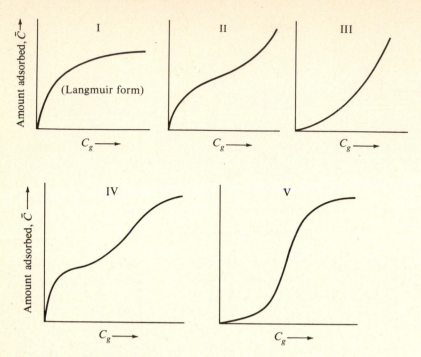

Figure 7-4 Types of adsorption isotherms.

where V and V_a are the volumes per mole of adsorbed component in the gas and on the surface, respectively. Neglecting the latter and assuming the ideal-gas law for V gives

$$\left[\frac{d(\ln p)}{dT}\right]_\theta = \frac{\Delta H_a}{R_g T^2} \tag{7-18}$$

For the Clausius-Clapeyron equation to be valid the process must be univariant. This means that Eq. (7-18) can be applied only for constant concentration of adsorbate on the surface, that is, constant θ. If adsorption-equilibrium data are available at different temperatures, the slopes of p-vs.-T curves at constant θ may be used with Eq. (7-18) to calculate ΔH_a. Figure 7-5, taken from Beeck,[†] shows such isosteric heats of adsorption as a function of θ for hydrogen on several metal films. These results are typical in showing a decrease in ΔH_a with increasing surface coverage. For physical adsorption, ΔH_a values are lower. For example, for nitrogen and n-butane the heats of adsorption (at essentially zero surface coverage) on a molecular sieve catalyst (synthetic zeolite, 5A type) were 4.3 and 10.3 kcal/g mol.[‡]

Two other well-known isotherms may be classified in terms of the $\Delta H_a - \theta$ dependency. The *Temkin isotherm* may be derived from the Langmuir isotherm

† O. Beeck, *Disc. Faraday Soc.*, **8**, 118 (1950).
‡ N. Hashimoto and J. M. Smith, *Ind. Eng. Chem. Fundam.*, **12**, 353 (1973).

by assuming that the heat of adsorption drops linearly with increasing θ.[†] The result is

$$\theta = k_1 \ln k_2 p \tag{7-19}$$

where k_1 and k_2 are constants at a given temperature. Figure 7-5 shows that a linear decrease in ΔH_a with θ fits some data for H_2 on metal films. Similar agreement has been obtained for N_2 chemisorbed on a promoted iron catalyst.[‡]

The *Freundlich isotherm* can be derived[§] by assuming a logarithmic decrease in ΔH_a with θ; that is,

$$\Delta H_a = -\Delta H_0 \ln \theta \tag{7-20}$$

The isotherm itself has the form

$$\theta = c(p)^{1/n} \tag{7-21}$$

where n has a value greater than unity. Owing to its flexibility, the Freundlich isotherm usually fits experimental data over a reasonable range of concentrations. For example, hydrogen gas on tungsten[¶] and sulfur dioxide (in aqueous solution) on activated carbon.[††] In the latter system, the concentration form of Eq. (7-21) at 23°C was found to be

$$\bar{C}_{SO_2} = 0.180[(C_L)_{SO_2}]^{1/1.88} \tag{7-22}$$

† A. Frumkin and A. Slygin, *Acta Physico chim.*, **3**, 791 (1935); S. Brunauer, K. S. Love, and R. G. Keenan, *J. Am. Chem. Soc.*, **64**, 751 (1942).
‡ P. H. Emmett and S. Brunauer, *J. Am. Chem. Soc.*, **56**, 35 (1934).
§ G. Halsey and H. S. Taylor, *J. Chem. Phys.*, **15**, 624 (1947).
¶ R. M. W. Trapnell, *Proc. Roy. Soc.* (*London*), **A206**, 39 (1951).
†† Hiroshi Komiyama and J. M. Smith, *AIChE J.*, **21**, 664 (1975).

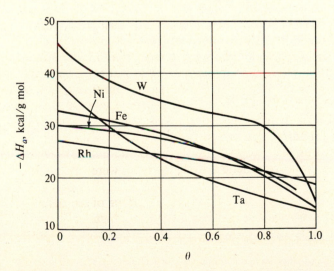

Figure 7-5 Heats of adsorption vs. surface coverage for hydrogen on metal films.

where

$$C_L = \text{concentration of } SO_2 \text{ in water, g mol/cm}^3$$

$$\bar{C} = \text{adsorbed concentration of } SO_2, \text{ g mol/(g of carbon)}$$

In conclusion, note two points. First, the Freundlich isotherm can be reduced to either the Langmuir or the Temkin form by proper simplication. It may be considered a general empirical form encompassing the other more specific types. Second, a single isotherm of the Langmuir or Temkin type cannot be expected to fit data over the entire range of θ.

7-7 Rates of Adsorption

As already noted we are going to use the Langmuir formulation [Eqs. (7-10) and (7-11)] for adsorption rates in Chap. 9 to derive rate expressions for catalytic reactions. Also as noted, the Langmuir formulation is seldom substantiated over a wide range of adsorbent concentrations. Therefore, it is useful to examine what is known about adsorption rates and compare this information with the Langmuir postulates.

The rate at which molecules of a gas strike a surface, in molecules/(s)(cm^2 surface), is $p/(2\pi m k_B T)^{1/2}$. If s is the fraction of the collisions which result in chemisorption, that is, the *sticking probability*, the rate of adsorption is

$$\mathbf{r}_a = \frac{sp}{(2\pi m k_B T)^{1/2}} \tag{7-23}$$

where p is the gas pressure and m is the mass of the molecule. As Hayward and Trapnell point out,[†] $s < 1$ for several reasons. Two are particularly important: only those molecules possessing the required activation energy can be chemisorbed, and molecules possessing the necessary energy may not be chemisorbed because the configuration of molecule and surface site may not allow the activated complex to be traversed. The fraction of the molecules possessing the required energy will be $e^{-E/R_g T}$, where E is the activation energy for chemisorption. The configuration probability for a molecule occupying a single site[‡] will be proportional to the fraction of the surface that is unoccupied, that is, $1 - \theta$. Thus

$$s = \alpha(1 - \theta)e^{-E/R_g T} \tag{7-24}$$

where α is the proportionality constant, often called the *condensation coefficient*. Combining Eqs. (7-23) and (7-24) yields

$$\mathbf{r}_a = \frac{\alpha e^{-E/R_g T}}{(2\pi m k_B T)^{1/2}} p(1 - \theta) \tag{7-25}$$

[†] D. O. Hayward and B. M. W. Trapnell, "Chemisorption," 2d ed., Butterworth & Co. (Publishers), London, 1964.

[‡] If a molecule (for example, O$_2$) dissociates upon adsorption and occupies two adjacent sites, the configuration probability will be $(1 - \theta)^2$, at least for small values of θ.

This ideal expression for the rate of adsorption does not usually agree with experimental data. Observed rates decrease so rapidly with increasing coverage θ that they can be explained only if the activation energy increases with θ. Also, the condensation coefficient may vary with θ. These variations may be caused by surface heterogeneity; that is, the activity of the sites varies, so that different sites possess different values of α and E. The most active sites would have the lowest activation energy and would be occupied first. In addition, interaction forces between occupied and unoccupied sites can cause deviations. In any event, it is necessary to rewrite Eq. (7-25) showing α and E as functions of θ:

$$\mathbf{r}_a = \frac{\alpha(\theta)e^{-E(\theta)/R_g T}}{(2\pi m k_B T)^{1/2}} p(1 - \theta) \tag{7-26}$$

Comparing the Langmuir equation, Eq. (7-10), with Eq. (7-26) shows that k in Eq. (7-10) is a function of surface coverage. However, the reason α and E are functions of θ is that the first two postulates of the Langmuir treatment (Sec. 7-6, A) are not satisfied experimentally; that is, in real surfaces all sites do not have the same activity, and interactions do exist.

For many cases of chemisorption the variation in rate with surface coverage can be accounted for entirely in the exponential term of Eq. (7-26), because $\alpha(\theta)$ and $1 - \theta$ are so much weaker functions. This leads to the result (for constant temperature)

$$\mathbf{r}_a = \beta p e^{-\gamma\theta} \tag{7-27}$$

commonly known as the *Elovich equation*.† Equation (7-27) can be obtained‡ from Eq. (7-26) by assuming that α is constant and that E is a linear function of θ. It can also be derived by supposing that E is constant but that the number of sites is a function of the coverage and of the temperature.§

Experimental data for rates of *chemisorption* are meager.¶ By pulse-response measurements†† adsorption rates of hydrogen‡‡ on commercial nickel and cobalt (on Kieselguhr) catalysts at temperatures of about 297 K were found to be 1.0×10^{-2} and 1.6×10^{-3} g mol/(g catalyst)(s). With a similar technique rates of *physical* adsorption of ethane propane and *n*-butane on silica gel have been measured at 50°C.§§ The rate constants, $(k_c \bar{C}_m)$, in the concentration range where the rate was first-order [$\bar{C} \to 0$ in Eq. (7-13)] were: 167 cm³/(s)(g silica gel) for

† S. Yu. Elovich and G. M. Zhabrova, *Zh. Fiz. Khim.*, **13**, 1761 (1939); S. Roginsky and Ya. Zeldovich, *Acta Physicochim.*, **1**, 554, 595 (1934).

‡ D. O. Hayward and B. M. W. Trapnell, "Chemisorption," 2d ed., Butterworth & Co. (Publishers), London, 1964.

§ H. A. Taylor and N. J. Thon, *J. Am. Chem. Soc.*, **74**, 4169 (1952); M. J. D. Low, *Chem. Rev.*, **60**, 267 (1960).

¶ For example, see Taylor and Thon, Low, and Hayward and Trapnell for summaries of chemisorption rate measurements.

†† G. Padberg and J. M. Smith, *J. Catalysis*, **12**, 172 (1968); J. C. Adrian and J. M. Smith, *J. Catalysis*, **18**, 57 (1970).

‡‡ Actually exchange rates were measured for deuterium in an atmosphere of hydrogen so that the catalyst surface was saturated with hydrogen.

§§ P. Schneider and J. M. Smith, *AIChE J.*, **14**, 762 (1968).

C_2H_6, 255 for C_3H_8, and about 1500 for n-C_4H_{10}. The increase in rate constant with increasing molecular weight is expected since heavier hydrocarbons are more easily and rapidly adsorbed. Transition-state theory (presented in Sec. 2-6 for homogeneous gas-phase reactions) has been applied to the prediction of the condensation coefficient. Lack of precise knowledge about the structure of the activated complex on the solid surface hinders these calculations and the evaluation of numerical values for the rates.

Before we leave the topic of rates it is important to mention the desorption step. Since desorption is normally an endothermic process,† it will have a significant activation energy (see Fig. 2-1). Thus desorption will be an activated process even when adsorption is not. If ΔH_a is the heat of adsorption on the site, the activation energies are related by

$$E_d = E_a - \Delta H_a \qquad (7\text{-}28)$$

Rates of chemisorption of hydrogen on a given nickel catalyst were found to be nonactivated ($E_a \sim 0$) but the desorption activation energy was 13.8 kcal/g mol, about the same magnitude as the heat of adsorption (-13.5 kcal/g mol).‡ For hydrogen adsorption on a cobalt/Kieselguhr catalyst, the chemisorption also was nonactivated with $E_d = 5.2$ kcal/(g mol).§

PROBLEMS

7-1 In a sketch similar to Fig. 7-1 show schematically the concentration profiles for a first-order *reversible* catalytic reaction. Consider three cases: (*a*) reaction control, (*b*) diffusion control, and (*c*) an intermediate case.

7-2 Explain why mass-transfer resistance reduces the global rate more at higher temperatures than at lower temperatures. Assume no heat-transfer resistance.

7-3 For endothermic reactions, do mass- and heat-transfer resistances have complementary or counterbalancing effects on the global rate?

7-4 A gaseous reaction with a solid catalyst is carried out in a flow reactor. The system is isothermal, but it is believed that mass-transfer resistances are important. (*a*) Would increasing the turbulence in the gas region next to the catalyst surface increase or decrease the global rate? (*b*) If the system is not isothermal and the reaction is exothermic, would increasing the turbulence increase or decrease the global rate?

7-5 There is a growing need to obtain a better understanding of reactions involving bacteria. For example, most all-sewage disposal plants for purifying municipal wastewater are based upon bacterial oxidation (activated sludge process). A special characteristic of bacterial oxidation is that the bacteria not only catalyze the oxidation of carbonaceous compounds, but also the carbon compound provides "fuel" for growth of the bacteria. As an illustration, consider a stirred-tank reactor operating continuously at *steady state* and at constant temperature. The feed consists of an aqueous solution of glucose

---

† A few chemisorption processes appear to be endothermic, so that desorption in these instances would be exothermic. See J. M. Thomas and W. J. Thomas, "Introduction to the Principles of Heterogeneous Catalysis," p. 30, Academic Press Inc., New York, 1967.

‡ G. Padberg and J. M. Smith, loc. cit.

§ J. C. Adrian and J. M. Smith, loc. cit.

(the carbonaceous material) which does not contain bacteria. The concentration of glucose in the feed is S_0, and its volumetric flow rate into the reactor is Q. The mixture in the tank contains bacteria in a concentration B, and the product stream has a flow rate equal to Q. This product stream contains bacteria and glucose in concentrations B and S (both in mg/liter).

The two major reactions may be written (in a highly simplified manner) as follows:

1. $C_6H_{12}O_6 + B \rightarrow nB + \text{inert products}$ $(n > 1)$

2. $C_6H_{12}O_6 + O_2 \xrightarrow{B} H_2O + CO_2 + \text{inert products}$

The Michaelis-Menten† rate equations have been found to fit some types of bacterial oxidations. For conditions where the oxygen supply is adequate, the equations can be written for reactions 1 and 2 as

$$R_x = \frac{\mu_m S}{K + S}(B)$$

$$R_S = \left(\frac{1}{Y}\right)\frac{\mu_m S}{K + S}(B)$$

where μ_m and K are rate constants.

R_x = rate of formation of bacteria (cell mass), mg/(liter)(h)

R_S = rate of disappearance of glucose (substrate), mg/(liter)(h)

R_x represents the growth of bacteria according to reaction 1, while R_S is the total rate of disappearance of glucose by *both* reactions.

Notice that reaction 2 is catalyzed by the bacteria. The symbol Y represents the yield, or ratio of rates of production of bacteria to the total rate of consumption of glucose. Other symbols are defined in part D.

A. For steady-state operation, derive a relationship between Q/V (called the dilution rate, D) and the concentration S of glucose in the product stream.

B. If S is very large with respect to K, corresponding to a very large excess of glucose in the feed stream, how is the relationship of part A simplified?

C. The result obtained in part B gives a specific relation between the dilution rate D and the rate constant, or constants. If the actual dilution rate is increased above the value given by this specific relation, for example by increasing the flow rate, what would happen to the bacteria in the reactor?

D. For a particular case, the following data apply:

$Q = 1.0$ liters/h

$V = $ volume of stirred tank reactor $= 4$ liters

$\mu_m = $ rate constant $= 0.5$ (h)$^{-1}$

$K_S = $ constant in the rate equation $= 15$ mg/liter

$S_0 = 80$ mg/liter $= $ concn of glucose in feed

$Y = $ yield (assume constant) $= 0.5$

Calculate the following (for steady-state operation):

1. the concentration of glucose in the product stream leaving the reactor
2. the concentration of bacteria in the product stream leaving the reactor

E. For the conditions of part D, what would be the maximum volumetric flow rate that could be used and yet avoid the problem referred to in part C?

† L. Michaelis and M. L. Menten, *Biochem. Z.*, **49**, 333 (1913).

7-6 The following data were obtained at 70°C for the equilibrium adsorption of n-hexane on silica gel particles.

Partial pressure of C_6H_{14} in gas, atm	C_6H_{14} adsorbed, g mol/(g gel)
0.0020	10.5×10^{-5}
0.0040	16.0×10^{-5}
0.0080	27.2×10^{-5}
0.0113	34.6×10^{-5}
0.0156	43.0×10^{-5}
0.0206	47.3×10^{-5}

(a) Determine how well the Langmuir isotherm fits these data. Establish the values of the constants $\bar{C}_m$ and K_c by least-mean-squares analysis of the linearized form of Eq. (7-15):

$$\frac{C_g}{\bar{C}} = \frac{1}{K_c \bar{C}_m} + \frac{C_g}{\bar{C}_m}$$

Note that a straight line should be obtained if $C_g/\bar{C}$ is plotted against C_g.

(b) Up to what gas concentration is the isotherm linear?

7-7 Repeat Prob. 7-6 for the following data for the equilibrium adsorption of C_6H_6 on the same silica gel at 110°C.

Partial pressure of C_6H_6 in gas, atm	C_6H_6 adsorbed, g mol/(g gel)
5.0×10^{-4}	2.6×10^{-5}
1.0×10^{-3}	4.5×10^{-5}
2.0×10^{-3}	7.8×10^{-5}
5.0×10^{-3}	17.0×10^{-5}
1.0×10^{-2}	27.0×10^{-5}
2.0×10^{-2}	40.0×10^{-5}

7-8 Can the Freundlich isotherm, Eq. (7-21), represent the physical-adsorption data for n-hexane given in Prob. 7-6? What are the appropriate values of c and n?

7-9 Will the Temkin isotherm fit the n-hexane data shown in Prob. 7-6?

7-10 Ward† studied the chemisorption of hydrogen on copper powder and found the heat of adsorption to be independent of surface coverage. Equilibrium data from experiments at 25°C are as follows:

Hydrogen pressure, mm Hg	1.05	2.95	5.40	10.65	21.5	45.1	95.8	204.8
Volume adsorbed, cm³ at 0°C and 1 atm	0.239	0.564	0.659	0.800	0.995	1.160	1.300	1.471

What kind of isotherm fits these results?

† A. F. H. Ward, *Proc. Roy. Soc.* (*London*), **A133**, 506 (1931).

7-11 Brunauer et al.† concluded that the Temkin isotherm, Eq. (7-19), explained equilibrium chemisorption data for N_2 on an ammonia-synthesis catalyst (promoted iron powder). Data taken at 396°C are as follows:

Nitrogen pressure, kPa	3.33	7.06	20.0	52.9	102
Volume adsorbed, $\times 10^6$, m^3 at 273 K and 101.3 kPa	2.83	3.22	3.69	4.14	4.55

Can Brunauer's conclusion be reached by plotting the data according to linearized forms of the Langmuir, Temkin, and Freundlich isotherms? Explain your answer.

7-12 Plot the isosteric heat of adsorption vs. the amount adsorbed from the following data of Shen and Smith‡ for benzene on silica gel.

Partial pressure of benzene, atm	Moles adsorbed/g gel $\times 10^5$			
	70°C	90°C	110°C	130°C
5.0×10^{-4}	14.0	6.7	2.6	1.13
1.0×10^{-3}	22.0	11.2	4.5	2.0
2.0×10^{-3}	34.0	18.0	7.8	3.9
5.0×10^{-3}	68.0	33.0	17.0	8.6
1.0×10^{-2}	88.0	51.0	27.0	16.0
2.0×10^{-2}	.....	78.0	42.0	26.0

7-13 Show that the Elovich equation for the rate of chemisorption can be integrated with respect to surface coverage (at constant pressure of the adsorbing gas) to give

$$\theta = a \ln (1 + bt)$$

where t is the time from the start of the adsorption process and a and b are constants. If the activation energy for chemisorption is a linear function of surface coverage, according to $E = E_0 + c\theta$, derive the integrated form of the Elovich equation.

---

† S. Brunauer, K. S. Love, and R. G. Keenan, *J. Am. Chem. Soc.*, **64**, 751 (1942).
‡ John Shen and J. M. Smith, *Ind. Eng. Chem. Fundam.*, **7**, 100 (1968).

7-14 Ward† studied the chemisorption of hydrogen on small copper particles at 25°C. His data for one run are:

Hydrogen pressure cm of Hg	Vol. of hydrogen adsorbed, cm^3 at 25°C and 1 atm
0.019	0.042
0.066	0.138
0.097	0.163
0.101	0.161
0.110	0.171
0.190	0.221
0.265	0.256
0.405	0.321
0.555	0.371
0.750	0.411
0.815	0.421
1.19	0.471
1.75	0.550

Do these data fit the Langmuir isotherm or the Freundlich isotherm?

† A. F. H. Ward, *Proc. Roy. Soc. (London)*, **A133**, 506 (1931).

EIGHT

SOLID CATALYSTS

In Chap. 7 general concepts of catalysis were discussed. Now we want to examine solid catalysts specifically. Such catalysts depend for their activity in part, at least, on the extent of surface area. It is difficult to obtain outer surface areas of more than 1 m^2/g by subdividing nonporous solids into small particles (see Example 8-1). To be effective, most catalytic solids must have surface areas in the 5 to 1000 m^2/g range. Hence, solid catalysts are normally porous. For such materials the geometrical properties of the pores can affect the global reaction rate. The objective of the first part of this chapter is to present methods of measuring the pertinent physical (geometric) properties. Following this is a discussion of theories of the catalytic behavior of solids along with a listing of typical catalysts for various types of reactions. Finally, practical aspects, such as catalyst preparation, poisons, and parameters, are treated.

The importance of surface area for catalytic activity is immediately evident by considering nickel. This metal is an active catalyst under certain conditions for oxidation and hydrogenation because it absorbs oxygen and hydrogen. The surface area of a solid has a pronounced effect on the amount of gas adsorbed and on its activity as a catalyst. For example, if a sample of fresh Raney nickel, which is highly porous and has a large surface, is held in the hand, the heat due to adsorption of oxygen can be felt immediately. No evidence of heat is apparent when a single piece of nonporous nickel of the same mass is held. This relationship between surface area and extent of adsorption has led to the development of highly porous materials with areas as high as 1500 m^2/g. Sometimes the catalytic material itself can be prepared in a form with large surface area. When this is not possible, materials which can be so prepared may be used as a *carrier* or *support* on which the catalytic substance is dispersed. Silica gel and alumina are widely used as supports.

The dependence of rates of adsorption and catalytic reactions on surface makes it imperative to have a reliable method of measuring surface area. For surface areas in the range of hundreds of square meters per gram a porous material with equivalent cylindrical pore radii (see Sec. 8-1) in the range of 10 to 100 Å is needed. The following example shows that such areas are not possible with nonporous particles of the size which can be economically manufactured.

Example 8-1 Spray drying and other procedures for manufacturing small particles can produce particles as small as 2 to 5 microns. Calculate the external surface area of nonporous spherical particles of 2 microns diameter. What size particles would be necessary if the external surface is to be 100 m^2/g (10^5 m^2/kg)? The density of the particles is 2.0 g/cm^3 (2.0 × 10^3 kg/m^3).

SOLUTION The external surface area per unit volume of a spherical particle of diameter d_p is

$$\frac{\pi d_p^2}{\pi d_p^3/6} = \frac{6}{d_p}$$

If the particle density is ρ_P, the surface area, per gram of particles, would be

$$S_g = \frac{6}{\rho_p d_p}$$

For $d_p = 2$ microns (2 × 10^{-4} cm) and $\rho_p = 2.0$ g/cm^3

$$S_g = \frac{6}{2(2 \times 10^{-4})} = 1.5 \times 10^4 \text{ cm}^2/\text{g} \ (1.5 \times 10^3 \text{ m}^2/\text{kg})$$

or

$$S_g = 1.5 \text{ m}^2/\text{g} \ (1.5 \times 10^3 \text{ m}^2/\text{kg})$$

This is about the largest surface area to be expected for nonporous particles. If a surface of 100 m^2/g were required, the spherical particles would have a diameter of

$$d_p = \frac{6}{\rho_p S_g} = \frac{6}{2.0(100 \times 10^4)} = 0.03 \times 10^{-4} \text{ cm} \ (0.03 \times 10^{-6} \text{ m})$$

or

$$d_p = 0.03 \text{ micron}$$

Particles as small as this cannot, at present, be produced economically. It may be noted that the smaller particles found in a fluidized-bed reactor are retained on 400 mesh screen, which has a sieve opening of 37 microns.

The quantitative effects of intraparticle mass and energy transfer on the rate when a reaction occurs on the interior pore surface of a catalyst particle is treated in Chap. 11. The method of predicting their effects requires a geometric model for the extent and distribution of void spaces within the complex porous structure of the particle. It would be best to know the size and shape of each void space in the particle. In the absence of this information the parameters in the model should be evaluated from reliable and readily obtainable average properties. In addition to the surface area, three other properties fall into this classification: void volume, the density of the solid material in the particle, and the distribution of void volume according to void size (pore-volume distribution). The methods of measurement of these four properties are considered in Secs. 8-1 to 8-3.

8-1 Determination of Surface Area

The standard method for measuring catalyst areas is based on the physical adsorption of a gas on the solid surface. Usually the amount of nitrogen adsorbed at equilibrium at the normal boiling point ($-195.8°C$) is measured over a range of nitrogen pressures below 1 atm. Under these conditions several layers of molecules may be adsorbed on top of each other on the surface. The amount adsorbed when one molecular layer is attained must be identified in order to determine the area. The historical steps in the development of the Brunauer-Emmett-Teller method[†] are clearly explained by Emmett.[‡] There may be some uncertainty as to whether the values given by this method correspond exactly to the surface area. However, this is relatively unimportant, since the procedure is standardized and the results are reproducible. It should be noted that the surface area so measured may not be the area effective for catalysis. Only certain parts of the surface, the active centers, may be active for chemisorption of a reactant, while nitrogen may be physically adsorbed on much more of the surface. When the catalyst is dispersed on a large-area support, only part of the support area may be covered by catalytically active atoms and this area may be several atoms in depth. Thus, the active atoms may be together in clusters so that the catalytic surface is less than if the atoms were more completely dispersed or separated. For example a nickel-on-kieselguhr catalyst was found to have a surface of 205 m^2/g as measured by nitrogen adsorption.[§] To determine the area covered by nickel atoms, hydrogen was chemisorbed on the catalyst at 25°C. From the amount of hydrogen chemisorbed, the surface area of nickel atoms was calculated to be about 40 m^2/g. It would be useful to know surface areas for chemisorption of the reactant at reaction conditions. However, this would require measurement of relatively small amounts of chemisorption at different, and often troublesome, conditions (high temperature and/or pressure), for each reaction system. In contrast, nitrogen can be adsorbed easily and rapidly in a routine fashion with standard equipment.

In the classical method of determining surface area an all-glass apparatus is used to measure the volume of gas adsorbed on a sample of the solid material.[¶]

---

† S. Brunauer, P. H. Emmett, and E. Teller, *J. Am. Chem. Soc.*, **60**, 309 (1938).

‡ P. H. Emmett (ed.), "Catalysis," vol. I, chap. 2. Reinhold Publishing Corporation, New York, 1954.

§ G. Padberg and J. M. Smith, *J. Catalysis*, **12**, 111 (1968).

¶ For a complete description of apparatus and techniques see L. G. Joyner, "Scientific and Industrial Glass Blowing and Laboratory Techniques," Instruments Publishing Company, Pittsburgh, 1949; see also S. Brunauer, "The Adsorption of Gases and Vapors," vol. 1, Princeton University Press, Princeton, N.J., 1943.

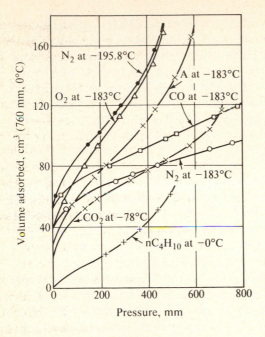

Figure 8-1 Adsorption isotherms for various gases on a 0.606-g sample of silica gel [by permission from P. H. Emmett (ed.), "Catalysis," vol. I, Reinhold Publishing Corporation, New York, 1954].

The apparatus operates at a low pressure which can be varied from near zero up to about 1 atm. The operating temperature is in the range of the normal boiling point. The data obtained are gas volumes at a series of pressures in the adsorption chamber. The observed volumes are normally corrected to cubic centimeters at 0°C and 1 atm (standard temperature and pressure) and plotted against the pressure in millimeters, or as the ratio of the pressure to the vapor pressure at the operating temperature. Typical results from Brunauer and Emmett's work† are shown in Fig. 8-1 for the adsorption of several gases on a 0.606-g sample of silica gel. To simplify the classical experimental procedure a flow method has been developed in which a mixture of helium (or other nonadsorbed gas) and the gas to be adsorbed is passed continuously over the sample of solid.‡ The operating total pressure is constant, and the partial pressure of adsorbable gas is varied by changing the composition of the mixture. The procedure§ is to pass a mixture of known composition over the sample until equilibrium is reached, that is, until the solid has adsorbed an amount of adsorbable component corresponding to equilibrium at its partial pressure in the mixture. Then the gas is desorbed by heating the sample while a stream of pure helium flows over it. The amount desorbed is measured with a thermal-conductivity cell or other detector. This gives one point on an isotherm, such as shown in Fig. 8-1. Then the process is repeated at successively different compositions of the mixture until the whole isotherm is obtained.

† S. Brunauer and P. H. Emmett, *J. Am. Chem. Soc.*, **59**, 2682 (1937).

‡ F. M. Nelson and F. T. Eggertsen, *Anal. Chem.*, **30**, 1387 (1958).

§ A description of the operating procedure and the data obtained are given by S. Masamune and J. M. Smith [*AIChE J.*, **10**, 246 (1964)] for the adsorption of nitrogen on Vycor (porous glass).

The curves in Fig. 8-1 are similar to the extent that at low pressures they rise more or less steeply and then flatten out for a linear section at intermediate pressures. After careful analysis of much data it was concluded that the lower part of the linear region corresponded to complete monomolecular adsorption. If this point could be located with precision, the volume of one monomolecular layer of gas, v_m, could then be read from the curve and the surface area evaluated. The Brunauer-Emmett-Teller method locates this point from an equation obtained by extending the Langmuir isotherm to apply to multilayer adsorption. The development is briefly summarized as follows: Equation (7-12) can be rearranged to the form

$$\frac{p}{v} = \frac{1}{Kv_m} + \frac{p}{v_m} \tag{8-1}$$

Brunauer, Emmett, and Teller adapted this equation for multilayer adsorption and arrived at the result

$$\frac{p}{v(p_0 - p)} = \frac{1}{v_m c} + \frac{(c-1)p}{cv_m p_0} \tag{8-2}$$

where p_0 is the saturation or vapor pressure and c is a constant for the particular temperature and gas-solid system.

According to Eq. (8-2), a plot of $p/v(p_0 - p)$ vs. p/p_0 should give a straight line. The data of Fig. 8-1 are replotted in this fashion in Fig. 8-2. Of additional significance is the fact that such straight lines can be accurately extrapolated to

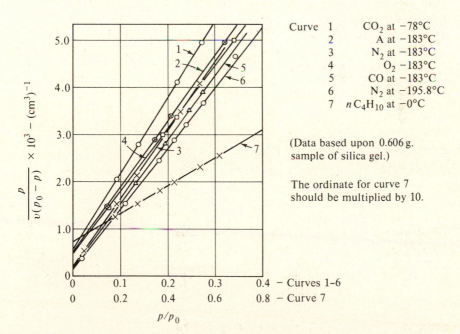

Curve 1 CO_2 at $-78°C$
2 A at $-183°C$
3 N_2 at $-183°C$
4 O_2 at $-183°C$
5 CO at $-183°C$
6 N_2 at $-195.8°C$
7 $n\,C_4H_{10}$ at $-0°C$

(Data based upon 0.606 g. sample of silica gel.)

The ordinate for curve 7 should be multiplied by 10.

Figure 8-2 Plot of Brunauer-Emmett-Teller equation [Eq. (8-2)] for data of Fig. 8-1 [*by permission from P. H. Emmett (ed.), "Catalysis," vol. I, Reinhold Publishing Corporation, New York, 1954*].

$p/p_0 = 0$. The intercept I obtained from this extrapolation, along with the slope s of the straight line, gives two equations from which v_m can be obtained,

$$I = \frac{1}{v_m c} \quad \text{at } p/p_0 = 0 \tag{8-3}$$

$$s = \frac{c-1}{v_m c} \tag{8-4}$$

Solving these equations for the volume of adsorbed gas corresponding to a monomolecular layer gives

$$v_m = \frac{1}{I+s} \tag{8-5}$$

The volume v_m can be readily converted to the number of molecules adsorbed. However, to determine the surface area it is necessary to select a value for the area covered by one adsorbed molecule. If this is α, the total surface area is given by

$$S_g = \left[\frac{v_m N_0}{V}\right]\alpha \tag{8-6}$$

where N_0 is Avogadro's number, 6.02×10^{23} molecules/mol, and V is the volume per mole of gas at conditions of v_m. Since v_m is recorded at standard temperature and pressure, $V = 22,400$ cm^3/g mol. The term in brackets represents the number of molecules adsorbed. If v_m is based on a 1.0 g sample, then S_g is the total surface per gram of solid adsorbent.

Emmett and Brunauer[†] proposed that α is the projected area of a molecule on the surface when the molecules are arranged in close two-dimensional packing. This value is slightly larger than that obtained by assuming that the adsorbed molecules are spherical and their projected area on the surface is circular. The proposed equation is

$$\alpha = 1.09 \left[\frac{M}{N_0 \rho}\right]^{2/3} \tag{8-7}$$

where M is molecular weight and ρ is the density of the adsorbed molecules. The term in brackets represents the volume of one adsorbed molecule. The density is normally taken as that of the pure liquid at the temperature of the adsorption experiment. For example, for N_2 at $-195.8°C$, $\rho = 0.808$ g/cm^3 and the area per molecule from Eq. (8-7) is 16.2×10^{-16} cm^2, or 16.2 Å^2. If this result is used in Eq. (8-6), along with the known values of N_0 and V, the surface area per gram is

$$S_g = 4.35 \times 10^4 v_m \quad \text{cm}^2/\text{g solid adsorbent} \tag{8-8}$$

Table 8-1 shows surface areas determined by the Brunauer-Emmett-Teller method for a number of common catalysts and carriers. Calculations of surface areas from adsorption data are illustrated in Examples 8-2 and 8-3.

---

† P. H. Emmett and S. Brunauer, *J. Am. Chem. Soc.*, **59**, 1553 (1937).

Table 8-1 Surface area, pore volume, and mean pore radii for typical solid catalysts

Catalyst	Surface area, m^2/g	Pure volume, cm^3/g	Mean pore radius, Å
Activated carbons	500–1,500	0.6–0.8	10–20
Silica gels	200–600	0.4	15–100
SiO-Al$_2$O$_3$ cracking			
catalysts	200–500	0.2–0.7	33–150
Activated clays	150–225	0.4–0.52	100
Activated alumina	175	0.39	45
Celite (Kieselguhr)	4.2	1.1	11,000
Synthetic ammonia			
catalysts, Fe	....	0.12	200–1,000
Pumice	0.38		
Fused copper	0.23		

Source: In part from A. Wheeler, "Advances in Catalysis," vol. III, pp. 250–326, Academic Press, Inc., New York, 1950.

Example 8-2 From the Brunauer-Emmett-Teller plot in Fig. 8-2 estimate the surface area per gram of the silica gel. Use the data for adsorption of nitrogen at $-195.8°C$.

SOLUTION From curve 6 of Fig. 8-2, the intercept on the ordinate is

$$I = 0.1 \times 10^{-3} \text{ cm}^{-3}$$

The slope of the curve is

$$s = \frac{(5.3 - 0.1) \times 10^{-3}}{0.4 - 0} = 13 \times 10^{-3} \text{ cm}^{-3}$$

These values of s and I may be substituted in Eq. (8-5) to obtain v_m,

$$v_m = \frac{10^3}{0.1 + 13} \frac{1}{(0.606)} = 126 \text{ cm}^3/g \text{ catalyst}$$

The factor 0.606 is introduced because the data in Fig. 8-2 are for a silica gel sample of 0.606 g, and v_m is the monomolecular volume per gram. For nitrogen at $-195.8°C$, the application of Eq. (8-8) yields

$$S_g = 4.35(126) = 550 \text{ m}^2/g$$

Example 8-3 For comparison, estimate the surface area of the silica gel by using the adsorption data for oxygen at $-183°C$. The density of the liquefied oxygen at $-183°C$, from the International Critical Tables, is 1.14 g/cm.

SOLUTION First the area of an adsorbed molecule of O_2 must be calculated from Eq. (8-7):

$$\alpha = 1.09 \left[\frac{32}{(6.02 \times 10^{23})1.14} \right]^{2/3} = 14.2 \times 10^{-16} \text{ cm}^2$$

With this value of α Eq. (8-6) gives

$$S_g = \frac{v_m(6.02 \times 10^{23})}{22,400} \, 14.2 \times 10^{-16} = 3.8 \times 10^4 v_m \qquad cm^2/g$$

From curve 4 of Fig. (8-2),

$$I = 0.40 \times 10^{-3} \, cm^{-3}$$

$$s = \frac{(5.4 - 0.4) \times 10^{-3}}{0.38 - 0} = 13.2 \times 10^{-3} \, cm^{-3}$$

Then the monomolecular volume per gram of silica gel is, from Eq. (8-5),

$$v_m = \frac{10^3}{0.4 + 13.2} \frac{1}{(0.606)} = 122 \, cm^3/g \text{ catalyst}$$

Finally, substituting this value of v_m in the area expression gives

$$S_g = 3.8 \times 10^4 (122) = 465 \times 10^4 \, cm^2/g \qquad or \quad 465 \, m^2/g$$

The difference in area determined from the N_2 and O_2 data is somewhat larger than expected for these gases. The adsorption curve for N_2 at $-183°C$ gives a value in closer agreement with 550 m^2/g (see Prob. 8-4).

8-2 Void Volume and Solid Density

The void volume, or pore volume, of a catalyst particle can be estimated by boiling a weighed sample immersed in a liquid such as water. After the air in the pores has been displaced, the sample is superficially dried and weighed. The increase in weight divided by the density of the liquid gives the pore volume.

A more accurate procedure is the *helium-mercury method*. The volume of helium displaced by a sample of catalyst is measured; then the helium is removed, and the volume of mercury that is displaced is measured. Since mercury will not fill the pores of most catalysts at atmospheric pressure, the difference in volumes gives the pore volume of the catalyst sample. The volume of helium displaced is a measure of the volume occupied by the solid material. From this and the weight of the sample, the density of the solid phase, ρ_S, can be obtained. Then the void fraction, or porosity, of the particle, ε_p, may be calculated from the equation

$$\varepsilon_p = \frac{\text{void (pore) volume of particle}}{\text{total volume of particle}} = \frac{m_p V_g}{m_p V_g + m_p(1/\rho_S)}$$

$$= \frac{V_g \rho_S}{V_g \rho_S + 1} \tag{8-9}$$

where m_p is the mass of the particle and V_g is the void volume per gram of particles. If the sample of particles is weighed, the mass divided by the mercury volume gives the density ρ_p of the porous particles. Note that the porosity is also obtainable from this density by the expression

$$\varepsilon_p = \frac{\text{void volume}}{\text{total volume}} = \frac{V_g}{1/\rho_p} = \rho_p V_g \tag{8-10}$$

From the helium-mercury measurements the pore volume, the solid density, and the porosity of the catalyst particle can be determined. Values of ε_p are of the order of 0.5, indicating that the particle is about half void space and half solid material. Since overall void fractions in packed beds are about 0.4, a rule of thumb for a packed-bed catalytic reactor is that about 30% of the volume is pore space, 30% is solid catalyst and carrier, and 40% is void space between catalyst particles. Individual catalysts may show results considerably different from these average values, as indicated in Examples 8-4 and 8-5 and Table 8-2.

Example 8-4 In an experiment to determine the pore volume and catalyst-particle porosity the following data were obtained on a sample of activated silica (granular, 4 to 12 mesh size):

Mass of catalyst sample placed in chamber = 101.5 g
Volume of helium displaced by sample   = 45.1 cm³
Volume of mercury displaced by sample   = 82.7 cm³

Calculate the required properties.

SOLUTION The volume of mercury displaced, minus the helium-displacement volume, is the pore volume. Hence

$$V_g = \frac{82.7 - 45.1}{101.5} = 0.371 \text{ cm}^3/\text{g}$$

The helium volume is also a measure of the density of the solid material in the catalyst; that is,

$$\rho_S = \frac{101.5}{45.1} = 2.25 \text{ g/cm}^3$$

Substituting the values of V_g and ρ_S in Eq. (8-9) gives the porosity of the silica gel particles,

$$\varepsilon_p = \frac{0.371(2.25)}{0.371(2.25) + 1} = 0.455$$

Catalyst particles used in packed-bed reactors normally are pelleted, or agglomerated by other means, to sizes of $\frac{1}{16}$ in. to $\frac{1}{2}$ in. in order to avoid excessive pressure drop and to provide mechanical strength. Usually the pellets are cylindrical or granular. Agglomeration of *porous* particles gives a pellet containing two void regions: small void spaces within the individual particles and larger spaces between particles. Hence such materials are said to contain *bidisperse pore systems*. Although the shape and nature of these two void regions may vary from thin cracks to a continuous region surrounding a group of particles, it has been customary to designate both regions as *pores*. The void spaces within the particles are commonly termed *micropores*, and the void regions between particles are called *macropores*. *Particle* refers only to the small individual unit from which the pellet is produced.

Perhaps the most widely used pellets are those of alumina. Porous alumina particles (20 to 200 microns diameter) containing micropores of 10 to 200 Å

Table 8-2 Physical properties of alumina pellets

Density, g/cm³		Pore volume, cm³/g		Void fraction		
Particle	Pellet	Micro	Macro	Total	Macro	Micro
1.292	1.121	0.365	0.120	0.543	0.134	0.409
1.264	1.010	0.383	0.198	0.587	0.200	0.387
1.238	0.896	0.400	0.308	0.634	0.275	0.359
1.212	0.785	0.416	0.451	0.680	0.353	0.327
1.188	0.672	0.434	0.670	0.725	0.450	0.275

Notes: All properties are based on Al_2O_3. Micro refers to pore radii less than 100 Å, and macro refers to radii greater than 100 Å.

Source: R. A. Mischke and J. M. Smith, *Ind. Eng. Chem., Fund. Quart.*, **1**, 288 (1962).

diameter are readily prepared by spray drying. These somewhat soft particles are easily made into pellets. The macroporosity and macropore diameter depend on the pelleting pressure and can be varied over a wide range. Table 8-2 shows macro and micro properties of five alumina pellets, each prepared at a different pelleting pressure. The pellet density listed in the second column is approximately proportional to the pressure used. Comparison of the least and greatest pellet density shows that the macropore volume has decreased from 0.670 to 0.120 with increased pelleting pressure, while the micropore volume has decreased only from 0.434 to 0.365.

The external surface area of even very fine particles has been shown (Example 8-1) to be small with respect to the internal surface of the pores. Hence, in a catalyst pellet the surface resides predominantly in the small pores within the particles. The external surface of the particles, and of course the external area of the pellets, is negligible.

Macro- and micropore volumes and porosities for bidisperse catalyst pellets are calculated by the same methods as used for monodisperse pore systems. Example 8-5 illustrates the procedure.

Example 8-5 A hydrogenation catalyst is prepared by soaking alumina particles (100 to 150 mesh size) in aqueous $NiNO_3$ solution. After drying and reduction, the particles contain about 7 wt % NiO. This catalyst is then made into large cylindrical pellets for rate studies. The gross measurements for one pellet are

$$
\begin{aligned}
\text{Mass} &= 3.15 \text{ g} \\
\text{Diameter} &= 1.00 \text{ in.} \\
\text{Thickness} &= \tfrac{1}{4} \text{ in.} \\
\text{Volume} &= 3.22 \text{ cm}^3
\end{aligned}
$$

The Al_2O_3 particles contain micropores, and the pelleting process introduces macropores surrounding the particles. From the experimental methods already described, the macropore volume of the pellet is 0.645 cm³ and the micropore volume is 0.40 cm³/g of particles. From this information calculate:

(a) The density of the *pellet*
(b) The macropore volume in cubic centimeters per gram
(c) The macropore void fraction in the *pellet*
(d) The micropore void fraction in the *pellet*
(e) The solid fraction
(f) The density of the *particles*
(g) The density of the solid phase
(h) The void fraction of the *particles*

SOLUTION
(a) The density of the pellet is

$$\rho_P = \frac{3.15}{3.22} = 0.978 \text{ g/cm}^3$$

(b) The macropore volume per gram is

$$(V_g)_M = \frac{0.645}{3.15} = 0.205 \text{ cm}^3/\text{g}$$

(c) The macropore void fraction ε_M is obtained by applying Eq. (8-10) to the pellet. Thus

$$\varepsilon_M = \frac{\text{macropore volume}}{\text{total volume}} = \frac{(V_g)_M}{1/\rho_P} = \frac{0.205}{1/0.978} = 0.200$$

(d) Since

$$(V_g)_\mu = 0.40 \text{ cm}^3/\text{g}$$

the micropore void fraction ε_μ in the pellet is

$$\varepsilon_\mu = \frac{(V_g)_\mu}{1/\rho_P} = \frac{0.40}{1/0.978} = 0.391$$

(e) The solids fraction ε_S is given by

$$1 = \varepsilon_M + \varepsilon_\mu + \varepsilon_S$$

$$\varepsilon_S = 1 - 0.200 - 0.391 = 0.409$$

(f) The density ρ_p of the *particles* can be calculated by correcting the total volume of the *pellet* for the macropore volume. Thus

$$\rho_p = \frac{3.15}{3.22 - 0.645} = 1.22 \text{ g/cm}^3$$

or, in terms of 1 g of pellet,

$$\rho_p = \frac{1}{1/\rho_P - (V_g)_M} = \frac{\rho_P}{1 - (V_g)_M \rho_P}$$

$$\rho_p = \frac{0.978}{1 - 0.205(0.978)} = 1.22 \text{ g/cm}^3$$

(g) The density of the solid phase is

$$\rho_S = \frac{\text{mass of pellet}}{(\text{volume of pellet}) \, \varepsilon_S}$$

$$= \frac{\rho_P}{\varepsilon_S} = \frac{0.978}{0.409} = 2.39 \text{ g/cm}^3$$

(h) The void fraction of the *particles* is given by

$$\varepsilon_p = \frac{(V_g)_\mu}{1/\rho_p} = \rho_p (V_g)_\mu$$

$$= 1.22(0.40) = 0.49$$

For this pellet a fraction equal to $\varepsilon_M + \varepsilon_\mu = 0.591$ is void and 0.409 is solid. Of the individual particles, a fraction 0.49 is void. Note that all these results were calculated from the mass and volume of the pellet and the measurements of macro- and micropore volumes.

8-3 Pore-Volume Distribution

We shall see in Chap. 11 that the effectiveness of the internal surface for catalytic reactions can depend not only on the volume of the void spaces (V_g), but also on the radius of the void regions. Therefore it is desirable to know the distribution of void volume in a catalyst according to size of the pore. This is a difficult problem because the void spaces in a given particle are nonuniform in size, shape, and length, and normally are interconnected. Further, these characteristics can change from one type of catalyst particle to another. Figure 8-3 shows electron-microscope (scanning type) photographs of porous silver particles ($S_g = 19.7$ m²/g). The material was prepared by reducing a precipitate of silver fumarate by heating at 350°C in a stream of nitrogen. The larger darker regions probably represent void space between individual particles, and the smaller dark spaces are intraparticle voids. The light portions are solid silver. The complex and random geometry shows that it is not realistic to describe the void spaces as pores. It is anticipated that other highly porous materials such as alumina and silica would have similar continuous and complex void phases. For a material such as Vycor, with its relatively low porosity (0.3) and continuous solid phase, the concept of void spaces as pores is more reasonable.

In view of evidence such as that in Fig. 8-3, it is unlikely that detailed quantitative descriptions of the void structure of solid catalysts will become available. Therefore, to account quantitatively for the variations in rate of reaction with location within a porous catalyst particle, a simplified model of the pore structure is necessary. The model must be such that diffusion rates of reactants through the void spaces into the interior surface can be evaluated. More is said about these models in Chap. 11. It is sufficient here to note that in all the widely used models the void spaces are simulated as cylindrical pores. Hence the size of the void space is interpreted as a radius a of a cylindrical pore, and the distribution of void volume is defined in terms of this variable. However, as the example of the silver

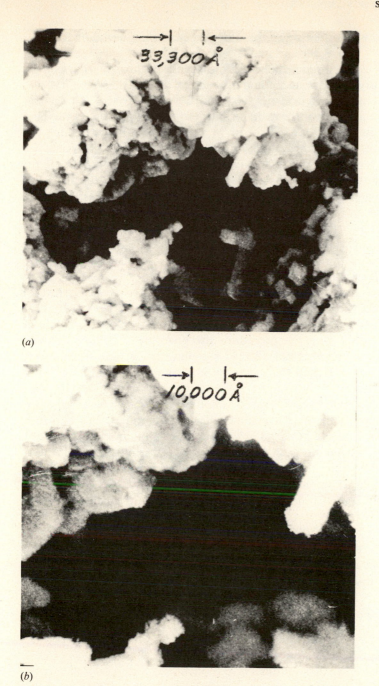

(a)

(b)

Figure 8-3 Electron micrographs of porous silver particles (approximate surface area = 19.7 m²/g): (a) magnification = 3000 (1 cm = 33000 Å); (b) magnification = 10,000 (1 cm = 10,000 Å).

catalyst indicates, this does not mean that the void spaces are well-defined cylindrical pores.

There are two established methods for measuring the distribution of pore volumes. The mercury-penetration method depends on the fact that mercury has a significant surface tension and does not wet most catalytic surfaces. This means that the pressure required to force mercury into the pores depends on the pore radius. The pressure varies inversely with a; 100 lb/in.² (approximately) is required to fill pores for which $a = 10,000$ Å, and 10,000 lb/in.² is needed for $a = 100$ Å. Simple techniques and equipment are satisfactory for evaluating the pore-volume distribution down to 100 to 200 Å, but special high-pressure apparatus is necessary to go below $a = 100$ Å, where much of the surface resides. In the second method, the nitrogen-adsorption experiment (described in Sec. 8-1 for surface area measurement) is continued until the nitrogen pressure approaches the saturation value (1 atm at the normal boiling point). At $p/p_0 \rightarrow 1.0$, where p_0 is the saturation pressure, all the void volume is filled with adsorbed and condensed nitrogen. Then a desorption isotherm is established by lowering the pressure in increments and measuring the amount of nitrogen evaporated and desorbed for each increment. Since the vapor pressure of a liquid evaporating from a capillary depends on the radius of the capillary, these data can be plotted as volume desorbed vs. pore radius. Thus this procedure also gives the distribution of pore volumes. The vapor pressure is not affected significantly by radii of curvature greater than about 200 Å. Hence, this method is not suitable for pores larger than 200 Å.

If a high-pressure mercury porosimeter is not available a combination of the two methods is necessary to cover the entire range of pore radii (10 to 10,000 Å) which may exist in a bidisperse catalyst or support, such as alumina pellets. For a monodisperse pore distribution, such as that in silica gel, the nitrogen-desorption experiment is sufficient, since there are few pores of radius greater than 200 Å. In a bidisperse pore system the predominant part of the catalytic reaction occurs in pores less than about 200 Å (the micropore region), since that is where the bulk of the surface resides. However, the transport of reactants to these small pores occurs primarily in pores of 200 to 10,000 Å (the macropore region). Hence the complete distribution of pore volume is required in order to establish the effectiveness of the interior surface, that is, the global rate of reaction. Calculation procedures and typical results are discussed briefly in the following paragraphs.

Mercury-penetration method By equating the force due to surface tension (which tends to keep mercury out of a pore) to the applied force, Ritter and Drake[†] obtained

$$\pi a^2 p = -2\pi a \sigma \cos \theta$$

or

$$a = \frac{-2\sigma \cos \theta}{p} \tag{8-11}$$

[†] H. L. Ritter and L. C. Drake, *Ind. Eng. Chem., Anal. Ed.*, **17**, 787 (1945).

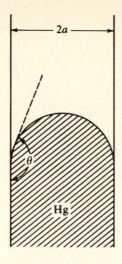

Figure 8-4 Mercury penetration in a pore of radius a.

where θ is the contact angle between the mercury and pore wall (Fig. 8-4). While θ probably varies somewhat with the nature of the solid surface, 140° appears to be a good average value. Then the working equation for evaluating the radius corresponding to a given pressure is

$$a(\text{Å}) = \frac{8.75 \times 10^5}{p \; (\text{lb/in.}^2)} \tag{8-12}$$

Calculation of the pore-size distribution by this method is illustrated in Example 8-6.

Example 8-6 The mercury-penetration data given in Table 8-3 were obtained on a 0.624-g sample of a uranium dioxide pellet formed by sintering particles at 1000°C for 2 h. Since the particles were nonporous, the void space was entirely between the particles (that is, in macropores). At the beginning of the experiment (when the pressure was 1.77 lb/in.² abs) the amount of mercury displaced by the sample was found to be 0.190 cm³. Calculate the porosity and pore-volume distribution of the pellet.

SOLUTION According to Eq. (8-12), at $p = 1.77$ lb/in.² abs only pores larger than about 500,000 Å (50 microns) would be filled with mercury. No pores larger than this are likely. Hence 0.190 cm³ is the total volume of the sample. At the highest pressure, 3,500 lb/in.² abs, only pores less than $a = 250$ Å would remain unfilled. Since the pores were entirely of the macro type, few pores smaller than 250 Å are expected. Neglecting these pores, the porosity can be calculated from the porosimeter measurements alone. Thus

$$\varepsilon_P = \frac{0.125}{0.190} = 0.66$$

Table 8-3 Mercury porosimeter data for uranium dioxide pellet (mass of sample 0.624 g)

Pressure lb/in.^{2}	Mercury concentration, cm^3	Pentration, cm^3/g	
116	0.002	0.003	0.196
310	0.006	0.010	0.189
344	0.010	0.016	0.183
364	0.014	0.022	0.177
410	0.020	0.032	0.167
456	0.026	0.042	0.157
484	0.030	0.048	0.151
540	0.038	0.061	0.138
620	0.050	0.080	0.119
710	0.064	0.102	0.097
800	0.076	0.122	0.077
830	0.080	0.128	0.071
900	0.088	0.141	0.058
1,050	0.110	0.160	0.039
1,300	0.112	0.179	0.020
1,540	0.118	0.189	0.010
1,900	0.122	0.196	0.003
2,320	0.124	0.198	0.001
3,500	0.125	0.199	0

A check on this result is available from air-pycnometer† data, which gave a solid volume of 0.0565 cm^3. Thus the total porosity is

$$(\varepsilon_P)_t = \frac{0.190 - 0.0565}{0.190} = 0.70$$

The difference between values suggests that there were a few pores smaller than 250 Å in the sample, although the comparison also includes experimental errors in the two methods.

To calculate the pore-volume distribution the penetration data are first corrected to a basis of 1 g of sample, as given in the third column of Table 8-3. If we neglect the pores smaller than 250 Å, the penetration data can be reversed, starting with $V = 0$ at 3500 lb/in.^{2} abs (250 Å). The last column shows the figures. Then, from Eq. (8-12) and the pressure, the radius corresponding to each penetration value can be established. This gives the penetration curve shown in Fig. 8-5. The penetration volume at any pore radius a is the volume of pores larger than a. The derivative of this curve, $\Delta V/\Delta a$, is the volume of pores between a and $a + \Delta a$ divided by Δa; that is, it is the distribution function for the pore volume according to pore radius. It is customary to

† A device which uses air at two pressures to measure void volumes of porous materials. It provides the same data as the helium measurement described in Sec. 8-2.

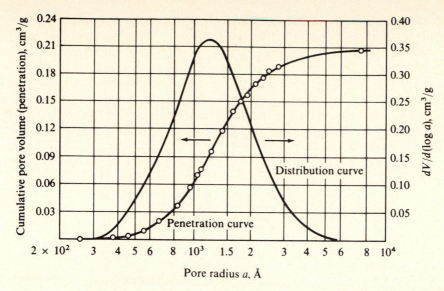

Figure 8-5 Pore-volume distribution in a UO_2 pellet.

plot the pore radius on a logarithmic coordinate as shown. Hence the distribution function is taken as the derivative of the curve so plotted, that is, dV/d (log a). The distribution function is also shown plotted against a in Fig. 8-5.

For this pellet the distribution is seen to be reasonably symmetrical with most of the volume in pores from 300 to 8000 Å and with a most probable pore radius of 1200 Å. Note that the flatness of the penetration curve at low pore radii justifies neglecting the pores smaller than 250 Å.

Wheeler[†] has summarized the assumptions and accuracy of the mercury-penetration method. It is important to note that erroneous results would be obtained if the porous particle contains large void spaces that are connected only to smaller void spaces. Such large "bottleneck" pores would fill with mercury at the higher pressure corresponding to the connecting smaller pores. For accurate results each porous region must be connected to at least one larger pore.

Nitrogen-desorption method As the low-temperature nitrogen-adsorption experiment (Sec. 8-1) is continued to higher pressures multilayer adsorption occurs, and ultimately the adsorbed films are thick enough to bridge the pore.[‡] Then further uptake of nitrogen will result in capillary condensation. Since the vapor pressure decreases as the capillary size decreases, such condensation will occur first in the smaller pores. Condensation will be complete, as $p/p_0 \rightarrow 1.0$, when the entire void

---

[†] A. Wheeler, in P. H. Emmett (ed.), "Catalysis," vol. II, p. 123, Reinhold Publishing Corporation, New York, 1955.

[‡] L. H. Cohan, *J. Am. Chem. Soc.*, **60**, 433 (1938); A. G. Foster, *J. Phys. Colloid Chem.*, **55**, 638 (1951).

region is filled with condensed nitrogen. Now, if the pressure is reduced by a small increment, a small amount of nitrogen will evaporate from the meniscus formed at the ends of the largest pores. Pores which are emptied of condensate in this way will be those in which the vapor pressure of nitrogen is greater than the chosen pressure. The Kelvin equation gives the relationship between vapor pressure and radius of the concave surface of the meniscus of the liquid. Since some of the nitrogen is adsorbed on the surface, and therefore not present because of capillary condensation, the Kelvin relationship must be corrected for the thickness δ of the adsorbed layers. With this correction, the pore radius is related to the saturation-pressure ratio (vapor pressure p in the pore divided by the normal vapor pressure p_0) by

$$a - \delta = \frac{-2\sigma V_1 \cos \theta}{R_g T \ln (p/p_0)} \tag{8-13}$$

where V_l = molal volume of the condensed liquid
σ = surface tension
θ = contact angle between surface and condensate

Since nitrogen completely wets the surface covered with adsorbed nitrogen, $\theta = 0°$ and $\cos \theta = 1$. The thickness δ depends on p/p_0. The exact relationship has been the subject of considerable study,[†] but Wheeler's form

$$\delta(\text{Å}) = 9.52 \left(\log \frac{p_0}{P} \right)^{-1/n} \tag{8-14}$$

is generally used.

For nitrogen at $-195.8°C$ (normal boiling point) Eq. (8-13), for $a - \delta$ in angstroms, becomes

$$a - \delta = 9.52 \left(\log \frac{p_0}{p} \right)^{-1} \tag{8-15}$$

with δ determined from Eq. (8-14).

For a chosen value of p/p_0, Eqs. (8-15) and (8-14) give the pore radius above which all pores will be empty of capillary condensate. Hence, if the amount of desorption is measured for various p/p_0, the pore volume corresponding to various radii can be evaluated. Differentiation of the curve for cumulative pore volume vs. radius gives the distribution of volume as described in Example 8-6. Descriptions of the method of computation are given by several investigators.[‡] As in the mercury-penetration method, errors will result unless each pore is connected to at least one larger pore.

Figure 8-6 shows the result of applying the method to a sample of Vycor

† A. Wheeler, in P. H. Emmett (ed.), "Catalysis," vol. II, chap. 2, Reinhold Publishing Corporation, New York, 1955; G. D. Halsey, *J. Chem. Phys.*, **16**, 931 (1948); C. G. Shull, *J. Am. Chem. Soc.*, **70**, 1405 (1948); J. O. Mingle and J. M. Smith, *Chem. Eng. Sci.*, **16**, 31 (1961).

‡ E. P. Barrett, L. G. Joyner, and P. P. Halenda, *J. Am. Chem. Soc.*, **73**, 373 (1951); C. J. Pierce, *J. Phys. Chem.*, **57**, 149 (1953); R. B. Anderson, *J. Catalysis*, **3**, 50 (1964).

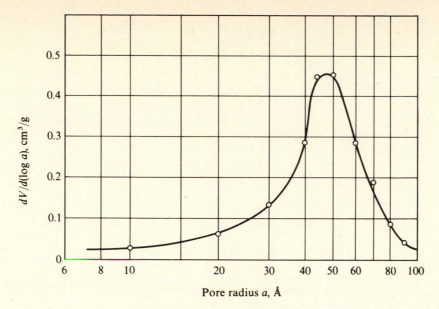

Figure 8-6 Pore-volume distribution in Vycor; $\rho_p = 1.46$ g/cm³, $V_g = 0.208$ cm³/g, $S_g = 90$ m²/g.

(porous glass).† This material, which contained only micropores, had the properties

$$\rho_p = 1.46 \text{ g/cm}^3$$

$$V_g = 0.208 \text{ cm}^3/\text{g}$$

$$\varepsilon_p = 0.304$$

$$S_g = 90 \text{ m}^2/\text{g}$$

The surface area was determined from nitrogen-adsorption data in the low p/p_0 range, as described in Sec. 8-1, while the distribution results in Fig. 8-6 were established from the desorption curve in the capillary-condensation (high p/p_0) region.

By combining mercury-penetration and nitrogen-desorption measurements, pore-volume information can be obtained over the complete range of radii in a pelleted catalyst containing both macro- and micropores. Figure 8-7 shows the cumulative pore volume for two alumina pellets, each prepared by compressing porous particles of boehmite ($Al_2O_3 \cdot H_2O$). The properties‡ of the two pellets are given in Table 8-4. The only difference in the two is the pelleting pressure. Increasing this pressure causes drastic reductions in the space between particles (macropore volume) but does not greatly change the void volume within the particles or

† M. R. Rao and J. M. Smith, *AIChE J.*, **10**, 293 (1964).

‡ The properties and pore-volume distribution were determined by M. F. L. Johnson, Sinclair Research Laboratories, Harvey, Ill., by the methods described in Secs. 8-1 to 8-3. They were orginally reported in J. L. Robertson and J. M. Smith, *AIChE J.*, **9**, 344 (1963).

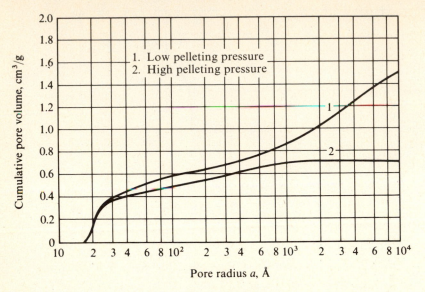

Figure 8-7 Pore volume in alumina (boehmite) pellets.

the surface area. The derivative of the volume curves in Fig. 8-7 gives the pore-volume distribution, and these results are shown in Fig. 8-8. In this figure the bidisperse pore system, characteristic of alumina pellets, is clearly indicated. The micropore range within the particles is narrow, with a most probable radius of 20 Å. The macropores cover a much wider range of radii and show the effect of pelleting pressure. For the high-pressure pellet all volume with pores greater than 2000 Å has been squeezed out, while the most probable radius for the low-pressure pellet is 8000 Å. Pelleting pressure seems to have little effect on the micropores, which suggests that the particles themselves are not crushed significantly during the pelleting process.

Some models (see Chap. 11) for quantitative treatment of the effectiveness of the internal catalyst surface require only the average pore radius $\bar{a}$, rather than the distribution of pore volumes. Wheeler† has developed a simple equation for $\bar{a}$

---

† A. Wheeler, in P. H. Emmett (ed.), "Catalysis," vol. II, chap. 2, Reinhold Publishing Corporation, New York, 1955.

Table 8-4 Properties of boehmite ($Al_2O_3 \cdot H_2O$) pellets

Pelleting pressure	Macropore volume, cm^3/g	Micropore volume, cm^3/g	Surface area, m^2/g
Low	1.08	0.56	389
High	0.265	0.49	381

Notes: Volume and surface area refer to mass of Al_2O_3 obtained by ignition of boehmite. The pore-volume distribution is given in Fig. 8-8. Average particle size from which pellets were made was 85 microns.

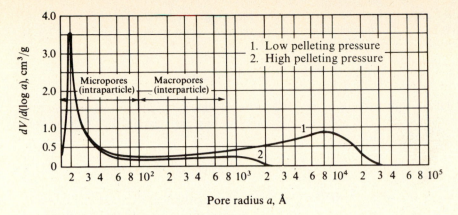

Figure 8-8 Pore-volume distribution in alumina pellets.

which requires only surface-area and pore-volume measurements. Suppose all the pores in a hypothetical particle are straight, cylindrical, not interconnected, and have the same radius $\bar{a}$ and length $\bar{L}$. The average pore radius may be found by writing equations for the total surface and volume in the hypothetical particle and equating these quantities to the surface $m_p S_g$ and volume $m_p V_g$ in the actual particle; i.e.

$$m_p S_g = (2\pi \bar{a} \bar{L})n \tag{8-16}$$

$$m_p V_g = (\pi \bar{a}^2 \bar{L})n \tag{8-17}$$

where m_p and n are the mass and number of pores per particle. Dividing the two equations gives the average pore radius,

$$\bar{a} = \frac{2V_g}{S_g} \tag{8-18}$$

This expression agrees well with volume-average values obtained from the distribution curve for monodisperse pore systems. For example, from the data for the Vycor sample (Fig. 8-6) Eq. (8-18) gives

$$\bar{a} = \frac{2(0.208)}{90 \times 10^4} = 46 \times 10^{-8} \text{ cm} \qquad \text{or } 46 \text{ Å}$$

The volume-average value is calculated from the pore volume data used to obtain the distribution curve (Fig. 8-6) and the expression

$$\bar{a} = \frac{\int_0^{V_g} a \, dV}{V_g}$$

By this method $\bar{a} = 45$ Å.

Accurate values of the small areas existing in macropore systems make it difficult to use Eq. (8-18) to calculate $\bar{a}$ for interparticle pores. Hence the average radius for systems such as the UO_2 pellets discussed in Example 8-6 should be obtained by integrating under the cumulative-volume-vs.-$\bar{a}$ curve shown in

Fig. 8-5. Also a single value of $\bar{a}$ has no meaning for a bidisperse pore system such as that in an alumina pellet. Thus using the *total* pore volume in Table 8-4 for the low-pressure pellet gives

$$\bar{a} = \frac{2(1.64)}{389 \times 10^4} = 84 \times 10^{-8} \text{ cm} \qquad \text{or } 84 \text{ Å}$$

As Fig. 8-8 shows, there is very little pore volume (very few pores) in this radius region. If Eq. (8-18) is applied to the micropore region, arbitrarily taken as pores smaller than $a = 100$ Å,

$$\bar{a}_\mu = \frac{2(0.56)}{389 \times 10^4} = 29 \times 10^{-8} \text{ cm} \qquad \text{or } 29 \text{ Å}$$

The curve in Fig. 8-8 indicates that this is an approximate value for the *average* pore radius in the micropore region. Note that the micropore distribution is asymmetrical in such a way that the average radius is greater than the most probable value (20 Å).

Summary In concluding the treatment of physical properties of catalysts, let us review the purpose for studying properties and structure of porous solids. Heterogeneous reactions with solid catalysts occur on parts of the surface active for chemisorption. The number of these active sites and the rate of reaction is, to a first approximation, proportional to the extent of the surface. Hence it is necessary to know the surface area. This is evaluated by low-temperature-adsorption experiments in the pressure range where a monomolecular layer of gas (usually nitrogen) is physically adsorbed on the catalyst surface. The effectiveness of the interior surface of a particle (and essentially all of the surface is in the interior) depends on the volume and size of the void spaces. The pore volume (and porosity) can be obtained by simple pycnometer-type measurements (see Examples 8-4 and 8-5). The average size (pore radius) can be estimated by Eq. (8-18) from the surface area and pore volume in some *monodisperse* systems. Determination of the complete distribution of pore volume according to pore radius requires either mercury-penetration measurements or nitrogen-adsorption data at pressures where capillary condensation occurs, or both. Accurate values of the average pore radius can be evaluated from such pore-volume-vs.-radius data. Note also that a measurement of the complete nitrogen-adsorption-desorption isotherm is sufficient to calculate surface area and pore volume, and distribution of pore sizes, in the range 10 Å $< a <$ 200 Å.

8-4 Theories of Heterogeneous Catalysis

Understanding how solid catalysts work has been an elusive and challenging problem. Several theories have gained prominence, only to lose acceptance as *general* explanations of catalytic activity as additional experimental evidence becomes available. In chronological order, Sabatier[†] suggested that a mechanism for

---

[†] P. Sabatier, "Catalysis in Organic Chemistry," trans. by E. E. Reid, in "Catalysis, Then and Now," Franklin Publishing Company, Englewood, N.J., 1965.

the activity of nickel as a hydrogenation catalyst might involve formation of a chemical compound, nickel hydride. Since then, the *chemical factor* continues to be recognized as important. Later Taylor,* Balandin,† and Beeck‡ provided evidence for the significance of *geometric properties*. According to this concept the catalytic activity of a solid surface depends on the spacing of atoms so as to facilitate adsorption of reactant molecules. Over the years most of the evidence for the geometric theory has proved suspect, except that for metallic films. The work of Boudart§ and Beeck changed the emphasis from geometric considerations to *electronic properties*. In 1948 Dowden and coworkers¶ proposed that catalysts might be classified, on the basis of electron mobility, as conductors, semiconductors, and insulators.

The conductor catalysts are the metals (silver, platinum, vanadium, iron, etc.) and have the property of chemisorption by electron transfer. The semiconductor catalysts are the oxides, such as NiO, Cu_2O, and ZnO. These materials have the capability of interchanging electrons from the filled valence bands in a compound when sufficient energy is provided, for example, by heating. Upon this electron transfer the semiconductor becomes a conductor.†† The insulator catalysts include such widely used substances as silica gel, alumina, and their combinations. Even at high temperatures, electrons are not supposed to be able to move through these two solids freely enough to justify their being called conductors. These substances are also known to be strong acids. Their activity in the many hydrocarbon reactions which they catalyze is presumably due to the formation of carbonium ions at the acid sites on the surface. Carbonium-ion mechanisms are well described in the original work of Whitmore‡‡ and the later work of Greensfelder.§§ It should be emphasized that the electronic theory is not without uncertainties and at present should be considered as a concept in transition.¶¶ However, it does provide a convenient, and probably helpful, method of classifying solid catalysts.

In order fully to understand the behavior of solid catalysts the chemical structure of the adsorbed species must be known. Up to now this has been an

---

* H. S. Taylor, *Proc. Roy. Soc. (London)*, **A108**, 105 (1925).

† A. A. Balandin, "Advances in Catalysis," vol. X, p. 96, Academic Press, Inc., New York, 1958.

‡ O. Beeck, *Disc. Faraday Soc.*, **8**, 118 (1950).

§ M. Boudart, *J. Am. Chem. Soc.*, **72**, 1040 (1950).

¶ D. A. Dowden, *Research*, **1**, 239 (1948); D. A. Dowden and P. W. Reynolds, *Disc. Faraday Soc.*, **8**, 187 (1950).

†† Semiconductors are classified as *p* type if they tend to attract electrons from the chemisorbed species, or as *n* type if they donate electrons to this species. The *p* type are normally the compounds, such as NiO. The *n* type are substances which contain small amounts of impurities, or the oxide is present in nonstoichiometric amounts (as, for example, when some of the zinc in ZnO has been reduced). Reviews of semiconductors as catalysts are given by P. H. Emmett ("New Approaches to the Study of Catalysis," 36th Annual Priestly Lectures, Pennsylvania State University, April 9–13, 1962) and by P. G. Ashmore ("Catalysis and Inhibition of Chemical Reactions," Butterworth & Co. (Publishers), London, 1963.

‡‡ F. C. Whitmore, *J. Am. Chem. Soc.*, **54**, 3274 (1932).

§§ B. S. Greensfelder, "Chemistry of Petroleum Hydrocarbons," vol. II, chap. 27, Reinhold Publishing Corporation, New York, 1955.

¶¶ An analysis of the electronic theory is given by Th. Volkenstein, in "Advances in Catalysis," vol. XII, p. 189, Academic Press Inc., New York, 1960.

unsolved problem, except for a few special cases. However, recent progress in developing and improving spectroscopic techniques for examing surface and bulk properties of solids provides encouragement. These techniques include x-ray scattering, nuclear magnetic resonance, Raman spectroscopy, high resolution electron microscopy, extended x-ray absorption, fine structure spectroscopy, and photoacoustic spectroscopy. It seems unlikely that a general theory of catalysis, applicable to a wide variety of reactions, will emerge. Nevertheless, these new instrumental methods may provide explanations for the behavior of many important catalytic processes.

8-5 Classification of Catalysts

From an accumulation of practical experience it is possible to narrow the range of solids that are likely to be catalysts for a type of chemical reaction. We discuss a few cases here and give information for more reactions in Table 8-5.

Metals chemisorb oxygen and hydrogen and therefore are usually effective catalysts for oxidation-reduction and hydrogenation-dehydrogenation reactions. Thus platinum is a successful catalyst for the oxidation of SO_2, and Ni is used effectively for hydrogenation of hydrocarbons. The metals oxides, as semiconductors, catalyze the same kinds of reactions, but often higher temperatures are required. Because of the relative strength of the chemisorption bond with which such gases as O_2 and CO are attached to metals, these gases are poisons when metals are employed as hydrogenation catalysts. The semiconductor oxides are less susceptible to such poisoning. Oxides of the transition metals, such as MoO_3 and Cr_2O_3, are good catalysts for polymerization of olefins. Also aluminum alkyl-titanium chloride [for example, $Al(C_2H_5)_3 + TiCl_4$] constitutes an excellent catalyst for producing isotactic polymers from olefins.[†] Alumina and silica catalysts are widely used for alkylation, isomerization, polymerization, and particularly for cracking of hydrocarbons. In each case the mechanism presumably involves carbonium ions formed at the acid sites on the catalyst. While the emphasis here has been on solid catalysts, liquid and gaseous acids, particularly H_2SO_4 and HF, are well-known alkylation and isomerization catalysts.

Often catalysts are specific. An important example is the effectiveness of iron for producing hydrocarbons from hydrogen and carbon monoxide (the Fischer-Tropsch synthesis). Dual-function catalysts for isomerization and reforming reactions consist of two active substances in close proximity to each other. For example, Ciapetta and Hunter[‡] found that a silica-alumina catalyst upon which nickel was dispersed was much more effective in isomerizing *n*-hexane than silica alumina alone. The explanation depends on the fact that olefins are more readily isomerized than paraffin hydrocarbons. Nickel presumably acts as a dehydrogenating agent, producing hexene, after which the silica alumina isomerizes the hexene to isohexene. Finally, the nickel is effective in hydrogenating hexene back to isohexane.

[†] K. Ziegler, *Angew. Chem.*, **64**, 323 (1952); G. Natta and I. Pasquon, in "Advances in Catalysis," vol. XI, p. 1, Academic Press Inc., New York, 1959.

[‡] F. G. Ciapetta and J. B. Hunter, *Ind. Eng. Chem.*, **45**, 155 (1953).

Table 8-5 Catalysts for Commercial Processes†

Process	Typical catalysts	Poisons
Alkylation of hydrocarbons	$H_2SO_4(l)$, $HF(l)$, $AlCl_3 + HCl$ H_3PO_4/kieselguhr	Substances which reduce acidity
Hydrocarbon cracking	Crystalline synthetic SiO_2-Al_2O_3 (zeolites)	Nitrogen compounds, metals (Ni, V, Cu) coke deposition
Chlorination of hydrocarbons	$CuCl_2$/Al_2O_3	
Dehydration	γ-Al_2O_3, SiO_2-Al_2O_3, WO_3	Coke deposition
Dehydrogenation	Cr_2O_3/Al_2O_3, Fe, Ni, Co, ZnO, Fe_2O_3	H_2O
Desulfurization of petroleum fractions	Sulfided Co-Mo/Al_2O_3	
Fischer-Tropsch process	Ni/kieselguhr, $Fe + Fe_2C + Fe_3O_4$	
Hydrogen from naphtha, coal	Ni/refractory	Sulfur, arsenic, coke deposition
Hydrogenation	Ni/kieselguhr, NiO, Ni-Al (Raney nickel), Pt/Al_2O_3, Pd/Al_2O_3, Ru/Al_2O_3	Sulfur, chlorine compounds
Hydrocracking of coal, heavy oil	NiS, Co_2O_3-MoO_3/Al_2O_3 W_2O_3, $ZnCl_2$	
Isomerization	$AlCl_3 + HCl$, Pt/Al_2O_3	
Oxidation, inorganic	Pt, V_2O_5, Rh, $CuCl_2$(HCl to Cl_2)	Arsenic, chlorine compounds
Oxidation, organic (liquid phase)	$CuCl_2$(aq) + $PdCl_2$, Pd/Al_2O_3, Co + Cu acetates	
Oxidation, organic (gas phase)	V_2O_5/Al_2O_3, Ag-AgO, CuO bismuth molybdate	
Polymerization	$Al(C_2H_5)_3$, P_2O_5/kieselguhr, MoO_3 -CoO/Al_2O_3, CrO_3/(SiO_2-Al_2O_3), $TiCl_3$-$Al(C_2H_5)_3$	H_2O, O_2, sulfur compounds CO, CO_2

† Condensed from material in "Catalytic Processes and Proven Catalysts" by Charles L. Thomas, Academic Press, New York, 1970.

Much has and is being written on solid catalysts, and helpful sources of summary information are available.††

8-6 Catalyst Preparation

Experimental methods and techniques for catalyst manufacture are particularly important because chemical composition is not enough by itself to determine activity. The physical properties of surface area, pore size, particle size, and particle structure also have an influence. These properties are determined to a large

† P. H. Emmett (ed.), "Catalysis," Reinhold Publishing Corporation, New York, 1954–; "Advances in Catalysis," Academic Press Inc., New York, 1949–; *J. Catalysis*, **1**, 1962–; A. A. Balandin, "Scientific Selection of Catalysts," trans. by A. Aledjem, Davey Publishing Company, Hartford, Conn., 1968.
‡ J. H. Sinfelt, *AIChE J.*, **19**, 673 (1973).

extent by the preparation procedure. To begin with, a distinction should be drawn between preparations in which the entire material constitues the catalyst and those in which the active ingredient is dispersed on a *support* or *carrier* having a large surface area. The first kind of catalyst is usually made by precipitation, gel formation, or simple mixing of the components.

Precipitation provides a method of obtaining the solid material in a porous form. It consists of adding a precipitating agent to solutions of the desired components. Washing, drying, and usually calcination and activation (or pretreatment) are subsequent steps in the process. For example, a magnesium oxide catalyst can be prepared by precipitating the magnesium from nitrate solution by adding sodium carbonate. The precipitate of $MgCO_3$ is washed, dried, and calcined to obtain the oxide. Such variables as concentration of the aqueous solutions, temperature, and time of the drying and calcining steps may influence the surface area, pore structure, and intrinsic activity of the final product. This illustrates the difficulty in reproducing catalysts and indicates the necessity of carefully following tested recipes. Of particular importance is the washing step to remove all traces of impurities, which may act as poisons.

A special case of the precipitation method is the formation of a colloidal precipitate which gels. The steps in the process are essentially the same as for the usual precipitation procedure. Catalysts containing silica and alumina are especially suitable for preparation by gel formation, since their precipitates are of a colloidal nature. Detailed techniques for producing catalysts through gel formation or ordinary precipitation are given by Ciapetta and Plank.† ‡

In some instances a porous material can be obtained by mixing the components with water, milling to the desired grain size, drying, and calcining. Such materials must be ground and sieved to obtain the proper particle size. A mixed magnesium and calcium oxide catalyst can be prepared in this fashion. The carbonates are milled wet in a ball machine, extruded, dried, and reduced by heating in an oven.

Catalyst *carriers* provide a means of obtaining a large surface area with a small amount of active material. This is important when expensive agents such as platinum, palladium, ruthenium, and silver are used.§ Berkman *et al.*¶ have treated the subject of carriers in some detail.

The steps in the preparation of a catalyst impregnated on a carrier may include (1) evacuating the carrier, (2) contacting the carrier with the impregnating solution, (3) removing the excess solution, (4) drying, (5) calcination and activation. For example, a nickel hydrogenation catalyst can be prepared on alumina by soaking the evacuated alumina particles with nickel nitrate solution, draining to remove the excess solution, and heating in an oven to decompose the nitrate to nickel oxide. The final step (activation), reduction of the oxide to metallic nickel, is

† F. G. Ciapetta and C. J. Plank, in P. H. Emmett (ed.), "Catalysis," vol. I, chap. 7, Reinhold Publishing Corporation, New York, 1954.

‡ J. R. Anderson "Structure of Metallic Catalysts", Academic Press, New York, 1975.

§ Such metals are normally deposited only on the outer region of the particle.

¶ S. Berkman, J. C. Morrell, and G. Egloff, "Catalysis," Reinhold Publishing Corporation, New York, 1940.

best carried out with the particles in place in the reactor by passing hydrogen through the equipment. Activation *in situ* prevents contamination with air and other gases which might poison the reactive nickel. In this case no precipitation was required. This is a desirable method of preparation, since thorough impregnation of all the interior surface of the carrier particles is relatively simple. However, if the solution used to soak the carrier contains potential poisons such as chlorides or sulfates, it may be necessary to precipitate the required constituent and wash out the possible poison.

The nature of the support can affect catalyst activity and selectivity. This effect presumably arises because the support can influence the surface structure of the atoms of dispersed catalytic agent. For example, changing from a silica to alumina carrier may change the electronic structure of deposited platinum atoms. This question is related to the optimum amount of catalyst that should be deposited on a carrier. When only a small fraction of a monomolecular layer is added, increases in amount of catalyst should increase the rate. However, it may not be helpful to add large amounts to the carrier. For example, the conversion rate of the ortho to para hydrogen with a NiO catalyst deposited on alumina was found† to be less for 5.0 wt % NiO than for 0.5 wt % NiO. The dispersion of the catalyst on the carrier may also be an important factor in such cases. The nickel atoms were deposited from a much more concentrated $NiNO_3$ solution to make the catalyst containing 5.0 wt % NiO. This may have led to larger clusters of nickel atoms. That is, many more nickel atoms were deposited on top of each other, so that the dispersion of nickel on the surface was less uniform than with the 0.5 wt % catalyst. It is interesting to note that a 5.0 wt % NiO catalyst prepared by 10 individual depositions of 0.5 wt % was much more active (by a factor of 11) than the 5.0 wt % added in a single treatment. The multiple deposition method presumably gave a much larger active nickel surface, because of better dispersion of the nickel atoms on the Al_2O_3 surface. Since the total amount of nickel was the same for the two preparations, one would conclude that the individual particles of nickel were smaller in the 10-application catalyst. These kinds of data indicate the importance of measuring surface areas for chemisorption of the reactants involved. A technique based on the chemisorption of H_2 and CO has been developed‡ to study the effect of dispersion of a catalyst on its activity and the effect of interaction between catalyst and support on activity.

8-7 Promoters and Inhibitors

As normally used, the term "catalyst" designates the composite product used in a reactor. Components of the catalyst include the catalytically active substance itself and may also include a carrier, promoters, and inhibitors.

Innes§ has defined a *promoter* as a substance added during the preparation of

---

† N. Wakao, J. M. Smith, and P. W. Selwood, *J. Catalysis*, **1**, 62 (1962).

‡ G. K. Boreskov and A. P. Karnaukov, *Zh. Fiz. Khim.*, **26**, 1814 (1952); L. Spenadel and M. Boudart, *J. Phys. Chem.*, **64**, 204 (1960).

§ W. B. Innes, in P. H. Emmett (ed.), "Catalysis," vol. 1, chap. 7, Reinhold Publishing Corporation, New York, 1954.

a catalyst which improves activity or selectivity or stabilizes the catalytic agent so as to prolong its life. The promoter is present in a small amount and by itself has little activity. There are various types, depending on how they act to improve the catalyst. Perhaps the most extensive studies of promoters has been in connection with iron catalysts for the ammonia synthesis reaction.† It was found that adding Al_2O_3 (other promoters are CaO, K_2O) prevented reduction (by sintering) in surface area during catalyst use and gave an increased activity over a longer period of time. Chlorides are sometimes added as promoters to hydrogenation and isomerization catalysts, and sulfiding improves hydro-desulfurization (Co-Mo) catalysts. Some promoters are also believed to increase the number of active centers and so make the existing catalyst surface more active. The published information on promoters is largely in the patent literature.

An *inhibitor* is the opposite of a promoter. When added in small amounts during catalyst manufacture, it lessens activity, stability, or selectivity. Inhibitors are useful for reducing the activity of a catalyst for an undesirable side reaction. For example, silver supported on alumina is an excellent oxidation catalyst. In particular, it is used widely in the production of ethylene oxide from ethylene. However, at the same conditions complete oxidation to carbon dioxide and water also occurs, so that selectivity to C_2H_4O is poor. It has been found that adding halogen compounds to the catalyst inhibits the complete oxidation and results in satisfactory selectivity.

8-8 Catalyst Deactivation (Poisoning)

The activity of a catalyst normally decreases with time. In the development of a new catalytic process, the life of the catalyst is usually a major economic consideration. Shutting down a process, and its associated separation and preparation units, for regenerating or replacing catalyst is prohibitive except at infrequent intervals. In many instances, very active catalytic substances have been discovered only to be discarded because activity cannot be maintained and regeneration was not practical. Hence, an understanding of how catalysts lose activity is important. In some systems the catalyst activity decreases so slowly that exchange for new material, or regeneration, is required only at intervals measured in months or years. Examples are promoted catalysts for synthetic ammonia and catalysts containing metals such as platinum and silver. Catalysts for cracking and some other hydrocarbon reactions, however, lose much of their activity in seconds. The decrease in activity is due to *poisons*, which will be defined here as substances, either in the reactants stream or produced by the reaction, which lower the activity of the catalyst. The continuous (or frequent) regeneration of cracking catalysts is necessary because of the deposition of one of the products, carbon, on the surface.

The slow decrease is usually due to chemisorption of reactants, products, or impurities in the liquid stream. Rapid deactivation is caused by physically depositing a substance which blocks the active sites of the catalyst. We will use the term *poisoning* to describe both processes, although rapid deactivation is also called

---

† P. H. Emmett and S. Brunauer, *J. Am. Chem. Soc.*, **62**, 1732 (1940).

fouling. Deactivation can also be caused by a change in the surface structure of the catalyst (e.g., sintering) due to prolonged exposure to elevated temperature in the reacting atmosphere. In this section we list and briefly describe poisons. Quantitative treatment and the effect of deactivation on intrinsic rates of catalytic reactions are considered in Sec. 9-6. The effect on the global rate, which includes the influence of intraparticle diffusion, is taken up in Sec. 11-14.

Poisons can be differentiated in terms of the way in which they operate. Many summaries listing specific poisons and fouling reactions are available.† The following arrangement has been taken in part from Innes.

Deposited poisons Carbon deposition on catalysts used in the petroleum industry falls into this category. The carbon covers the active sites of the catalyst and may also partially plug the pore entrances. This type of poisoning is at least partially reversible, and regeneration can be accomplished by burning to CO and CO_2 with air and/or steam. The regeneration process itself is a heterogeneous reaction, a gas-solid noncatalytic one. In the design of the reactor, attention must be given to the regeneration as well as to the reaction parts of the cycle.‡

Chemisorbed poisons Compounds of sulfur and other materials are frequently chemisorbed on nickel, copper, and platinum catalysts. The decline in activity stops when equilibrium is reached between the poison in the reactant stream and that on the catalyst surface. If the strength of the adsorption bond is not great, the activity will be regained when the poison is removed from the reactants. If the adsorbed material is tightly held, the poisoning is more permanent. The mechanism appears to be one of covering the active sites, which could otherwise adsorb reactant molecules.

Selectivity poisons The selectivity of a solid surface for catalyzing one reaction with respect to another is not well understood. However, it is known that some materials in the reactant stream will adsorb on the surface and then catalyze other undesirable reactions, thus lowering the selectivity. The very small quantities of nickel, vanadium, iron, etc., in petroleum stocks may act as poisons in this way. When such stocks are cracked, the metals deposit on the catalyst and act as dehydrogenation catalysts. This results in increased yields of hydrogen and coke and lower yields of gasoline.

Stability poisons When water vapor is present in the sulfur dioxide–air mixture supplied to a platinum-alumina catalyst, a decrease in oxidation activity occurs. This type of poisoning is due to the effect of water on the structure of the alumina

† R. H. Griffith, "The Mechanism of Contact Catalysis," p. 93, Oxford University Press, New York, 1936; E. B. Maxted, _J. Soc. Chem. Ind._ (_London_), **67**, 93 (1948); P. H. Emmett (ed.), "Catalysis," vol. 1, chap. 6, Reinhold Publishing Corporation, New York, 1954. An extensive review by J. B. Butt, Chemical Reaction Engineering, Adv. Chemistry Series, **109**, 259 (1972); B. W. Wojciechowski, _Catal. Rev._ **9**, 79 (1974).

‡ Typical references are M. Sagara, S. Masamune, and J. M. Smith, _AIChE J._, **13**, 1226 (1967); G. F. Froment and K. B. Bischoff, _Chem. Eng. Sci._, **16**, 189 (1961); P. B. Weisz and R. D. Goodwin, _J. Catalysis_, **2**, 397 (1963).

Table 8-6 Poisons for various catalysts

Catalyst	Reaction	Types of poisoning	Poisons
Silica, alumina	Cracking	Chemisorption	Organic bases
		Deposition	Carbon
		Stability	Water
		Selectivity	Heavy metals
Nickel, platinum, copper	Hydrogenation Dehydrogenation	Chemisorption	Compounds of S, Se, Te, P, As, Zn, halides, Hg, Pb, NH_3, C_2H_2. H_2S, Fe_2O_3, etc.
Cobalt	Hydrocracking	Chemisorption	NH_3, S, Se, Te, P
Silver	$C_2H_4 + O \rightarrow C_2H_4O$	Selectivity	CH_4, C_2H_6
Vanadium oxide	Oxidation	Chemisorption	As
Iron	Ammonia synthesis	Chemisorption	O_2, H_2O, CO, S, C_2H_2
	Hydrogenation	Chemisorption	Bi, Se, Te, P, H_2O
	Oxidation	Chemisorption	VSO_4, Bi

Source: In part from W. B. Innes in P. H. Emmett (ed), "Catalysis," vol. I, chap. 7, p. 306, Reinhold Publishing Corporation, New York, 1954.

carrier. Temperature has a pronounced effect on stability poisoning. Sintering and localized melting may occur as the temperature is increased, and this, of course, changes the catalyst structure.

Diffusion poisons This kind of poisoning has already been mentioned in connection with carbon deposition on cracking catalysts. Blocking the pore mouths prevents the reactants from diffusing into the inner surface. Entrained solids in the reactants, or fluids which can react with the catalyst to form a solid residue, can cause this type of poisoning.

Tables 8-5 and 8-6 list poisons for various catalysts and reactions. The materials that are added to reactant streams to improve the performance of a catalyst are called *accelerators*. They are the counterparts of poisons. For example, steam added to the butene feed of a dehydrogenation reactor appeared to reduce the amount of coke formed and increase the yield of butadiene. The catalyst in this case was iron.†

PROBLEMS

8-1 The silica gel for *n*-hexane adsorption mentioned in Prob. 7-6 has the following properties: $S_g = 832$ m^2/g, $\varepsilon_p = 0.486$, $\rho_p = 1.13$ g/cm^3, $V_g = 0.43$ cm^3/g. Using the adsorption data given in Prob. 7-6: (*a*) Estimate the fraction of the surface covered with an adsorbed monomolecular layer at each partial pressure of *n*-hexane. The surface occupied by one molecule of hexane at 70°C is estimated to be 58.5×10^{-16} cm^2; (*b*) Calculate the value of $\bar{C}_m$ from the total surface area and the area

† K. K. Kearly, *Ind. Eng. Chem.*, **42**, 295 (1950).

occupied by one molecule of hexane. What conclusions can be drawn from the comparison of this value of $\bar{C}_m$ with that obtained in Prob. 7-6?

8-2 Repeat Prob. 8-1 for benzene adsorption on the same silica gel but at 110°C. The surface occupied by one benzene molecule is estimated to be 34.8×10^{-16} cm^2. Use the adsorption data for benzene given in Prob. 7-7.

8-3 Refer to Prob. 7-12. Using a constant value of 34.8×10^{-16} cm^2 for the surface area occupied by one molecule of benzene, determine the fractional surface coverage at each partial pressure of benzene given in Prob. 7-12.

8-4 Curve 3 of Fig. 8-2 is a Brunauer-Emmett-Teller plot for the adsorption data of N$_2$ at -183°C on the sample of silica gel. The density of liquid N$_2$ at this temperature is 0.751 g/cm^3. Estimate the area of the silica gel from these data, in square meters per gram, and compare with the results of Example 8-2.

8-5 The "point-B method" of estimating surface areas was frequently used prior to the development of the Brunauer-Emmett-Teller approach. It entailed choosing from an adsorption diagram such as Fig. 8-1 the point at which the central linear section begins. This procedure worked well for some systems, but it was extremely difficult, if not impossible, to select a reliable point B on an isotherm such as that shown for n-butane in Fig. 8-1. In contrast, the Brunauer-Emmett-Teller method was found to be reasonably satisfactory for this type of isotherm. Demonstrate this by estimating the surface area of the silica gel sample from the n-butane curve in Fig. 8-2 (multiply the ordinate of the n-butane curve by 10). The density of liquid butane at 0°C is 0.601 g/cm^3.

8-6 An 8.01-g sample of Glaucosil is studied with N$_2$ adsorption at -195.8°C. The following data are obtained:

Pressure, mmHg	6	25	140	230	285	320	430	505
Volume adsorbed, cm^3 (at 0°C and 1 atm)	61	127	170	197	215	230	277	335

The vapor pressure of N$_2$ at -195.8°C is 1 atm. Estimate the surface area (square meters per gram) of the sample.

8-7 Low-temperature (-195.8°C) nitrogen-adsorption data were obtained for an Fe-Al$_2$O$_3$ ammonia catalyst. The results for a 50.4-g sample were:

Pressure, mmHg	8	30	50	102	130	148	233	258	330	442	480	507	550
Volume adsorbed, cm^3 (at 0°C and 1 atm)	103	116	130	148	159	163	188	198	221	270	294	316	365

Estimate the surface area for this catalyst.

8-8 Ritter and Drake† give the true density of the solid material in an activated alumina particle as 3.675 g/cm^3. The density of the particle determined by mercury displacement is 1.547. The surface area by adsorption measurement is 175 m^2/g. From this information compute the pore volume per gram, the porosity of the particles, and the mean pore radius. The bulk density of a bed of the alumina particles in a 250-cm^3 graduate is 0.81 g/cm^3. What fraction of the total volume of the bed is void space between the particles and what fraction is void space within the particles?

8-9 Two samples of silica-alumina cracking catalysts have particle densities of 1.126 and 0.962 g/cm^3, respectively, as determined by mercury displacement. The true density of the solid material in each case

† H. L. Ritter and L. C. Drake, *Ind. Eng. Chem., Anal. Ed.*, **17**, 787 (1945).

is 2.37 g/cm^3. The surface area of the first sample is 467 m^2/g and that of the second is 372 m^2/g. Which sample has the larger mean pore radius?

8-10 Mercury porosimeter data are tabulated below for a 0.400-g sample of UO$_2$ pellet. At the beginning of the measurements ($p = 1.77$ lb/in.^{2} abs) the mercury displaced by the sample was 0.125 cm^3. At this low pressure no pores were penetrated. Data obtained with a pycnometer gave a true density of the solid phase of $\rho_S = 7.57$ g/cm^3.

Calculate the total porosity of the pellet and the porosity due to pores of larger than 250 Å radius. Also plot the pore-volume distribution for the pores larger than 250 Å radius, using the coordinates of Fig. 8-5.

Pressure, lb/in.^{2}	196	296	396	500	600	700	800	900
Mercury penetration, cm^3	0.002	0.004	0.008	0.014	0.020	0.026	0.032	0.038

	1000	1200	1400	1800	2400	2800	3400	5000
	0.044	0.052	0.057	0.062	0.066	0.066	0.067	0.068

RATE EQUATIONS FOR
FLUID-SOLID CATALYTIC REACTIONS

The nonuniformity of catalyst surfaces and lack of accurate knowledge of the structure of chemisorbed species and their concentrations have been emphasized in Chaps. 7 and 8. In view of these uncertainties it is debatable how much detail should be postulated in formulating equations for rates of reaction. The most simple approach, wholly empirical, would be to use the power-law form of the rate equation, employed for homogeneous reactions, that is, Eq. (2-9). The values of the exponents on the concentrations, the apparent orders of the reaction, are determined by fitting the equation to the data. This approach ignores all the problems associated with adsorption and catalytic surfaces and provides no information about how the reaction occurs. Frequently, but not always, such an equation can correlate experimental rates just as accurately, and with fewer adjustable parameters, than more elaborate methods. When the objective is reactor design, where calculations of engineering accuracy are going to be made with the rate equation, the simplicity of the power-law form is advantageous. Hence, it has been used widely in industrial reactor design. As examples, such equations have been used to represent kinetics data for the water-gas shift reaction using iron oxide catalysts[†]

$$CO + H_2O \rightleftarrows H_2 + CO_2$$

and for the methanation reaction

$$CO_2 + 4H_2 \rightleftarrows CH_4 + 2H_2O$$

on a Ru/Al_2O_3 catalyst.[‡]

[†] H. Bohlbro, "An Investigation on the Kinetics of the Conversion of Carbon Monoxide with Water Vapor Over Iron Oxide Catalysts", 2d ed. Gjellerup, Copenhagen, 1969.
[‡] P. J. Lunde and F. L. Kester, *J. Catal.*, **30**, 423 (1973).

At the other extreme, we might separate the adsorption and surface reaction steps. Then use rate equations like Eq. (7-25) for the adsorption rate, and postulate forms of the adsorbed species and reaction mechanism to formulate the rate of the surface reaction. The necessity for the occurrence of chemisorption as an explanation of catalytic activity provides strong support for separating the adsorption-desorption and surface processes. However, the θ function in Eq. (7-25) and many aspects of the surface reaction are unknown. Therefore, the resultant equations are not only extremely complex but dependent upon so many assumptions that the results may be no more meaningful than the power-law approach.

An intermediate procedure, the Langmuir-Hinshelwood formulation† will be followed, for the most part, in this book.‡ The fundamental concepts on which the Langmuir rate and isotherm expressions are based appear sound, although, as we have seen, some of the assumptions are incorrect. The first-order relationship between gas concentration C_g, adsorbed concentration $\bar{C}$, and total concentration on the surface $\bar{C}_m$ in the Langmuir approach [Eqs. (7-13) to (7-15)] permit great simplification in formulating rate equations. Therefore these equations for adsorption and desorption along with simple first- and second-order expressions for the surface reaction, will be used in the sections that follow. Because of these assumptions, agreement of the resultant rate equation does not mean that the reaction mechanism resulting from the assumptions has been proven. Establishing a mechanism, as distinguished from developing a rate equation, requires information about the adsorbed species and how they react. Sufficient information of this type cannot normally be obtained from measurements of reaction rates alone. The advantages of the Langmuir-Hinshelwood method are: (1) the resultant rate equation may be extrapolated more accurately to concentrations beyond the range of experimental measurements used, and (2) the method does take into account adsorption and surface reactions (which must occur) in a consistent manner.

9-1 Rates of Adsorption, Desorption, Surface Reaction

In this section equations are presented for the rates of adsorption, surface reaction, and desorption. In Sec. 9-2 these equations will be combined to give expressions for the rate in terms of fluid concentrations; that is, concentrations on the surface (denoted by $\bar{C}$) will be eliminated.

---

† This approach to kinetics of fluid-solid catalytic reactions was proposed by C. N. Hinshelwood ("Kinetics of Chemical Change," Oxford University Press, London, 1940) and developed in detail by O. A. Hougen and K. M. Watson ("Chemical Process Principles," pt. 3, "Kinetics and Catalysis," John Wiley & Sons, Inc., New York, 1947).

‡ For critical evaluations of various methods of formulating rate equations see the reviews: Sol W. Weller, "Chemical Reaction Engineering Reviews," *Adv. Chem. Ser.*, 148, p. 26, Amer. Chem. Soc., Washington D.C. (1945) and M. Boudart, *AIChE J.* **18**, 465 (1972). Also the engineering viewpoint toward catalysis and rate equations is well presented by O. A. Hougen, *Ind. Eng. Chem.* **53** (7), 509 (1961).

Adsorption The net rate of adsorption of a component A is given by the difference between Eqs. (7-13) and (7-14), written as†

$$\mathbf{r}_a = k_a C_A (\bar{C}_m - \bar{C}) - k_a' \bar{C}_A = k_a \left[C_A (\bar{C}_m - \bar{C}) - \frac{1}{K_A} \bar{C}_A \right]$$

where $\bar{C}_m - \bar{C}$ represents the concentration of vacant sites, $\bar{C}_v$. If only A were adsorbed, then $\bar{C}_m - \bar{C}$ would equal $\bar{C}_m - \bar{C}_A$. However, other components of the reaction may be adsorbed, so that it is necessary to write $\bar{C}_m - \bar{C} = \bar{C}_v$, where $\bar{C}_v$ is the concentration of vacant sites per unit mass of catalyst. Then r_a is given by

$$\mathbf{r}_a = k_a \left(C_A \bar{C}_v - \frac{1}{K_A} \bar{C}_A \right) \tag{9-1}$$

In this equation C_A is the concentration of A in the gas phase at the catalyst surface. If the intrinsic adsorption rate (k_a) is large with respect to other steps in the overall conversion process, the concentration of A on the catalyst surface will be in equilibrium with the concentration of A in the gas phase. The net rate of adsorption, from Eq. (9-1), approaches zero, and the equilibrium concentration of A is given by the expression

$$(\bar{C}_A)_{eq} = K_A C_A \bar{C}_v \tag{9-2}$$

where K_A designates the adsorption equilibrium constant for A.

Note that this result would reduce to the Langmuir isotherm, Eq. (7-15), if only A were adsorbed. Equations (9-1) and (9-2) are applicable when A occupies one site. Often in chemisorption a diatomic molecule, such as oxygen, will dissociate upon adsorption with each atom occupying one site. Formally, dissociative adsorption may be written

$$A_2 + 2X \rightarrow 2A \cdot X$$

For this case the net rate of adsorption is

$$\mathbf{r}_a = k_a \left(C_{A_2} \bar{C}_v^2 - \frac{1}{K_{A_2}} \bar{C}_A^2 \right) \tag{9-3}$$

At equilibrium the concentration of atomically adsorbed A is‡

$$(\bar{C}_A)_{eq} = K_{A_2}^{1/2} C_{A_2}^{1/2} \bar{C}_v \tag{9-4}$$

---

† The gas phase concentration C_g is now written as C_A to denote component A. Also, the rate constant k_c is written as k_a to denote adsorption.

‡ If A_2 is the only adsorbable gas, then $\bar{C}_v = \bar{C}^m - \bar{C}_A$, and Eq. (9-4) becomes

$$(\bar{C}_A)_{eq} = \frac{K_{A_2}^{1/2} C_{A_2}^{1/2} \bar{C}_m}{1 + K_{A_2}^{1/2} C_{A_2}^{1/2}} \qquad \text{or} \qquad \theta_A = \frac{K_A^{1/2} C_{A_2}^{1/2}}{1 + K_{A_2}^{1/2} C_{A_2}^{1/2}}$$

in comparison with Eqs. (7-15) and (7-16) for molecularly adsorbed A_2.

Surface reaction The mechanism assumed for the surface process will depend on the nature of the reaction. Suppose that the overall reaction is of the type

$$A + B \rightleftharpoons C \tag{9-5}$$

An immediate question concerning the surface process is whether the reaction is between an adsorbed molecule of A and a gaseous molecule of B at the surface or between adsorbed molecules of both A and B on adjacent active centers. In the former case the process might be represented by the expression

$$A \cdot X + B \rightleftharpoons C \cdot X \tag{9-6}$$

If the concentration of adsorbed product C on the surface is $\bar{C}_C$, in moles per unit mass of catalyst (analogous to $\bar{C}_A$), the net rate of this step would be

$$r_s = k_s \bar{C}_A C_B - k_s' \bar{C}_C = k_s \left(\bar{C}_A C_B - \frac{1}{K_s} \bar{C}_C \right) \tag{9-7}$$

In this equation we suppose that the rate of the forward reaction is first order in A on the solid surface and first order in B in the gas phase. Similarly, the rate of the reverse process is first order in C on the surface.

If the mechanism is a reaction between adsorbed A and adsorbed B, the process may be represented by the expression

$$A \cdot X + B \cdot X \rightleftharpoons C \cdot X + X \tag{9-8}$$

Here only those A molecules will react which are adsorbed on sites immediately adjacent to adsorbed B molecules. Hence the rate of the forward reaction should be proportional to the concentration of the pairs of adjacent sites occupied by A and B. The concentration of these pairs will be equal to $\bar{C}_A$ multiplied by the fraction of the adjacent sites occupied by B molecules. This fraction is proportional to the fraction of the total surface occupied by B molecules, i.e., to θ_B.† If $\bar{C}_m$ is defined as the molal concentration of total sites, then $\theta_B = \bar{C}_B / \bar{C}_m$. The rate of the forward reaction, according to Eq. (9-8), will be

$$r = k_s \bar{C}_A \frac{\bar{C}_B}{\bar{C}_m}$$

The reverse rate is proportional to the pairs of centers formed by adsorbed product C molecules and adjacent vacant centers,

$$r' = k_s' \bar{C}_C \frac{\bar{C}_v}{\bar{C}_m}$$

Combining these two expressions gives the net surface rate by the mechanism of Eq. (9-8),

$$r_s = \frac{1}{\bar{C}_m} (k_s \bar{C}_A \bar{C}_B - k_s' \bar{C}_C \bar{C}_v) = \frac{k_s}{\bar{C}_m} \left(\bar{C}_A \bar{C}_B - \frac{1}{K_s} \bar{C}_C \bar{C}_v \right) \tag{9-9}$$

---

† This is correct only if the fraction of the surface occupied by A molecules is small. It would be more accurate to postulate that the fraction of the adjacent centers occupied by B is proportional to $\theta_B/(1 - \theta_A)$. For small values of θ_A the two results are nearly the same.

If the surface step is intrinsically fast with respect to the others, the process would occur at equilibrium, and Eq. (9-7) or Eq. (9-9) could be used to relate the concentrations of the reactants and products on the catalyst surface. For example, if the chosen mechanism is Eq. (9-8), the concentration of product C is given by Eq. (9-9) with $r_s = 0$; that is,

$$K_s = \left(\frac{\bar{C}_v \bar{C}_C}{\bar{C}_A \bar{C}_B} \right)_{eq} \tag{9-10}$$

where K_s is the equilibrium constant for the surface reaction.

Desorption The mechanism for the desorption of the product C may be represented by the expression

$$C \cdot X \rightarrow C + X$$

The rate of desorption will be analogous to Eq. (9-1) for the adsorption of A,

$$\mathbf{r}_d = k_d' \bar{C}_C - k_d C_C \bar{C}_v = -k_d \left(C_C \bar{C}_v - \frac{1}{K_C} \bar{C}_C \right) \tag{9-11}$$

9-2 Rate Equations in terms of Fluid-Phase Concentrations at the Catalyst Surface

At steady state the rates of adsorption $\mathbf{r}_a$, surface reaction $\mathbf{r}_s$, and desorption $\mathbf{r}_d$ are equal. To express the rate solely in terms of fluid concentrations, the adsorbed concentrations $\bar{C}_A$, $\bar{C}_B$, $\bar{C}_C$, and $\bar{C}_v$ must be eliminated from Eqs. (9-2) to (9-11). In principle, this can be done for any reaction, but the resultant rate equation involves all the rate constants k_i and the equilibrium constants K_i. Normally neither type of constant can be evaluated independently.† Both must be determined from measurements of the rate of conversion from fluid reactants to fluid products. However, there are far too many constants, even for simple reactions, to obtain meaningful values from such overall rate data. The problem can be simplified by supposing that one step in the overall reaction controls the rate. Then the other two steps occur at near-equilibrium conditions. This greatly simplifies the rate expression and reduces the number of rate and equilibrium constants that must be determined from experiment. To illustrate the procedure

---

† It might be supposed that rate and equilibrium constants for adsorption or desorption could be established from pure-component adsorption data on the components involved. However, such results rarely agree with the values of the constants determined from rate data for the reaction. Interaction effects between components, other inadequacies of the Langmuir theory, and the assumption of a single controlling step, explain the deviation. An exception to this conclusion is provided by Kabel and Johanson [AIChE J. **8**, 621 (1962)] in their study of the vapor-phase dehydration of ethanol to diethyl ether over an acid catalyst (the acid form of Dowex 50, a sulfonated styrene-divinylbenzene copolymer). Evaluation of the adsorption equilibrium constants from fitting rate data and from independent, adsorption-equilibrium measurements gave nearly the same values.

equations for the rate will be developed, for various controlling steps, for the reaction system

$$A + X \rightleftharpoons A \cdot X \qquad \text{adsorption}$$

$$B + X \rightleftharpoons B \cdot X$$

$$A \cdot X + B \cdot X \rightleftharpoons C \cdot X + X \qquad \text{surface reaction}$$

$$C \cdot X \rightleftharpoons C + X \qquad \text{desorption}$$

$$\overline{\qquad\qquad\qquad\qquad\qquad\qquad\qquad\qquad\qquad\qquad}$$

$$A + B \rightleftharpoons C \qquad \text{overall reaction}$$

Surface reaction controlling The concentrations $\bar{C}_A$, $\bar{C}_B$, and $\bar{C}_C$ will be those corresponding to equilibrium for the adsorption and desorption steps. Equation (9-2) gives the equilibrium value for $\bar{C}_A$. Similar results for $\bar{C}_B$ and $\bar{C}_C$ are

$$(\bar{C}_B)_{eq} = K_B C_B \bar{C}_v \tag{9-12}$$

$$(\bar{C}_C)_{eq} = K_C C_C \bar{C}_v \tag{9-13}$$

Substituting these results in Eq. (9-9) for the surface rate gives

$$\mathbf{r} = \frac{k_s}{\bar{C}_m}\left(K_A K_B C_A C_B \bar{C}_v^2 - \frac{K_C}{K_s}C_C \bar{C}_v^2\right) \tag{9-14}$$

The concentration of vacant sites can be expressed in terms of the total concentration of sites $\bar{C}_m$,

$$\bar{C}_m = \bar{C}_v + \bar{C}_A + \bar{C}_B + \bar{C}_C \tag{9-15}$$

Since $\bar{C}_A$ (or $\bar{C}_B$ and $\bar{C}_C$) corresponds to the equilibrium value for adsorption, Eqs. (9-2), (9-12), and (9-13) can be combined with Eq. (9-15) to yield

$$\bar{C}_v = \frac{\bar{C}_m}{1 + K_A C_A + K_B C_B + K_C C_C} \tag{9-16}$$

Now Eqs. (9-14) and (9-16) can be combined to give a relatively simple expression for the rate in terms of fluid-phase concentrations,

$$\mathbf{r} = k_s \bar{C}_m \frac{K_A K_B C_A C_B - (K_C/K_s)C_C}{(1 + K_A C_A + K_B C_B + K_C C_C)^2} \tag{9-17}$$

This result can be further reduced by using the relationship between the several equilibrium constants K_A, K_B, K_C, and K_s. If the equilibrium constant for the overall reaction is denoted by K, then

$$K = \left(\frac{C_C}{C_A C_B}\right)_{eq} \tag{9-18}$$

This is the conventional K for a homogeneous reaction computed from thermodynamic data, as outlined in Chap. 1. It may be related to the adsorption and surface-reaction equilibrium constants by the equilibrium equations for each of these processes, Eqs. (9-2), (9-12), and (9-13). Thus

$$K = \frac{\bar{C}_C/K_C\bar{C}_v}{(\bar{C}_A/K_A\bar{C}_v)(\bar{C}_B/K_B\bar{C}_v)} = \frac{K_A K_B}{K_C}\left(\frac{\bar{C}_v\bar{C}_C}{\bar{C}_A\bar{C}_B}\right)_{eq} \tag{9-19}$$

According to Eq. (9-10), the last group of surface concentrations is K_s, and so

$$K = \frac{K_A K_B}{K_C} K_s \tag{9-20}$$

Substituting this relationship in Eq. (9-17) gives the final expression for the rate in terms of fluid-phase concentrations,

$$\mathbf{r} = k_s \bar{C}_m K_A K_B \frac{C_A C_B - (1/K)C_C}{(1 + K_A C_A + K_B C_B + K_C C_C)^2} \tag{9-21}$$

It is well at this point to review the chief premises involved in Eq. (9-21):

1. It is supposed that the surface reaction controls the rate of the three steps.
2. The equation applies to the simple reaction $A + B \rightarrow C$. Furthermore, it is assumed that the surface step is an elementary reaction between an adsorbed molecule of A and an adsorbed molecule of B.
3. The adsorption rates are given by the Langmuir-Hinshelwood theories referred to earlier.

An application of Eq. (9-21) is the study of ethanol dehydration in the vapor phase using a sulfonated styrene-divinylbenzene polymer as a catalyst.† The reaction is

$$2C_2H_5OH \rightleftharpoons (C_2H_5)_2O + H_2O$$

Experimental rate data agreed well with the Langmuir-Hinshelwood concept based upon a rate-controlling, reversible, surface reaction between two adsorbed molecules of ethanol. For this reaction Eq. (9-21) takes the form:

$$\mathbf{r} = k_s \bar{C}_m K_A^2 \frac{C_A^2 - (1/K)C_W C_E}{(1 + K_A C_A + K_W C_W + K_E C_E)^2}$$

where subscripts A, W, and E refer to ethanol, water, and ether.

Consider a reaction for which the adsorption is weak for all components. The denominator of Eq. (9-21) approaches unity, and the rate expression reduces to the homogeneous form

$$\mathbf{r} = k_s \bar{C}_m K_A K_B \left(C_A C_B - \frac{1}{K} C_C \right)$$

or

$$\mathbf{r} = k \left(C_A C_B - \frac{1}{K} C_C \right) \tag{9-22}$$

The decomposition of formic acid on several catalytic surfaces follows such behavior.‡ Here there is only one reactant, so that the rate becomes first order in concentration of formic acid.

† R. L. Kabel and L. N. Johanson, *AIChE J.* **8**, 621 (1962).
‡ C. N. Hinshelwood and B. Topley, *J. Chem. Soc.*, **123**, 1014 (1923).

If the product of a reaction is strongly adsorbed and the reactant adsorption is weak, the term $K_C C_C$ is much larger than all others in the denominator. If the reaction is also irreversible, Eq. (9-21) becomes

$$\mathbf{r} = k_s \bar{C}_m K_A K_B \frac{C_A C_B}{(K_C C_C)^2}$$

Suppose there was only one reactant; that is, the reaction was of the form $A \rightarrow C$. The denominator term would be raised to a power of unity instead of being squared, so that

$$\mathbf{r} = k_s \bar{C}_m K_A \frac{C_A}{K_C C_C} \tag{9-23}$$

This result shows the retarding effect that a strongly adsorbed product can have on the rate. The decomposition of ammonia on a platinum wire has been found in one instance to follow this form of rate equation; i.e., hydrogen is strongly adsorbed and ammonia is only weakly adsorbed,† so that

$$\mathbf{r} = k \frac{C_{NH_3}}{C_{H_2}}$$

Again consider a reaction of the form $A \rightarrow C$, but suppose that C is only weakly adsorbed, while A is very strongly adsorbed. Then the rate expression, in analogy with Eq. (9-23), would become zero order in A; that is

$$\mathbf{r} = k_s \bar{C}_m K_A \frac{C_A}{K_A C_A} = k \tag{9-24}$$

These special cases for the surface reaction controlling the rate all follow from the form of the Langmuir isotherm. Weak adsorption corresponds to small values of θ or $\bar{C}_A$, and Eq. (7-16) shows that $\bar{C}_A$ is first order in C_A, as predicted by Eq. (9-22). When the adsorption is very strong, the critical parameter is the concentration of vacant sites remaining for adsorption of reactant. Equation (9-16) shows that this is inversely proportional to C_C, in agreement with Eq. (9-23).

Adsorption or Desorption Controlling Still retaining the simple reaction $A + B \rightleftharpoons C$, let us now suppose that the adsorption of A is the slow step. Then the adsorption of B, the surface reaction, and the desorption of C will occur at equilibrium. The rate can be formulated from the adsorption equation (9-1). The adsorbed concentration $\bar{C}_A$ in this expression is obtained from the equilibrium equations for the surface rate [Eq. (9-10)], adsorption of B [Eq. (9-12)], and desorption of C [Eq. (9-13)]. Thus

$$\bar{C}_A = \frac{\bar{C}_v \bar{C}_C}{K_s \bar{C}_B} = \frac{\bar{C}_v (K_C C_C \bar{C}_v)}{K_s (K_B C_B \bar{C}_v)} = \frac{\bar{C}_v K_C C_C}{K_s K_B C_B}$$

---

† C. N. Hinshelwood and R. E. Burk, *J. Chem. Soc.*, **127**, 1105 (1925). Although there are two products in the decomposition, the nitrogen is only slightly adsorbed and does not appear in the rate equation.

From the relationship of the several equilibrium constants, Eq. (9-20), the expression for $\bar{C}_A$ may be simplified to

$$\bar{C}_A = \frac{K_A \bar{C}_v C_C}{K C_B} \tag{9-25}$$

Substituting this value of $\bar{C}_A$ in the rate-controlling equation (9-1) gives

$$\mathbf{r} = k_a \bar{C}_v \left(C_A - \frac{1}{K} \frac{C_C}{C_B} \right) \tag{9-26}$$

The expression for $\bar{C}_v$ can be formulated from Eq. (9-15), the equilibrium values of $\bar{C}_B$ and $\bar{C}_C$ from Eqs. (9-12) and (9-13), and $\bar{C}_A$ from Eq. (9-25). With this expression for $\bar{C}_v$ substituted in Eq. (9-26), the final rate equation for adsorption of A controlling the process is

$$\mathbf{r} = \frac{k_a \bar{C}_m [C_A - (1/K)(C_C/C_B)]}{1 + K_B C_B + (K_A/K)(C_C/C_B) + K_C C_C} \tag{9-27}$$

If, instead of adsorption, the rate of desorption of product C controls the whole reaction, the expression for $\mathbf{r}$ should be formulated from Eq. (9-11). The adsorption and surface steps will occur at equilibrium conditions. Substituting the equilibrium values of $\bar{C}_C$ and $\bar{C}_v$ in Eq. (9-11) leads to the result

$$\mathbf{r} = k_d \bar{C}_m K \frac{C_A C_B - (1/K) C_C}{1 + K_A C_A + K_B C_B + K_C K C_A C_B} \tag{9-28}$$

9-3 Qualitative Analysis of Rate Equations

The procedure for developing rate expressions in terms of fluid properties according to the Langmuir concepts has been illustrated for a single reaction. Yang and Hougen† have considered various kinds of reactions and mechanisms and examined the results when adsorption, desorption, or surface reaction controls the rate. By dividing the final equation into a kinetic coefficient [for example, $k_d \bar{C}_m K$ in Eq. (9-28)], a driving force $[C_A C_B - (1/K)C_C]$, and an adsorption term $(1 + K_A C_A + K_B C_B + K_C K C_A C_B)$, they were able to prepare tables from which the rate equation for a specific situation could be quickly assembled.‡

Equations such as (9-21), (9-27), and (9-28) will have value if they can be used to predict the rate over a wide range of conditions and hence be suitable for designing reactors. To be useful as working expressions the various constants (kinetics and equilibrium) must be given numerical values.§ It has not proved possible, except in isolated instances, to obtain reliable K values from separate

---

† K. H. Yang and O. A. Hougen, *Chem. Eng. Progr.*, **46**, 146 (1950).

‡ See also J. M. Thomas and W. J. Thomas, in "Introduction to the Principles of Heterogeneous Catalysis," pp. 458–459, Academic Press Inc., New York, 1967, for rate equations assembled in tabular form for various controlling mechanisms for the two reactions $A \rightleftharpoons B$ and $A + B \rightleftharpoons C$.

§ Equations (9-21), (9-27), and (9-28) were developed for constant-temperature conditions. Hence the various specific rates k and equilibrium constants K are termed constants. They are, in theory, constant with respect to pressure (or concentration) and conversion but change with temperature.

adsorption measurements. Hence all the constants must be determined from experimental kinetic data. This means, for example, that $k_d \bar{C}_m$, K_A, K_B, and K_C in Eq. (9-28) would be obtained from rate measurements. Since four-constant equations offer considerable flexibility, it is frequently possible to fit experimental data with several equations, each based on a different mechanism and assumption regarding the step which is controlling. The advantage of this method of formulating rate reactions is the systematic way in which the pertinent parameters are introduced. It is a mechanized procedure which permits little flexibility and provides little insight into the actual mechanism of heterogeneous catalytic reactions. Progress in mechanism studies for specific reactions has been summarized by Thomas and Thomas.† Interesting comparisons of concepts and correlations for homogeneous and heterogeneous catalytic reactions are presented by Boudart.‡

The quantitative interpretation of kinetic data in terms of this type of rate equation is illustrated in Sec. 9-4. Before we proceed to the evaluation of constants in rate equations, let us consider some of the implications of the rate expressions from a qualitative standpoint.

The adsorption terms in the denominator are all different. For surface rate controlling, adsorption equilibrium groups $(K_i C_i)$ are included for each component, and the entire term is squared because reaction is between adsorbed A and adsorbed B. For adsorption of A controlling, no group for the adsorption of A is present, and the entire term is to the first power. For desorption controlling, no group for product C is in the adsorption term. These differences result in separate relationships between the pressure (for a gaseous reaction) and the rate. Hence the treatment of data as a function of pressure provides a useful method for distinguishing between equations. Temperature is not so useful a variable as pressure because all the constants are strong functions of temperature, and many of the same ones are part of each equation. The primary value of temperature is to determine the energies of activation for the adsorption and surface processes. Measurements at different conversion levels, obtained by varying the flow rate in a continuous reactor, can be used to evaluate various equations. However, as the conversion changes, the various concentrations (or partial pressures for a gaseous reaction) do not change independently, but in a specific way determined by the reaction stoichiometry. As with homogeneous reactions, initial rate data can be helpful. The reactants and their concentrations are known accurately, and the reverse reaction (and product concentrations) can be ignored. In a continuous-flow tubular reactor, initial rates would be evaluated from runs at low conversions (e.g., in a differential reactor) for various feed concentrations.

Quantitative treatment of conversion-vs.-rate data is illustrated in Examples 9-2 and 9-3. The importance of total pressure as a variable to evaluate forms of rate equations is shown in the following example.

Example 9-1 A solid-catalyzed gaseous reaction has the form

$$A + B \rightarrow C$$

---

† See also J. M. Thomas and W. J. Thomas, in "Introduction to the Principles of Heterogeneous Catalysis," pp. 458–459, Academic Press Inc., New York, 1967, for rate equations assembled in tabular form for various controlling mechanisms for the two reactions $A \rightleftharpoons B$ and $A + B \rightleftharpoons C$.

‡ Michel Boudart, "Kinetics of Chemical Processes," chaps. 8 and 9, Prentice-Hall, Inc., Englewood Cliffs, N.J., 1968.

Sketch curves of the *initial* rate (rate at zero conversion) vs. the total pressure for the following cases:

(a) The mechanism is the reaction between adsorbed A and adsorbed B molecules on the catalyst. The controlling step is the surface reaction.

(b) The mechanism is the same as (a), but adsorption of A is controlling.

(c) The mechanism is the same as (b), but desorption of C is controlling. Assume that the overall equilibrium constant is large with respect to the adsorption equilibrium constants.

(d) The mechanism is a reaction between adsorbed A and B in the gas phase. The controlling step is the surface reaction.

In each instance suppose that the reactants are present in an equimolal mixture.

SOLUTION For a gaseous reaction the concentration C_i of any component i is proportional to its partial pressure; for an ideal-gas mixture $C_i = p_i/R_g T$. Hence at constant temperature partial pressures may be substituted for C_i in the rate equations, causing a change only in the value of the constants. At zero conversion the pressure of product C is zero, and for an equimolal mixture

$$p_A = p_B = \tfrac{1}{2}p_t$$

For the first three cases Eqs. (9-21), (9-27), and (9-28) are the appropriate rate equations.

(a) At initial conditions Eq. (9-21) simplifies to

$$\mathbf{r}_0 = k_s \bar{C}_m K_A K_B \frac{\tfrac{1}{4}p_t^2/(R_g T)^2}{[1 + \tfrac{1}{2}(K_A + K_B)p_t/R_g T]^2}$$

By combining constants we may write this expression as

$$\mathbf{r}_0 = \frac{a p_t^2}{(1 + b p_t)^2} \tag{A}$$

where a and b are the resulting overall constants.

(b) In a similar manner Eq. (9-27) for the adsorption of A controlling may be reduced to the form

$$\mathbf{r}_0 = k_a \bar{C}_m \frac{\tfrac{1}{2}p_t/R_g T}{1 + \tfrac{1}{2}K_B p_t/R_g T} = \frac{a' p_t}{1 + b' p_t} \tag{B}$$

(c) Equation (9-28), for the case where desorption of C is controlling, may be written

$$\mathbf{r}_0 = k_d \bar{C}_m K \frac{\tfrac{1}{4}p_t^2/(R_g T)^2}{1 + \tfrac{1}{2}(K_A + K_B)p_t/R_g T + \tfrac{1}{4}K_C K p_t^2/(R_g T)^2}$$

If the equilibrium constant K is large with respect to K_A, K_B, and K_C, only the last term in the denominator is important, and the result is

$$\mathbf{r}_0 = \frac{k_d \bar{C}_m}{K_C} = a'' \tag{C}$$

Equation (A) for surface reaction controlling shows that the initial rate will be proportional to the square of the pressure at low pressures and will approach a constant value at high pressures. This type of relation is shown in Fig. 9-1a. The case for adsorption controlling is indicated in Fig. 9-1b, and that for desorption controlling is shown in Fig. 9-1c. If the equilibrium constant were not very large for case (c), the initial rate equation would be as shown in Fig. 9-1a.

(d) For this case the rate equation can be obtained from Eq. (9-7). Combining this with Eqs. (9-2) and (9-13) for the equilibrium values for $\bar{C}_A$ and $\bar{C}_C$ gives

$$\mathbf{r} = k_s \bar{C}_v \left(K_A C_A C_B - \frac{K_C}{K_s} C_C \right)$$

Since there is no adsorption of B in this instance, Eq. (9-16) for $\bar{C}_v$ becomes

$$\bar{C}_v = \frac{\bar{C}_m}{1 + K_A C_A + K_C C_C}$$

The relationship of the equilibrium constants [Eq. (9-20)] is

$$K = \frac{K_A K_s}{K_C}$$

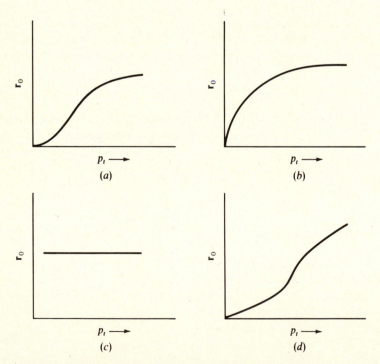

Figure 9-1 Initial rate vs. total pressure for reaction $A + B \rightarrow C$.

Substituting these two expressions into the rate equation yields

$$\mathbf{r} = k_s K_A \bar{C}_m \frac{C_A C_B - (1/K)C_C}{1 + K_A C_A + K_C C_C} \tag{D}$$

Equation (D) is the appropriate expression for the surface rate controlling, with no adsorption of B. At initial conditions, and replacing concentrations with partial pressures, it becomes

$$\mathbf{r}_0 = k_s K_A \bar{C}_m \frac{\frac{1}{4}p_t^2/(R_g T)^2}{1 + \frac{1}{2}K_A p_t/R_g T} = \frac{a'''p_t^2}{1 + b'''p_t} \tag{E}$$

A schematic diagram of Eq. (E) shows the rate proportional to p_t at high pressures (Fig. 9-1d).

Suppose that experimental rate data for the reaction $A + B \rightarrow C$ were obtained over a wide range of total pressures, all at the same temperature. Comparison of a plot of the observed results with curves such as shown in Fig. 9-1 would be of value in establishing the most accurate rate equation. However, it is sometimes difficult to cover a wide enough range of pressures to observe all the changes in shape of the curves.

9-4 Quantitative Interpretation of Kinetics Data

To evaluate the rate and adsorption equilibrium constants in equations such as (9-21) rate data are needed as a function of concentrations in the fluid phase. Data are required at a series of temperatures in order to establish the temperature dependency of these constants. The proper concentrations to employ are those directly adjacent to the site. In the treatment that follows we shall suppose that these local concentrations have been established from the measurable concentrations in the bulk stream by the methods to be given in Chaps. 10 and 11. Our objective here is to find the most appropriate rate equation for the chemical steps at a catalyst site.

Statistical methods are required to obtain the best fit of the equation to the kinetic data. Minimizing the deviations between the observed rate and that predicted from the equation is straightforward as long as the constants are linearly related in the rate equation.† When nonlinearities exist the analysis is more complicated, but the general procedure is the same and consists of the following steps:

1. Assume various mechanisms and controlling steps for each mechanism. Develop a rate equation for each combination of mechanism and controlling step.
2. Determine the numerical values of the constants which give the best fit of each equation to the observed rate data.
3. Choose the equation which best fits the data and agrees with available independent information about the reaction.

---

† O. A. Hougen and K. M. Watson, "Chemical Process Principles," vol. III, "Kinetics and Catalysis," John Wiley & Sons, Inc., New York, 1948.

Nonlinearities in the relations for the constants complicate step 2. However, powerful iterative methods have been developed† for finding the best values of the constants, and numerous applications are available.‡ Kinetics data§ for the hydrogenation of ethylene and propylene with a Pt/Al_2O_3 catalyst is an example of a dual-site mechanism analyzed many times§¶†† by the Langmuir-Hinshelwood approach.

The method of evaluating the various constants is partially illustrated for a linear case in Example 9-2.

Example 9-2 Olson and Schuler‡‡ determined reaction rates for the oxidation of sulfur dioxide, using a packed bed of platinum-on-alumina catalyst pellets. A differential reactor was employed, and the partial pressures as measured from bulk-stream compositions were corrected to fluid-phase values at the catalyst surface by the methods described in Chap. 10 (see Example 10-1). The total pressure was about 790 mmHg.

From previous studies§§ and the qualitative nature of the rate data, a suitable combination appeared to be a controlling surface reaction between adsorbed atomic oxygen and adsorbed sulfur dioxide. In order to determine all the constants in the rate equation for this mechanism, it is necessary to vary each partial pressure independently in the experimental work. Thus measuring the rate of reaction at different total pressures but at constant composition is not sufficient to determine all the adsorption equilibrium constants. Similarly, if the data are obtained at constant composition of initial reactants but varying conversions, the partial pressures of the individual components do not vary independently. However, in these cases it is possible to verify the validity of the rate equation even though values of the separate adsorption equilibrium constants cannot be ascertained. Olson and Schuler studied the effect of conversion alone. Some of their data at 480°C (interpolated for even intervals of reaction rates) are given in Table 9-1.

Derive the rate equation suggested from the assumed mechanism and test how well it fits the experimental rate data.

SOLUTION To develop an expression for the rate of reaction we must postulate the method of obtaining adsorbed atomic oxygen. If it is supposed that molecular oxygen is first adsorbed on a pair of vacant centers and that this

† G. E. P. Box and W. G. Hunter, *Technometrics*, **3**, 311 (1962); W. G. Hunter and R. Mezaki, *AIChE J.*, **10**, 315 (1964); J. R. Kittrell, W. G. Hunter, and R. Mezaki, *AIChE J.*, **10**, 1014 (1964); J. R. Kittrell, W. G. Hunter, and C. C. Watson, *AIChE. J.*, **11**, 105 (1965) and **12**, 5 (1966).

‡ See Sol W. Weller, "Chemical Reaction Engineering Reviews," *Adv. Chem. Ser.*, 148, p. 26, Amer. Chem. Soc. Washington D.C. (1975).

§ G. B. Rogers, M. M. Lih, and O. A. Hougen, *AIChE. J.*, **12**, 369 (1966).

¶ R. Mezaki, *J. Catal.*, **10**, 238 (1966).

†† S. Kolboe, *J. Catal.*, **24**, 40 (1972).

‡‡ R. W. Olson, R. W. Schuler, and J. M. Smith, *Chem. Eng. Progr.*, **46**, 614 (1950).

§§ O. Uyehara and K. M. Watson, *Ind. Eng. Chem.*, **35**, 541 (1943).

Table 9-1

r, g mol/(h)(g catalyst)	Partial pressure (atm) at catalyst surface		
	SO_3	SO_2	O_2
0.02	0.0428	0.0255	0.186
0.04	0.0331	0.0352	0.190
0.06	0.0272	0.0409	0.193
0.08	0.0236	0.0443	0.195
0.10	0.0214	0.0464	0.196
0.12	0.0201	0.0476	0.197

product then dissociates into two adsorbed atoms, the process may be written

$$O_2 + 2X \rightarrow \begin{matrix} O-O \\ | \quad | \\ X \quad X \end{matrix} \rightarrow 2O \cdot X$$

Since the surface reaction is controlling, the adsorption of oxygen must be at equilibrium. Then the concentration of adsorbed atomic oxygen is given by the equilibrium equation

$$K_O = \frac{\bar{C}_O^2}{p_{O_2}\bar{C}_v^2}$$

where $\bar{C}_O$ represents the concentration of adsorbed atomic oxygen and $\bar{C}_v$ represents the concentration of vacant centers. Solving this expression for $\bar{C}_O$ yields

$$\bar{C}_O = K_O^{1/2}p_{O_2}^{1/2}\bar{C}_v \tag{A}$$

The surface-reaction step is represented by

$$SO_2 \cdot X + O \cdot X \rightarrow SO_3 \cdot X + X$$

$$\mathbf{r}_s = k_s\left(\bar{C}_{SO_2}\bar{C}_O - \frac{1}{K_s}\bar{C}_{SO_3}\bar{C}_v\right) \tag{B}$$

The concentrations of SO_3 and SO_2 adsorbed on the catalyst are given by the conventional equilibrium expressions

$$\bar{C}_{SO_3} = K_{SO_3}p_{SO_3}\bar{C}_v \qquad \bar{C}_{SO_2} = K_{SO_2}p_{SO_2}\bar{C}_v \tag{C}$$

Substituting the values of $\bar{C}_O$, $\bar{C}_{SO_2}$, and $\bar{C}_{SO_3}$ in Eq. (B) gives

$$\mathbf{r}_s = k_s\left(K_O^{1/2}K_{SO_2}p_{SO_2}p_{O_2}^{1/2} - \frac{K_{SO_3}}{K_s}p_{SO_3}\right)\bar{C}_v^2 \tag{D}$$

The total concentration of centers is the summation

$$\bar{C}_m = \bar{C}_v + K_O^{1/2}p_{O_2}^{1/2}\bar{C}_v + K_{SO_2}p_{SO_2}\bar{C}_v + K_{SO_3}p_{SO_3}\bar{C}_v + K_{N_2}p_{N_2}\bar{C}_v \tag{E}$$

The last term is included to take into account the possibility that N_2 may be adsorbed on the catalyst. Eliminating $\bar{C}_v$ from Eq. (D) by introducing Eq. (E) and noting that $K_O^{1/2} K_{SO_2} K_s / K_{SO_3} = K$, we may write Eq. (D) as

$$r = \frac{k_s \bar{C}_m^2 K_O^{1/2} K_{SO_2}[p_{SO_2} p_{O_2}^{1/2} - (1/K)p_{SO_3}]}{[1 + K_{SO_2} p_{SO_2} + K_O^{1/2} p_{O_2}^{1/2} + K_{SO_3} p_{SO_3} + K_{N_2} p_{N_2}]^2} \tag{F}$$

Because the partial pressures of sulfur dioxide and sulfur trioxide are both small, the value of p_{N_2} will not vary significantly with conversion. Hence $K_{N_2} p_{N_2}$ may be regarded as a constant in Eq. (F). Since conversion was the only variable causing the composition to change, p_{O_2} and p_{SO_3} and p_{SO_2} are not independent, but are related to the initial, constant composition and the conversion. Hence p_{O_2} in the denominator of Eq. (F) can be expressed in terms of conversion or, more conveniently, in terms of p_{SO_3}. Thus Eq. (F) may be simplified to

$$r = \frac{k_s \bar{C}_m^2 K_O^{1/2} K_{SO_2}[p_{SO_2} p_{O_2}^{1/2} - (1/K)p_{SO_3}]}{[A' + B' p_{SO_3}]^2}$$

Combining the constants and rearranging leads to the linear form

$$R = A + Bp_{SO_3} \tag{G}$$

where

$$R = \left[\frac{p_{SO_2} p_{O_2}^{1/2} - (1/K)p_{SO_3}}{r}\right]^{1/2} \tag{H}$$

Since conversion was the sole variable, individual adsorption equilibrium constants K_O, K_{SO_2}, K_{SO_3}, K_{N_2} cannot be determined. However, rate-vs.-composition data such as that in Table 9-1 can be used to evaluate the constants A and B in rate Equation (G). This is done by calculating a value of R for each of the n sets of data given in Table 9-1. Then, A and B can be determined by minimizing the square of the deviations between R and $A + Bp_{SO_3}$. The equations for A and B are

$$B = \frac{\sum_i R_i p_{SO_3,i} - (\sum_i p_{SO_3,i} \sum_i R_i)/n}{\sum_i p_{SO_3,i}^2 - (\sum_i p_{SO_3,i})^2/n} \tag{I}$$

$$A = \frac{\sum_i R_i - B \sum_i p_{SO_3,i}}{n} \tag{J}$$

The summations include all n values of R and p_{SO_3}. In order to evaluate R it is necessary to know the overall equilibrium constant K. At 480°C K is estimated from the equation

$$\ln K = \frac{22{,}200}{R_g T} - 10.5$$

to be 73. Following this approach, and using the 6 ($n = 6$) sets of data in Table 9-1, it is found that

$$B = 12.9 \text{ (atm)}^{1/2}/[\text{g mol}/(\text{h})(\text{g catalyst})]$$

$$A = 0.176 \text{ (atm)}^{3/2}/[\text{g mol}/(\text{h})(\text{g catalyst})]$$

With these values the rate equation [from Eqs. (G) and (H)] is

$$r = \frac{p_{SO_2} p_{O_2}^{1/2} - (1/K)p_{SO_3}}{[0.176 + 12.9 \; p_{SO_3}]^2} \qquad \text{g mol}/(\text{h})(\text{g catalyst}) \qquad \text{(K)}$$

This expression may be used to calculate the rate at any conversion, or p_{SO_3}. For example, for the first set of data in Table 9-1

$$r = \frac{0.0255(0.186)^{1/2} - (1/73)(0.0428)}{[0.176 + 12.9(0.0428)]^2}$$

$$= 0.0196 \text{ g mol}/(\text{h})(\text{g catalyst})[\text{or } 5.44 \times 10^{-6} \text{ kg mol}/(\text{s})(\text{kg catalyst})]$$

vs. the experimental rate of 0.02 in Table 9-1

Results for the other sets of data show similar agreement, indicating that Eq. (K) represents the rate data well.

9-5 Redox Rate Equations

In some instances it is useful to formulate rate equations from known behavior of catalysts for a class of reactions. Oxidation on metal oxides is an example. It has long been demonstrated† that flowing the substance to be oxidized, without air or oxygen, over the catalyst produces some oxidized product and extracts oxygen from the metal oxide. The catalyst can be reactivated by exposure to oxygen or air. This two-step process, whereby the substance to be oxidized reduces the catalyst and then is reoxidized, is known as the redox mechanism. It leads to rate equations of the same general form as those based upon the Langmuir-Hinshelwood approach and, indeed, is based upon the same assumptions in formulating rates. According to the Langmuir idea, the rate of oxidation of a component is proportional to its concentration C in the fluid and to the concentration $\bar{C}_O$ of the oxidized sites in the catalyst. Thus

$$r = k_1 C \bar{C}_O \qquad (9\text{-}29)$$

Suppose that the rate r_0 of reoxidation of the catalyst surface is directly proportional to the concentration of oxygen C_{O_2} in the fluid and to the concentration of unoxidized sites. The latter will be $\bar{C}_M - \bar{C}_O$, assuming that the extent of adsorption of other substances than oxygen is small.

$$r_0 = k_2 C_{O_2}(\bar{C}_M - \bar{C}_O) \qquad (9\text{-}30)$$

---

† A review of the literature on catalytic oxidations with extensive references is given by A. Cappelli in "Chemical Reaction Engineering Reviews," *Adv. Chem. Ser.* 148, p. 212, Amer. Chem. Soc., Washington, D.C. (1975).

For a stationary-state condition on the catalyst, and if one mole of oxygen is required for the oxidation, the two rates may be set equal to each other to give

$$\bar{C}_O = \frac{(k_2 \bar{C}_M)C_{O_2}}{k_1 C + k_2 C_{O_2}} \tag{9-31}$$

Substituting this expression for $\bar{C}_O$ in Eq. (9-29) gives the rate of oxidation:

$$\mathbf{r} = \frac{(k_1 k_2 \bar{C}_M)C_{O_2} C}{k_1 C + k_2 C_{O_2}} = \frac{k C_{O_2} C}{k_1 C + k_2 C_{O_2}} \tag{9-32}$$

A more general form of Eq. (9-32), in which the rate of reoxidation is proportional to any power of the oxygen concentration† and any number of moles of oxygen needed for the oxidation, has been developed.‡ Equation (9-32) suggests that the rate is first order in the substance to be oxidized at low concentrations (where $k_1 C$ in the denominator would be much less than $k_2 C_{O_2}$) and zero order at high concentrations. Example 9-3 is an illustration of catalytic oxidation in the liquid phase.

Example 9-3 Reaction rates have been measured for the liquid-phase oxidation of dilute aqueous solutions of acetic acid in water, at 68 atm total pressure and at temperatures from 250°C to 280°C, in a differential, packed-bed catalytic reactor.§ Oxygen was predissolved in the solution of acetic acid in water so that only a liquid phase flowed over the bed of catalyst particles (no gas phase was present). The oxidation reaction is

$$CH_3COOH + 2O_2 \rightarrow 2CO_2 + 2H_2O$$

The catalyst consisted of Mn, Co, and La oxides on a zinc aluminate carrier. Three catalyst particle sizes were studied. The equivalent spherical diameters were: $d_p = 0.038$ cm, $d_p = 0.054$ cm, and $d_p = 0.18$ cm.

The effect of oxygen concentration on the global rate was measured at 260°C, for $d_p = 0.054$ cm and for an average acetic acid concentration of $C_{HA} = 33.3 \times 10^{-7}$ g mol/cm³. The rates of oxidation [g mol/(s)(g catalyst)], expressed as the rate (r_{CO_2}) of production of carbon dioxide, are as follows:

Average oxygen concn. C_{O_2}, g mol/cm³	(r_{CO_2}/C_{HA}), cm³/(g)(s)
1.23×10^{-7}	2.70×10^{-2}
3.68×10^{-7}	4.29×10^{-2}
5.69×10^{-7}	5.19×10^{-2}
8.70×10^{-7}	6.30×10^{-2}

---

† For example, if the oxygen dissociates on adsorption $\mathbf{r}_0$ would be proportional to $C_{O_2}^{1/2}$ (see footnote p. 361).

‡ P. Mars and D. W. van Krevelan, *Chem. Eng. Sci. Spec. Suppl.* **3**, 41 (1949) were amongst the first to quantify the redox mechanism. They developed more general equations than (9-32).

§ Janez Levec and J. M. Smith, *AIChE J.*, **22**, 159 (1976).

The effect of acetic acid concentration on the global rate was measured at three temperatures and for a constant, feed concentration of oxygen of 10.4×10^{-7} g mol/cm^3. The data for particle size $d_p = 0.054$ cm are:

$1/C_{HA}$ cm^3/g mol	$(C_{O_2}^{1/2}/r_{CO_2})$, (g)(s)/(mol)$^{1/2}$(cm)$^{1.5}$		
	260°C	270°C	280°C
1.21×10^5	3.48×10^3	2.47×10^3	1.90×10^3
3.10×10^5	4.70×10^3	3.30×10^3	2.55×10^3
6.30×10^5	6.00×10^3	4.05×10^3	3.00×10^3
13.0×10^5	10.2×10^3	5.75×10^3	4.40×10^3

The liquid-phase concentrations in the two preceding tables are for average values in the *differential* reactor.

Finally, global rates were measured for the three particle sizes over a temperature range at constant feed concentrations of acetic acid and oxygen of 33.3×10^{-7} g mol/cm^3 and 10.4×10^{-7} g mol/cm^3, respectively. The data are as follows:

$1000/T$ (K)$^{-1}$	$r_{CO_2} \times 10^7$, g mol/(g)(s)		
	$d_p = 0.038$ cm	$d_p = 0.054$ cm	$d_p = 0.18$ cm
1.807	3.50	3.55	2.10
......	.....	3.45	.....
1.842	2.50	2.60	1.70
1.876	.....	2.00	1.40
1.912	.....	1.42	1.13

Preliminary runs at constant temperature (280°C) and feed concentrations demonstrated that the liquid flow rate through the reactor did not affect the reaction rate, r_{CO_2}. The porosity of the catalyst particles was 0.55. For all the runs the reactor operated isothermally. Derive an equation for the *intrinsic* rate (i.e., the rate with no diffusion resistance) of the oxidation reaction. Obtain numerical values for the constants in the equation for each temperature.

SOLUTION The experimental rates are global values. Before we can develop an equation for the intrinsic rate at a catalyst site, the potential significance of external and internal (intraparticle) diffusion effects must be considered. The statement that the liquid flow rate did not influence the results tells us that external mass transfer is not affecting the reaction rate. This is because the liquid rate affects the mass transfer coefficient from fluid to particle and changes the concentration of reactants at the external surface of the catalyst particle (see Chap. 10). The data in the third table of the problem statement show that the rate is independent of particle size for $d_p = 0.054$ cm or less, but decreases for larger particles. This means that intraparticle diffusion does not affect the rate for $d_p = 0.054$ cm. Particle size is a pertinent variable to use in analyzing intraparticle diffusion effects since the diffusion path length is determined by d_p (see Example 11-10). Accordingly, the global rates for $d_p = 0.054$ cm may be considered to be intrinsic rates corresponding to the bulk-fluid values of the concentrations.†

The next step is to propose various rate equations to see if they correctly represent the effects of oxygen and acetic acid concentrations, as given by the data in the first two tables. The reaction order with respect to oxygen is about one-half, as noted by plotting r_{CO_2} vs. $C_{O_2}^{1/2}$ using the values given in the first table (which are for a constant acetic acid concentration). The one-half order plot is shown in Fig. 9-2. At constant oxygen concentration, there is a linear relationship between $(C_{O_2})^{1/2}/r_{CO_2}$ and $1/C_{HA}$. This is evident by the straight line obtained when the data in the second table are plotted with these coordinates, as shown in Fig. 9-3. The equation for the straight lines in Fig. 9-3 is of the form

$$\frac{(C_{O_2})^{1/2}}{r_{CO_2}} = a\left(\frac{1}{C_{HA}}\right) + b \tag{A}$$

† Intraparticle transport effects are treated in detail in chap. 11.

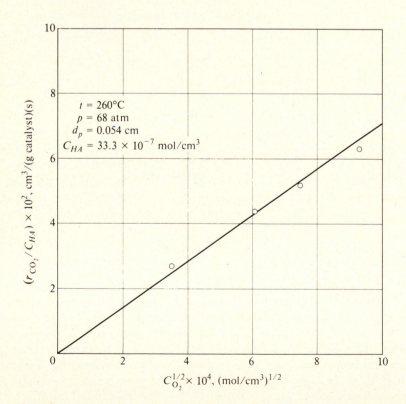

Figure 9-2 Effect of oxygen concentration on catalytic oxidation rate of aqueous acetic acid.

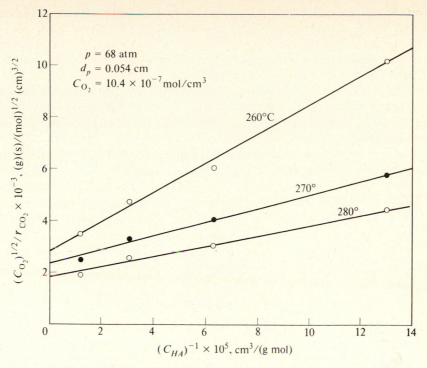

Figure 9-3 Effect of acetic acid concentration on catalytic oxidation rate of aqueous acetic acid.

which may be solved for the rate to give

$$r_{CO_2} = \frac{(1/a)C_{O_2}^{1/2}C_{HA}}{1 + (b/a)C_{HA}} = \frac{k_A C_{O_2}^{1/2}C_{HA}}{1 + k_B C_{HA}} \tag{B}$$

where k_A and k_B are functions of temperature. The values of k_A and k_B can be found from the slope $(a = 1/k_A)$ and intercept $(b = ak_B)$ of the lines in Fig. 9-3. The results for each temperature are:

t	k_A	k_B
°C	cm$^{4.5}$/(mol)$^{1/2}$(g)(s)	cm^3/(g mol)
260	1.9×10^2	5.2×10^5
270	3.2×10^2	7.1×10^5
280	5.4×10^2	9.8×10^5

So far we have drawn no conclusions about the reaction mechanism and actually very little can be said from the rate data alone. What has been done is to find that Eq. (B) correctly correlates the experimental data. If we wish to speculate, we can choose a mechanism (probably several) which leads to Eq. (B). As an illustration of the redox mechanism for oxidation, let's do just that. Suppose that the reaction occurs by the following series of elementary steps:

1. Reversible adsorption of oxygen by a dissociative mechanism forming an oxidized site on the catalyst surface
2. Reaction of this oxidized site with acetic acid from the liquid phase to form a complex $HA \cdot O$
3. Decomposition of $HA \cdot O$ into intermediate products, according to a rate constant k_3
4. Further reaction with more oxygen to give CO_2 and H_2O by steps which are intrinsically rapid with respect to steps 1 to 3

The individual reactions according to this mechanism are

$$\tfrac{1}{2}O_2(aq) + X \overset{k_1}{\rightleftharpoons} O \cdot X \tag{C}$$

$$O \cdot X + HA(aq) \overset{k_2}{\rightarrow} HA \cdot O + X \tag{D}$$

$$HA \cdot O \overset{k_3}{\rightarrow} \text{intermediates} + O_2 \overset{\text{fast}}{\rightarrow} CO_2 + H_2O \tag{E}$$

where X refers to a reduced catalyst site. At steady state, assume that the stationary-state hypothesis (Sec. 2-4) applies to the complexes $O \cdot X$ and $HA \cdot O$. Then

$$\frac{d\bar{C}_{O \cdot X}}{dt} = 0 = k_1 C_{O_2}^{1/2} \bar{C}_X - k_1' \bar{C}_{O \cdot X} - k_2 \bar{C}_{O \cdot X} C_{HA} \tag{F}$$

and

$$\frac{d\bar{C}_{HA \cdot O}}{dt} = 0 = k_2 \bar{C}_{O \cdot X} C_{HA} - k_3 \bar{C}_{HA \cdot O} \tag{G}$$

Equations (F) and (G) may be solved for $\bar{C}_{O \cdot X}$ and $\bar{C}_{HA \cdot O}$ to yield

$$\bar{C}_{O \cdot X} = \frac{k_1 C_{O_2} \bar{C}_X}{k_1' + K_2 C_{HA}} \tag{H}$$

$$\bar{C}_{HA \cdot O} = \frac{k_1 k_2 C_{O_2}^{1/2} C_{HA} \bar{C}_X}{k_3(k_1' + k_2 C_{HA})} \tag{I}$$

Then the rate of CO_2 formation is given by reaction (E)

$$r_{CO_2} = k_3 \bar{C}_{HA \cdot O} = \frac{(k_1 k_2 \bar{C}_X) C_{HA} C_{O_2}^{1/2}}{k_1' + k_2 C_{HA}}$$

$$= \frac{(k_1 k_2 \bar{C}_X / k_1') C_{O_2}^{1/2} C_{HA}}{1 + (k_2/k_1) C_{HA}} \tag{J}$$

For the very low concentrations of oxygen and acetic acid in these dilute aqueous solutions, the fraction of the surface covered by the $X \cdot O$ complex could be low. Then, $\bar{C}_X$ in Eq. (J) would be nearly constant and the result is a rate expression in agreement with the experimental data, Eq. (B).

9-6 Kinetics of Catalyst Deactivation

We mentioned in Sec. 8-8 that loss in catalyst activity with time was a frequent occurrence in commercial reactors. Actually, the need to account for deactivation in *designing* a reactor is not as significant as might be supposed. This is because (1) if a catalyst loses activity rapidly and cannot be regenerated in place, it probably will not be economical to use it, and (2) if regeneration is possible, the regeneration process would be carried out continuously (as in a fluidized-bed reactor and regenerator unit) so that the reactor operates with a constant intermediate activity. However, in some instances the loss in activity is gradual so that operation for an economical length of time is possible. Also, dual reactors are sometimes used with the catalyst in one unit being regenerated (in place) while the reaction is carried out in the other unit. In cases such as these it is important to know how the conversion of the desired reaction changes with time on stream. For this to be determined, a quantitative description must be known for the effect of deactivation on rates. We consider here the effect on the intrinsic rate at a catalyst site, ignoring the transport influences listed in Chap. 7. Then in Sec. 11-14 the effect of deactivation on the *global* rate is discussed.

Formulating mechanisms to explain rates of poisoning and the effect on the desired (or main) reaction is likely to be as difficult as establishing the mechanism of the main reaction itself. Hence, there are compelling reasons to adopt the empirical power-law rate equations for treating deactivation. This procedure has been well developed by Levenspiel[†] for various types of deactivation reactions.

We illustrate how to determine the rate as a function of time for a simple case for which deactivation is assumed to occur by blocking the active sites and where the rate of the main reaction is assumed to be directly proportional to the unblocked sites.[‡] Suppose that q represents the concentration of blocked sites (or concentration of poison on the catalyst) at any time and q_0 is the maximum concentration corresponding to completely deactivated catalyst. Then, if the main reaction is

$$A \rightarrow B \qquad (9\text{-}33)$$

its rate will be proportional to the fraction of the sites that are not poisoned, that is, to $1 - q/q_0$:

$$\mathbf{r}_B = kC_A\left(1 - \frac{q}{q_0}\right) = kC_A(1 - \phi) \qquad (9\text{-}34)$$

where ϕ represents the fraction, q/q_0, of poisoned sites.

The rate of the poisoning reaction determines how ϕ varies with time. If the sites are blocked by a product, C, formed by reaction of B, the process is called *series* poisoning, or fouling, and the deactivation reaction is written

$$B \rightarrow C(s) \qquad (9\text{-}35)$$

[†] O. Levenspiel, *J. Catal.*, **25**, 265 (1972).

[‡] This case has been considered by S. Masamune and J. M. Smith, *AIChE J.*, **12**, 384 (1966). The Levenspiel treatment is more general in that it allows for other than first-order processes.

Its rate will be proportional to the concentration of B and the fraction of un-poisoned sites, that is,

$$\mathbf{r}_c = \frac{dq}{dt} = \frac{q_0 \, d\phi}{dt} = k_{B,p} C_B(1 - \phi) \tag{9-36}$$

Carbon deposition in hydrocarbon cracking, where some of the primary products react further to hydrogen and coke, is an example of series deactivation.

Suppose the poison is an impurity in the feed stream. The poison will deposit independently of the main reaction, according to the reaction:

$$P \rightarrow C(s) \tag{9-37}$$

If the deactivation reaction (or adsorption) occurs only on the active sites, its rate is given by

$$\mathbf{r}_c = \frac{q_0 \, d\phi}{dt} = k_p C_p(1 - \phi) \tag{9-38}$$

where C_p is the concentration of the poison. This is termed *independent* poisoning.

A third type, *parallel* deactivation, occurs when A can react in two ways: (1) to deposit poison C on the sites, and (2) to form desired product B by Eq. (9-33). The poisoning reaction and its rate are then written

$$A \rightarrow C(s) \tag{9-39}$$

$$\mathbf{r}_c = q_0 \frac{d\phi}{dt} = k_{A,p} C_A(1 - \phi) \tag{9-40}$$

For any of these types of poisoning, the rate of production of B as a function of time is obtained by integrating Eq. (9-36), (9-38), or (9-40) and combining the result with Eq. (9-34). This is easily done if the concentration in Eqs. (9-36), (9-38), and (9-40) is constant. We conclude that if rates are measured as a function of time, with the proper concentration held constant, an appropriate expression for predicting deactivation effects can be established.

For example, if *independent* poisoning is suspected, Eq. (9-38) can be integrated, for a fixed poison concentration C_p to yield

$$\int_0^\phi \frac{d\phi}{1 - \phi} = \left(\frac{k_p}{q_0} C_p\right) \int_0^t dt$$

or

$$1 - \phi = \exp\left[-\left(\frac{k_p}{q_0} C_p\right) t\right] \tag{9-41}$$

With this result for $1 - \phi$, the rate of the main reaction, Eq. (9-34), becomes

$$\mathbf{r}_B = k C_A \exp\left[-\left(\frac{k_p}{q_0} C_p\right) t\right] \tag{9-42}$$

Suppose that $\mathbf{r}_B$ was measured experimentally with the same feed (in order that C_p remain constant) but for different reactant concentrations. If the results showed a first-order dependency on C_A and an exponential decay with time, independent poisoning could be a possibility. If the measurements were made at constant reactant concentrations, an exponential decay with time would still be expected for both independent and parallel poisoning. Thus integration of Eq. (9-40) with C_A constant, and substitution in Eq. (9-34) gives

$$\mathbf{r}_B = kC_A \exp\left[-\left(\frac{k_{A,P}}{q_0}C_A\right)t\right] = a_1 \exp\left[-a_2 t\right] \text{ (for constant } C_A) \quad (9\text{-}43)$$

where a_1 and a_2 are constants. Hence, experiments at different reactant concentrations are necessary to distinguish between parallel and independent poisoning. Rate equations for all types of poisoning can be developed and compared with experimental rate data (see Prob. 9-12).

PROBLEMS

9-1 It has been proposed† that the gas-phase catalytic hydrochlorination of acetylene occurs by the following steps

1. Adsorption of HCl
2. Surface reaction between absorbed HCl and acetylene in the gas phase to produce adsorbed vinyl chloride which desorbs into the gas phase.

   Preliminary experimental studies indicate that both external and internal diffusional resistances are negligible. Also, the equilibrium constant for the homogeneous reaction is very large.
   The rate of reaction per gram of catalyst is measured at varying total pressures, but constant composition of components. The results show a linear relationship between $\mathbf{r}$ and p_t all the way down to pressures approaching zero. What conclusions may be drawn from these data concerning the controlling step in the reaction?

9-2 Two gas-solid catalytic reactions, (1) and (2), are studied in fixed-bed reactors. Rates of reaction per unit mass of catalyst, at constant composition and total pressure, indicate the variations with mass velocity and temperature shown in Figure 9-4. The interior pore surface in each case is fully effective. What do the results shown suggest about the two reactions?

† S. Shankar, Ph.D. Thesis, Monash University, Australia, 1976.

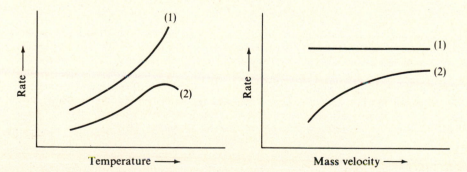

Figure 9-4 Rate vs. temperature and mass velocity.

9-3 In a study of the kinetics of two different gas-solid catalytic reactions it is found that all diffusional resistances are negligible. Also, both reactions are irreversible. As an aid in establishing the mechanism of the reactions the rate is measured at a constant composition over a wide range of temperature. For the first reaction, (1), the rate increases exponentially over the complete temperature range. For the second reaction, (2), the rate first increases and then decreases as the temperature continues to rise. What does this information mean with regard to the controlling step in each of the reactions?

9-4 An isomerization reaction has the simple form $A \rightarrow B$. Assuming that the reaction is isothermal and reversible and that the external- and internal-diffusion steps have negligible concentration gradients, propose intrinsic rate equations for the following cases:

(a) The adsorption of A on the catalyst is controlling.
(b) The surface interaction between adsorbed A and an adjacent vacant center is controlling.
(c) The desorption of B from the surface is controlling.

In all cases the mechanism is adsorption of A, reaction on the surface to form adsorbed B, and desorption of B into the gas phase. Sketch the rate of reaction (per unit mass of catalyst) vs. total pressure (at constant composition) in each of the above three cases. Also, for comparison, include a sketch of the rate of the homogeneous reaction, assuming that it is first order. Sketches should be for constant composition.

9-5 Thodos and Stutzman[†] studied the formation of ethyl chloride, using a zirconium oxide catalyst (on silica gel) in the presence of inert methane.

$$C_2H_4 + HCl \rightleftharpoons C_2H_5Cl$$

If the surface reaction between adsorbed ethylene and adsorbed HCl controls the overall kinetics, derive an expression for the rate. Neglecting external and internal transport resistances, evaluate the constants in the rate equation at 350°F from the following data:

r × 10⁴, lb mol/(h)(lb catalyst)	Partial pressures, atm			
	CH_4	C_2H_4	HCl	C_2H_5Cl
2.71	7.005	0.300	0.370	0.149
2.63	7.090	0.416	0.215	0.102
2.44	7.001	0.343	0.289	0.181
2.58	9.889	0.511	0.489	0.334
2.69	10.169	0.420	0.460	0.175

The equilibrium constant for the overall reaction at 350°F is 35.

9-6 Potter and Baron[‡] studied the reaction

$$CO + Cl_2 \rightarrow COCl_2$$

at atmospheric pressure, using an activated carbon catalyst. Preliminary studies showed that the rate of reaction did not depend on the mass velocity of gases through the reactor. Analysis of the rate data indicated that the reaction occurred by adsorption of Cl_2 and CO on the catalyst and surface reaction between the adsorbed complexes. It appeared that the surface reaction, rather than the adsorption or desorption steps, was controlling the overall reaction rate. Furthermore, preliminary adsorption measurements indicated that chlorine and phosgene were readily adsorbed on the catalyst, while carbon monoxide was not. Hence the adsorption equilibrium constant of carbon monoxide, although it was not zero, was considered negligible with respect to those for Cl_2 and $COCl_2$.

† G. Thodos and L. F. Stutzman, *Ind. Eng. Chem.*, **50**, 413 (1958).
‡ C. Potter and S. Baron, *Chem. Eng. Progr.*, **47**, 473 (1951).

(a) On the basis of this information, develop an expression for the intrinsic rate of reaction in terms of the bulk partial pressures in the gas phase. The reaction is irreversible. (b) Determine the best values for the adsorption equilibrium constants for Cl_2 and $COCl_2$ and the product $\bar{C}_m k_s K_{CO}$ from the following experimental data, where the temperature was 30.6°C, the catalyst size was 6 to 8 mesh, and

$$\bar{C}_m = \text{total concentration of active centers, in moles per gram of catalyst}$$
$$k_s = \text{specific reaction-rate constant for surface reaction}$$
$$K_{CO} = \text{adsorption equilibrium constant for CO}$$

r,	Partial pressure, atm		
g mol/(h)(g catalyst)	CO	Cl_2	$COCl_2$
0.00414	0.406	0.352	0.226
0.00440	0.396	0.363	0.231
0.00241	0.310	0.320	0.356
0.00245	0.287	0.333	0.376
0.00157	0.253	0.218	0.522
0.00390	0.610	0.113	0.231
0.00200	0.179	0.608	0.206

Assume that the 6- to 8-mesh catalyst particles are small enough that the pore surface was fully effective.

9-7 Potter and Baron also made rate measurements at other temperatures, and their results at 42.7, 52.5, and 64.0°C are as shown below. Assume that the adsorption equilibrium constants and the rate constant k_s follow an equation of the form $y = ae^{-b/R_g T}$, where a and b are constant and y is K_{Cl_2}, K_{COCl_2}, or $\bar{C}_m k_s K_{CO}$.

T, °C	r × 10³, g mol/(h)(g catalyst)	Partial pressure, atm		
		CO	Cl_2	$COCl_2$
42.7	4.83	0.206	0.578	0.219
42.7	10.73	0.569	0.194	0.226
42.7	1.34	0.128	0.128	0.845
42.7	9.18	0.397	0.370	0.209
42.7	9.10	0.394	0.373	0.213
52.5	14.28	0.380	0.386	0.234
52.5	15.46	0.410	0.380	0.210
52.5	6.00	0.139	0.742	0.118
52.5	3.68	0.218	0.122	0.660
64.0	25.74	0.412	0.372	0.216
64.0	24.46	0.392	0.374	0.234
64.0	13.78	0.185	0.697	0.118
64.0	8.29	0.264	0.131	0.605

Determine values of a and b for each case.

9-8 Reinicker and Gates† investigated the liquid phase reaction between phenol and acetone to bisphenol A and water at 364 K. The catalyst was a sulfonated styrene-divinylbenzene copolymer. The reaction is:

$$2C_6H_5OH + (CH_3)_2CO \rightarrow (OH)C_6H_4 - C(CH_3)_2 - C_6H_4OH + H_2O$$

† *AIChE J.*, **20**, 933 (1974).

Derive a rate equation based upon the following elementary steps (A = acetone, —SO_3H = active group in the polymer, P = phenol, I = tertiary alcohol intermediate):

(a) reversible adsorption of acetone
(b) reversible reaction of adsorbed acetone and phenol (not adsorbed) to give the adsorbed tertiary alcohol intermediate
(c) irreversible surface reaction between adsorbed phenol and adsorbed tertiary alcohol to give adsorbed water and bisphenol A (not adsorbed).

Assume that step (c) is rate controlling. This mechanism is an approximation to the more careful development of Reinicker and Gates.

9-9 The Deacon process for production of chlorine involves the gas-phase catalytic reaction

$$HCl + \tfrac{1}{4}O_2 \rightleftarrows \tfrac{1}{2}Cl_2 + \tfrac{1}{2}H_2O$$

Furusaki[†] studied the kinetics of this reversible reaction, with a $CuCl_2$, KCl, $SnCl_2$-on-silica catalyst, in differential and integral reactors. The rate of disappearance of HCl could be correlated by the equation:

$$r = \frac{k[C_{HCl}C_{O_2}^{1/4} - (1/K)C_{Cl_2}^{1/2}C_{H_2O}^{1/2}]}{[1 + K_1 C_{HCl} + K_2 C_{Cl_2}]^2}$$

(a) Devise a series of fundamental steps of adsorption and surface reaction which will give the above rate expression.
(b) The rate data at 350°C ($p_t = 1$ atm), obtained in the differential reactor with a feed of HCl and air (no Cl_2 or H_2O), is as follows:

Rate $\times 10^6$ g mol/(g catalyst)(s)	$C_{HCl} \times 10^6$ g mol/cm^3
10.5	0.24
11.2	0.27
10.3	0.33
13.5	0.44
12.8	0.45
15.2	0.68
15.3	0.74
15.7	0.89

Using these data evaluate as many of the constants, k, K_1, and K_2, as possible.

9-10 Intrinsic rate data for the hydrogenolysis of ethane[‡]

$$C_2H_6 + H_2 \rightarrow 2CH_4$$

over metal catalysts (such as Ni, CO) obeys the empirical equation

$$r = kC_{C_2H_6}^a C_{H_2}^b$$

† Shintaro Furusaki, *AIChE J.*, **19**, 1009 (1973).
‡ J. H. Sinfelt, *Cat. Rev.*, **3**, 175 (1970).

The parameters k, a, and b depend upon the catalyst. Suppose that the mechanism is:

1. reversible adsorption of C_2H_6 on a catalyst site

$$C_2H_6 + 2X \underset{k_1'}{\overset{k_1}{\rightleftarrows}} C_2H_5 \cdot X + H \cdot X$$

2. Hydrogen extraction

$$C_2H_5 \cdot X + H \cdot X \underset{k_2'}{\overset{k_2}{\rightleftarrows}} C_2H_x \cdot X + nH_2 + X$$

3. Reaction of H_2 to break the C—C bond

$$X + C_2H_x \cdot X + H_2 \overset{k_3}{\rightarrow} CH_y \cdot X + CH \cdot X$$

4. Further reaction with H_2 to give CH_4

$$CH_y \cdot X + CH \cdot X + nH_2 \rightarrow 2CH_4$$

    A. Derive a rate equation from the postulated mechanism by assuming that the rate is controlled by step 3 (which is irreversible) and that the stationary-state hypothesis may be used to obtain the concentration of adsorbed C_2H_x.

    B. Will the empirical equation be good approximation for the result derived in A?

9-11 Rates of oxidation of aqueous solutions of formic acid with dissolved oxygen over a commercial CuO—ZnO catalyst at 200° to 240°C suggest† a first-order dependency for both molecular oxygen and formic acid. Preliminary data showed that the catalyst could be reduced in the presence of formic acid solutions containing no oxygen and then reoxidized when oxygen was added to the solution flowing over the catalyst bed. Hence, a redox mechanism is indicated.

    A. Suggest a mechanism and note the assumptions that are necessary to explain the experimental data (that is, the first-order effects of both oxygen and formic acid).

    B. At 225°C, some of the measured rate data are as follows:

$C_{O_2} \times 10^7$ g mol/cm³	(Rate/C_{FA}) cm³/(g)(min)	$C_{FA} \times 10^7$ g mol/cm³	Rate/C_{O_2} cm³/(g)(min)
2.5	3.2	12	14
6.0	6.3	18	20
7.2	7.4	20	20
10.0	10.6	23	23
		30	32
($C_{FA} = 28.7 \times 10^{-7}$ g mol/cm³)		63	74
		96	110
FA = formic acid		[$C_{O_2} = 10 \times 10^{-7}$ g mol/cm³]	

For $C_{O_2} = 10.9 \times 10^{-7}$ and $C_{FA} = 28.7 \times 10^{-7}$ g mol/cm³ data for the effect of temperature on the rate are:

t, °C	240	232	225	215	208
$k \times 10^{-5}$ cm⁶/(g mol)(g)(s)	5.7	3.9	1.8	0.88	0.53

    † G. Baldi, S. Goto, C.-K. Chow, and J. M. Smith, *Ind. Eng. Chem., Proc. Design Develop.*, **13**, 447 (1974).

From this information determine the constants k_0 and E in the following expression for the second-order rate constant

$$k_2 = k_0 \exp{(-E/R_A T)}$$

9-12 Derive an expression for the rate as a function of time for first-order main and poisoning reactions in a slurry reactor. The reactor operates batchwise for both solid catalyst particles and the liquid reaction mixture. Deactivation occurs by blocking of active sites with poison C, which is produced from reactant A in *parallel* with the production of desired product B. (The solution is simplified by dividing the rate equations for production of B and for C to eliminate time as a variable. Note that for a batch reactor the deactivation time is equal to the reaction time.)

9-13 Reconsider Prob. 9-12 for *series* poisoning where the reactions are:

$$A \rightarrow B \rightarrow C \text{ (poison)}$$

For this case assume that the rate of production of C is much less than the rate of formation of B from A (slow poisoning). This means that the concentration of B is equal to the initial concentration of A minus the concentration of A, at any time.

EXTERNAL TRANSPORT PROCESSES IN HETEROGENEOUS REACTIONS

No matter how active a catalyst particle is, it can be effective only if the reactants can reach the catalytic surface. The transfer of reactant from the bulk fluid to the outer surface of the catalyst particle requires a driving force, the concentration difference. Whether this difference in concentration between bulk fluid and particle surface is significant or negligible depends on the velocity pattern in the fluid near the surface, on the physical properties of the fluid, and on the intrinsic rate of the chemical reactions at the catalyst; that is, it depends on the mass-transfer coefficient between fluid and surface and the rate constant for the catalytic reaction. The concentration of reactant is less at the surface than in the bulk fluid. Hence the observed rate, the *global* rate, is less than the intrinsic rate evaluated at the concentration of reactant in the *bulk* fluid.

The same reasoning suggests that there will be a temperature difference between bulk fluid and catalyst surface. Its magnitude will depend on the heat-transfer coefficient between fluid and catalyst surface, the reaction-rate constant, and the heat of reaction. If the reaction is endothermic, the temperature of the catalyst surface will be less than that in the bulk fluid (Fig. 10-1), and the observed rate will be less than that corresponding to the bulk-fluid temperature; the resistance to mass and energy transfer supplement each other. If the reaction is exothermic, the temperature of the catalyst surface will be greater than that of the bulk fluid. Now the global rate may be higher or lower than that corresponding to bulk-fluid conditions; it is increased because of the temperature rise and reduced because of the drop in reactants concentration.

In this chapter our objective is to study quantitatively how these external physical processes affect the global rate, which is the rate that we need in order to design heterogeneous reactors. Such processes are designated as *external* to signify that they are completely *separated* from, and in series with, the chemical reaction

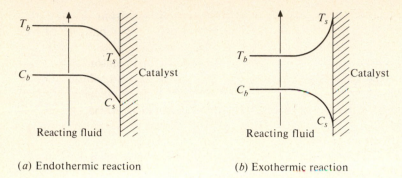

(a) Endothermic reaction   (b) Exothermic reaction

Figure 10-1 External concentration and temperature profiles for fluid–solid catalytic reactions.

on the catalyst surface. For porous catalysts both reaction and heat and mass transfer occur at the same internal location *within* the catalyst pellet. The quantitative analysis in this case requires *simultaneous* treatment of the physical and chemical steps. The effect of these internal physical processes will be considered in Chap. 11.

Mass- and heat-transfer coefficients between fluid and catalyst depend upon the flow and geometric arrangement of fluid and catalyst particle. In other words, they depend upon the type of reactor. For example, such coefficients are different in a fixed-bed arrangement, where the fluid flows around stationary particles, from those in a fluidized bed where both particles and fluid are in motion. In this chapter the fixed-bed case is considered first and this is followed by fluidized beds and 3-phase (slurry and trickle-bed) reactors.

In designing a reactor we suppose that an equation is known for the *intrinsic* rate, that is, the rate in terms of concentrations and temperature at the surface of the catalyst. With respect to external transport effects, the problem is to evaluate a *global* rate for any set of bulk fluid concentrations and temperatures. This is accomplished by using equations for heat and mass transfer to evaluate concentrations and temperature at the catalyst surface. Then the surface C and T are used in the intrinsic rate equation to determine the global rate. The other form of the problem is the interpretation of laboratory measurements to obtain an intrinsic rate equation. Here the global rate is known for a set of bulk concentrations and temperature. The heat- and mass-transfer equations are now used to evaluate surface concentrations and temperature. These values can then be associated with the measured rate to seek an intrinsic rate equation, according to the methods of Chap. 9. Both types of problems are considered in the examples in this chapter.

FIXED-BED REACTORS

Fixed- (or packed-) bed reactors refer to two-phase systems in which the reacting fluid flows through a tube filled with stationary catalyst particles or pellets. In such reactors there will be regions near the outer surface of the particles where the fluid velocity is very low. In these regions, which are particularly evident near points of contact between particles, mass and energy transfer between bulk fluid and pellet surface will be primarily by conduction. Away from the surface a convective mechanism will be dominant. The complexity of flow patterns around an individual pellet suspended in a fluid stream is considerable. When this is combined with interactions between pellets, as in a fixed-bed reactor, the problem of predicting or correlating *local* velocities is, at present, beyond solution. Therefore transport rates are normally defined in terms of an *average* heat- or mass-transfer coefficient. Even though experimental data show that variations exist, it will be assumed that the average coefficient can be applied to *all* the outer surface of a pellet. With these assumptions a single value of the heat- or mass-transfer coefficient can be used to describe the rates of transfer between bulk fluid and pellet surface.

The error introduced in using an average coefficient is not as serious as might be expected, since the correlations for the mass-transfer coefficient k_m and heat-transfer coefficient h are based on experimental data for beds of particles (see Sec. 10-2). That is, the experimental results are, in general, for average values of the coefficients. Gillespie et al.,[†] however, have carried out an interesting study of the variation of heat-transfer coefficients with respect to the position of the pellet in the bed and the location on the surface of a single pellet. The results showed that h values were lower for the first two layers of pellets (measured from the entrance) than for the remainder of the bed. In addition, the coefficient was higher near the wall of the bed than in the central section of the cylindrical tube, presumably because of the higher velocity about one pellet diameter from the wall.[‡] The local coefficients also varied with position; the highest values were obtained for the surface perpendicular to the direction of bulk flow in the bed. The status of analytical attempts to predict local variations in h are described by Petersen.[§]

In reactor design the rate at the interior catalyst site is presumed to be known from the rate equation (see Chap. 9). Also the temperature and concentrations in the bulk fluid are known. The global rate is then calculated by simultaneous solution of the equations for the rate of reaction and mass and energy transfer, and in the process the unknown concentrations at the surface of the catalyst pellet are eliminated. If all the equations are linear in the surface concentrations and temperatures, an analytical solution is possible, as illustrated by Eqs. (10-5) and (10-6). If some equations are nonlinear, numerical solution is necessary. When the objective is an intrinsic rate equation, the global rate is directly obtained from the experimental measurements; for example, from differential or integral reactor data, as explained in Sec. 4-3, or from other types of laboratory reactors, as described in Chap. 12. Such rates are then used with the equations for mass and energy transfer to determine the concentrations and temperature at the surface of the catalyst pellet. These surface values are correlated with the observed rate in order to establish the form of the rate equation by the procedure described in Chap. 9.

† B. M. Gillespie, E. D. Crandall, and J. J. Carberry, *AIChE J.*, **14**, 483 (1968).

‡ C. E. Schwartz and J. M. Smith, *Ind. Eng. Chem.*, **45**, 1209 (1953).

§ E. E. Petersen, "Chemical Reaction Analysis," pp. 129–164, Prentice-Hall, Inc., Englewood Cliffs, N.J., 1965.

Before we proceed with quantitative illustrations, let us consider the qualitative effect of external resistances on reaction rates (Sec. 10-1) and then summarize the available information for mass- and heat-transfer coefficients (Sec. 10-2).

10-1 The Effect of Physical Processes on Observed Rates of Reaction

Suppose an irreversible gaseous reaction on a *solid* catalyst pellet is of order n. At steady state the rate, expressed per unit mass of pellet, may be written either in terms of the diffusion rate from the bulk gas to the surface or in terms of the rate on the surface. These expressions (C = reactant concentration) are:

$$\mathbf{r}_P = k_m a_m (C_b - C_s) \tag{10-1}$$

$$\mathbf{r}_P = kC_s^n \tag{10-2}$$

where C_b and C_s are the concentrations in the bulk gas and at the surface, respectively. In the first expression k_m (for example, in centimeters per second) is the mass-transfer coefficient between bulk gas and solid surface, and a_m is the external surface area per unit mass of the pellet. Suppose that the reaction-rate constant k (per unit mass of catalyst) is very much greater than $k_m a_m$. Under these conditions C_s approaches zero and Eq. (10-1) shows that the rate per pellet is

$$\mathbf{r}_P = k_m a_m C_b \tag{10-3}$$

At the other extreme, k is very much less than $k_m a_m$. Then C_s approaches C_b, and the rate, according to Eq. (10-2), is

$$\mathbf{r}_P = kC_b^n \tag{10-4}$$

Equation (10-3) represents the case when diffusion controls the overall process. The rate is determined by $k_m a_m$; the kinetics of the chemical step at the catalyst surface are unimportant. Equation (10-4) gives the rate when the mass-transfer resistance is negligible with respect to that of the surface step; i.e., the kinetics of the surface reaction control the rate.

Consider a situation where the true order of the surface reaction is 2 [according to Eq. (10-2)] but the rate is diffusion controlled, so that Eq. (10-3) is applicable. Experimental data plotted as rate vs. C_b would yield a straight line. If diffusion were not considered, and Eq. (10-2) were used to interpret the data, the order would be identified as unity—a false conclusion. This simple example illustrates how erroneous conclusions can be reached about kinetics of a catalytic reaction if external mass transfer is neglected.

When both diffusion and reaction resistance are significant, and for a first-order reaction, Eqs. (10-1) and (10-2) can be easily solved for the unknown surface concentration. This result for C_s can be substituted in Eq. (10-1) to obtain an expression for the rate in terms of the bulk concentration C_b. This was done in Chap. 7 (see Sec. 7-1 and Fig. 7-1) with the results

$$C_s = \frac{k_m a_m}{k + k_m a_m} C_b \tag{7-3}$$

$$r_p = k_0 C_b = \frac{1}{1/k + 1/k_m a_m} C_b \tag{7-4} \quad \text{or} \quad (10-5)$$

where

$$\frac{1}{k_0} = \frac{1}{k} + \frac{1}{k_m a_m} \tag{10-6}$$

Equations (10-5) and (10-6) show that for this intermediate case the observed rate is a function of both the rate-of-reaction constant k and the mass-transfer coefficient k_m. In a design problem k and k_m would be known, so that Eqs. (10-5) and (10-6) give the global rate in terms of C_b. Alternately, in interpreting laboratory kinetic data k_o would be measured. If k_m is known, k can be calculated from Eq. (10-6). In the event that the reaction is not first order Eqs. (10-1) and (10-2) cannot be combined easily to eliminate C_s. The preferred approach is to utilize the mass-transfer coefficient to evaluate C_s and then apply Eq. (10-2) to determine the order of the reaction n and the numerical value of k.[†] One example of this approach is described by Olson et al.[‡]

Diffusion effects can also lead to a false activation energy. Suppose global reaction rates are measured for a nonporous catalyst pellet at different temperatures and that the surface rate is first order. The observed rates can be used to calculate the overall rate constant k_o from Eq. (10-5). If external diffusion is arbitrarily neglected, an apparent activation energy E' could then be calculated from the Arrhenius equation.

$$k_o = A'e^{-E'/R_g T} \tag{10-7}$$

where A' is the apparent frequency factor. This result would give an erroneous value for E if external diffusion were a significant resistance. In fact, the data points for different temperatures would not form a straight line, but would give a curve, as indicated by the solid line in Fig. 10-2. This can be seen by writing Eq. (10-6) as

$$\frac{1}{k_o} = \frac{1}{k} + \frac{1}{k_m a_m} = \frac{e^{E/R_g T}}{A} + \frac{1}{k_m a_m}.$$

or

$$k_o = \frac{A k_m a_m e^{-E/R_g T}}{k_m a_m + A e^{-E/R_g T}} \tag{10-8}$$

where $A e^{-E/R_g T}$ has been substituted for the rate constant k of the surface step. Here E is the true activation energy of the surface reaction. Since the mass-transfer coefficient is relatively insensitive to temperature, Eq. (10-8) shows that k_o approaches a nearly constant value equal to $k_m a_m$ at high temperatures. At low temperatures $k_o \approx A e^{-E/R_g T}$, since $k_m a_m$ is the dominant term in the denominator, and a straight line is obtained on the Arrhenius plot. Figure 10-2 illustrates these results. At low temperatures the slope of the straight line gives the correct activation energy E of the surface reaction. As the temperature increases, a curve which

---

[†] Note that the kinetics of the surface reaction need not be expressed as a simple case of order n, as in Eq. (10-2). The procedure is equally applicable to the form of rate equation developed in Chap. 9 [e.g., Eq. (9-21)].

[‡] R. W. Olson, R. W. Schuler, and J. M. Smith, *Chem. Eng. Progr.*, **46**, 614 (1950).

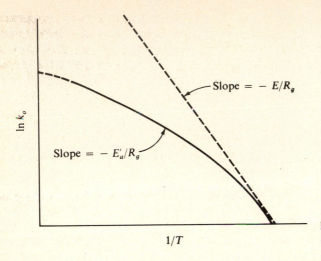

Figure 10-2 True (E) and apparent (E') activation energies.

ultimately flattens to a nearly horizontal line is obtained. In this region E' would vary with the temperature. When experimental rate data for fluid-solid catalytic reactions show a curved line, as in Fig. 10-2, it is possible that external diffusion resistances are important.

These illustrations of the significance of external mass-transfer resistances have been based on isothermal conditions. Temperature differences due to external heat-transfer resistance can also be important. Nonisothermal situations will be considered in Sec. 10-3.

10-2 Mass- and Heat-Transfer Coefficients (Fluid-Particle) in Packed Beds

Average transport coefficients between the bulk stream and particle surface can be correlated in terms of dimensionless groups which characterize the flow conditions. For mass transfer the Stanton number, $k_m \rho / G$, is an empirical function of the Reynolds number $d_p G / \mu$, and the Schmidt number $\mu / \rho \mathscr{D}$. It is common practice to correlate experimental data in terms of j-factors which are defined as the following functions of the Stanton and Schmidt numbers:

$$j_D = \frac{k_m \rho}{G} \left(\frac{a_m}{a_t} \right) \left(\frac{\mu}{\rho \mathscr{D}} \right)^{2/3}$$ (10-9)

The ratio (a_m / a_t) allows for the possibility that the effective mass-transfer area, a_m, may be less than the total external area, a_t, of the particles.†

---

† In experimental studies of mass transfer only the product $k_m a_m$ can be directly measured. However, it is this product that is needed to evaluate mass transfer effects in reactors, as indicated by Eqs. (10-5) or (10-8).

There have been many experimental studies*–¶ of mass transfer in fixed beds along with critical summaries and analyses of the results.**–†† For Reynolds numbers greater than 10, the following relationship†† between j_D and the Reynolds number well represents available data:

$$j_D = \frac{0.455}{\varepsilon_B} \left(\frac{d_p G}{\mu}\right)^{-0.407} \tag{10-10}$$

G = mass velocity (superficial) based upon cross-sectional area of empty reactor ($G = u\rho$)
d_p = diameter of catalyst particle for spheres‡‡
μ = viscosity of fluid
ρ = density of fluid
$\mathscr{D}$ = molecular diffusivity of component being transferred
ε_B = void fraction of the interparticle space (void fraction of the bed)

Heat transfer between a fluid and particle surface in a packed bed occurs by the same molecular and convective processes as describe mass transfer. For heat transfer the equation analogous to Eq. (10-9) is

$$j_H = \frac{h}{c_p G}\left(\frac{a_m}{a_t}\right)\left(\frac{c_p \mu}{k_f}\right)^{2/3} \tag{10-11}$$

The heat-transfer coefficient h is defined in terms of the temperature difference and particle surface

$$Q'_L = h a_m (T_s - T_b) \tag{10-12}$$

where Q' is the heat-transfer rate per unit mass of catalyst. While literature correlations sometimes make a distinction between j_H and j_D, the validity of the difference (in the absence of radiation) is uncertain. Hence, Eq. (10-10) may also be used for j_H, in accordance with the original analogy proposed by Chilton and Colburn.§§ At temperatures above about 400°C and for large ($\geq \frac{1}{4}''$) particles heat transfer by radiation may be significant.¶¶

The relationship between the temperature and concentration differences between fluid and pellet surface can be established by combining the correlations for k_m and h. An energy balance on the pellet requires, for steady state,

$$k_m a_m (C_b - C_s)(-\Delta H) = h a_m (T_s - T_b) \tag{10-13}$$

* D. Thoenes and H. Kramers, *Chem. Eng. Sci.*, **8**, 271 (1958).
† R. D. Bradshaw and C. O. Bennett, *AIChE J.*, **7**, 48 (1961).
‡ J. J. Carberry, *AIChE J.*, **6**, 460 (1960).
§ L. J. Petrovic and G. Thodos, *Ind. Eng. Chem. Fundam.*, **7**, 274 (1968).
¶ E. J. Wilson and C. J. Geankoplis, *Ind. Eng. Chem.*, **5**, 9 (1966).
** S. Whitaker, *AIChE J.*, **18**, 361 (1972).
†† P. N. Dwivedi and S. N. Upadhay, *Ind. Eng. Chem. Proc. Des. Dev.*, **16**, 157 (1977).
‡‡ For other shapes, an approximate value of d_p is that of a sphere with the same external area as the nonspherical particle.
§§ T. C. Chilton and A. P. Colburn, *Ind. Eng. Chem.*, **26**, 1183 (1934).
§§ W. B. Argo and J. M. Smith, *Chem. Engr. Prog.*, **49**, 443 (1953).

Using Eqs. (10-9) and (10-11) for k_m and h yields

$$T_s - T_b = (C_b - C_s)\frac{-\Delta H}{c_p\rho}\left(\frac{c_p\mu/k_f}{\mu/\rho\mathscr{D}}\right)^{2/3}\left(\frac{j_D}{j_H}\right) \tag{10-14}$$

This expression can be used to evaluate the temperature difference from $C_b - C_s$. For many *gases* the *Lewis number*, the ratio of the Prandtl and Schmidt numbers, is about 1.0 and also $j_H \sim j_D$. Hence for most *gases* Eq. (10-14) reduces approximately to

$$T_s - T_b = \frac{-\Delta H}{c_p\rho}(C_b - C_s) \tag{10-15}$$

If the global rate is influenced by diffusion, so that $C_b - C_s$ is appreciable, Eq. (10-15) shows that significant temperature differences between fluid and pellet are possible. In fact the arrangement of properties in Eq. (10-14) is such that $T_s - T_b$ may be appreciable when $C_b - C_s$ is very small.† Hence, external heat transfer can have an effect on the global rate when the external mass-transfer resistance is negligible. For example, experiments‡ on the oxidation of hydrogen $(H_2 + \frac{1}{2}O_2 \rightarrow H_2O)$ with a platinum-on-alumina catalyst gave $T_s - T_b$ values of as high as 115°C while the fractional concentration difference $(C_b - C_s)/C_b$ was less than 5%. When mass-transfer resistance is not negligible, even larger $T_s - T_b$ are observed.§ In the extreme case where external mass-transfer controls the global rate $[C_s \rightarrow 0]$, Eq. (10-15) shows that the maximum $T_s - T_b$ would be that correspond to the adiabatic temperature rise, $(-\Delta H)C_b/c_p\rho$, for the reaction.

A variation of the fixed-bed reactor is an assembly of screens or gauze of catalytic solid over which the reacting fluid flows. Such an arrangement is sometimes employed for oxidation reactions; for example, the oxidations of acetaldehyde to acetic acid using silver screens as a catalyst. These reactions are frequently characterized by a high rate constant, so that mass transfer of reactants from fluid to solid surface can be a significant part of the total resistance. Data on mass transfer from single screens has been reported by Gay and Maughan.¶ Their correlation is of the form

$$j_D = \frac{\varepsilon k_m\rho}{G}\left(\frac{\mu}{\rho\mathscr{D}}\right)^{2/3} = C\left(\frac{4G}{\beta\mu}\right)^{-m} \tag{10-16}$$

where ε is the porosity of the single screen and β is the external-heat-transfer area of the screen per unit volume, i.e., the reciprocal of the hydraulic radius of the screen, and the other symbols are defined as in Eq. (10-9). The coefficients C and m are given in Table 10-1 for the four screen sizes investigated. Heat-transfer results were obtained for similar screens by Coppage and London.††

---

† John Hutchings and J. J. Carberry, *AIChE J.*, **12**, 20 (1966).

‡ J. Maymo and J. M. Smith, *AIChE J.*, **12**, 845 (1966).

§ C. N. Satterfield and H. Resnick, *Chem. Eng. Prog.*, **50**, 504 (1954); F. Yoshida, D. Ramaswami and O. A. Hougen, *AIChE J.*, **8**, 5 (1962).

¶ B. Gay and R. Maughan, *Intern. J. Heat Mass Transfer*, **6**, 277 (1963).

†† J. E. Coppage and A. L. London, *Chem. Eng. Progr.*, **52**, 57 (1956).

Table 10-1 Coefficients C and m in Eq. (10-16)

Screen mesh size	C	m	ε	β, ft^{-1}
10	2.62	0.73	0.817	390
16	4.26	0.85	0.795	535
24	2.80	0.81	0.763	858
60	1.46	0.77	0.690	2,030

For a matrix of 20 or more screens heat-transfer data were found† to follow the j_H lines of Fig. 10-3. The curves given are for extremes of porosity. For intermediate porosities the curves were located between the two shown in the figure. The Reynolds number for Fig. 10-3 is defined as

$$\text{Re}' = \frac{4r_H G}{\mu} \tag{10-17}$$

where r_H, the hydraulic radius, is

$$r_H = L\frac{A_f}{A_h} \tag{10-18}$$

L = length of matrix of screens in the direction of flow
A_f = total cross-sectional area of screen multiplied by porosity ε_m
       of matrix of screens
A_h = heat-transfer area of matrix of screens
G = mass velocity based on A_f

† A. L. London, J. W. Mitchell, and W. A. Sutherland, *ASME J. Heat Transfer*, **82**, 199 (1960).

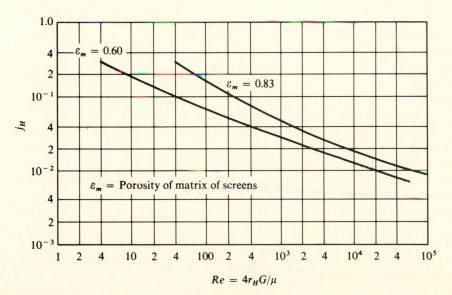

Figure 10-3 Heat-transfer correlation for matrix of screens.

10-3 Quantitative Treatment of External Transport Effects

Four equations are needed to treat external transport effects: one each for the rate of mass and energy *transfer* to the catalyst surface, and two rate of *reaction* equations, one for the mass converted and one for the energy involved due to reaction. We will assume first-order kinetics for the reaction equations, for simplicity, but note that the method of using the equations is the same for nonlinear kinetics (see Example 10-2). We also limit the discussion in this section to single reactions; selectivity in multiple reaction systems is considered in Sec. 10-5.

The rates of mass and energy transfer are Eqs. (10-1) and (10-12) and the mass converted by reaction is given by Eq. (10-2) with $n = 1$. The corresponding energy evolved by reaction is

$$Q'_R = (-\Delta H)\mathbf{r}_p = (-\Delta H)A(e^{-E/R_g T_s})C_s \tag{10-19}$$

where the second equality is obtained by expressing k in terms of the Arrhenius function of temperature.

A. Isothermal behavior Applications of these equations to the design problem of calculating the global rate and to the problem of interpreting laboratory kinetics data are illustrated in Examples (10-2) and (10-1) for isothermal behavior. For this treatment only equations such as (10-1) and (10-2) for transfer and conversion of mass of reactant are needed. Then in Examples (10-3) and (10-4) nonisothermal behavior is analyzed.

> **Example 10-1** Rates of oxidation of SO_2 with air have been measured† in a differential, fixed-bed reactor. The platinum catalyst was deposited only in the outer surface‡ of the $\frac{1}{8}$-in. $\times \frac{1}{8}$-in. cylindrical Al_2O_3 pellets so that intraparticle transport effects were negligible. Bulk gas compositions entering and leaving the reactor and flow rate were measured. Hence, the global rate could be calculated from the usual equation for differential reactors [Eq. (4-5) with the reactor volume replaced by the mass of catalyst pellets]. The objective of the study was to determine the significance of external diffusion resistance by calculating the magnitude of $C_b - C_s$. If this difference is appreciable, then the C_s values must be used in developing an equation for the intrinsic rate. Experimental global rates are given in Table 10-2 for two levels of conversion of SO_2 to SO_3. Evaluate the concentration difference for SO_2 between bulk gas and pellet surface and comment on the significance of external diffusion. Neglect possible temperature differences. The gases passed through the reactor at a superficial mass velocity of 147 lb/(h)(ft)^2, or 0.199 kg/(m)^2(s), and at a pressure of 790 mmHg, or 105×10^3 pascals (Pa). The temperature of the catalyst pellets was 480°C (753 K), and the feed gas contained 6.42 mole %

---

† R. W. Olson and J. M. Smith, *Chem. Eng. Prog.*, **42**, 614 (1950).

‡ Noble metal catalysts are normally manufactured so that the active metal penetrates only a small distance into the pores of the inert carrier (such as alumina). Hence, intraparticle diffusion does not reduce the effectiveness of the catalyst (see Chap. 11).

Table 10-2

Mean conversion of SO_2	r_p, g moles SO_2/(h) (g catalyst)	p_b, atm		
		SO_2	SO_3	O_2
0.1	0.0956	0.0603	0.0067	0.201
0.6	0.0189	0.0273	0.0409	0.187

SO_2 and 93.58 mole % air. Assume that the gas properties are those of air. The external area of the catalyst pellets was 5.12 ft^2/(lb of pellets), and the void fraction of the bed was 0.43.

SOLUTION For interpretation of laboratory data where the rate r_p has been measured, only Eq. (10-1) is needed to evaluate $C_b - C_s$. If $k_m a_m$ is eliminated from this equation by using the correlation for mass-transfer data, Eq. (10-9), the result is

$$C_b - C_s = \frac{r_p(\mu/\rho\mathscr{D})^{2/3}}{a_t(G/\rho)j_D} \tag{10-20}$$

To calculate $C_b - C_s$ for SO_2 from this equation we need to evaluate the Reynolds and Schmidt numbers. At 480°C the viscosity of air is about 0.09 lb/(h)(ft) or 3.72×10^{-5} (Pa)(s). The particle diameter to employ is the diameter of the sphere with the same area as that of the cylindrical pellets. Hence πd_p^2 will equal the sum of the areas of the lateral and end surfaces of the cylinder:

$$\pi d_p^2 = \pi dL + 2\frac{\pi d^2}{4} = \pi \frac{1}{96}\left(\frac{1}{96}\right) + \frac{2\pi}{4}\left(\frac{1}{96}\right)^2$$

$$d_p^2 = \frac{3}{2}\left(\frac{1}{96}\right)^2$$

$$d_p = 0.0128 \text{ ft or } 0.00390 \text{ m.}$$

The Reynolds number is

$$\frac{d_p G}{\mu} = \frac{0.0128(147)}{0.09} = 21$$

or, in SI units,

$$\frac{d_p G}{\mu} = \frac{0.0039(0.199)}{3.72 \times 10^{-5}} = 21$$

From Eq. (10-10)

$$j_D = \frac{0.458}{0.43}(21)^{-0.407} = 0.31$$

In the Schmidt group the correct value for $\mathscr{D}$ would be the molecular diffusivity of sulfur dioxide in a mixture of nitrogen, oxygen, and sulfur trioxide, in which O_2 and SO_3 would also be diffusing. Procedures are available†† for evaluating diffusivities in such complex systems. However, in this instance little error will be introduced by considering $\mathscr{D}$ the binary diffusivity of SO_2 in air. This may be estimated from the Chapman-Enskog kinetic theory. The equation for $\mathscr{D}$ and illustrations of its use are given in Sec. 11-1. From Example 11-1, the molecular diffusivity of SO_2-air is 0.629 cm²/s, or 2.44 ft²/h (6.30×10^{-5} m²/s).

The density of air will be

$$\frac{28.9}{359}\left(\frac{273}{480 + 273}\right)\left(\frac{790}{760}\right) = 0.0304 \text{ lb/ft}^3(0.487 \text{ kg/m}^3)$$

Then the Schmidt group is

$$\frac{\mu}{\rho\mathscr{D}} = \frac{0.09}{0.0304(2.44)} = 1.21 \quad \text{or} \quad \frac{3.72 \times 10^{-5}}{0.487(6.30 \times 10^{-5})} = 1.21$$

For 10% conversion $\mathbf{r}_p = 0.0956$ g mol/(h)(g), or lb mol/(h)(lb), and $a_m = 5.12$ ft²/lb. Substituting all these results in Eq. (10-20) yields

$$C_b - C_s = \frac{0.0956}{5.12}\frac{(1.21)^{2/3}}{0.31(147/0.0.0304)}$$

$$= 1.40 \times 10^{-5} \text{ lb mol/ft}^3(16.0 \text{ kg mol/m}^3)$$

The numerical results are more meaningful if they are converted to partial pressures. In atmospheres, the difference between bulk and surface pressures of sulfur dioxide is

$$(p_b - p_s)_{SO_2} = R_g T(C_b - C_s) = 0.73[1.8(480 + 273)](1.40 \times 10^{-5})$$

$$= 0.0139 \text{ atm}$$

$$p_s = 0.0603 - 0.0139 = 0.0464 \text{ atm (or } 4.7 \times 10^3 \text{ Pa)}$$

Thus p_s is about 23% less than p_b.

If the Δp for 60% conversion is calculated in the same manner, the results at the two conversions can be summarized as in Table 10-3. The relatively large values of $p_b - p_s$ indicate that external-diffusion resistance is significant, although the effect is less at the higher conversion because the rate is less.

If external diffusion were neglected in Example 10-1 and C_b values were used to relate the surface rate and composition (i.e., to obtain an equation for the chemical step at the surface), serious errors would result, particularly at low conversions. The high temperature, low mass velocity, and relatively fast reaction, lead to large external-diffusion resistances in this case. At lower temperatures the surface rate would be lower, and at higher velocities k_m would increase; both

---

†† R. C. Reid, J. M. Prausnitz, and T. K. Sherwood, "The Properties of Gases and Liquids," 3d ed., Chap. 11, McGraw-Hill Book Company, New York, 1977.

Table 10-3

Conversion level	Partial pressures of SO_2, atm			
	p_b	$p_b - p_s$	p_s	$(p_b - p_s)/p_b$
0.1	0.0603	1.39×10^{-2}	0.0464	0.23
0.6	0.0273	0.27×10^{-2}	0.0246	0.10

effects would reduce $C_b - C_s$. In laboratory investigations it is customary to make preliminary measurements at a series of increasing mass velocities. If all other conditions are constant during this series of runs, the importance of external diffusion will decrease with increasing mass velocity. When the experimental rate no longer increases as G is increased, external diffusion is negligible and $C_s \approx C_b$. If rate measurements are made at this mass velocity, bulk concentrations can be used to obtain an expression for the chemical step at the surface. The critical velocity must be determined at the maximum temperature if diffusion resistance is to be negligible at all other temperatures. Other methods of eliminating external resistances in laboratory reactors are described in Chap. 12. Instead of eliminating these resistances by choice of experimental conditions, we may account for them by calculating the surface values of concentration (and temperature), as illustrated in Example 10-1. However, this introduces uncertainty in the results to the extent that the correlations for k_m (and h) are in error.

> **Example 10-2** Suppose that a fixed-bed reactor is to be designed for the oxidation of SO_2 using a platinum/Al_2O_3 catalyst. The *intrinsic* rate equation is that developed in Example 9-2 from experimental rate data. Intraparticle transport effects are negligible, but external concentration and temperature differences may be important. We will assume isothermal conditions at 480°C in order to simplify the calculations, even though this assumption is unrealistic (see Example 10-3). The superficial mass velocity of gas through the bed will be 147 lb/(h)(ft)^2. To demonstrate the procedure for accounting for external mass-transport limitation, calculate the global rate at a location in the reactor where the bulk partial pressures are those given in Table 10-2 (corresponding to 10% conversion).
>
> SOLUTION Equation (F) of Example 9-2 gives the Langmuir-Hinshelwood, intrinsic rate equation. Since the feed composition is fixed, the several partial pressures are uniquely related to each other by the stoichiometry of the reaction. Hence, Eq. (F) can be expressed as Eq. (K) of Example 9-2, which is written
>
> $$r = \left[\frac{p_{SO_2} p_{O_2}^{1/2} - (1/K)p_{SO_3}}{(0.176 + 12.9p_{SO_3})^2}\right]_s, \text{ g mol/(h)(g catalyst)} \qquad (A)$$
>
> The partial pressures (in atm) in this expression are at the catalyst surface. We must use Eqs. (10-1) and (A) to evaluate these surface pressures from the

known bulk values (in Table 10-2). Nonlinear Eq. (A) replaces Eq. (10-2) for this problem. Hence a numerical solution is necessary. One procedure is to assume a value of the global rate, and then calculate p_s for each component from Eq. (10-1). In this particular problem there is no uncertainty about the initial assumption for the rate since the experimental value [$r = 0.0956$ g mol SO_2/(h)(g catalyst) or 2.66×10^{-5} kg mol/(s)(kg catalyst)] is available from Table 10-2. Equation (10-1) with Eq. (10-9) for $k_m a_m$ becomes Eq. (10-20). We have already used this equation [in Example (10-1)] to calculate $(p_b - p_s)$ for SO_2. The result is

$$(p_b - p_s) = 0.0139 \text{ atm } (1.408 \text{ Pa})$$

or

$$(p_s)_{SO_2} = 0.0603 - 0.0139 = 0.0464 \text{ atm } (4.70 \text{ Pa})$$

For oxygen the rate will be $0.0956/2$ g mol/(h)(g catalyst) so that Eq. (10-20) becomes

$$(C_b - C_s)_{O_2} = \frac{(0.0956)/2}{5.12} \frac{(0.81)^{2/3}}{0.31(147/0.0304)} = 5.4 \times 10^{-6} \text{ lb mol/ft}^3$$

or

$$(p_b - p_s)_{O_2} = R_g T(C_b - C_s)_{O_2} = 0.73[1.8(753)]5.4 \times 10^{-6}$$
$$= 0.0054 \text{ atm (or 550 Pa)}$$
$$(p_s)_{O_2} = 0.201 - 0.005 = 0.196 \text{ atm}$$

In SI units

$$(C_b - C_s)_{O_2} = \frac{2.66 \times 10^{-5}}{1.05} \frac{(0.81)^{2/3}}{0.31(0.199/0.427)} = 8.7 \times 10^{-5} \text{ kg mol/m}^3$$

and

$$(p_b - p_s)_{O_2} = R_g T(C_b - C_s)_{O_2} = 8.314(1000)(753)(8.7 \times 10^{-5})$$
$$= 550 \text{ Pa}$$

A similar calculation for SO_3, gives

$$(p_b - p_s)_{SO_3} = 0.0130 \text{ atm}$$

and

$$(p_s)_{SO_3} = 0.0067 + 0.0150 = 0.0217 \text{ atm}$$

With these values for the partial pressures at the catalyst surface we can employ Eq. (A) to check the assumed rate.

$$r = \frac{0.0464(0.196)^{1/2} - (1/73)0.0217}{[0.176 + 12.9(0.0217)]^2}$$
$$= 0.097 \text{ lb mol/(h)(g catalyst)}$$

Since this result is close to the assumed value, it is the desired global rate for a location in the reactor where the conversion is 10%.

B. Nonisothermal behavior At steady state the heat loss Q'_L from a catalyst, given by Eq. (10-12), is equal to the heat Q'_R evolved due to reaction as given by Eq. (10-19). This equality establishes the temperature difference $T_s - T_b$ in terms of the heat- and mass-transfer coefficients and the properties (ΔH, k, E) of the reaction system. Thus, if we equate Q'_L and Q'_R, and also equate r_p from Eqs. (10-1) and (10-2), two expressions are obtained. For a *first-order* reaction they are:

$$\underbrace{h\, a_m(T_s - T_b)}_{\text{Eq. (10-12) for } Q'_L} = \underbrace{r_p(-\Delta H) = (-\Delta H)A(e^{-E/R_gT_s})C_s}_{\text{Eq. (10-19) for } Q'_R} \tag{10-21}$$

$$\underbrace{k_m a_m(C_b - C_s)}_{\text{Eq. (10-1)}} = \underbrace{r_p = A(e^{-E/R_gT_s})C_s}_{\text{Eq. (10-2)}} \tag{10-22}$$

In a design problem, A, E, ΔH, h, k_m and a_m would be known. The global rate is to be calculated for a given bulk concentration and temperature. Equations (10-21) and (10-22) can be solved numerically for the unknown C_s and T_s. Then the rate is given by either Eq. (10-1) or (10-2). The problem of interpretating laboratory data to determine an intrinsic rate equation is more direct. Here, A and E are unknown, the global rate and C_b and T_b are measured, and T_s and C_s are to be found. We calculate C_s and T_s directly from the left-hand side equalities in Eqs. (10-22) and (10-21). Examples (10-3) and (10-4) illustrate the calculations.

Example 10-3 The temperature reported in Example 10-1 was measured by inserting thermocouples in the catalyst pellets, thus giving T_s. Calculate the temperature difference $T_s - T_b$ for the conditions of Example 10-1. The heat of reaction for

$$SO_2 + \tfrac{1}{2}O_2 \rightarrow SO_3$$

is approximately $-23{,}000$ cal/g mol at 480°C, and the activation energy may be taken as 20,000 cal/g mol.

SOLUTION $T_s - T_b$ can be obtained from Eq. (10-21). Thus

$$T_s - T_b = \frac{r_p(-\Delta H)}{h a_m} \tag{A}$$

Expressing h in terms of j_H and using Eq. (10-11) gives

$$T_s - T_b = \frac{r_p}{a_t}\frac{(-\Delta H)(c_p\mu/k_f)^{2/3}}{j_H c_p G} \tag{B}$$

For air at 480°C the *Prandtl number* is $c_p\mu/k = 0.70$ and $c_p = 0.26$ Btu/lb°F. From Example 10-1, the Reynolds number is 21. Then from Eq. (10-10) $j_H = j_D = 0.31$. Using the data given in Example 10-1, we find for the 0.1 conversion level

$$T_s - T_b = \frac{0.0956}{5.12}\frac{23{,}000(1.8)(0.70)^{2/3}}{0.31(0.26)(147)} = 51°\text{F or } 28°\text{C}$$

$$T_b = T_s - 28 = (480 + 273) - 28 = 725 \text{ K or } 452°\text{C}$$

The temperature difference could also have been obtained from the concentration difference (calculated in Example 10-1) by using the energy balance of Eq. (10-14).

$$T_s - T_b = 1.40 \times 10^{-5} \frac{23{,}000(1.8)}{0.26(0.0304)} \left(\frac{0.70}{1.21}\right)^{2/3} = 51°F \text{ or } 28°C$$

With an activation energy of 20,000 cal/g mol, a 28°C temperature drop would cause a 40% decrease in rate at a temperature level of 480°C. Hence, a significant error in rate would be made by neglecting the temperature difference between bulk fluid and catalyst surface.

According to Eq. (B) the temperature difference is proportional to the rate. Hence, at the 60% conversion level

$$T_s - T_b = \frac{0.0189}{0.0956} 28 = 5°C.$$

Because of the low rate, the external temperature difference is now of little significance. Note that in Example 10-1 we found that the external concentration difference was also small at 60% conversion.

Example 10-4 The general treatment of the problem of predicting the global rate at given bulk conditions accounts for both mass- and heat-transfer effects. Let us treat this generalization of Example 10-2 by supposing that we know $t_b = 452°C$ ($T_b = 725$ K) and the bulk concentrations given for the 10% conversion level in Table 10-2. We also presume that the intrinsic rate equation [Eq. (K) of Example 9-2] is available. The objective is to calculate a global rate. It is known from the data in Table 10-2 that the answer is $r_p = 0.0956$ g mol/(h) g catalyst). However, to illustrate this type of problem, we will pretend that we don't know the rate, just as we did in Example 10-2.

SOLUTION As in Example (10-2), one method of solution is to assume a rate. Then calculate T_s and C_s (for each component) from the left-hand-side equalities in Eqs. (10-21) and (10-22). Finally, check the assumed rate by using these values of T_s and C_s in the expression for the intrinsic rate.

We assume $r = 0.0956$ and evaluate $k_m a_m$ and $h a_m$ from Eqs. (10-9) and (10-11), using Eq. (10-10) with $j_D = j_H$. Substituting these results in Eqs. (10-21) and (10-22) we obtain

$$T_s - T_b = \frac{r_p(-\Delta H)}{h a_m} = 28°C$$

$$T_s = T_b + 28 = (452 + 273) + 28 = 753 \text{ K}$$

$$(C_b - C_s)_{SO_2} = 1.40 \times 10^{-5} \text{ lb mol/ft}^3$$

or

$$(p_b - p_s)_{SO_2} = 0.0139 \text{ atm}$$

The numerical values shown are those already determined in Examples (10-3) for ΔT and (10-1) for ΔC. Similar ΔC values for oxygen and sulfur trioxide are also available from Example 10-2. In summary, the corresponding p_s values are

$$(p_s)_{SO_2} = 0.0603 - 0.0139 = 0.0464 \text{ atm}$$

$$(p_s)_{O_2} = 0.201 - 0.0045 = 0.196 \text{ atm}$$

$$(p_s)_{SO_3} = 0.0067 + 0.0150 = 0.0217 \text{ atm}$$

We check the assumed rate by using these surface partial pressures and T_s in the intrinsic rate equation

$$\mathbf{r}_p = \frac{p_{SO_2} p_{O_2}^{1/2} - (1/K)p_{SO_3}}{[A + Bp_{SO_2}]^2} = \frac{0.0464(0.196)^{1/2} - (1/73)0.0217}{[0.176 + 12.9(0.0217)]^2} \tag{A}$$

$$= 0.097 \text{ lb mol/(h)(g catalyst)}$$

This result agrees well with the assumed value. In the normal case where a good initial estimate for $\mathbf{r}_p$ is not available, iterative calculations would probably be required. In such cases the *initial* estimate for $\mathbf{r}_p$ could be that value calculated from the intrinsic rate equation assuming $p_s = p_b$ and $T_s = T_b$. Note also that the two "constants" in the intrinsic rate equation are temperature dependent. Thus, in the normal case it is necessary to know how A and B in Eq. (A) vary with temperature and to use the values corresponding to the calculated surface temperature. In our calculations this was unnecessary because $A = 0.176$ and $B = 12.9$ were the correct values for $T_s = 753$ K (480°C).

For first-order reactions another method of calculation of T_s and C_s values may be used. This is to employ the equality of the left-hand and right-hand terms in Eqs. (10-21) and (10-22). In this method a rate is not assumed; instead Eqs. (10-21) and (10-22) are solved simultaneously for T_s and C_s.

Hougen† has summarized many aspects of the problem of external-mass- and energy-transfer resistances and has included figures for predicting when such resistances need to be considered. This was done for mass transfer by eliminating k_m from Eqs. (10-9) and (10-1), thus establishing $(C_b - C_s)C_b$ in terms of the rate of reaction, the Reynolds number, and the Schmidt number, $\mu\rho/\mathscr{D}$. If this concentration ratio is less than, say, 0.1, diffusion resistance is unimportant. In a similar way, Eqs. (10-11), (10-12), and (10-19) were used to prepare a plot of $T_s - T_b$ as a function of the rate, the Reynolds number, and the Prandtl number. The exponential effect of temperature on the rate means that small values of $T_s - T_b$ can have a large effect on the rate. Thus care must be taken when external heat-transfer resistances are neglected. The safest procedure is to evaluate the rate at T_s and at T_b and note whether the difference is significant, as was done in Example 10-2. These methods for evaluating external resistances are applicable for any form of rate equation, since they utilize an experimental rate of reaction.

† O. A. Hougen, *Ind. Eng. Chem.*, **53**, 509 (1961).

10-4 Stable Operating Conditions

The requirement that $Q'_R = Q'_L$ at steady state introduces interesting questions about stable operating conditions. The problem is similar to the situation in stirred-tank reactors, discussed in Sec. 5-5. We consider this problem for two cases: negligible and finite external-diffusion resistance.

Negligible mass transfer resistance The equality of Q'_L and Q'_R given by Eq. (10-21), and equality of rates of mass transfer and reaction given by Eq. (10-22) establish steady-state values of C_s and T_s. While an analytical solution cannot be achieved, the intersection of curves of Q'_L and Q'_R vs. $T_s - T_b$, or vs. the dimensionless temperature rise $\theta = (T_s - T_b)/T_b$, determines T_s. When the external mass transfer resistance is negligible, $C_s = C_b$, and Eq. (10-21) alone gives T_s for a given T_b and C_b. According to Eq. (10-19), Q'_R will be an exponential curve in θ, as shown in Fig. 10-4 for an exothermic reaction. Equation (10-12) for Q'_L is linear in θ. It could intersect Q'_R as indicated by curves Q'_{L_1} or Q'_{L_2}, or it could be below Q'_R, as shown by curve Q'_{L_3}, depending on the magnitude of h and the location of Q'_R.

The curves in Fig. 10-4 were originally proposed by Frank-Kamenetski† and are useful for describing regions where multiple values of T_s are possible. If the relative position is as described by curves Q'_R and Q'_{L_1}, stable operation occurs at point A_1. If the rate of heat transfer from solid to fluid is less, so that the system is given by Q'_R and Q'_{L_2}, there can be two temperatures for stable operation, given by A_2 and B_2. The region between A_2 and B_2 is unstable in that the heat loss is greater than the heat of reaction. An *initial* condition in this region would stabilize

† D. A. Frank-Kamenetski, "Diffusion and Heat Exchange in Chemical Kinetics," chap. IX, Princeton University Press, Princeton, N.J., 1955.

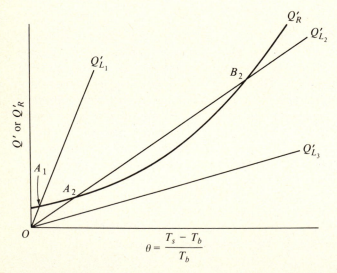

Figure 10-4 Temperature difference between bulk fluid and surface of catalyst pellet (negligible diffusion resistance).

at A_2. An *initial* condition above B_2 would be unstable, with no temperature of stabilization, because $Q'_R > Q'_{L_2}$. Similarly, the Q'_R and Q'_{L_3} system is not usually stable since Q'_R is greater than Q'_{L_3} at any reasonable value of $T_s - T_b$.

Finite mass transfer resistance In this case $C_s < C_b$ so that Eqs. (10-21) and (10-22) must be solved simultaneously. Solving Eq. (10-22) for C_s and substituting the result for C_s in Eq. (10-19) for Q'_R gives

$$Q'_R = \frac{(-\Delta H)C_b}{(1/A)e^{E/R_g T_s} + (1/k_m a_m)} \tag{10-23}$$

The steady-state value of T_s is determined by the intersection of the curve of Q'_R vs. $(T_s - T_b)/T_b$, from Eq. (10-23), and the curve for Q'_L given by Eq. (10-12).

The curve for Q'_R tends to flatten as $T_s - T_b$ increases to a high value, because the $(1/k_m a_m)$ term in the denominator of Eq. (10-23) becomes dominant. Physically this means that the rate is so high that $C_s \to 0$ and mass transfer determines the global rate. In comparison with the Q'_R curves in Fig. 10-4, we see that mass transfer tends to quench the reaction; i.e., the rate reaches a plateau corresponding to the flat portion of the Q'_R curve in Fig. 10-5. Both Q'_R and Q'_L curves are shown in Fig. 10-5. The relative position of the Q'_R and Q'_L curves is illustrated by drawing several lines for Q'_L. For the set $Q'_{L_1} - Q'_R$ the stable operating point is at A_1. If the initial temperature is on either side of this point, the pellet will cool or heat until A_1 is reached. For the set $Q'_R - Q'_{L_2}$ initial T_s values below point A_2 will increase to A_2, and initial T_s values above B_2 will fall to B_2. Intersection C_2 is, in theory, another stable condition. However, it is pseudostable, because small perturbations below C_2 force the temperature to A_2, while slight deviations above force the temperature to B_2. In practice, the entire region between A_2 and B_2 is

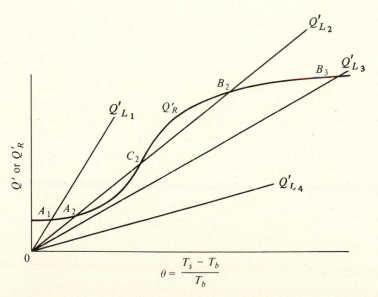

Figure 10-5 Temperature differences between bulk fluid and surface of catalyst pellet (general case).

unstable. For $Q'_R - Q'_{L_3}$ stable operation occurs only at a high value of $T_s - T_b$ corresponding to B_3. Here the Q'_R curve is nearly flat and is determined by the diffusion term $1/k_m$ in Eq. (10-23). Frank-Kamenetski termed this region the *diffusion-controlled regime* and the region at A_1 and A_2 the *reaction-controlled regime*. For the set $Q'_R - Q'_{L_4}$ there is usually no stable region, because the heat transfer away from the pellet is normally less than the heat of reaction. Regions of large $(T_s - T_b)/T_b$ are frequently encountered for combustion, since $(-\Delta H)$ for these systems is high. However, hydrogenations and other reactions can exhibit both reaction and diffusion regimes at feasible operating conditions.

The same type of stability behavior is observed for gas-solid *noncatalytic* reactions, such as the combustion of carbon pellets in air. This problem is a dynamic one, however, because the carbon reactant and size of pellet changes with time. This class of heterogeneous reactions is discussed in Chap. 14.

10-5 Effect of External Transport Processes on Selectivity

When more than one reaction occurs, the effect of external transport on selectivity is important. The local, or point, selectivity, S_p, at any location in a reactor has been defined (Sec. 2-11) as the ratio of the rates of the desirable to the undesirable product. To illustrate the effect of transport processes we consider parallel and consecutive reactions for isothermal and nonisothermal conditions.

Consecutive reactions—isothermal Consider the production of high-octane, branched chain hydrocarbons by the isomerization reactions:[†]

$$n\text{-}C_6H_{14} \underset{k'_1}{\overset{k_1}{\rightleftarrows}} 2\text{-methyl pentane} \underset{k'_2}{\overset{k_2}{\rightleftarrows}} 3\text{-methyl pentane}$$
$$A \qquad\qquad\qquad B \qquad\qquad\qquad C$$

where 2-methyl pentane is the most desirable product. While these catalytic reactions are first order, they are also reversible. The reversible case is considered in Problem 10-9 but in this simple treatment we will neglect the reverse rates.[‡]

The effect of external mass transfer will be to reduce the surface concentration of *n*-hexane (A) below the bulk value. This reduces the rate of formation of 2-methyl pentane (B) by the first reaction. Conversely, the concentration of B will be greater at the catalyst surface than in the bulk fluid stream. This increases the rate of disappearance of B by the second reaction. From this qualitative reasoning we expect that the selectivity S_p of B with respect to C will be reduced by external mass transfer. Quantitatively, the selectivity may be written as the ratio of the rates of production of B and C:

$$S_p = \frac{\mathbf{r}_B}{\mathbf{r}_C} = \frac{k_1(C_A)_s - k_2(C_B)_s}{k_2(C_B)_s} \tag{10-24}$$

[†] A suitable catalyst is Pt on Al_2O_3.

[‡] For a more realistic treatment of hexane isomerization see A. Voorhies, Jr., and R. G. Beecher, paper 20B, 61st Annual Meeting of A.I.Ch.E., Los Angeles, Calif., Dec. 1–5, 1968.

The surface concentrations can be expressed in terms of bulk values by writing mass balances for A and B. For A,

$$(k_m a_m)_A (C_b - C_s)_A = k_1 C_{A_s}$$

or

$$C_{A_s} = \frac{(k_m a_m)_A}{(k_m a_m)_A + k_1} (C_A)_b \qquad (10\text{-}25)$$

Component B is produced by reaction 1 and disappears by reaction 2. Hence, the mass balance for B is

$$(k_m a_m)_B (C_s - C_b)_B = k_1 (C_A)_s - k_2 (C_B)_s$$

Solving this expression for $(C_B)_s$ gives

$$(C_B)_s = \frac{k_1 (C_A)_s + (k_m a_m)_B (C_B)_b}{(k_m a_m)_B + k_2}$$

With Eq. (10-25) for $(C_A)_s$ this expression for $(C_B)_s$ becomes:

$$(C_B)_s = \frac{\dfrac{k_1 (k_m a_m)_A (C_A)_b}{(k_m a_m)_A + k_1} + (k_m a_m)_B (C_B)_b}{(k_m a_m)_B + k_2} \qquad (10\text{-}26)$$

Equations (10-25) and (10-26) can be substituted in Eq. (10-24) to yield an expression for S_p in terms of bulk concentrations. The result is

$$S_p = \frac{k_1}{k_2} \left(\frac{C_A}{C_B}\right)_b \left[\frac{1 + k_2/(k_m a_m)_B}{[k_1/(k_m a_m)_B](C_A/C_B)_b + [1/(k_m a_m)_A][(k_m a_m)_A + k_1]]} \right] - 1 \qquad (10\text{-}27)$$

If there were no mass-transfer resistance $C_s = C_b$ for both A and B, then Eq. (10-24) becomes

$$S'_p = \frac{k_1}{k_2} \left(\frac{C_A}{C_B}\right)_b - 1 \qquad (10\text{-}28)$$

Comparison of Eqs. (10-27) and (10-28) indicates that the selectivity is reduced by mass-transport effects to the extent that the term in brackets in Eq. (10-27) is less than unity. As the mass-transfer coefficients approach infinity (no mass-transfer resistance) the term in brackets approaches 1.0, in agreement with Eq. (10-28).

For our *isomerization* example $(k_m a_m)_A \approx (k_m a_m)_B = k_m a_m$. Then the term in brackets becomes

$$\left[\frac{k_m a_m + k_2}{k_1 (C_A/C_B)_b + (k_m a_m) + k_1} \right]$$

This quantity will be less than unity for $k_1 > k_2$ at every location in the reactor, that is, regardless of the value of $(C_A/C_B)_b$.

Parallel reactions—isothermal Consider parallel reactions with a common reactant (A):

$$A \xrightarrow{k_1} B$$

$$A \xrightarrow{k_2} C$$

where B is the desired product. The selectivity of B with respect to C is given by the ratio of the rates

$$S_p = \frac{\mathbf{r}_B}{\mathbf{r}_C} = \frac{k_1(C_A)_s}{k_2(C_A)_s} = \frac{k_1}{k_2} \tag{10-29}$$

Regardless of how much the surface concentration is reduced by mass transfer Eq. (10-29) shows that the selectivity is unaffected. Thus, Eq. (10-29) is identical with Eq. (2-74), which ignored mass-transfer effects. For this type of parallel reactions the *rate* is reduced by mass transfer but the *selectivity* is unchanged.

If the parallel reactions are completely independent:

$$A \xrightarrow{k_1} B \text{ (desired)}$$

$$R \xrightarrow{k_2} S$$

Equation (10-25) can be used to express the surface concentration of both A and R in terms of bulk values. Then the selectivity is

$$S_p = \frac{\mathbf{r}_B}{\mathbf{r}_s} = \frac{k_1(C_A)_s}{k_2(C_R)_s} = \left[\frac{(1/k_m a_m)_R + 1/k_2}{(1/k_m a_m)_A + 1/k_1} \right] \frac{(C_A)_b}{(C_R)_b} \tag{10-29a}$$

For no external resistances, the selectivity S'_p is

$$S'_p = \frac{k_1}{k_2} \frac{(C_A)_b}{(C_R)_b} \tag{10-29b}$$

Hence the effect of external mass transport on selectivity is given by the ratio

$$\frac{(S_p)}{(S'_p)} = \left[\frac{(k_m a_m)_R + k_2}{(k_m a_m)_A + k_1} \right] \frac{(k_m a_m)_A}{(k_m a_m)_R} \tag{10-29c}$$

The two mass-transfer coefficients will be nearly the same for most reaction systems. Hence, the selectivity will be reduced by external mass transfer if $k_1 > k_2$. There will be little effect for low mass-transfer resistances since $k_m a_m$ becomes large with respect to k_1 or k_2 and the ratio in Eq. (10-29c) approaches unity.

Nonisothermal conditions Equations (10-27) and (10-29) are applicable when external temperature differences are significant, but the rate constants k_1 and k_2 must be evaluated at the surface temperature. The effect of heat transfer on selectivity will then depend upon the activation energies of the reactions. In general, the selectivity is increased by heat-transfer resistances for an exothermic reaction when the activation energy, E_1, for the desired reaction is larger than E_2 for the by-product reaction. For example, for parallel reactions, Eq. (10-29) gives

$$S_p = \frac{A_1 e^{-E_1/R_g T_s}}{A_2 e^{-E_2/R_g T_s}} \tag{10-30}$$

If there were no heat-transfer resistance $T_s = T_b$. Hence the ratio of selectivities with and without considering external temperature differences would be

$$\frac{(S_p)_{T_s}}{(S_p)_{T_b}} = \frac{\exp\left[\dfrac{E_1}{R_g}\left(\dfrac{T_s - T_b}{T_s T_b}\right)\right]}{\exp\left[\dfrac{E_2}{R_g}\left(\dfrac{T_s - T_b}{T_s T_b}\right)\right]} \tag{10-31}$$

When the net heat of the reactions is exothermic, $T_s - T_b$ is positive. Hence, if $E_1 > E_2$, the selectivity at T_s will be greater than that at T_b. For a net endothermic heat effect the effect of heat transfer would be to reduce the selectivity when $E_1 > E_2$.

For consecutive reactions the same conclusions apply. Suppose that mass transfer is important as well as heat transfer. Then the selectivity is reduced by the mass-transfer resistance and increased by the heat-transfer resistance, provided $E_1 > E_2$ and the reaction is exothermic. The temperature effect will be the dominant factor for moderate or high heats of reaction. As an illustration, imagine that the conditions (heats of reaction, heat-transfer coefficient) are such that $(T_s - T_b) = 25°C$ for a set of *consecutive* reactions. Suppose that the mass-transfer resistance is relatively low. Typical values for the outer conditions might be:

$$(k_m a_m)_A = (k_m a_m)_B = 50 \text{ cm}^3/(\text{s})(\text{g catalyst})$$

$$E_1 = 20 \text{ kcal/g mol}$$

$$E_2 = 15 \text{ kcal/g mol}$$

$$A_1 = 6.2 \times 10^8 \text{ cm}^3/(\text{s})(\text{g catalyst})$$

$$A_2 = 1.9 \times 10^6 \text{ cm}^3/(\text{s})(\text{g catalyst})$$

$$T_b = 500 \text{ K}$$

$$(C_A/C_B)_b = 2.$$

If external mass and heat transfer are both neglected, the selectivity is, from Eq. (10-28),

$$S_p = \left(\frac{k_1}{k_2}\right)_b \left(\frac{C_A}{C_B}\right)_b - 1$$

$$= \frac{6.2 \times 10^8 \exp[-20,000/R_g(500)]}{1.4 \times 10^6 \exp[-15,000/R_g(500)]} (2) - 1$$

$$= 4.2 - 1 = 3.2$$

The combined effects of both mass and heat transfer are given by Eq. (10-27) with k_1 and k_2 evaluated at $T_s = 525$ K. These values are $k_1 = 2.87$ and $k_2 = 1.07$ cm^3/(s)(g catalyst). Hence

$$S_p = \frac{2.87}{1.07}\,(2)\left[\frac{1 + (1.07)/50}{(2.87/50)(2) + (50 + 2.87)/50}\right] - 1$$

$$= 5.36[0.87] - 1 = 3.7$$

The term in brackets indicates that the effect of mass transfer reduces the selectivity about 13% [1 vs. 0.87]. However, the effect of temperature is to increase S_p by $(5.36 - 4.2)/4.2$ or 30%. The net result is an increase from 3.2 to 3.7.

FLUIDIZED-BED REACTORS

In the fixed-bed reactor the catalyst particles are relatively large and stationary. In contrast, in a fluidized-bed reactor, Fig. 1-8, the small particles (50 to 250 microns) move about in a manner dependent on the velocity of the reacting fluid. Figure 10-6† shows the range of behavior of particles in a vertical tube through which fluid flows. At very low velocities the particles are not disturbed so that, in essence, fixed-bed behavior is observed. At the other extreme the velocity is so high that the particles are carried out the top of the reactor with the fluid.‡ In this

---

† Adapted from Fig. 1, Chap. 1, "Fluidization Engineering," Daizo Kunii and Octave Levenspiel, John Wiley & Sons, New York, 1969.

‡ F. A. Zenz and D. F. Othmer, "Fluidization and Fluid-Particle Systems," Reinhold Publishing Corporation, New York, 1960.

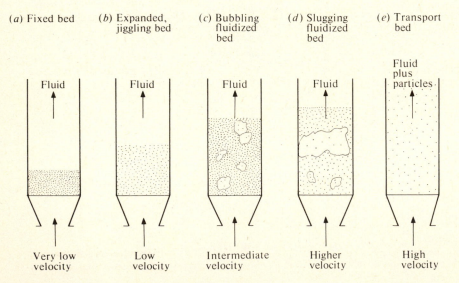

Figure 10-6 Particle motion in tubular-flow reactor.

condition the system operates as a transport, or transfer-line reactor. In most *fluidized-bed* catalytic reactors, the fluid is a gas and the normal operating condition is in the bubbling regime. In this condition the gas moves through the reactor in two ways: as "bubbles" containing relatively few solid particles and moving at above the average velocity, and as a continuous "dense" or "emulsion" phase where the particle concentration is high.† Kunii and Levenspiel‡ have described this behavior in terms of a "bubbling-bed model" and used the model for design of reactors.§ We will consider the design problem later in Chap. 13. Here we are interested in mass- and heat-transfer rates between the fluid and the solid particles.

10-6 Particle-Fluid Mass and Heat Transfer

An important characteristic of fluidized beds is the near uniformity of temperature throughout the reactor. This is due to the mixing in the emulsion phase, caused by the gas bubbles, and to the very high heat-transfer area, per unit mass of catalyst, that is associated with the small particles. The same large area is available for mass transfer. Therefore, the temperature and concentration differences between fluid and particle surface are usually negligible. This is in direct contrast to the result for fixed-bed reactors even though the mass- and heat-transfer *coefficients* in fluidized beds are less than those in fixed beds. As a result, external concentration and temperature effects are not very significant in design of fluidized-bed reactors. The more important problem is accounting for the bypassing of the catalyst by the gas "bubbles" and the mass transfer between bubble and dense phases as discussed in Chap. 13.

Kunii and Levenspiel¶ have summarized available mass- and heat-transfer data in the form of graphs of Sherwood and Nusselt numbers vs. Reynolds number. Typical of the results are those of Chu et al.,†† which may be expressed in terms of j factors as

$$j_D \text{ or } j_H = 1.77 \left[\frac{d_p G}{\mu(1 - \varepsilon_B)} \right]^{-0.44} \tag{10-32}$$

for the range $30 < d_p G/\mu(1 - \varepsilon_B) < 5,000$. Here j_D and j_H are as defined in Eqs. (10-9) and (10-11) and

$$\varepsilon_B = \text{void fraction in the bed}$$

$$G = \text{superficial mass velocity}$$

$$d_p = \text{average particle diameter}$$

This correlation is based on data for both liquid-solid and gas-solid beds.

---

† J. J. van Deempter, *Chem. Eng. Sci.*, **13**, 343 (1961).

‡ Op. cit.

§ J. F. Davidson and D. Harrison, in "Fluidized Solids," Cambridge University Press, London, 1963, also proposed a bubble and emulsion phase model to describe fluidized-bed reactor behavior.

¶ Op. cit.

†† J. C. Chu, J. Kaiil, and W. A. Wetterath, *Chem. Eng. Prog.*, **49**, 141 (1953).

Example 10-5 Estimate the ratio of mass-transfer rates, per unit volume, for fixed-bed and fluidized-bed reactors at reasonable operating conditions. Suppose these conditions are:

	Fluidized bed	Fixed bed
Particle size, d_p, cm	0.0063	0.635
	(250 mesh)	($\frac{1}{4}$ in.)
Void fraction of the bed, ε_B	0.90	0.40
Fluid mass velocity, G g/cm^2)(s)	0.02	0.10

SOLUTION The mass transfer rate per unit *volume* of reactor is, from Eq. 10-1):

$$r_v = \rho_p(1 - \varepsilon_B)r_p = \rho_p(1 - \varepsilon_B)k_m a_m(C_b - C_s) \tag{A}$$

where ρ_p is the density of the particles themselves. For spherical particles of diameter d_p, the external area per unit mass, a_m, is $6/d_p\rho_p$. Hence Eq. (A) becomes

$$r_v = \frac{6(1 - \varepsilon_B)}{d_p} k_m(C_b - C_s) \tag{B}$$

The ratio of the mass-transfer rates can be obtained by using Eq. (10-9) for k_m in Eq. (B). For the same fluid and the same concentration difference, the result is

$$\frac{r_{\text{fluid}}}{r_{\text{fixed}}} = \frac{[(1 - \varepsilon_B)/d_p]_{\text{fluid}}(Gj_D)_{\text{fluid}}}{[(1 - \varepsilon_B)/d_p]_{\text{fixed}}(Gj_D)_{\text{fixed}}} \tag{C}$$

Substituting Eqs. (10-32) and (10-10) for j_D,

$$\frac{r_{\text{fluid}}}{r_{\text{fixed}}} = \frac{[1.77(1 - \varepsilon_B)^{1.44}G^{0.56}d_p^{-1.44}]_{\text{fluid}}}{[(0.458/\varepsilon_B)(1 - \varepsilon_B)G^{0.593}d_p^{-1.407}]_{\text{fixed}}} \tag{D}$$

Substituting numerical values

$$\frac{r_{\text{fluid}}}{r_{\text{fixed}}} = 32$$

Thus the mass-transfer rate between particle and fluid for a fluidized bed can be an order of magnitude greater than that for a fixed bed. With this large transport rate it is evident that $C_b - C_s$, calculated by the methods described in Sec. 10-3, will be negligible. A similar result applies for external temperature differences.

SLURRY REACTORS

Heretofore in this chapter external transport has been analyzed for two-phase systems, fluid and solid catalyst. When there are both volatile and nonvolatile reactants, or when a liquid solvent is necessary with all gaseous reactants, three-phase reactors are needed. Examples are hydrogenation of oils with a nickel catalyst, oxidation of liquids (or pollutants dissolved in liquids) with metal oxide catalysts, and polymerization of ethylene or propylene in a slurry of solid catalyst particles in liquid cyclohexane. External transport effects can be especially important in three-phase reactors because more interphase steps exist; a gaseous reactant must be transferred from gas to liquid and then from liquid to solid catalyst before reaction occurs.

The two common forms of three-phase reactors are the slurry† and trickle-bed‡ types described briefly in Chap. 1 (Fig. 1-6). The design of three-phase reactors consists of the usual two steps associated with heterogeneous systems: formulation of an expression for the global rate of reaction, applicable at any location, followed by using this expression to predict overall reactor performance. By using the global rate, the second step can be carried out with the procedures for homogeneous reactors presented in Chaps. 3 to 5. In this chapter we discuss only the first step of accounting for external transport in formulating a global rate equation, first for slurry reactors and then for trickle beds. Then in Chap. 13 the prediction of reactor performance is considered.

The distinguishing feature of a slurry reactor is that *small* particles (~ 100 microns) of catalyst are suspended in a liquid. In a three-phase system bubbles of gas rise through the agitated slurry, as indicated in Fig. 10-7. Unlike the fluidized bed, there is little relative movement between particles and fluid, even though the liquid is agitated mechanically. The particles tend to move with the liquid. The small particle size, low diffusivities in liquids, and low relative velocity all reduce the mass transfer coefficient. Hence external mass transfer can significantly retard the global rate. However, the relatively high thermal conductivity of liquids increases the heat transfer coefficient. This coupled with the small heat of reaction per unit *volume* of the slurry means that there is little temperature difference between particle and liquid. Thus, external temperature differences can normally be neglected in slurry reactors. Note that the relative importance of heat- and mass-transfer effects in slurries is the opposite of that in fixed-bed reactors.§

A similar situation exists for heat and mass transfer between gas bubble and liquid. The bubble velocity with respect to the liquid may be large, but the area for

---

† Another form of the slurry reactor is the two-phase type where all the reactants are liquid.

‡ The term trickle bed means cocurrent downflow of gas and liquid over a fixed bed of catalyst particles (see Fig. 1-6*b*). Downward flow of liquid and countercurrent, upflow of gas, and cocurrent upflow of both gas and liquid are possible alternatives of three-phase reactors.

§ For other treatments of slurry reactors see C. N. Satterfield and T. K. Sherwood, "The Role of Diffusion in Catalysis," pp. 43–55, Addison-Wesley Publishing Co., Inc., Reading, Mass., 1963; C. N. Satterfield, "Mass Transfer in Heterogeneous Catalysis," pp. 107–122, Massachusetts Institute of Technology Press, Cambridge, Mass., 1970; Y. T. Shah, "Gas-Liquid-Solid Reactor Design," p. 133–135, McGraw-Hill Book Company, New York, 1979.

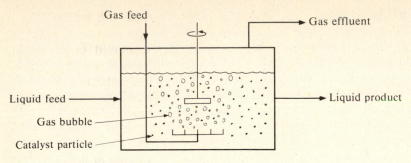

Figure 10-7 Three-phase slurry reactor.

mass transfer with respect to that of the particles is often small.† Hence, bubble-liquid mass transfer can also significantly retard the global rate.

It is reasonable to assume that the liquid and catalyst particles in a slurry reactor are well mixed (as in a stirred-tank). However, the gas bubbles rise through the liquid without complete mixing. Therefore, the concentration of gaseous reactant in the bubble will change with position (unless the gas is a pure component, as in hydrogenations). As a result, the global rate can change with vertical position in the reactor.‡ *At one position*, the overall reaction consists of the following processes in series:§

1. Mass transfer from the bulk concentration in the gas bubble to the bubble-liquid interface
2. Mass transfer from the bubble interface to the bulk-liquid phase
3. Mixing and diffusion in the bulk liquid
4. Mass transfer to the external surface of the catalyst particles
5. Reaction at the catalyst surface¶

As mentioned, the rise of the bubbles through the liquid, along with mechanical agitation is normally sufficient to achieve uniform conditions in the bulk liquid. Hence the resistance of step 3 can be neglected.†† The golbal rate can be expressed in terms of the known bulk concentrations by writing rate equations for

† For a wide range of operating conditions the vertical velocity of bubbles rising through a liquid is about 20 cm/s. Also, the bubble diameter is 1–3 mm (much larger than the particle size) and varies but little [D. Misic and J. M. Smith, *Ind. Eng. Chem. Fundam.*, **10**, 380 (1971)]. Hence the bubble surface per unit volume of slurry can be much smaller than the particle surface.

‡ When there is no gas phase, that is, there are only liquid reactants, and the slurry is well mixed, the global rate is the same throughout the reactor.

§ For an irreversible reaction. For a reversible case, the reverse transport processes for the limiting product would be added.

¶ Since our concern in this chapter is only with external (to the catalyst particle) resistances, the possibility of resistances within a porous catalyst particle is not included. The particles are inherently small, so that intraparticle mass-transfer resistance may be unimportant. The conditions for establishing the significance of such internal resistances are given in Chap. 11.

†† This conclusion has been verified experimentally by H. Kolbel and W. Siemes [*Umschau*, **24**, 746 (1957)] and by W. Siemes and W. Weiss [*Dechema Monographien*, **32**, 451 (1959)].

each of the four remaining steps. The rates of all the steps will be identical at steady state, and this equality permits elimination of the unknown interfacial concentrations. The procedure is the same as that employed in developing Eq. (10-5) for a single mass-transfer step. As in that situation, a simple explicit equation for the global rate can be written only for a first-order reaction at the catalyst surface.§ If we make the assumption of a first-order irreversible catalytic reaction, the rate per unit volume of bubble-free slurry may be written

$$\mathbf{r}_v = ka_c C_s \qquad \text{reaction at surface} \qquad (10\text{-}33)$$

where a_c = external area of catalyst particles per unit volume of liquid (bubble free)

k = first-order rate constant

C_s = concentration of reactant (hydrogen) at the outer surface of the catalyst particle

The rates of the three mass-transfer processes may be expressed as

$$\mathbf{r}_v = k_g a_g (C_g - C_{i_g}) \qquad \text{bulk gas to bubble interface} \qquad (10\text{-}34)$$

$$\mathbf{r}_v = k_L a_g (C_{i_L} - C_L) \qquad \text{bubble interface to bulk liquid} \qquad (10\text{-}35)$$

$$\mathbf{r}_v = k_c a_c (C_L - C_s) \qquad \text{bulk liquid to catalyst surface} \qquad (10\text{-}36)$$

where a_g is the gas bubble-liquid interfacial area per unit volume of bubble-free liquid and k_g, k_L, and k_c are the appropriate mass-transfer coefficients. The various concentrations are shown schematically in Fig. 10-8 for a single bubble-

§ For hydrogenation reactions the low solubility of hydrogen usually means that it is the limiting component. Then first-order behavior is expected.

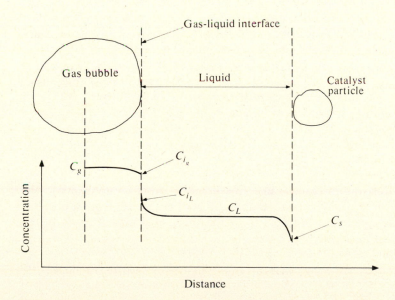

Figure 10-8 Concentration profiles in a slurry reactor.

particle combination. If equilibrium exists at the bubble-liquid interface, C_{i_g} and C_{i_L} are related by Henry's law,†

$$C_{i_g} = HC_{i_L} \tag{10-37}$$

These five equations can be combined to eliminate C_{i_g}, C_{i_L}, C_L, and C_s. Then the rate can be expressed solely in terms of the concentration of reactant in the gas:

$$\mathbf{r}_v = k_0 \, a_c \, C_g \tag{10-38}$$

and

$$\frac{1}{k_0} = \frac{a_c}{a_g} \frac{1}{k_g} + \frac{a_c}{a_g} \frac{H}{k_L} + H\left(\frac{1}{k_c} + \frac{1}{k}\right) \tag{10-39}$$

Equation (10-39) shows that the *global* k_0 is a function of the three mass-transfer coefficients, the specific reaction rate k, and the area ratio a_c/a_g. The rate will increase as this ratio falls, which corresponds to increasing the concentration of gas bubbles in the slurry. The rate is enhanced by increasing the concentration of catalyst particles in the slurry, because this increases a_c in Eq. (10-38). For low concentration of catalyst there is a sufficient supply of dissolved reactant for mass transfer to each particle to be independent of the number of particles. Under these circumstances the global rate is linear in catalyst concentration (particles per unit volume of reactor), since a_c in Eq. (10-38) is directly proportional to this concentration. As the catalyst concentration increases to higher values, there is competition between particles for reactant, and the rate ultimately approaches a constant upper limit.

The rate constant k is sensitive to temperature and, in principle, should be associated with the temperature of the catalyst particle. However, as mentioned, the catalyst temperature is essentially the same as that of the liquid. Hence, external temperature differences are not normally important in slurry reactors.

Not all of the four resistances indicated in Eq. (10-39) are significant in every case. For example, in hydrogenations pure hydrogen is normally used as reactant. Then there is no resistance to diffusion from bulk gas (in the bubble) to bubble-liquid interface. Also, for slightly soluble gases the resistance to mass transfer on the liquid side of the interface is predominant. For these conditions $C_g = C_{i_g}$ and Eq. (10-39) reduces to

$$\frac{1}{k_0 H} = \frac{a_c}{a_g} \frac{1}{k_L} + \frac{1}{k_c} + \frac{1}{k} \tag{10-40}$$

Even when the gaseous reactant is in a mixture with other components in the bubbles, k_g appears to be much larger than k_L/H, so that Eq. (10-40) is applicable. In this case it is sometimes desirable to define the global rate in terms of the

---

† For a linear solubility relationship. Sometimes a Freudlich-type equation [similar to Eq. (7-21)] is indicated. This is the case for sulfur dioxide solubility in water, even at low concentrations [Hiroshi Komiyama and J. M. Smith, *AIChE J.*, **21**, 664 (1975)].

liquid-phase concentration in equilibrium with C_g, that is, $(C_L)_{eq}$. Since $C_{i_g} = C_g = H(C_L)_{eq}$, Eq. (10-38) may be written

$$\mathbf{r}_v = k_0 H a_c (C_L)_{eq} \tag{10-41}$$

where $k_0 H$ is given by Eq. (10-40).

If the catalyst is very active,[†] k will be much greater than k_c or k_L/H. Then the global rate is determined by the mass-transfer coefficients k_L and k_c. In contrast, for some chemical systems and for common bubble and particle concentrations, all three rate coefficients, k_L, k_c, and k may be significant.[‡] In any event, the important mass-transport parameters are k_L and k_c. Available data for these coefficients are summarized in Secs. 10-7 and 10-8.

10-7 Mass-Transfer Coefficients: Gas Bubble to Liquid (k_L)

Data for the coefficient k_g in Eq. (10-39) are meager. Fortunately, accurate values are seldom required. Even when the bubble is a gas mixture, the major resistance to transport for slightly soluble gases is in the liquid. Hence k_L is usually the important coefficient in gas bubble-to-liquid mass transfer. Several experimental studies and correlations[§]–[§§] for k_L are available. Its value depends somewhat on the geometry of the reactor system (especially the dimensions of the stirrer and baffles) and the stirrer speed. In one correlation[†] these effects were accounted for in terms of a mild dependence on the energy dissipation rate using the following dimensional equation.

$$k_L = 0.592 \, \mathscr{D}_A^{1/2} (\sigma/v)^{1/4} \tag{10-42}$$

The energy dissipation rate σ per unit of mass of liquid [in erg/(s)(g)] is given by

$$\sigma = \frac{N_p \rho_L N^3 D_I^5}{W} \phi \tag{10-43}$$

where D_I = impeller diameter, cm
$\mathscr{D}_A$ = molecular diffusivity of reactant in liquid, cm^2/s
v = kinematic viscosity, cm^2/s
k_L = liquid-side mass-transfer coefficient, cm/s
N = impeller speed in rps
W = mass of liquid in the slurry, g

[†] For example, nickel is an extremely active catalyst, when used in finely divided form, for many hydrogenation reactions.

[‡] For example, in the catalytic oxidation of SO_2 to SO_3 in an aqueous slurry of activated carbon at 25°C [Hiroshi Komiyama and J. M. Smith, *AIChE J.*, **21**, 670 (1975)].

[§] P. H. Calderbank and M. B. Moo-Young, *Chem. Eng. Sci.*, **16**, 39 (1961).

[¶] B. D. Prasher and G. B. Wills, *Ind. Eng. Chem. Proc. Des. Dev.*, **12**, 351 (1973).

[††] P. H. Calderbank, *Trans. Inst. Chem. Eng.* (London), **37**, 173 (1959).

[‡‡] D. Misic and J. M. Smith, *Ind. Eng. Chem. Fundam.*, **10**, 380 (1971).

[§§] V. A. Juvekar and M. M. Sharma, *Chem. Eng. Sci.*, **28**, 825 (1973).

N_p = power number, defined by the expression

$$N_p = \frac{P}{\rho_L N^3 D_I^5} \tag{10-44}$$

This number depends upon the design (number and size) of baffles, geometry of baffles and vessel, but frequently is about 10. P is the power input, erg/s.

ϕ = a correction factor ($0 < \phi < 1$) to account for the decrease in energy dissipation rate due to gas bubbles. For $Q/ND_I^3 < 0.035$, Calderbank gives†

$$\phi = 1 - 12.6Q/ND_I^3 \tag{10-45}$$

where Q is the gas flow rate (cm³/s).

This method of correlation seems to account for variations in impeller speed, baffles, and impeller geometry but requires a measurement of the energy dissipation rate by determining the torque of the agitator.

In the *absence* of mechanical agitation and for bubbles whose diameter is less than 2.5 mm (the usual size range for slurry reactors), the following correlation is available

$$k_L \left(\frac{\mu_L}{\rho_L \mathscr{D}} \right)^{2/3} = 0.31 \left(\frac{\Delta\rho\mu_L g}{\rho_L^2} \right)^{1/3} \tag{10-46}‡$$

where $\Delta\rho$ = difference in density between liquid phase and gas bubbles, g/(cm)³
μ_L = viscosity of liquid phase, g/(cm)(s)
g = acceleration of gravity, cm/(s)²
ρ_L = density of liquid phase, g/(cm)³
k_L = mass-transfer coefficient, cm/s

This correlation is for bubbles rising through the liquid phase because of gravitational force. Equation (10-46) has been applied with apparent success to transport from bubbles to liquid in systems containing catalyst particles in slurry form.§

In the absence of gravitational force, mass transfer from a stagnant bubble would be by molecular diffusion through the surrounding stagnant liquid. Then the *Sherwood number* is Sh = $d_b k_L/\mathscr{D}$ = 2, where d_b is the diameter of the bubble. Other correlations have been developed by modifying this relation to include the effect of turbulence, in the liquid phase, induced by mechanical stirring or gravitational force. Hughmark¶ proposed this type of correlation for the Sherwood number when the turbulence was due to gravitational force. Both Eq. (10-46) and Hughmark's correlation are for column-type reactors, where gas bubbles rise through an appreciable height.

---

† P. H. Calderbank, *Trans. Inst. Chem. Eng.*, **36**, 443 (1958).
‡ P. H. Calderbank, *Trans. Inst. Chem. Engrs.* (*London*), **37**, 173 (1959).
§ T. Matsuura, doctoral dissertation, Technische Universitat Berlin, Berlin, 1965.
¶ G. A. Hughmark, *Ind. Eng. Chem., Process Design Develop. Quart.*, **6**, 218 (1967).

Example 10-6 One method of determining the gas bubble-to-liquid mass-transfer coefficients, k_L, is to introduce a step function of transferable component, A, in the gas stream fed to the slurry reactor (Fig. 10-7). The variation with time of the partial pressure p_L of the component in the effluent gas stream is a function of the rate and equilibrium properties of the system. If no catalyst particles are present, so that the gas bubbles rise through only the liquid, the p_L vs. time relation depends on k_L and the solubility of A.

A. Derive an equation for the fraction, $(p_0 - p_L)/p_0$, of A removed from the gas bubbles as a function of t for a well-stirred liquid of height L. Assume that the gas bubbles, which contain a low concentration of A, do not coalesce and are in plug flow. Assume also that A is slightly soluble in the liquid and that its solubility follows Henry's law, $p = H_p C_L$.† In the time required for a bubble to move through the liquid the change in concentration of A in the bubble is large, but the change in concentration in the liquid is negligible. There is no liquid feed or effluent (batch liquid).
B. This dynamic method was used‡ to determine k_L for the transfer of benzene from gas bubbles to water at 24°C and one atmosphere pressure. At $t = 0$ the helium stream bubbling through the water was switched to a He - C_6H_6 stream. Then the concentration of benzene in the effluent gas was measured continuously as a function of time.

Experimental results for one run are given in Table 10-4. The ratio, V_B, (the gas holdup) is a key parameter and is obtained from the gas flow rate Q and the bubble velocity v_B. The total gas holdup in the liquid of volume V_L is

† Note that this definition of Henry's law is in terms of the partial pressure in the gas, not the concentration as in Eq. (10-37).
‡ D. M. Misic and J. M. Smith, op. cit.

Table 10-4 Absorption of benzene from gas bubbles into liquid water (24°C)

Time, min	$(p_0 - p_L)/p_0$
1.5	0.68
3.5	0.47
6.5	0.30
8.0	0.23
12	0.15
16	0.092
18	0.060
23	0.039
25	0.019
33	0.015
38	0.011

$V_B V_L$. This total holdup is also equal to the product of the gas flow rate and the bubble residence time L/v_B:

$$V_B V_L = QL/v_B$$

or

$$V_B = \frac{QL}{v_B V_L}$$

For the given run, $V_L = 50$ cm^3 (50 $\times$ 10^{-6} m^3) and $Q = 50.7$ cm^3/min (8.45 $\times$ 10^{-7} m^3/s). The vertical bubble velocity $v_B = 22.5$ cm/s (0.225 m/s). And $L = 6.0$ cm (0.06 m). The average bubble diameter was 1.8 mm (1.8 $\times$ 10^{-3} m). Calculate k_L for this run.

SOLUTION

A. For slightly soluble gas A, the concentration C_{i_L} (see Fig. 10-7) at the bubble-liquid interface will be that in equilibrium with the pressure p_g in the bulk gas of the bubble. Hence, $C_{i_L} = p_g/H_p$. Then Eq. (10-35) for the rate of mass transfer from gas bubble to liquid, per unit volume of bubble-free liquid is

$$\mathbf{r} = k_L a_g \left(\frac{p_g}{H_p} - C_L \right) \tag{A}$$

A mass balance of A in the gas phase† according to Eq. (3-1), is

$$-v_B V_B \frac{1}{RT} \left(\frac{\partial p_g}{\partial z} \right) - \mathbf{r} = \frac{\partial}{\partial t} \left(\frac{V_B p_g}{RT} \right) \tag{B}$$

The first term represents the net molal flux in and out of the volume element and the last term is the accumulation of A in the gas bubbles within the volume element. The vertical bubble velocity, v_B, and the volume of bubbles per unit volume of liquid, V_B, will be constant for dilute concentrations and small bubble sizes. Also the temperature does not vary. Since the liquid concentration does not change significantly with the residence time of the bubble, but the change in p_g is large, the accumulation term in Eq. (B) may be neglected. Then with Eq. (A) for $\mathbf{r}$, we have

$$-\left(\frac{v_B V_B}{RT} \right) \frac{dp_g}{dz} = k_L a_g \left(\frac{p_g}{H_p} - C_L \right) \tag{C}$$

Since C_L is constant with respect to z (well-stirred liquid) Eq. (C) can be integrated from the bubble entrance, $z = 0$, $p_g = p_0$, to any liquid height z where the partial pressure is $p(z)$. The result is

$$p(z) = H_p C_L + (p_0 - H_p C_L) \exp\left[-\frac{k_L a_g RT z}{H_p V_B v_B} \right] \tag{D}$$

† In a volume element corresponding to a vessel height dz and unit cross-sectional area.

In order to express Eq. (D) in terms of time we need to know how the concentration of A in the liquid varies with time. To do this a mass balance is written for the total volume of bubble-free liquid. According to Eq. (3-1),

$$\frac{V_L}{L}\int_0^L k_L a_g\left(\frac{p(z)}{H_p} - C_L\right)dz = V_L\frac{\partial C_L}{\partial t} \tag{E}$$

The first term represents the total rate of mass transfer from gas bubble to liquid, and the second term is the accumulation of A in the liquid; where V_L is the total liquid volume. Substituting $p(z)$ from Eq. (D) in the left-hand side of Eq. (E), and integrating with respect to z, gives an expression for dC_L/dt as a function of C_L. This differential equation can be integrated from $C_L = 0$ at $t = 0$, to yield the following expression for $C_L(t)$

$$C_L(t) = \frac{p_0}{H}\left[1 - \exp\left\{-k_L a_g\left(\frac{1 - e^{-\alpha}}{\alpha}\right)t\right\}\right] \tag{F}$$

where α is the dimensionless parameter

$$\alpha = \frac{k_L a_g RTL}{H_p V_B v_B} \tag{G}$$

To obtain p_L Eq. (D) is applied at $z = L$ with C_L expressed in terms of time using Eq. (F). The result may be rearranged to give an equation for the fraction of A removed from the gas bubbles by the liquid:

$$\frac{p_0 - p_L}{p_0} = (1 - e^{-\alpha})\exp\left[-k_L a_g\frac{1 - e^{-\alpha}}{\alpha}t\right] \tag{H}$$

Equation (H) is the desired result; expressing $p_L(t)$ in terms of the solubility constant H_p and mass-transfer coefficient k_L.

B. Equation (H) suggests that p_L vs. time data plotted as $\ln[(p_0 - p_L)/p_0]$ vs. t should yield a straight line with a slope equal to

$$-k_L a_g(1 - e^{-\alpha})/\alpha.$$

Hence, from experimental data for $(p_0 - p_L)/p_0$ vs. t, a value of α can be determined. Then k_L is calculated from Eq. (G), provided H_p is known. The data plotted in this manner are shown in Fig. 10-9. The slope is -2.8×10^{-3} $(s)^{-1}$.

Solubility measurements† for C_6H_6 in water at 24°C gave $H_p = 4.85 \times 10^3$ atm(cm)^3/(g mol) [0.491(kPa)(m^3)/kg mol)]. The ratio of interfacial area to bubble volume for the 1.8 mm bubbles is

$$\frac{a_g}{V_B} = \frac{\pi d_b^2}{\pi d_b^3/6} = 33.4 \text{ cm}^{-1} (33.4 \times 10^2 \text{ m}^{-1})$$

† D. M. Misic and J. M. Smith, op. cit.

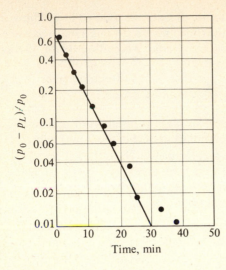

Figure 10-9 Fraction of benzene absorbed in water from helium–benzene gas bubbles.

The bubble volume itself is

$$V_B = \frac{QL}{v_B V_L} = \frac{50.7(1/60)6.0}{22.5(50)} = 4.51 \times 10^{-3}$$

Then

$$a_g = 33.4(4.51 \times 10^{-3}) = 0.151 \text{ cm}^{-1} \text{ (15.1 m}^{-1})$$

With these values, Eq. (G) gives for α

$$\alpha = \frac{0.151(82.05)(273 + 24)6.0}{4.85 \times 10^3(4.51 \times 10^{-3})(22.5)} k_L$$

$$= 44.9 k_L$$

From the known slope of the line in Fig. 10-9

$$\text{Slope} = -2.8 \times 10^{-3} = -\frac{k_L a_g(1 - e^{-\alpha})}{\alpha}$$

$$= -\frac{0.151 k_L(1 - e^{-44.9 k_L})}{44.9 k_L}$$

or

$$1 - e^{-44.9 k_L} = \frac{44.9(2.8 \times 10^{-3})}{0.151} = 0.833$$

$$k_L = 4.0 \times 10^{-2} \text{ cm/s} \text{ (4.0} \times 10^{-4} \text{ m/s)}$$

For comparison, from Eq. (10-46)

$$k_L = 0.31 \frac{[(1.0(0.01)980/1^2)]^{1/3}}{[(0.01/2 \times 10^{-5})]^{2/3}}$$

$$= 1.0 \times 10^{-2} \text{ cm/s} \text{ (1.0} \times 10^{-4} \text{ m/s)}$$

where the diffusivity of benzene in liquid water is estimated to be 2.0×10^{-5} cm²/s $(2.0 \times 10^{-9}$ m²/s$)$. Equation (10-46) is based upon data in an unstirred liquid which may explain part of the difference between the two values. However, the difference also indicates the difficulty in obtaining accurate values of mass-transfer coefficients in difficult-to-characterize physical situations such as bubbles rising through a liquid phase.

10-8 Mass-Transfer Coefficients: Liquid to Particle (k_c)

The relative velocity between the particles and the liquid determines the extent to which convection increases the Sherwood number above that for stagnant conditions, i.e., above 2. As mentioned, this relative velocity is low in slurries, because the particles are so small that they tend to move with the liquid. In agitated slurries the relative velocity is due primarily to the shearing action induced by the agitator. The basis for correlating k_c as a function of agitation speed and particle size is Kolmogoroffs' theory† of isotropic turbulence. According to this theory the Reynolds' number (Re) is defined in terms of the energy dissipation rate [see Eq. (10-43)]. If the eddy size ζ is greater than the particle diameter

$$\text{Re} = \left(\frac{\sigma d_p^4}{v^3}\right)^{1/2} \qquad \zeta > d_p \qquad (10\text{-}47)$$

† D. M. Levins and J. R. Glastonbury, *Chem. Eng. Sci.,* **27**, 537 (1972).

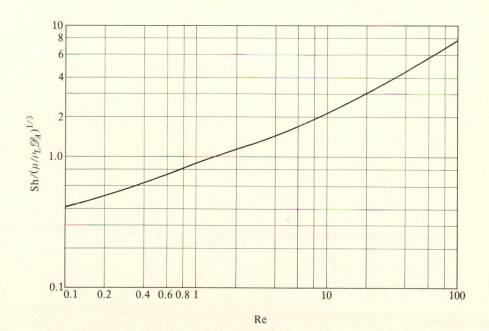

Figure 10-10 Correlation of fluid-particle mass-transfer coefficients in slurries.

or otherwise,

$$Re = \left(\frac{\sigma d_p^4}{v^3} \right)^{1/3} \qquad \zeta < d_p \qquad (10\text{-}48)$$

where ζ is a function of σ and the kinematic viscoscity v. Thus

$$\zeta = \left(\frac{v^3}{\sigma} \right)^{1/4} \qquad (10\text{-}49)$$

The experimental data† ‡ for k_c are well correlated by plotting the Sherwood number $(k_c d_p / \mathcal{D}_A)$ vs. the Reynolds number, defined by Eqs. (10-47) or (10-48). Figure 10-10 shows the results.

In three-phase slurries, both bubble-liquid and liquid-particle mass transfer can influence the global rate, as indicated in the following Example.

Example 10-7 The reaction

$$SO_2(g) + \tfrac{1}{2}O_2(g) \to SO_3(g) + H_2O(l) \to H_2SO_4(aq)$$

is catalyzed by activated carbon. Hence, it provides a means of studying mass transfer in a three-phase aqueous slurry of carbon particles. During the steady-state period the rates are calculated [Eq. (4-5)] from the measured concentrations of SO_2 in the gas streams entering (C_{gi}) and leaving (C_{go}) the reactor. Thus, the rate of disappearance of oxygen is $(Q = $ volumetric gas flow rate; $V_L = $ volume of liquid):

$$r_{O_2} = \frac{1}{2} r_{SO_2} = \frac{1}{2} \frac{Q}{V_L} (C_{gi} - C_{go})_{SO_2} \qquad (A)$$

Such measurements have been made§ at 25°C and 1 atm pressure with gas feeds containing 1.2 to 2.3% SO_2 and 21% O_2 in helium. Since oxygen is slightly soluble in water, it was the limiting reactant. However, the O_2 concentration in the bubbles is greatly in excess of the stoichiometric requirement, and, therefore, is essentially constant as the bubbles rise through the liquid. The results showed that at these conditions the rate was first-order in oxygen and zero order in SO_2. Since the oxygen concentration does not change as the bubbles rise through the liquid, the rate is constant throughout the slurry. Then Eqs. (10-40) and (10-41) which were developed for one vertical position, may be applied to the whole slurry.

Rates of disappearance of oxygen calculated from Eq. (A) are given in Table 10-5 for two sizes (d_p) and for several concentrations (m_s) of carbon particles. The gas bubble size was about 3 mm.

Rates were also measured by enclosing the carbon particles in stationary baskets made from stainless steel screen. In this case the relative velocity between particles and liquid could be made very high by operating the stirrer

---

† P. Harriott, *AIChE J.*, **8**, 93 (1962).
‡ P. L. T. Brian, H. B. Hales, and T. K. Sherwood, *AIChE J.*, **15**, 419, 727 (1969).
§ Hiroshi Komiyama and J. M. Smith, *AIChE J.*, **21**, 670 (1975).

Table 10-5 Rates of SO_2 oxidation in aqueous slurries of activated carbon at 25°C

Carbon particle size, d_p, mm	Particle concentration, m_s g/(cm³ of water)	Rate, $r_{O_2} \times 10^9$ mol/(s)(cm³ of liquid)
0.099	0.0131	8.4
0.099	0.00560	4.22
0.099	0.00222	1.78
0.030	0.0370	21.0
0.030	0.0111	10.4
0.030	0.00560	7.44
0.030	0.00278	4.11
0.030	0.00139	2.33

at a high speed. Under these conditions the resistance to mass transfer between liquid and particle approaches zero. For $m_s = 0.0333$ g/(cm³ of water) and for $d_p = 0.542$ mm, the rates without and with particles held in the baskets were:

$$r_{O_2} = 5.05 \times 10^{-9} \text{ g mol/(s)(cm}^3 \text{ of water)}$$
$$\text{(no baskets)}$$

$$r_{O2,b} = 6.08 \times 10^{-9} \text{ g mol/(s)(cm}^3 \text{ of water)}$$
$$\text{(particles in baskets)}$$

A. Calculate the mass-transfer coefficients k_L and k_c
B. For the conditions given, what fraction of the total resistance in the global rate is due to (a) gas-liquid mass transfer, and (b) liquid-particle mass transfer?

SOLUTION

A. The two sets of rate data are sufficient to evaluate separately the three contributions to the global rate indicated by Eq. (10-40). Combining Eqs. (10-40) and (10-41) to eliminate $k_0 H$ gives

$$\frac{(C_L)_{eq}}{r_{O_2}} = \frac{1}{a_g k_L} + \frac{1}{a_c}\left[\frac{1}{k_c} + \frac{1}{k}\right] \tag{B}$$

where a_c is the outer surface area of the particles per unit volume of liquid. For spherical particles

$$a_c = \frac{\pi d_p^2}{\pi d_p^3/6}\left(\frac{1}{\rho_p}\right)m_s = \frac{6m_s}{d_p \rho_p} \tag{C}$$

where ρ_p is the density of the carbon particles ($\rho_p = 0.80$ g/cm³). Substituting Eq. (C) in (B) yields

$$\frac{(C_L)_{eq}}{r_{O_2}} = \frac{1}{a_g k_L} + \frac{d_p \rho_p}{6m_s}\left[\frac{1}{k_c} + \frac{1}{k}\right] \tag{D}$$

Since $(C_g)_{O_2}$ is constant, $\mathbf{r}_{O_2}$ is the same throughout the slurry. Hence, Eqs. (10-40) and (D), which were derived for a fixed position in the slurry, are applicable throughout the liquid. Henry's law constant for oxygen in water at 25°C is $H = 35.4$ [g mol/(cm^3 of gas)]/[g mol/cm^3 of liquid]. Then

$$(C_L)_{eq} = \left(\frac{C_g}{H}\right)_{O_2} = \frac{0.21}{(82)(298)(35.4)} = 2.43 \times 10^{-7} \text{ g mol/cm}^3$$

Equation (D) indicates that a plot of $1/r_{O_2}$ vs. $1/m_s$ for constant $(C_L)_{eq}$ and d_p should be a straight line with an intercept equal to $1/(C_L)_{eq}a_g k_L$. As Fig. 10-11 shows, the data in Table 10-5 when so plotted do establish straight lines with an intercept of about 3.6×10^7 (cm^3(s)/(g mol)). Then from Eq. (D)

$$\left(\frac{1}{\mathbf{r}_{O_2}}\right)_{\text{intercept}} = 3.6 \times 10^7 = [a_g k_L (C_L)_{eq}]^{-1} \tag{E}$$

For spherical bubbles of diameter d_b

$$a_g = \frac{\pi d_b^2}{\pi d_b^3/6} V_B = \frac{6}{d_b} V_B \tag{10-50}$$

where V_B is the gas holdup, as bubble volume per unit volume of liquid. Holdup measurements (measuring the increase in volume when gas is

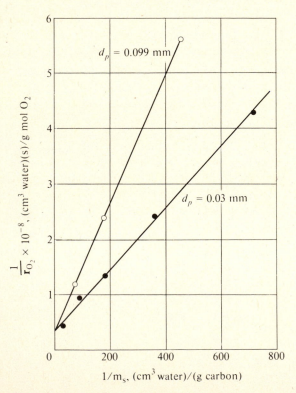

Figure 10-11.

bubbled through the slurry) gave $V_B = 0.070$ cm^3 gas/(cm^3 of liquid). Hence

$$a_g = \frac{6}{0.3}(0.07) = 1.4 \text{ cm}^{-1}$$

Substituting the values of $(C_L)_{eq}$ and a_g in Eq. (E) gives

$$k_L = \frac{1}{(3.6 \times 10^7)(2.43 \times 10^{-7})1.4}$$

$$= 0.08 \text{ cm/s}$$

The rate (r_{O_2}) values for runs in which the particles were stationary (held in baskets) can be used to evaluate the liquid-to-particle coefficient k_c. If $k_c a_c$, in Eq. (D), is very high when the particles were held in baskets,

$$\frac{(C_L)_{eq}}{r_{O_2, b}} = \frac{1}{a_g k_L} + \frac{d_p \rho_p}{6 m_s}\left(\frac{1}{k}\right) \tag{F}$$

With the particles free to move, Eq. (D) is applicable. Subtracting Eq. (F) from (D) yields

$$(C_L)_{eq}\left(\frac{1}{r_{O_2}} - \frac{1}{r_{O_2, b}}\right) = \frac{d_p \rho_p}{6 m_s}\left(\frac{1}{k_c}\right) \tag{G}$$

Using the rate data for the $d_p = 0.542$ mm particles, Eq. (G) gives

$$\frac{1}{k_c} = \frac{6 m_s (C_L)_{eq}}{d_p \rho_p}\left(\frac{1}{r_{O_2}} - \frac{1}{r_{O_2, b}}\right)$$

$$= \frac{6(0.0167)(2.43 \times 10^{-7})}{0.0542(0.80)}\left(\frac{1}{5.05 \times 10^{-9}} - \frac{1}{6.08 \times 10^{-9}}\right)$$

$$= 37 \text{ s/cm}$$

or

$$k_c = 0.027 \text{ cm/s}$$

For comparison, an estimate can be made of k_c from Fig. 10-10. For the particular bubble and stirrer arrangement used by Komiyama† the power number was 9.5 [from Eq. (10-44)]. Then σ calculated from Eq. (10-43) was 2.43×10^4 erg/(s)(g); ϕ was 0.67. From Eq. (10-49) $\zeta = 0.024$ mm. Since $\zeta < d_p$, the Reynolds number is given by Eq. (10-48).

$$\text{Re} = \left[\frac{2.43 \times 10^4 (0.0542)^4}{(0.01)^3}\right]^{1/3} = 58$$

From Fig. 10-10

$$\text{Sh} = \frac{k_c d_p}{\mathscr{D}_A} = 5.6\left(\frac{\mu_L}{\rho_L \mathscr{D}_A}\right)^{1/3}$$

† Op. cit.

The diffusivity of oxygen in liquid water at 25°C is about 2.6×10^{-5} cm^2/s, which corresponds to a Schmidt number, $\mu_L/\rho_L \mathscr{D}_A$, of about 340. Then

$$k_c = 5.6(340)^{1/3} \frac{2.6 \times 10^{-5}}{0.0542} = 0.019 \text{ cm/s}$$

which is reasonably close to the value of 0.027 cm/s obtained from the rate data.

B. The resistances to the global rate can be estimated from the values of the three terms in Eq. (D). If it is assumed that the particle size does not affect bubble-to-liquid transport (as indicated by the common intercept in Fig. 10-11) the first term is (from Part A)

$$\frac{1}{a_g k_L} = \frac{1}{1.4(0.08)} = 8.9 \text{ s}$$

The liquid-to-particle resistance is

$$\frac{d_p \rho_p}{6 m_s} \left(\frac{1}{k_c} \right) = \frac{0.0542(0.80)}{6(0.0333)(0.027)} = 8.0 \text{ s}$$

The reaction resistance is obtained from the rate and Eq. (D):

$$\frac{d_p \rho_p}{6 m_s} \left(\frac{1}{k} \right) = \frac{(C_L)_{eq}}{r_{O_2}} - 8.9 - 8.0$$

$$= \frac{2.43 \times 10^{-7}}{5.05 \times 10^{-9}} - 8.9 - 8.0 = 31.2 \text{ s}$$

Hence, the fractional resistances are: mass transfer, bubble-to-liquid = $(8.9/48.1) \times 100 = 18\%$, mass transfer, liquid-to-particle = $(8.0/48.1) \times 100 = 17\%$, reaction resistance = $(31.2/48.1) \times 100 = 65\%$.

10-9 The Effect of Mass Transfer on Observed Rates

Equation (10-40) shows that the relative importance of diffusion resistances from bubbles to bulk liquid and from liquid to catalyst particle depends on the area ratio. If a large concentration of very small particles is used, a_c/a_g is so large that only the first term is important. Then Eq. (10-40) reduces to

$$\frac{1}{k_o H} = \frac{a_c}{a_g} \frac{1}{k_L}$$

and the rate is, according to Eq. (10-41)

$$r_v = k_L a_g (C_L)_{eq} \tag{10-51}$$

Calderbank et al.† studied the hydrogenation of ethylene using a large concentration of Raney nickel particles in a slurry reactor in order to approach these

---

† P. H. Calderbank, F. Evans, R. Farley, G. Jepson, and A. Poll, "Proceedings of a Symposium on Catalysis in Practice," p. 66, Institution of Chemical Engineers (London), 1963.

conditions. Analysis of the data† indicated that the controlling step was the mass transfer of hydrogen from gas bubble to bulk liquid.

Equation (10-51) indicates that the important parameters for this case are k_L and the gas-liquid interfacial area a_g. The latter depends on the size and concentration of bubbles in the slurry. Alternately, it may be expressed in terms of the bubble diameter and gas holdup by Eq. (10-50) (see Example 10-7).

Koide et al.‡ give the following equation for the diameter of bubbles formed over a porous plate:

$$d_b = 1.35 \frac{u/g\,\delta}{(\delta u^2 \rho_L/\sigma_L)^{1/2}} \left(\frac{\sigma_L\,\delta}{g\rho_L}\right)^{1/3} \tag{10-52}$$

where d_b = bubble diameter, cm

$\quad u$ = gas velocity through the porous plate

$\quad \delta$ = pore diameter of porous plate

$\quad \sigma_L$ = surface tension of liquid

$\quad \rho_L$ = density of liquid

Massimilla, Calderbank, and others also have reported experimental data on bubble sizes.§

At the other extreme, a large concentration of small gas bubbles (large a_g) might be combined with a low concentration of relatively large active catalyst particles (low a_c) and poor agitation (low k_c). Then Eq. (10-40) reduces to

$$\frac{1}{k_o H} = \frac{1}{k_c} + \frac{1}{k} \approx \frac{1}{k_c} \qquad \text{for very active catalyst} \tag{10-53}$$

The rate is, from Eq. (10-41),

$$\mathbf{r}_v = k_c a_c (C_L)_{eq} \tag{10-54}$$

Here the surface area of the catalyst particles is of primary importance, along with k_c. For spherical particles a_c is given by Eq. (C) of Example (10-7); that is,

$$a_c = \frac{6m_s}{d_p \rho_p} \tag{10-55}$$

Under many conditions a_c/a_g would be greater than unity. This follows from the fact that bubble diameters will be of the order of 1 mm, while catalyst particles will be in the order of 0.1 mm (100 microns or 140 mesh). Then for spherical shapes a_c/a_g would be equal to $d_b/d_p = 10$, *for equal volumes of bubbles and particles in the slurry.* If the gas holdup were 10 times larger than the catalyst volume, a_c/a_g would be reduced to unity, but this ratio of volumes is infrequent. For example, consider the conditions of Example 10-7. For $m_s = 0.0333$ g/cm^3 and

† D. MacRae, doctoral dissertation, University of Edinburgh, Edinburgh, 1956.

‡ K. T. Koide, T. Hirahara, and H. Kubata, *Chem. Eng.* (Japan), **30**, 712 (1966).

§ L. Massimilla, A. Solimando, and E. Squillace, *Brit. Chem. Eng.*, **6**, 232 (1961); P. H. Calderbank, F. Evans, and J. Rennie, *Proc. Intern. Symp. Distn., Suppl. Trans. Inst. Chem. Engrs. (London),* p. 51 (1960).

$d_p = 0.0542$ mm, which are very large particles for a slurry, Eqs. (10-50) and (10-55) give

$$\frac{a_c}{a_g} = \frac{m_s/d_p\rho_p}{V_B/d_b} = \frac{0.0333/0.0542(0.8)}{0.07/0.3} = 3.3$$

Except for large particles at low concentrations (low m_s), the effect of k_L on the global rate will be greater than that of k_c. The relative importance of the total mass-transfer process, from gas bubble to particle, on the global rate also will depend on the intrinsic rate, as illustrated in Example 10-8.

Example 10-8 Coenan† has reported rates of hydrogenation of sesame seed oil with a nickel-on-silica catalyst in a slurry reactor. Hydrogen was added at the bottom of a small cylindrical vessel equipped with stator and stirrer blades. Initial rates of reaction were measured as function of catalyst concentration at 180°C, a stirrer speed of 750 rpm, atmospheric pressure, and a hydrogen rate of 60 liters/h. The data, converted to global rates in terms of g mol/(min)(cm³ oil), are given in Table 10-6 (based on an oil density of 0.9 g/cm³). Estimate $k_L a_g/H$ from these data. Comment on the importance of the resistance of hydrogen to solution in the oil and estimate what the reaction rate would be if this resistance could be eliminated for a catalyst concentration of 0.07% Ni in oil.

SOLUTION Equation (10-40) is applicable. Combining it with Eq. (10-38) gives

$$r_v = \frac{a_c C_g}{H[(a_c/a_g)(1/k_L) + 1/k_c + 1/k]} = \frac{C_g}{H[(1/a_g k_L) + (1/a_c)(1/k_c + 1/k)]} \quad \text{(A)}$$

Since the gas rate and stirrer speed are constant, k_c, k_L, and a_g will not vary from run to run. Because the temperature is constant, k will also be invariant. The catalyst consists of small particles of silica containing nickel, and its concentration is changed by adding more such particles. The activity per particle is constant. If we assume that the particles do not agglomerate, a_c will

† J. W. E. Coenan, The Mechanism of the Selective Hydrogenation of Fatty Oils, in J. H. deBoer (ed.), "The Mechanism of Heterogeneous Catalysis," p. 126, Elsevier Publishing Company, New York, 1960.

Table 10-6

Catalyst particle concn, % Ni in oil	$r_v \times 10^5$	C_g/r_v, min
0.018	5.2	0.52
0.038	8.5	0.32
0.07	10.0	0.27
0.14	12.0	0.22
0.28	13.6	0.20
1.0	14.6	0.18

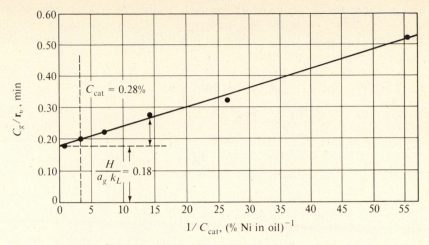

Figure 10-12 Effect of catalyst concentration on rate in a slurry reactor.

be directly proportional to the catalyst concentration, and Eq. (A) may be written

$$\frac{C_g}{r_v} = \frac{H}{a_g k_L} + \frac{AH}{C_{cat}}\left(\frac{1}{k_c} + \frac{1}{k}\right) \tag{B}$$

where A is the proportionality constant. This expression suggests that the data plotted as C_g/r_v vs. $1/C_{cat}$ would yield a straight line. Assuming the ideal-gas law, we have

$$C_g = \frac{p}{R_g T} = \frac{1}{82(273 + 180)} = 2.7 \times 10^{-5} \text{ g mol/cm}^3$$

The data in Table 10-6, when plotted in this way (Fig. 10-12), do yield a linear result. Extrapolation to $1/C_{cat} = 0$ gives an intercept value of 0.18 min. According to Eq. (B),

$$0.18 = \frac{H}{a_g k_L}$$

or

$$\frac{k_L a_g}{H} = 5.7 \text{ min}^{-1}$$

In Fig. 10-12 the resistance to solution of hydrogen in the slurry is represented by the horizontal dashed line. The total resistance is the ordinate of the solid line. At low catalyst concentration the combined resistance of diffusion to the particle and chemical reaction on the catalytic surface is large, although it does not determine the rate by itself. At high concentrations the resistance to transfer of hydrogen from bubble to the bulk liquid dominates the rate. In fact at $C_{cat} = 0.28\%$, 0.18/0.20, or 90%, of the total resistance is for this step. The results show that a_c/a_g increases enough, as the catalyst concentration increases to 0.28%, that the first term in Eq. (10-40) dominates the whole quantity. The results also indicate that increases in C_{cat} beyond 0.28% would do little to speed up the reaction.

If the resistance to dissolving hydrogen could be eliminated at $C_{cat} = 0.07\%$ then $C_g/\mathbf{r}_v$ would be, from Fig. 10-12,

$$\frac{C_g}{\mathbf{r}_v} = \frac{AH}{C_{cat}}\left(\frac{1}{k_c} + \frac{1}{k}\right) = 0.27 - 0.18 = 0.09 \text{ min}$$

$$\mathbf{r}_v = \frac{2.7 \times 10^{-5}}{0.09} = 30 \times 10^{-5} \text{ g mol/(min)(cm}^3 \text{ oil)}$$

This value is threefold that of the observed (global) rate. Decreasing the bubble size would be a step in this direction. Increasing the hydrogen-flow rate at constant bubble size would have the same effect, since this would increase gas volume and a_g. Better agitation would also reduce the bubble-to-liquid resistance.

TRICKLE-BED REACTORS†

In downflow of liquid and gas over a fixed bed of catalyst particles (see Figs. 1-6b and 10-16), the nature of the flow depends in a complex way upon the liquid and gas flow rates. Figure 10-13 shows *approximate* boundaries for the different flow regimes. At low liquid and gas mass velocities the gas phase is continuous and the liquid falls in rivulets from one particle to the next (trickle-flow regime). At high liquid rates and low gas rates the liquid phase is continuous and the gas moves in bubbles (dispersed bubble regime). With very high gas rates and low liquid rates, the liquid falls in droplets through the gas (spray regime). Finally, with high rates of both liquid and gas, the two phases fall in slugs (pulsed-flow regime). The boundary curves in Fig. 10-13 are approximate and more detailed information and discussion about factors that affect flow behavior have been published.‡–††

Mass-transfer rates are available primarily in the trickle-flow regime and our discussion is limited to this region. For trickle flow the concentration profile for a reactant in the gas phase is sketched in Fig. 10-14. This profile is similar to that in Fig. 10-8 for a slurry reactor. However, in Fig. 10-14 a small part of the particle (labeled gas-covered) is shown without a liquid rivulet. Reaction studies in trickle

---

† For reviews of literature on trickle-bed reactors see: Satterfield, C. N., *AIChE J.*, **21**, 209 (1975); S. Goto, J. Levec, and J. M. Smith, *Cat. Rev.—Sci. Eng.*, **15**, 187 (1977); Y. T. Shah, "Gas-Liquid-Solid Reactor Design," pp. 180–229, McGraw-Hill Book Company, New York, 1979.

‡ E. Talmor, *AIChE J.*, **23**, 868 (1977).

§ Y. T. Sato, H. Hirose, F. Takahashi, M. Toda, and Y. Hashigughi, *J. Chem. Eng. (Japan)*, **6**, 315 (1973).

¶ J. C. Charpentier and M. Favier, *AIChE J.*, **21**, 1213 (1975).

†† T. S. Chou, F. L. Worley, Jr., and D. Luss, *Ind. Eng. Chem. Proc. Des. Dev.*, **16**, 424 (1977).

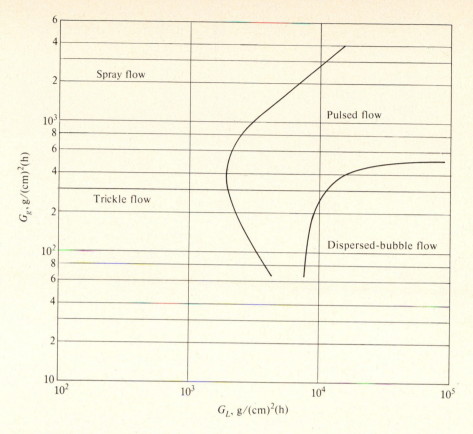

Figure 10-13 Flow regimes in trickle-bed reactors (cocurrent downflow).

beds have demonstrated that such surfaces exist †‡§ at low liquid rates. On this surface there would be much less mass-transfer resistance for a gaseous reactant. Accounting for the effect of "gas-covered" surface complicates the design of trickle beds. Here and in Chap. 13 it will be supposed that the liquid flow rate is sufficient to cover all of the outer surface. In Secs. 10-10 to 10-12 that follow, we examine mass-transfer effects. Then in Chap. 13 the objective is to combine these transport resistances with the intrinsic rate, and a reactor model, to predict the conversion in the whole reactor.

In trickle beds the external mass-transfer limitations are determined by the two volumetric coefficients: gas-to-liquid, $k_L a_g$, and liquid-to-particle, $k_c a_c$. The areas a_g, a_c refer to the effective mass-transfer surface per unit volume of empty reactor. Since they are difficult to determine experimentally correlations are normally reported in terms of the products $k_L a_g$ or $k_c a_c$.

† M. Herskowitz, R. G. Carbonell, and J. M. Smith, *AIChE J.*, **25**, 272 (1979).
‡ W. Sedriks and C. N. Kenny, *Chem. Eng. Sci.*, **28**, 559 (1973).
§ M. Hartman and Robert W. Coughlin, *Chem. Eng. Sci.*, **27**, 867 (1972).

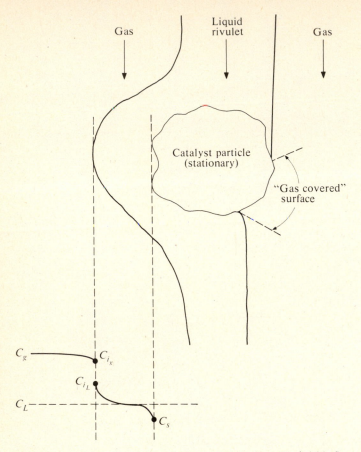

Gas   Liquid rivulet   Gas

Catalyst particle (stationary)

"Gas covered" surface

C_g C_{i_g}

C_{i_L}

C_L

C_s

Figure 10-14 Concentration profiles in a trickle-bed reactor (trickle-flow regime).

10-10 Mass-Transfer Coefficients: Gas to Liquid $(k_L a_g)$

In trickle beds the gas phase is often either a nearly pure component (e.g., hydrogenations) or a slightly soluble gas (oxidations). Hence, the most important coefficient for gas-liquid transport is the liquid side value $k_L a_g$. However, correlations are available for the gas-side coefficient $k_g a_g$.[†][‡][§]

Two types of correlations have been used for $k_L a_g$: one[†][‡][¶] relates $k_L a_g$ to the pressure drop for two-phase flow in the reactor, and the other is in terms of the flow velocities. Illustrative of the latter type is the *dimensional* equation[††]

$$\frac{k_L a_g}{\mathscr{D}_A} = \alpha_L \left(\frac{G_L}{\mu_L}\right)^{\eta_L} \left(\frac{\mu_L}{\rho_L \mathscr{D}_A}\right)^{1/2} \tag{10-56}$$

[†] L. P. Reiss, *Ind. Eng. Chem. Proc. Des. Dev.*, **6**, 486 (1967).

[‡] A. Gianetto, V. Specchia, and G. Baldi, *AIChE J.*, **19**, 916 (1973).

[§] S. Goto, J. Levec, and J. M. Smith, *Ind. Eng. Chem. Proc. Des. Dev.*, **14**, 473 (1975).

[¶] J. C. Charpentier, *Chem. Eng. J.*, **11**, 161 (1976).

[††] S. Goto and J. M. Smith, *AIChE J.*, **21**, 706 (1975).

where α_L is about 7 $(cm)^{n_L-2}$, and $\eta_L = 0.40$ for granular catalyst particles
(0.054 and 0.29 cm in diameter)

$\mathscr{D}_A$ = molecular diffusivity of the diffusing component (cm^2/s),

G_L = superficial mass velocity of the liquid, g/(cm^2)(s)

μ_L = liquid viscosity, g/(cm)(s)

$k_L a_g$ = volumetric liquid-side mass-transfer coefficient (gas-liquid), s^{-1}

This correlation applies for the trickle-flow regime; much higher values can be obtained at higher liquid and gas rates. Note that Eq. (10-56) does not involve the gas rate.

10-11 Mass-Transfer Coefficients: Liquid to Particle $(k_c a_c)$

Data for mass transfer between fluid and particle have been determined primarily by measuring rates of dissolution of nonporous particles such as benzoic acid and β-naphthol.[†]–[¶] A correlation for the trickle-flow regime which includes much of the data is [††]

$$j_D = 1.64(\text{Re}_L)^{-0.331} \tag{10-57}$$

(for $0.2 < \text{Re}_L < 2400$)

where

$$j_D = \frac{k_c a_c}{u_L a_t}\left(\frac{\mu_L}{\rho_L \mathscr{D}_A}\right)^{2/3} \tag{10-58}$$

and u_L = superficial velocity of liquid

a_t = total external area of particles per unit volume of reactor

$\text{Re}_L = (d_p u_L \rho_L/\mu_L)$

Satterfield et al.[‡‡] has obtained data and proposed correlations for $k_c a_c$ for both the trickle- and pulse-flow regimes. In agreement with Eq. (10-57), their results for trickle-flow were nearly independent of gas flow rate.

10-12 Calculation of Global Rate

In Sec. 10-8 (e.g., Example 10-7 involving a slurry reactor) a method was illustrated for extracting mass-transfer coefficients and the intrinsic reaction rate from laboratory data. In the present section we examine the reverse process for a trickle-bed reactor. That is, we want to calculate what is occurring at a particular location in a trickle-bed from intrinsic kinetics and mass-transfer coefficients. The concept of a *global* rate is not particularly helpful in reactor design when there are

[†] D. W. van Krevelan and J. T. C. Krekels, *Rec. Trav. Chim.*, **67**, 512 (1948).

[‡] Y. Sato, T. Hirose, F. Takahashi, and M. Toda, *PACHEC*, Paper 8.3, p. 1187 (1972).

[§] T. Hirose, M. Toda, and Y. Sato, *J. Chem. Eng. (Japan)*, **7**, 187 (1974).

[¶] S. Goto, J. Levec, and J. M. Smith, *Ind. Eng. Chem. Proc. Des. Dev.*, **14**, 473 (1975).

[††] A. Dharwadkar and N. D. Sylvester, *AIChE J.*, **23**, 376 (1977).

[‡‡] C. N. Satterfield, M. W. van Eek, and G. S. Bliss, *AIChE J.*, **24**, 709 (1978).

two or more phases in continuous flow. This is because a global rate expressed in terms of bulk properties of one phase does not describe what is occurring in other moving phases. For design, it is most direct to write conservation equations for each flowing phase and this procedure, which does not involve the global rate, is followed in Chap. 13 for trickle-bed reactors. However, in a trickle-bed where the liquid completely covers the catalyst particles, a global rate can be evaluated in terms of bulk liquid properties. This rate may be compared with the mass-transfer rate between gas and liquid streams. It is helpful to do this since the comparison tells us whether the reactant is being depleted or augmented in the liquid and gas streams. The global rate is obtained by equating, at steady state, the rate of mass transfer of reactant from the liquid to the catalyst particle to the rate of reaction. For a unit volume of reactor the equality may be written

$$\mathbf{r}_v = (k_c a_c)(C_L - C_s) = \rho_B \mathbf{r}(C_s, T_s) \tag{10-59}$$

where C_L is the bulk concentration in the liquid and $\mathbf{r}(C_s, T_s)$ is the intrinsic rate,[†] per unit mass of catalyst, evaluated at the reactant concentration and temperature at the catalyst surface. When the temperature T_s is known (or when T_b is known and there is no heat-transfer resistance so that $T_s = T_b$) as well as C_L, Eq. (10-59) can be solved for C_s. Then this result, substituted in the function $\mathbf{r}(C_s, T_s)$, gives the global rate corresponding to the bulk values C_L and T_b. The procedure is analogous to that illustrated in Examples 10-2 and 10-4 for fixed-bed (two-phase) reactors.

Note that the global rate is obtained without considering the gas phase. This is because the assumption is made that the liquid completely covers the outer surface of the particle. No reaction can occur without transfer of reactant from liquid to particle. In order to evaluate the transfer rate of a gaseous reactant from gas to liquid, mass-transfer coefficient $(k_L a_g)$ would be necessary, as indicated in Example 10-9, part B.

In Example 10-9, part A, the global rate is calculated for a system in which one reactant can exist in both gas and liquid phases while the other is only in the liquid. The kinetics are not first order but of the Langmuir-Hinshelwood form.

Example 10-9 The oxidation of dilute aqueous solutions of acetic acid was studied[‡] in a trickle-bed reactor using a commercial, iron oxide catalyst. Air and solutions of acetic acid saturated with oxygen (by bubbling air through the solution at atmospheric pressure) flowed at 252°C and 67 atm downward over 0.0541 cm catalyst particles (packed to a depth of 2.2 cm) in a 2.54 cm ID reactor. The flow rates are $Q_L = 0.66$ cm^3/s and $Q_g = 3.5$ cm^3/s. At the reactor entrance the concentration C_{L, O_2} of oxygen in the liquid is 2.40×10^{-7} g mol/cm^3, that for acetic acid is $C_{L, HA} = 33.7 \times 10^{-7}$ g mol/cm^3, and the gas is air, saturated with water at 252°C. The bulk density of the catalyst bed is $\rho_B = 1.17$ g/cm^3, while the particle density is, $\rho_p = 2.05$ g/cm^3.

---

† For first-order kinetics, $\mathbf{r}(C_s, T_s) = k(T_s)C_s$.
‡ Janez Levec and J. M. Smith, *AIChE J.*, **22**, 159 (1976).

For the reaction, $CH_3COOH(aq) + 2O_2(aq) \rightarrow 2CO_2(aq) + 2H_2O$, the intrinsic rate for the disappearance of oxygen is given by

$$r_{O_2}(\text{g mol/(g catalyst)(s)}) = \frac{9.9 \times 10^9[\exp(-21{,}000/R_g T)]C_{s,HA}C_{s,O_2}^{1/2}}{1 + 7.2 \times 10^5 \, C_{s,HA}} \quad \text{(A)}$$

where the concentrations are those at the outer surface of the catalyst particle and T is given in degrees Kelvin. Henry's law constant for oxygen in liquid water at 252°C is 0.89 [g mol/(cm^3 gas at 25°C, 1 atm)]/[g mol/cm^3 liquid]. Other physical properties at 252°C and 67 atm for liquid water are estimated to be:

$$\rho_L = 0.81 \text{ g/cm}^3$$

$$\mu_L = 1.18 \times 10^{-3} \text{ g/(cm)(s)}$$

$$(\mathscr{D}_{O_2}) = 2.1 \times 10^{-4} \text{ cm}^2/\text{s}$$

$$(\mathscr{D}_{HA})_L = 1.2 \times 10^{-4} \text{ cm}^2/\text{s}$$

A. From this information calculate the global rate at the reactor entrance.
B. Also calculate the rate of mass transfer of oxygen from the gas to the liquid.

SOLUTION The reaction occurs by oxygen and acetic acid in the liquid diffusing to the catalyst particles and reacting. Simultaneously, oxygen (but not nonvolatile acetic acid) is transferred from the air stream to replenish the oxygen in the liquid. Since oxygen is but slightly soluble in water, the overall transfer from gas to liquid may be safely assumed to be governed by the liquid-side coefficient $(k_L a_g)_{O_2}$. Then the rate of oxygen transfer, per unit volume of reactor, from *gas to liquid* is

$$N_{O_2}\left[\frac{\text{g mol}}{\text{(s)(cm)}^3}\right] = (k_L a_g)_{O_2}(C_{g,O_2}/H - C_{L,O_2}) \quad \text{(B)}$$

The concentration difference in this expression is large since C_{L,O_2} in the feed corresponds to saturation with air at atmospheric pressure, while C_{g,O_2} in the feed is for air at 67 atm.

Similarly, the rates of oxygen and acetic acid transfer from *liquid to particle* are given by the expressions

$$(r_v)_{O_2} = (k_c a_c)_{O_2}(C_{L,O_2} - C_{s,O_2}) \quad \text{(C)}$$

$$(r_v)_{HA} = (k_c a_c)_{HA}(C_{L,HA} - C_{s,HA}) \quad \text{(D)}$$

A. At steady state the rates of mass transfer from liquid to particle are equal to the rates of reaction. These expressions relate surface and bulk liquid concentrations. Equating Eqs. (A) and (C), and (A), and (D), yields

$$r_{O_2} = (k_c a_c)_{O_2}(C_{L,O_2} - C_{s,O_2})$$

$$= \rho_B \frac{9.9 \times 10^9[\exp(-21{,}000/R_g T)]C_{s,HA}C_{s,O_2}^{1/2}}{1 + 7.2 \times 10^5 C_{s,HA}} \quad \text{(E)}$$

$$r_{HA} = (k_c a_c)_{HA}(C_{L, HA} - C_{s, HA})$$

$$= \frac{1}{2} \rho_B \frac{9.9 \times 10^9 [\exp(-21,000/R_g T)] C_{s, HA} C_{O_2}^{1/2}}{1 + 7.2 \times 10^5 C_{s, HA}} \quad (F)$$

Equations (E) and (F) may be solved for the two surface concentrations after $(k_c a_c)_{O_2}$ and $(k_c a_c)_{HA}$ are estimated.

From the data given

$$G_L = \frac{0.66(0.81)}{(\pi/4)(2.54)^2} = 0.11 \text{ g/(cm)}^2(s)$$

$$u_L = \frac{0.66}{(\pi/4)(2.54)^2} = 0.13 \text{ cm/s}$$

$$\text{Re}_L = \frac{d_p u_L \rho}{\mu_L} = \frac{0.0541(0.13)(0.81)}{1.18 \times 10^{-3}} = 4.8$$

$$\mu_L/\rho_L \mathscr{D}_{O_2} = \frac{1.18 \times 10^{-3}}{0.81(2.10 \times 10^{-4})} = 6.9$$

$$\mu_L/\rho_L \mathscr{D}_{HA} = \frac{1.18 \times 10^{-3}}{0.81(1.2 \times 10^{-4})} = 12.1$$

for spherical particles,

$$a_t = \frac{\pi d_p^2 \rho_B}{(\pi d_p^3/6)\rho_p} = \frac{6}{0.0541}\left(\frac{1.17}{2.05}\right) = 63 \text{ cm}^2/\text{cm}^3$$

Substituting these values in Eqs. (10-57) and (10-58) gives

$$(k_c a_c)_{O_2} = j_D(u_L a_t)\left(\frac{\mu_L}{\rho_L D_{O_2}}\right)^{-2/3}$$

$$= 0.97(0.13)(63)(0.28) = 2.2 \text{ s}^{-1}$$

$$(k_c a_c)_{HA} = 0.97(0.13)(63)(0.19) = 1.5 \text{ s}^{-1}$$

With these mass-transfer coefficients, $T = 252 + 273 = 525$ K, and $C_{L, O_2} = 2.40 \times 10^{-7}$ g mol/cm^3 and $C_{L, HA} = 33.7 \times 10^{-7}$, all quantities in Eqs. (E) and (F) are known except $C_{s, HA}$ and C_{s, O_2}. Numerical solution yields

$$C_{s, O_2} = 2.35 \times 10^{-7} \text{g mol/cm}^3$$

$$C_{s, HA} = 33.6 \times 10^{-7} \text{g mol/cm}^3$$

The global rate is then obtained by using these surface concentrations in Eq. (A):

$$r_{O_2} = \frac{9.9 \times 10^9 [\exp(-21,000/525R)](33.6 \times 10^{-7})(2.35 \times 10^{-7})^{1/2}}{1 + 7.2 \times 10^5(33.7 \times 10^{-7})}$$

$$= 8.4 \times 10^{-9} \text{ g mol/(s)(g catalyst)}$$

In this example, the external liquid-particle mass-transfer resistance is relatively small; $(C_{L, O_2} - C_{s, O_2})/C_{L, O_2} = 0.05/2.4 = 0.02$ or 2%. For this particular case bulk concentrations could have been used in Eq. (A), to calculate the rate. This is because the intrinsic rate is slow.

B. Equation (B) is applicable for evaluating the oxygen-transfer rate from gas to liquid. The coefficient $(k_L a_g)_{O_2}$ is determined from Eq. (10-56).

$$(k_L a_g)_{O_2} = 2.1 \times 10^{-4}(7)\left(\frac{0.11}{1.18 \times 10^{-3}}\right)^{0.40}(6.9)^{1/2}$$

$$= 0.024 \text{ s}^{-1}$$

Note that this coefficient is about two orders of magnitude less than $k_c a_c$. Both $k_L a_g$ and $k_c a_c$ would be much higher at the higher liquid rates normally found in industrial trickle-bed reactors. The concentration of oxygen in the gas at 67 atm and 252°C is[†]

$$C_g = \frac{p_{O_2}}{R_g T} = \frac{0.21(26.2)}{82(525)} = 1.3 \times 10^{-4} \text{ g mol/cm}^3$$

Henry's law constant is $H_{O_2} = 0.89$ when the gas volume of pure oxygen is measured at 25°C and 1 atm. At $p_{O_2} = 0.21(26.2) = 5.50$ atm and $t = 252$°C

$$H_{O_2} = 0.89\frac{5.5}{1}\left(\frac{298}{252 + 273}\right)$$

$$= 2.78 \frac{\text{g mol/(cm}^3 \text{ gas at 5.5 atm, 252°C)}}{\text{g mol/cm}^3 \text{ of liquid}}$$

Then, from Eq. (B)

$$N_{O_2} = 0.024\left[\frac{1.3 \times 10^{-4}}{2.78} - 2.4 \times 10^{-7}\right]$$

$$= 0.024(4.59 - 0.02) \times 10^{-5} = 1.1 \times 10^{-6} \text{ g mol/(s)(cm)}^3$$

Per unit mass of catalyst this is $1.1 \times 10^{-6}/\rho_B = 9.4 \times 10^{-7}$ g mol/(s)(g catalyst). Since this rate is much greater than the rate of reaction, r_{O_2}, at the particle surface, the oxygen concentration in the liquid would tend to be augmented at this location in the reactor. This was observed by Leve[‡]; the oxygen concentration in the liquid increased in the direction of flow through the reactor. In a similar study[§] for the more rapid oxidation of formic acid, the intrinsic rate was much higher. For that situation the rate of depletion of oxygen from the liquid due to reaction was greater than the rate of transfer from gas to liquid in some parts of the reactor. It was observed that C_{L, O_2} decreased in the direction of flow through the reactor. Also in that case the concentration difference $(C_{L, O_2} - C_{s, O_2})$ was significant, in contrast to the results of this example for acetic acid oxidation.

[†] The vapor pressure of water at 252°C is 40.8 atm. Hence, the pressure of air at 67 atm total pressure (assuming saturation) is $67 - 40.8 = 26.2$ atm.

[‡] Op. cit.

[§] S. Goto and J. M. Smith, *AIChE J.*, **21**, 706 (1975).

Since the pertinent parameter that influences external mass transfer in trickle beds is the product of an area as well as a coefficient, for example, $k_c a_c$, the numerous flow and geometric characteristics are important as well as physical properties of the system. Hence, it is difficult to state firm conclusions about the significance of external mass transfer. At low gas and liquid rates with small catalyst particles, $k_L a_g$ seems to be less than $k_c a_c$ so that gas-liquid transport has more of an effect on the global rate than liquid-particle transport (see Example 10-9). For large particles where a_c is reduced, gas-liquid and liquid-particle transport may be of about equal influence.†

Countercurrent and cocurrent upflow of gas and liquid are alternate arrangements to trickle-beds operation. Measurements, at similar rates and for the same reactor and particles, for *cocurrent* flow show about the same $(k_c a_c)$ values at low flow rates. For upflow the coefficients increase with gas flow rate; for downflow gas rate seems to have little effect. While data at comparable conditions are meager, it appears that gas-liquid coefficients are higher for upflow than for downflow, particularly at high-flow rates.‡ For countercurrent flow $k_L a_g$ is less than in cocurrent operation, except at high flow rates.

We have not considered the effect of temperature differences between gas and liquid, and liquid and solid, on the global rate in trickle beds. In many cases these differences are negligible because the heat effect is small. In Example 10-9 and in related examples for oxidation of pollutants, the concentrations are so low that the heat evolved due to reaction is negligible, even though ΔH_c (heats of combustion) are high. Similarly, in hydrodesulfurization processes, the concentrations of sulfur components in the liquid are so low that near-isothermal operation results. However, unlike the situation in slurry reactors, a_c is small for large particles. Hence, for high concentrations and heats of reaction (e.g., hydro-cracking of petroleum fractions) significant temperature differences may exist between gas and particle.

PROBLEMS

10-1 To illustrate the effect of mass velocity on external diffusion in the oxidation of sulfur dioxide with a platinum catalyst, consider the following data, all at 480°C.

Mass velocity, lb/(h)(ft²)	Bulk partial pressures, p_b, atm			Global rate, r_p, g mol SO_2/(h)(g catalyst)
	SO_2	SO_3	O_2	
514	0.0601	0.00668	0.201	0.1346
350	0.0599	0.00666	0.201	0.1278
245	0.0603	0.00668	0.201	0.1215
147	0.0603	0.00670	0.201	0.0956

---

† V. Specchia, G. Baldi, and A. Gianetto, *Proc. 4th Int. Symp. Chem. React. Eng.*, p. 390, Heidelberg, April, 1976.

‡ S. Goto, J. Levec, and J. M. Smith, *Ind. Eng. Chem. Proc. Des. Dev.*, **14**, 473 (1975).

The reactor consisted of a fixed bed of $\frac{1}{8} \times \frac{1}{8}$-in cylindrical pellets. The pressure was 790 mmHg. The external area of catalyst particles was 5.12 ft^2/lb, and the platinum did not penetrate into the interior of the alumina particles. Calculate the partial-pressure difference between the bulk-gas phase and the surface of the catalyst for SO_2 at each mass velocity. What conclusions may be stated with regard to the importance of external diffusion? Neglect temperature differences.

10-2 The global rates of SO_2 oxidation have been measured with a platinum catalyst impregnated on the outer surface of $\frac{1}{8} \times \frac{1}{8}$-in. cylindrical pellets of Al_2O_3. The data were obtained in a differential reactor consisting of a 2-in.-ID tube packed with the catalyst pellets. The superficial mass velocity of the reaction mixture was 350 lb/(h)(ft^2). At constant conversion of 20% of a feed consisting of 6.5 mol % SO_2 and 93.5 mol % air, the rates are as shown below. The total pressure was 790 mmHg. Consider the properties of the mixture to be those of air, except for the Schmidt number for SO_2, which is 1.21.

t, °C	350	360	380	400	420	440	460	480	500
r_p, g mol SO_2/ (h)(g catalyst)	0.0049	0.00788	0.01433	0.02397	0.0344	0.0514	0.0674	0.0898	0.122

(a) If only the data at 460 to 500°C are used and external diffusion is neglected, what is the *apparent* activation energy? (b) By calculating $(C_b - C_s)_{SO_2}$ at various temperatures and using all the data above, estimate the true activation energy for the combined adsorption and reaction processes at the catalyst surface. Neglect temperature differences between bulk gas and catalyst surface.

10-3 Predict the global rate of reaction for the oxidation of SO_2 at bulk-gas conditions of 20% conversion at 480°C. Other conditions are as given in Prob. 10-2. The rate at the catalyst surface is to be calculated from Eqs. (G) and (H) of Example 9-2. Assume isothermal conditions. The constants in Eqs. (G) and (H) at 480°C are

$$A = -0.127 \ (\text{atm})^{3/2}(\text{h})(\text{g catalyst})/\text{g mol}$$
$$B = 15.3 \ (\text{atm})^{1/2}(\text{h})(\text{g catalyst})/\text{g mol}$$
$$K = K_p = 73 \ (\text{atm})^{-1/2}$$

Also calculate the ratio γ of the global rate and the rate evaluated at bulk conditions.

10-4 For the conditions of Prob. 10-2 predict the temperature difference between bulk gas and pellet surface at 350 and at 500°C. Comment on the validity of the isothermal assumptions made in Probs. 10-1 to 10-3.

10-5 (a) Estimate the *maximum* temperature difference $T_s - T_b$ for a gas-solid catalytic reaction for which

$$\Delta H = -20,000 \ \text{cal/g mol}$$
$$c_p = 8.0 \ \text{cal/g mole (K)}$$
$$p_t = 2 \ \text{atm}$$
$$T_b = 473 \ \text{K} (200°C)$$

The mole fraction of reactant in the bulk gas is 0.25. (b) If external diffusion resistance is not controlling, but $C_b - C_s = C_b/2$, what will be the value of $T_s - T_b$?

10-6 The reaction $H_2 + \frac{1}{2}O_2 \rightarrow H_2O$ occurs with a platinum-on-alumina catalyst at low temperatures. Following the work of Maymo and Smith,† suppose that the rate, in g mol O_2/(s)(g catalyst), at the catalyst surface is given by

$$\mathbf{r} = 0.327(p_{O_2})_s^{0.804} e^{-5,230/R_g T_s}$$

where p_{O_2} is in atmospheres and T is in degrees Kelvin.

† J. A. Maymo and J. M. Smith, *AIChE J.*, **12**, 845 (1966).

(a) Calculate the global rate of reaction at a location in a packed-bed reactor where the bulk conditions are

$$T_b = 373 \text{ K}(100°\text{C})$$
$$p_t = 1 \text{ atm}$$
$$(p_{O_2})_b = 0.060 \text{ atm}$$
$$(p_{H_2})_b = 0.94 \text{ atm}$$

Are external mass- and heat-transfer resistances negligible?

The diameter and mass of the nonporous catalyst pellet are 1.86 cm and 2.0 g, and the superficial mass velocity of the flowing gas is 250 lb/(h)(ft²). At these conditions the molecular diffusivity of oxygen is about 1.15 cm²/s. The properties of the gas mixture are essentially the same as those of hydrogen.

(b) Determine the ratio γ of the global rate to the rate evaluated at bulk-gas conditions.

10-7 The catalytic isomerization of n-butane,

$$n\text{-}C_4H_{10}(g) \rightarrow i\text{-}C_4H_{10}(g),$$

is studied in a plug-flow laboratory reactor in which the catalyst is deposited on the walls of the cylindrical tube. Pure n-butane (as a gas) is fed to the reactor at a steady rate of F (lb mol/h), and the whole system operates isothermally. Assume that the reaction rate at the catalyst surface is first order and irreversible, according to the expression

$$\mathbf{r} = kC_s$$

where $\mathbf{r}$ is the rate of formation of iso-butane, lb mol/(h)(ft² tube surface) and C_s is the surface concentration of n-butane, lb mol/ft³.

(a) Suppose that the rate of mass transfer radially from the bulk gas, where C_g is the concentration of n-C₄H₁₀, to the catalyst wall, where C_s is the concentration, is proportional to the mass-transfer coefficient k_m, in feet per hour. Derive an expression for the conversion of n-butane in terms of the reactor length L and other necessary quantities. The constant temperature is $T°\text{R}$ and the pressure is p atmospheres.

(b) Suppose the resistance to mass transfer radially is negligible. What form would the equation for the conversion take?

(c) Suppose the plug-flow model is not valid because axial diffusion is important. Let the effective diffusivity in the axial direction be D_L. Other conditions are as in part (a). Derive an equation whose solution will give the conversion as a function of reactor length and other necessary quantities.

(d) Calculate the conversion of n-butane leaving the reactor from the equation derived in part (a) and the conditions

$$k = 446 \text{ ft/h}$$
$$d = 6 \text{ in.}$$
$$t = 500°\text{F}$$
$$p_t = 1 \text{ atm}$$
$$F = 20 \text{ lb mol/h}$$
$$\text{Reactor length } L = 10 \text{ ft}$$

Also calculate the conversion from the equation derived in part (c), using the above data, and $D_L/uL = 0.05$. This value of D_L/uL corresponds to an intermediate amount of dispersion.

10-8 Reconsider parts (a) to (c) of Prob. 10-7 for a reversible first-order reaction. It is suggested that the rate equation be written in terms of the equilibrium concentration of n-butane. Also show how this concentration is related to the equilibrium constant K.

10-9 The catalytic isomerization of normal hexane to 2-methyl pentane and 3-methylpentane was discussed in Sec. 10-5. The effect of external mass transfer on selectivity was evaluated by assuming that the consecutive reactions were irreversible and first order. In this problem treat the realistic situation of reversible, first-order kinetics. Derive an expression for the selectivity of 2-methylpentane with respect to 3-methylpentane under isothermal conditions and for bulk-fluid concentrations of $(C_A)_b$, $(C_B)_b$, and $(C_C)_b$. Express the result in terms of the two first-order, forward-direction rate constants k_1 and k_2, the corresponding equilibrium constants K_1 and K_2, and the necessary mass-transfer coefficients between bulk fluid and catalyst surface.

10-10 Imagine that the catalytic (V_2O_5) oxidation of naphthalene can be approximated by the following irreversible, parallel (not balanced) reactions:

$$C_{10}H_8 \xrightarrow{k_1} C_8H_4O_3 \text{ (phthalic anhydride)}$$

$$C_{10}H_8 \xrightarrow{k_2} 10CO_2 + 4H_2O$$

Suppose the activation energies for the two first-order reactions are $E_1 = 117$ kJ/mol and $E_2 = 167$ kJ/mol. For reaction 1, $\Delta H_1 = -3770$ kJ/mol and for reaction 2, $\Delta H_2 = -4770$ kJ/mol. At one location in a fixed-bed reactor, the temperature difference between catalyst particle and gas phase is 15 K, and the gas temperature is 620 K.

Estimate how much error would be made in the selectivity of phthalic anhydride with respect to carbon dioxide, if the effect of the heat-transfer resistance between gas and particle is neglected.

10-11 Suppose that a temperature difference exists between bulk gas and catalyst surface. How would the equation derived in Prob. 10-9 be modified? What are the additional quantities that would have to be known to evaluate the effect of nonisothermal behavior?

10-12 A. Estimate gas-to-particle mass- and heat-transfer coefficients for a gas (whose properties are those of air) at 373 K and 1 atm flowing through a fixed bed (void fraction = 0.40) of 0.0064 m(¼-in.) spherical particles. The superficial mass velocity of the gas is 1.47 kg/(s)(m)², or 0.3 lb/(s)(ft)².Compare these coefficients with those for a fluidized bed (void fraction = 0.95) of particles with an average diameter of 8×10^{-5} m (170-200 mesh size). The superficial mass velocity is 0.29 kg/(s)(m)² [or 0.06 lb/(s)(ft)²].

B. Also compare the mass- and heat-transfer rates, per unit volume of the bed per unit driving force (ΔC or ΔT), for the two beds.

C. Suppose the particles are a catalyst for a reaction that occurs on the outer surface at a rate of 3.0×10^{-5} (g mol of product produced)/(s)(g catalyst). The heat of reaction is $\Delta H = -20,000$ cal/(g mol of product). Estimate temperature and concentration differences ($C_b - C_s$ and $T_b - T_s$) for the two beds. Catalyst particle density = 1.0 g/cm³.

D. Comment on the influence of external mass- and heat-transfer effects on the global rate of reaction for the two beds.

10-13 In Example 10-6, part A, an equation was derived for the fraction of component A removed from gas bubbles rising through a height L of liquid. No solid phase was present. Now derive the more general equation applicable when the gas bubbles rise through a slurry of solid particles on which A is absorbed. That is, derive an expression for $(p_0 - p_L)/p_0$ as a function of time. Assume that very rapid physical adsorption occurs on the solid particles. The adsorption follows a linear isotherm, $n = KC_s$, where n is the adsorbed concentration of A, g mol/(g particle). Also assume that rate coefficient $k_c a_c$ is much greater than $k_L a_g$, so that $C_L - C_s \rightarrow 0$. The mass of particles in the slurry is m_s (g/cm³ of liquid).

All other conditions are the same as in Example 10-6. The slurry liquid is well mixed and the bubbles rise separately in plug flow through the slurry. A is slightly soluble in the liquid and the change in C_L is negligible during the residence time of one bubble.

Show that the derived expression reduces to Eq. (H) of Example 10-6 when there are no solid particles present.

10-14 Estimate the mass-transfer coefficient k_L for bubbles ($d_b = 1.0$ mm) of pure hydrogen rising through a liquid ($\mu_L = 2.0$ centipoises, $\rho_L = 0.9$ g/cm³). The system is at 150°C and 5 atm pressure.

Also estimate the value of k_c if the liquid contains catalyst particles of 100 mesh size. The density of the particles is 1.8 g/cm³ and the diffusivity of hydrogen in the liquid is 6×10^{-5} cm²/s. The energy dissipation rate evaluated from Eq. (10-43) is 2.0×10^4 erg/(s)(g of liquid).

10-15 If the intrinsic rate of hydrogenation at the catalyst site is very large for the conditions of Prob. 10-14, calculate the global rate of hydrogenation r_v for the following cases, corresponding to a constant concentration of gas bubbles ($a_g = 3.0$ cm^{-1}) and the following catalyst-particle concentrations:

(a) $a_c/a_g = 100$ (high-particle concentration)
(b) $a_c/a_g = 10$
(c) $a_c/a_g = 2.0$ (low-particle concentration)

The solubility of hydrogen in the liquid is such that $C_g/(C_L)_{eq} = 10$ at 150°C.

10-16 If the rate of hydrogenation at the catalyst can be represented by Eq. (10-33) with $k = 0.0020$ cm/s, what are the global rates of reaction for the three cases of Prob. 10-15?

10-17 A continuous, stirred-tank reactor and separator (see Fig. 10-15) are used to study the heterogeneous reaction between pure liquid A (phase 1) and reactant B, which is dissolved in phase 2 (also liquid). Phase 2 contains the solvent, a catalyst, and reactant B, all three substances being completely miscible. However, neither solvent, catalyst, reactant B, nor the products of reaction are soluble in phase 1. The reactor operates isothermally at 25°C and at this temperature A has a limited solubility in phase 2 of 2.7×10^{-5} g mol/liter. Phase 2 is present in the reactor as bubbles dispersed in phase 1 (continuous phase). The two phases are separated in the separator and phase 1 is recycled. Neglect the reaction that occurs in the separator.

There is good mixing in the reactor but motion within the bubbles of phase 2 is insufficient to eliminate all the mass-transfer resistance within the bubbles. Assume that this mass-transfer resistance is concentrated in a thin film of liquid near the interface between phases 1 and 2, and that this mass-transfer resistance is governed by a mass-transfer coefficient k_g (based on interfacial area of bubbles). From independent mass-transfer measurements it is estimated that the reaction resistance within the bubbles of phase 2 is 75% of the total resistance (mass transfer plus reaction resistances). In a practical example, such a system could represent an alkylation reaction catalyzed by an acid phase.

A. Derive a relationship between the concentration, C_{B_0}, of B in the feed stream and in the product (from the separator) stream (concn = C_B) of phase 2.

B. In an alkylation study using this reactor system, the following data were obtained:

Feed rate of phase 2 = 0.2 cm³/s
Total liquid volume in reactor = 1500 cm³
Volume fraction of phase 2 in reactor = 0.24
C_{B_0} = 0.02 g/liter (feed stream)
C_B = 0.0125 g/liter (product stream)

Calculate the value of the reaction-rate constant, k_2, from these data.

Notes

1. Assume that the reaction is second order and irreversible

$$A + B \rightarrow \text{products}$$

2. The liquid density of either phase does not change significantly with reaction.
3. Assume that equilibrium exists at the interface between phases 1 and 2.

10-18 As an example of external mass transfer in catalytic reactors, consider a continuous slurry (two-phase) reactor (well mixed) in which benzaldehyde is adsorbed from an aqueous solution by activated carbon particles at 25°C.

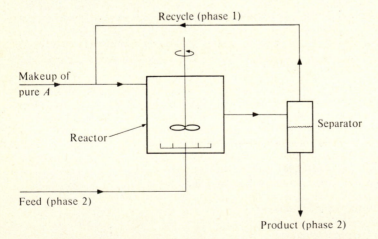

Figure 10-15 Liquid–liquid reaction (alkylation) system (for Prob. 10-17).

One method for evaluating the mass-transfer coefficient, k_c, from liquid to particle surface is by measuring the response curve when a pulse [Dirac-delta function $\delta(t)$] of benzaldehyde solution is introduced into the feed to the reactor. The system is illustrated in the following sketch:

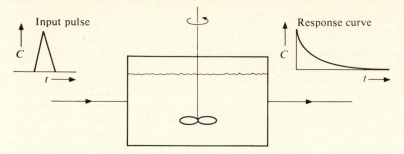

For this chemical system, and the particle sizes studied, it may be assumed that the adsorption process is irreversible and first order and that intraparticle diffusion resistance is negligible. That is, the overall adsorption process depends only on the external mass-transfer process (k_c, cm/s) and the intrinsic rate of adsorption (k, cm³/(g)s). No gas phase is present.

A. Using the nomenclature given later, derive equations for the first (μ_1) and second (μ_2) moments of the response curve. These moments are defined as:

$$\mu_1 = \frac{m_1}{m_0}$$

$$\mu_2 = \frac{m_2}{m_0}$$

where

$$m_n = (-1)^n \lim_{s \to 0} \frac{d^n \bar{C}(s)}{ds^n}; \qquad n = 0, 1, 2, \ldots$$

and $\bar{C}(s)$ is the Laplace transform of $C(t)$. For $n = 0$, $d^n \bar{C}(s)/ds^n = \bar{C}(s)$.

B. Experimental values for the first moment were determined from the measured response curve using the equation

$$\mu_1 = \frac{\displaystyle\int_0^\infty Ct\,dt}{\displaystyle\int_0^\infty C\,dt}$$

These experimental results for different particle sizes of an activated carbon are given in the following table.

Experimental data at 25°C

r_s microns	$m_s \times 10^3$ g/cm³	ρ_p g/cm³	$(1/\mu_1 - 1/\tau) \times 10^3$ (s)⁻¹
80	0.471	0.956	2.09
114	0.940	0.956	2.94
192	0.940	0.956	1.49
271	0.940	0.956	1.14
322	0.940	0.956	0.818

From these data and the derivation in part A, calculate a numerical value for k_c.

    C. Sketch the shape of the response curves (for comparison) for:

1. Irreversible adsorption
2. Reversible adsorption

Use the following nomenclature

A = strength of input pulse. Input pulse is defined as $C_0 = A\,\delta(t)$ where $\delta(t)$ is Dirac-delta
    function; C_0 = concentration in inlet pulse.
C = concentration of benzaldehyde in bulk liquid of slurry, g/cm^3; C_s = concn at outer
    surface of particle.
m_s = concentration of carbon particles in slurry, g/(cm^3 of particle-free liquid).
n = concentration of benzaldehyde adsorbed on carbon particles, mol/g.
s = transform variable, s
Q = volumetric flow rate of liquid to adsorber, cm^3/s (assume constant).
r_s = radius of carbon particle (assume spherical).
t = time, s
V = volume of particle-free liquid in reactor, cm^3
ρ_p = density of carbon particles, g/cm^3
τ = average residence time of liquid in reactor, V/Q, s
S_c = outer surface area of particles per unit volume of particle-free liquid in slurry, cm^{-1}
W = mass of carbon particles in slurry adsorber, g

10-19 Mass transfer between gas and liquid in trickle beds is studied by measuring the *desorption* of
oxygen from water into nitrogen as both liquid and gas streams flowed down through a packed bed of
particles. The rate of oxygen transfer was determined by measuring the oxygen concentration $C_{L,f}$ in the
liquid feed and in the liquid stream leaving the reactor ($C_{L,e}$) when the gas feed was pure nitrogen.
However, some mass transfer occurs in the entrance region between the liquid distributor (see
Fig. 10-16) and the top of the bed, and also between the bottom of the bed (where the screen holding
the particles is located) and the effluent liquid. To account for these end-effects the concentration in the

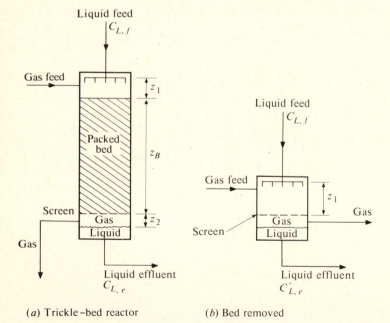

(a) Trickle–bed reactor   (b) Bed removed

Figure 10-16 Trickle-bed reactor.

effluent is also measured (at the same flow rates and for the same $C_{L, f}$) when the bed of particles was removed and the liquid distributor lowered to the distance z_1 above the screen. If this latter concentration is $C'_{L, e}$, derive an equation for $(k_L a_g)$ for the bed itself in terms of the measured concentrations. The height of the packed bed is z_B and the superficial mass velocity of the liquid is G_L (mass)/(area)(time).

The gas rate through the bed is large enough that the concentration of oxygen in the gas stream was essentially zero. Also neglect axial dispersion.

It is convenient to imagine the apparatus to consist of three sections: the region between liquid distributor and top of bed with mass-transfer coefficient $(k_L a)_{top}$ and length z_1, the bed proper with coefficient $k_L a_g$ (the desired quantity) and length z_B, and the region between the bottom of the bed and the liquid withdrawal with mass-transfer coefficient $(k_L a)_{btm}$ and length z_2. Then assume that the mass-transfer coefficients $(k_L a)_{top}$ and $(k_L a)_{btm}$ and length z_1 and z_2 are the same in the runs with and without the bed of particles.

10-20 Repeat Prob. 10-19 for the *adsorption* of oxygen. In this case the gas stream entering the reactor is pure oxygen (designate the concentration of oxygen in water in equilibrium with pure oxygen as $(C_L)_{eq}$). The liquid feed is oxygen-free water.

10-21 Consider a trickle-bed reactor for the catalytic oxidation of formic acid (in dilute aqueous solution) with air. The temperature and pressure are 240°C and 40 atm.

A. Calculate the global rate for the reaction

$$HCOOH(aq) + \tfrac{1}{2}O_2(aq) \rightarrow H_2O + CO_2(aq)$$

at a location in the reactor where the bulk liquid concentrations are:

$$C_{L, O_2} = 2.70 \times 10^{-7} \text{ g mol/cm}^3$$

$$C_{L, FA} = 28.7 \times 10^{-7} \text{ g mol/cm}^3$$

and the gas phase is air saturated with water vapor at 240°C. The gas and liquid flow rates are: $Q_g = 4.0 \text{ cm}^3/\text{s}$ (at 25°C, and 1 atm) and $Q_L = 2.0 \text{ cm}^3/\text{s}$ in a 1-in. ID reactor. The bed is packed with commercial CuO·ZnO catalyst particles ($d_p = 0.29/\text{cm}$). The particle density is 1.90 g/cm³ and the bed density, $\rho_B = 1.05 \text{ g/cm}^3$.

Assume that the liquid covers all the surface of the particles. Properties of the system at 240°C, and 40 atm are estimated to be:

$$H_{O_2} = 3.82 \text{ [g mol/(cm}^3 \text{ of oxygen gas at 25°C, 1 atm)]/(g mol/cm}^3 \text{ of water)}$$
$$\rho_L = \text{density of water} = 0.79 \text{ g/cm}^3$$
$$\mu_L = \text{viscosity of water } 0.92 \text{ g/(cm)(s)}$$
$$(\mathscr{D}_{O_2})_L = \text{diffusivity of O}_2 \text{ in water} = 3.4 \times 10^{-4} \text{ cm}^2/\text{s}$$
$$(\mathscr{D}_{FA})_L = \text{diffusivity of formic acid in water } 2.7 \times 10^{-4} \text{ cm}^2/\text{s}$$

The intrinsic rate of disappearance of oxygen at the catalyst surface is second order;[†] at 240°C

$$r_{O_2}(\text{g mol/(g catalyst)(s)}) = 4.37 \times 10^{-6} C_{s, O_2} C_{s, FA}$$

where the concentrations are in the liquid at the outer surface of the catalyst particles.

B. Also calculate the mass transfer rate of oxygen from gas to liquid at the same location in the reactor. Comment on the significance of the values of the global rate (part A) and oxygen transfer rate (part B).

---

† G. Baldi, S. Goto, C. K. Chow, and J. M. Smith, *Ind. Eng. Chem. Proc. Des. Dev.*, **13**, 447 (1974).

ELEVEN

INTERNAL TRANSPORT PROCESSES—
REACTION AND DIFFUSION
IN POROUS CATALYSTS

Essentially all of the active surface of porous catalyst pellets is internal (see p. 327). The reaction that occurs within the pellet consumes reactant and evolves (or absorbs) the heat of reaction. This, in turn, induces *internal* concentration and temperature gradients which can be large enough to cause a significant variation in rate of reaction with position inside the pellet.

At steady state the average rate for the whole pellet will be equal to the global rate at the location of the pellet in the reactor. The concentration and temperature of the bulk fluid at this location may not be equal to the values at the outer surface of the pellet. In Chap. 10 we studied how to account for these *external* transport resistances, but we did not consider the influence of the internal concentration and temperature gradients. This latter effect is the objective of the present chapter. That is to say, we want to evaluate the *average* rate for the whole pellet in terms of the concentration and temperature at the outer surface. Then the results of Chaps. 10 and 11 can be combined to express the rate in terms of the known bulk properties for use in design.

Because there is a continuous variation in concentration and temperature with radius of the pellet, *differential* conservation equations are required to describe the concentration and temperature profiles. These profiles are used with the intrinsic rate equation to integrate through the pellet and so obtain the average rate for the pellet. The differential equations involve the effective† diffusivity

---

† In the remainder of the text the term "effective" is used to indicate that a transport coefficient applies to a porous material, as distinguished from a homogeneous region. Effective diffusivities D_e and thermal conductivities k_e are based on a unit of *total area* (void plus nonvoid) perpendicular to the direction of transport. For example, for diffusion in a spherical catalyst pellet at radius r, D_e is based on the area $4\pi r^2$.

and thermal conductivity of the porous pellet. Data and theories for these quantities are given in Secs. 11-1 to 11-6, and in Secs. 11-7 to 11-11 the results are used to establish the rate for the whole pellet. The method of combining the effects of *internal* and *external* transport resistance to give the global rate in terms of bulk fluid properties is discussed in Example 11-8 and also in Chap. 12.

The effect of intrapellet mass transfer is to reduce the reactant concentration within the pellet. Hence, the average rate will be less than what it would be if there were no internal concentration gradient. The effect of the temperature gradient is to increase the rate for an exothermic reaction, the opposite of the mass transfer effect. This is because intrapellet temperatures will be greater than surface values. For endothermic reactions temperature and concentration gradients both reduce the rate below that evaluated at outer-surface conditions.

For a gaseous reaction accompanied by a change in number of moles, an intrapellet gradient in total pressure will also develop (at steady state). If the moles decrease, there will be a flow of reactant toward the center of the pellet due to this total-pressure gradient. This augments the diffusion of reactant inward toward the center of the pellet and retards the diffusion of product outward. Unless the decrease in number of moles is very large, which is unlikely, the effect on the average rate is small.† Also, it should be remembered that all intrapellet-transport effects will become less important as the pellet size decreases. For fluidized-bed and slurry reactors intraparticle transport processes can usually be neglected.

INTRAPELLET MASS TRANSFER

It is seldom possible to predict diffusion rates in porous materials by simply correcting bulk-phase diffusivities for the reduction in cross-sectional area due to the solid phase. There are several reasons for this:

1. The tortuous, random, and interconnected arrangement of the porous regions makes the length of the diffusion path unknown.
2. The diffusion in the pore volume itself will be influenced by the pore walls, provided the diffusing molecule is likely to collide with a wall rather than another molecule. This often is the case for gaseous systems. Then the pore-volume contribution to total mass transport is not dependent solely on the *bulk diffusivity*, but may be affected (or determined) by the *Knudsen diffusivity*.
3. A catalyst is characterized as a substance which adsorbs reactant. When it is adsorbed it may be transported either by desorption into the pore space or by migration to an adjacent site on the surface. The contribution of surface diffusion probably is small in most cases, but it must be added to the diffusion in the pore volume to obtain the total mass transport.

---

† See A. Wheeler in P. H. Emmett (ed.), "Catalysis," vol. II, p. 143, Reinhold Publishing Corporation, New York, 1955; Von E. Wicke and Peter Hugo, *Z. Phys. Chem.*, **28**, 26 (1961); S. Otani, N. Wakao, and J. M. Smith, *AIChE J.*, **10**, 135 (1965); N. Wakao, S. Otani, and J. M. Smith, *AIChE J.*, **11**, 435 (1965).

Surface diffusion is not well understood, and rates of diffusion by this mechanism cannot yet be predicted. We must be content with summarizing in Sec. 11-4 the available data for surface migration. Volume diffusion in a straight cylindrical pore is amenable to analysis. Hence the procedure used (see Sec. 11-3) to predict an effective diffusivity is to combine the established equations for diffusion in a single cylindrical pore with a geometric model of the pore structure of the catalyst pellet. If the surface diffusivity is known for the surface of a cylindrical pore, it can be used with the same model to obtain the total mass-transport rate (in the absence of flow due to a total-pressure gradient). The nature of gaseous diffusion in cylindrical pores is discussed in Sec. 11-1. When the reaction mixture in the pore is liquid Knudsen diffusion does not occur, but uncertainty as to how the bulk diffusivity varies with concentration is a complicating feature. The effective length of the diffusion path is determined by the pore structure of the pellet. Hence the path length is intimately connected with the model chosen to represent the porous catalyst. Such models are considered in Sec. 11-3.

It is evident from the foregoing discussion that the effective diffusivity cannot be predicted accurately for use under reaction conditions unless surface diffusion is negligible and a valid model for the pore structure is available. The prediction of an effective thermal conductivity is also difficult. Hence sizable errors are frequent in predicting the global rate from the rate equation for the chemical step on the interior catalyst surface. This is not to imply that for certain special cases accuracy is not possible (see Sec. 11-9). It does mean that heavy reliance must be placed on experimental measurements for effective diffusivities and thermal conductivities. Also note from some of the examples and data mentioned later that intrapellet resistances can greatly affect the rate. Hence the problem is significant.

Only a brief account will be given of the considerable theory on *special* aspects of isothermal mass transfer. More emphasis will be placed on interpretation of laboratory-reactor data with respect to the importance of intrapellet resistances.

11-1 Gaseous Diffusion in Single Cylindrical Pores

Basic Equations For many catalysts and reaction conditions (especially pressure) both bulk and Knudsen diffusion contribute to the mass-transport rate within the pore volume. For some years the proper combination of the two mechanisms was in doubt. About 1961 three independent investigations† proposed identical equations for the rate of diffusion (in a binary gaseous mixture of A and B) in terms of the bulk diffusivity $\mathscr{D}_{AB}$ and Knudsen diffusivity $\mathscr{D}_K$. If N_A is the molal flux of A, it is convenient to represent the result as

$$N_A = -\frac{p_t}{R_g T} D \frac{dy_A}{dx} \tag{11-1}$$

---

† R. B. Evans, III, G. M. Watson, and E. A. Mason, *J. Chem. Phys.*, **35**, 2076 (1961); L. B. Rothfeld, *AIChE J.*, **9**, 19 (1963); D. S. Scott and F. A. L. Dullien, *AIChE J.*, **8**, 29 (1962).

where y_A is the mole fraction of A, x is the coordinate in the direction of diffusion, and D is a *combined* diffusivity given by

$$D = \frac{1}{(1 - \alpha y_A)/\mathscr{D}_{AB} + 1/(\mathscr{D}_K)_A} \tag{11-2}$$

The quantity α is related to the ratio of the diffusion rates of A and B by

$$\alpha = 1 + \frac{N_B}{N_A} \tag{11-3}$$

For reactions at steady state, α is determined by the stoichiometry of the reaction. For example, for the reaction $A \to B$, reaction and diffusion in a pore would require equimolal counterdiffusion; that is, $N_B = -N_A$. Then $\alpha = 0$, and the effective diffusivity is

$$D = \frac{1}{1/\mathscr{D}_{AB} + 1/(\mathscr{D}_K)_A} \tag{11-4}$$

When the pore radius is large, Eqs. (11-1) and (11-2) reduce to the conventional constant-pressure form for bulk diffusion. For this condition $(\mathscr{D}_K)_A \to \infty$. Then combining Eqs. (11-1) to (11-3) gives

$$N_A = -\frac{p_t}{R_g T} \mathscr{D}_{AB} \frac{dy_A}{dx} + y_A(N_A + N_B) \tag{11-5}$$

If, in addition, the diffusion is equimolal, $N_B = -N_A$, Eq. (11-5) may be written

$$N_A = \frac{-p_t}{R_g T} \mathscr{D}_{AB} \frac{dy_A}{dx} \tag{11-6}$$

If the pore radius is very small, collisions will occur primarily between gas molecules and pore wall, rather than between molecules. Then the Knudsen diffusivity becomes very low, and Eqs. (11-1) and (11-2) reduce to

$$N_A = -\frac{p_t}{R_g T} (\mathscr{D}_K)_A \frac{dy_A}{dx} \tag{11-7}$$

This equation is the usual one† expressing Knudsen diffusion in a long capillary.

Although Eq. (11-2) is the proper one to use for regions where both Knudsen and bulk diffusion are important, it has a serious disadvantage: the combined diffusivity D is a function of gas composition y_A in the pore. This dependency on composition carries over to the effective diffusivity in porous catalysts (see Sec. 11-3) and makes it difficult later to integrate the combined diffusion and transport equations. The variation of D with y_A is usually not strong (see Example 11-3). Therefore it has been almost universal,‡ in assessing the importance of

---

† M. Knudsen, *Ann. Physik*, **28**, 75 (1909); N. Wakao, S. Otani, and J. M. Smith, *AIChE J.*, **11**, 435 (1965).

‡ N. Wakao and J. M. Smith [*Ind. Eng. Chem., Fund. Quart.*, **3**, 123 (1964)] used the general equations (11-1) and (11-2) in developing expressions for intrapellet effects for isothermal first-order reactions.

intrapellet resistances, to use a composition-independent form for D, for example, Eq. (11-4). In fact, the concept of a single effective diffusivity loses its value if the composition dependency must be retained. Note also that the problem disappears when the reaction stoichiometry is such that diffusion is countercurrent and equimolal.

Effective diffusivities in porous catalysts are usually measured under conditions where the *pressure is maintained constant by external means* (experimental methods are described in Sec. 11-3). Under this condition, and for a binary counterdiffusing system, the ratio N_B/N_A is the same regardless of the extent of Knudsen and bulk diffusion. Evans et al† have shown this constant ratio to be (at constant pressure)

$$\frac{N_B}{N_A} = -\sqrt{\frac{M_A}{M_B}} \tag{11-8}$$

or

$$\alpha = 1 - \sqrt{\frac{M_A}{M_B}}$$

where M represents the molecular weight. Equation (11-8) applies for nonreacting conditions. When reaction occurs, stoichiometry determines α (see Example 11-3).

Calculation of Diffusivities In analyzing Knudsen and bulk diffusivities the important parameter is the size of the pore with respect to the mean free path. The bulk diffusivity is a function of the molecular velocity and the mean free path; that is, it is a function of temperature and pressure. The Knudsen diffusivity depends on the molecular velocity v and the pore radius a. In terms of simple kinetic theory, these two diffusivities may be described by the equations

$$\mathscr{D}_{AB} = \tfrac{1}{3}\bar{v}\lambda \qquad \left(\lambda \sim \frac{1}{p}\right) \tag{11-9}$$

$$(\mathscr{D}_K)_A = \tfrac{2}{3}a\bar{v} \tag{11-10}$$

where λ is the mean free path. Since λ is of the order of 1000 Å for gases at atmospheric pressure, diffusion in micropores of a catalyst pellet will be predominantly by the Knudsen mechanism. This would be the case for a material such

---

† R. B. Evans, III, G. M. Watson, and E. A. Mason, *J. Chem. Phys.*, **35**, 2076 (1961).

---

† Values of σ and ε/k_B are from J. O. Hirschfelder, C. F. Curtiss, and R. B. Bird, "Molecular Theory of Gases and Liquids," pp. 1110–1112, John Wiley & Sons, Inc., New York, 1954 (also Addenda and Corrigenda, p. 11). The above values are computed from viscosity data and are applicable for temperatures above 100 K.

‡ Values of T_c, p_c, and V_c are from K. A. Kobe and R. E. Lynn, Jr., *Chem. Rev.*, **52**, 117–236 (1952); F. D. Rosinni (ed.), *Am. Petrol. Inst. Res. Proj.*, Carnegie Institute of Technology, **44** (1952).

Source: By permission from R. B. Bird, W. E. Stewart, and E. N. Lightfoot, "Transport Phenomena," John Wiley & Sons, Inc., New York, 1960.

Table 11-1 Lennard-Jones constants and critical properties

Substance	Molecular weight	Lennard-Jones parameters†		Critical constants‡		
		σ, Å	ε/k_B, K	T_c, K	p_c, atm	V_c, cm³/g mole
Light elements						
H_2	2.016	2.915	38.0	33.3	12.80	65.0
He	4.003	2.576	10.2	5.26	2.26	57.8
Noble gases						
Ne	20.183	2.789	35.7	44.5	26.9	41.7
Ar	39.944	3.418	124.0	151.0	48.0	75.2
Kr	83.80	3.61	190.0	209.4	54.3	92.2
Xe	131.3	4.055	229.0	289.8	58.0	118.8
Simple polyatomic substances						
Air	28.97	3.617	97.0	132.0	36.4	86.6
N_2	28.02	3.681	91.5	126.2	33.5	90.1
O_2	32.00	3.433	113.0	154.4	49.7	74.4
O_3	48.00	......	......	268.0	67.0	89.4
CO	28.01	3.590	110.0	133.0	34.5	93.1
CO_2	44.01	3.996	190.0	304.2	72.9	94.0
NO	30.01	3.470	119.0	180.0	64.0	57.0
N_2O	44.02	3.879	220.0	309.7	71.7	96.3
SO_2	64.07	4.290	252.0	430.7	77.8	122.0
F_2	38.00	3.653	112.0			
Cl_2	70.91	4.115	357.0	417.0	76.1	124.0
Br_2	159.83	4.268	520.0	584.0	102.0	144.0
I_2	253.82	4.982	550.0	800.0		
Hydrocarbons						
CH_4	16.04	3.822	137.0	190.7	45.8	99.3
C_2H_2	26.04	4.221	185.0	309.5	61.6	113.0
C_2H_4	28.05	4.232	205.0	282.4	50.0	124.0
C_2H_6	30.07	4.418	230.0	305.4	48.2	148.0
C_3H_6	42.08	......	......	365.0	45.5	181.0
C_3H_8	44.09	5.061	254.0	370.0	42.0	200.0
$n\text{-}C_4H_{10}$	58.12	......	......	425.2	37.5	255.0
$i\text{-}C_4H_{10}$	58.12	5.341	313.0	408.1	36.0	263.0
$n\text{-}C_5H_{12}$	72.15	5.769	345.0	469.8	33.3	311.0
$n\text{-}C_6H_{14}$	86.17	5.909	413.0	507.9	29.9	368.0
$n\text{-}C_7H_{16}$	100.20	......	......	540.2	27.0	426.0
$n\text{-}C_8H_{18}$	114.22	7.451	320.0	569.4	24.6	485.0
$n\text{-}C_9H_{20}$	128.25	......	......	595.0	22.5	543.0
Cyclohexane	84.16	6.093	324.0	553.0	40.0	308.0
C_6H_6	78.11	5.270	440.0	562.6	48.6	260.0
Other organic compounds						
CH_4	16.04	3.822	137.0	190.7	45.8	99.3
CH_3Cl	50.49	3.375	855.0	416.3	65.9	143.0
CH_2Cl_2	84.94	4.759	406.0	510.0	60.0	
$CHCl_3$	119.39	5.430	327.0	536.6	54.0	240.0
CCl_4	153.84	5.881	327.0	445.4	45.0	276.0
C_2N_2	52.04	4.38	339.0	400.0	59.0	
COS	60.08	4.13	335.0	378.0	61.0	
CS_2	76.14	4.438	488.0	552.0	78.0	170.0

as silica gel, where the mean pore radius is from 15 to 100 Å (see Table 8-1). For a pelleted catalyst of alumina with the pore-volume distribution shown in Fig. 8-8, the mean macropore radius is about 8000 Å. At atmospheric pressure bulk diffusion would prevail in these pores. In the micropores of the same pellet, where $\bar{a} = 20$ Å, diffusion would be by the Knudsen process. Since the mean free path is inversely proportional to pressure, bulk diffusivity becomes more important as the pressure increases.

For more accurate calculations the *Chapman-Enskog formula*† has been found suitable for evaluating the bulk diffusivity at moderate temperatures and pressures. The equation is (for the binary gas mixture A, B)

$$\mathscr{D}_{AB} = 0.0018583 \frac{T^{3/2}(1/M_A + 1/M_B)^{1/2}}{p_t \sigma_{AB}^2 \Omega_{AB}} \tag{11-11}$$

where $\mathscr{D}_{AB}$ = bulk diffusivity, cm^2/s
T = temperature, K
M_A, M_B = molecular weights of gases A and B
p_t = total pressure of the gas mixture, atm
σ_{AB}, ε_{AB} = constants in the Lennard-Jones potential-energy function for the molecular pair AB; σ_{AB} is in Å.
Ω_{AB} = collision integral, which would be unity if the molecules were rigid spheres and is a function of $k_B T/\varepsilon_{AB}$ for real gases (k_B = Boltzmann's constant)

Since the Lennard-Jones potential-energy function is used, the equation is strictly valid only for nonpolar gases. The Lennard-Jones constants for the unlike molecular pair AB can be estimated from the constants for like pairs AA and BB:

$$\sigma_{AB} = \tfrac{1}{2}(\sigma_A + \sigma_B) \tag{11-12}$$

$$\varepsilon_{AB} = (\varepsilon_A \varepsilon_B)^{1/2} \tag{11-13}$$

The force constants for many gases are given in the literature and are summarized in Table 11-1. Those that are not available otherwise may be approximated by the expressions

$$\sigma = 1.18 \, V_b^{1/3} \tag{11-14}$$

$$\frac{k_B T}{\varepsilon} = 1.30 \frac{T}{T_c} \tag{11-15}$$

where k_B = Boltzmann's constant
T_c = critical temperature
V_b = volume per mole (cm^3/g mole) at normal boiling point

If necessary, V_b may be estimated by adding the increments of volume for the atoms making up the molecule (Kopp's law). Such increments are given in Table

---

† See J. O. Hirschfelder, C. F. Curtiss, and R. B. Bird, "Molecular Theory of Gases and Liquids," pp. 539, 578, John Wiley & Sons, Inc., New York, 1954.

Table 11-2 Volume increments for estimating molecular volume at normal boiling point

Kind of atom in molecule	Volume increment, cm^3/g mol
Carbon	14.8
Chlorine, terminal as R—Cl	21.6
Chlorine, medial as —CHCl—	24.6
Fluorine	8.7
Helium	1.0
Hydrogen	3.7
Mercury	15.7
Nitrogen in primary amines	10.5
Nitrogen in secondary amines	12.0
Oxygen in ketones and aldehydes	7.4
Oxygen in methyl esters and ethers	9.1
Oxygen in ethyl esters and ethers	9.9
Oxygen in higher esters and ethers	11.0
Oxygen in acids	12.0
Oxygen bonded to S, P, or N	8.3
Phosphorus	27.0
Sulfur	25.6
For organic cyclic compounds	
3-membered ring	−6.0
4-membered ring	−8.5
5-membered ring	−11.5
6-membered ring	−15.0
Naphthalene	−30.0
Anthracene	−47.5

Source: In part from C. N. Satterfield and T. K. Sherwood, "The Role of Diffusion in Catalysis," p. 9, Addison-Wesley Publishing Company, Reading, Mass., 1963.

11-2. The collision integral Ω_{AB} is given as a function of $k_B T/\varepsilon_{AB}$ in Table 11-3. From these data and the equations, binary diffusivities may be estimated for any gas. For polar gases, or for pressures above 0.5 critical pressure, the errors may be greater than 10%. The effects of composition on $\mathscr{D}$ are small for gases at moderate conditions, so that the same procedure may be used as an approximation for multicomponent mixtures. An improved result for mixtures may be obtained from empirical correlations.†

For evaluating the Knudsen diffusivity we may use the following equation for the average molecular velocity $\bar{v}$ for a component of gas in a mixture:

$$\bar{v}_A = \left(\frac{8R_g T}{\pi M_A}\right)^{1/2} \tag{11-16}$$

† See R. C. Reid, J. M. Prausnitz, and T. K. Sherwood, "The Properties of Gases and Liquids", 3d ed., chap. 11, McGraw-Hill Book Company, New York, 1977.

Table 11-3 Values of Ω_{AB} for diffusivity calculations (Lennard-Jones model)

$k_B T/\varepsilon_{AB}$	Ω_{AB}	$k_B T/\varepsilon_{AB}$	Ω_{AB}
0.30	2.662	2.0	1.075
0.35	2.476	2.5	1.000
0.40	2.318	3.0	0.949
0.45	2.184	3.5	0.912
0.50	2.066	4.0	0.884
0.55	1.966	5.0	0.842
0.60	1.877	7.0	0.790
0.65	1.798	10.0	0.742
0.70	1.729	20.0	0.664
0.75	1.667	30.0	0.623
0.80	1.612	40.0	0.596
0.85	1.562	50.0	0.576
0.90	1.517	60.0	0.560
0.95	1.476	70.0	0.546
1.00	1.439	80.0	0.535
1.10	1.375	90.0	0.526
1.20	1.320	100.0	0.513
1.30	1.273	200.0	0.464
1.40	1.233	300.0	0.436
1.50	1.198	400.0	0.417
1.75	1.128		

Source: By permission from J. O. Hirshfelder, C. F. Curtiss, and R. B. Bird, "Molecular Theory of Gases and Liquids," pp. 1126, 1127, John Wiley & Sons, Inc., New York, 1954.

Combining this with Eq. (11-10) gives a working expression for $(\mathscr{D}_K)_A$ in a circular pore of radius a,

$$(\mathscr{D}_K)_A = 9.70 \times 10^3 a \left(\frac{T}{M_A}\right)^{1/2} \tag{11-17}$$

where $(\mathscr{D}_K)_A$ is in square centimeters per second, a is in centimeters, and T is in degrees Kelvin.

Example 11-1 Estimate the diffusivity of SO_2 for the conditions of Example 10-1.

SOLUTION The gas composition is about 94% air, with the remainder SO_2 and SO_3. Hence a satisfactory simplification is to consider the system as a binary mixture of air and SO_2.

From Table 11-1, for air

$$\frac{\varepsilon}{k_B} = 97°K \qquad \sigma = 3.617 \text{ Å or } 3.617 \times 10^{-10} \text{ m, or } 0.3617 \text{ nanometers}$$

and for SO_2

$$\frac{\varepsilon}{k_B} = 252°K \qquad \sigma = 4.290 \text{ Å } (0.429 \text{ nm})$$

From Eqs. (11-12) and (11-13),

$$\sigma_{AB} = \tfrac{1}{2}(3.617 + 4.290) = 3.953 \text{ Å } (0.3953 \text{ nm})$$

$$\varepsilon_{AB} = k_B[97(252)]^{1/2}$$

At the temperature of 480°C,

$$\frac{k_B T}{\varepsilon_{AB}} = \frac{k_B(753)}{k_B[97(252)]^{1/2}} = 4.8$$

and so, from Table 11-3,

$$\Omega_{AB} = 0.85$$

Substituting all these values in Eq. (11-11) gives

$$\mathscr{D}_{SO_2\text{-air}} = 0.0018583 \frac{753^{3/2}(1/64.1 + 1/28.9)^{1/2}}{(790/760)(3.953)^2(0.85)}$$

$$= 0.629 \text{ cm}^2/\text{s } (0.629 \times 10^{-4} \text{ m}^2/\text{s})$$

Example 11-2 A nickel catalyst for the hydrogenation of ethylene has a mean pore radius of 50 Å. Calculate the bulk and Knudsen diffusivities of hydrogen for this catalyst at 100°C, and 1 and 10 atm pressures, in a hydrogen-ethane mixture.

SOLUTION From Table 11-1, for H_2

$$\frac{\varepsilon}{k_B} = 38 \text{ K} \qquad \sigma = 2.915 \text{ Å}$$

and for C_2H_6

$$\frac{\varepsilon}{k_B} = 230 \text{ K} \qquad \sigma = 4.418 \text{ Å}$$

Then for the mixture, from Eqs. (11-12) and (11-13),

$$\sigma_{AB} = \tfrac{1}{2}(2.915 + 4.418) = 3.67 \text{ Å}$$

$$\varepsilon_{AB} = k_B[38(230)]^{1/2}$$

and

$$\frac{k_B T}{\varepsilon_{AB}} = \frac{273 + 100}{[38(230)]^{1/2}} = 4.00$$

From Table 11-3,

$$\Omega_{AB} = 0.884$$

Substituting these values in the Chapman-Enskog equation (11-11) gives the bulk diffusivity,

$$\mathscr{D}_{H_2, C_2H_6} = 0.001858 \frac{373^{3/2}(1/2.016 + 1/30.05)^{1/2}}{p_t(3.67)^2(0.884)} = \frac{0.86}{p_t}$$

This gives 0.86 cm^2/s at 1 atm, or 0.086 cm^2/s at 10 atm.

The Knudsen diffusivity, which is independent of pressure, is obtained from Eq. (11-17) as

$$(\mathscr{D}_K)_{H_2} = 9.70 \times 10^3 (50 \times 10^{-8}) \left(\frac{373}{2.016}\right)^{1/2} = 0.065 \text{ cm}^2/\text{s}$$

These results show that 1 atm pressure the Knudsen diffusivity is much less than the bulk value. Hence Knudsen diffusion controls the diffusion rate. At 10 atm both bulk and Knudsen diffusivities are important.

Example 11-3 (a) Calculate the combined diffusivity of hydrogen in a mixture of ethane, ethylene, and hydrogen in a pore of radius 50 Å (5.0 nm) at two total pressures, corresponding to 1 and 10 atm. Suppose that the pore is closed at one end, and that the open end is exposed to a mixture of ethylene and hydrogen. The pore wall is a catalyst for the reaction

$$C_2H_4 + H_2 \rightarrow C_2H_6$$

The temperature is 100°C.
(b) For comparison calculate the combined diffusivity of hydrogen for diffusion through a noncatalytic capillary of radius 50 Å (5.0 nm). Hydrogen is supplied at one end and ethane at the other. The pressure is maintained the same at both ends of the capillary. Make the calculations for two compositions, $y_{H_2} = 0.5$ and 0.8.

SOLUTION (a) Assume that the binary $H_2-C_2H_6$ will be satisfactory for representing the diffusion of hydrogen in the three-component system. From the reaction stoichiometry, the molal diffusion rates of H_2 and C_2H_6 will be equal and in opposite directions. Hence $\alpha = 0$, and Eq. (11-4) is applicable. From the results of Example 11-2,

$$D = \begin{cases} \dfrac{1}{1/0.86 + 1/0.065} = 0.060 \text{ cm}^2/\text{s } (0.060 \times 10^{-4} \text{ m}^2/\text{s}) & \text{at 1 atm} \\[2ex] \dfrac{1}{1/0.086 + 1/0.065} = 0.037 \text{ cm}^2/\text{s } (0.037 \times 10^{-4} \text{ m}^2/\text{s}) & \text{at 10 atm} \end{cases}$$

(b) For constant-pressure diffusion Eq. (11-2) should be used. From Eq. (11-3)

$$\alpha = 1 + \frac{N_{C_2H_6}}{N_{H_2}} = 1 - \sqrt{\frac{M_{H_2}}{M_{C_2H_6}}} = 1 - \sqrt{\frac{2.016}{30.05}} = 0.741$$

Then at $y_{H_2} = 0.5$, using Eq. (11-2),

$$D = \frac{1}{[1 - 0.741(0.5)]/\mathscr{D}_{H_2-C_2H_6} + 1/(\mathscr{D}_K)_{H_2}} = \frac{1}{0.630/\mathscr{D}_{H_2-C_2H_6} + 1/(\mathscr{D}_K)_{H_2}}$$

For the two pressures this expression gives

$$D = \begin{cases} \dfrac{1}{0.630/0.86 + 1/0.065} = 0.062 \text{ cm}^2/\text{s} \ (0.062 \times 10^{-4} \text{ m}^2/\text{s}) \\ \hspace{7cm} \text{at 1 atm} \\[4pt] \dfrac{1}{0.630/0.086 + 1/0.065} = 0.044 \text{ cm}^2/\text{s} \ (0.044 \times 10^{-4} \text{ m}^2/\text{s}) \\ \hspace{7cm} \text{at 10 atm} \end{cases}$$

At $y_{H_2} = 0.8$ the two results are

$$D = \begin{cases} 0.063 \text{ cm}^2/\text{s} \ (0.063 \times 10^{-4} \text{ m}^2/\text{s}) & \text{at 1 atm} \\ 0.050 \text{ cm}^2/\text{s} \ (0.050 \times 10^{-4} \text{ m}^2/\text{s}) & \text{at 10 atm} \end{cases}$$

This example illustrates the following point. The variation of D with y_A depends on the importance of bulk diffusion. At the extreme where the Knudsen mechanism controls, the composition has no effect on D. When bulk diffusion is significant, the effect is a function of α. For equimolal counterdiffusion, $\alpha = 0$ and y_A has no influence on D. In our example, where $\alpha = 0.741$, and at 10 atm pressure, D increased only from 0.044 to 0.050 cm^2/s as y_{H_2} increased from 0.5 to 0.8.

Also note that under reaction conditions in a pore, as in part (a), the ratio of the diffusion rates of the species is determined by stoichiometry. In contrast, for nonreacting systems at constant pressure, Eq. (11-8) is applicable.

11-2 Diffusion in Liquids

The mean free path in liquids is so small that Knudsen diffusion is not significant. Thus the diffusion rate is unaffected by the pore diameter and pressure (in the absence of surface diffusion). The *effective* diffusivity is determined by the molecular diffusivity and the pore structure of the catalyst pellet. Since the molecules in liquids are close together, the diffusion of one component is strongly affected by the force fields of nearby molecules and at the pore wall. As a result, diffusivities are concentration dependent and difficult to predict. As an approximation, we may express the diffusion flux of component A in a single cylindrical pore as:

$$N_A = -\mathscr{D}_{AB} \frac{dC_A}{dx} \tag{11-18}$$

where $\mathscr{D}_{AB}$ is the molecular diffusivity of liquid A in a solution of A and B and dC_A/dx is the concentration gradient in the direction of diffusion.

Values of $\mathscr{D}_{AB}$ are much less than those for gases and are, in general, of the order of 1×10^{-5} cm^2/s. For gases (for example, hydrogen) dissolved in liquids the diffusivity can be an order of magnitude larger, as noted in Examples 10-9 and 11-8. Several correlations are available† for estimating diffusivities in liquids at low concentrations (infinite dilution). These may be used as approximate values in reactor problems but the uncertain and sometimes large effect of concentration on $\mathscr{D}_{AB}$ should not be forgotten.

The major need for liquid-phase diffusivities is in problems involving slurry or trickle-bed reactors. Even though a gas phase is present, the wetting of the catalyst particles by the liquid means that the pores will be essentially filled with liquid. Since diffusivities in liquids are lower than those in gases, internal transport resistances can have a larger effect on the global rate for trickle-bed reactors than for gas-solid (two-phase), fixed-bed reactors. This effect is illustrated in Example 11-8. Even in slurry reactors employing catalyst particles of the order of 100 microns diameter, intraparticle diffusion can affect the global rate.‡

11-3 Diffusion in Porous Catalysts

Considerable experimental data has been accumulated for effective diffusivities in gas-filled pores.§ Since reactors normally are operated at steady state and nearly constant pressure, diffusivities have also been measured under these restraints. The usual apparatus¶ is of the steady-flow type, illustrated in Fig. 11-1 for studying diffusion rates of H_2 and N_2. The effective diffusivity is defined in terms of such rates (per unit of total cross-sectional area) by the equation

$$(N_A)_e = -D_e \frac{dC_A}{dr} = -\frac{p}{R_g T} D_e \frac{dy_A}{dr} \tag{11-19}$$

where the subscript e on N_A emphasizes that this is a diffusion flux in a porous catalyst rather than for a single pore, as given by Eq. (11-1). If we use the concentration-independent diffusivity D given by Eq. (11-4), D_e for a porous pellet will also be constant. Then Eq. (11-19) can be integrated to give

$$(N_A)_e = -\frac{p}{R_g T} D_e \frac{(y_A)_2 - (y_A)_1}{\Delta r} \tag{11-20}$$

where Δr is the length of the pellet. If the flow rates and concentrations are measured for the experiment pictured in Fig. 11-1, $(N_A)_e$ can be calculated. Then this flux and the measured concentrations and pellet length are substituted in Eq. (11-20) to obtain an experimental effective diffusivity.

Dynamic methods have also been used to measure D_e. For example,†† a pulse input of diffusing component A can be inserted into a stream of helium flowing

---

† See R. C. Reid, J. M. Prausnitz, and T. K. Sherwood, "The Properties of Gases and Liquids," 3d ed., chap. 11, McGraw-Hill Book Company, New York, 1977.

‡ Takehiko Furusawa and J. M. Smith, *Ind. Eng. Chem. Fundam.*, **12**, 197 (1979).

§ A summary up to 1969 is available in C. N. Satterfield, "Mass Transfer in Heterogeneous Catalysis," pp. 56–77, Massachusetts Institute of Technology Press, Cambridge, Mass., 1970. See also C. N. Satterfield and P. J. Cadle, *Ind. Eng. Chem., Fund. Quart.*, **7**, 202 (1968); *Process Design Develop.*, **7**, 256 (1968); L. F. Brown, H. W. Haynes, and W. H. Manogue, *J. Catalysis*, **14**, 220 (1969).

¶ Originally proposed by E. Wicke and R. Kallenbach [*Kolloid-Z.*, **17**, 135 (1941)].

†† Gulsen Dogu and J. M. Smith, *AIChE J.*, **21**, 58 (1975).

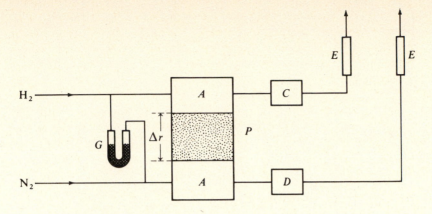

A — Mixing chambers
C — Detector for determining composition of N_2 in H_2 stream
D — Detector for determining composition of H_2 in N_2 stream
E — Flow meters
P — Catalyst pellet
G — Pressure equalization gauge

Figure 11-1 Constant-pressure apparatus for measuring diffusion rates in porous catalysts.

through the upper chamber, as indicated in Fig. 11-2. Pure helium also flows through the lower chamber. Some of the pulse of A diffuses through the pellet and is measured as a response pulse at the detector in the lower stream. For high flow rates of a nonadsorbing component, the effective diffusivity is given by

$$D_e = \frac{(\Delta r)^2 \varepsilon_p}{6\mu_1} \tag{11-21}$$

Here, μ_1 is the first moment (retention time) of the diffusing component in the pellet. It is obtained from the measured response pulse $C_A(t)$ by the equation

$$\mu_1 = \frac{\int_0^\infty C_A(t)t\, dt}{\int_0^\infty C_A(t)\, dt} \tag{11-22}$$

where $C_A(t)$ = concentration (function of time, t) in the response pulse
ε_p = porosity of pellet
Δr = length of pellet

Other equations have been developed,† both for adsorbing components and for lower flow rates across the pellet faces (see Prob. 11-5).

Comparison of diffusivities from steady-state and dynamic experiments can, in principle, provide information about pore structure. For example, dead-end pores should not affect diffusion at steady state, but would influence the results in dynamic experiments. Actually, uncertainties in accuracy of experimental results

---

† A. Burghardt and J. M. Smith, *Chem. Eng. Sci.,* **34**, 367 (1979) and Gulsen Dogu and J. M. Smith, *Chem. Eng. Sci.,* **23**, 123 (1976).

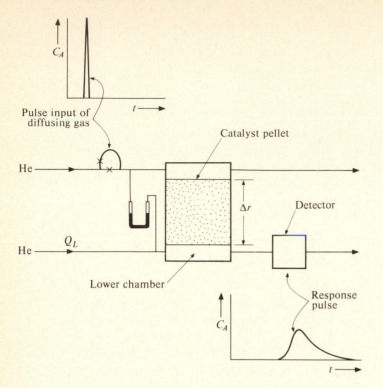

Figure 11-2 Dynamic method for measuring diffusivities in porous catalysts.

has prevented, as yet, meaningful comparisons. Another method of evaluating effective diffusivities, provided the rate equation for the chemical step is known, is comparison of the observed and predicted global rates by combining the equation for the chemical step with the intraparticle mass-transport resistance (see Example 11-9 for a variation of this method).

In the absence of experimental data it is necessary to estimate D_e from the physical properties of the catalyst. In this case the first step is to evaluate the diffusivity for a single cylindrical pore, that is, to evaluate D from Eq. (11-4). Then a geometric *model* of the pore system is used to convert D to D_e for the porous pellet. A model is necessary because of the complexity of the geometry of the void spaces. The optimum model is a realistic representation of the geometry of the voids (with tractable mathematics) that can be described in terms of easily measurable physical properties of the catalyst pellet. As noted in Chap. 8, these properties are the surface area and pore volume per gram, the density of the solid phase, and the distribution of void volume according to pore size.

The Parallel-pore Model Wheeler† proposed a model, based on the first three of these properties, to represent the monodisperse pore-size distribution in a catalyst

---

† A. Wheeler, in P. H. Emmett (ed.), "Catalysis," vol. II, chap. 2, Reinhold Publishing Corporation, New York, 1955; A. Wheeler, "Advances in Catalysis," vol. III, p. 250, Academic Press, Inc., New York, 1951.

pellet. From ρ_s and V_g the porosity ε_p is obtained from Eq. (8-9). Then a mean pore radius $\bar{a}$ is evaluated† by writing equations for the total pore volume and total pore surface in a pellet. The result, developed as Eq. (8-18), is

$$\bar{a} = \frac{2V_g}{S_g} \tag{11-23}$$

By using V_g, S_g, and ρ_s, Wheeler replaced the complex porous pellet with an assembly (having a porosity ε_p) of cylindrical pores of radius $\bar{a}$. To predict D_e from the model the only other property necessary is the length x_L of the diffusion path. If we assume that, on the average, the pore makes an angle of 45° with the coordinate r in the resultant direction of diffusion (for example, the radial direction in a spherical pellet), $x_L = \sqrt{2}\,r$. Owing to pore interconnections and non-cylindrical shape, this value of x_L is not very satisfactory. Hence, it is customary to define x_L in terms of an adjustable parameter, the *tortuosity factor* δ, as follows:

$$x_L = \delta r \tag{11-24}$$

An effective diffusivity can now be predicted by combining Eq. (11-1) for a single pore with this parallel-pore mode. To convert D, which is based on the cross-sectional area of the pore, to a diffusivity based upon the total area perpendicular to the direction of diffusion, D should be multiplied by the porosity. In Eq. (11-1), x is the length of a single, straight cylindrical pore. To convert this length to the diffusion path in a porous pellet, x_L from Eq. (11-24) should be substituted for x. With these modifications the diffusive flux in the porous pellet will be

$$(N_A)_e = -\frac{p}{R_g T}\frac{\varepsilon D}{\delta}\frac{dy_A}{dr} \tag{11-25}$$

Comparison with Eq. (11-19) shows that the effective diffusivity is

$$D_e = \frac{\varepsilon D}{\delta} \tag{11-26}$$

where D is given (in the absence of surface diffusion) by Eq. (11-4). The use of Eq. (11-26) to predict D_e is somewhat limited because of the uncertainty about δ. Comparison of D_e from Eq. (11-26) with values obtained from experimental data for various catalysts using Eq. (11-20) shows that δ varies from less than unity to more than 6.‡

The Random-pore Model This model was originally developed for pellets containing a bidisperse pore system, such as the alumina described in Chap. 8 (Fig. 8-8). It is supposed that the pellet consists of an assembly of small particles. When the

---

† Recall from Sec. 8-3 that regardless of the complexity of the void structure, it is assumed that the void spaces may be regarded as cylindrical pores.

‡ Tortuosity factors less than unity can occur when the surface diffusion is significant. This is because D_e is increased, while D as defined in Eq. (11-4) does not include this contribution. See Sec. 11-4 for a corrected D to include surface diffusion. C. N. Satterfield ("Mass Transfer in Heterogeneous Catalysis," Massachusetts Institute of Technology Press, Cambridge, Mass., 1970) has summarized data from the literature and recommended the use of $\delta = 4$ when surface diffusion is insignificant. See also P. Carman, *Trans. Inst. Chem. Engrs.*, **15**, 150 (1937).

particles themselves contain pores (micropores), there exists both a macro and a micro void-volume distribution. The voids are not imagined as capillaries, but more as an assembly of short void regions surrounding and between individual particles, as indicated in Fig. 11-3. The nature of the interconnection of macro and micro void regions is the essence of the model. Transport in the pellet is assumed to occur by a combination of diffusion through the macro regions (of void fraction ε_M), the micro regions (of void fraction ε_μ), and a series contribution involving both regions. It is supposed that both micro and macro regions can be represented as straight, short cylindrical pores of average radii $\bar{a}_M$ and $\bar{a}_\mu$. The magnitude of the individual contributions is dependent on their effective cross-sectional areas (perpendicular to the direction of diffusion). The details of the development are given elsewhere,† but in general these areas are evaluated from the probability of pore interconnections. The resultant expression for D_e may be written

$$D_e = \bar{D}_M \varepsilon_M^2 + \frac{\varepsilon_\mu^2 (1 + 3\varepsilon_M)}{1 - \varepsilon_M} \bar{D}_\mu \tag{11-27}$$

Here $\bar{D}_M$ and $\bar{D}_\mu$ are obtained by applying Eq. (11-4) to macro and micro regions. Thus

$$\frac{1}{\bar{D}_M} = \frac{1}{\mathscr{D}_{AB}} + \frac{1}{(\bar{\mathscr{D}}_K)_M} \tag{11-28}$$

$$\frac{1}{\bar{D}_\mu} = \frac{1}{\mathscr{D}_{AB}} + \frac{1}{(\bar{\mathscr{D}}_K)_\mu} \tag{11-29}$$

No tortuosity factor is involved in this model.‡ The actual path length is equal to the distance coordinate in the direction of diffusion. To apply Eq. (11-27)

---

† N. Wakao and J. M. Smith, *Chem. Eng. Sci.*, **17**, 825 (1962); *Ind. Eng. Chem., Fund. Quart.*, **3**, 123 (1964).

‡ Thus the random-pore model does not involve an adjustable parameter.

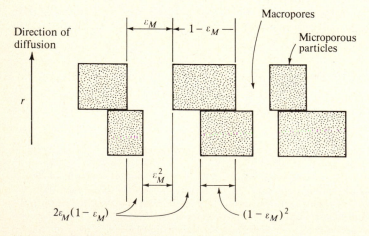

Figure 11-3 Random-pore model.

requires void fractions and mean pore radii for both macro and micro regions. The mean pore radii can be evaluated for the micro region by applying Eq. (11-23) to this region. However, $\bar{a}_M$ must be obtained from the pore-volume distribution, as described in Sec. 8-3. The mean pore radii are necessary in order to calculate $(\mathscr{D}_K)_M$ and $(\mathscr{D}_K)_\mu$ from Eqs. (11-28) and (11-29).

The random-pore model can also be applied to monodisperse systems. For a pellet containing only macropores, $\varepsilon_\mu = 0$ and Eq. (11-27) becomes

$$D_e = \bar{D}_M \varepsilon_M^2 \tag{11-30}$$

Similarly, for a material such as silica gel, where $\varepsilon_M = 0$, the effective diffusivity is

$$D_e = \bar{D}_\mu \varepsilon_\mu^2 \tag{11-31}$$

Comparison of these last two equations with Eq. (11-26) indicates that $\delta = 1/\varepsilon$. The significance of the random-pore model is that the effective diffusivity is proportional to the square of the porosity. This has also been proposed by Weisz and Schwartz.[†] Johnson and Stewart[‡] have developed another method for predicting D_e that utilizes the pore-volume distribution. Evaluation of their model and the random-pore model with extensive experimental data has been carried out by Satterfield and Cadle[§] and Brown et al.[¶]

The examples that follow illustrate the methods that have been outlined for evaluating D_e from experimental data and predicting it from a pore model.

Example 11-4 Rothfeld [††] has measured diffusion rates for isobutane, in the isobutane-helium system, through a $\frac{1}{8}$-in.-long pelleted cylinder of alumina (diameter $\frac{1}{8}$ in.). The measurements were at 750 mmHg total pressure and 25°C, and the diffusion direction was through the pellet parallel to the central axis. The following data are available:

$$S_g = 76 \text{ m}^2/\text{g}$$

$$\varepsilon_M = 0.18 \qquad \varepsilon_\mu = 0.34$$

$$\bar{a}_M = 4800 \text{ Å} \qquad \bar{a}_\mu = 84 \text{ Å}$$

The mole fraction of isobutane is 1.0 on one face of the pellet and zero on the other face. The experimental results gave

$$\frac{(N_A)_e R_g T \Delta r}{\mathscr{D}_{AB} p_t (y_2 - y_1)} = -0.023$$

where N_A is the diffusion flux of isobutane and $\mathscr{D}_{AB}$ is the bulk diffusivity in the isobutane-helium system.

† P. B. Weisz and A. B. Schwartz, *J. Catalysis*, **1**, 399 (1962).

‡ M. F. L. Johnson and W. E. Stewart, *J. Catalysis*, **4**, 248 (1965).

§ C. N. Satterfield and P. J. Cadle, *Ind. Eng. Chem., Fund. Quart.*, **7**, 202 (1968); *Process Design Develop.*, **7**, 256 (1968).

¶ L. F. Brown, H. W. Haynes, and W. H. Manogue, *J. Catalysis*, **14**, 220 (1969).

†† L. B. Rothfeld, *AIChE J.*, **9**, 19 (1963).

(a) Calculate the experimental value of D_e. (b) What macropore tortuosity factor is indicated by the data? What is δ_M predicted by the random pore model?

SOLUTION (a) According to the procedure of Example 11-1 (the Chapman-Enskog equation), the bulk diffusivity in the isobutane-helium system at 750 mmHg pressure and 25°C is

$$\mathscr{D}_{AB} = 0.313 \text{ cm}^2/\text{s}$$

Thus the measured diffusion flux of isobutane is

$$(N_A)_e = \frac{-0.023(0.313)(\frac{750}{760})(0 - 1.0)}{82(25 + 273)\frac{1}{8}(2.54)} = 9.1 \times 10^{-7} \text{ g mol/(s)(cm}^2)$$

This value of $(N_A)_e$ can be used in Eq. (11-20) to calculate the experimental D_e,

$$D_e = -\frac{(N_A)_e R_g T \Delta r}{p(y_2 - y_1)} = \frac{-9.1 \times 10^{-7}(82)(298)(\frac{1}{8})(2.54)}{\frac{750}{760}(0 - 1.0)} = 0.0072 \text{ cm}^2/\text{s}$$

(b) The *parallel-pore* model is designed for a *monodisperse* pore system and hence is not directly applicable to this catalyst. Since the macropores are much larger than the micropores, an approximate approach is to neglect the contribution of the micropores to the mass transport. Using the macro properties in Eq. (11-26) and the experimental D_e, we have

$$D_e = 0.0072 = \frac{\varepsilon_M \bar{D}_M}{\delta_M} = \frac{0.18 \bar{D}_M}{\delta_M}$$

or

$$\delta_M = \frac{0.18 \, \bar{D}_M}{0.0072}$$

The combined diffusivity in the macropores is given by Eq. (11-28)

$$\frac{1}{\bar{D}_M} = \frac{1}{\mathscr{D}_{AB}} + \frac{1}{(\bar{D}_K)_M}$$

From Eq. (11-17),

$$(\mathscr{D}_K)_M = 9.70 \times 10^3 (4{,}800 \times 10^{-8})\left(\frac{298}{58.1}\right)^{1/2} = 1.10 \text{ cm}^2/\text{s}$$

$$\frac{1}{\bar{D}_M} = \frac{1}{0.313} + \frac{1}{1.10}$$

$$\bar{D}_M = 0.243 \text{ cm}^2/\text{s}$$

Then the tortuosity factor suggested by the data is

$$\delta_M = \frac{0.18(0.243)}{0.0072} = 6.1$$

If we assume that all the diffusion is in the macropores, Eq. (11-30) gives D_e for the *random-pore model*. Combining Eq. (11-30) with Eq. (11-26) yields

$$\frac{\varepsilon_M \bar{D}_M}{\delta_M} = D_e = \bar{D}_M \varepsilon_M^2$$

or

$$\delta_M = \frac{1}{\varepsilon_M} = \frac{1}{0.18} = 5.5$$

in reasonable agreement with the experimental result. Actually, this pellet is relatively dense (low ε_M) and probably was prepared with a high pelleting pressure. Hence the diffusion likely is affected by the micropores.†

Example 11-5 Vycor (porous silica) appears to have a pore system with fewer interconnections than alumina. The pore system is monodisperse, with the somewhat unusual combination of a small mean pore radius (45 Å) and a low porosity 0.31. Vycor may be much closer to an assembly of individual voids than to an assembly of particles surrounded by void spaces. Since the random-pore model is based on the assembly-of-particles concept, it is instructive to see how it applies to Vycor. Rao and Smith‡ measured an effective diffusivity for hydrogen of 0.0029 cm²/s in Vycor. The apparatus was similar to that shown in Fig. 11-1, and data were obtained using an $H_2 - N_2$ system at 25°C and 1 atm. Predict the effective diffusivity by the random-pore model.

SOLUTION Only micropores are present, and so D_e should be predicted by means of Eq. (11-31). Furthermore, mass transport in the small pores will be predominately by Knudsen diffusion. Hence $\bar{D}_\mu = (\bar{\mathscr{D}}_K)_\mu$, from Eq. (11-29), and

$$D_e = (\bar{\mathscr{D}}_K)_\mu \varepsilon_\mu^2$$

$$(\bar{\mathscr{D}}_K)_\mu = 9.7 \times 10^3 (45 \times 10^{-8}) \left(\frac{298}{2.02}\right)^{1/2} = 0.053 \text{ cm}^2/\text{s}$$

Then the predicted D_e is

$$D_e = 0.053(0.31)^2 = 0.0050 \text{ cm}^2/\text{s}$$

This value is 70% greater than the experimental result—evidence that the random-pore model is not very suitable for Vycor.

---

† W. Wakao and J. M. Smith [*Chem. Eng. Sci.*, **17**, 825 (1962)] analyzed the diffusion data of Rothfeld. These data were obtained in an apparatus of the type shown in Fig. 11-1. For the butane-helium system this means that N_{He}/N_{C_4} is 3.80. Diffusion is far from equimolal, suggesting that Eqs. (11-28) and (11-29) for D values are not exact. For this particular case Eq. (11-2) should be used. In most reaction systems the counterdiffusion of reactants and products is much closer to equimolal, so that Eqs. (11-28) and (11-29) are better approximations.

‡ M. J. Rao and J. M. Smith, *AIChE J.*, **10**, 243 (1964).

The experimental tortuosity, from Eq. (11-26), is

$$\delta = \frac{0.31\bar{D}_\mu}{D_e} = \frac{0.31(\bar{\mathscr{D}}_K)_\mu}{D_e} = \frac{0.31(0.053)}{0.0029} = 5.6$$

In contrast, the tortuosity predicted by the random-pore model would be $1/\varepsilon = 3.2$.

Silica gel of large surface area has even smaller pores than Vycor, but a larger void fraction; for example, for one grade of silica gel $\varepsilon = 0.486$ and $\bar{a} = 11$ Å. Schneider† studied the diffusion of ethane in this material at 200°C (at this temperature surface diffusion was negligible) and found an effective diffusivity such that δ from Eq. (11-26) was 3.34. The value predicted from the random-pore model is $1/\varepsilon = 2.1$. Note that the Wheeler model would predict $\delta = \sqrt{2}$ for Vycor, silica gel, or other porous material.

A final word should be said about the variety of porous materials. Porous catalysts cover a rather narrow range of possibilities. Perhaps the largest variation is between monodisperse and bidisperse pellets, but even these differences are small in comparison with materials such as freeze-dried beef, which is like an assembly of solid fibers, and freeze-dried fruit, which appears to have a structure like an assembly of ping-pong balls with holes in the surface to permit a continuous void phase.‡ Also for sintered pellets, where pore interconnections have been greatly reduced by heating, tortuosity factors as high as 100 have been measured.§

11-4 Surface Diffusion

Surface migration is pertinent to a study of intrapellet mass transfer if its contribution is significant with respect to diffusion in the pore space at reaction temperature. When multimolecular-layer adsorption occurs, surface diffusion has been explained as a flow of the outer layers as a condensed phase.¶ However, surface transport of interest in relation to catalytic reaction occurs in the monomolecular layer. It is more appropriate to consider, as proposed by deBoer,†† that such transport is an activated process, dependent on surface characteristics as well as those of the adsorbed molecules. Imagine that a molecule in the gas phase strikes the pore wall and is adsorbed. Then two alternatives are possible: desorption into the gas or movement to an adjacent active site on the pore wall (surface diffusion). If desorption occurs, the molecule can continue its journey in the void space of the pore or be readsorbed by again striking the wall. In moving along the wall, the same molecule would be transported sometimes on the surface and sometimes in the gas phase. If this view is correct, the relative contribution of surface migration

† Petr Schneider and J. M. Smith, *AIChE J.*, **14**, 886 (1968).

‡ J. C. Harper, *AIChE J.*, **8**, 298 (1962).

§ J. H. Krasuk and J. M. Smith, *AIChE J.*, **18**, 506 (1972).

¶ D. H. Everett in F. S. Stone (ed.), *Structure and Properties of Porous Materials, Proc. Tenth Symp. Colston Res. Soc.*, Butterworth, London, p. 178, 1958.

†† J. H. deBoer, in "Advances in Catalysis," vol. VIII, p. 18, Academic Press, Inc., New York, 1959.

would increase as the surface area increases (or the pore size decreases). There is evidence to indicate that such is the case.[†]

Experimental verification of surface diffusion is usually indirect, since concentrations of adsorbed molecules on a surface are difficult to measure. When gas concentrations are obtained, the problem arises of separating the surface and pore-volume transport rates. One solution is to measure both N_A and N_B in the apparatus shown in Fig. 11-1, using a nonadsorbable gas for A. If the diffusion rate of B is greater than that calculated from N_A by Eq. (11-8), the excess is attributable to surface migration. Barrer and Barrie[‡] used this procedure with Vycor at room temperature and found surface migration significant for such gases as CO_2, CH_4, and C_2H_4 and negligible for helium and hydrogen. Rivarola[§] found that for CO_2 the surface contribution on an alumina pellet increased from 3.5 to 54% of the total mass-transfer rate as the macropore properties changed from $\bar{a}_M = 1,710$ Å and $\varepsilon_M = 0.33$ to $\bar{a}_M = 348$ Å and $\varepsilon_M = 0.12$.

By analogy to Fick's law, a surface diffusivity $\mathscr{D}_s$ may be defined in terms of a surface concentration C_s in moles of adsorbate per square centimeter of surface:

$$N_s = -\mathscr{D}_s \frac{dC_s}{dx} \tag{11-32}$$

where N_s is the molal rate per unit perimeter of pore surface. In order to combine surface and pore-volume contributions in a *catalyst pellet*, the surface flux should be based on the total area of the catalyst perpendicular to diffusion and on the coordinate r. If this flux is $(N_s)_e$, then

$$(N_s)_e = -D_s \rho_P \frac{d\bar{C}}{dr} \tag{11-33}$$

where ρ_P is the density of the catalyst and $\bar{C}$ now represents the moles adsorbed per gram of catalyst.[§] To be useful, Eq. (11-33) must be expressed in terms of the concentration in the gas phase. If the adsorption step is fast with respect to surface transfer from site to site on the catalyst, we may safely assume equilibrium between gas and surface concentration. Otherwise the relation between the two concentrations depends on the intrinsic rates of the two processes.[¶] If equilibrium is assumed, and if the isotherm is linear, then

$$\bar{C}_A = (K_A \bar{C}_m) C_g = K'_A \frac{p y_A}{R_g T} \tag{11-34}$$

where the subscript A refers to the adsorbate and K' is the linear form of the equilibrium constant, in cubic centimeters per gram. The latter is obtained from

[†] J. B. Rivarola and J. M. Smith, *Ind. Eng. Chem., Fund. Quart.*, **3**, 308 (1964).

[‡] R. M. Barrer and J. M. Barrie, *Proc. Roy. Soc. (London)*, **A213**, 250 (1952).

[§] To relate $\mathscr{D}_s$ and D_s requires a model for the porous structure. The parallel-pore and random-pore models have been applied to surface diffusion by J. H. Krasuk and J. M. Smith [*Ind. Eng. Chem., Fund. Quart.*, **4**, 102 (1965)] and J. B. Rivarola and J. M. Smith [*Ind. Eng. Chem., Fund. Quart.*, **3**, 308 (1964)].

[¶] J. A. Moulijn, et al. [*Ind. Eng. Chem. Fundam.* **16**, 301 (1077)] have extended the theory of intraparticle diffusion to allow for finite rates of surface diffusion.

the Langmuir isotherm, Eq. (7-15), when adsorption is small enough that the linear form is valid. Now, applying Eq. (11-33) to component A,

$$(N_s)_e = -\frac{p}{R_g T}\rho_P K'_A D_s \frac{dy_A}{dr} \tag{11-35}$$

Equation (11-35) gives the surface diffusion of A in the same form as Eq. (11-19) applied earlier to transport in the gas phase of the pores. From these equations the total flux and total effective diffusivity are given by

$$(N_A)_t = -\frac{p}{R_g T}(D_e + \rho_P K'_A D_s)\frac{dy_A}{dr} \tag{11-36}$$

$$(D_e)_t = D_e + \rho_P K'_A D_s \tag{11-37}$$

If the density of the catalyst and the adsorption equilibrium constant are known, Eq. (11-37) permits the evaluation of a total effective diffusivity from D_s. Data for D_s have been reported in the literature in a variety of ways, depending on the definitions of adsorbed concentrations and equilibrium constants. Schneider† has summarized much of the information for light hydrocarbons on various catalysts in the form defined by Eq. (11-35). Such D_s values ranged from 10^{-3} to 10^{-6}, depending on the nature of the adsorbent and the amount adsorbed. Most values were in the interval 10^{-4} to 10^{-5} cm^2/s. Data for other adsorbates‡ have similar magnitudes and show that the variation with adsorbed concentration can be large.

The effect of temperature on D_s, given an activated process, is described by an Arrhenius-type expression,

$$D_s = \mathbf{A}e^{-E_s/R_g T}$$

where E_s is the activation energy for surface diffusion. The variation of K'_A with temperature is given by van't Hoff equation

$$\frac{d \ln K'_A}{dT} = \frac{\Delta H}{R_g T^2}$$

or

$$K'_A = \mathbf{A}'e^{-\Delta H/R_g T}$$

The observed rate of surface diffusion, according to Eq. (11-35), will be proportional to the product $K'_A D_s$. Substituting the above expressions for K'_A and D_s in Eq. (11-35),

$$(N_s)_e = -\frac{p}{R_g T}\rho_P(\mathbf{A}\mathbf{A}')e^{-(1/R_g T)(\Delta H + E_s)}\frac{dy_A}{dr}$$

† Petr Schneider and J. M. Smith, *AIChE J.*, **14**, 886 (1968).

‡ R. A. W. Haul, *Angew. Chem.*, **62**, 10 (1950); P. S. Carman and F. A. Raal, *Proc. Roy. Soc. (London)*, **209A**, 38 (1951), *Trans. Faraday Soc.*, **50**, 842 (1954); D. H. Everett in F. S. Stone (ed.), Structure and Properties of Porous Materials, *Proc. Tenth Symp. Colston Res. Soc.*, Butterworth, London, p. 178, 1958.

The exponential term expresses a much stronger temperature dependency than the coefficient $1/T$. If we neglect the latter, we may express the *temperature* effect on the rate as

$$(N_s)_e = -A'' e^{-(1/R_g T)(\Delta H + E_s)}$$

where

$$A'' = AA' \frac{dy_A}{dr} \frac{p}{R_g T} \rho_p$$

This equation shows that the observed or apparent activation energy for surface diffusion is related to E_s by

$$E' = \Delta H + E_s \tag{11-38}$$

From available data† it appears that E_s is only a few kilocalories per mole. The heat of adsorption ΔH is generally greater than this, particularly for chemisorption, and is always negative. Therefore the observed effect is a decrease in rate of surface diffusion with increase in temperature. Note that ΔH is negative.

From the assumptions and approximations presented, it is clear that surface diffusion is not well understood. It is hoped that improved interpretations of surface migration will permit a more accurate assessment of its effect on global rates of reaction. When we consider the effect of intraparticle resistances in Secs. 11-7 to 11-12 we shall suppose that the D_e used is the most appropriate value and includes, if necessary, a surface contribution.

INTRAPELLET HEAT TRANSFER

11-5 Concept of Effective Thermal Conductivity

The effective thermal conductivities of catalyst pellets are surprisingly low. Therefore significant intrapellet temperature gradients can exist, and the global rate may be influenced by thermal effects. The effective conductivity is the energy transferred per unit of *total* area of pellet (perpendicular to the direction of heat transfer). The defining equation, analogous to Eq. (11-19) for mass transfer, may be written

$$Q_e = -k_e \frac{dT}{dr} \tag{11-39}$$

where Q_e is the rate of energy transfer per unit of total area.

A major factor contributing to small values of k_e is the numerous void spaces that hinder the transport of energy. In addition, the path through the solid phase offers considerable thermal resistance for many porous materials, particularly pellets made by compressing microporous particles. Such behavior is readily understood if these materials are viewed as an assembly of particles which contact each other only through adjacent points. There is strong experimental evidence that such point contacts are regions of high thermal resistance. For example, the

† Petr Schneider and J. M. Smith, *AIChE J.*, **14**, 886 (1968).

thermal conductivity of the bulk solid (zero porosity) from which the particles are prepared does not have a large effect on k_e. Masamune† found that the effective thermal conductivity of pellets of microporous particles of silver was only two to four times that of alumina pellets, at the same macropore void fraction, pressure, and temperature. In contrast, the thermal conductivity of solid silver is about 200 times as large as that of solid alumina. Furthermore, k_e is a strong function of the void fraction, increasing as ε_p decreases. Materials such as alumina pellets may be regarded as a porous assembly within a second porous system. Each particle from which the pellet is made contains a microporous region. These particles are in point contact with like particles and are surrounded by macroporous regions. When viewed in this way, the thermal conductivity of the bulk solid should have little influence on k_e.

The pressure and nature of the fluid in the pores has an effect on the effective thermal conductivity. With liquids the effect of pressure is negligible and k_e is of the same magnitude as the true conductivity of the liquid. For gases at low pressures, where the mean free path is the same or larger than the pore size, free-molecule conduction controls the energy transfer. In this region k_e increases with pressure. At higher pressures k_e is about independent of pressure. The transition pressure depends on the gas as well as on the pore size. For air the transition pressure is about 470 mm in a silver pellet with mean pore diameter of 1500 Å. For helium the value would be above 760 mm.† For alumina pellets, at 120°F and a macropore void fraction of $\varepsilon_M = 0.40$, k_e was 0.050 in vacuum, 0.082 with pores filled with air, and 0.104 Btu/(h)(ft)(°F) with helium, at atmospheric pressure.‡ Temperature does not have a strong influence. The effect is about what would be expected for the combination of variations of thermal conductivity with temperature for the solid and fluid phases.

11-6 Effective Thermal-Conductivity Data

Most of the experimental information on k_e for catalyst pellets is described by Masamune and Smith.† Mischke and Smith‡ and Sehr.§ Sehr gives single values for commonly used catalysts. The other two works present k_e as a function of pressure, temperature, and void fraction for silver and alumina pellets. Both transient and steady-state methods have been employed. Figure 11-4 shows the variation of k_e with pellet density and temperature for alumina (boehmite, $Al_2O_3 \cdot H_2O$) pellets. Different densities were obtained by increasing the pressure used to pellet the microporous particles. The data are at vacuum conditions and therefore represent the conduction of the solid matrix of the pellet. Note how low k_e is in comparison with the thermal conductivity of solid alumina [about 1.0 Btu/(h)(ft)(°F)]. The low value is due to the small heat-transfer areas at the point-to-point contacts between particles. As the pelleting pressure increases (lower macropore void fraction), these contact areas increase, and so does k_e. Figure 11-5 shows the effect of macropore void fraction on k_e for pellets of microporous silver

† S. Masamune and J. M. Smith, *J. Chem. Eng. Data*, **8**, 54 (1963).

‡ R. A. Mischke and J. M. Smith, *Ind. Eng. Chem.*, *Fund. Quart.*, **1**, 288 (1962).

§ R. A. Sehr, *Chem. Eng. Sci.*, **2**, 145 (1958).

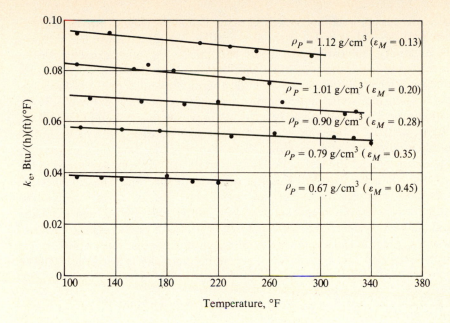

Figure 11-4 Effective thermal conductivity of alumina (boehmite) catalyst pellets at 10 to 25 microns Hg pressure.

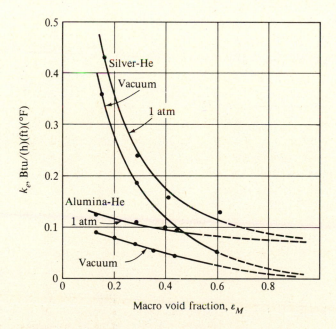

Figure 11-5 Effect of macropore void fraction on k_e at 34°C.

and microporous alumina (boehmite) particles. Data are given for two conditions: vacuum and pores filled with helium at 1 atm, both at 34°C. Since helium has a high conductivity, the two curves would encompass the values expected for any reaction mixture. As 1 atm is at or beyond the transition pressure, increasing the pressure would have little effect (until the thermodynamic critical point is approached). Similarly, silver and boehmite represent close to the extremes of conductivity for the solids that might be expected in porous catalysts. Figure 11-6 is a similar plot of data for three cases: vacuum, air-filled pores at 1 atm pressure, and helium-filled pores at 1 atm pressure. These results are for boehmite pellets.†

The theory of heat transfer in porous materials has not been developed to the level of that for mass transfer. The contribution of the solid phase makes the problem more complex. It is not yet possible to predict k_e accurately from the properties of the fluid and solid phases. Butt‡ has applied and extended the random-pore model to develop a valuable method of predicting the effects of void fraction, pressure, and temperature on k_e. A different and more approximate approach§ proposes that the effective thermal conductivity is a function only of

† R. A. Mischke and J. M. Smith, *Ind. Eng. Chem., Fund. Quart.*, **1**, 288 (1962).
‡ J. B. Butt, *AIChE J.*, **11**, 106 (1965).
§ W. Woodside and J. H. Messmer, *J. Appl. Phys.*, **32**, 1688 (1961).

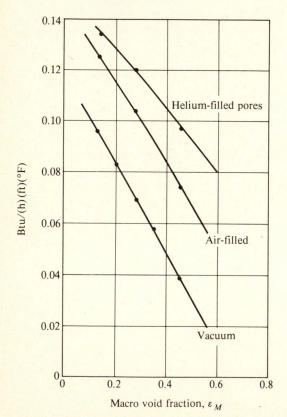

Figure 11-6 Effective thermal conductivity of alumina (boehmite) pellets vs. void fraction at 120°F.

the volume fraction of the void phase and the thermal conductivities of the bulk-fluid and solid phases, k_f and K_s. The relationship is

$$k_e = k_s\left(\frac{k_f}{k_s}\right)^{1-\varepsilon} \tag{11-40}$$

In spite of the difficulties in predicting k_e, it is still possible to choose a value which will be reasonably correct, because the possible range of values (excluding vacuum conditions) is only from about 0.1 to 0.4 Btu/(h)(ft)(°F). Furthermore, the nature of the variations within this range that are due to void fraction, temperature, and pressure are known approximately from Figs. 11-4 to 11-6.

MASS TRANSFER WITH REACTION

After discussing the effective transport coefficients D_e and k_e, we now turn to the main objective of the chapter: an expression of the rate of reaction for the whole catalyst pellet,† $\mathbf{r}_P$, in terms of the temperature and concentrations existing at the *outer* surface. We start in a formal way by defining an effectiveness factor η as follows:

$$\eta = \frac{\text{actual rate for the whole pellet}}{\text{rate evaluated at outer surface conditions}} = \frac{\mathbf{r}_P}{\mathbf{r}_s} \tag{11-41}$$

The equation for the local rate (per unit mass of catalyst) as developed in Chap. 9 may be expressed functionally as $\mathbf{r} = f(C, T)$, where C represents, symbolically, the concentrations of all the involved components.‡ Then Eq. (11-41) gives for $\mathbf{r}_P$

$$\mathbf{r}_P = \eta \mathbf{r}_s = \eta f(C_s, T_s) \tag{11-42}$$

With the formulation of Eq. (11-42) the objective becomes the evaluation of η rather than $\mathbf{r}_P$. Once η is known, Eq. (11-42) gives the rate for the whole pellet in terms of the temperature and concentration at the outer surface. Then the rate can be expressed in terms of the temperature and concentration of the bulk fluid by accounting for external resistances (Chap. 10). The effectiveness factor is a function of k_e, D_e, and the rate constants associated with the chemical step at the site, i.e., the constants in the rate equations developed in Chap. 9. In the remainder of this chapter we shall develop the relationship between η and these rate parameters. In Secs. 11-7 to 11-10 isothermal conditions are assumed. With this restraint k_e is not involved, and Eq. (11-42) becomes

$$\mathbf{r}_P = \eta f(C_s) \tag{11-43}$$

The nonisothermal problem is considered in Secs. 11-11 and 11-12.

---

† $\mathbf{r}_P$ is the rate for the whole pellet, but based on a unit mass of catalyst. The description "whole pellet" is used to denote that intrapellet effects are accounted for in $\mathbf{r}_P$. If the rate measured for one or more pellets is $\mathbf{r}$ and the mass of the pellets is m, then $\mathbf{r}_P = \mathbf{r}/m$.

‡ For example, for the reaction $A + B \rightleftharpoons C$, Eq. (9-21) indicates that $f(C, T)$ would be

$$f(C, T) = k_s \bar{C}_m K_A K_B \frac{C_A C_B - (1/K)(C_C)}{(1 + K_A C_A + K_B C_B + K_C C_C)^2}$$

where k and all the K values are functions of temperature.

11-7 Effectiveness Factors

Suppose that an irreversible reaction $A \to B$ is first order, so that for isothermal conditions $\mathbf{r} = f(C_A) = k_1 C_A$. Then Eq. (11-43) becomes

$$\mathbf{r}_P = \eta k_1 (C_A)_s \tag{11-44}$$

We want to evaluate η in terms of D_e and k_1. The first step is to determine the concentration profile of A in the pellet. This is shown schematically in Fig. 11-7 for a spherical pellet (also shown is the external profile from C_b to C_s). The differential equation expressing C_A vs. r is obtained.† by writing a mass balance over the spherical-shell volume of thickness Δr (Fig. 11-7). According to Eq. (3-1) at steady state the rate of diffusion into the element‡ less the rate of diffusion out will equal the rate of disappearance of reactant within the element. This rate will be $\rho_p k_1 C_A$ per unit volume, where ρ_P is the density of the pellet. Hence the balance may be written, omitting subscript A on C,

$$\left(-4\pi r^2 D_e \frac{dC}{dr} \right)_r - \left(-4\pi r^2 D_e \frac{dC}{dr} \right)_{r+\Delta r} = 4\pi r^2\, \Delta r \rho_p k_1 C \tag{11-45}$$

---

† This development was first presented by A. Wheeler in W. G. Frankenburg, V. I. Komarewsky, and E. K. Rideal (eds.), "Advances in Catalysis," vol. III, p. 297, Academic Press, Inc., New York, 1951; see also P. H. Emmett (ed.), "Catalysis," vol. II, p. 133, Reinhold Publishing Corporation, New York, 1955.

‡ The diffusive flux into the element is given by Eq. (11-19). Note that for $A \to B$ there is equimolal counterdiffusion of A and B ($\alpha = 0$). The rate of diffusion is the product of the flux and the area, $4\pi r^2$.

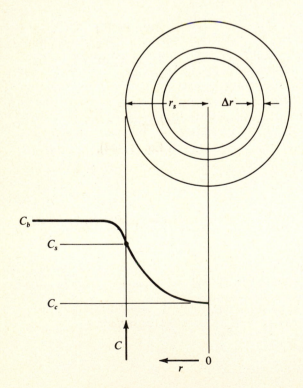

Figure 11-7 Reactant (A) concentration vs. position for first-order reaction on a spherical catalyst pellet.

If we take the limit as $\Delta r \to 0$ and assume that the effective diffusivity is independent of the concentration of reactant (see Sec. 11-2), this difference equation becomes

$$\frac{d^2 C}{dr^2} + \frac{2}{r}\frac{dC}{dr} - \frac{k_1 \rho_P}{D_e} C = 0 \tag{11-46}$$

At the center of the pellet symmetry requires

$$\frac{dC}{dr} = 0 \qquad \text{at } r = 0 \tag{11-47}$$

and at the outer surface

$$C = C_s \qquad \text{at } r = r_s \tag{11-48}$$

Linear differential equation (11-46) with boundary conditions (11-47) and (11-48) may be solved by conventional methods to yield

$$\frac{C}{C_s} = \frac{r_s}{r}\frac{\sinh(3\Phi_s r/r_s)}{\sinh 3\Phi_s} \tag{11-49}$$

where Φ_s is a dimensionless group (a Thiele-type modulus for a *spherical* pellet) defined by

$$\Phi_s = \frac{r_s}{3}\sqrt{\frac{k_1 \rho_P}{D_e}} \tag{11-50}$$

The second step is to use the concentration profile given by Eq. (11-49), to evaluate the rate of reaction r_P for the whole pellet. We have two choices for doing this: calculating the diffusion rate of reactant *into* the pellet at r_s, or integrating the local rate over the whole pellet. Choosing the first approach,

$$r_P = \frac{1}{m_P} 4\pi r_s^2 D_e \left(\frac{dC}{dr}\right)_{r=r_s} = \frac{3}{r_s \rho_P} D_e \left(\frac{dC}{dr}\right)_{r=r_s}$$

where the mass of the pellet is $m_P = \frac{4}{3}\pi r_s^3 \rho_P$. Then, from Eq. (11-44),

$$\eta = \frac{3 D_e}{r_s \rho_P k_1 C_s}\left(\frac{dC}{dr}\right)_{r=r_s} \tag{11-51}$$

Differentiating Eq. (11-49), evaluating the derivative at $r = r_s$, and substituting this into Eq. (11-51) gives

$$\eta = \frac{1}{\Phi_s}\left(\frac{1}{\tanh 3\Phi_s} - \frac{1}{3\Phi_s}\right) \tag{11-52}$$

If this equation for the effectiveness factor is used in Eq. (11-44), the desired rate for the whole pellet in terms of the *concentration at the outer surface* is

$$r_P = \frac{1}{\Phi_s}\left(\frac{1}{\tanh 3\Phi_s} - \frac{1}{3\Phi_s}\right)k_1 C_s \tag{11-53}$$

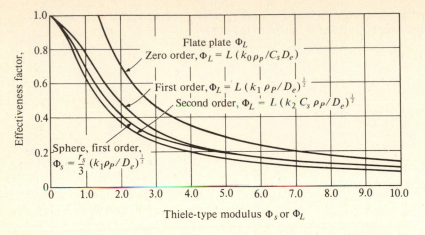

Figure 11-8 Effectiveness factor for various pellet shapes and kinetic equations.

Both D_e and k_1 are necessary to use Eq. (11-53). The relative importance of diffusion and chemical-reaction processes is evident from Eq. (11-52), which is represented by the second-lowest curve in Fig. 11-8. This curve shows that for small values of Φ_s, $\eta \to 1$. Then intraparticle mass transport has no effect on the rate per pellet; the chemical step controls the rate. From Eq. (11-50), small values of Φ_s are obtained when the pellets are small, the diffusivity is large, or the reaction is intrinsically slow (catalyst of low activity). For $\Phi_s > 5$ a good approximation for Eq. (11-52) is

$$\eta = \frac{1}{\Phi_s} \tag{11-54}$$

For such large Φ_s intraparticle diffusion has a large effect on the rate. Practically, these conditions mean that diffusion into the pellet is *relatively* slow, so that reaction occurs before the reactant has diffused far into the pellet. In fact, an alternate definition for η is the fraction of the whole surface that is as active as the external surface. If $\eta \to 1$, Eq. (11-43) shows that the rate for the whole pellet is the same as the rate if all the surface were available to reactant at concentration C_s; i.e., the rate at the center is the same as the rate at the outer surface—all the surface is fully effective. In this special case the concentration profile shown in Fig. 11-6 would be horizontal, with $C = C_s$. In contrast, if $\eta \ll 1$, only the surface near the outer periphery of the pellet is effective; the concentration drops from C_s to nearly zero in a narrow region near r_s. In this case the catalyst in the central portion of the pellet is not utilized. Note that such a situation is caused by large pellets, low D_e, or high k_1. The latter factor shows that low effectiveness factors are more likely with a very active catalyst. Thus the more active the catalyst, the more likely it is that intrapellet diffusion resistance will reduce the global rate.

Equations (11-52) and (11-53) provide a method for accounting for intrapellet mass transport for one case, a spherical pellet, and first-order irreversible reaction. The effect of shape on the η-vs.-Φ relationship has been examined by several investigators.† For a flat plate (slab) of catalyst *sealed from reactants on one side* and all ends,

$$\eta = \frac{\tanh \Phi_L}{\Phi_L} \tag{11-55}$$

$$\Phi_L = L \sqrt{\frac{k_1 \rho_P}{D_e}} \tag{11-56}$$

where L is the thickness of the plate. The curve for Eq. (11-55), as well as flat plate curves for zero and second-order kinetics, are also shown in Fig. 11-8. The first-order curves for spherical and flat-plate geometry show little deviation—less than the error involved in evaluating k_1 and D_e. Hence the shape of the catalyst pellet is not very significant,‡ provided the different definitions of Φ_L and Φ_s are taken into account. For example, suppose the intrinsic rate is second order for a spherical catalyst particle. We may obtain a reasonably accurate η value by using the second-order curve for a plate in Fig. 11-8. However, in using this curve we would take L equal to $r_s/3$ for evaluating Φ_L. The curve for the flat plate can also be used when both sides of the plate are exposed to reactants, provided L is one-half the plate thickness.

First-order kinetics was chosen in writing Eq. (11-46), so that an analytical solution could be obtained. Numerical solutions for η vs. Φ have been developed for many other forms of rate equations.§ Solutions include those for Langmuir-Hinshelwood equations¶ with denominator terms, as derived in Chap. 9 [e.g., Eq. (9-21)]. To illustrate the extreme effects of reaction, Wheeler†† obtained solutions for zero- and second-order kinetics for a flat plate of catalyst, and these results are also shown in Fig. 11-8. For many catalytic reactions the rate equation is approximately represented by a first-, second-, or intermediate-order kinetics. In these cases the curves for first- and second-order kinetics in Fig. 11-8 define a region within which the effectiveness factor will lie. For unusual situations—for example, when the desorption of a reaction product limits the rate—η values outside of this region may exist. For unusual cases the previous references of Hougen, Schneider, Satterfield, and their colleagues should be consulted. Also,

---

† R. Aris, *Chem. Eng. Sci.*, **6**, 262 (1957); A Wheeler in W. G. Frankenburg, V. I. Komarewsky, and E. K. Rideal (eds.), "Advances in Catalysis," vol. III, Academic Press, Inc., New York, 1951.

‡ The curve for η vs. Φ for a cylinder lies between the curves for the sphere and flat plate in Fig. 11-8. For first-order kinetics, Φ for the cylinder (no axial diffusion) is defined

$$\Phi_c = (r_s/2)\sqrt{k_1 \rho_P/D_e}$$

§ O. A. Hougen and Chu, *Chem. Eng. Sci.*, **17**, 167 (1962); P. Schneider and R. A. Mitschka, *Collection Czecho. Chem. Commun.*, **30**, 146 (1965), **31**, 1205, 3677 (1966); P. Schneider and R. A. Mitschka, *Chem. Eng. Sci.*, **21**, 455 (1965).

¶ G. W. Roberts and C. N. Satterfield, *Ind. Eng. Chem., Fund. Quart.*, **4**, 288 (1965), **5**, 317, 325 (1966).

†† A. Wheeler in W. G. Frankenburg, V. I. Komarewsky, and E. K. Rideal (eds.), "Advances in Catalysis," vol. III, Academic Press, Inc., New York, 1951. The definition of Φ_L for the second-order curve is $L(k_2 C_s \rho_P/D_e)^{1/2}$, where k_2 is the conventional second-order rate constant. For the zero-order case $\Phi_L = (k_0 \rho_P/C_s D_e)^{1/2}$.

Aris[†] has presented a comprehensive mathematical treatment of the effectiveness-factor problem, including the effects of particle shape, multiple reactions, and nonisothermal operation, as well as various types of kinetics. Bischoff[‡] has developed a procedure for evaluating η for any form of kinetics by defining a generalized Thiele-type modulus.

The earliest studies of diffusion and reaction in catalysts were by Thiele,[§] Damkoehler,[¶] and Zeldowitsch.[††] Thiele considered the problem from the standpoint of a single cylindrical pore (see Prob. 11-12). Since the catalytic area per unit length of diffusion path does not change in a straight cylindrical pore whose walls are catalytic, the results are of the form of those for a flat plate [Eqs. (11-55) and (11-56)].

Nothing has been said as yet about reversible reactions. For the first-order case the irreversible result can be used, with some modification, for reversible reactions (Example 11-6).

Example 11-6 Derive equations for the effectiveness factor for a first-order reversible reaction $A \rightleftharpoons B$ at isothermal conditions, for a spherical catalyst pellet.

SOLUTION It has been shown [Eqs. (2-62) and (2-64)] that the rate equation for a reversible first-order reaction can be written

$$\mathbf{r} = k_R(C - C_{eq}) \tag{A}$$

where C_{eq} is the equilibrium concentration of reactant at the temperature involved, and k_R is related to the forward-rate constant k_1 and the equilibrium constant K by

$$k_R = \frac{k_1(K + 1)}{K} \tag{B}$$

Since Eq. (A) is to be applied to a catalytic reaction, the rate is expressed as g mol/(s)(g catalyst) and k_R has the dimensions $cm^3/(s)(g\ catalyst)$.

The mass balance of reactant on a spherical shell (see Fig. 11-7) will be the same as Eq. (11-45) except for the reaction term; that is,

$$\left(-4\pi r^2 D_e \frac{dC}{dr}\right)_r - \left(-4\pi r^2 D_e \frac{dC}{dr}\right)_{r+\Delta r} = 4\pi r^2\,\Delta r \rho_p k_R(C - C_{eq})$$

Taking the limit as $\Delta r \to 0$ and replacing C with the variable $C' = C - C_{eq}$,

$$\frac{d^2C'}{dr^2} + \frac{2}{r}\frac{dC'}{dr} - \frac{k_R \rho_p}{D_e}C' = 0 \tag{C}$$

† Rutherford Aris, "The Mathematical Theory of Diffusion and Reaction in Permeable Catalysts," chaps. 3–5, vol. I, Clarendon Press, Oxford, 1975.
‡ K. B. Bischoff, *AIChE J.*, **11**, 351 (1965).
§ E. W. Thiele, *Ind. Eng. Chem.*, **31**, 916 (1939).
¶ G. Damkoehler, *Chem. Eng.*, **3**, 430 (1937).
†† J. B. Zeldowitsch, *Acta Physicochim. U.R.S.S.*, **10**, 583 (1939).

The boundary conditions are

$$\frac{dC'}{dr} = 0 \qquad\qquad \text{at } r = 0 \qquad\qquad \text{(D)}$$

$$C' = C_s - C_{eq} = C'_s \qquad \text{at r} = r_s \qquad\qquad \text{(E)}$$

Equations (C) to (E) are the same as Eqs. (11-46) to (11-48), with C' replacing C and k_R replacing k_1. Hence the solution for the effectiveness factor will be identical with Eq. (11-52), but Eq. (11-50) for Φ_s becomes

$$\Phi'_s = \frac{r_s}{3} \sqrt{\frac{k_1(K + 1)\rho_P}{K D_e}} \qquad\qquad (11\text{-}57)$$

These results show that the first-order curves in Fig. 11-8 can be used for reversible as well as irreversible reactions, provided k_1 in the definition of Φ_s or Φ_L is replaced by $k_1(K + 1)/K$. Since $(K + 1)/K$ is greater than unity, Φ will be greater, and η less, for reversible reactions than for irreversible ones.

11-8 The Significance of Intrapellet Diffusion: Evaluation of the Effectiveness Factor

Isothermal effectiveness factors for practical reactions cover a wide range. With normal pellet sizes ($\frac{1}{8}$ to $\frac{1}{2}$ in.) η is 0.7 to 1.0 for intrinsically slow reactions, such as the ammonia synthesis, and of the order of $\eta = 0.1$ for fast reactions, such as some hydrogenations of unsaturated hydrocarbons. Satterfield[†] and Sherwood[‡] have summarized much of the experimental data for effectiveness factors for various reactions, temperatures, and pellet sizes. For reactor design it is important to be able to answer these questions:

1. Should intrapellet diffusion resistance be considered in evaluating the global rate? That is, is η significantly less than unity?
2. If $\eta < 1$, how can η be evaluated from a minimum of experimental data?

We use the results of Sec. 11-7 to formulate answers to these questions.

Suppose that the rate $\mathbf{r}_P$ is measured at a given bulk concentration of reactant. Suppose also that either the external resistance is negligible, or the surface concentration C_s has been evaluated from the bulk value by the methods discussed in Chap. 10. Weisz[§] has provided a criterion for deciding, from these measurements and D_e, whether intrapellet diffusion may be disregarded. The basic premise is that if $\Phi_s \leq \frac{1}{3}$, then η is not much less than unity (Fig. 11-8 indicates that η will be

---

[†] C. N. Satterfield, "Mass Transfer in Heterogeneous Catalysis," pp. 152–156, Massachusetts Institute of Technology Press, Cambridge, Mass., 1970.

[‡] C. N. Satterfield and T. K. Sherwood, "The Role of Diffusion in Catalysis," pp. 72–75, Addison-Wesley Publishing Company, Reading, Mass., 1963.

[§] P. B. Weisz, *Z. Phys. Chem.*, **11**, 1 (1957).

greater than 0.9 for $\Phi_s \leq \frac{1}{3}$). Equation (11-50) shows that the criterion may be written

$$r_s \sqrt{\frac{k_1 \rho_P}{D_e}} \leq 1$$

or

$$r_s^2 \frac{k_1 \rho_P}{D_e} \leq 1 \tag{11-58}$$

The unknown rate constant k_1 can be eliminated in favor of the measured rate $\mathbf{r}_P$ from Eq. (11-44), noting that $\eta \to 1.0$. In terms of $\mathbf{r}_P$ Eq. (11-58) becomes

$$r_s^2 \frac{\mathbf{r}_P \rho_P}{C_s D_e} \leq 1 \tag{11-59}$$

The usefulness of Eq. (11-59) stems from the fact that the curves for η vs. Φ for first- and higher-order reactions in Fig. 11-8 would be nearly coincident at $\Phi_s \leq \frac{1}{3}$. Hence Eq. (11-59) is satisfactory as an approximate criterion for most catalytic kinetics, even though it was derived for a first-order case.

Example 11-7 The rate of isomerization of n-butane with a silica-alumina catalyst is measured at 5 atm (507 kPa) and 50°C (323 K) in a laboratory reactor with high turbulence in the gas phase surrounding the catalyst pellets. Turbulence ensures that external-diffusion resistances are negligible, so that $C_s = C_b$. Kinetic studies indicate that the rate is first order and reversible. At 50°C the equilibrium conversion is 85%. The effective diffusivity is 0.08 cm^2/s at reaction conditions, and the density of the catalyst pellets is 1.0 g/cm^3, regardless of size. The measured, global rates when pure n-butane surrounds the pellets are as follows:

d_P, in.	1/8	1/4	3/8
$\mathbf{r}_P$, g mol/(s)(g catalyst) or kg/mol/(s)(kg catalyst)	4.85×10^{-4}	4.01×10^{-4}	3.54×10^{-4}

(a) To reduce pressure drop in the proposed fixed-bed reactor it is desirable to use the maximum pellet size for which there will be little reduction in the global rate due to intrapellet resistances. The heat of isomerization is low enough that the whole pellet is at 50°C. What is the largest size pellet that may be used? (b) Calculate the effectiveness factor for each size.

SOLUTION (a) In Example 11-6 the proper definition of Φ_s' for a reversible first-order reaction was shown to be given by Eq. (11-57). From this definition, the criterion for $\eta \to 1$ is

$$\left(\frac{r_s}{3}\right)^2 \frac{k_1(K+1)\rho_P}{KD_e} \leq \left(\frac{1}{3}\right)^2$$

or

$$r_s^2 \frac{k_1(K+1)\rho_P}{KD_e} \leq 1 \tag{A}$$

The rate for the whole pellet (per unit mass of catalyst) for the first-order reversible case is the same as Eq. (11-44), with $k_R(C_s - C_{eq})$ replacing $k_1 C_s$:

$$\mathbf{r}_P = \eta k_R(C_s - C_{eq}) = \eta \left[\frac{k_1(K + 1)}{K} \right](C_s - C_{eq}) \qquad (11\text{-}60)$$

Combining Eqs. (A) and (11-60) to eliminate k_1, and noting that $\eta \to 1$, we obtain

$$r_s^2 \frac{\mathbf{r}_P \rho_P}{D_e(C_s - C_{eq})} \leq 1 \qquad (11\text{-}61)$$

Equation (11-61) is the proper criterion to use for reversible reactions in place of Eq. (11-59).

For an equilibrium conversion of 85%,

$$\frac{C_s - C_{eq}}{C_s} = 0.85$$

At 5 atm and 50°C

$$C_s - C_{eq} = 0.85 \frac{p}{R_g T} = 0.85 \frac{5}{82(323)}$$

$$= 1.60 \times 10^{-4} \text{ g mole/cm}^3 \ (1.60 \times 10^{-1} \text{ kg/mol/m}^3)$$

With numerical values, Eq. (11-61) becomes

$$r_s^2 \mathbf{r}_P \frac{1.0}{0.08(1.60 \times 10^{-4})} = 7.82 \times 10^4 r_s^2 \mathbf{r}_P \leq 1 \qquad (B)$$

For $\frac{1}{8}$-in. pellets, $r_s = \frac{1}{8}(\frac{1}{2})(2.54) = 0.159$ cm $(0.159 \times 10^{-2}$ m). Hence

$$r_s^2 \frac{\mathbf{r}_P \rho_P}{D_e(C_s - C_{eq})} = 7.82 \times 10^4 (0.159)^2 (4.85 \times 10^{-4}) = 0.95$$

For the other two sizes

$$\frac{r_s^2 \mathbf{r}_P \rho_P}{D_e(C_s - C_{eq})} = 3.2 \qquad \frac{1}{4}\text{-in. pellets}$$

$$\frac{r_s^2 \mathbf{r}_P \rho_P}{D_e(C_s - C_{eq})} = 6.3 \qquad \frac{3}{8}\text{-in. pellets}$$

The $\frac{1}{8}$-in. pellets are the largest for which intrapellet diffusion has a negligible effect on the rate.

(b) To calculate η, one relation between Φ_s' and η is obtained from Eq. (11-57) of Example 11-6 and a second is Eq. (11-52), or the Φ_s curve in Fig. 11-8. From Eqs. (11-57) and (11-60) k_1 can be eliminated, giving

$$\Phi_s' = \frac{r_s}{3} \sqrt{\frac{\mathbf{r}_P \rho_P}{\eta D_e(C_s - C_{eq})}} \qquad (11\text{-}62)$$

Table 11-4

d_P, in.	Φ'_s	η
1/8	0.33	0.93
1/4	0.67	0.77
3/8	1.00	0.68

The only unknowns in Eq. (11-62) are η and Φ'_s for a given pellet size. For the $\frac{3}{8}$-in. pellets

$$\Phi'_s = \frac{0.476}{3}\sqrt{\frac{3.54 \times 10^{-4}(1.0)}{0.08(1.60 \times 10^{-4})\eta}} = 0.158\sqrt{\frac{27.6}{\eta}}$$

Simultaneous solution of this expression and Eq. (11-52) yields

$$\Phi'_s = 1.00 \qquad \eta = 0.68$$

Results for the other sizes, obtained the same way, are shown in Table 11-4.

Example 11-7 illustrates one of the problems in scale-up of catalytic reactors. The results showed that for all but $\frac{1}{8}$-in. pellets intrapellet diffusion significantly reduced the global rate of reaction. If this reduction were not considered, erroneous design could result. For example, suppose the laboratory kinetic studies to determine a rate equation were made with $\frac{1}{8}$-in. pellets. Then suppose it was decided to use $\frac{3}{8}$-in. pellets in the commercial reactor to reduce the pressure drop through the bed. If the rate equation were used for the $\frac{3}{8}$-in. pellets without modification, the rate could be erroneously high. At the conditions of part (b) of Example 11-7 the correct r_P would be only 0.68/0.93, or 73% of the rate measured with $\frac{1}{8}$-in. pellets.

With respect to the second question of this section, part (b) of Example 11-7 illustrated one method of evaluating η when intrapellet mass transfer is important. In addition to an experimentally determined rate, it was necessary to know the effective diffusivity of the pellet. The need for D_e may be eliminated by making rate measurements for two or more sizes of pellets, provided D_e is the same for all sizes. To show this, note from Eq. (11-44) that the ratio of the rates for two sizes 1 and 2 is

$$\frac{(r_P)_2}{(r_P)_1} = \frac{\eta_2}{\eta_1} \qquad (11\text{-}63)$$

Also, from Eq. (11-50),

$$\frac{(\Phi_s)_2}{(\Phi_s)_1} = \frac{(r_s)_2}{(r_s)_1} \qquad (11\text{-}64)$$

Furthermore, Eq. (11-52) gives a relation between η and Φ_s which must be satisfied for both pellets. These are the four relations between the four unknowns, η_1, $(\Phi_s)_1$,

η_2, $(\Phi_s)_2$. Solution by trial is easiest. One procedure† is to assume a value of η_2 and calculate η_1 from Eq. (11-63). Determine $(\Phi_s)_1$ from Eq. (11-52); then $(\Phi_s)_2$ can be calculated from Eq. (11-64). Finally, applying Eq. (11-52) for pellet 2 gives a revised value of η_2. The calculations are continued until the initial and calculated values of η_2 agree. This method is illustrated in Example 11-8 where data including the effect of external mass transport are analyzed. The method is not valid for large Φ_s, for then Eq. (11-52) reduces to Eq. (11-54), and combination with Eqs. (11-63) and (11-64) shows that the rate is inversely proportional to r_s *for all sizes.*

If one of the pellets for which $\mathbf{r}_P$ is measured is very small (e.g., powder form), η equals 1.0 for that size. Then the ratio of Eq. (11-43) for the small pellet, $(\mathbf{r}_P)_1$, and a larger size pellet, $(\mathbf{r}_P)_2$, gives, for any reaction order,

$$\frac{(\mathbf{r}_P)_2}{(\mathbf{r}_P)_1} = \frac{\eta_2\, f(C_s)}{1.0 f(C_s)}$$

or

$$\eta_2 = \frac{(\mathbf{r}_P)_2}{(\mathbf{r}_P)_1} \tag{11-65}$$

Hence the effectiveness factor for a given pellet can be obtained by measuring the rate for the pellet and for a small particle size of the same catalyst at the same concentration of reactant.

The rate measurement for the small particles determines the rate of the chemical reaction at the catalyst site, i.e., without intrapellet-diffusion resistance. Then the rate constant k can be calculated directly. Thus it is not necessary that the measurements for the small particles and the pellet be at the same concentration of reactant, as required by Eq. (11-65). Consider an irreversible first-order reaction. From Eq. (11-44),

$$k_1 = \frac{(\mathbf{r}_P)_1}{(C_s)_1} \tag{11-66}$$

since $\eta_1 = 1.0$. Using this value of k_1 in Eq. (11-44), applied to the pellet, we have

$$\eta_2 = \frac{(\mathbf{r}_P)_2}{(C_s)_2 k_1} = \frac{(\mathbf{r}_P/C_s)_2}{(\mathbf{r}_P/C_s)_1} \tag{11-67}$$

Example 11-8 To illustrate the analysis of laboratory-reactor data to account for the combined effects of external and internal mass transfer, consider the following information.‡

A differential, packed-bed, catalytic reactor (ID = 0.95 cm) is used to study hydrogenation of α-methyl styrene to cumene. Liquid styrene containing only dissolved hydrogen is pumped through a short bed of Pd/Al$_2$O$_3$ catalyst particles. The concentration of H$_2$ in the flowing liquid stream is

† C. N. Satterfield and T. K. Sherwood, loc. cit.
‡ M. Herskowitz, Ph.D. Thesis, University of California, Davis, 1978.

approximately constant throughout the reactor and equal to 2.6×10^{-6} g mol/cm^3. The reactor operates at a constant temperature of 40.6°C and at steady state. Catalyst and reactor-bed properties are:

Catalyst = 0.2 wt% Pd on Al$_2$O$_3$ (granular)
Particle density, $\rho_p = 1.53$ g/cm^3
Bed void fraction, $\varepsilon_B = 0.48$
Particle porosity $\varepsilon_p = 0.50$

Table 11-5 gives the reaction rate, **r** (determined by measuring the production of cumene), as a function of liquid flow rate, Q, for two sizes (d_p = equivalent spherical diameter) of catalyst particles. Both sizes were taken from the same batch of catalyst. At these conditions the reaction rate is first order with respect to hydrogen.

From the data given calculate effectiveness factors and the effective diffusivity for hydrogen in the liquid-filled pores of the catalyst particles.

SOLUTION In this study of the reaction,

$$C_6H_5—CH{=}CH_2 + H_2 \rightarrow C_6H_5—CH_2CH_3$$

there is no gas phase, and the pores of the catalyst are filled with liquid. Even though the particle sizes are small, the low diffusivity in liquids may mean that intraparticle diffusion retards the rate. Also, the increase in reaction rate with liquid flow rate shown in Table 11-5 indicates that external mass transfer retards the rate. Under these conditions the global rate is given by Eq. (11-44) and also by the external mass transfer expression, Eq. (10-1). Equating these rate equations:

$$\mathbf{r}_p = \eta k_1 (C_{H_2})_s = k_m a_m (C_b - C_s)_{H_2} \tag{A}$$

Table 11-5 Global rate data for the hydrogenation of α-methyl styrene

Q cm^3/s	$r \times 10^6$ g mol/(g catalyst)(s)	
	$d_p = 0.054$ cm	$d_p = 0.162$ cm
2.5	.....	0.65
3.0	1.49	
5.0	1.56	0.72
8.0	1.66	0.80
10.0	1.70	0.82
11.5	.....	0.85
12.5	1.80	
15.0	1.90	0.95
25.0	1.94	1.02
30.0	.....	1.01

where k_m is the mass transfer coefficient and a_m is the mass transfer area per unit mass of catalyst. The last equality can be solved for C_s and then substituted in the first equality to obtain the rate in terms of the bulk liquid concentration. Doing this, and omitting the H_2 subscript, gives

$$r = \frac{C_b}{1/k_1\eta + 1/k_m a_m}$$

or

$$\frac{C_b}{r} = \frac{1}{k_1\eta} + \frac{1}{k_m a_m} \tag{B}$$

Experimental data are available for r vs. Q. Since $k\eta$ is not a function of liquid flow rate and $k_m a_m$ is, Eq. (B) can be used to separate and evaluate $k\eta$ and $k_m a_m$. Then the values for $k\eta$ for the two particle sizes permit determining k and η separately by using Eqs. (11-63) and (11-64).

An alternate procedure would be to use the correlation of Eq. (10-10) with Eq. (10-9) to evaluate $k_m a_m$. Then $k\eta$ would be obtained from Eq. (B). It is probably more accurate to use the experimental data for the effect of Q, rather than the correlation, and we will follow this procedure.

Combination of Eqs. (10-9) and (10-10) to eliminate j_D indicates that $k_m a_m$ is proportional to $(d_p G/\mu)^b$, or Q^b, at constant d_p. Then Eq. (B) becomes

$$\frac{C_b}{r} = \frac{1}{k_1\eta} + A\frac{1}{Q^b} \tag{C}$$

where A is a constant for data at different Q and the same catalyst particle size. Equation (C) shows that a graph of known values of C_b/r vs. Q^{-b} should be a straight line with an intercept, at $Q^{-b} = 0$, equal to $1/k_1\eta$. The power b is given as 0.407 in the correlation of Eq. (10-10). However, we can establish b by trial, noting the value which leads to the best straight line. This b would supposedly account for the characteristics of the particular bed (uniformity of packing, etc.) not accounted for in Eq. (10-10).

Figure 11-9 shows that with $b = 0.3$ the data establish reasonably good straight lines.† The intercepts are

$$(k_1\eta_1)^{-1} = 0.77 \quad \text{for} \quad (d_p)_1 = 0.054 \text{ cm} \tag{D}$$

$$(k_1\eta_2)^{-1} = 1.32 \quad \text{for} \quad (d_p)_2 = 0.162 \text{ cm} \tag{E}$$

Since k_1 is the same for both particle sizes, the ratio of the intercepts gives

$$\frac{\eta_1}{\eta_2} = \frac{(k_1\eta_2)^{-1}}{(k_1\eta_1)^{-1}} = \frac{1.32}{0.77} = 1.71$$

† A range of b values (from 0.25 to 0.45) give good straight lines for all of which the intercepts are about the same. Since there are sufficient data points, a linear mean square fit of the data might be a logical way to establish $k_1\eta$.

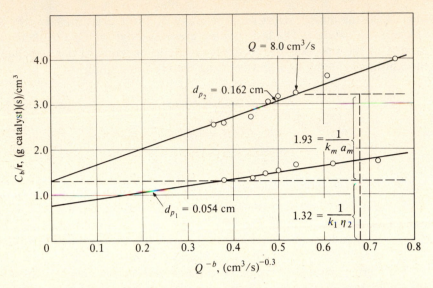

Figure 11-9 Effect of external mass transfer on global rate for hydrogenation of α-methyl styrene.

This equation is analogous to Eq. (11-63). Equation (11-64) provides the second relation between η_1, η_2, $(\phi_s)_1$ and $(\phi_s)_2$:

$$\frac{(\Phi_s)_1}{(\Phi_s)_2} = \frac{(d_p)_1}{(d_p)_2} = \frac{0.054}{0.162} = 0.333$$

The other two relations (one for each particle size) are provided by Eq. (11-52) or by the curve for spherical particles in Fig. 11-8. Solving by trial

$$\eta_1 = 0.88 \qquad (\Phi_s)_1 = 0.58$$
$$\eta_2 = 0.51 \qquad (\Phi_s)_2 = 1.63$$

Then from Eq. (C)

$$\frac{1}{k_1(0.88)} = 0.77$$

and

$$k_1 = 1.5 \text{ cm}^3/(g)(s)$$

Now the diffusivity can be found from the definition of Φ_s. Thus, from Eq. (11-50) for d_{p_1}

$$(\Phi_s)_1 = 0.58 = \frac{0.054/2}{3}\left[\frac{1.5(1.53)}{D_e}\right]^{1/2}$$

$$D_e = 5.5 \times 10^{-4} \text{ cm}^2/s$$

This diffusivity is relatively high for liquids (see Sec. 11-3) but is typical of values for dissolved hydrogen. Note that for the larger particles the effect of both external and internal transport was significant for this relatively fast reaction. For example, at $Q = 8.0$ cm^3/s from Fig. 11-9, the values of the terms in Eq. (B) are, for $(d_p)_2 = 0.162$ cm

$$\frac{2.6 \times 10^{-6}}{0.80 \times 10^{-6}} = 3.25 = \frac{1}{1.5(0.51)} + 1.93$$

$$= 1.32 + 1.93$$

At these conditions the resistance to external mass transfer (1.93) is greater than the combined resistance (1.32) of reaction and intraparticle diffusion. Also, the η_2 of 0.51 means that the effect of intraparticle diffusion was to reduce the intrinsic rate by 50%.

11-9 Experimental and Calculated Effectiveness Factors

When the rate is measured for a catalyst pellet and for small particles, and the diffusivity is also measured or predicted, it is possible to obtain both an experimental and a calculated result for η. For example, for a first-order reaction Eq. (11-67) gives η_{exp} directly. Then the rate measured for the small particles can be used in Eq. (11-66) to obtain k_1. Provided D_e is known, Φ can be evaluated from Eq. (11-50) for a spherical pellet or from Eq. (11-56) for a flat plate of catalyst. Then η_{calc} is obtained from the proper curve in Fig. 11-7. Comparison of the experimental and calculated values is an overall measure of the accuracy of the rate data, effective diffusivity, and the assumption that the intrinsic rate of reaction (or catalyst activity) is the same for the pellet and the small particles.

Example 11-9 illustrates the calculations and results for a flat-pellet of NiO catalyst, on an alumina carrier, used for the ortho-para-hydrogen conversion.

> **Example 11-9**† The pellet reactor is shown in Fig. 11-10. The reaction gases were exposed to one face of the cylindrical disk (1 in. in diameter and $\frac{1}{4}$ in. in depth), while the other face and the cylindrical surface were sealed. Vigorous turbulence near the exposed face ensured uniform composition in the gas region and eliminated external diffusion resistance. The reaction
>
> $$o\text{-H}_2 \rightleftharpoons p\text{-H}_2$$
>
> has a very low heat of reaction, and the whole reactor was enclosed in a liquid nitrogen bath. Thus isothermal conditions obtain at $-196°C$.
>
> Rates were measured at $-196°C$ and 1 atm pressure for pellets of three densities. The rate of reaction was also measured for the catalyst in the form of 60-micron (average size) particles. With this small size, $\Phi \ll 1$ so that $\eta = 1.0$. The rate data and pellet properties are given in Table 11-6.

---

† This example is taken from the data of M. R. Rao, N. Wakao, and J. M. Smith, *Ind. Eng. Chem.*, 3, 127 (1964).

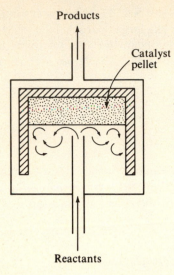

Products

Catalyst pellet

Reactants

Figure 11-10 Reactor for a single, flat-plate catalyst pellet.

Table 11-6 Catalyst and rate data for ortho-hydrogen conversion on NiO/Al_2O_3

Catalyst particles

$d_P = 60$ microns
$S_g = 278$ m^2/g
$V_g = 0.44$ cm^3/g
$\rho_S = 2.63$ g/cm^3
$\rho_P = 2.24$ g/cm^3
$\bar{a}_\mu = 29$ Å (from integration of pore-volume distribution)
$k_R = 0.688$ cm^3/(s)(g), calculated from the measured rate $(r_P)_1$, for the
   small particles ($\eta = 1.0$) using Eq. (A) of Example 11-6.

Catalyst pellets			
ρ_P, g/cm^3	ε_M	Macropore radius $\bar{a}_M$, Å	$(r_P)_2/(C_s - C_{eq})$ cm^3/(s)(g)
1.09	0.48	2,100	0.186
1.33	0.37	1,690	0.129
1.58	0.33	1,270	0.109

Note: The physical-property data were obtained from pore-size distribution and surface-area measurements, as described in Chap. 8.

SOLUTION Equation (11-67) is for an irreversible reaction, and so it cannot be used to calculate η_{exp} for the ortho-hydrogen reaction. The proper equation is obtained by applying Eq. (11-60) to the pellet. Solving for η gives

$$\eta_{exp} = \frac{(r_P)_2}{k_R(C_s - C_{eq})_2}$$

From this equation and the data given in Table 11-6, η_{exp} can be immediately evaluated. The results are:

ρ_P	η_{exp}
1.09	$\dfrac{0.186}{0.688} = 0.27$
1.33	$\dfrac{0.129}{0.688} = 0.19$
1.58	$\dfrac{0.109}{0.688} = 0.16$

To obtain η_{calc} we must estimate the effective diffusivity. Since the macropores are much larger than the micropores (see $\bar{a}_M$ and $\bar{a}_\mu$ in Table 11-6), it is safe to assume that diffusion is predominantly through the macropores. Then, according to the random-pore model [Eqs. (11-27) and (11-28)],

$$D_e = \varepsilon_M^2 \frac{1}{1/\mathscr{D}_{H_2} + 1/(\bar{\mathscr{D}}_K)_{H_2}} \tag{A}$$

For hydrogen at $-196°C$ and 1 atm, $\mathscr{D}_{H_2} = 0.14$ cm^2/s. Using Eq. (11-17) and the $\bar{a}_M$ values from Table 11-5, we obtain:

ρ_P	1.09	1.33	1.58
$(\bar{\mathscr{D}}_K)_{H_2}$, cm^2/s	1.28	1.03	0.77

From Eq. (A), the effective diffusivity is, for $\rho_P = 1.09$,

$$D_e = 0.48^2 \frac{1}{1/0.14 + 1/1.28} = 0.029 \text{ cm}^2/\text{s}$$

The results for the other pellet densities are given in the second column of Table 11-7. The Thiele-type modulus is calculated from Eq. (11-56), modified for a reversible reaction by substituting k_R for k_1:

$$\Phi'_L = L\sqrt{\frac{k_R \rho_P}{D_e}} = \frac{1}{4}(2.54)\sqrt{\frac{0.688(1.09)}{0.029}} = 3.2$$

From the first-order Φ_L-vs.-η curve in Fig. 11-8 for a flat plate, $\eta_{calc} = 0.30$. The results for all three pellets are given in the last columns of the table.

Table 11-7

ρ_P	D_e, cm^2/s	Φ'_L	η_{calc}	η_{exp}
1.09	0.029	3.2	0.30	0.27
1.33	0.017	4.6	0.21	0.19
1.58	0.013	5.7	0.17	0.16

The calculated and experimental effectiveness factors agreed well with each other in Example 11-9. Note that the calculated η required the rate for the small particles as the only experimental data. Hence the method offers an attractive procedure for predicting the rate for any size of pellet. However, there are data for other catalysts and reactions which suggest less favorable agreement. Results for the same reaction with a catalyst of NiO supported on silica gel showed good coincidence between η_{exp} and η_{calc}.[†] With Vycor as a carrier the agreement for effectiveness factors is reasonably good, even though calculated (by the random-pore model) and experimentally measured D_e were in poor accord[‡] Relatively large deviations in D_e cause small differences in η. Otani and Smith[§] applied the method to the reaction

$$CO + \tfrac{1}{2}O_2 \rightarrow CO_2$$

using spherical catalyst pellets of NiO on Al$_2$O$_3$. The calculated η were 50 to 100% higher than the experimental results, depending on the temperature. The reasons for the large deviation were obscure, but a decrease in catalyst activity may have been caused by plugging of some of the pores in the particles when the pellets were formed. Such results emphasize that the method of predicting η depends on assuming that the intrinsic rate of reaction is the same for the surface in the small particles as for the surface in the pellet. Another possible error in the method is the variation in effective diffusivity within a pellet. It has been found[¶] that D_e can vary with position near the surface of a pellet, presumably because of skin effects caused by nonuniform stresses in the pelleting process.

11-10 The Effect of Intrapellet Mass Transfer on Observed Kinetics

In Sec. 10-1 we saw that neglecting external resistances could lead to misleading conclusions about reaction order and activation energy. Similar errors may occur when intrapellet mass transfer is neglected. Consider the situation where $\Phi_s > 5$. In this region intrapellet transport has a strong effect on the rate; Fig. 11-8 shows

† M. R. Rao and J. M. Smith, *AIChE J.*, **9**, 445 (1963).
‡ M. R. Rao and J. M. Smith, *AIChE J.*, **10**, 293 (1964).
§ Seiya Otani and J. M. Smith, *J. Catalysis*, **5**, 332 (1966).
¶ C. N. Satterfield and S. K. Saraf, *Ind. Eng. Chem., Fund, Quart.*, **4**, 451 (1965).

that η is less than about 0.2. From Eqs. (11-54) and (11-50) for a first-order reaction

$$\eta = \frac{3}{r_s} \sqrt{\frac{D_e}{k_1 \rho_P}}$$

and for second-order kinetics

$$\eta = \frac{3}{r_s} \sqrt{\frac{D_e}{k_2 \rho_P C_s}}$$

Using these expressions for η in the equations for the rates for the whole pellet [for example, Eq. (11-44)], we obtain

$$\mathbf{r}_P = \frac{3C_s}{r_s} \sqrt{\frac{D_e k_1}{\rho_P}} \qquad \text{first order}$$

$$\mathbf{r}_P = \frac{3C_s^{3/2}}{r_s} \sqrt{\frac{D_e k_2}{\rho_P}} \qquad \text{second order}$$

If the rate constants are expressed as Arrhenius functions of temperature, $k = Ae^{-E/R_g T}$, then

$$\mathbf{r}_P = \frac{3A_1^{1/2} C_s}{r_s} \left(\frac{D_e}{\rho_P}\right)^{1/2} e^{-E/2 R_g T} \qquad \text{first order} \qquad (11\text{-}68)$$

$$\mathbf{r}_P = \frac{3A_2^{1/2} C_s^{3/2}}{r_s} \left(\frac{D_e}{\rho_P}\right)^{1/2} e^{-E/2 R_g T} \qquad \text{second order} \qquad (11\text{-}69)$$

These equations give the correct influence of concentration and temperature when intrapellet diffusion is important.

Now suppose intrapellet resistance is neglected. The rate for a first-order reaction would be correlated in terms of the apparent activation energy E_a by the expression

$$\mathbf{r}_P = A_1 e^{-E_a/R_g T} C_s \qquad (11\text{-}70)$$

Comparison of Eqs. (11-70) and (11-68) indicates that the apparent activation energy determined from Eq. (11-70) would be one-half the true value, E. The measured rate, when plotted in Arrhenius coordinates, would appear as in Fig. 11-11. At low enough temperatures the data would determine a line with a slope equal to $-E/R_g$, because η would approach unity. However, at high enough temperatures intrapellet diffusion would be important; Eq. (11-68) would be applicable, and a line with a slope of $-\frac{1}{2}E/R_g$ would result. These conclusions would be the same regardless of the reaction order.

Equations (11-68) and (11-69) show that the rate $\mathbf{r}_P$ (recall that it is for the whole pellet, but per unit mass of catalyst) is inversely proportional to pellet size r_s. If intrapellet diffusion is neglected, the rate is independent of pellet size, as illustrated by Eq. (11-70).

If diffusion in the pores is of the Knudsen type, D_e is independent of pressure, and therefore of concentration. Then Eq. (11-68) indicates that first-order kinetics would be observed, even though intrapellet diffusion is important. However, a

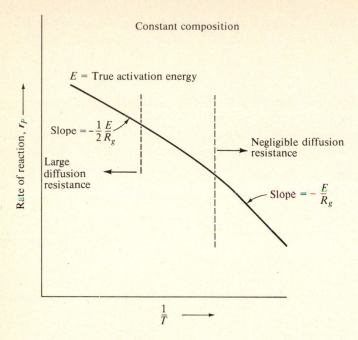

Figure 11-11 Effect of intrapellet diffusion on apparent activation energy.

second-order reaction would appear to be of order $\frac{3}{2}$. If bulk diffusion were involved, $D_e \approx 1/p$. Hence when concentration is varied by changing the pressure, a first-order reaction would appear to be of order $\frac{1}{2}$. Similarly, Eq. (11-69) shows that a second-order reaction would appear to be first order.

These effects have been observed in many instances. In particular, the flattening of the line on an Arrhenius plot (as in Fig. 11-11) is frequently found when heterogeneous reactions are studied over a wide temperature range (see Example 11-10).

Example 11-10 In Example 11-8 the reaction was first order. Now we examine the same type of problem for nonlinear kinetics using experimental global rate data for the catalytic oxidation of dilute solutions of acetic acid with dissolved oxygen. The data were given in Example 9-3 and the results for the small particle sizes were analyzed there to determine an equation for the intrinsic rate. The result was Eq. (B) of Example 9-3. Our objective here is to use the data for the largest particles to establish the importance of intraparticle diffusion in the liquid-filled pores.

(a) Using the data given in Example 9-3 calculate effectiveness factors at each temperature for the 0.18 cm (diameter) particles.
(b) Estimate the intraparticle diffusivity and tortuosity factor at 280°C. The molecular diffusivities of oxygen and of acetic acid in water at 68 atm and 280°C are estimated to be 2.0×10^{-4} cm²/s and 1.2×10^{-4} cm²/s, respectively.

(c) Calculate an activation energy for the reaction and also an apparent value, E_a, from the global rate data given for the 0.18 cm particles.

SOLUTION

(a) The rate data for various temperatures (given in the third table of the problem statement for Example 9-3) are for constant concentrations of both oxygen and acetic acid. Also, the effectiveness factor for 0.054 cm particles is 1.0, as explained in Example 9-3. Hence, Eq. (11-65) is applicable for evaluating the effectiveness factors. At 260°C ($10^3/T = 1.876$),

$$(\eta)_{d_p=0.18} = \frac{(r_p)_{d_p=0.18}}{(r_p)_{d_p=0.054}} = \frac{1.40}{2.00} = 0.70$$

For all the temperatures (using rate values from the table) the results are:

t, °C	280	270	260	250
$(\eta)_{d_p=0.18}$	0.59	0.64	0.70	0.79

(b) Since the intrinsic rate equation is available, and we know η, the effective diffusivity can be calculated. If the kinetics were first order, this would be simple. From η at a particular temperature we evaluate Φ_s from Fig. 11-8. Then from Eq. (11-50), using the known k and particle size (r_s), D_e could be calculated. For the nonlinear kinetics of Eq. (B) of Example 9-3 another approach must be used. An exact method would be to write a differential mass balance like Eq. (11-46) but using Eq. (B) for the intrinsic rate. This balance could be integrated numerically to find $C = f(r)$, and then to evaluate an effectiveness factor, as is done in Sec. 11-7. Alternately, we could search the literature[†] to see if such a solution for the kinetics of Eq. (B) was already available. Let us use a third approach (already mentioned), that of defining a generalized Thiele modulus suitable for any type of kinetics.[‡] According to this method η is still given by Eq. (11-52), but the generalized Thiele modulus is, for spherical pellets and for oxygen,

$$\Phi_s = \frac{r_s}{3}(r_{O_2})_s \left[2 \int_0^{(C_{O_2})_s} r_{O_2} D_e \, dC_{O_2} \right]^{-1/2} \tag{A}$$

where $(r_{O_2})_s$ is the rate of disappearance of oxygen evaluated at the surface concentration (which is equal to the bulk value in this case).

The quantity r_{O_2} is equal to r_{CO_2} so that Eq. (B) of Example 9-3 gives r_{O_2}. To carry out the integration indicated in Eq. (A), C_{HA} in Eq. (B) must be expressed in terms of C_{O_2}. This is done by noting that stoichiometry

---

[†] For example, in G. W. Roberts and C. N. Satterfield, *Ind. Eng. Chem. Fundam.*, **4**, 288 (1965); **5**, 317 (1966).

[‡] K. B. Bischoff, *AIChE J.*, **11**, 351 (1965).

requires that the mass transfer rate of oxygen be twice that of acetic acid. Hence,

$$(D_e)_{O_2} \frac{dC_{O_2}}{dr} = 2(D_e)_{HA} \frac{dC_{HA}}{dr} \tag{B}$$

Integrating from the outer surface to any intraparticle radius r yields

$$(D_e)_{O_2}[(C_{O_2})_s - C_{O_2}] = 2(D_e)_{HA}[(C_{HA})_s - C_{HA}]$$

or

$$C_{HA} = (C_{HA})_s + \frac{1}{2}\left(\frac{D_{O_2}}{D_{HA}}\right)[C_{O_2} - (C_{O_2})_s] \tag{C}$$

Equation (C) gives C_{HA} in terms of C_{O_2} and the known surface concentrations. The effective diffusivity ratio, $(D_e)_{O_2}/(D_e)_{HA}$, is equal to the molecular diffusivity ratio by Eq. (11-26), since the porosity ε_p and tortuosity factor δ are properties of the geometry of the catalyst particle and should be the same for oxygen and acetic acid.

Equation (C) can be substituted for C_{HA} in Eq. (B) of Example 9-3 to obtain an expression for r_{O_2} with C_{O_2} as the only variable. With this expression the term in brackets in Eq. (A) can be integrated directly to yield

$$\left[2(D_e)_{O_2}\int_0^{(C_{O_2})_s} r_{O_2} \, dC_{O_2}\right] = 2(D_e)_{O_2}k_A B \int_0^{(C_{O_2})_s} \frac{C_{O_2}^{3/2} \, dC_{O_2}}{1 + k_B(A + BC_{O_2})}$$

$$+ \, 2(D_e)_{O_2}k_A A \int_0^{(C_{O_2})_s} \frac{C_{O_2}^{1/2} \, dC_{O_2}}{1 + k_B(A + BC_{O_2})} \tag{D}$$

where

$$A = (C_{HA})_s - \frac{1}{2}\left(\frac{D_{O_2}}{D_{HA}}\right)(C_{O_2})_s \tag{E}$$

$$B = \frac{1}{2}\frac{D_{O_2}}{D_{HA}} \tag{F}$$

Now all quantities on the right-hand side of Eq. (A) are known except D_e. Surface concentrations for oxygen and for acetic acid are equal to their bulk values. The molecular diffusivities are known at 280°C, as are k_A and k_B in the intrinsic rate equation. The rate at the *surface* is, from Eq. (B) of Example 9-3,

$$(r_{O_2})_s = (r_{CO_2})_s = \frac{k_A(C_{O_2})_s^{1/2}(C_{HA})_s}{1 + k_B(C_{HA})_s} \tag{G}$$

A solution procedure is: first evaluate Φ_s from Eq. (11-52) at 280°C using the η values found in part (b). Then taking $r_s = d_p/2 = 0.18$ cm, find the value of $(D_e)_{O_2}$ from the right-hand side of Eq. (A) which agrees with this value of Φ_s. With $\eta = 0.59$, $\Phi_s = 1.25$ and, by trial, $(D_e)_{O_2} = 2 \times 10^{-5}$ cm²/s.

Finally, the tortuosity factor is obtained from $(D_e)_{O_2}$ using Eq. (11-26).

$$\delta = \frac{\varepsilon_p D_{O_2}}{(D_e)_{O_2}} = \frac{0.55(2 \times 10^{-4})}{2 \times 10^{-5}} = 5.5$$

In contrast to the first-order curve in Fig. 11-8, the effectiveness factor for nonlinear kinetics will depend upon the surface concentrations. For example, for reactor design the objective is to determine η and a global rate for known intrinsic kinetics. In this case we would use Eq. (A) to calculate Φ_s, and then evaluate η from Eq. (11-52). The result would depend upon surface concentrations through Eq. (D) and $(r_{O_2})_s$. This means that the effectiveness factor would vary with location in an integral reactor.

(c) Suppose that the effect of temperature on the intrinsic rate, given by Eq. (B) and the values of k_A and k_B in Example 9-3, can be represented by the Arrhenius equation

$$\mathbf{r}_{CO_2} = Ae^{-E/R_g T} f(C_{O_2}, C_{HA}) \tag{H}$$

Then the slope of a plot of $\mathbf{r}_{CO_2}$ vs. $1/T$ at constant composition is equal to $-E/R_g$. To obtain the correct activation energy for the chemical reaction, the rate values must exclude mass transport effects. This was shown to be true in Example 9-3 for the 0.038 and 0.054 cm particles. Such a plot is shown in Fig. 11-12 for the data in the third table of Example 9-3, which

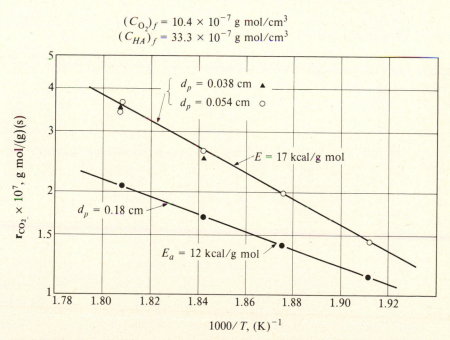

Figure 11-12 Effects of temperature and particle size on oxidation rate of acetic acid.

are for constant composition. The slope of the line corresponds to $E = 17,000$ cal/g mol.

The Arrhenius plot for the data $d_p = 0.18$ cm is also shown in Fig. 11-12. The slope of this line gives an apparent activation energy, $E_a = 12,000$ cal/g mol. This low value in comparison with E is expected, since intraparticle diffusion is influencing the rate for the 0.18 cm particles. Reference to Fig. 11-11 suggests that there is an intermediate amount of intraparticle diffusion resistance, since E_a lies between the true activation energy and $\frac{1}{2}E$, corresponding to large intraparticle diffusion resistance.

MASS AND HEAT TRANSFER WITH REACTION

When the heat of reaction is large, intrapellet temperature gradients *may* have a larger effect on the rate per pellet than concentration gradients. Even when ΔH is low, the center and surface temperatures may differ appreciably, because catalyst pellets have low thermal conductivities (Sec. 11-6). The combined effect of mass and heat transfer on $\mathbf{r}_P$ can still be represented by the general definition of the effectiveness factor, according to Eq. (11-41). Hence Eq. (11-42) may be used to find $\mathbf{r}_P$, provided η is the nonisothermal effectiveness factor. The nonisothermal η may be calculated in the same way as the isothermal η, except that an energy balance must be combined with the mass balance.

11-11 Nonisothermal Effectiveness Factors

Consider the same irreversible first-order reaction $A \rightarrow B$ used in Sec. 11-7 to obtain the isothermal η. If the effect of temperature on D_e is neglected, the differential mass balance and boundary conditions, Eqs. (11-46) to (11-48), are still applicable. The differential energy balance over the spherical shell of thickness Δr (see Fig. 11-7) is

$$\left(-4\pi r^2 k_e \frac{dT}{dr}\right)_r - \left(-4\pi r^2 k_e \frac{dT}{dr}\right)_{r+\Delta r} = (4\pi r^2 \, \Delta r)\rho_P k_1 C \, \Delta H \quad (11\text{-}71)$$

Taking the limit as $\Delta r \rightarrow 0$ and assuming that k_e is independent of temperature, we find

$$\frac{d^2 T}{dr^2} + \frac{2}{r}\frac{dT}{dr} - \frac{k_1 \rho_P C}{k_e} \Delta H = 0 \quad (11\text{-}72)$$

with boundary conditions

$$\frac{dT}{dr} = 0 \qquad \text{at } r = 0 \quad (11\text{-}73)$$

$$T = T_s \qquad \text{at } r = r_s \quad (11\text{-}74)$$

The solution of Eqs. (11-46) to (11-48) and (11-72) to (11-74) gives the concentration and temperature profiles within the pellet. A numerical solution is necessary because Eqs. (11-46) and (11-72) are coupled through the nonlinear dependence of k_1 on temperature; $k_1 = Ae^{-E/R_g T}$. Nevertheless, the similarity of the nonreaction terms in the two differential equations does permit an analytical relation between concentration of reactant and temperature at any point in the pellet. Thus eliminating $k_1 \rho_P C$ from the two equations yields

$$D_e\left(\frac{d^2C}{dr^2} + \frac{2}{r}\frac{dC}{dr}\right) = \frac{k_e}{\Delta H}\left(\frac{d^2T}{dr^2} + \frac{2}{r}\frac{dT}{dr}\right)$$

or

$$D_e\frac{d}{dr}\left(r^2\frac{dC}{dr}\right) = \frac{k_e}{\Delta H}\frac{d}{dr}\left(r^2\frac{dT}{dr}\right) \tag{11-75}$$

If this equation is integrated once, using Eqs. (11-47) and (11-73) for the boundary conditions, and then integrated a second time, using Eqs. (11-48) and (11-74),

$$T - T_s = \frac{(\Delta H)D_e}{k_e}(C - C_s) \tag{11-76}$$

This result, originally derived by Damkoehler,[†] is not restricted to first-order kinetics, but is valid for any form of rate expression, since the rate term was eliminated in forming Eq. (11-75). The maximum temperature rise in a pellet would occur when the reactant has been consumed by the time it diffuses to the center. Applying Eq. (11-76) for $C = 0$ gives

$$(T_c - T_s)_{max} = -\frac{(\Delta H)D_e}{k_e}C_s \tag{11-77}$$

Equation (11-77) shows that the maximum temperature rise depends on the heat of reaction, transport properties of the pellet, and the surface concentration of reactant. It permits a simple method of estimating whether intrapellet temperature differences are significant (see Example 11-11).

Let us return to the nonisothermal effectiveness factor. Weisz and Hicks solved Eqs. (11-46) and (11-72) numerically[‡] to determine the concentration profile within the pellet. Then η was obtained from Eq. (11-51), which is not limited to isothermal conditions, provided k_1 is evaluated at the surface temperature. The results expressed η as a function of three dimensionless parameters:

1. The Thiele-type modulus,

$$3(\Phi_s)_s = r_s\sqrt{\frac{(k_1)_s\rho_P}{D_e}} \tag{11-78}$$

Note that Φ_s is evaluated at the surface temperature; that is, $(k_1)_s$ in Eq. (11-78) is the rate constant at T_s.

[†] G. Damkoehler, *Z. Phys. Chem.*, **A193**, 16 (1943).

[‡] P. B. Weisz and J. S. Hicks, *Chem. Eng. Sci.*, **17**, 265 (1962). Actually, Eq. (11-76) permitted expression of k_1 as a function of concentration rather than of temperature ($k_1 = Ae^{-E/R_g T}$). This uncoupled Eqs. (11-46) and (11-72), so that only Eq. (11-46) needed to be solved.

2. The Arrhenius number,

$$\gamma = \frac{E}{R_g T_s} \tag{11-79}$$

3. A heat-of-reaction parameter,

$$\beta = \frac{(-\Delta H) D_e C_s}{k_e T_s} \tag{11-80}$$

Figure 11-13 shows η as a function of $(\Phi_s)_s$ and β for $\gamma = 20$, which is in the middle of the practical range of γ. Weisz and Hicks give similar figures for $\gamma = 10, 30$, and 40. The curve for $\beta = 0$ corresponds to isothermal operation ($\Delta H = 0$) and is identical with the curve for a spherical pellet in Fig. 11-8.

For an exothermic reaction (positive β) the temperature rises going into the pellet. The increase in rate of reaction accompanying the temperature rise can more than offset the decrease in rate due to drop in reactant concentration. Then η values are greater than unity. While $\eta > 1$ increases the rate per pellet, and therefore the production per unit mass of catalyst, there may also be some disadvantages. With large η there will be a large increase in temperature toward the center of the pellet, resulting in sintering and catalyst deactivation. The desired product may be subject to further reaction to unwanted product, or undesirable side reactions may occur. If these reactions have higher activation energies than the desired reaction, the rise in temperature would reduce selectivity.

For an endothermic reaction there is a decrease in temperature and rate into the pellet. Hence η is always less than unity. Since the rate decreases with drop in temperature, the effect of heat-transfer resistance is diminished. Therefore the curves for various β are closer together for the endothermic case. In fact, the decrease in rate going into the pellet for endothermic reactions means that mass

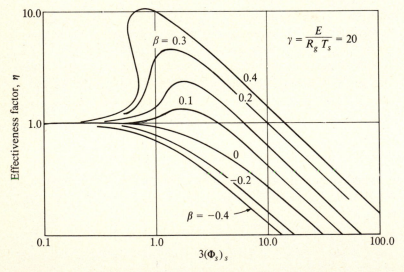

Figure 11-13 Nonisothermal effectiveness factors for first-order reactions in spherical catalyst pellets.

transfer is of little importance. It has been shown† that in many endothermic cases it is satisfactory to use a *thermal* effectiveness factor. Such thermal η neglects intrapellet mass transport; that is, η is obtained by solution of Eq. (11-72), taking $C = C_s$ (see Prob. 11-14).

Figure 11-13 indicates that for $\beta > 0.3$ and $3(\Phi_s)_s < 1.0$, up to three values of η exist for a single set of γ, $3(\Phi_s)_s$, and β. Such behavior can be explained by noting that the equality of heat evolved due to reaction and heat transferred by conduction in the pellet (such equality must exist for steady-state operation) can occur with different temperature profiles. Thus the solution giving the highest η for a given β curve in Fig. 11-13 would correspond to a steep temperature profile in the pellet; physical processes dominate the rate for the whole pellet. The solution for the lowest η, which is near unity, corresponds to small temperature gradients in the pellet; the rate is controlled by the chemical-reaction step. The intermediate η is identified as a metastable state, similar to the metastable conditions described in Sec. 10-4.

Carberry‡ has shown that for $(\Phi_s)_s > 2.5$ the η-vs.-$(\Phi_s)_s$ relationship could be characterized approximately by the single parameter $\beta\gamma$, rather than by γ and β separately. For example, for a first-order irreversible reaction he found

$$\eta = \frac{1}{(\Phi_s)_s} e^{\beta\gamma/5} \qquad \text{for } (\Phi_s)_s > 2.5 \tag{11-81}$$

where

$$\beta\gamma = \frac{(-\Delta H)D_e C_s}{k_e T_s} \frac{E}{R_g T_s} \tag{11-82}$$

Carberry also obtained results for a second-order rate equation.

In Sec. 11-8, Eq. (11-59) was developed as a criterion for deciding if intrapellet mass transport had a significant effect on the rate. Weisz and Hicks§ have extended the analysis to the problem of combined mass and energy transport. For a first-order irreversible reaction the criterion may be expressed as

$$\Theta = r_s^2 \frac{\mathbf{r}_P \rho_P}{C_s D_e} e^{\gamma\beta/(1+\beta)} \leq 1 \tag{11-83}$$

where γ and β are defined by Eqs. (11-79) and (11-80). When $\beta = 0$ (isothermal conditions) this criterion reduces to Eq. (11-59). As in that equation, $\mathbf{r}_P$ is the measured rate. Thus the significance of intrapellet gradients can be evaluated from measurable or known properties. If $\Theta \leq 1$, the nonisothermal effectiveness factor will be near unity, so that intrapellet gradients can be neglected.

11-12 Experimental Nonisothermal Effectiveness Factors

The magnitude of intrapellet temperature differences and nonisothermal η are illustrated by the following examples, based on experimental measurements for specific systems.

---

† J. A. Maymo, R. E. Cunningham, and J. M. Smith, *Ind. Eng. Chem., Fund Quart.*, **5**, 280 (1966).
‡ J. J. Carberry, *AIChE J.*, **7**, 350 (1961).
§ P. B. Weisz and J. S. Hicks, *Chem. Eng. Sci.*, **17**, 265 (1962).

Example 11-11 Rate data for the reaction $H_2 + \tfrac{1}{2}O_2 \to H_2O$ have been measured[†] for single catalyst pellets (1.86-cm diameter) of platinum on Al_2O_3. Catalyst properties, k_e, D_e, and center and surface temperatures were also evaluated. The rate was obtained in a stirred-tank reactor in which the pellet was surrounded by well-mixed reaction gases. In one run the data were as follows:

$$\text{Gas temperature } t_g = 90°C$$
$$\text{Average surface temperature } t_s = 101°C$$
$$\text{Pellet-center temperature } t_c = 148°C$$
$$\text{Density of catalyst particles in pellet[‡]} = 0.0602 \text{ g catalyst/cm}^3$$
$$\text{Mole fraction oxygen at pellet surface} = 0.0527$$
$$\text{Effective diffusivity of pellet} = 0.166 \text{ cm}^2/\text{s}$$
$$\text{Effective thermal conductivity} = 6.2 \times 10^{-4} \text{ cal/(s)(cm)(°C)}$$
$$\text{Pellet rate of reaction, } \mathbf{r}_P = 2.49 \times 10^{-5} \text{ g mol } O_2/$$
$$\text{(g catalyst)(s)}$$
$$\text{Total pressure } p_t = 1 \text{ atm}$$

Rate data were also obtained for the small (80 to 250 mesh) particles from which the pellets were prepared. The results, expressed as the rate of oxygen consumption, g mol/(g catalyst)(s), were correlated by

$$\mathbf{r}_{\text{part}} = 0.327 p_{O_2}^{0.804} e^{-5.230/R_g T}$$

where p_{O_2} is in atmospheres. Do internal concentration and temperature gradients have a significant effect on the pellet rate? Estimate the maximum temperature difference, $T_c - T_s$.

SOLUTION Equation (11-83) will be satisfactory as a criterion, since the Φ_s-vs.-η curves for a first-order reaction will be nearly the same as those for 0.8 order:

$$(C_{O_2})_s = \frac{(p_{O_2})_s}{R_g T} = \frac{1(0.0527)}{82(101 + 273)} = 1.72 \times 10^{-6} \text{ mole/cm}^3$$

$$r_s^2 \frac{\mathbf{r}_P \rho_P}{C_s D_e} = \left(\frac{1.86}{2}\right)^2 \frac{2.49 \times 10^{-5}(0.0602)}{1.72 \times 10^{-6}(0.166)} = 4.4$$

From Eqs. (11-79) and (11-80),

$$\gamma = \frac{5,230}{2(374)} = 7.0$$

$$\beta = \frac{115,400(0.166)(1.72 \times 10^{-6})}{6.2 \times 10^{-4}(374)} = 0.14$$

[†] J. A. Maymo and J. M. Smith, *AIChE J.*, **12**, 845 (1966).

[‡] The spherical pellet was prepared by compressing a mixture of inert Al_2O_3 particles and Al_2O_3 particles containing 0.005 wt % platinum. The ratio of active to total particles was 0.1. The density of the pellet was 0.602 g/cm³, so that the density of catalyst particles in the pellet was 0.0602.

Note that the heat of reaction per mole of oxygen is $-57,700(2) = -115,400$ cal/g mol.

Introducing these results in Eq. (11-83) gives

$$\Theta = (4.4)e^{7(0.14)/1.14} = 4.4(2.36) = 10.4$$

Since the value of Θ far exceeds unity, intrapellet resistances do affect the rate. Equation (11-77) gives the maximum value for $T_c - T_s$,

$$T_c - T_s = -\frac{(-115,400)(0.166)}{6.2 \times 10^{-4}}(1.72 \times 10^{-6}) = 52°C$$

The observed temperatures gives $148 - 101 = 47°C$. Comparison between the maximum and observed $T_c - T_s$ suggests that the oxygen is nearly all consumed when it reaches the center of the pellet. Hence the rate also will vary with radial position due to the drop in reactant (O_2) concentration.

This preliminary analysis suggests that both intrapellet temperature and concentration gradients will significantly affect the rate. Since the reaction is exothermic, the two factors have opposite effects on the rate per pellet. The dominant effect can be ascertained by calculating the resultant effectiveness factor.

Example 11-12 From the experimental data given in Example 11-11 evaluate the effectiveness factor for the catalyst pellet. Also predict η, using the results of Weisz and Hicks.†

SOLUTION The *particles* for which rates were measured are small enough that η is unity. Hence an experimental effectiveness factor can be obtained from Eq. (11-65). This equation was derived for isothermal operation but will be applicable for nonisothermal conditions as long as the two rates are evaluated at the same surface temperature as well as reactant concentration. Applying Eq. (11-65)

$$\eta_{\exp} = \frac{r_P(\text{at } T_s, C_s)}{r_{\text{part}}(\text{at } T_s, C_s)}$$

The rate for the particles at T_s and C_s is obtainable from the correlation of particle rates given in Example 11-10:

$$r_{\text{part}} = 0.327(0.0527^{0.804})e^{-5.230/2(374)} = 2.77 \times 10^{-5} \text{ g mol } O_2/(\text{g catalyst})$$

Then

$$\eta_{\exp} = \frac{2.49 \times 10^{-5}}{2.77 \times 10^{-5}} = 0.9$$

To calculate η we first obtain $3(\Phi_s)_s$, using Eq. (11-78). The reaction is not first order, and so $(k_1)_s$ is not known. However, it may be replaced by its equivalent, r/C, where r is the rate of the chemical step, that is, the rate

---

† P. B. Weisz and J. S. Hicks, *Chem. Eng. Sci.*, **17**, 265 (1962).

uninfluenced by intrapellet resistances. This is the rate measured for the small catalyst particles, 2.77×10^{-5} g mol/(g catalyst)(s). Hence

$$3(\Phi_s)_s = r_s \sqrt{\frac{r_{part}\, \rho_P}{C_s\, D_e}} = \frac{1.86}{2} \sqrt{\frac{2.77 \times 10^{-5}(0.0602)}{1.72 \times 10^{-6}(0.166)}} = 2.20$$

With this value for $3(\Phi_s)_s$, and $\gamma = 7.0$ and $\beta = 0.14$, a calculated η can be found from the Weisz and Hicks charts [similar to Fig. (11-13)]. The value is difficult to estimate from the available charts, but is approximately

$$\eta_{calc} = 0.94$$

A more accurate η can be obtained by solving equations such as (11-46) and (11-72) but avoiding the first-order restriction. This can be done by replacing $k_1 C$ in the reaction term of each equation with the more correct relationship $k p_{O_2}^{0.804}$. Maymo did this, solving the differential equations numerically by means of a high-speed digital computer. The calculated η for $3(\Phi_s)_s = 2.20$, $\gamma = 7.0$, and $\beta = 0.14$ is $\eta_{calc} = 0.96$. The calculated η for a first-order reaction agrees somewhat better with the experimental result. Probably this is not significant, because the calculated values rest on the assumption that the intrinsic rate is the same for the particles as for the pellets. There is evidence† to indicate that when the particles are made into pellets there is a slight reduction in intrinsic rate owing to partial blocking of the micropores in the particles. Regardless of the reasons for the small deviation (7%), the agreement between calculated and experimental η is quite adequate, particularly in view of the variety of independent data $(D_e, k_e, \text{rates, etc.})$ that are required to calculate the effectiveness factor.

Since η is less than unity, the effect of the concentration gradient is more important than the effect of the temperature gradient in this instance.

The significance of the thermal effect on the pellet rate is established primarily by β and secondarily by γ. For example, Otani and Smith‡ studied the reaction $CO + \frac{1}{2}O_2 \rightarrow CO_2$, making measurements similar to the $H_2 + O_2$ reaction, but at temperatures from 250 to 370°C. Here β was of the order of 0.05 and $\gamma = 9.7$. Intrapellet temperature gradients were much lower than in Example 11-11; the largest measured value for $T_c - T_s$ was 7°C. In contrast, hydrogenation reactions have large heats of reaction; D_e is relatively high for hydrogen, and often T_s is low. All these factors tend to increase β. Cunningham et al.§ found β to be of the order of 0.5 (and γ about 25) for the hydrogenation of ethylene, and thermal effects were large. Similarly, in the work of Prater¶ for the dehydrogenation of cyclohexane, β and γ were sufficient to reduce η significantly.

† J. A. Maymo and J. M. Smith, *AIChE J.*, **12**, 845 (1966).

‡ S. Otani and J. M. Smith, *J. Catalysis*, **5**, 332 (1966).

§ R. A. Cunningham, J. J. Carberry, and J. M. Smith, *AIChE J.*, **11**, 636 (1965).

¶ C. D. Prater, *Chem. Eng. Sci.*, **8**, 284 (1958), as analyzed in C. N. Satterfield and T. K. Sherwood "The Role of Diffusion in Catalysis," p. 90, Addison-Wesley Publishing Company, Reading, Mass, 1963.

Increasing reactant concentration increases β and η. When the partial pressure of oxygen was increased to 0.11 atm in the $H_2 + O_2$ reaction, with other conditions approximately the same as in Example 11-11, β became 0.34, and the experimental η was 1.10. At these conditions $T_c - T_s$ increased to $219 - 118 = 101°C$.

EFFECT OF INTERNAL TRANSPORT ON SELECTIVITY AND POISONING

In many catalytic systems, multiple reactions occur, so that selectivity becomes important. In Sec. 2-11 point and overall selectivities were evaluated for homogeneous, well-mixed systems of parallel and consecutive reactions. In Sec. 10-5 we found that external diffusion and heat-transfer resistances in consecutive reactions affected the selectivity. Here we shall examine the influence of intrapellet resistances. Systems with first-order kinetics at isothermal conditions are analyzed analytically in Sec. 11-13 for parallel and consecutive reactions. Results for other kinetics, or for nonisothermal conditions, can be developed in a similar way but require numerical solution.†

11-13 Selectivities for Porous Catalysts

The selectivity at a location in a fluid-solid catalytic reactor is equal to the ratio of the global rates at that point. The combined effect of both external and internal diffusion resistance can be displayed easily for parallel irreversible, first-order reactions. We shall do this first and then consider how internal resistance influences the selectivity for other reaction sequences.

Consider two parallel reactions of the independent form

$$1. \quad A \xrightarrow{k_1} B + C \quad \text{(desired)}$$

$$2. \quad R \xrightarrow{k_2} S + W$$

where $k_1 > k_2$.

An example might be the dehydrogenation of mixed feed of propane and n-butane, where the desired catalyst is selective for the n-butane dehydrogenation. Suppose that the temperature is constant and that both external and internal diffusion resistances affect the rate. At steady state, the rate (for the pellet, expressed per unit mass of catalyst) may be written in terms of either Eq. (10-1) or Eq. (11-44),

$$r_P = k_m a_m (C_b - C_s) \tag{11-84}$$

or

$$r_P = \eta k_1 C_s \tag{11-85}$$

---

† Selectivity for Langmuir-Hinshelwood kinetics of the form of Eqs. (9-21) and (9-27), have been evaluated for isothermal conditions by G. Roberts and C. N. Satterfield [*Ind. Eng. Chem., Fund. Quart.*, **4**, 288 (1965)] and J. Hutchings and J. J. Carberry [*AIChE J.*, **12**, 20 (1966)].

where a_m is the external area for mass transfer per unit mass of catalyst. The surface concentration can be eliminated between (11-84) and (11-85) to give

$$\mathbf{r}_p = \frac{1}{1/k_m a_m + 1/\eta k_1} C_b \tag{11-86}$$

This expression is equivalent to Eq. (B) developed in Example 11-8 and expresses the combined effect of external and internal mass-transport resistances. Note that the reduction in rate due to internal diffusion (through η) is combined with the rate constant k for the chemical step, while the external effect is separate. If the external resistance is negligible, then $k_m a_m \gg \eta k_1$. If internal transport is insignificant, then $\eta \to 1$. If both conditions are satisfied, the rate is determined solely by the chemical step; that is, Eq. (11-85) reduces to $\mathbf{r}_P = k_1 C_b$.

The selectivity of product B with respect to product S for a pellet in the reactor is obtained by applying Eq. (11-86) to the two reactions; thus the pellet selectivity S_P is

$$S_P = \frac{(\mathbf{r}_P)_1}{(\mathbf{r}_P)_2} = \frac{[1/(k_m)_R a_m + 1/\eta_2 k_2](C_A)_b}{[1/(k_m)_A a_m + 1/\eta_1 k_1](C_R)_b} \tag{11-87}$$

If there were neither external nor internal resistances, the pellet selectivity would be

$$S_P = \frac{k_1 (C_A)_b}{k_2 (C_R)_b} \tag{11-88}$$

Since $(k_m)_R$ and $(k_m)_A$ will not differ greatly, comparison of Eqs. (11-87) and (11-88) shows that external-diffusion resistance reduces the selectivity. The same conclusion was reached in Sec. 10-5 for parallel independent reactions (see Eq. (10-29c)). To evaluate qualitatively the effect of internal diffusion note that k_1 is presumably greater than k_2 (B is the desired product). Since Φ increases, and, therefore, η decreases as k increases (see Fig. 11-8), η_1 will be less than η_2. Then Eq. (11-87) indicates that internal diffusion also decreases the selectivity.

Equation (11-86) shows that the external effect on the rate can always be treated as a separate, additive resistance. From here on we shall focus on internal effects, using Eq. (11-85) for the global rate and taking bulk and surface concentrations to be equal. The internal problem can be expressed analytically if we are satisfied with examining the extreme case of large intrapellet resistance characterized by $\Phi_s \geq 5$ ($\eta \leq 0.2$) where Eq. (11-54) is valid. Then Eq. (11-85) becomes

$$\mathbf{r}_P = \frac{1}{\Phi_s} k_1 C_b = \frac{3}{r_s} \sqrt{\frac{k_1 D_e}{\rho_P}} C_b \tag{11-89}$$

If Eq. (11-89) is applied to the stated, independent parallel reactions, the selectivity is

$$S_P = \frac{(\mathbf{r}_P)_1}{(\mathbf{r}_P)_2} = \frac{\sqrt{k_1 (D_A)_e} (C_A)_b}{\sqrt{k_2 (D_R)_e} (C_R)_b} \tag{11-90}$$

Neglecting differences in diffusivities of A and R, we have

$$S_P = \left(\frac{k_1}{k_2}\right)^{1/2} \frac{(C_A)_b}{(C_R)_b} \tag{11-91}$$

Comparing Eqs. (11-88) and (11-91) shows that the effect of *strong* intrapellet diffusion resistance is to reduce the selectivity to the square root of its intrinsic value.

Wheeler† characterized the *independent parallel* reactions $A \to B$ and $R \to S$ as *type I selectivity*. *Type II selectivity* refers to *parallel* reactions with a common reactant:

$$1. \quad A \xrightarrow{k_1} B$$

$$2. \quad A \xrightarrow{k_2} C$$

An example would be the dehydration of ethanol to ethylene and its dehydrogenation to acetaldehyde. If both reactions are first order and irreversible, selectivity is unaffected by internal mass transport; the ratio of the rates of reactions 1 and 2 is k_1/k_2 at any position within the pellet. This same conclusion applied for external transport [Sec. 10-5, Eq. (10-29)]. Equation (11-89) cannot be applied separately to the two reactions because of the common reactant A. The development of the effectiveness-factor function would require writing a differential equation analogous to Eq. (11-45) for the *total* consumption of A by both reactions. Hence k in Eq. (11-89) would be $k_1 + k_2$ and $\mathbf{r}_P$ would be $(\mathbf{r}_P)_1 + (\mathbf{r}_P)_2$. Such a development would shed no light on selectivity.

If the kinetics of the two reactions are different, diffusion has an effect on selectivity. Suppose reaction 2 is second order in A and reaction 1 first order. The reduction in concentration of A due to diffusion resistance would lower the rate of reaction 2 more than that of reaction 1. For this case the selectivity of B would be improved by diffusion resistance.

Type III selectivity applies to *successive* reactions of the form

$$A \xrightarrow{k_1} B \xrightarrow{k_3} D$$

where B is the desired product. Examples include successive oxidations, successive hydrogenations, and other sequences, such as dehydrogenation of butylenes to butadiene, followed by polymerization of the butadiene formed. In Chap. 2, Eq. (2-87) was developed for the intrinsic selectivity of B with respect to A,‡ that is, the point selectivity determined by the kinetics of the reactions,

$$S = \frac{\text{net rate of production of } B}{\text{rate of disappearance of } A} = \frac{dC_B}{-dC_A} = 1 - \frac{k_3}{k_1}\frac{C_B}{C_A} \tag{11-92}$$

---

† The treatment in this section follows in part that developed by A. Wheeler in W. G. Frankenburg, V. I. Komarewsky, and E. K. Rideal (eds.), "Advances in Catalysis," vol. III, p. 313, Academic Press, Inc., New York, 1951; and P. P. Weisz and C. D. Prater in "Advances in Catalysis," vol. VII, Academic Press, Inc., New York, 1954.

‡ In Sec. 10-5 the same successive reactions were analyzed. There the selectivity of B with respect to final product D was considered, rather than the selectivity with respect total disappearance of reactant A.

Equation (11-92) is also applicable for the whole catalyst pellet when diffusion resistance is negligible and gives the selectivity at any location within the pellet. The selectivity will vary with position in the pellet as C_B/C_A changes. Diffusion resistance causes C_A to decrease going from the outer surface toward the center of the pellet. Since B is formed within the pellet and must diffuse outward in order to enter the bulk stream, C_B increases toward the pellet center. Equation (11-92) shows qualitatively that these variations in C_A and C_B both act to reduce the global, or pellet, selectivity for B.

A quantitative interpretation can also be developed. For first-order reactions the variation of C_A with radial position in a spherical pellet is given by Eq. (11-49). The concentration profile for B can be obtained in a similar manner by writing a differential mass balance for B. The expression will be like Eq. (11-45), except that the term on the right must represent the *net* disappearance of B; that is, it must account for the conversion of B to D and the production of B from A. Wheeler† solved such a differential equation to obtain $C_B = f(r)$. The solution, combined with Eq. (11-49) for C_A, can be used to obtain the selectivity of B with respect to A as a function of radial position. Integrating across the radius gives the selectivity for the whole pellet. For strong diffusion resistance ($\eta \leq 0.2$) and equal effective diffusivities the result is

$$S_P = \frac{(\mathbf{r}_P)_B}{-(\mathbf{r}_P)_A} = \frac{(k_1/k_3)^{1/2}}{1 + (k_1/k_3)^{1/2}} - \left(\frac{k_3}{k_1}\right)^{1/2} \frac{(C_B)_b}{(C_A)_b} \tag{11-93}$$

Comparison of Eqs. (11-92) and (11-93) demonstrates that the selectivity is significantly reduced when diffusion resistances are large. The magnitude of the reduction depends on $(C_B/C_A)_b$, which, if external-diffusion resistance is neglected, is equal to C_B/C_A at any position in the reactor. At the entrance to a *reactor* $(C_B/C_A)_b$ is a minimum. If no B is present in the feed, $C_B = 0$ at this point. From Eq. (11-92) the selectivity is 1.0. Equation (11-93) shows that strong diffusion resistance reduces this to

$$S_P = \frac{(k_1/k_3)^{1/2}}{1 + (k_1/k_3)^{1/2}} \tag{11-94}$$

The reduction is most severe for small values of k_1/k_3.

Suppose the pellet selectivity for B is observed to be low in the $A \rightarrow B \rightarrow D$ series of reactions, and diffusion resistance is significant. Either preparing the catalyst with larger pores or reducing the pellet size may improve the selectivity. However, these changes would be most effective if they increased η from a low value to near unity. The changes would be wholly ineffective if η remained ≤ 0.2, since Eq. (11-93) is valid for all η below 0.2. The first change would probably reduce the capacity of a fixed mass, or volume, of catalyst, since the pore surface available for reaction would decrease as pore radius increased. The second change would increase the pressure drop in a fixed-bed reactor.

---

† A. Wheeler in W. G. Frankenburg, V. I. Komarewsky, and E. K. Rideal (eds.), "Advances in Catalysis," vol. III, Academic Press, Inc., New York, 1951.

Example 11-13 An ethylene stream is fed to a polymerization reactor in which the catalyst is poisoned by acetylene. The ethylene is prepared by catalytically dehydrogenating ethane. Hence it is important that the dehydrogenation catalyst be selective for dehydrogenating C_2H_6 rather than C_2H_4. The first-order reactions are

$$\overset{(A)}{} \qquad \overset{(B)}{} \qquad \overset{(D)}{}$$

$$C_2H_6 \overset{k_1}{\rightarrow} C_2H_4 \overset{k_3}{\rightarrow} C_2H_2$$

For one catalyst $k_1/k_3 = 16$. It is suspected that intrapellet diffusion strongly retards both dehydrogenations. Estimate the potential improvement in selectivity if diffusion resistance could be eliminated. Make the estimate for a concentration ratio $(C_B/C_A)_b = 1.0$. Neglect differences in D_e between ethane and ethylene.

SOLUTION Equation (11-93) is applicable for the *diffusion-limited* situation. The pellet selectivity for ethylene formation with respect to ethane disappearance is

$$S_P = \frac{16^{1/2}}{1 + 16^{1/2}} - \frac{1}{16^{1/2}}(1) = 0.55$$

If diffusion resistance is eliminated, Eq. (11-92) gives the point as well as the pellet selectivity

$$S_P = 1 - \tfrac{1}{16}(1) = 0.94$$

These selectivities give the ratio of rates of formation of ethylene and total disappearance of ethane (ethylene plus acetylene). If we start with pure C_2H_6, a mass balance requires that the rate of formation of acetylene with respect to disappearance of ethane be $1 - S_P$. Hence eliminating diffusion resistance reduces the rate of formation of undesirable C_2H_2 from 0.45 to 0.06.

The discussion has been limited to selectivities which exist at one conversion (set of concentrations) in a reactor. Wheeler has extended the development to show cumulative selectivities that would exist at any level of conversion in the reactor effluent.

We have seen that when intrapellet temperature gradients exist the rate per pellet (or η) cannot be expressed analytically. Hence selectivities for nonisothermal conditions must be evaluated numerically. Examples are available for several forms of rate equations.†

---

† John Beek, *AIChE J.*, 7, 337 (1961); John Hutchings and J. J. Carberry, *AIChE J.*, 12, 20 (1966).

11-14 Rates for Poisoned Porous Catalysts

As mentioned in Chap. 8, the rates of fluid-solid catalytic reactions are frequently reduced by poisoning. It is of interest to know how the reduction in rate varies with the extent of catalytic surface which has been poisoned and with on-stream time. Experimental data show that the rate may drop linearly with the fraction of surface poisoned, or it may fall much more rapidly. The second form of behavior may be caused by selective adsorption of the poisoning substance on those catalyst sites which are active for the main reaction. Such adsorption on a relatively small part of the total catalyst would cause a large reduction in rate. An alternate explanation is possible if the main reaction occurs primarily on the outer part of the catalyst pellet (low effectiveness factor) and the poison is also adsorbed primarily in this region. Then a relatively small part of the total surface will be deactivated by poison, but this small part is where the main reaction occurs.

As Wheeler[†] has shown, the interaction of intrapellet diffusion on the rate of the main and poisoning reactions can lead to a variety of relations between activity and extent of poisoning. We examine some of these here as a further illustration of the effect of intrapellet mass transfer upon the rate of catalytic reactions. It is assumed that the activity of the unpoisoned catalytic surface is the same throughout the pellet and that the main reaction is first order.

The discussion will be for *parallel* and independent poisoning such as caused by deposition of an impurity in the feed stream. Poisoning also commonly occurs by a *successive* reaction sequence. Thus, an intermediate may be the desired product, but reaction continues to end products which are deposited on the catalytic sites. Poisoning by carbon deposition in refinery (e.g., cracking) and petrochemical (e.g., dehydrogenation, etc.) processing are examples. Quantitative treatment and reviews are available[‡][§] for the reduction in rate due to successive (series) poisoning processes. Also, the treatment here will be for the extremes of uniform distribution of poison or of pore-mouth poisoning, both at isothermal conditions. Examples of more general analysis encompassing the regions in between these extremes are available.[†] Nonisothermal poisoning has also been investigated.[¶][††]

Uniform distribution of poison Suppose the rate of the adsorption (or reaction) process which poisons the catalytic site is slow with respect to intrapellet diffusion. Then the surface will be deactivated uniformly through the pellet. If α is the fraction of the surface so poisoned, the rate constant for the *main* reaction will become $k_1(1 - \alpha)$. The rate per pellet, according to Eq. (11-44), is

$$\mathbf{r}_P = \eta k_1 (1 - \alpha) C_s \tag{11-95}$$

Consider first the case where the diffusion resistance for the main reaction

[†] A. Wheeler in W. G. Frankenburg, V. I. Komarewsky, and E. K. Rideal (eds.), "Advances in Catalysis," vol. III, p. 307, Academic Press, Inc., New York, 1951.

[‡] Shinobu Masamume and J. M. Smith, *AIChE J.*, **12**, 384 (1966).

[§] J. B. Butt, "Catalyst Deactivation," *Adv. Chem. Ser.*, vol. 109, p. 259 (1972).

[¶] M. Sagara, S. Masamume, and J. M. Smith, *AIChE J.*, **13**, 1226 (1967).

[††] E. K. T. Kam and Ronald Hughes, *AIChE J.*, **25**, 359 (1979).

also is low; that is, consider a slow main reaction. Then $\eta \to 1$, and Eq. (11-95) shows that the rate drops linearly with α ($F = 1 - \alpha$). At the other extreme of large intrapellet resistance ($\Phi_s > 5$), $\eta = 1/\Phi_s$, and Eq. (11-95) becomes

$$
\mathbf{r}_P = \frac{3}{r_s} \sqrt{\frac{D_e}{k_1(1 - \alpha)\rho_P}} \, k_1(1 - \alpha)C_s
$$

$$
= \frac{3}{r_s} \sqrt{\frac{D_e}{\rho_P}} \, k_1(1 - \alpha)C_s \tag{11-96}
$$

Hence the effect of poisoning on the rate of the main reaction is proportional to $\sqrt{1 - \alpha}$; that is, it is less than a linear effect. This situation, corresponding to a slow poisoning reaction and a fast (large Φ_s) main reaction, was termed *antiselective poisoning* by Wheeler.

The two cases are shown in Fig. 11-14 by curves A and B, where the ratio F of poisoned to unpoisoned rate is plotted against α. Note that the ratio is the quotient of Eq. (11-95) evaluated at any α and at $\alpha = 0$. Thus for curve B, from Eq. (11-96),

$$
F = \frac{(3/r_s)\sqrt{D_e k_1(1 - \alpha)/\rho_P}\,C_s}{(3/r_s)\sqrt{D_e k_1/\rho_P}\,C_s} = \sqrt{1 - \alpha} \tag{11-97}
$$

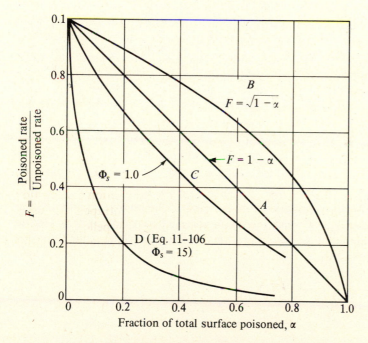

A – Uniform poisoning and slow main reaction $\eta \to 1$
B – Uniform poisoning and fast main reaction $\eta \leqslant 0.2$
C – Pore-mouth poisoning and slow main reaction $\Phi_s = 1.0$
D – Pore-mouth poisoning and fast main reaction $\Phi_s = 15$

Figure 11-14 Intrapellet diffusion and poisoning in catalytic reactions.

For intermediate cases of diffusion resistance for the main reaction, the curves would fall between A and B in the figure.

Pore-mouth (shell) poisoning If the adsorption (or reaction) causing poisoning is very fast, the outer part of a catalyst pellet will be completely deactivated, while the central portion retains its unpoisoned activity. An example is sulfur poisoning of a platinum-on-Al$_2$O$_3$ catalyst, where the sulfur-containing molecules diffuse only a short distance into the pellet before they are adsorbed on the platinum surface. In this type of poisoning a layer of poisoned catalyst will start to grow at the outer surface and will continue to increase in thickness with time† until all the pellet is deactivated. In the extreme case the boundary between the deactivated and active catalyst will remain sharply defined during the deactivation process. This type of poisoning, described by the *progressive shell model*, will be discussed in Chap. 14 in connection with noncatalytic reactions.

The effect of pore-mouth poisoning can be obtained by equating the rate of diffusion through the outer, deactivated layer to the rate of reaction on the inner fully active part of the pellet. Figure 11-15 depicts a spherical catalyst pellet at a *time* when the radius of the unpoisoned central portion is r_c, corresponding to a thickness of completely poisoned catalyst $r_s - r_c$. Consider a first-order reaction where the concentration of reactant at the outer surface is C_s. The rate of diffusion into one pellet will be

$$\mathbf{r} = D_e 4\pi r^2 \frac{dC}{dr} \tag{11-98}$$

This equation is applicable between r_s and r_c. Integrating‡ to express the rate in terms of the concentrations C_s and C_c (at $r = r_c$) gives

$$\mathbf{r} = \frac{4\pi D_e r_s r_c}{r_s - r_c}(C_s - C_c) \tag{11-99}$$

This diffusion rate is equal to the rate of reaction in the inner core, or

$$\mathbf{r} = \frac{4\pi D_e r_s r_c}{r_s - r_c}(C_s - C_c) = m_P \eta k_1 C_c \tag{11-100}$$

where m_P is the mass of the active core of the pellet. The second equality in Eq. (11-100) may be solved for C_c and the result inserted into Eq. (11-99). Taking $m_P = \frac{4}{3}\pi r_c^3 \rho_P$ and dividing by $\frac{4}{3}\pi r_s^3 \rho_P$ to give the rate for the whole pellet, but on a unit mass basis, gives

$$\mathbf{r}_P = \frac{\mathbf{r}}{\frac{4}{3}\pi r_s^3 \rho_P} = \frac{k_1 C_s}{r_s^3/\eta r_c^3 + (k_1 \rho_P r_s^2/3D_e)[(r_s - r_c)/r_c]} \tag{11-101}$$

---

† For example, suppose that the poison is brought into contact with the catalyst pellet as a contaminant in the reactant stream flowing steadily in the reactor. Then the amount of poisoning material available is directly proportional to time.

‡ It is assumed here that the thickness of the poisoned layer is not changing in the time required for diffusion. Then **r** can be regarded as a constant during this short time interval. Such a pseudo-steady state is discussed in Chap. 14.

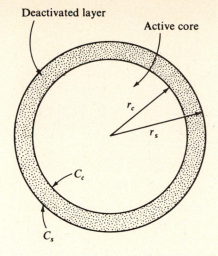

Deactivated layer

Active core

C_c

C_s

Figure 11-15 Shell model of catalyst poisoning.

The fraction of the total surface unpoisoned is

$$1 - \alpha = \frac{\frac{4}{3}\pi r_c^3 \rho_P S_g}{\frac{4}{3}\pi r_s^3 \rho_P S_g} = \frac{r_c^3}{r_s^3} \tag{11-102}$$

Using this relation for r_c/r_s in Eq. (11-101), the rate in terms of α becomes

$$\mathbf{r}_P = \frac{k_1 C_s}{1/\eta(1 - \alpha) + 3\Phi_s^2[1 - (1 - \alpha)^{1/3}]/(1 - \alpha)^{1/3}} \tag{11-103}$$

where Eq. (11-50) has been used to introduce Φ_s. If there is no poisoning, then $\alpha = 0$, and Eq. (11-103) reduces to $\mathbf{r}_P = \eta k_1 C_s$ [i.e., Eq. (11-44)]. Hence the ratio of poisoned and unpoisoned rates is

$$F = \frac{1}{1/(1 - \alpha) + 3\eta(\Phi_s)^2[1 - (1 - \alpha)^{1/3}]/(1 - \alpha)^{1/3}} \tag{11-104}$$

Again consider two extremes of the main reaction. If its intrinsic rate is very slow, diffusion resistance in the inner core of active catalyst will be negligible and $\eta \to 1$. At these conditions Φ_s will also be low, but not as small as would be expected. This is because η is a function of $\Phi_c = (r_c/3)\sqrt{k_1 \rho_P/D_e}$, and, therefore, of α, and not of Φ_s. The second term in the denominator of Eq. (11-104), which is positive, tends to reduce F. When this term is negligible, $F = 1 - \alpha$. This is represented by curve A in Fig. 11-14. For these conditions of $\eta \to 1$, Φ_s will be of the order of 1.0 and $\Phi_c = \frac{1}{3}$. Then the second term in the denominator is small and F is not much less than that given by curve A. Curve C in the figure corresponds to these conditions.

For the opposite situation of a very fast main reaction ($\Phi_c > 5$), the effectiveness factor is given by an expression similar to Eq. (11-54). However, it is written for the active core, so that

$$\eta = \frac{1}{\Phi_c} = \frac{r_s}{r_c}\frac{1}{\Phi_s} = \frac{1}{(1 - \alpha)^{1/3}}\frac{1}{\Phi_s} \tag{11-105}$$

Then Eq. (11-104) becomes

$$F = \frac{1}{1/(1 - \alpha) + 3\Phi_s[1 - (1 - \alpha)^{1/3}]/(1 - \alpha)^{2/3}} \tag{11-106}$$

For $\Phi_c \geq 5$, Φ_s will be even larger. For $r_c/r_s = \frac{1}{3}$, $\Phi_s \geq 15$. Curve D in Fig. 11-14 is a plot of Eq. (11-106) with $\Phi_s = 15$. In this case a sharp drop in activity is observed. Physically, the situation corresponds to a large diffusion resistance or a fast main reaction. The reactant molecules cannot penetrate far into the active part of the catalyst before reaction occurs. Thus both the poisoning and main reaction compete for sites on the same (outer) surface. As the outer surface is poisoned, the rate falls dramatically.

Our discussion has expressed the effect of poisoning (F) in terms of the fraction, α, of the catalytic surface that has been poisoned. Perhaps a more important practical problem is how F varies with time. To obtain this information the relation between α and time is required. For *pore-mouth* poisoning this means finding the relation between (r_c/r_s) and t for the shell model. Such problems are identical to those considered in Chap. 14 for gas-solid noncatalytic reactions where a product of the reaction is also a solid. Quantitative relationships between (r_c/r_s) and t, based upon the shell model, are developed in Sec. 14-3. For *uniform distribution of poison*, the relation between α and time does not involve intraparticle diffusion for the poisoning reaction but only its intrinsic kinetics. A simple example of the type of equation to be solved is

$$\frac{dn}{dt} = k_p C_p(1 - \alpha) \tag{11-107}$$

where n and C_p are the adsorbed and fluid phase concentrations of the poison. In this expression it is assumed that the rate of deposition of the poison is first order in C_p and in the fraction, $1 - \alpha$, of unpoisoned surface. If monomolecular adsorption is assumed, n is proportional to α; that is $\alpha = n/n_0$, where n_0 is the adsorbed concentration corresponding to complete deactivation (a monomolecular layer of deposited poison). Then Eq. (11-107) can be written

$$\frac{d\alpha}{dt} = (k_p/n_0)C_p(1 - \alpha) \tag{11-108}$$

Integration of this expression gives the required relationship between α and t. It may be combined with curves A or B in Fig. 11-14 to determine how F varies with time (see Prob. 11-16).

We should again mention that this discussion has been limited to *parallel and independent poisoning*, caused, for example, by deposition of an impurity in the reaction stream. For *series fouling* the equations relating F and α, and α and t are coupled. They become partial differential equations and their simultaneous solution becomes more complex† when intraparticle diffusion is involved (see Prob. 11-17 for a special, simple case).

---

† For a more complete analysis of the effects of both parallel and series poisoning see Shinobu Masamune and J. M. Smith, *AIChE J.*, **12**, 384 (1966).

PROBLEMS

11-1 In a cylindrical pore of 30 Å radius at what pressure would the bulk diffusivity in a $H_2-C_2H_6$ mixture be equal to the Knudsen diffusivity for H_2? The temperature is 100°C.

11-2 At 10 atm pressure and 100°C what would be the pore radius for which bulk and Knudsen diffusivities would be equal for H_2 in the hydrogen-ethane system. Locate a curve for equal values of bulk and Knudsen diffusivities on coordinates of pressure-vs.-pore radius. Mark the regions on the figure where Knudsen and where bulk diffusivity would be dominant.

11-3 Constant-pressure diffusion experiments are carried out in the apparatus shown in Fig. 11-1 with the H_2-N_2 binary system. In one experiment the nitrogen diffusion rate through the porous pellet was 0.49×10^{-5} g mol/s. What would be the counterdiffusion rate of hydrogen?

11-4 Diffusion rates for the H_2-N_2 system were measured by Rao and Smith[†] for a cylindrical Vycor (porous-glass) pellet 0.25 in. long and 0.56 in. in diameter, at 25°C and 1 atm pressure. A constant-pressure apparatus such as that shown in Fig. 11-1 was used. The Vycor has a mean pore radius of 45 Å, so that diffusion was by the Knudsen mechanism. The diffusion rates were small with respect to the flow rates of the pure gases on either side of the pellet. The average diffusion rate of hydrogen for a number of runs was 0.44 cm³/min (25°C, 1 atm). The porosity of the Vycor was 0.304.

(a) Calculate the effective diffusivity D_e of hydrogen in the pellet. (b) Calculate the tortuosity using the parallel-pore model. (c) Predict what the counterdiffusion rate of nitrogen would be.

11-5 Figure 11-2 displays a dynamic method for measuring the effective diffusivity in a catalyst pellet. Equation (11-21) relates D_e to the first moment μ_1 of the response peak measured at the detector. This expression is applicable when the volumetric flow rate Q_L in the chamber on the lower side of the pellet (Fig. 11-2) is very large.

(a) Derive a relationship between D_e, μ_1, pellet length (Δr), pellet porosity ε_p, and Q_L when Q_L is *not* especially large.

(b) Show that the equation obtained in (a) reduces the Eq. (11-21) when $Q_L \to \infty$.

Notes:

1. Neglect the accumulation of diffusing component in the small volume of the lower chamber.
2. Assume that turbulence in the lower chamber is such that there is negligible mass-transfer resistance between the bottom face of the catalyst pellet and the gas in the lower chamber.
3. Assume that the composition of gas throughout the lower chamber is uniform at any instant.

11-6 Rao and Smith[††] also studied the first-order (at constant total pressure) reversible reaction

$$o\text{-}H_2 \rightleftharpoons p\text{-}H_2$$

at $-196°C$ and 1 atm pressure using a NiO-on-Vycor catalyst. For particles with an average diameter of 58 microns, rate measurements gave $r_p/(y_s - y_{eq})_{O-H_2} = 5.29 \times 10^{-5}$ g mol/(s)(g catalyst). For *pellets* $\frac{1}{8}$-in. long and $\frac{1}{2}$-in. in diameter $r_p/(y_s - y_{eq})_{O-H_2} = 2.18 \times 10^{-5}$. The catalyst pellets were encased on the cylindrical surface and on one end, so that hydrogen was available only to the other face of the cylinder, as illustrated in Fig. 11-10. The density of the pellet was 1.46 g/cm³.

(a) From the experimental rate data evaluate the effectiveness factor for the pellet. (b) Using the random-pore model to estimate D_e, predict an effectiveness factor for comparison with the answer to part (a). Only micropores ($\bar{a}_\mu = 45$ Å) exist in Vycor, and the porosity of the pellet was $\varepsilon_\mu = 0.304$.

11-7 Cunningham et al.[‡] measured global rates of ethylene hydrogenation on copper–magnesium oxide catalysts of two types: particles of 100 mesh size and $\frac{1}{2}$-in. spherical pellets of three pellet densities. For both materials external concentration and temperature differences were negligible. The

[†] M. R. Rao and J. M. Smith, *AIChE J.*, **10**, 293 (1964).
[††] Loc. cit.
[‡] R. A. Cunningham, J. J. Carberry, and J. M. Smith, *AIChE J.*, **11**, 636 (1965).

rate data at about the same surface concentrations of ethylene and hydrogen at the outer surface of the particles and pellets are as shown below.

		Rate r, g mol/(s)(g catalyst)		
			Pellets	
t_s, °C	Particles	$\rho_P = 0.72$	$\rho_P = 0.95$	$\rho_P = 1.18$
124	1.45×10^{-5}	6.8×10^{-6}	4.3×10^{-6}	2.2×10^{-6}
112	6.8×10^{-6}	6.7×10^{-6}	4.2×10^{-6}	2.1×10^{-6}
97	2.9×10^{-6}	6.4×10^{-6}	4.0×10^{-6}	2.0×10^{-6}
84	1.2×10^{-6}	6.0×10^{-6}	3.7×10^{-6}	1.9×10^{-6}
72	....	5.5×10^{-6}	3.4×10^{-6}	1.7×10^{-6}
50	....	4.3×10^{-6}	2.6×10^{-6}	1.3×10^{-6}

(a) From Arrhenius plots of these data, what is the true activation energy of the chemical steps at the catalytic sites? (b) Calculate effectiveness factors for pellets of each density at all the listed temperatures. (c) Explain why some effectiveness factors are greater than unity and why some are less than unity. (d) Why do the rate and the effectiveness factor increase with decreasing pellet density? (e) Suggest reasons for the very low apparent activation energies suggested by the pellet data. Note that even for the maximum effect of intrapellet gradients the apparent activation energy would still be one-half the true value.

11-8 Wheeler[†] has summarized the work on internal diffusion for catalytic cracking of gas-oil. At 500°C the rate data for fixed-bed operation, with relatively large ($\frac{1}{8}$-in.) catalyst particles and that for fluidized-bed reactors (very small particle size) are about the same. This suggests that the effectiveness factor for the large particles is high. Confirm this by estimating η for the $\frac{1}{8}$-in. catalyst if the mean pore radius is 30 Å, the particle diameter (spherical) is 0.31 cm, and the pore volume is 0.35 cm³/g catalyst. Molecular weight of oil is 120.

At atmospheric pressure with 30-Å pores the diffusion will be of the Knudsen type. The rate data, interpreted in terms of a first-order rate equation, indicate (at atmospheric pressure) that $(k_1)_{exp} = 0.25$ cm³/(s)(g catalyst). Assume that a tortuosity factor of 3.0 is applicable.

11-9 Blue et al.[‡] have studied the dehydrogenation of butane at atmospheric pressure, using a chromia-alumina catalyst at 530°C. For a spherical catalyst size of $d_p = 0.32$ cm the experimental data suggest a first-order rate constant of about 0.94 cm³/(s)(g catalyst). The pore radius is given as 110 Å. Assuming Knudsen diffusivity at this low pressure and estimating the pore volume as 0.35 cm³/g, predict an effectiveness factor for the catalyst. Use the parallel-pore model with a tortuosity factor of 3.0.

11-10 Rate data for the pyrolysis of normal octane (C_8H_{18}) at 450°C give an apparent, first-order, irreversible rate constant, k_a, of 0.25 cm³/(s)(g). This apparent rate constant is defined:

$$r \text{ (mol/(s)(g catalyst)} = k_a C$$

where C_s = concn of hydrocarbon gas at outer surface of catalyst pellet, mol/cm³.

The data were obtained at 1 atm pressure with a monodisperse, silica catalyst whose average pore radius was 30 Å. At these conditions Knudsen diffusion predominates in the pores. Other properties of the $\frac{1}{8}$-in., spherical catalyst pellets are as follows: surface area = 230 (m²)/(g catalyst), pore volume = 0.35 cm³/g, and tortuosity factor = 2.0. Using the parallel pore model, determine the effectiveness factor for this reaction-catalyst system.

---

† A. Wheeler in W. G. Frankenburg, V. I. Komarewsky, and E. K. Rideal (eds.), "Advances in Catalysis," vol. III, pp. 250–326, Academic Press, Inc., New York, 1950.

‡ R. W. Blue, V. C. F. Holm, R. B. Reiger, E. Fast, and L. Heckelsberg, *Ind. Eng. Chem.*, **44**, 2710 (1952).

11-11 In Example 11-8 effectiveness factors were calculated from global rate data for the liquid-phase hydrogenation of α-methyl styrene. The dissolved hydrogen concentration was constant in the differential reactor. The global rate was calculated from the cumene concentration C_c in the effluent liquid, the volumetric liquid flow rate Q, and the mass m of catalyst, by the conventional equation

$$r = \frac{QC_c}{m}$$

Explain, with appropriate equations, how Eqs. (B), (C) would need to be modified if C_{H_2} in the liquid varied from entrance to exit of the reactor. (Retain the differential reactor concept.)

11-12 (a) Develop an expression for the effectiveness factor for a straight cylindrical pore of length $2L$. Both ends of the pore are open to reactant gas of concentration $C = C_s$. A first-order irreversible reaction $A \rightarrow B$ occurs on the pore walls. Express the result as $\eta = f(\Phi_p)$, where Φ_p is the Thiele modulus for a single pore. Take the rate constant for the reaction as k_s, expressed as $\text{mol}/(\text{cm}^2)(\text{s}) \times (\text{g mol}/\text{cm}^3)$. The pore radius is a and the diffusivity of A in the pore is D.

(b) How does the relationship $\eta = f(\Phi_p)$ compare with Eq. (11-55)? What is the relationship between Φ_p and Φ_L?

(c) Comparison of the definitions of Φ_p [derived in part (a)] and Φ_L [from Eq. (11-56)] leads to what relationship between the effective diffusivity D_e in a flat plate of catalyst and the diffusivity D for a single pore? Note that L, the actual thickness of the slab in Eq. (11-56), is equal to one-half the pore length. Also note that k_1 in Eq. (11-56) is equal to $k_s S_g$.

(d) What simple model of a porous catalyst will explain the relation between D_e and D? Assume that the Wheeler equation relating the pore radius, the pore volume, and the surface area [Eq. (11-23)] is applicable.

11-13 Consider the effect of surface diffusion on the effectiveness factor for a first-order, irreversible, gaseous reaction on a porous catalyst. Assume that the intrinsic rates of adsorption and desorption of reactant on the surface are rapid with respect to the rate of surface diffusion. Hence equilibrium is established between reactant in the gas in the pore and reactant adsorbed on the surface. Assume further that the equilibrium expression for the concentration is a linear one. Derive an equation for the effectiveness factor for each of the following two cases:

(a) A porous slab of catalyst (thickness L) in which diffusion is in one direction only, perpendicular to the face of the slab
(b) A spherical catalyst pellet of radius r.

Use the expression for total effective diffusivity given by Eq. (11-37) and follow the development in Sec. 11-7.

11-14 A limiting case of intrapellet transport resistances is that of the *thermal effectiveness factor*.† In this situation of zero mass-transfer resistance, the resistance to intrapellet heat transfer alone establishes the effectiveness of the pellet. Assume that the temperature effect on the rate can be represented by the Arrhenius function, so that the rate at any location is given by

$$r = Ae^{-E/R_g T} f(C_s)$$

where $f(C_s)$ represents the concentration dependency of the rate, evaluated at the outer surface of the pellet.

(a) Derive a dimensionless form of the differential equation for the temperature profile within the pellet, using the variables

$$T^* = \frac{T}{T_s} \qquad r^* = \frac{r}{r_s}$$

† J. A. Maymo, R. E. Cunningham, and J. M. Smith, *Ind. Eng. Chem., Fund. Quart.*, **5**, 280 (1966).

where T_s is the temperature at the outer surface of the pellet (radius r_s). What are the dimensionless coefficients in the equation? One of the parameters in the coefficient should be the rate of reaction evaluated at the outer surface; that is,

$$r_s = Ae^{-E/R_g T_s} f(C_s)$$

(b) Derive an integral equation for the effectiveness factor, where the integral is a function of the dimensionless temperature profile.

(c) Are the results in parts (a) and (b) restricted to a specific form for $f(C)$, such as a first-order reaction? (d) Would you expect the thermal effectiveness factor, given by the solution of the integral equation of part (b), to be more applicable to exothermic or to endothermic reactions? (e) Determine the effectiveness factor as a function of a dimensionless coefficient involving r_s, where the other coefficient is $E/R_g T_s = 20$.

A digital computer will facilitate the numerical solution of the equations derived in parts (a) and (b).

11-15 When the intrinsic rate of poison deposition on a porous catalyst is very fast with respect to intraparticle diffusion *pore-mouth* poisoning occurs. For these conditions a shell model (Fig. 11-15) can be used to approximate the effect of poisoning on the rate of a catalytic reaction. Equation (11-104) was presented to show the effect of the Thiele modules Φ_s for the main reaction and the effect of the size (α) of the poisoned shell on the rate. Show how Eq. (11-104) can be derived starting with the equations (11-98) and

$$r = m_p \eta k_1 C_c$$

11-16 The activity of a catalyst is known to decrease with time because of deposition of a poison in the feed stream. The poison-deposition reaction is intrinsically slow with respect to the rate of intraparticle diffusion of the poison. The rate of decrease of active catalyst surface is governed by Eq. (11-108). The main reaction is first order and irreversible. Also it is intrinsically fast so that intraparticle diffusion resistance is large. Derive an expression showing how the rate of the main reaction *for a catalyst particle* decreases with time for a constant $(C_A)_s$ and poison concentration C_p.

11-17 As a simple illustration of series poisoning (for example, by carbon deposition in hydrocarbon cracking processes) consider the following reactions

$$A(g) \rightarrow B(g) \text{ main reaction } (B \text{ is desired product})$$

$$B(g) \rightarrow C(s) \text{ poisoning reaction (deposition of } C \text{ on catalyst)}$$

The main reaction is irreversible, first order in A, and its rate is directly proportional to the fraction of unpoisoned catalytic surface. The kinetics of the poisoning reaction obey Eq. (11-108) and its rate is slow with respect to the rate of intraparticle mass transfer.

(a) For a slow main reaction (no intraparticle diffusion) derive an equation showing how the rate of the main reaction decreases with time for a catalyst particle at constant $(C_A)_s$, $(C_B)_s$.

(b) Can a similar expression for an intrinsically fast main reaction (large intraparticle diffusion resistance) be derived?

11-18 The "wetting efficiency" question in trickle-bed reactors introduces the problem of effectiveness factors in catalyst pellets with nonuniform boundary conditions. Consider a simple example of this problem for a catalyst pellet with slab geometry as sketched below. One face of the porous slab has a uniform concentration $C_{s, g}$ and the other face has a concentration $C_{s, L}$. Define the effectiveness factor η for a first-order reaction on the pore walls as:

$$\text{Rate } [\text{k mol}/(\text{kg})(\text{s})] = \eta k C_{s, g}$$
$$k = \text{first-order rate constant, } m^3/(\text{kg})(\text{s})$$
$$\rho_p = \text{density of porous catalyst slab, kg}/(m)_3$$
$$C = \text{concentration of reactant in pores, k mol}/(m)^3$$

A. Derive an expression for η in terms of the Thiele modules,

$$\Phi = L(k/\rho_p D_e)^{1/2}$$

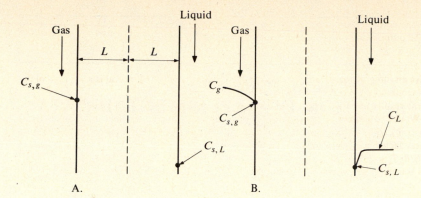

and the concentration ratio

$$C_{s,g}/C_{s,L}$$

where $2L$ = thickness of the slab, m

D_e = effective diffusivity of the reactant in the catalyst slab, m^2/s

B. How could external mass-transfer resistances on both faces of the slab be introduced in order to express the rate of reaction in terms of *bulk* concentrations C_g (see sketch) and C_L.

TWELVE

LABORATORY REACTORS—
INTERPRETATION OF EXPERIMENTAL DATA

It was noted in Chap. 7 that if global rates of reaction are used, design equations for heterogeneous reactors are similar to those for homogeneous reactors. We shall take advantage of this simplification in Chap. 13. The effects of external and internal transport processes on the global rate were analyzed in detail in Chaps. 10 and 11. In preparation for reactor design let us summarize the application of these results to relate the global and intrinsic rate.

As intrinsic rate equations cannot yet be predicted (Chap. 9), they must be evaluated from laboratory kinetics data. Such data are measurements of the global rate of reaction. The problem is to extract the equation for the intrinsic rate from the global rate data. Since laboratory reactors are much less costly than commercial-scale units, there are fewer economic restraints in designing laboratory equipment. Construction and operating conditions can be chosen to reduce or eliminate the differences between the global and intrinsic rates so that more accurate equations for the intrinsic rate can be extracted from the experimental data. Several types of laboratory reactors have been developed, and these will be discussed in Secs. 12-2 and 12-3.

After the rate equation at a catalyst site has been obtained, the relations between global and intrinsic rates must be used again, this time to obtain a global rate for a range of operating conditions. Such information is necessary for designing the commercial-scale reactor. The type of commercial unit will be dictated by economics, and whatever the effects of physical processes, they must be included in evaluating the global rate. This aspect of the problem will be discussed in Sec. 12.4.

Since the objectives are different, operating conditions and reactor types are usually different for laboratory and large-scale units. Hence, the global rate in the laboratory reactor normally cannot be used directly for designing large-scale equipment. For example, economic considerations may require a different (usually greater) velocity in the commercial reactor, resulting in different external mass- and heat-transfer coefficients between catalyst pellet and reacting fluid. If external resistances are significant, there will be a different global rate in the commercial-scale reactor for the same intrinsic rate. Similarly, different sizes of catalyst pellets in the commercial and laboratory reactors will lead to different global rates if internal resistances are significant. Thus, to obtain a global rate useful for design, the measurements in the laboratory should be reduced to the intrinsic rate at the catalyst site and then recombined with physical resistances corresponding to the conditions of commercial operation.

The relation between global and intrinsic rates is not always important. As a general rule, the global rate will be the same as the intrinsic rate† when the chemical step at the catalyst site is very slow. The difference between the two rates increases as the intrinsic rate increases, and for systems where the chemical step is extremely rapid a careful accounting of mass- and heat-transfer processes is important. In fixed-bed catalytic reactors external resistances are small for normal operating conditions. Internal resistances, however, may be important when external ones are negligible. It is often true that external resistances are significant only when intrapellet resistances are more significant.‡ Care should be taken to distinguish between mass- and heat-transfer resistances. We have seen (Sec. 10-2) that large external temperature differences can exist when the external concentration difference is small. In fluidized-bed and slurry reactors, where the catalyst particles are small, a situation opposite to that in fixed-bed reactors exists. Here internal resistances are usually negligible, but external resistances, particularly for mass transfer, may be significant. As we have seen in Chap. 11 (Example 11-8), both external and internal mass-transfer resistances can be important in fixed-bed reactors. Various types of transport effects are examined again in Chap. 13 where models are formulated for reactor design.

12-1 Interpretation of Laboratory Kinetics Data

Suppose that the global rate has been measured in a laboratory reactor over a range of concentration and temperature for a single pellet size. The first step is to determine the surface concentration C_s and temperature T_s from measured bulk values by means of Eqs. (10-21) and (10-22), as in Example 10-4. Next the global rate and C_s and T_s are used to establish the rate equation. In this second step the effect of the internal resistances, the *effectiveness factor*, is obtained simultaneously. The second step is a complicated one for the general case when there are significant temperature gradients within the catalyst pellet. For *isothermal* conditions Eq. (11-43) and the appropriate curve in Fig. 11-8 can be used to establish

---

† When comparisons between global and intrinsic rates are made, it is understood that the comparison is for the same temperature and concentration. Note that in a reactor operating at steady state the two rates are always the same, but the temperature and composition in the bulk fluid (i.e., the global conditions) are different from those at a site within the catalyst pellet (intrinsic conditions).

‡ For porous pellets with uniform distribution of catalytic material. If the catalyst is deposited only on the outer layer of the pellet, this is not true.

both the intrinsic rate $[f(C)$ in Eq. (11-43)] and the effectiveness factor. A trial procedure is usually necessary. The following steps describe a method of solution:

1. Assume a form for $f(C)$ (first order, second order, etc.).
2. For a given set of experimental data $(C_s, T_s,$ and $\mathbf{r}_P)$ assume a value of η.
3. From the appropriate curve in Fig. 11-8 obtain Φ and then evaluate k from the defining expression for Φ [for example, Eq. (11-50) for a first-order rate on spherical catalyst pellets].
4. Check the assumed value of η with Eq. (11-43).
5. Repeat steps 2 to 4 for all sets of data to see whether the assumed form of $f(C)$ is valid (for a first-order rate equation the evaluation would be to see if k_1 is constant for data at various concentrations but constant temperature).
6. Finally, the values of k for different temperatures would be plotted as $\ln k$ vs. $1/T$ to obtain $\mathbf{A}$ and E in the Arrhenius equation, $k = \mathbf{A}e^{-E/R_g T}$.

The resultant values of $\mathbf{A}$ and E and the nature of $f(C)$ establish the desired equation for the intrinsic rate. Many aspects of this procedure were illustrated for a fixed-bed reactor in Example 11-8. Note for step 3 we must know, or estimate, the effective diffusivity. In Example 11-8, D_e was obtained from experimental data since rates were measured for two sizes of catalyst pellets. Alternately, the global rate may be measured for *small* catalyst particles for which $\eta \to 1.0$. Then internal (and usually external) transport resistances are negligible; the bulk temperature and concentrations may be taken as equal to those at a catalyst site, and the observed rate and bulk temperature and concentration data can thus be used directly to obtain $f(C)$ and $\mathbf{A}$ and E. This procedure was illustrated in Example 11-10.

If intrapellet temperature gradients are significant, an analytic procedure similar to that for the isothermal case can be employed, in principle, to evaluate an equation for the intrinsic rate. However, step 1 now must include an assumption for the activation energy as well as for $f(C)$. The subsequent steps for checking the assumptions involve β and γ, defined by Eqs. (11-79) and (11-80), since η is now a function of these parameters as well as of Φ. In addition to the effective diffusivity, we need the effective thermal conductivity. The uncertainties in these quantities and the complex analysis required to obtain them severely limits the usefulness of the procedure for nonisothermal conditions. Hence it is preferable to use measured rates for small particles, or laboratory reactors in which nonisothermal conditions are eliminated.

Integral, or differential, tubular reactors or stirred-tank reactors are used for laboratory investigations. There are advantages in using stirred-tanks for kinetic studies. Steady-state operation with well-defined residence-time conditions and uniform concentrations in the fluid and on the solid catalyst are achieved. Isothermal behavior in the fluid phase is attainable. Stirred tanks have long been used for homogeneous liquid-phase reactors and slurry reactors and recently reactors of this type have been developed for large catalyst pellets. Recycle operation (see Secs. 4-9 and 4-10) of tubular flow units combine some of the advantages of stirred-tank and differential-tubular reactors.

When either a stirred-tank or a differential reactor is employed, the global rate is obtained directly, and the analysis procedure described above can be initiated immediately. This was the situation in Examples 11-8 and 11-10. However, if *integral* reactor data are measured, the data must first be differentiated in order to obtain the global rate. Alternately, the assumed rate equation may be integrated and the results compared with the observed data. For measurements made in a tubular-flow reactor packed with catalyst pellets, the mass balances of reactant, if we assume plug flow, is

$$\mathbf{r}_P \, dW = F \, dx \qquad (12\text{-}1)$$

where $\mathbf{r}_P$ = global rate of reaction per unit mass of catalyst
  W = mass of catalyst
  F = feed rate of reactant
  x = conversion of reactant

Equation (12-1) is the same as that developed for homogeneous reactors [Eq. (3-17)], except that W has replaced the reactor volume and $\mathbf{r}_P$ is based on a unit mass of catalyst. If the volume of the reaction mixture is constant (constant temperature, pressure, and number of moles) and C_0 is the feed concentration, then $dx = -(1/C_0) \, dC_b$, and Eq. (12-1) becomes

$$\frac{W}{F} = \int \frac{dx}{\mathbf{r}_P} = -\frac{1}{C_0} \int \frac{dC_b}{\mathbf{r}_P} \qquad (12\text{-}2)$$

Note that C_b is the concentration of *reactant* in the *bulk* stream.

The interpretation procedure for integral reactor data is illustrated in Example 12-1 for the reaction

$$o\text{-}H_2 \rightleftharpoons p\text{-}H_2$$

on Ni—Al_2O_3 catalyst pellets. Integral reactor data were measured under isothermal conditions. The analysis is simplified because the intrinsic rate equation is first order (at the fixed pressure existing in the reactor). Hence external and internal mass-transfer resistances can be combined in the simple way shown in Eq. (11-86). Both resistances are significant at some conditions, and the analysis permits quantitative evaluation of these resistances.

Example 12-1 Wakao et al.† studied the conversion of ortho hydrogen to para hydrogen in a fixed-bed tubular-flow reactor (0.50 in. ID) at isothermal conditions of $-196°C$ (liquid nitrogen temperature). The feed contained a mole fraction p-H_2 of $y_{b_1} = 0.250$. The equilibrium value at $-196°C$ is $y_{eq} = 0.5026$. The catalyst is Ni on Al_2O_3 and has a surface area of 155 m^2/g. The mole fraction p-H_2 in the exit stream from the reactor was measured for different flow rates and pressures and for three sizes of catalyst: granular particles of equivalent spherical diameter, 0.127 mm, granular particles 0.505 mm, and nominal $\frac{1}{8} \times \frac{1}{8}$-in. cylindrical pellets. The flow-rate, pressure, and composition measurements are given in Table 12-1.

The reaction is reversible and believed to be first order. Also, strong

† Noriaki Wakao, P. W. Selwood, and J. M. Smith, *AIChE J.*, **8**, 478 (1962).

Table 12-1 Experimental data

Type of catalyst	Mass of catalyst in reactor, g	Pressure, lb/in.^{2} gauge	Flow rate, cm^3/s (measured at 0°C, 1 atm)	Mole fraction p-H$_2$ in exit gas, $(y_b)_2$
Cylindrical pellets, $\frac{1}{8}$ in.	2.55	40	30.1	0.3790
			67.7	0.3226
			141.0	0.2886
		100	26.7	0.4031
			58.6	0.3416
		400	28.7	0.4286
			78.5	0.3480
0.505-mm particles	1.13	100	24.2	0.4261
			147.0	0.2965
		400	35.8	0.4300
			77.1	0.3485
0.127-mm particles	0.739	100	40.6	0.3378
		400	68.5	0.3381

adsorption of hydrogen reduces the rate. The concepts of Chap. 9 suggest that a logical assumption for the form of the rate equation at a catalyst site (an intrinsic rate equation) would be

$$\mathbf{r} = f(C) = \frac{k_c(C_o - C_p/K)}{1 + K_a(C_o + C_p)} \tag{A}$$

where $\mathbf{r}$ = rate of reaction, g mol/(s)(g catalyst)

C_o, C_p = concentrations of ortho and para hydrogen in the gas phase at a catalyst site, g mol/cm^3

k_c = reaction rate constant, cm^3/(g catalyst)(s)

K = equilibrium constant for the reaction o-H$_2 \rightleftharpoons p$-H$_2$

K_a = adsorption constant in the rate equation

Test the suitability of the proposed rate equation and evaluate the constants k_c and K_a. Also calculate effectiveness factors for the $\frac{1}{8}$-in. pellets. The following additional data are available:

Viscosity of H$_2$ at -196°C $\quad = 348 \times 10^{-7}$ poises

Diffusivity of H$_2$ at -196°C $\quad = \begin{cases} 0.0376 \text{ cm}^2/\text{s} & \text{at 40 lb/in.}^2 \text{ gauge} \\ 0.0180 \text{ cm}^2/\text{s} & \text{at 100 lb/in.}^2 \text{ gauge} \\ 0.00496 \text{ cm}^2/\text{s} & \text{at 400 lb/in.}^2 \text{ gauge} \end{cases}$

Schmidt number for H$_2$ at -196°C $\quad = \mu/\rho \, \mathscr{D}_{H_2} = 0.78$

Density of catalyst (particles or pellet) $\quad = 1.91$ g/cm^3

Void fraction of catalyst beds, $\varepsilon_B = 0.33$

SOLUTION The volume of the reaction mixture is constant for any one run. If plug-flow behavior is assumed, Eq. (12-2) is applicable. The data in Table 12-1 represent integral reactor results. To use such information Eq. (12-2) must be integrated. This requires expressing the rate in terms of C_b. When external composition differences are significant and the rate equation is not first order, this is difficult. The difference $C_b - C_s$ will vary in the axial direction. Hence, a trial-and-error stepwise procedure is necessary to accomplish the integration. In this case the intrinsic rate is first order, so that the integration of Eq. (12-2) is simple and analytical.

For any one run the total pressure and total concentration are uniform in the reactor, so that Eq. (A) may be written

$$\mathbf{r} = k_1 \left(C_o - \frac{C_p}{K} \right) = k_1 C_o - k_1' C_p \tag{B}$$

where

$$k_1 = \frac{k_c}{1 + K_a C_t} \tag{C}$$

We saw in Sec. 2-10 that for first-order reversible reactions Eqs. (2-62) and (2-63) show that Eq. (B) may be written

$$\mathbf{r} = \frac{k_1(K + 1)}{K} [C_o - (C_o)_{eq}] \tag{D}$$

where $(C_o)_{eq}$ is the equilibrium concentration of ortho hydrogen at the temperature involved. Since the total concentration of hydrogen is constant $(C_t = C_o + C_p)$,

$$C_o - (C_o)_{eq} = C_t - C_p - [C_t - (C_p)_{eq}] = (C_p)_{eq} - C_p \tag{E}$$

Then Eq. (D) may be written

$$\mathbf{r} = \frac{k_1(K + 1)}{K} [(C_p)_{eq} - C_p] = \frac{k_1(K + 1)}{K} (C_{eq} - C)\dagger \tag{F}$$

Equation (F) is for the intrinsic rate at a catalyst site, and so C is the $p\text{-}H_2$ concentration at a site within the catalyst pellet. To convert this to a rate for the pellet, $\mathbf{r}_P$, that is, to account for internal mass-transfer resistance, we use Eq. (11-43) to obtain

$$\mathbf{r}_P = \eta \frac{k_1(K + 1)}{K} (C_{eq} - C_s) \tag{G}$$

where the concentration C_s at the outer surface of the pellet has replaced C. Next, the relation between C_s and C_b is obtained by equating the mass-transfer rate (per unit mass) from Eq. (10-1) to the rate of reaction,

$$\mathbf{r}_P = k_m a_m (C_s - C_b) = \eta \frac{k_1(K + 1)}{K} (C_{eq} - C_s) \tag{H}$$

---

† Since only $p\text{-}H_2$ concentrations will be used from here on, the subscript p will be omitted.

where a_m is the external surface per unit mass, or $6/d_p \rho_p$ for a spherical pellet. Eliminating C_s from the two forms of Eq. (H) gives

$$r_P = \frac{C_{eq} - C_b}{d_p \rho_p/6k_m + K/[\eta k_1(K + 1)]} = K(C_{eq} - C_b) \tag{I}$$

where

$$\frac{1}{K} = \frac{d_p \rho_p}{6k_m} + \frac{K}{\eta k_1(K + 1)} \tag{J}$$

Equation (I) is the required expression for the rate in terms of the bulk concentration of p-H_2. It is the form of Eq. (11-86) applicable to a reversible first-order reaction.

Parameter K is constant throughout the reactor for a given run. Hence when Eq. (I) is substituted in Eq. (12-2) the integration is simple. We note that $dC_p = -dC_o$, so that Eq. (12-2) becomes, for p-H_2,

$$\frac{W}{F_p} = \frac{1}{(C_b)_1} \int_1^2 \frac{dC_b}{K(C_{eq} - C_b)} = \frac{-1}{(C_b)_1 K} \ln \frac{C_{eq} - (C_b)_2}{C_{eq} - (C_b)_1} \tag{K}$$

where subscripts 1 and 2 represent entrance and exit conditions, respectively. Assuming ideal-gas behavior, we have

$$(C_b)_1 = \frac{(y_b)_1}{R_g T} p_t$$

The feed rate of p-H_2 is $F_t(y_b)_1$. In terms of *total* feed rate and mole fractions, Eq. (K) may be written

$$\frac{W}{F_t} = -\frac{R_g T}{p_t K} \ln \frac{y_{eq} - (y_b)_2}{y_{eq} - (y_b)_1} \tag{L}$$

The data in Table 12-1 give $(y_b)_2$ for various values of flow rate. Also, $y_{eq} = 0.5026$ and $(y_b)_1 = 0.250$. Hence values of K can be calculated for each run. For example, for the first run at 40 lb/in.^{2} gauge with the $\frac{1}{8}$-in. catalyst pellets

$$\frac{W}{F_t} = \frac{2.55}{30.1/22,400} = 1,890 \text{ (g catalyst)(s)/(g mol)}$$

Then, from Eq. (L),

$$1,890 = -\frac{R_g T}{p_t K} \ln \frac{0.5026 - 0.3790}{0.5026 - 0.250}$$

$$\frac{R_g T}{p_t K} = 2,640 \text{ (s)(g catalyst)/(g mol)}$$

$$\frac{1}{K} = 2,640 \frac{(40 + 14.7)/14.7}{82(273 - 196)} = 1.56 \text{ (g catalyst)(s)/cm}^3$$

The results for other runs are given in column 5 of Table 12-2.

Table 12-2 Calculated results for the reaction $o\text{-}H_2 \rightleftharpoons p\text{-}H_2$

Type of catalyst	Mass of catalyst, g	p_t, lb/in.² gauge	W/F (g catalyst) (g mol)	$1/K$, (g catalyst) (s)/cm³	$d_P \rho_P / 6 k_m$, (g catalyst) (s)/cm³	ηk_1, cm³/(s) (g catalyst)
Cylindrical pellets, $\frac{1}{8}$ in.	2.55	40	1,890	1.56	0.12	0.348
		40	839	1.47	0.08	0.361
		40	403	1.46	0.06	0.359
		100	2,120	2.80	0.25	0.197
		100	969	2.65	0.18	0.203
		400	1,980	7.18	0.90	0.080
		400	724	6.60	0.60	0.084
0.505-mm particles	1.13	100	1,040	1.07	0.008	0.472
		100	171	1.04	0.005	0.485
		400	701	2.50	0.020	0.202
		400	325	2.93	0.024	0.173
0.127-mm particles	0.739	100	404	1.06	0.0006	0.474
		400	240	2.51	0.0017	0.200

If the form of the intrinsic rate equation is satisfactory, **K** should be constant at a given pressure. The values of $1/K$ in Table 12-2 are about constant for a given catalyst and pressure. Results for different catalyst sizes may be different, because **K** includes the effects of external and internal mass-transfer resistances.

The next step is to evaluate and separate the external mass-transport resistance by means of Eq. (J). The mass-transfer coefficient k_m can be estimated from Eq. (10-10). With the first run as an example,

$$\text{Area of reactor tube} = \frac{\pi}{4} d_t^2 = \frac{\pi}{4}\left(\frac{1}{2}\right)(2.54)^2 = 1.27 \text{ cm}^2$$

$$G = \frac{30.1(2.016)}{22,400(1.27)} = 2.12 \times 10^{-3} \text{ g/(s)(cm}^2)$$

$$d_P = \tfrac{1}{8}(2.54) = 0.318 \text{ cm}$$

$$\text{Re} = \frac{d_P G}{\mu} = \frac{0.318(2.12 \times 10^{-3})}{348 \times 10^{-7}} = 20$$

From Eq. (10-10) at this Reynolds number, we have

$$j_D = \frac{0.458}{0.33}(20)^{-0.407} = 0.41$$

Using this result in Eq. (10-9) gives

$$k_m = \frac{j_D G}{\rho}\left(\frac{\mu}{\rho D}\right)^{-2/3} = \frac{0.41(2.12 \times 10^{-3})}{1.18 \times 10^{-3}}(0.78^{-2/3}) = 0.87 \text{ cm/s}$$

where 1.18×10^{-3} g/cm^3 is the density of hydrogen at 40 lb/in.^{2} gauge and $-196°C$. The external resistance is given by the first term on the right-hand side of Eq. (J),

$$\frac{d_P \rho_P}{6 k_m} = \frac{0.318(1.91)}{6(0.87)} = 0.12 \text{ (g catalyst)(s)/cm}^3$$

Similar calculations for the other runs give the results in column 6 of Table 12-2. Comparison of $1/K$ and $d_P \rho_P / 6 k_m$ shows that external resistance is important only for the $\frac{1}{8}$-in. pellets. Even then, the external contribution to the total resistance is less than 10% for all but the 400-lb/in.^{2} gauge runs. In the most severe case the external contribution is 0.90/7.18, or 13% of the total. For the two smaller particle sizes such a value would be very small. These results are typical and illustrate the previous statement about the un-importance of external resistances for small particles in fixed-bed reactors.

Since the external resistance is significant for some of the runs with $\frac{1}{8}$-in. pellets, it is expected that the internal resistance will also be important. This can be established when η is evaluated. Equation (J) can be used to separate the external resistance from $\mathbf{K}$. Using the data for the first run,

$$\frac{K}{\eta k_1 (K + 1)} = \frac{1}{\mathbf{K}} - \frac{d_P \rho_P}{6 k_m} = 1.56 - 0.12 = 1.44 \text{ (g catalyst)(s)/cm}^3$$

The equilibrium constant K is

$$K = \frac{(C_p)_{eq}}{(C_o)_{eq}} = \frac{P_t(y_p)_{eq}}{P_t(y_o)_{eq}} = \frac{0.5026}{1 - 0.5026} = 1.01$$

Hence

$$\eta k_1 = \frac{K}{K + 1}\left(\frac{1}{1.44}\right) = \frac{1.01}{2.01}\left(\frac{1}{1.44}\right) = 0.348 \text{ cm}^3/\text{(s)(g catalyst)}$$

Values calculated in this way for all the runs are given in the last column of Table 12-2.

If ηk_1 were available only for particles for which $\eta < 1$, we would have to follow the trial procedure outlined earlier (and used in Example 11-8), assuming an effective diffusivity and then using Fig. 11-8 and the ηk_1 values together to find k_1. However, the data for the small particles enable us to use the alternate procedure based on $\eta = 1$ for these particles. Since ηk_1 is known for two particle sizes, we can substantiate that η actually is unity for the *particles*. For a fourfold increase in particle diameter (0.505 vs. 0.127 mm) the data in Table 12-2 show that ηk_1 is essentially unchanged at constant pressure. Since Φ_s, and hence η, vary with diameter [see Eq. (11-50)] for $\eta < 1$, it is clear that $\eta = 1$ for both sizes of small particles. Therefore, the values of ηk_1 for the particles in Table 12-2 are equal to k_1. We can use these results to evaluate k_c

and K_a, employing Eq. (C). Averaging the three values for 100 lb/in.^{2} gauge we have

$$\frac{0.472 + 0.485 + 0.474}{3} = \frac{k_c}{1 + K_a C_t}$$

$$C_t = \frac{p_t}{R_g T} = \frac{114.7/14.7}{82(77)} = 1.23 \times 10^{-3} \text{ g mol/cm}^3$$

and so

$$0.477 = \frac{k_c}{1 + 1.23 \times 10^{-3} K_a} \tag{M}$$

Similarly, at 400 lb/in.^{2} gauge

$$0.192 = \frac{k_c}{1 + 4.46 \times 10^{-3} K_a} \tag{N}$$

Solution of Eqs. (M) and (N), for k_c and K_a gives

$$K_a = 1.06 \times 10^3 \text{ cm}^3/\text{g mol}$$

$$k_c = 1.1 \text{ cm}^3/(\text{g catalyst})(\text{s})$$

These two constants depend only on temperature. If data had been available at a series of temperatures, $\ln k_c$ could have been plotted against $1/T$ to obtain the activation energy E and the frequency factor A, according to the Arrhenius equation. The adsorption equilibrium constant would also be expected to be an exponential function of temperature, according to a van't Hoff type of equation,

$$\frac{d(\ln K_a)}{dT} = \frac{\Delta H}{R_g T^2}$$

The analysis has now been completed for the assumed form for the intrinsic rate [Eq. (A)]. The results should be regarded as illustrative only. For accurate evaluation of k_c and K_a many more data points would be desired at several pressures and flow rates (W/F values). Then a statistical analysis could be used to obtain the best values of k_c and K_a.

While ηk_1 was the same for the two small particle sizes, Table 12-2 shows that ηk_1 is significantly less for the $\frac{1}{8}$-in. pellets. (The comparison must be made at the same pressure because k_1 and η both vary with pressure.) Therefore, considerable internal mass-transfer resistance exists. We can evaluate η from the ηk_1 data using Eq. (C). For example, at 40 lb/in.^{2} gauge.

$$k_1 = \frac{k_c}{1 + K_a C_t} = \frac{1.1}{1 + 1.06 \times 10^3 (0.59 \times 10^{-3})} = 0.68$$

Table 12-3 Effectiveness factors for $\frac{1}{8}$-in. pellets

p, lb/in.^{2} gauge	C_t, mol/cm^3	k_1, cm^3/ (g catalyst)(s)	$\bar{\eta}$
40	0.59×10^{-3}	0.68	0.53
100	1.23×10^{-3}	0.48	0.42
400	4.46×10^{-3}	0.19	0.42

Then

$$\bar{\eta} = \frac{\bar{\eta} k_1}{k_1} = \frac{(0.348 + 0.361 + 0.359)(\frac{1}{3})}{0.68} = 0.53$$

This is an average value for the effectiveness factor at 40 lb/in.^{2} gauge. The results for the other pressures are shown in Table 12-3.

Conclusions from this example may be summarized as follows:

1. The first-order form of rate equation with an adsorption term in the denominator fits the data.
2. For $\frac{1}{8}$-in. pellets both external and internal mass transport retard the rate. The internal mass-transfer resistance is sizable, as indicated by η values of the order of 0.5. The external resistance seldom exceeded 10% of the total.
3. For small catalyst particles both internal and external resistances were negligible, so that rate data for these particles could be used directly to calculate the constants in the intrinsic-rate equation.

12-2 Homogeneous Laboratory Reactors

Although our major interest in this chapter is heterogeneous laboratory reactors, for the sake of completeness we shall briefly review homogeneous systems. Experimental measurements in homogeneous reactors represent the influence of the intrinsic rate and the effects of mixing. The problem of extracting the intrinsic rate from the observed data is less difficult than for heterogeneous reactions. The methods of analysis for the two extremes of mixing—ideal tubular flow and ideal stirred tank—and for intermediate cases were considered in Chaps. 4 and 6. Isothermal operation is desirable for accurate results. Ideal stirred-tank performance can be achieved easily in most laboratory reactors. Ideal tubular-flow behavior is more difficult to realize. However, with differential reactors non ideal flow is not troublesome, because at low conversions the effect of a distribution of residence times is small.† Both stirred-tank and tubular-flow reactors are common laboratory forms. Gaseous reactions are usually studied in tubular-flow reactors, and both types are used for liquid systems. Integral tubular-flow reactors are not

---

† K. G. Denbigh, "Chemical Reactor Theory," p. 61, Cambridge University Press, Cambridge, 1965.

convenient for studying complex reactions because of the difficulty in extracting rate equations from the observed data. This difficulty was illustrated in Sec. 4-5, where we saw that analytic relations could be obtained for polymerization reactions only for a stirred-tank reactor.

The laboratory studies analyzed in Examples 4-3 and 4-5 for the methane and sulfur vapor reactions illustrate typical apparatus and procedures for differential and integral tubular-flow reactors. An example of a laboratory differential reactor for a homogeneous catalytic reaction has been given by Matsuura et al.†

As mentioned, the *batch* recycle reactor (shown in Fig. 12-1) combines some of the advantages of stirred-tank and tubular-flow operation. With high circulation rates and a small reactor volume with respect to total volume of system, the conversion per pass is very small. Differential operation is achieved in each pass through the reactor, yet over a period of time the conversion becomes significant. This is an advantage when it is difficult to make accurate measurements of the small concentration changes required in a differential tubular-flow reactor. Also, the recycle increases the heat capacity of the stream in the reactor thus reducing temperature gradients. The reaction system is charged with feed at the proper conditions, and then samples are withdrawn periodically for analysis. Suppose that the course of the reaction is followed by analyzing for a product. The data would give a curve like that in Fig. 12-2. If all the system is at the same composition, Eq. (4-32) written for the product, is applicable. That is

$$\mathbf{r} = \frac{V_t}{V_R}\frac{dC}{dt} \tag{12-3}$$

where V_R = volume of reactor
V_t = total volume (reactor + lines + reservoir)
C = concentration of product

† Matsuura, A. E. Cassano, and J. M. Smith, *AIChE J.*, **15**, 495 (1969).

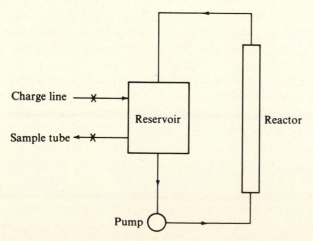

Charge line

Sample tube

Reservoir

Reactor

Pump

Figure 12-1 Batch recycle reactor of tubular-flow type.

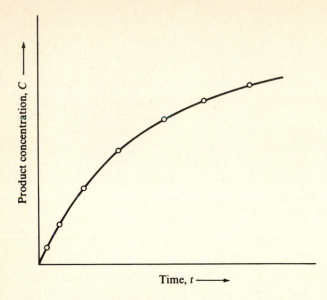

Figure 12-2 Experimental results in a batch recycle reactor.

The slope of the curve of C vs. t in Fig. 12-2 gives the rate at any time (or composition) during the run. By analyzing data for runs at different initial compositions and other runs at different temperatures, a rate equation can be obtained that includes the activation energy. A description of the apparatus and analysis procedure for a gaseous reaction (decomposition of acetone) is given by Cassano et al.† Examples of liquid-phase batch recycle reactors are available.‡ The validity of the assumption that the composition is uniform throughout all the system depends on the degree of mixing in the reservoir, the volume of the reservoir with respect to that of the whole system, and the conversion per pass in the reactor (see Sec. 4-9). If measurements at different circulation rates give the same rates of reaction, these conditions are closely approached. The apparatus can be designed so that the volumes of the reactor and connecting lines are small with respect to that of the reservoir. In the acetone study the reactor and lines had a volume of about 63 cm³ and that of the reservoir was 5500 cm³. At the circulation rates employed the average residence time in the reactor was about 0.20 s and that in the reservoir was 17 s. Under these conditions the change in concentration of reactant per pass was only 0.0055% for a concentration change of 1%/h in the reservoir.

As we saw in Sec. 4-10, the recycle reactor may be operated on a *steady-state*, flow basis. Then ideal stirred-tank behavior is approached as the recycle rate is increased. Recycle ratios (recycle rate to feed rate) of 15 or more are usually sufficient to achieve results close to stirred-tank operation.

---

† A. E. Cassano, T. Matsuura, and J. M. Smith, *Ind. Eng. Chem., Fund. Quart.*, **7**, 655 (1968).

‡ The homogeneous photo-polymerization of styrene was studied in a batch recycle reactor by Ibarra, *AIChE J.*, **20**, 404 (1974). Homogeneous photo decomposition kinetics of 2,4-dichlorophenoxy-acetic acid was investigated by Boval [*Chem. Eng. Sci.*, **28**, 1661 (1973)] in a similar reactor.

12-3 Heterogeneous Laboratory Reactors

Tubular-flow reactors operated in a differential, integral, or recycle manner are frequently used for investigating the kinetics of gas-solid catalytic reactions. A case of integral operation was discussed in Example 12-1. Another example is a study of hexane isomerization.† Illustrations of differential reactors which include experimental details are those for SO_2 oxidation,‡ C_2H_4 hydrogenation,§ hydrogen oxidation,¶ and carbon monoxide oxidation.†† Biskis and Smith‡‡ studied the hydrogenation of α-methyl styrene in a batch-type recycle reactor. The liquid stream (with dissolved hydrogen) was *pulsed* as it flowed over the catalyst particles in order to reduce the external mass-transfer resistance. Trickle-bed reactors have been operated in a combination mode for the hydrogenation of α-methyl styrene: a batch of liquid was recycled between the catalyst bed and a reservoir while the hydrogen gas flowed continuously in one pass through the bed.§§

Fixed-bed reactors have the disadvantage that external and, particularly, internal transport resistances may be significant. Special laboratory reactors have been designed to minimize or clarify these resistances. Carberry¶¶ and Weekman††† have discussed several types. The recycle form can be used to reduce external transport resistances by increasing the recycle rate, thus increasing turbulence in the fluid around the catalyst pellets. The same result can be obtained by rotating a basket containing the catalyst pellets in a tank. Figure 12-3 shows a recycle type (the Berty reactor) in which the impeller speed provides control of the recycle rate independent of the feed rate. Figure 12-4 illustrates the rotating basket form.‡‡‡ In both types increasing rotation speed reduces external transport resistances and achieves ideal stirred-tank behavior. Bennett et al.§§§ have reviewed such reactors and discussed experiments for evaluating their characteristics.

Instead of rotating the catalyst, Ford and Perlmutter¶¶¶ inserted a cylinder containing a deposited catalyst in a stirred vessel. Studies on the vapor-phase dehydrogenation of sec-butyl alcohol indicated that stirred-tank performance was achieved.

A number of single-pellet reactors have been employed in laboratory kinetic

---

† A. Voorhies, Jr., and R. G. Beecher, presented at Sixty-First Annual Meeting AIChE, Los Angeles, Dec. 1–5, 1968.

‡ R. W. Olson, R. W. Schuler, and J. M. Smith, *Chem. Eng. Progr.*, **42**, 614 (1950).

§ A. C. Pauls, E. W. Comings, and J. M. Smith, *AIChE J.*, **5**, 453 (1959).

¶ J. A. Maymo and J. M. Smith, *AIChE J.*, **12**, 845 (1966).

†† S. Otani and J. M. Smith, *J. Catalysis*, **5**, 332 (1966).

‡‡ E. R. Biskis and J. M. Smith, *AIChE J.*, **9**, 677 (1963).

§§ Mordechay Herskowitz, R. G. Carbonell, and J. M. Smith, *AIChE J.*, **25**, 272 (1979).

¶¶ J. J. Carberry, *Ind. Eng. Chem.*, **56**, 39 (1964).

††† V. W. Weekman, Jr., *AIChE J.*, **20**, 833 (1974).

‡‡‡ D. G. Tajbl, J. B. Simmons, and J. J. Carberry, *Ind. Eng. Chem. Fundam.*, **5**, 17 (1866).

§§§ C. O. Bennett, M. B. Cutlip, and C. C. Yang, *Chem. Eng. Sci.*, **27**, 2255 (1972).

¶¶¶ F. E. Ford and D. D. Perlmutter, presented at Fifty-fifth Annual Meeting AIChE, Chicago, Dec. 2–6, 1962.

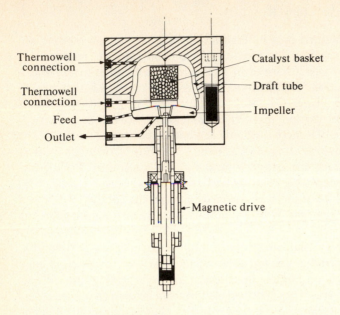

Thermowell connection

Thermowell connection

Feed

Outlet

Catalyst basket

Draft tube

Impeller

Magnetic drive

Figure 12-3 Flow recycle reactor (Berty type). (*Reproduced by permission from J. M. Berty, Chem. Eng. Prog.* **70**(*5*), *78, May 1974.*)

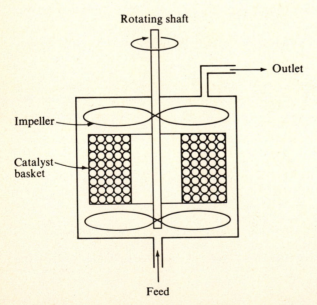

Rotating shaft

Outlet

Impeller

Catalyst basket

Feed

Figure 12-4 Rotating basket catalytic reactor [*from D. G. Tajbl, J. B. Simons, and J. J. Carberry, Ind. Eng. Chem., Fund. Quart.,* **5,** *17 (1966)*].

studies.† Generally, large single pellets are used to facilitate measurements of intrapellet temperature and concentration profiles when the purpose is to evaluate the effects of internal transport resistances. Such reactors are not convenient for establishing the intrinsic rate equation. In a single-pellet apparatus the surrounding fluid is normally well mixed, so that ideal stirred-tank behavior can be used for interpreting experimental data. The effect of flow rate on the global reaction rate in a spherical vessel containing a single, large catalyst pellet has been studied.‡

Many specialized laboratory reactors and operating conditions have been used. Sinfelt§ has *alternately* passed reactants and inert materials through a tubular-flow reactor. This mode of operation is advantageous when the activity of the fixed bed of catalyst pellets changes with time. A system in which the reactants flow through a porous semiconductor catalyst, heated inductively, has been proposed for studying the kinetics of high-temperature (500 to 2000°C) reactions.¶ An automated microreactor for investigating catalytic reactions is described in detail by Harrison et al.††; equipment specifications and control procedures are included.

12-4 Calculation of the Global Rate

For reactor design the global rate needs to be calculated, from bulk concentrations, temperature, and flow conditions, at all locations in the reactor. The procedure is essentially the reverse of that outlined in Sec. 12-1. The extensive trial calculations are not necessary since the intrinsic rate equation $[r = f(C) \exp(-E/RT)]$ is known. If intraparticle temperature gradients, are significant an effective thermal conductivity k_e, as well as the effective diffusivity, is required. The accounting for *external* mass and heat transfer was illustrated in Examples 10-4 and 10-9 for fixed-bed and trickle-bed reactors. In this section we want to account for *both internal* and *external* effects in calculating the global rate.‡‡ For an isothermal case we need Eq. (10-1) for external diffusion

$$\mathbf{r}_p = k_m a_m (C_b - C_s),$$

and Eq. (11-43) for internal diffusion

$$\mathbf{r}_p = \eta f(C_s).$$

In addition, the relationship between η and Φ and the defining expression for Φ are needed. These relationships are given in Fig. 11-8 for zero-, first-, and second-order intrinsic rate equations. The calculations are illustrated in Example 12-2.

† J. R. Balder and E. E. Petersen, *Chem. Eng. Sci.*, **23**, 1287 (1968); S. Otani and J. M. Smith, *J. Catalysis*, **5**, 332 (1966); J. A. Maymo and J. M. Smith, *AIChE J.*, **12**, 845 (1966); N. Wakao, M. R. Rao, and J. M. Smith, *Ind. Eng. Chem., Fund. Quart.*, **3**, 127 (1964).

‡ K. Patel and J. M. Smith, *J. Catal.*, **40**, 383 (1975).

§ J. H. Sinfelt, *Chem. Eng. Sci.*, **23**, 1181 (1968).

¶ W. E. Ranz and B. A. Bydal, presented at Fifty-fifth Annual Meeting AIChE, Chicago, Dec. 2–6, 1968.

†† D. P. Harrison, J. W. Hall, and H. F. Rase, *Ind. Eng. Chem.*, **57**, (1), 18 (1965).

‡‡ We did just this in Example 11-8, and Example 12-1 of this chapter, for the reverse problem of evaluating an intrinsic rate equation.

Example 12-2 Using the intrinsic rate equation obtained in Example 12-1, calculate the global rate of the reaction o-$H_2 \rightleftharpoons p$-H_2 at 400 lb/in.^{2} gauge $(2.86 \times 10^3$ kPa abs) and $-196°C$ (77 K), at a location where the mole fraction of ortho hydrogen in the bulk-gas stream is 0.65. The reactor is the same as described in Example 12-1; that is, it is a fixed-bed type with tube of 0.50 in. ID and with $\frac{1}{8} \times \frac{1}{8}$-in. cylindrical catalyst pellets of Ni on Al_2O_3. The superficial mass velocity of gas in the reactor is 15 lb/(h)(ft^2) [or 0.0204 kg/(m^2)(s)]. The effective diffusivity can be estimated from the random-pore model if we assume that diffusion is predominately in the macropores where Knudsen diffusion is insignificant. The macroporosity of the pellets is 0.36. Other properties and conditions are those given in Example 12-1.

SOLUTION The intrinsic rate equation developed in Example 12-1 is

$$r = \frac{1.1(C_o - C_p/K)}{1 + 1.06 \times 10^3 C_t} \qquad \text{mol/(g catalyst)(s)} \qquad (A)$$

It was shown that this expression could be written in terms of p-H_2 concentrations [Eq. (F) of Example 12-1] as

$$r = \frac{k_1(K + 1)}{K}(C_{eq} - C)_p$$

Then, the particular form for Eq. (11-43) for the rate of this reaction is [Eq. (G) of Example 12-1]

$$r_P = \eta \frac{k_1(K + 1)}{K}(C_{eq} - C_s)_p$$

where

$$k_1 = \frac{k_c}{1 + K_a C_t} = \frac{1.1}{1 + 1.06 \times 10^3 C_t}$$

Next this equation can be combined with Eq. (10-1), expressed in p-H_2 concentrations, to eliminate the surface concentration C_s. Then the rate can be written in terms of the bulk concentration C_b. In Example 12-1 [Eq. (I)] this resulted in

$$r_P = \frac{C_{eq} - (C_b)_p}{d_P \rho_P/6k_m + K/[\eta k_1(K + 1)]}$$

Now the problem is reduced to evaluating η, k_1, and k_m and substituting in Eq. (I) to obtain the global rate. We first calculate k_m, using properties from Example 12-1:

$$G = 15 \frac{454}{3,600(2.54)^2(12)^2} = 2.03 \times 10^{-3} \text{ g/(s)(cm}^2)$$

$$[\text{or } 2.03 \times 10^{-2} \text{ kg/(s)(m}^2)]$$

$$\text{Re} = \frac{0.318(2.03 \times 10^{-3})}{348 \times 10^{-7}} = 19$$

From Eq. (10-10), at $Re = 19$ we have

$$j_D = 0.41$$

Then k_m is given by Eq. (10-9) as

$$k_m = \frac{j_D G}{\rho} \left(\frac{\mu}{\rho D} \right)^{-2/3} = \frac{0.41(2.03 \times 10^{-3})}{8.95 \times 10^{-3}} (0.78^{-2/3})$$

$$= 0.11 \text{ cm/s } (0.11 \times 10^{-2} \text{ m/s})$$

Here 8.95×10^{-3} g/cm^3 is the density of H_2 at 400 lb/in.^{2} gauge and $-196°C$. The total concentration of H_2 at 400 lb/in.^{2} gauge is

$$C_t = \frac{p_t}{R_g T} = \frac{414.7/(14.7)}{82(77)} = 4.46 \times 10^{-3} \text{ g mol/cm}^3 \text{ (4.46 kg mol/m}^3)$$

At this total concentration

$$k_1 = \frac{1.1}{1 + (1.06 \times 10^3)(4.46 \times 10^{-3})}$$

$$= 0.19 \text{ cm}^3/\text{(g catalyst)(s) [or } 8.63 \times 10^{-5} \text{ m}^3/\text{(kg catalyst)(s)]}$$

According to the random-pore model, the effective diffusivity for macro-pore diffusion is given by Eq. (11-30). If the mass transfer is solely by bulk diffusion, Eq. (11-28) shows that $\bar{D}_M = \mathscr{D}_{AB}$, so that

$$D_e = \mathscr{D}_{AB} \varepsilon_M^2 = 0.00496(0.36)^2 = 6.4 \times 10^{-4} \text{ cm}^2/\text{s } (6.4 \times 10^{-8} \text{ m}^2/\text{s})$$

It was shown in Example 11-6 that the curve for first-order kinetics in Fig. 11-8 could be used for a reversible first-order reaction, provided

$$\Phi_s' = \frac{r_s}{3} \sqrt{\frac{k_1(K + 1)\rho_P}{K D_e}}$$

Using the values just calculated for k_1 and D_e, we find

$$\Phi_s' = \frac{0.318}{2(3)} \sqrt{\frac{0.19(1.01 + 1)(1.91)}{1.01(6.4 \times 10^{-4})}} = 1.8$$

Then, from Fig. 11-8, $\eta = 0.43$.

The global rate can now be evaluated from Eq. (I) of Example 12-1. For $(y_o)_b = 0.65$

$$(C_{eq} - C_b)_p = C_t(y_{eq} - y_b)_p = 4.46 \times 10^{-3} [0.5026 - (1 - 0.65)]$$

$$= 6.8 \times 10^{-4} \text{ g mol/cm}^3 \text{ (6.81} \times 10^{-1} \text{ kg mol/m}^3)$$

Hence

$$r_P = \frac{6.8 \times 10^{-4}}{0.318(1.91)/[6(0.11)] + 1.01/[0.43(0.19)(1.01 + 1)]} = \frac{6.8 \times 10^{-4}}{0.92 + 6.15}$$

$$= 0.96 \times 10^{-4} \text{ g mol/(g catalyst)(s) [or 0.96 kg mol/(kg catalyst)(s)]}$$

The same procedure could be used to calculate the global rate for any other location in the reactor. Because the reaction is first order, the effectiveness factor is independent of concentrations, and therefore independent of location. For nonlinear intrinsic rate equations, Φ and η are dependent on the reactant concentration, as illustrated in Fig. 11-8. For these cases the effectiveness factor will vary from point to point in the reactor (see Prob. 12-6).

Also, because the reaction is first order, C_s could be easily eliminated between Eqs. (10-1) and (11-43), so that the global rate could be expressed explicitly in terms of C_b [Eq. (I) of Example 12-1]. For other kinetics a trial solution would be simplest.

Finally, the first-order kinetics allowed a direct display of the relative importance of the diffusion resistances. The quantities 0.92 and 6.15 in the denominator of the previous equation measure the resistances of external diffusion and of internal diffusion plus reaction. The value of η divides the latter into internal-diffusion resistance and the resistance of the intrinsic reaction at the interior catalyst sites.

When external and internal temperature differences are also significant, trial solution is required to evaluate the global rate for a given C_b and T_b, regardless of the order of the intrinsic-rate equation. The effective thermal conductivity k_e is needed, as well as D_e. Equation (10-1) is applicable, and it is necessary to relate T_b and T_s by Eq. (10-13), written as

$$\mathbf{r}_P(-\Delta H) = ha_m(T_s - T_b)$$

To account for possible internal-temperature gradients, Eq. (11-42) is used instead of Eq. (11-43). Since the temperature effect on the intrinsic rate is expressed by the Arrhenius function, Eq. (11-42) may be written

$$\mathbf{r}_P = \eta f(C_s)e^{-E/R_gT_s} \tag{12-4}$$

where η is a function of Φ, β, and γ, as described in Sec. 11-11. These relations are sufficient to calculate the global rate at a given T_b and C_b. One procedure is to assume values of T_s and C_s and then calculate η from Φ, β, and γ as outlined in Sec. 11-11. Next $\mathbf{r}_P$ is obtained from Eq. (12-4), and finally, C_s and T_s are evaluated from Eqs. (10-1) and (10-13). If the calculated values of C_s and T_s do not agree with the assumed values, the process is repeated. When agreement is obtained, Eq. (12-4) gives the global rate. The procedure requires that the relation of η to Φ, β, and γ be known for the intrinsic rate of reaction. Such a relationship is shown in Fig. 11-13 for a first-order rate.

At the beginning of this chapter it was pointed out that both external and internal resistances are frequently, but not always, important, and general criteria were discussed. Quantitative methods of evaluating $C_b - C_s$ and $T_s - T_b$ were developed in Sec. 10-3. Quantitative criteria for the significance of internal mass- and heat-transfer resistances were given in Chap. 11 by Eqs. (11-59) and (11-83). If preliminary calculations suggest that some of the internal or external resistances are negligible, the calculation of the global rate is simplified. For example, if external mass-transfer resistance is negligible, then $C_b \rightarrow C_s$, and Eq. (10-1) is not needed.

12-5 The Structure of Reactor Design

Before we take up quantitative aspects of heterogeneous reactor design in Chap. 13, let us survey the problem. It is appropriate to present an overview here because of the interaction between laboratory and large-scale reactors; i.e., the purpose of the laboratory study is to obtain a rate equation useful for designing the large reactor.

Suppose a new chemical reaction, or new example of a known type of reaction, has been discovered. The product has promising economic possibilities. Samples of product have been made and tested, and a preliminary economic analysis has been completed. It is decided to proceed with calculations and experiment work necessary to design a large reactor to produce the product. The overall structure of the problem is represented schematically in Fig. 12-5. The items at the right indicate the scientific disciplines involved in the steps between the discovery of the reaction and the large-scale reactor. The *vertical* path from discovery to large reactor is an ideal, and usually a hypothetical, one. It implies that the intrinsic rate and selectivity equations can be obtained from analysis of data obtained in the discovery reactor. Usually this is impossible because the discovery reactor has not been designed according to the concepts discussed in Secs. 12-2 and 12-3. Therefore a laboratory (sometimes called bench-scale) reactor should be built with the objective of obtaining intrinsic rate and selectivity equations. The calculations for this step were illustrated in Example 12-1.

After the intrinsic rate equation has been established, it is used with a mathematical model for design of the large-scale reactor. The first step is to obtain the global rate at any location in the reactor, as illustrated in Example 12-2. To

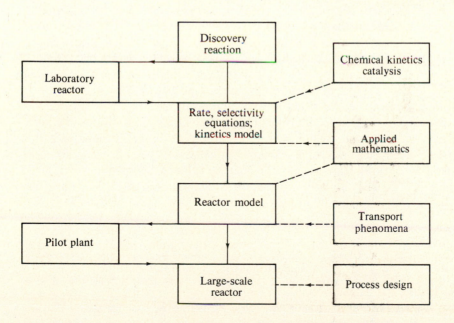

Figure 12-5 Structure of reactor design [*from J. M. Smith, Chem. Eng. Progr.,* **64**, *78 (1968)*].

complete the model, equations are derived to give the conversion and selectivities in the product stream leaving the reactor in terms of the proposed operating conditions. These equations and their solution are the subject of Chap. 13.

In principle it is possible to go directly from reactor model to large-scale reactor by the last vertical step in Fig. 12-5. This also is usually inadequate, and a pilot plant is necessary. The pilot plant may be omitted if the treatment of the physical processes in the model for the reactor as a whole (velocity distribution, mixing, etc.) is adequate to predict their influence on conversion and selectivity. Usually this is not the case. Hence one function of the pilot plant is to evaluate the reactor model. The pilot plant is also important for investigating questions about construction materials, corrosion, instrumentation, operating procedures, and control—all important in the large-scale unit. The pilot plant does have limitations. It is not an economical or technically appropriate apparatus for obtaining intrinsic rate equations, developing a catalyst, studying the effects of contaminated feed streams, or evaluating catalyst life. It is used most efficiently for its primary purpose: helping the chemical engineer to derive an adequate model for the large-scale reactor. It serves as a testing ground for how well macro effects such as temperature and velocity variations can be combined with rate equations to predict the production rates of products. The pilot plant also can be used to produce small quantities of product for marketing studies.

Scale-up of a reaction process involves direct application of the results from a laboratory unit to determine the size and operating conditions for a large-scale unit. It avoids the two-step process advocated in this chapter of evaluating rate equations from laboratory data and then using them to design the large-scale unit. The simplicity of scale-up is attractive, but unfortunately it is not often useful for chemical-reactor design. For example, to maintain geometric and dynamic similarity in a fixed-bed catalytic reactor, a larger catalyst pellet would be indicated for the large-scale reactor than was used in the laboratory reactor. However, using larger pellets could introduce larger intrapellet resistances and seriously change the global rate, which would destroy the initial similarity. Beek[†] has discussed these and other scale-up problems for fixed-bed catalytic reactors. Design and scale-up for several types of reactors have been reported by Rase.[‡]

In homogeneous reactors that are not isothermal, scale-up is dangerous because it is difficult to allow for the differences in heat-transfer conditions in laboratory and large-scale units. When precautions are taken to obtain the same global rate and the same heat-transfer conditions (for example, by requiring adiabatic or isothermal operation), scale-up concepts may be valuable. An illustration is given in Example 12-3. The similarity criteria on which scale-up methods depend are described by Johnstone and Thring.[§]

[†] John Beek, in T. B. Drew, J. W. Hoopes, Jr., and Theodore Vermeulen (eds.), "Advances in Chemical Engineering," vol. 3, p. 259, Academic Press, Inc., New York, 1962.

[‡] H. F. Rase, "Chemical Reactor Design for Process Plants," vols. I and II, John Wiley & Sons, New York, 1977.

[§] R. E. Johnstone and M. W. Thring, "Pilot Plants, Models, and Scale-up Methods in Chemical Engineering," McGraw-Hill Book Company, New York, 1957. Chapter 15 presents scale-up methods for chemical reactors and, in particular, describes the problems in scaling up heterogeneous types.

It is a simple matter to demonstrate that it is the interaction of reaction, geometric, and heat-transfer requirements that limits the value of scale-up. Consider a fixed-bed catalytic reactor. Suppose that the pressure drop has no effect on the rate and that plug flow exists. The mass balance for reactant is given by Eq. (12-1) as

$$\frac{W}{F} = \int_0^{x_e} \frac{dx}{\mathbf{r}_P}$$

where x_e is the conversion in the effluent from the reactor and $\mathbf{r}_P$ the global rate. Suppose plug-flow reactors of different geometries are being considered for the same chemical reaction. Equation (12-1) states that W/F will be the same for each reactor, provided that the value of the integral is the same. For the integral to be constant for a specified feed composition and conversion in the effluent, the global rate must be a function only of conversion. This is a necessary and sufficient criterion for scale-up. If this condition is met, the same conversion can be obtained in reactors of any geometry by fixing the magnitude of W/F.

The restriction that the global rate be a function only of conversion is the key to this simple scale-up procedure. Let us examine what it means. First, it rules out the possibility of intrapellet resistances, for then the catalyst-pellet size would influence the global rate. For example, the weight of catalyst W can be kept the same by reducing the size of the pellets and increasing their number. If intrapellet resistances were important, $\mathbf{r}_P$ would change. Similarly, external resistances are excluded, for if they were significant, the rate would become a function of velocity. Hence Eq. (12-1) is suitable for scale-up only when the global rate is determined by the intrinsic rate of the chemical step at the interior site in the catalyst pellet. In addition, the whole reactor must operate isothermally or adiabatically. Otherwise the rate becomes a function of temperature in addition to conversion. The adiabatic condition is permissible because an energy balance provides a unique relation between temperature and conversion, so that the rate can be expressed solely in terms of conversion [see Eq. (5-17)].

Example 12-3 A pilot plant for a liquid-solid catalytic reaction consists of a cylindrical bed 5 cm in radius and packed to a depth of 30 cm with 0.5-cm catalyst pellets. When the liquid feed rate is 0.2 liter/s the conversion of reactant to desirable product is 80%. To reduce pressure drop, a radial-flow reactor (Fig. 12-6) is proposed for the commercial-scale unit. The feed will be at a rate of 5 ft^3/s and will have the same composition as that used in the pilot plant. The inside radius of the annular bed is to be 2 ft, and its length is also 2 ft. What must the outer radius of the bed be to achieve a conversion of 80%?

The same catalyst pellets will be employed in both reactors. Laboratory studies have shown that external resistances are negligible for this system. The heat of reaction is small, so that isothermal operation is achievable. Assume plug-flow behavior and that the density of the reaction liquid does not change with conversion. The bulk density of catalyst in either reactor is 1.0 g/cm^3.

SOLUTION Under the stated conditions the global rate is a function only of conversion. Since the same conversion is desired for both reactors, the integral

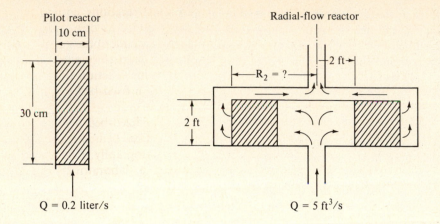

Figure 12-6 Scale-up from axial to radial flow reactor.

in Eq. (12-2) must be the same. Hence

$$\left(\frac{W}{F}\right)_{pilot} = \left(\frac{W}{F}\right)_{radial} \qquad (A)$$

The modal feed rate F of reactant is equal to QC_0, where Q is the volumetric flow rate and C_0 is the concentration of reactant in the feed. Then, if V is the volume of the catalyst bed and ρ_B is its density, Eq. (A) becomes

$$\frac{\rho_B}{C_0}\left(\frac{V}{Q}\right)_{pilot} = \frac{\rho_B}{C_0}\left(\frac{V}{Q}\right)_{radial}$$

Substituting numbers, we have

$$\left(\frac{V}{Q}\right)_{radial} = \frac{\pi(5^2)(30)}{0.2(1,000)} = 3.75\pi, \text{ s}$$

$$V_{radial} = 3.75\pi(5) = 18.75\pi, \text{ ft}^3$$

To obtain this volume in the radial-flow reactor the outer radius r_2 is given by

$$\pi L(r_2^2 - r_1^2) = \pi(2)(r_2^2 - 4) = 18.75\pi$$

$$r_2 = 3.7 \text{ ft}$$

This result shows that the outer diameter of the radial-flow reactor should be about 7.5 ft for an inner diameter of 4 ft.

In undergraduate courses in chemical engineering kinetics it is often advantageous to illustrate principles with laboratory experiments on suitable reactions.

Anderson† provides a detailed description of chemical reactions, equipment, experimental procedures, and methods of analyzing results, as developed by one educational institution. Of particular interest are actual data obtained by student groups. Experiments are described for measuring the performance of batch, stirred-tank, and tubular reactors with a homogeneous reaction (hydrolysis of acetic anhydride). Heterogeneous catalysis is studied in fixed-bed reactors, with the vapor-phase dehydrogenation and dehydration of isopropanol as the reactions. Other experiments have been conducted on fluidized-bed behavior, diffusion and reaction within catalyst pellets, diffusion in packed beds, reactor stability, biological reactions, and chemical analysis by chromatography. This report provides convenient source material for developing a laboratory course for chemical-reaction engineering.

PROBLEMS

12-1 A batch recycle reactor (Fig. 12-1), operated differentially and used for the gas-phase photolysis of acetone, has a volume of 62.8 cm^3, while the total volume of the system is 6620 cm^3. Runs at 97°C and 870 mmHg pressure were made at different initial concentrations of acetone in helium. Typical data for conversion to C$_2$H$_6$ are as given below for varying initial concentrations (gram moles per liter). The products of photolysis are primarily C$_2$H$_6$ and CO, according to the overall reaction

$$CH_3COCH_3 \rightarrow CO + C_2H_6$$

Therefore, the rate of acetone decomposition can be followed by measuring the conversion of acetone to ethane in the reaction mixture. Preliminary measurements showed that the composition of the system was essentially uniform, indicating good mixing in the reservoir (Fig. 12-1) and a recycle rate high enough that the conversion per pass through the reactor was very small.

(*a*) Calculate the rate of decomposition of acetone for each run. (*b*) Evaluate constants k_1 and k_2, assuming that the rate expression is of the form

$$\mathbf{r} = \frac{k_1 C}{1 + k_2 C}$$

where C is the concentration of acetone.

	Conversion to C$_2$H$_6$, %				
t, h	4.66×10^{-3}	4.62×10^{-3}	1.99×10^{-3}	1.33×10^{-3}	1.27×10^{-3}
0	0	0	0	0	0
0.5	0.6	0.5	0.8	1.0	1.0
1.0	0.9	0.75	1.4	1.35	1.45
2.0	1.85	1.65	3.0	3.30	3.55
3.0	2.80	2.30	4.50	5.0	5.45
4.0	3.65	3.15	6.05	6.65	7.20
5.0	4.60	3.90	7.50	8.25	8.95

Source: A. E. Cassano, T. Matsuura, and J. M. Smith, *Ind. Eng. Chem., Fund. Quart.*, 7, 655 (1968); 4.66×10^{-3}, etc. are initial concentrations of acetone.

† J. B. Anderson, "A Chemical Reactor Laboratory for Undergraduate Instruction," Princeton University Department of Chemical Engineering, Princeton, N.J., September, 1968.

12-2 The liquid-phase hydrogenation of α-methyl styrene to cumene,

$$H_2(\text{dissolved}) + C_6H_5C(CH_3) = CH_2(l) \rightarrow C_6H_5CH(CH_3)_2(l),$$

has been studied in a recycle batch reactor at 80 lb/in.^{2} abs and 55°C.† The reactor was packed with $\frac{1}{8}$-in. Al_2O_3 spheres containing 0.5 wt % palladium on the outer surface as a catalyst. The reaction was followed by analyzing microsamples of reaction liquid at various times. The recycle rate was such that the conversion per pass was very low. The reaction liquid was always saturated with hydrogen. Data for two runs with different compositions (styrene and cumene) in the initial mixture are as follows:

	Run C4	Run C5
Moles α-methyl styrene at $t = 0$	29.0	27.9
Mass of catalyst in reactor, g	30.7	30.7

t, h	Mole fraction cumene	
0	0.0264	0.1866
4	0.0530	0.2114 (4.5 h)
8	0.0810	0.2332
10.5	0.0995	0.2552 (11.5 h)
17.5	0.1518	
21.5	0.1866	

(a) For the integral changes in composition (during a run) shown, a plot of cumene mole fraction vs. time might be expected to be curved; yet the data show a linear relationship. What does this signify about the kinetics of the reaction? (b) Calculate rates of hydrogenation from the data given. The solubility of hydrogen in the liquid is very low.

12-3 A test procedure is to be developed for measuring the activity of various catalysts. A small, fixed-bed reactor is available in the laboratory. The same gaseous reaction will be used to evaluate each catalyst. The method of analysis of the product stream is fairly accurate but not good enough to justify differential reactor operation. Hence the sample of any catalyst tested will be sufficient in amount to obtain conversions above 30%. The feed rate, temperature, pressure, feed and effluent composition (or conversion), and mass of catalyst in the reactor will be routinely measured. The activity will be defined as the ratio of a measurable variable for any catalyst to the value of that variable for a standard catalyst. The variable chosen should be a true measure of the intrinsic rate of the reaction at a catalyst site.

(a) Propose a method of operating the reactor so as to obtain the activity of any catalyst. That is, describe what variables will be held constant during a run and within the integral reactor, as well as the variables which will be held the same for the given catalyst and the standard catalyst. Also give the variable whose value will be used in formulating the activity. Illustrative of the several variables to be considered are the size of the catalyst pellet, mass velocity of the gas through the reactor, and temperature.

(b) In part (a) no consideration was given to the experimental effort needed to carry out the proposed activity tests. Suppose that the ideal procedure described in (a) requires excessive effort. Propose an alternate, perhaps less desirable, but simpler method of operation.

12-4 In Example 10-7, mass transfer coefficients k_L and k_c in a slurry reactor were calculated from data for the catalytic oxidation of SO_2 in an aqueous slurry of activated carbon particles. Rates of reaction were given (Table 10-5) for two particle sizes.

A. In Example 10-7 the possibility of intraparticle diffusion resistance was not considered; that is, the apparent rate constant k in Eq. (D) of that example included intraparticle effects. To check the importance of intraparticle mass transfer, calculate the effectiveness factors for $d_p = 0.03$ and 0.099 mm

† E. G. Biskis and J. M. Smith, *AIChE J.*, **9**, 677 (1963).

particles. Do this by taking $k = k_{int}\eta$, where k_{int} is the intrinsic rate constant, and using the rate data given in Table 10-5.

B. Also calculate the effective diffusivity of oxygen in the liquid-filled pores of activated carbon.

12-5 It is often true that intraparticle mass transport has a negligible effect on the global rate for the small catalyst particles in slurries. However, the results of Prob. 12-4 indicate that the low diffusivities in the liquid-filled pores in slurries suggest caution in drawing general conclusions.

A. Using the results of Prob. 12-4, how small would the particles have to be for intraparticle diffusion to reduce the global rate by 5% or less?

B. What effectiveness factor would be expected for particles 1 mm in diameter? Such particles would be relatively large for a slurry reactor.

12-6 The vapor-phase dimerization of ethylene is to be carried out in a fixed-bed reactor operating isothermally and in plug flow. The $\frac{1}{4}$-in. nickel-aluminum catalyst pellets are packed in a 4-in ID reactor which will operate at 100°F and 200 lb/in.^{2} abs. The feed rate is to be 21 lb mol per hour of essentially pure ethylene. The packed-bed density of catalyst particles will be 1.1 g/cm^3.

While other reactions occur, consider only the dimerization (to butylenes) which may be taken as second order and irreversible at the operating conditions. External mass-transfer resistance may be neglected, but bench-scale studies indicate that intraparticle diffusion significantly retards the global rate. The bench-scale, differential reactor data give the following results at 100°F and 200 lb/in.^{2} abs:

Intrinsic, second-order rate contact, $k_2 = 5.0$ ft^6/(s)(mol)(lb of catalyst). Effective diffusivity, $D_e = 1.2 \times 10^{-3}$ cm^2/s. Catalyst particle density, $\rho_p = 1.9$ g/cm^3.

Calculate the effectiveness factor as a function of conversion of ethylene (use the second-order, plate curve in Fig. 11-8 with $L = r_s/3$).

12-7 A. For the conditions of Prob. 12-6, what depth of catalyst bed would be required for a conversion of 50%?

B. What catalyst-bed depth for a conversion of 50% would have been found if intraparticle diffusion had been neglected?

12-8 If external mass transfer had been important, describe a method of solution for Probs. 12-6 and 12-7A.

THIRTEEN

DESIGN OF HETEROGENEOUS CATALYTIC REACTORS

Our objective in this chapter is to predict the performance of large heterogeneous reactors using the kinetics and transport rates developed in Chaps. 9 to 11. It is not necessary to consider further the individual transport processes within and external to the catalyst pellet. Methods of combining these steps with the kinetics of the chemical reactions to obtain a global rate were summarized in Chap. 12. However, Example 13-2 illustrates again how these individual processes can be combined to predict the conversion in an entire fixed-bed reactor. The major goal now is to use the global rate information to evaluate the composition of the effluent from a reactor for a specified set of design conditions.† The design conditions that must be fixed are the temperature, pressure, and composition of the feed stream, the dimensions of the reactor and catalyst pellets, and enough information about the surroundings to evaluate the heat flux through the reactor walls.

As noted in Chap. 1, a common catalytic reactor is the fixed-bed type, in which the reaction mixture flows continuously through a tube filled with a stationary bed of catalyst pellets (Fig. 1-4a). Because of its importance, and because considerable information is available on its performance, most attention will be given to this reactor type. Fluidized-bed, trickle-bed, and slurry reactors are also

---

† An a priori design procedure, in contrast to a succession of larger experimental reactors, has been ably summarized by R. H. Wilhelm [*J. Pure Appl. Chem.*, **5**, 403 (1962)] for fixed beds. This review includes relations between the intrinsic and global rates and, as such, summarizes the effects of external and internal transport processes described in Chaps. 10 to 12. Also, an extensive review of more recent developments in reactor theory and design is available: "Chemical Reactor Theory," Leon Lapidus and Neal R. Amundson (eds.), Prentice-Hall, Englewood Cliffs, N.J., 1977. See particularly Chap. 6 on design of fixed-bed reactors and Chaps. 10 and 11 on fluidized-bed reactors.

considered later in the chapter. Some of the design methods given are applicable also to fluid-solid noncatalytic reactions. The global rate and integrated conversion-time relationships for noncatalytic gas-solid reactions will be considered in Chap. 14.

Only reactors operating at pseudo-steady state are discussed; that is, the design methods presented are applicable when conditions such as catalyst activity do not change significantly in time intervals of the order of the residence time in the reactor. Brief comments about transient conditions are included in Sec. 13-7, but these refer to changes from one stable state to another.

FIXED-BED REACTORS

Quantitative design methods of increasing complexity are considered in Secs. 13-3 to 13-6. First, however, let us summarize construction and operating characteristics of fixed-bed reactors.

13-1 Construction and Operation

Fixed-bed reactors consist of one or more tubes packed with catalyst particles and operated in a vertical position. The catalyst particles may be a variety of sizes and shapes: granular, pelleted, cylinders, spheres, etc. In some instances, particularly with metallic catalysts such as platinum, instead of using single particles, wires of the metal are made into screens. Multiple layers of these screens constitute the catalyst bed. Such screen or gauze catalysts are used in commercial processes for the oxidation of ammonia and the oxidation of acetaldehyde to acetic acid.

Because of the necessity of removing or adding heat, it may not be possible to use a single large-diameter tube packed with catalyst. In this event the reactor may be built up of a number of tubes encased in a single body, as illustrated in Fig. 13-1. The energy exchange with the surroundings is obtained by circulating, or perhaps boiling, a fluid in the space between the tubes. If the heat of reaction is large, each catalyst tube must be small (tubes as small as 1.0-in. diameter have been used) in order to prevent excessive temperatures within the reaction mixture (for an exothermic reaction). The problem of deciding how large the tube diameter should be, and thus how many tubes are necessary, to achieve a given production forms an important problem in the design of such reactors.

A disadvantage of this method of cooling is that the rate of heat transfer to the fluid surrounding the tubes is about the same all along the tube length, but the major share of the reaction usually takes place near the entrance. For example, in an exothermic reaction the rate will be relatively large at the entrance to the reactor tube owing to the high concentrations of reactants existing there. It will become even higher as the reaction mixture moves a short distance into the tube, because the heat liberated by the high rate of reaction is greater than that which can be transferred to the cooling fluid. Hence the temperature of the reaction mixture will rise, causing an increase in the rate of reaction. This continues as the mixture moves up the tube, until the disappearance of reactants has a larger effect on the rate than the increase in temperature. Farther along the tube the rate will

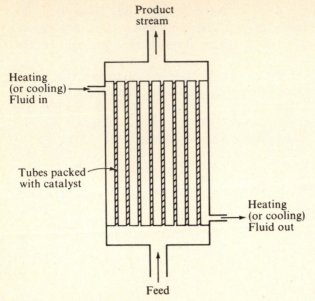

Figure 13-1 Multitube, fixed-bed reactor.

decrease. The smaller amount of heat can now be removed through the wall with the result that the temperature decreases. This situation leads to a maximum in the curve of temperature vs. reactor-tube length. An example is shown in Fig. 13-2 for a TVA ammonia-synthesis reactor. Such a maximum temperature (hot spot) is characteristic of an exothermic reaction in a tubular reactor (Chap. 5).

As mentioned in Chaps. 1 and 5, other means of cooling may be employed besides circulating a fluid around the catalyst tube. Dividing the reactor into parts with intercoolers between each part (see Fig. 13-3) is a common procedure.

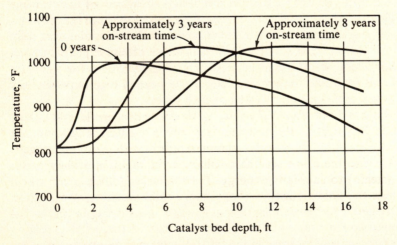

Figure 13-2 Variation in temperature profile with on-stream time in fixed-bed ammonia-synthesis reactor [*by permission from A. V. Slack, H. Y. Allgood, and H. E. Maune, Chem. Eng. Progr., 49, 393 (1953)*].

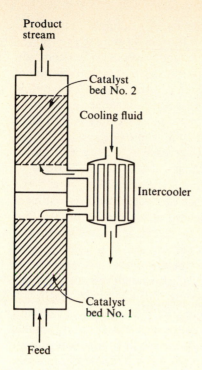

Figure 13-3 Divided reactor with inter-cooler between catalyst beds.

Another scheme which has been satisfactory for reactions of moderate heat of reaction, such as the dehydrogenation of butene, is to add a large quantity of an inert component (steam) to the reaction mixture.

The particular scheme employed for cooling (or heating) the fixed-bed reactor depends on a number of factors: cost of construction, cost of operation, maintenance, and special features of the reaction, such as the deactivation of the catalyst and the magnitude of ΔH. For example, the heat of reaction of naphthalene oxidation is so high that small externally cooled tubes provide about the only way to prevent excessive temperatures in fixed-bed equipment. In sulfur dioxide oxidation the much smaller heat of reaction permits the use of less expensive large-diameter adiabatic catalyst bins in series, with external intercoolers for removing the heat evolved. In dehydrogenation of butene the heat of reaction is not extremely high, so that small-diameter catalyst tubes are not required. Here the use of external heat exchangers is possible (the reaction is endothermic, and energy must be supplied to maintain the temperature), and a satisfactory system can be designed by alternating adiabatic reaction sections with heat exchangers. However, in this case there are several auxiliary advantages in adding a hot inert material (steam) to supply the energy. The blanketing effect of the steam molecules reduces the polymerization of the butadiene product. Also, the steam lowers the partial pressure of the hydrocarbons, and in so doing improves the equilibrium conversion.

All of the operating devices for energy exchange have the objective of preventing excessive temperatures or maintaining a required temperature level; i.e., they are attempts to achieve isothermal operation of the reactor. There are many advantages to operating at near-isothermal conditions. For example, in the naphthalene oxidation process it is necessary to control the temperature to prevent the oxidation from going all the way to carbon dioxide and water. This is a common situation in partial-oxidation reactions. A second example is the air oxidation of ethylene where ethylene oxide is the desired product. Another reason for avoiding excessive temperatures is to prevent loss of catalyst activity. Changes in structure of the solid catalyst particles as the temperature is increased may reduce their activity and shorten their useful life. For example, the iron oxide catalyst for the ammonia-synthesis reaction shows a more rapid decrease in activity with time if the synthesis unit is operated above the normal temperature range of 400 to 550°C.

The reason for limiting the temperature in sulfur dioxide oxidation is based on two factors: excessive temperatures decrease the catalyst activity, as just mentioned, and the equilibrium yield is adversely affected at high temperatures. This last point is the important one in explaining the need to maintain the temperature level in the dehydrogenation of butene. Still other factors, such as physical properties of the equipment, may require limiting the temperature level. For example, in reactors operated at very high temperatures, particularly under pressure, it may be necessary to cool the reactor-tube wall to preserve the life of the tube itself.

The problem of regenerating the catalyst to restore activity may be a serious one in the fixed-bed reactor. In a great many instances the catalyst is too valuable to discard. If the catalyst activity decreases rapidly with time, frequent regeneration may be necessary. Even when the cost is so low that regeneration is not required, shutting down the process and starting up again after new catalyst has been added is an expensive procedure. If this is necessary at frequent intervals, the entire process may become uneconomical. The exact economic limit on shutdown time depends on the particular process, but in general, if the activity cannot be maintained over a period of several months, the cost of shutdowns is likely to be prohibitive. Regeneration *in situ* is one way out of this difficulty. However, *in situ* regeneration requires two or more reactors if continuous operation is to be maintained, and hence increases the initial cost of the installation. The most successful fixed-bed reactor systems are those where the catalyst activity is sustained for long periods without regeneration. The fixed-bed reactor requires a minimum of auxiliary equipment and is particularly suitable for small commercial units.

In order to prolong the time between regenerations and shutdowns, the reactor tube may be made longer than required for the reaction. For example, suppose a 3-ft length of catalyst bed is necessary to approach the equilibrium conversion with fresh catalyst of high activity. The reactor may be built with tubes 10 ft long. Initially, the desired conversion will be obtained in the first 3 ft. As the catalyst activity decreases, the section of the bed in which the reaction is mainly accomplished will move through the bed, until finally all 10 ft are deactivated. This technique has been employed successfully in ammonia synthesis reactors.

13-2 Outline of the Design Problem

The global rate of reaction tells how much reaction is occurring at any location in the reactor, in terms of *bulk* concentrations and temperature. To evaluate temperatures and concentrations, energy- and mass-conservation equations are formulated according to Eqs. (3-1) and (5-1) for a fluid flowing through a bed of catalyst pellets. The solution of these equations gives the temperature and concentrations at any location, including the reactor exit. In fixed-bed reactor design it is assumed that all properties are constant in a volume element associated with a single catalyst pellet. This means that the global rate is the same within this volume element. Two methods have been used to solve the conservation equations. The most common approach is to assume that the volume element is small with respect to the reactor as a whole. Then the temperature and concentrations are regarded as continuous functions of reactor position, and the conservation expressions become differential equations. This is the method that will be used in the examples in this chapter. The second method† considers the volume element (associated with one catalyst pellet) to be an individual cell, or finite stage, within which mixing is complete so that the cell has uniform properties. The reactor is visualized as an interconnected assembly of these cells. Mass and energy transfer between cells is assumed to occur only by fluid flowing from one cell to adjacent cells. In this procedure the mass- and energy-conservation expressions become an assembly of algebraic (difference) equations.

When temperature gradients exist in the reactor, analytical solution of either the differential or difference equations is not possible. The design process requires numerical solution by a stepwise procedure. Machine computation is necessary. The procedure is illustrated in Secs. 13-4 to 13-6.

The complexity of the design problem depends primarily on the type (radial or axial) and magnitude of the temperature variation in the reactor. The reactants normally enter the catalyst bed at uniform temperature and composition, but as they pass through the bed and reaction occurs, the accompanying heat of reaction and heat exchange with the surroundings can cause both longitudinal and radial variations in temperature. The severity of these variations depends upon the heat of reaction and heat exchange with the surroundings. In the simplest case the entire reactor operates isothermally and there is no variation of axial velocity in the radial direction. The global rate is a function only of concentration. Further, the concentrations will change only in the axial direction. A *one-dimensional* model can be used in developing the mass-conservation equations as described in Sec. 13-3. An analytical solution for the conversion in the exit stream is sometimes possible (see Example 13-2).

It is infrequent in practice to achieve isothermal operation. Either the heat of reaction must be very low (as in isomerization reactions) or the reactant concentration must be very small (as in removal of pollutants from water or air by oxidation). However, in large-scale reactors, adiabatic operation is often closely approached. Again, a one-dimensional model may be employed, but both energy- and mass-conservation equations are needed to describe the conversion in the axial direction. The design procedure is discussed in Sec. 13-4 and illustrated in Example 13-3.

The most complex design problem arises when heat transfer through the reactor wall must be taken into account. This type of operation occurs when it is

---

† The mixing-cell or finite-stage model is described by H. A. Deans and L. Lapidus [*AIChE J.* **6**, 656, 663 (1960)] and by M. L. McGuire and L. Lapidus [*AIChE J.* **11**, 85 (1965)].

necessary to supply or remove heat through the wall, and the rate of energy transfer is not sufficient to approach isothermal operation. It is a frequent occurrence in commercial, fixed-bed reactors because of their size and because fluid velocities must be low enough to allow for the required residence time. The presence of the catalyst pellets prevents sufficient turbulence and mixing to obtain uniform concentration and temperature profiles. The concentration, temperature, and the global rate, will vary in both the radial and the axial direction. A *two-dimensional* model is needed for a correct formulation of the conservation equations. A relatively simple two-dimensional model is discussed and illustrated in Sec. 13-6. An approximate, simpler solution can be achieved if all the radial temperature variation is assumed to be concentrated in a thin layer of fluid at the reactor wall. Then with the plug-flow assumption the temperature would be uniform across the reactor radius except for a sharp change at the wall. Also, there would be no radial gradient in concentration. A one-dimensional model can be used for this approximate solution, as illustrated in Sec. 13-5. The computer time required to compute temperature and concentration profiles when radial gradients are accounted for depends upon the type of two-dimensional model employed. This subject is discussed briefly in Sec. 13-6. The approach followed in this introductory text is to illustrate the concepts with relatively simple models and provide references to more advanced methods.†

No mention has been made of the pressure drop in fixed-bed reactors. In most cases Δp is small with respect to the total pressure so that ignoring this effect is justified. However, for gaseous reactions at low pressures the change in pressure may affect the global rate significantly. Also, the Δp is needed for designing pumping equipment. For packed beds the pressure drop may be estimated from the Ergun equation.‡

ISOTHERMAL AND ADIABATIC FIXED-BED REACTORS

13-3 Isothermal Operation

In Chap. 4 the plug-flow model was used as a basis for designing homogeneous tubular-flow reactors. The equation employed to calculate the conversion in the effluent stream was Eq. (3-18). We shall find that the same equations and the same calculational procedure may be used for fixed-bed catalytic reactors, provided that plug-flow behavior is a valid assumption. All that is necessary is to replace the homogeneous rate of reaction in those equations with the global rate for the catalytic reaction, and replace the reactor volume with mass of catalyst. In

---

† Reviews of fixed-bed reactor design are available: V. Hlavacek and J. Votruba in "Chemical Reactor Theory," Leon Lapidus and Neal R. Amundson (eds.), chap. 6, Prentice-Hall, Englewood Cliffs, N.J. (1977); G. F. Froment, *Ind. Eng. Chem.* **59**(2), 18 (1967); J. Beek, *Adv. Chem. Eng.* **3**, 303 (1962).

‡ S. Ergun, *Chem. Eng. Prog.* **48**, 89 (1952); D. Mehta and M. C. Hawley, *Ind. Eng. Chem., Proc. Des. Dev.* **8**, 280 (1969).

the isothermal case deviations from plug-flow behavior arise from variations of axial velocity in the radial direction and from axial dispersion. The radial variation in velocity leads to a residence-time distribution, as does axial dispersion. The effects on the conversion of these deviations from plug-flow performance were discussed in Chap. 6. In isothermal fixed-bed reactors these effects are usually small for isothermal conditions, so that plug-flow equations are satisfactory. However, deviations are potentially large in nonisothermal reactors, and radial variations in particular must be taken into account. We consider first the two-dimensional form of the mass-conversion equation and then present the plug-flow, one-dimensional version normally applicable for isothermal conditions.

A section of a fixed-bed catalytic reactor is shown in Fig. 13-4. Consider a small volume element of radius r, width Δr, and height Δz, through which reaction mixture flows isothermally. Suppose that radial and axial mass transfer can be expressed by Fick's law, with $(D_e)_r$ and $(D_e)_L$ as *effective* diffusivities,[†] based on the total (void and nonvoid) area perpendicular to the direction of diffusion. The volume of the element, which is $2\pi r\,\Delta r\,\Delta z$, contains both solid catalyst pellets and surrounding fluid. The concentration in the fluid phase is constant within the element, and the global rate is known in terms of this bulk-fluid concentration. The axial velocity of the reacting fluid can vary in the radial direction. It will be described as a local superficial velocity $u(r)$ based on the total (void plus nonvoid) cross-sectional area.

We now apply Eq. (3-1) to obtain a mass-conservation expression for reactant in the volume element. For steady state, the result in differential form is

$$\frac{\partial}{\partial r}\left(r(D_e)_r\frac{\partial C}{\partial r}\right) + r\frac{\partial}{\partial z}\left(-uC + (D_e)_L\frac{\partial C}{\partial z}\right) - \mathbf{r}_p\rho_B r = 0 \qquad (13\text{-}1)$$

where r_p = *global* rate of disappearance of reactant per unit mass of catalyst
$\quad\rho_B$ = density of catalyst in the bed
$\quad u$ = superficial velocity in the axial direction

---

† These diffusivities include both molecular and turbulent contributions. In fixed beds some convection exists even at low velocities.

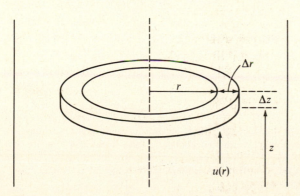

r

Δr

Δz

z

$u(r)$

Figure 13-4 Annular element in a fixed-bed catalytic reactor.

If the diffusivities are not sensitive to r or to z and the velocity is not a function of z,† Eq. (13-1) may be written

$$(D_e)_r \left(\frac{1}{r} \frac{\partial C}{\partial r} + \frac{\partial^2 C}{\partial r^2} \right) - u \frac{\partial C}{\partial z} + (D_e)_L \frac{\partial^2 C}{\partial z^2} - \mathbf{r}_p \rho_B = 0 \qquad (13\text{-}2)$$

If the velocity varies with z (due to changes in temperature or number of moles in a gaseous reaction), Eq. (13-1) should be used. If the concentration entering the reactor is C_0, and if there is no axial dispersion in the feed line, the boundary conditions for Eq. (13-2) are

$$\frac{dC}{dz} = 0 \qquad\qquad \text{at } z = L \qquad (13\text{-}3)$$

$$uC_0 = -(D_e)_L \left(\frac{\partial C}{\partial z} \right)_{>0} + u(C)_{>0} \qquad \text{at } z = 0 \text{ for all } r \qquad (13\text{-}4)$$

$$\frac{\partial C}{\partial r} = 0 \qquad\qquad \text{at } r = r_0 \text{ for all } z \qquad (13\text{-}5)$$

$$\frac{\partial C}{\partial r} = 0 \qquad\qquad \text{at } r = 0 \text{ for all } z \qquad (13\text{-}6)$$

It is instructive to write Eq. (13-2) in dimensionless form by introducing the conversion x and dimensionless coordinates $r*$ and $z*$ based on the diameter of the catalyst pellet:

$$x = \frac{C_0 - C}{C_0} \qquad (13\text{-}7)‡$$

$$r* = \frac{r}{d_P} \qquad (13\text{-}8)$$

$$z* = \frac{z}{d_P} \qquad (13\text{-}9)$$

In terms of these variables Eq. (13-2) becomes

$$-\frac{1}{\text{Pe}_r} \left[\frac{1}{r*} \frac{\partial x}{\partial r*} + \frac{\partial^2 x}{(\partial r*)^2} \right] + \frac{\partial x}{\partial z*} - \frac{1}{\text{Pe}_L} \frac{\partial^2 x}{(\partial z*)^2} - \frac{\mathbf{r}_p \rho_B \, d_P}{C_0 u} = 0 \qquad (13\text{-}10)$$

where

$$\text{Pe}_r = \frac{u d_P}{(D_e)_r} \qquad (13\text{-}11)$$

$$\text{Pe}_L = \frac{u d_P}{(D_e)_L} \qquad (13\text{-}12)$$

---

† This requirement is met for constant fluid density; that is, isothermal conditions for a liquid-phase reaction mixture, and isothermal conditions and no change in total molal flow rate for a gas-phase reaction mixture.

‡ For this equation to be valid for a gaseous reaction, there must be no change in total molal flow rate as well as isothermal operating conditions.

Equation (13-10) shows that the conversion depends on the dimensionless reaction-rate group $r_P \rho_B d_P / C_0 u$ and the radial and axial Peclet numbers, defined by Eqs. (13-11) and (13-12).

When the velocity u varies with *radial location*, a stepwise numerical solution of Eq. (13-10) or Eq. (13-2) would be needed. Axial velocities do vary with radial position in fixed beds. The typical profile† is flat in the center of the tube, increases slowly until a maximum velocity is reached about one pellet diameter from the wall, and then decreases sharply to zero at the wall. The radial gradients are a function of the ratio of tube to pellet diameter. Excluding the zero value at the wall, the deviation between the actual velocity at any radius and the average value for the whole tube is small when $d/d_P > 30$.

Radial Peclet numbers have been measured,‡ and some of the results are shown in Fig. 13-5. Above a modified Reynolds number $d_P G/\mu$ of about 40, Pe, is independent of flow rate and has a magnitude of about 10. As implied in Sec. 13-2 the two terms involving radial gradients in Eq. (13-10) are generally small for isothermal operation. The only way§ that concentration gradients can develop is through the variation of velocity with r. Also, the relatively large value of Pe$_r$ further reduces the magnitude of these two terms. If we neglect them, Eq. (13-10) reduces to

$$\frac{\partial x}{\partial z^*} - \frac{1}{Pe_L}\frac{\partial^2 x}{(\partial z^*)^2} - \frac{r_P \rho_B d_P}{C_0 u} = 0 \qquad (13\text{-}13)$$

† C. E. Schwartz and J. M. Smith, *Ind. Eng. Chem.*, **45**, 1209 (1953).

‡ R. W. Fahien and J. M. Smith, *AIChE J.*, **1**, 28 (1955); C. L. de Ligny, *Chem. Eng. Sci.*, **25**, 1175 (1970).

§ Note that this is not so if radial temperature gradients exist. Then the rate can vary significantly with r, and large concentration gradients develop. In such a case use is made of the data in Fig. 13-5 as described in Example 13-7.

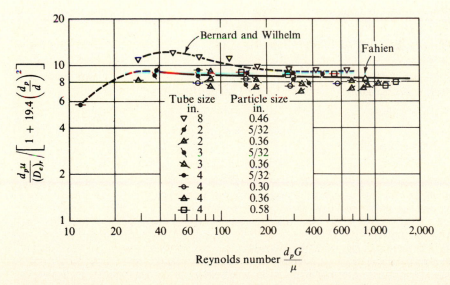

Figure 13-5 Correlation of average Peclet number, $d_P u/(D_e)_r$, with Reynolds number and d_P/d.

or, in dimensional form,

$$-u\frac{\partial C}{\partial z} + (D_e)_L \frac{\partial^2 C}{\partial z^2} - r_P \rho_B = 0 \tag{13-14}$$

This expression retains the effect of axial dispersion. It is identical to Eq. (6-43), except that the rate for a homogeneous reaction has been replaced with the global rate $r_p \rho_B$, per unit volume, for a heterogeneous catalytic reaction. In Sec. 6-9 Eq. (6-43) was solved analytically for first-order kinetics to give Eq. (6-47). That result can be adapted for fixed-bed catalytic reactors. The first-order *global* rate would be

$$r_p = k_o C \qquad \text{g mol/(s)(g catalyst)} \tag{13-15}$$

where k_o is an overall rate constant that includes the effects of fluid-to-particle and intraparticle mass-transfer effects. The solution of Eq. (13-14) is the same as Eq. (6-47), but instead of Eq. (6-48) we have

$$\beta = \left(1 + 4k_0 \rho_B \frac{(D_e)_L}{u^2}\right)^{1/2} \tag{13-16}$$

If the catalytic rate equation were not first order, numerical solution of Eq. (13-14) with Eqs. (13-3) and (13-4) would be necessary.

To utilize Eq. (6-47), or solutions of Eq. (13-14) for other kinetics, requires the axial diffusivity $(D_e)_L$ as well as the global rate. $(D_e)_L$ has been measured for both gases and liquids flowing through fixed beds. The experimental data of McHenry and Wilhelm† (for gases) and theoretical predictions‡ both indicate that $Pe_L = 2$ for Reynolds numbers above about 10. For liquids§ Pe_L is less, particularly at low Reynolds numbers. A more detailed correlation including bed porosity and Schmidt number is available¶.

The significance of the axial dispersion term in Eq. (13-14) depends on the length L of the reactor, the effective diffusivity, and the velocity. Its importance decreases as the velocity and length increase. For low velocities (Re < 1) in short reactors, longitudinal dispersion can be significant.†† In most other cases axial dispersion is negligible. Then Eq. (13-14) takes the form

$$-u\frac{dC}{dz} = r_P \rho_B \tag{13-17}$$

---

† K. W. McHenry, Jr., and R. H. Wilhelm, *AIChE J.*, **3**, 83 (1957).

‡ R. Aris and N. D. Amundson, *AIChE J.*, **3**, 280 (1957).

§ J. J. Carberry and R. H. Bretton, *AIChE J.*, **4**, 367 (1958); E. J. Cairns and J. M. Prausnitz, *Chem. Eng. Sci.*, **12**, 20 (1960).

¶ D. J. Gunn, *Chem. Eng.*, No. 219, 153 (1968).

†† Criteria for neglecting axial dispersion in isothermal and adiabatic reactors are given by Carberry and Wendel [*AIChE J.*, **9**, 129 (1963)], Hlavacek and Marek [*Chem. Eng. Sci.* **21**, 501 (1966)] and Levenspiel and Bischoff [*Adv. Chem. Eng.*, **4**, 95 (1963)].

Multiplying the numerator and the denominator by the cross-sectional area of the reactor gives

$$-Q\frac{dC}{dV} = \mathbf{r}_P\rho_B$$

or

$$\frac{V}{Q} = -\frac{1}{\rho_B}\int_{C_0}^{C}\frac{dC}{\mathbf{r}_P} = \frac{C_0}{\rho_B}\int\frac{dx}{\mathbf{r}_P} \tag{13-18}$$

Since $V\rho_B$ is the mass of catalyst in the reactor and $C_0 Q = F$, an alternate form, which is also applicable for variable fluid density, is Eq. (12-1):

$$\frac{W}{F} = \int\frac{dx}{\mathbf{r}_P} \tag{12-1}$$

Equation (12-1) is of the same form as Equation (3-18). The application of these equations to reactor design is the same as discussed in Chap. 4, for example, in Example 4-7.

Example 13-1 Derive Eq. (13-2) by starting with the point, mass-conservation equation (continuity equation), Eq. (3-21), for species j in a reacting system:

$$\frac{\partial C_j}{\partial t} - \nabla\cdot D_e(\nabla C_j) + \nabla\cdot(C_j\mathbf{v}) = \sum_i\mathbf{r}_{ij} \tag{A}\dagger$$

where $\mathbf{r}_i$ is the rate of *production* of species j, by reaction i, per unit volume of reactor. Note $\nabla\cdot\mathbf{N}_A = -\nabla\cdot D_e(\nabla C_j) + \nabla\cdot(C_j\mathbf{v})$

SOLUTION For steady state $\partial C_j/\partial t = 0$. In the cylindrical coordinates of a tubular reactor the radial component of the velocity will be zero and we can assume angular symmetry. Then the divergence of the convective flux of j is

$$\nabla\cdot(C_j\mathbf{v}) = \frac{\partial}{\partial z}(C_j v_z) \tag{B}$$

where v_z is now the velocity in the axial direction. In cylindrical coordinates the divergence of the gradient of the diffusion flux is

$$\nabla\cdot D_e(\nabla C_j) = \frac{\partial}{\partial z}\left[(D_e)_L\left(\frac{\partial C_j}{\partial z}\right)\right] + \frac{1}{r}\frac{\partial}{\partial r}\left[r(D_e)_r\frac{\partial C}{\partial r}\right] \tag{C}$$

For one reaction, and in terms of the global rate of disappearance per unit mass, $\mathbf{r}_p$,

$$\sum_i\mathbf{r}_{ij} = -\rho_B\mathbf{r}_p \tag{D}$$

---

† See books on transport phenomena for derivation of the species continuity equation; for example, R. B. Bird, W. E. Stewart, and E. N. Lightfoot, "Transport Phenomena," p. 556, John Wiley & Sons, New York (1960).

Substituting Eqs. (B) to (D) in Eq. (A) gives

$$-\frac{\partial}{\partial z}\left[(D_e)_L\left(\frac{\partial C_j}{\partial z}\right)\right] - \frac{1}{r}\frac{\partial}{\partial r}\left[r(D_e)_r\frac{\partial C}{\partial r}\right] + \frac{\partial}{\partial z}(C_j v_z) = -\rho_B r_p \qquad (E)$$

Finally, if the axial and radial diffusivities are not sensitive to r or to z, and if the velocity is not a function of z, Eq. (E) becomes

$$-(D_e)_r\left(\frac{1}{r}\frac{\partial C_j}{\partial r} + \frac{\partial^2 C_j}{\partial r^2}\right) + v_z\frac{\partial C_j}{\partial z} - (D_e)_L\frac{\partial^2 C_j}{\partial z^2} + \rho_B r_p = 0 \qquad (F)$$

This equation is identical with Eq. (13-2) where we have used u, rather than v_z, to be the velocity in the axial direction.

We have mentioned that in this chapter it is supposed that the global rate of reaction is known from the methods of Chaps. 9 to 11. That is, the intrinsic rate has been combined with external and intraparticle transport rates to obtain the global rate as a function of bulk concentrations and temperature. However, it is perhaps helpful to illustrate how all the individual transport effects are combined to predict reactor performance. For isothermal conditions and first-order kinetics, it is possible to obtain a simple analytical solution which clearly displays the influence of axial dispersion, fluid-to-particle mass transport, intraparticle diffusion, and intrinsic reaction. This is done in Example 13-2.

Example 13-2 An irreversible, first-order reaction occurs, at constant density, on the pore surface of catalyst particles in a fixed-bed. Assume isothermal operation and the dispersion model for the reaction mixture flowing through the bed. The reactant concentration in the feed is C_0. Derive an equation for the conversion in the reactor effluent. How is the equation simplified if:

(a) intraparticle diffusion resistance is unimportant?
(b) fluid-to-particle (or external) mass transfer resistance is unimportant?
(c) the global rate is controlled by the intrinsic reaction at an interior site?
(d) axial dispersion is unimportant (plug-flow conditions)?
(e) both conditions (c) and (d) apply?

SOLUTION
    A. *Intrinsic reaction and intraparticle diffusion.* The rate of the intrinsic reaction per unit mass of catalyst is

$$\mathbf{r} = k_1 C' \qquad (A)$$

where the prime on C designates the *intraparticle* concentration. The mass-conservation equation within the particle (assumed to be spherical) is given by Eq. (11-46). Its solution gives the concentration profile [Eq. (11-49)]. This, in turn, can be used to express the rate of reaction for the whole particle, $\mathbf{r}_p$, in terms of the concentration C_s' (which is equal to C_s) at the outer surface of the particle. The result for $\mathbf{r}_p$ is Eq. (11-53), which may be written

$$\mathbf{r}_p = \eta k_1 C_s = \frac{1}{\Phi_s}\left[\frac{1}{\tanh 3\Phi_s} - \frac{1}{3\Phi_s}\right]k_1 C_s \qquad (11\text{-}53)$$

This expression combines the effects of intrinsic reaction and intraparticle diffusion to give the rate for one particle (per unit mass of catalyst) in terms of the Thiele modulus Φ_s.

B. *External mass transfer.* The next step is to express $\mathbf{r}_p$ in terms of the bulk concentration; that is, convert Eq. (11-53) to a form involving C_b instead of C_s, thus giving an expression for the global rate. This is done by equating the mass-transfer rate from bulk gas to catalyst surface to the rate of reaction, as discussed in Chap. 10 (and illustrated in Example 11-8). Equating Eqs. (10-1) and (11-53),

$$k_m a_m (C_b - C_s) = \eta k_1 C_s \tag{B}$$

where a_m, the outer surface area per unit mass, is $6/d_p \rho_p$ for a spherical particle of diameter d_p and density ρ_p.

Eliminating C_s from Eqs. (11-53) and (B) in order to express $\mathbf{r}_p$ in terms of C_b gives

$$\mathbf{r}_p = \left[\frac{1}{d_p \rho_p / 6 k_m + 1/\eta k_1} \right] C_b \tag{C}$$

C. *Reactor Model.* For the dispersion model the mass-conservation expression is Eq. (13-14) with $\mathbf{r}_p$ given by Eq. (C). As noted, the solution of Eq. (13-14) with its boundary conditions is Eq. (6-47) with β given by Eq. (13-16). Thus the conversion x, in the effluent is

$$\left(\frac{C}{C_0} \right)_b = 1 - x$$

$$= \frac{4\beta}{(1 + \beta)^2 \exp \left[-\frac{1}{2} \frac{uL}{(D_e)_L} (1 - \beta) \right] - (1 - \beta)^2 \exp \left[-\frac{1}{2} \frac{uL}{(D_e)_L} (1 + \beta) \right]} \tag{D}$$

where

$$\beta = \left(1 + 4 k_o \rho_B \frac{(D_e)_L}{u^2} \right)^{1/2} \tag{E}$$

Comparison of Eqs. (13-15) and (C) shows that the overall rate constant is

$$k_o = \frac{1}{d_p \rho_p / 6 k_m + 1/\eta k_1} \tag{F}$$

Equation (F) can be written in dimensionless terms by multiplying by $d_p \rho_B / u$ to give

$$\frac{k_o \, d_p \rho_B}{u} = \left[\frac{1}{S} + \frac{1}{\eta \Lambda} \right]^{-1} \tag{G}$$

where $S = \dfrac{6 \rho_B k_m}{\rho_p u}$;   external mass-transfer parameter (H)

$\Lambda = \dfrac{k_1 \rho_B d_p}{u}$;   intrinsic reaction parameter (I)

Then β may be expressed in terms of these parameters and the Peclet number Pe_L:

$$\beta = \left[1 + 4\left(\frac{1}{S} + \frac{1}{\eta\Lambda}\right)^{-1}\left(\frac{u}{d_p}\right)\frac{(D_e)_L}{u^2}\right]^{1/2} = \left[1 + 4\left(\frac{1}{S} + \frac{1}{\eta\Lambda}\right)^{-1}\frac{1}{Pe_L}\right]^{1/2} \quad \text{(J)}$$

Finally, Eq. (D) can be written in terms of Pe_L:

$$1 - x = \frac{4\beta}{(1 + \beta)^2 \exp\left[-\frac{1}{2}(L/d_p)\,Pe_L(1 - \beta)\right] - (1 - \beta)^2 \exp\left[-\frac{1}{2}(L/d_p)\,Pe_L(1 + \beta)\right]} \quad \text{(K)}$$

Equation (K) is the required solution which expresses the conversion in terms of the axial dispersion parameter Pe_L and, through β, in terms of S, Λ, and the intraparticle diffusion parameter (the effectiveness factor η). Note that the product $L/d_p(Pe_L) = Lu/(D_e)_L$ is a Peclet number based upon L length. This indicates that the importance of axial dispersion depends upon the reactor length.

(a) If intraparticle diffusion is unimportant, η in Eq. (J) becomes unity.

(b) If external diffusion is unimportant, k_m becomes relatively large; then S is much larger than $\eta\Lambda$ and Eq. (J) for β becomes

$$\beta = (1 + 4\eta\Lambda/Pe_L)^{1/2} \quad \text{(L)}$$

(c) If intrinsic kinetics controls the global rate, $k_o = k_1$; this is the same as neglecting both external and internal mass transfer resistances. Then Eq. (J) becomes

$$\beta = (1 + 4\Lambda/Pe_L)^{1/2} \quad \text{(M)}$$

(d) If axial dispersion is unimportant, Pe_L is very large and $\beta \to 1.0$. Then Eq. (K) becomes indeterminate. However, Eq. (13-17) is applicable and may be integrated directly with r_p given by Eq. (13-15):

$$-u\frac{dC_b}{dz} = k_0 C_b \rho_B$$

or, in integrated form,

$$\left(\frac{C}{C_0}\right)_b = \exp\left(-k_0\rho_B L/u\right) \quad \text{(N)}$$

Substituting Eq. (F) for k_o,

$$\left(\frac{C}{C_0}\right)_b = 1 - x = \exp\left(-\frac{\rho_B L/u}{d_p\rho_p/6k_m + 1/\eta k_1}\right)$$

or, in terms of the dimensionless parameters

$$1 - x = \exp\left[-\frac{L/d_p}{(1/S) + (1/\eta\Lambda)}\right] \quad \text{(P)}$$

This result was also obtained in Example 12-1 where experimental conversion data were interpreted in terms of a plug-flow model which included external and intraparticle mass transfer. Equations (I) for the global rate and (K) for the effluent concentration in Example 12-1 are of the same form as Eqs. (C) and (P) of this example.

(e) If axial dispersion is unimportant and also intrinsic kinetics controls the global rate, Eq. (P) is applicable with $S \to \infty$ and $\eta \to 1$. Hence,

$$1 - x = \exp\left[-(\Lambda L/d_p)\right] = \exp\left[-\left(\frac{k_1 \rho_B L}{u}\right)\right] \qquad (Q)$$

In this equation the only rate constant is that for intrinsic kinetics. Equation (Q) is analogous to Eq. (4-18) developed for ideal-flow (plug flow) in a tubular reactor for a first-order homogeneous reaction.

13-4 Adiabatic Operation

Large-diameter or well-insulated reactors approach adiabatic operation more closely than isothermal operation. Therefore, an adiabatic, one-dimensional model may be a good representation of actual behavior. With this model, as for isothermal operation, radial gradients in concentrations and temperature are generally small because they are caused solely by radial variation in the axial velocity.

If axial dispersion of heat is included, the two-dimensional energy conservation equation is

$$(k_e)_r\left(\frac{1}{r}\frac{\partial T}{\partial r} + \frac{\partial^2 T}{\partial r^2}\right) - (u\rho)c_p\frac{\partial T}{\partial z} + (k_e)_L\frac{\partial^2 T}{\partial z^2} - \mathbf{r}_p\rho_B(\Delta H) = 0 \qquad (13\text{-}19)$$

This expression is analogous to Eq. (13-2) for mass transfer and involves the same kinds of assumptions. It is obtained from Eq. (5-1) by including terms for axial and radial transfer of heat. The quantities $(k_e)_r$ and $(k_e)_L$ are *effective* thermal conductivities in the radial and axial directions and are analogous to $(D_e)_r$ and $(D_e)_L$.

For the *adiabatic* model the radial dispersion term disappears so that Eq. (13-19) becomes

$$-u\rho c_p\frac{\partial T}{\partial z} + (k_e)_L\frac{\partial^2 T}{\partial z^2} - \mathbf{r}_p\rho_B(\Delta H) = 0 \qquad (13\text{-}20)$$

This is analogous to the mass-conservation expression [Eq. (13-14)].

The axial dispersion term† in the energy-conservation equation may be more important‡ than axial dispersion of mass. However, for the relatively high velocities and long bed depths encountered in commercial reactors it is normal that axial dispersion of both heat and mass can be neglected. If this is not so, the

---

† The axial effective thermal conductivity $(k_e)_L$ may be estimated from the correlation of J. Votruba, V. Hlavacek, and M. Marck, *Chem. Eng. Sci.* **27**, 1845 (1972).

‡ L. C. Young and B. A. Finlayson, *Ind. Eng. Chem. Fundam.* **12**, 412 (1973); Hlavacek, V., et al., *Chem. Eng. Sci.* **28**, 1897 (1973).

solution of Eqs. (13-14) and (13-20) becomes a boundary-value problem with boundary conditions applicable at the entrance and exit of the catalyst bed. The situation is analogous to the dispersion model considered for homogeneous tubular reactions in Sec. 6-9. However, now the energy-conservation equation is also involved and complex numerical solution techniques† are necessary. Even with high-speed computers considerable time is required.

If we neglect axial dispersion of heat and mass, Eq. (13-20) reduces to

$$u\rho C_p \frac{dT}{dz} = \mathbf{r}_p \rho_B(-\Delta H) \tag{13-21}$$

and the mass-conservation expression is Eq. (12-1). As done in Chap. 5 for adiabatic, homogeneous reactors [see Eqs. (5-15) and (5-16)], we can combine Eqs. (13-21) and (12-2) so as to integrate the energy equation. Thus, substituting $\mathbf{r}_p$ from the differential form [Eq. (12-2)] into Eq. (13-21) gives

$$u\rho c_p \frac{dT}{dz} = \rho_B(-\Delta H)F\frac{dx}{dW}$$

$$dT = \frac{\rho_B(-\Delta H)F}{u\rho c_p}\left(\frac{dz}{dW}\right)dx$$

Since $dW = (\rho_B A_c\, dz)$ and $u\rho A_c = F_t$,

$$dT = (-\Delta H)\frac{F}{F_t c_p}\, dx \tag{13-22}$$

The flow rates F and F_1 are the mass-feed rates of reactant and of total mixture. For constant, ΔH and c_p, integration yields

$$T - T_f = (-\Delta H)\frac{F}{F_t c_p}(x - x_f) \tag{13-23}$$

where x_f and T_f are the conversion and temperature of the feed.

With the temperature given by Eq. (13-23), the global rate can be evaluated at any conversion. Then only Eq. (12-1) need be integrated to determine the conversion for any catalyst bed depth or mass W. The calculations, which are similar to those for homogeneous reactors (Example 5-2), are illustrated in Example 13-3.

Example 13-3 Wenner and Dybdal‡ studied the catalytic dehydrogenation of ethyl benzene and found that with a certain catalyst the rate could be represented by the reaction

$$C_6H_5C_2H_5 \rightleftharpoons C_6H_5CH{=}CH_2 + H_2$$

---

† Orthogonal collocation [J. V. Villadsen and W. E. Stewart, *Chem. Eng. Sci.* **22**, 1483 (1967)] has proven to be an efficient procedure for solution of such equations [L. C. Young and B. A. Finlayson, loc. cit.].

‡ R. R. Wenner and F. C. Dybdal, *Chem. Eng. Progr.*, **44**, 275 (1948).

The global rate was given as

$$\mathbf{r}_p = k\left(p_E - \frac{1}{K} p_s p_H\right)$$

where p_E = partial pressure of ethyl benzene
p_s = partial pressure of styrene
p_H = partial pressure of hydrogen

The specific reaction rate and equilibrium constants are

$$\log k = -\frac{4770}{T} + 4.10$$

where k is the pound moles of styrene produced per hr(atm)(lb catalyst) and T is in degrees Kelvin; in SI units, $k_{SI} = 2.74 \times 10^{-6} k$, kg mol/(s)(kPa) (kg catalyst).

t, °C	K
400	1.7×10^{-3}
500	2.5×10^{-2}
600	2.3×10^{-1}
700	1.4

Estimate the amount of catalyst necessary to produce 15 tons (13,620 kg) of styrene a day, using vertical tubes 4 ft (1.22 m) in diameter, packed with catalyst pellets. Wenner and Dybdal considered this problem by taking into account the side reactions producing benzene and toluene. However, to simplify the calculations in this introductory example, suppose that the sole reaction is the dehydrogenation to styrene, and that there is no heat exchange between the reactor and the surroundings. Assume that under normal operation the exit conversion will be 45%. However, also prepare graphs of conversion and temperature vs. catalyst bed depth, up to equilibrium conditions. The feed rate per reactor tube is 13.5 lb mol/h (1.70×10^{-3} kg mol/s) for ethyl benzene and 270 lb mol/h (34.0×10^{-3} kg mol/s) for steam. In addition,

Temperature of
  mixed feed entering reactor = 625°C (898 K)

Bulk density of catalyst
  as packed = 90 lb/ft³ (1440 kg/m³)

Average pressure in
  reactor tubes = 1.2 atm (121 kPa)

Heat of reaction ΔH = 60,000 Btu/lb mol (1.39×10^5 kJ/(kg mol)

Surroundings temperature = 70°F (294 K)

SOLUTION The reaction is endothermic, so that heat must be supplied to maintain the temperature. In this problem energy is supplied by adding steam at 625°C to the feed. An alternate approach of transferring heat from the surroundings is utilized for the same system in Example 13-4.

The operation is adiabatic so that Eq. (13-23) is applicable with $x_f = 0$. For ethylbenzene $F = 13.5$ lb mol/h. As there is a large excess of steam, it will be satisfactory to take $c_p = 0.52$. Then the heat capacity of the reaction mixture will be

$$F_t c_p = (270 \times 18 + 13.5 \times 106)(0.52)$$

$$= 3270 \text{ Btu/°F(h) or } (1.72 \text{ kJ/(K)(s))} \tag{A}$$

Substituting numerical values in Eq. (13-23), we obtain

$$T - T_f = \frac{-60,000 \ (13.5)}{3270} x$$

$$T - 1616 = -248x \tag{B}$$

where T is in degrees Rankine, and 1616°R is the entering temperature of the feed.

Equation (12-1), written in differential form, with the weight of catalyst expressed as $dW. = \rho_B A_c dz$, where A_c is the cross-sectional area, is

$$F \, dx = \mathbf{r}_P \rho_B A_c \, dz$$

$$dz = \frac{F}{\mathbf{r}_P \rho_B A_c} \, dx = \frac{13.5 \, dx}{90(0.7854)(16)\mathbf{r}_P} = \frac{0.0119}{\mathbf{r}_P} \, dx \tag{C}$$

The partial pressures can be expressed in terms of the conversion as follows: At any conversion x the moles of each component are

$$\text{Steam} = 20$$
$$\text{Ethyl benzene} = 1 - x$$
$$\text{Styrene} = x$$
$$\text{Hydrogen} = x$$
$$\text{Total moles} = 21 + x$$

Then

$$p_E = \frac{1 - x}{21 + x} (1.2)$$

$$p_S = p_H = \frac{x}{21 + x} (1.2)$$

Then the rate equation becomes

$$\mathbf{r}_P = \frac{1.2}{21 + x} k \left[(1 - x) - \frac{1.2}{K} \frac{x^2}{21 + x} \right]$$

or, with the expression for k determined by Wenner and Dybdal,[†]

$$r_P = \frac{1.2}{21 + x}(12,600)e^{-19,800/T}\left[(1 - x) - \frac{1.2}{K}\frac{x^2}{21 + x}\right] \qquad \text{(D)}$$

Substituting this value of r_P in Eq. (C) gives an expression for the catalyst-bed depth in terms of the conversion and temperature,

$$\frac{dz}{dx} = \frac{21 + x}{1,270,000}e^{19,800/T}\left[(1 - x) - \frac{1.2}{K}\frac{x^2}{21 + x}\right]^{-1} \qquad \text{(E)}$$

With Eq. (B) to express T in terms of x, and the K vs. T data given in the problem statement, the entire right-hand side of Eq. (E) is in terms of x. Then we may write Eq. (E) as

$$\frac{dz}{dx} = f(x) \qquad \text{(F)}$$

This expression may be solved numerically for z as a function of x, starting at the feed ($z = 0$, $T = 1616°R$, or 898 K, for $x = 0$). Equation (F) is of the same form as Eq. (B) of Example 5-1 with one dependent variable, z. Hence, it may be solved by the Runge-Kutta procedure described in that example. The results of the computations obtained with an increment size $\Delta x = 0.01$ are given in Table 13-1.

The rate of reaction becomes zero at a conversion of about $x = 0.69$ and a temperature of 1445°R (803 K), as determined from Eqs. (B) and (D). From Fig. 13-6 it is found that a bed depth of 3.8 ft (1.16 m) is required for a

[†] In the exponential term T has been converted to degrees Rankine.

Table 13-1 Data for conversion of ethyl benzene to styrene in an adiabatic reactor

| Conversion | Temperature | | Catalyst-bed depth, ft |
	°R	°C	
0	1616	625	0
0.10	1591	611	0.40
0.20	1566	597	0.96
0.30	1542	584	1.75
0.40	1517	570	2.93
0.50	1492	556	4.95
0.55	1480	549	6.3
0.60	1467	542	10.0
0.62	1462	539	13.0
0.69	1445	530	∞

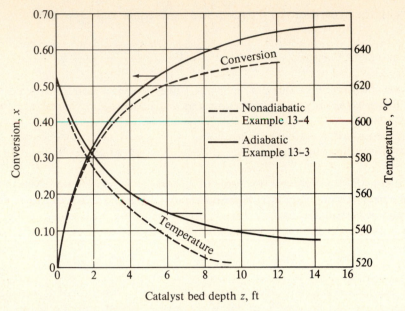

Figure 13-6 Conversion vs. catalyst-bed depth for the production of styrene from ethyl benzene.

conversion of 45%. The production of styrene from each reactor tube would be

$$\text{Production/tube} = 13.5(0.45)(104)(24)$$

$$= 15{,}200 \text{ lb/day} \qquad (7.6 \text{ tons/day, } 6900 \text{ kg/day})$$

Hence two 4-ft-diameter reactor tubes packed with catalyst to a depth of at least 3.8 ft would be required to produce 15 tons/day of crude styrene. The mass of catalyst required is

$$3.8(\pi d_t^2/4)90(15/7.6) = 8500 \text{ lb } (3860 \text{ kg})$$

NONISOTHERMAL, NONADIABATIC FIXED-BED REACTORS

In homogeneous-reactor design (Chap. 5) only heat exchange with the surroundings was considered. Radial mixing was supposed to be sufficiently good that all the resistance to energy transfer could be concentrated at the reactor wall. The temperature was assumed to be flat up to the wall, where a discontinuous change to the wall temperature occurred. The temperature of the reaction mixture changed only in the axial direction. This one-dimensional model is normally satisfactory for homogeneous reactors, because radial mixing is sufficient to give a reasonably flat profile. However, in fixed beds this is not as good an assumption, because the catalyst pellets hinder radial mixing. Hence, a two-dimensional model which accounts for radial temperature gradients within the catalyst bed may be necessary. The one- and two-dimensional models are considered separately in Secs. 13-5 and 13-6.

13-5 The One-Dimensional Model

The form of the radial temperature profile in a nonadiabatic fixed-bed reactor has been observed experimentally to have a parabolic shape. Data for the oxidation of sulfur dioxide with a platinum catalyst on $\frac{1}{8} \times \frac{1}{8}$-in. cylindrical pellets in a 2-in.-ID reactor are illustrated in Fig. 13-7. Results are shown for several catalyst-bed depths. The reactor wall was maintained at 197°C by a jacket of boiling glycol. This is an extreme case. The low wall temperature resulted in severe radial temperature gradients, more so than would exist in many commercial reactors. The longitudinal profiles are shown in Fig. 13-8 for the same experiment. These curves show the typical hot spots, or maxima, characteristic of exothermic reactions in a nonadiabatic reactor. The greatest increase above the feed temperature is at the center, $r/r_0 = 0$. This rise decreases as the wall is approached and actually disappears at a radial position of 0.9. The temperature is so low here, even in the entering stream, that very little reaction occurs. Hence the curve in Fig. 13-8 at $r/r_0 = 0.9$ is essentially a cooling curve, approaching 197°C as the bed depth increases.

A completely satisfactory design method for nonadiabatic reactors entails predicting the radial and longitudinal variations in temperature, such as those shown in Figs. 13-7 and 13-8, predicting the analogous concentration profiles, and the bulk mean conversion. To make such predictions we must know the effective

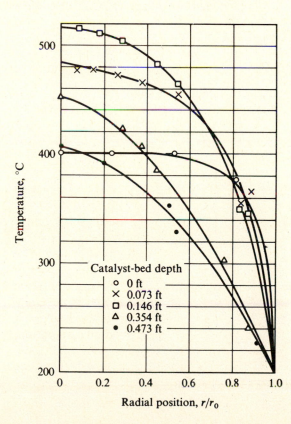

Figure **13-7** Radial temperature profiles in a fixed-bed reactor for the oxidation of SO_2 with air.

thermal conductivity and diffusivity for heat and mass transfer in the radial and axial directions since these qualities appear in the conservation expressions [Eqs. (13-2) and (13-19)]. However, it is worthwhile first to consider the one-dimensional model that eliminates the need for effective conductivities and diffusivities and gives an approximate prediction for the average temperature across the bed.

The parabolic shape of the radial temperature curves in Fig. 13-7 suggests that most of the resistance to heat transfer is near the wall of the reactor and only a small amount is in the central core. To carry this idea farther, if we assume that all the heat-transfer resistance is in a very thin layer next to the wall, the temperature profile will be as shown by the dashed lines in Fig. 13-9. The solid line is the 0.146-ft bed-depth curve of Fig. 13-7. The horizontal, dashed line represents the bulk-mean temperature obtained by integration of the data for the solid line. If the actual situation is replaced by this approximate model, the only information necessary to establish the energy exchange with the surroundings is the heat-transfer coefficient at the wall, h_w. To complete the definition of the one-dimensional model, axial dispersion of mass and energy is also neglected. With these conditions, the design procedure would be the same as for nonadiabatic, homogeneous tubular reactors, as illustrated in Example 5-2. To calculate

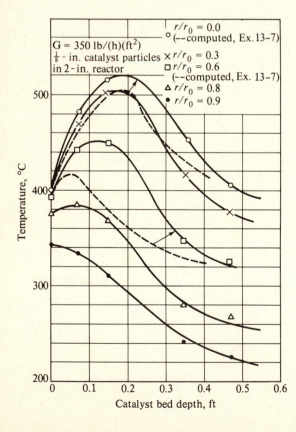

Figure 13-8 Longitudinal temperature profiles in SO_2 reactor.

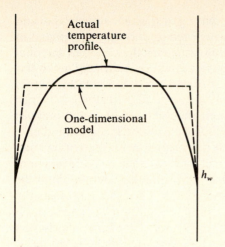

Figure 13-9 Comparison of model (one-dimensional) temperature profile with an actual profile.

the bulk conversion and temperature along the reactor length we need only plug-flow forms of the conservation equations: Eq. (12-2) for mass conservation and Eq. (5-15)† or (5-16) for energy conservation. However, before we apply the one-dimensional model it is necessary to review the available data for h_w.

Wall heat-transfer coefficients If T_b is the bulk mean temperature of the reaction fluid and T_w is the wall temperature, h_w is defined by

$$dQ = h_w(T_b - T_w)\, dA_h \tag{13-24}$$

where Q is the heat-transfer rate to the wall and A_h is the wall area. The presence of the solid particles increases the heat-transfer coefficient in a packed bed several times over that in an empty tube at the same gas flow rate. In the earliest experimental investigations of the subject,‡ the results were reported as ratios of the coefficient in the packed bed to that in the empty pipe. It was found that this ratio varied with the ratio of the pellet diameter to the tube diameter, reaching a maximum value at about $d_p/d = 0.15$. Colburn's results for the ratio of heat-transfer coefficients in packed and empty tubes, h_w/h, are:

d_p/d	0.05	0.10	0.15	0.20	0.25	0.30
h_w/h	5.5	7.0	7.8	7.5	7.0	6.6

Presumably the large increase in heat-transfer coefficient for a packed tube over that for an empty tube is due to the mixing, or turbulence, caused by the presence of the solid particles. This turbulence tends to prevent the buildup of a slow-moving layer of fluid next to the wall and also increases the radial transfer of heat

† Note that Eq. (13-21), which was derived as the plug-flow form of Eq. (13-19), is the adiabatic form of Eq. (5-15).

‡ A. P. Colburn, *Ind. Eng. Chem.*, **28**, 910 (1931); *Trans. AIChE*, **26**, 166 (1931); E. Singer and R. H. Wilhelm, *Chem. Eng. Progr.*, **46**, 343 (1950).

within the fluid in the tube. Up to a point, decreasing the particle size increases the importance of these factors, and the heat transfer coefficient continues to increase. However, the maximum at a certain d_p/d ratio suggests that another factor is involved. This is concerned with the size of the radial eddies in the fluid in the bed. As the particle size continues to decrease, the size of the eddies decreases, and the distance over which each mixing process occurs is decreased. Also, there is a larger number of more-or-less stagnant films between the fluid and the solid particles which the heat must cross in reaching the wall. The maximum in the heat-transfer coefficient would represent the point at which this second factor counterbalances favorable effects of the mixing process obtained with the smaller particles.

More recent correlations of wall coefficients are available.† The recommendation of Beek‡ for gases in fixed beds is

$$h_w \, d_p/k_f = 2.58(\text{Re})^{1/3}(\text{Pr})^{1/3} + 0.094(\text{Re})^{0.8}(\text{Pr})^{0.4} \qquad \text{cylin. particles} \qquad (13\text{-}25)$$

$$h_w \, d_p/k_f = 0.203(\text{Re})^{1/3}(\text{Pr})^{1/3}$$
$$+ 0.220(\text{Re})^{0.8}(\text{Pr})^{0.4} \qquad \text{for spherical particles} \qquad (13\text{-}26)$$

where $\text{Re} = d_p u\rho/\mu$ and $\text{Pr} = c_p \mu/k_f$. The molecular thermal conductivity of the fluid is k_f. These correlations apply only to the coefficient at the inside wall of the tube. If the wall temperature is not known, the resistance on the outside of the tube wall must be included in calculating the heat-exchange rate with the surroundings.

Application of one-dimensional model This model requires much less computational time than the two-dimensional procedure. Hence, it is particularly useful for a preliminary design. It provides a rapid procedure for estimating reactor size and the effect of such variables as tube diameter. As the tube diameter decreases, the ratio of the heat-transfer area to the reactor volume will increase. Therefore, the temperature rise of the reaction mixture as it passes through the bed will be less, and the radial temperature variation within the bed will also be less. Hence, where it is necessary not to exceed a certain temperature in the catalyst bed, small-diameter tubes are indicated. The approximate size necessary for a given temperature can be determined by means of this simplified procedure. Problem 13-7 illustrates calculations for a phthalic anhydride reactor. The oxidation of naphthalene has a high heat of reaction, so that the size of the catalyst tubes is a critical point in the design.

Examples 13-4 to 13-6 illustrate the simplified design method for different cases. The first is for the endothermic styrene reaction, where the temperature decreases continually with catalyst-bed depth. Example 13-5 is for an exothermic reaction carried out under conditions where radial temperature gradients are not large. Example 13-6 is also for an exothermic case, but here the gradients are severe, and the simplified solution is not satisfactory.

---

† G. F. Froment, *Adv. Chem. Ser.*, **109**, Amer. Chem. Society, Washington, D.C. (1972); R. E. Chow, R. A. Caban, and M. M. Irizarry, *Can. Chem. Eng.*, **51**, 67 (1973); S. Yagi and D. Kunii, *AIChE J.*, **6**, 97 (1960).

‡ J. Beek, *Adv. Chem. Eng.*, **3**, 303 (1962). Based on original correlations of T. J. Hanratty, *Chem. Eng. Sci.*, **3**, 209 (1954) and of D. Thoenes, Jr., and H. Kramers, *Chem. Eng. Sci.*, **8**, 271 (1958).

Example 13-4 Under actual conditions the reactor described in Example 13-3 would not be truly adiabatic. Suppose that with reasonable insulation the heat loss would correspond to a heat-transfer coefficient of $U = 1.6$ Btu/(h) (ft^2 inside tube area) (°F). This value is based on the difference in temperature between the reaction mixture and the surroundings at 70°F. Determine revised curves of temperature and conversion vs. catalysed-bed depth for this nonadiabatic operation.

SOLUTION The mass conservation equation (C) and the rate expression (D) of Example 13-3 are applicable, as is their combination, Eq. (E). The expression for conservation of energy for nonadiabatic operation will be Eq. (5-16); that is

$$U(T_s - T) \, dA_h = \Delta H_R F \, dx + F_t c_p \, dT$$

Substituting numerical values,

$$1.6(530 - T)(4\pi \, dz) = (60{,}000)13.5 \, dx + 3270 \, dT$$

or

$$dT = -248 \, dx - 0.00615(T - 530) \, dz \qquad (A)$$

Equation (E) of Example (13-3) and Eq. (A) determine the conversion and temperature as a function of catalyst bed depth. Equation (E) is of the form $dx/dz = f(x, T)$ and Eq. (A) of the form $dT/dz = f(x, T)$. Hence, the fourth order Runge-Kutta method with two dependent variables (T and x) can be used for the numerical solution. The procedure and the working equations are the same as those for Example 4-7. However, in order to gain an appreciation for the magnitudes of the numbers, let us illustrate for one increment the calculations by the Euler method (also first described in Example 4-7).

In applying the Euler method it is convenient to choose an increment of conversion. It is necessary to assume a temperature at the end of the increment in order to evaluate a reaction rate for use in Eq. (E) of Example 13-3. The assumption can be checked in Eq. (A). If $\Delta x = 0.1$, and T is assumed to be 1591°R, we have from Eq. (E):

at $z = 0$, $x = 0$ and $T = 1616$°R

$$\frac{dz}{dx} = \frac{21}{1{,}270{,}000} \, (e^{19{,}800/1616})[(1 - 0) - 0]^{-1} = 3.30$$

at $x = 0.1$, $T = 1591$°R

$$\frac{dz}{dx} = \frac{21 + 0.1}{1{,}270{,}000} \, (e^{19{,}800/1591}) \left[(1 - 0.1) - \frac{1.2}{0.28} \frac{0.1^2}{21 + 0.1} \right]^{-1}$$

$$= 4.65$$

Then, by the Euler method,

$$\Delta z = \left(\frac{dz}{dx} \right)_{\text{ave}} \Delta x = \frac{3.30 + 4.65}{2} (0.1) = 0.40 \text{ ft.}$$

Table 13-2 Conversion of ethylbenzene to styrene in a nonadiabatic reactor

Catalyst-bed depth, ft	Conversion x	Mean bulk temperature	
		°R	°C
0	0	1616	625
0.40†	0.10†	1589†	610†
0.50	0.12	1583	606
0.98†	0.20†	1560†	594†
1.0	0.20	1558	593
2.0	0.31	1525	574
3.0	0.39	1501	561
4.0	0.44	1483	551
5.0	0.47	1468	543
6.0	0.50	1457	536
7.0	0.51	1447	531
8.0	0.531	1438	526
9.0	0.534	1434	524
9.0†	0.55†	1426†	519†

† Calculated by the Euler method with an increment size $\Delta x = 0.1$.

Now, checking the assumed temperature in Eq. (A), we find

$$\Delta T = -248(0.1) = 0.00615\left(\frac{1{,}616 + 1{,}591}{2} - 530\right)(0.40)$$

$$= -24.8 - 2.6 = -27.4°F$$

$$T_1 = 1{,}616 - 27 = 1589°R$$

Since this temperature is close to the assumed value, the calculations need not be repeated.

The same kind of calculations can be made for subsequent increments to give T and x for successively larger bed depths. The results marked with a dagger in Table 13-2 were obtained by the Euler method. The other values in the table and in Fig. 13-6 are the more accurate results calculated by the fourth-order Runge-Kutta procedure, using an increment size $\Delta z = 0.5$ ft.

The bed depth for a given conversion is greater than for the adiabatic case, because the temperature is less. For example, for a conversion of 50%, 6.0 ft of catalyst is required, in comparison with 4.95 ft in Example 13-3.

Example 13-5 A bench-scale study of the hydrogenation of nitrobenzene was published by Wilson† in connection with reactor design studies. Nitrobenzene and hydrogen were fed at a rate of 65.9 g mol/h to a 3.0-cm-ID reactor

---

† K. B. Wilson, *Trans. Inst. Chem. Engrs.* (London), **24**, 77 (1946).

containing the granular catalyst. A thermocouple sheath, 0.9 cm in diameter, extended down the center of the tube. The void fraction was 0.424 and the pressure atmospheric. The feed entered the reactor at 427.5 K, and the tube was immersed in an oil bath maintained at the same temperature. The heat-transfer coefficient from the mean reaction temperature to the oil bath was determined experimentally to be 8.67 cal/(h)(cm²)(°C). A large excess of hydrogen was used, so that the specific heat of the reaction mixture may be taken equal to that for hydrogen and the change in total moles as a result of reaction may be neglected. The heat of reaction is approximately constant and equal to $-152,100$ cal/g mol.

The entering concentration of nitrobenzene was 5.0×10^{-7} g mol/cm³. The *global* rate of reaction was represented by the expression†

$$r_P = 5.79 \times 10^4 C^{0.578} e^{-2,958/T}$$

where $r_P =$ g mol nitrobenzene reacting/cm³(h), expressed in terms of void
volume in reactor
$C =$ concentration of nitrobenzene, g mol/cm³
$T =$ temperature, K

The experimental results for temperature vs. reactor length are shown in Fig. 13-10. From the data given, calculate temperatures up to a reactor length of 25 cm and compare them with the observed results.

† Includes effects of external and intraparticle resistances.

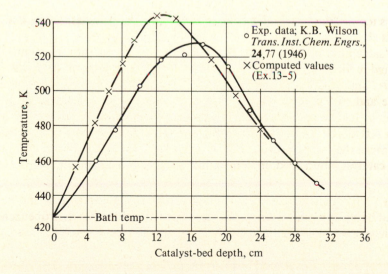

Figure 13-10 Longitudinal temperature profile in a reactor for hydrogenation of nitrobenzene.

SOLUTION The concentration depends on temperature as well as conversion. If Q is the volumetric flow rate at a point in the reactor where the concentration is C, and Q_0 is the value at the entrance, the conversion of nitrobenzene is

$$x = \frac{C_0 Q_0 - CQ}{C_0 Q_0} = 1 - \frac{C}{C_0} \frac{Q}{Q_0}$$

Since there is no change in pressure or number of moles, Q changes only because of temperature changes. Hence, assuming perfect-gas behavior, we have

$$x = 1 - \frac{C}{C_0} \frac{T}{T_0}$$

or

$$C = (1 - x) \frac{C_0 T_0}{T} = (5 \times 10^{-7}) \frac{427.5}{T} (1 - x) \tag{A}$$

Substituting this expression into the rate equation gives the global rate in terms of the temperature and conversion,

$$r_p = 439 \left(\frac{1 - x}{T} \right)^{0.578} e^{-2,958/T} \tag{B}$$

The mass-conservation expression, Eq. 12-1, should be written in terms of the void volume because of the form in which the rate data are reported. Thus

$$F \, dx = r_p (0.424) \, dV$$

The feed rate of nitrobenzene is

$$F = 65.9(22,400) \left(\frac{427.5}{273} \right) (5.0 \times 10^{-7}) = 1.15 \text{ g mol/h}$$

Hence,

$$1.15 \, dx = r_p (0.424) \frac{\pi}{4} (9 - 0.81) \, dz$$

or

$$\frac{dx}{dz} = \frac{r_p}{0.423} = 1040 \left(\frac{1 - x}{T} \right)^{0.578} e^{-2458/T} \tag{C}$$

The energy-conservation expression in Eq. 5-16. In numerical terms,

$$-(-152,100)F \, dx - 8.67\pi(3)(T - 427.5) \, dz = 65.9(6.9 \, dT)$$

where 6.9 cal/(g mol)(K) is the molal heat capacity of hydrogen at 427.5 K. This equation may be simplified to

$$\frac{dT}{dz} = 385 \left(\frac{dx}{dz} \right) - 0.180(T - 427.5) \tag{D}$$

If Eq. (C) is employed to eliminate dx/dz, Eq. (D) becomes

$$\frac{dT}{dz} = 4.0 \times 10^5 \left(\frac{1-x}{T}\right)^{-0.578} e^{-2958/T} - 0.180(T - 427.5) \qquad \text{(E)}$$

Equations (C) and (E) are of the form $dx/dz = f(x, T)$ and $dT/dz = f(x, T)$. Hence, they can be solved by the fourth-order Runge-Kutta method to give T and x as a function of z. The method of solution and equations are the same as those given for the similar, two-dependent variable problem of Example 4-7. The calculated results, using an increment $\Delta z = 0.5$ cm, are shown in Table 13-3 and Fig. 13-10.

As the bed depth increases the heat-loss term in Eq. (E) increases because T increases. At low conversions the heat-of-reaction term also increases. However, the decrease in reactant concentration (proportional to $1 - x$) ultimately slows the increase in the first term. The trend continues until the heat transferred to the oil bath is as large as that evolved as a result of reaction. The temperature reaches a maximum at this point, the so-called "hot spot." Figure 13-10 shows that the computed hot spot is reached at about 13 cm from the entrance to the reactor. This location is 4 cm before the experimental hot spot. Also, the experimental temperature is 20°C less than the computed value at the maximum. Three points should be mentioned in making this comparison. First, the measured temperature corresponds to the center of the tube, while the computed results are for a bulk mean temperature. Second, the thermocouples were contained in a metal sheath entering down the center of the reactor. The sheath would reduce the observed temperatures, because of longitudinal conduction, and make them more nearly comparable with the bulk mean computed values. Third, using a specific heat of hydrogen for the

Table 13-3 Temperatures and conversions in a nitrobenzene hydrogenation reactor

Catalyst bed depth cm	Conversion x	Bulk temperature °C
0	0	427.5
0.5	0.016	433
1.0	0.033	439
2.0	0.072	451
3.0	0.116	463
5.0	0.220	486
7.0	0.344	508
9.0	0.481	528
11.0	0.620	543
12.0	0.685	546
13.0	0.744	548
14.0	0.796	546
15.0	0.840	542
17.0	0.906	529
19.0	0.948	511
21.0	0.972	494

whole reaction mixture results in a value that is too low, causing the calculated temperatures to be high.

The agreement shown in Fig. 13-10 is good, in view of these three points and since the one-dimensional model neglects radial temperature gradients. The fact that a heat-transfer coefficient was determined experimentally in the same apparatus probably improved the agreement.

The simplified approach to the design of nonadiabatic reactors led to good results in Example 13-5 partly, at least, because radial temperature variations were not large. This is because the wall temperature was the same as the temperature of the entering reactants. At the hot spot the maximum temperature difference between the center and the wall was about 120°C. In Example 13-6 much larger radial temperature gradients exist, and the simplified method is not as suitable.

Example 13-6 Using the one-dimensional method, compute curves for temperature and conversion vs. catalyst-bed depth for comparison with the *experimental* data shown in Figs. 13-8 and 13-12 for the oxidation of sulfur dioxide. The reactor consisted of a cylindrical tube, 2.06 in. ID. The superficial gas mass velocity was 350 lb/(h)(ft^2), and its inlet composition was 6.5 mol % SO$_2$ and 93.5 mol % dry air. The catalyst was prepared from $\frac{1}{8}$-in. cylindrical pellets of alumina and contained a surface coating of platinum (0.2 wt % of the pellet). The measured global rates in this case were not fitted to a kinetic equation, but are shown as a function of temperature and conversion in Table 13-4 and Fig. 13-11. Since a fixed inlet gas composition was used, independent variations of the partial pressures of oxygen, sulfur dioxide, and sulfur trioxide were not possible. Instead these pressures are all related to one variable, the extent of conversion. Hence the rate data shown in Table 13-4 as a function of conversion are sufficient for the calculations. The total pressure was essen-

Table 13-4 Experimental global rates [g mol/(h)(g catalyst)] of oxidation of SO$_2$ with a 0.2 % Pt-on-Al$_2$O$_3$ catalyst

t, °C	% conversion† of SO$_2$						
	0	10	20	30	40	50	60
350	0.011	0.0080	0.0049	0.0031			
360	0.0175	0.0121	0.00788	0.00471	0.00276	0.00181	
380	0.0325	0.0214	0.01433	0.00942	0.00607	0.00410	
400	0.0570	0.0355	0.02397	0.01631	0.0110	0.00749	0.00488
420	0.0830	0.0518	0.0344	0.02368	0.0163	0.0110	0.00745
440	0.1080	0.0752	0.0514	0.03516	0.0236	0.0159	0.0102
460	0.146	0.1000	0.0674	0.04667	0.0319	0.0215	0.0138
480	....	0.1278	0.0898	0.0642	0.0440	0.0279	0.0189
500	....	0.167	0.122	0.0895	0.0632	0.0394	0.0263

† Conversion refers to a constant feed composition of 6.5 mole % SO$_2$ and 93.5 mole % air.

Source: R. W. Olson, R. W. Schuler, and J. M. Smith, *Chem. Eng. Progr.*, **46**, 614 (1950).

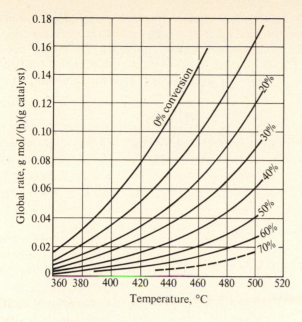

Figure 13-11 Rate of oxidation of SO_2 on $\frac{1}{8}$-in. catalyst particles containing 0.2% platinum [mass velocity 350 lb/(h)(ft^2)].

tially constant at 790 mmHg. The heat of reaction is nearly constant over a considerable temperature range and was equal to $-22,700$ cal/g mol of sulfur dioxide reacted. The gas mixture was predominantly air, so that its specific heat may be taken equal to that of air. The bulk density of the catalyst as packed in the reactor was 64 lb/ft^3.

From Fig. 13-7 it is apparent that the entering temperature across the diameter is not constant, but varies from a maximum value of about 400°C at the center down to 197°C at the wall. Since the one-dimensional method of solution to be used in this example is based on a uniform temperature radially, we use a mean value of 364°C. In the two-dimensional model, considered in Example 13-7, the actual entering-temperature profile can be taken into account.

For the heat-transfer coefficient from gas to tube wall take $h_w = 6.3$ Btu/(h)(ft)^2(°F). The tube wall temperature is 197°C.

SOLUTION The energy conservation equation for an element of reactor height dz is, according to Eq. (5-16)

$$-(-22,700)F(1.8)\,dx + 6.3\pi\frac{2.06}{12}(197 - T)1.8\,dz = F_t c_p\,dT(1.8) \quad \text{(A)}$$

In this expression the conversion factor of 1.8°F/°C has been introduced so that the temperature may be expressed in degrees centigrade

$$F = 350\pi\left(\frac{1.03}{12}\right)^2\frac{1}{31.2}(0.065) = 0.017 \text{ lb mol/h}$$

The number 31.2 is the molecular weight of the feed containing 6.5 mol% SO_2. The heat capacity of the reaction mixture, assumed to be that of air at an average temperature of 350°C, is 0.26 Btu/(lb)(°F). Hence

$$F_t c_p = 350\pi \left(\frac{1.03}{12}\right)^2 (0.26) = 2.11 \text{ Btu/°F}$$

Substituting these values in Eq. (A), we find

$$0.017(22,700) \, dx - 3.40(T - 197) \, dz = 2.11 \, dT$$

or

$$\frac{dT}{dz} = 182 \frac{dx}{dz} - 1.61(T - 197) \tag{B}$$

The mass-conservation expression [Eq. (12-1)] may be written

$$\pi \left(\frac{1.03}{12}\right)^2 \mathbf{r}_p \rho_B \, dz = F \, dx = 0.017 \, dx$$

The bulk density of the catalyst as packed is given as 64 lb/ft³. Hence Eq. (B) simplifies to the form

$$\frac{dx}{dz} = \frac{\mathbf{r}_p}{0.0112} \tag{C}$$

In contrast to the previous examples, the rate of reaction is not expressed in equation form, but as a tabulation of experimental rates as a function of x and T. A simple first- or second-order expression would not correlate the data for this reaction. Instead, a Langmuir type of equation, based on adsorption of oxygen on the catalyst, was necessary. This equation was developed and tested in Example 9-2. The tabulation of rates is more convenient to use in this problem than the rate equation. The units of $\mathbf{r}_P$ in Eq. (C) should be lb mol reacted/(h)(lb catalyst), but numbers in these units are numerically equivalent to g mol/(h)(g catalyst), so that the data in Table 13-4 can be used directly.

The method of solution is the same as in Example 13-5.

Table 13-5 Conversion and temperature in an SO_2 reactor calculated by the one-dimensional model

Conversion	t, °C	Catalyst-bed depth, ft
0	364	0
0.05	365	0.029
0.10	365	0.065
0.15	362	0.112
0.18	357	0.149
0.21	351	0.020
0.24	(300)	(0.5)

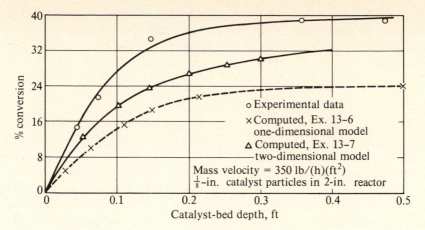

Figure 13-12 Comparison of calculated and experimental conversion values in an SO_2 reactor.

Equation (C) with $\mathbf{r}_p$ data from Table 13-4 is an expression of the form $dx/dz = f(x, T)$. Likewise, Eq. (B) is of the form $dT/dz = f(x, T)$. Hence, the numerical solution can again be carried out by the Runge-Kutta method. There are two dependent variables so that the method and equations for the calculations are the same as those used for Example 4-7. The calculations are started at the feed, $z = 0$, $x = 0$, $t = 364°C$. Increments, Δz, of catalyst bed depth are chosen and x and T at the end of each increment obtained from Eqs. (C) and (B).

The results obtained with $\Delta z = 0.001$ ft are summarized in Table 13-5 and shown in Fig. 13-12. It is apparent that the mean computed temperature never rises appreciably above the entering value of 364°C. Indeed, when a bed depth of about 0.2 ft is reached, the temperature is so low that the rate is no longer high enough to give a significant increase in conversion.

In the experimental reactor used by Schuler et al.† the low wall temperature results in severe radial temperature gradients. Hence the mean bulk temperature may be low, but the temperatures near the center of the bed may be high enough to cause a significant amount of reaction. This is a major factor in the large difference between the experimental conversion curve and the results computed in this example, both shown in Fig. 13-12. A critical quantity in the simplified design method applied to this problem is the wall heat-transfer coefficient. Small changes in h_w can cause large differences in the conversion. In the next section this example is repeated, this time with radial temperature and concentration gradients taken into account.

The one-dimensional model is particularly useful for investigating complex reactor systems. For example, Lucas and Gelbin‡ have used the approach to design multistage reactors for ammonia synthesis. Design calculations which

---

† R. W. Schuler, V. P. Stallings, and J. M. Smith, *Chem. Eng. Progr., Symp. Ser. 4,* **48**, 19 (1952).
‡ K. Lucas and D. Gelbin, *Brit. Chem. Eng.,* **7**, 336 (1962).

included radial gradients would be time consuming, even with large digital computers. With the one-dimensional method, approximate results and the effects of changes in operating conditions can be evaluated for a series of reactors with a modest amount of computer time.

13-6 The Two-Dimensional Model

In this section the design procedure is made more accurate by taking into account radial variations in temperature and concentrations. A thorough treatment would consider the radial distribution of velocity, would account for radial concentration and temperature gradients by using Peclet numbers which themselves varied with radial position, and would allow for axial dispersion of mass and energy. Such sets of elliptical partial differential equations are difficult to solve numerically since they constitute boundary-value problems. Further, transport data adequate to allow for radial variation in velocity and the Peclet numbers are not available. Also, the boundary conditions at the entrance to the bed are complex when the temperature varies radially. Young and Finlayson† have presented the solution to this problem for the case of uniform velocity and Peclet numbers in the radial direction. For these conditions the conservation expressions are Eqs. (13-2) and (13-19). Their results indicated that, for nonisothermal nonadiabatic conditions, axial dispersion could be important even for long catalyst beds. The numerical methods required for solution of Eqs. (13-2) and (13-19) are rather involved and will not be discussed here, except to note that the orthogonal-collocation technique‡ has proven to be an efficient solution method.

To illustrate the two-dimensional model we will neglect axial dispersion of mass and energy. This means omitting the terms involving $\partial^2/\partial z^2$ in Eqs. (13-2) and (13-19). If Eq. (13-2) is expressed in terms of conversion, the two conservation expressions become (G = mass velocity)

$$\frac{\partial x}{\partial z} - \frac{d_p}{\text{Pe}_r}\left(\frac{1}{r}\frac{\partial x}{\partial r} + \frac{\partial^2 x}{\partial r^2}\right) - \frac{\mathbf{r}_p \rho_B}{(G/\bar{M})y_0} = 0 \tag{13-27}$$

and

$$-Gc_p\frac{\partial T}{\partial z} + (k_e)_r\left(\frac{1}{r}\frac{\partial T}{\partial r} + \frac{\partial T}{\partial r^2}\right) - \mathbf{r}_p \rho_B \,\Delta H = 0 \tag{13-28}$$

where $(G/\bar{M})y_0 = u_0 C_0$ = molal feed rate of reactant per unit area of reactor
$\qquad\qquad \bar{M}$ = average molecular weight of the feed stream
$\qquad\qquad y_0$ = reactant mole fraction in the feed

The objective is to solve Eqs. (13-27) and (13-28) for the temperature and conversion at any point in the bed. The solution of these parabolic equations is an initial value problem where the only boundary conditions (in the axial direction) needed are at the entrance of the bed. Thus, we must know the feed temperature

† L. C. Young and B. A. Finlayson, *Ind. Eng. Chem.*, **12**, 412 (1973).

‡ B. A. Finlayson, *Chem. Eng. Sci.*, **26**, 1081 (1971); J. V. Villadsen and W. E. Stewart, *Chem. Eng. Sci.*, **22**, 1483 (1967).

and conversion profiles across the diameter of the reactor. Further boundary conditions applicable at any axial location are that the conversion is flat ($\partial x/\partial r = 0$) at both the centerline and at the wall of the tube. The temperature gradient at the centerline is zero, but the condition at the wall is determined by the heat-transfer characteristics. In the illustration that follows, the wall temperature is maintained constant by a boiling liquid (ethylene glycol) in a jacket surrounding the reactor. In other cases the wall temperature may vary with z. For example, if the fluid in the jacket or surroundings is at T_s and the heat-transfer coefficient between the inside wall surface and the surroundings is U, the proper boundary condition is

$$U(T_w - T_s) = -k_e\left(\frac{\partial T}{\partial r}\right)_w \tag{13-29}$$

where $(\partial T/\partial r)_w$ is the temperature gradient in the reactor at the wall.

The transport information needed, in addition to that for the one-dimensional model, are the radial values for D_e (or Pe) and k_e. Radial diffusivities have been presented in Sec. 13-3 (Fig. 13-5). Next, available information on effective thermal conductivities is summarized.

Effective thermal conductivities In a bed of solid particles through which fluid flows, heat can be transferred in the radial direction by several mechanisms: through the particles by conduction, through the fluid by conduction and convection, and by radiation. In writing Eqs. (13-19) and (13-28), all this heat transfer has been assumed to occur by conduction according to an *effective* thermal conductivity $(k_e)_r$. This is equivalent to supposing that the bed is replaced by a solid of thermal conductivity $(k_e)_r$. In view of the several mechanisms that are involved, $(k_e)_r$ is a property of the bed. Its value depends on a large number of variables, such as fluid flow rate, particle diameter, porosity, molecular thermal conductivity of the fluid and of the solid phases, and temperature level. Therefore, the most logical method of correlating data is to divide $(k_e)_r$ into separate contributions, each of which corresponds to a mechanism of heat transfer. Many correlations† have been developed on this basis and may be used to estimate $(k_e)_r$. For gaseous reaction mixtures the numerical range for $(k_e)_r$ is about 0.1 to 0.3 Btu/(h)(ft)(°R) [or 0.17 to 0.52 J/(m)(s)(K)] at temperatures of less than 500°F. At higher temperatures, radiation from particle to particle can lead to higher values.

Application of two-dimensional model A number of numerical methods are available‡ for solving Eqs. (13-27) and (13-28) with the appropriate boundary conditions. For simplicity, we will describe an unsophisticated, explicit method

---

† E. Singer and R. H. Wilhelm, *Chem. Eng. Prog.*, **46**, 343 (1950); W. B. Argo and J. M. Smith, *Chem. Eng. Prog.*, **49**, 443 (1953); John Beek, Design of Packed Catalytic Reactors, in "Advances in Chemical Engineering," vol. 3, p. 229, Academic Press, Inc., New York, 1962; S. Yagi and N. Wakao, *AIChE J.*, **5**, 71 (1960); A. P. deWasch and G. F. Froment, *Chem. Eng. Sci.*, **27**, 567 (1972).

‡ See Leon Lapidus, "Digital Computation for Chemical Engineers," McGraw-Hill Book Company, New York, 1960, and V. Hlavacek and J. Votruba, chap. 6, pp. 355–359, "Chemical Reactor Theory," ed. by Leon Lapidus and N. R. Amundson, Prentice-Hall, Englewood Cliffs, N.J., 1977.

based on writing the differential equations in difference form. The solution is carried out by starting at the reactor entrance and proceeding stepwise, first radially and then axially, to the desired catalyst bed depth.

Let n and L represent the number of increments in the radial and axial directions, respectively, and Δr and Δz their magnitude, so that

$$r = n\,\Delta r \tag{13-30}$$

$$z = L\,\Delta z \tag{13-31}$$

The temperature at any point in the bed can be represented by $T_{n,\,L}$, that is, the temperature at $r = n\,\Delta r$ and $z = L\,\Delta z$. Note that r is measured from the center of the bed and z from the feed entrance.

The first forward difference in temperature in the r direction can be written

$$\Delta_r T = T_{n+1,\,L} - T_{n,\,L} \tag{13-32}$$

Similarly, the first forward difference in the z direction may be written

$$\Delta_z T = T_{n,\,L+1} - T_{n,\,L} \tag{13-33}$$

The second central difference in the r direction is

$$\Delta_r^2 T = (T_{n+1,\,L} - T_{n,\,L}) - (T_{n,\,L} - T_{n-1,\,L}) \tag{13-34}$$

With these definitions, the difference form of Eq. (13-28) is

$$T_{n,\,L+1} = T_{n,\,L} + \frac{\Delta z}{(\Delta r)^2}\frac{(k_e)_r}{Gc_p}$$
$$\times \left[\frac{1}{n}(T_{n+1,\,L} - T_{n,\,L}) + T_{n+1,\,L} - 2T_{n,\,L} + T_{n-1,\,L}\right] - \frac{\Delta H \bar{r}_P \rho_B\,\Delta z}{Gc_P} \tag{13-35}$$

Similarly, Eq. (13-27) may be written in difference form as

$$x_{n,\,L+1} = x_{n,\,L} + \frac{\Delta z}{(\Delta r)^2}\frac{d_P}{\text{Pe}_r}$$
$$\times \left[\frac{1}{n}(x_{n+1,\,L} - x_{n,\,L}) + x_{n+1,\,L} - 2x_{n,\,L} + x_{n-1,\,L}\right] + \frac{\bar{r}_P \rho_B \bar{M}\,\Delta z}{Gy_0} \tag{13-36}$$

Provided the magnitude of the reaction terms involving r_P can be estimated, Eqs. (13-35) and (13-36) can be solved step by step to obtain the conversion. The first step is to compute values of T and x across the diameter, at $z = 1\,\Delta z$, or $L = 1$, from known values at $L = 0$. Then continue to the next longitudinal increment, $L = 2$, etc. The indeterminate form of the equations at $n = 0$ can be avoided by using the special expressions

$$T_{0,\,L+1} = T_{0,\,L} + \frac{2\,\Delta z}{(\Delta r)^2}\frac{(k_e)_r}{c_p G}(2T_{1,\,L} - 2T_{0,\,L}) - \frac{\Delta H \bar{r}_P \rho_B\,\Delta z}{Gc_p} \tag{13-37}$$

$$x_{0,\,L+1} = x_{0,\,L} + \frac{2\,\Delta z}{(\Delta r)^2}\frac{d_P}{\text{Pe}_r}(2x_{1,\,L} - 2x_{0,\,L}) + \frac{\bar{r}_P \rho_B \bar{M}\,\Delta z}{Gy_0} \tag{13-38}$$

derived from L'Hôpital's rule.

The effect of the reaction terms in Eqs. (13-35) and (13-36) is, for an exothermic reaction, to increase both the temperature and the conversion. Since the rate depends on the temperature and composition, and the average value for the increment L to $L + 1$ is not known until Eqs. (13-35) and (13-36) are solved, a trial-and-error procedure is indicated. The calculations are illustrated in Example 13-7, where the SO_2 reactor problem of Example 13-6 is recomputed with radial variations taken into account. In this example experimental data for the global rate are available. These data were obtained for the same size of catalyst pellets and for the same gas velocity as for the integral reactor. Hence external and internal transport effects on the global rate were the same for the laboratory conditions in which r_P was measured as for the integral reactor to be designed. In this way the calculations discussed in Chap. 12 for evaluating a global rate to fit the transport situation in the large-scale reactor were avoided. In most cases global rates have to be evaluated from intrinsic rates, effectiveness factors, and external transport coefficients.

Example 13-7 Recompute the conversion-vs.-bed-depth curve for the SO_2 reactor of Example 13-6 by the two-dimensional method, with the assumption that $(k_e)_r$, $(D_e)_r$, and G are constant. The temperature profile at the entrance to the reactor is given in the following table and also plotted in Fig. 13-13:

Feed temperature, °C	400.1	399.5	400.1	400.4	376.5	376.1	197.0
Radial position	0.023	0.233	0.474	0.534	0.797	0.819	1.000

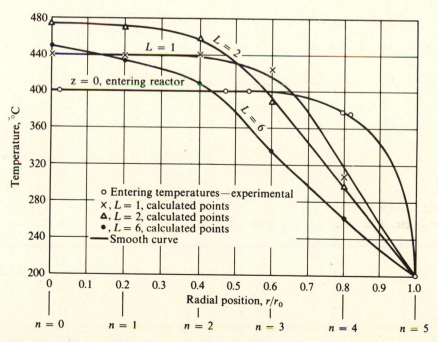

Figure 13-13 Calculated temperatures in an SO_2 reactor.

The reactants composition across the diameter may be assumed to be uniform.

SOLUTION The effective thermal conductivity calculated from the Argo† correlation is 0.216 Btu/(h)(ft)(°F). The Peclet number from Fig. 13-5 is 9.6 for the conditions of this example. The quantities needed in Eqs. (13-35) and (13-36) are

$$\frac{k_e}{c_p G} = \frac{0.216}{0.26(350)} = 0.00238 \text{ ft}$$

$$\frac{d_P}{Pe_r} = \frac{1/8}{12(9.6)} = 0.00109 \text{ ft}$$

Temperature and conversion equations It is convenient to divide the radius of the bed into five increments, so that

$$\Delta r = 0.2 r_0 = 0.2 \left(\frac{1.03}{12} \right) = 0.0172 \text{ ft}$$

If Δz is chosen to be 0.05 ft,

$$\frac{\Delta z}{(\Delta r)^2} = \frac{0.05}{0.0172^2} = 170 \text{ ft}^{-1}$$

Then the coefficients in Eqs. (13-35) and (13-36) are

$$\frac{k_e}{c_p G} \frac{\Delta z}{(\Delta r)^2} = 0.00238(170) = 0.404 \qquad \text{dimensionless}$$

$$\frac{d_P}{Pe_r} \frac{\Delta z}{(\Delta r)^2} = 0.00109(170) = 0.185 \qquad \text{dimensionless}$$

$$\frac{\Delta H \bar{r}_P \rho_B \Delta z}{G c_p} = \frac{-22{,}700(64)(0.05)\bar{r}_P}{350(0.26)} = -798\bar{r}_P \qquad °\text{C}$$

$$\frac{\bar{r}_P \rho_B \bar{M} \Delta z}{G y_0} = \frac{64(31.2)(0.05)\bar{r}_P}{350(0.065)} = 4.38\bar{r}_P \qquad \text{dimensionless}$$

Substituting these values in Eqs. (13-35) and (13-36) gives working expressions for calculating the temperature and conversion at a bed depth $L + 1$, from data at the previous bed depth L. Thus

$$T_{n,\,L+1} = T_{n,\,L} + 0.404$$

$$\times \left[\frac{1}{n}(T_{n+1,\,L} - T_{n,\,L}) + T_{n+1,\,L} - 2T_{n,\,L} + T_{n-1,\,L} \right] + 798\bar{r}_P \quad \text{(A)}$$

---

† W. B. Argo and J. M. Smith, *Chem. Eng. Progr.* **49**, 443 (1953).

$$x_{n, L+1} = x_{n, L} + 0.185$$

$$\times \left[\frac{1}{n} (x_{n+1, L} - x_{n, L}) + x_{n+1, L} - 2x_{n, L} + x_{n-1, L} \right] + 4.38\bar{r}_P \quad (B)$$

Calculations for the first bed-depth increment $(L = 1)$ The entering-temperature distribution is known (see Fig. 13-13), and the entering conversion will be zero at all radial positions. Starting at $n = 1$, the temperatures $T_{0, 0}$, $T_{1, 0}$, and $T_{2, 0}$, as read from Fig. 13-13, are all 400°C. Substituting these values into Eq. (A) gives $T_{1, 1}$ in terms of the average rate $\bar{r}_P$ over the increment of bed depth from 0 to $L = 1$ ($z = 0.05$ ft):

$$T_{1, 1} = T_{1, 0} + 0.404 \left[\frac{1}{1} (T_{2, 0} - T_{1, 0}) + T_{2, 0} - 2T_{1, 0} + T_{0, 0} \right] + 798\bar{r}_P$$

$$= 400 + 0.404(0) + 798\bar{r}_P = 400 + 798\bar{r}_P \quad (C)$$

Since $x_{0, 0}$, $x_{1, 0}$, and $x_{2, 0}$ are all 0, Eq. (B) gives, for the conversion at $n = 1$ and $L = 1$,

$$x_{1, 1} = 0 + 0.185(0) + 4.38\bar{r}_P = 4.38\bar{r}_P \quad (D)$$

Equations (C) and (D) and the rate data, Table 13-4 or Fig. 13-11, constitute three relationships of the unknown quantities $T_{1, 1}$, $x_{1, 1}$, and $\bar{r}_P$. One method of solution is the following four-step process:

1. Assume a value of $\bar{r}_P$, after obtaining $r_{1, 0}$ from Fig. 13-11.
2. Compute $T_{1, 1}$ and $x_{1, 1}$ from Eqs. (C) and (D).
3. Evaluate the rate $r_{1, 1}$ at the end of the increment from Fig. 13-11.
4. Average $r_{1, 1}$ and $r_{1, 0}$, and compare the result with the assumed $\bar{r}_P$. If agreement is not obtained, repeat the sequence with a revised value of $\bar{r}_P$.

Let us carry out these steps. At 400°C and zero conversion we have

$$r_{1, 0} = 0.055$$

We assume $\bar{r}_P = 0.051$. Then from Eq. (C) and Eq. (D),

$$T_{1, 1} = 400 + 798(0.051) = 441°C$$

$$x_{1, 1} = 4.38(0.051) = 0.223$$

From Fig. 13-11 at 441°C and 22.3% conversion, $r_{1, 1} = 0.046$. Hence

$$\bar{r}_P = \frac{0.055 + 0.046}{2} = 0.0505$$

This result is close to the assumed value of 0.051, so the calculated temperature and conversion at $n = 1$ and $L = 1$ may be taken as 441°C and 22.3%.

The same result would apply at $n = 0$ and 2, because the entering temperature is 400°C up to a radial position of $n = 3(r/r_0 = 0.6)$, as noted in

Fig. 13-13. At $n = 3$ the situation changes because $T_{4,0} = 376°C$. Let us make the stepwise calculations at this radial position. Starting with

$$\bar{r}_{3,0} = 0.055$$

we assume $\bar{r}_P = 0.046$. Then, from Eq. (A),

$$T_{3,1} = 400 + 0.404[\tfrac{1}{3}(376 - 400) + 376 - 2(400) + 400] + 798(0.046)$$

$$= 400 - 13 + 37 = 424°C$$

$$x_{3,1} = 0 + 4.38(0.046) = 0.201$$

From Fig. 13-11 at 424°C and 20.1% conversion, $r_{3,1} = 0.037$, and so

$$\bar{r}_P = \frac{0.055 + 0.037}{2} = 0.046$$

Continuing the calculations at $n = 4(r/r_0 = 0.8)$, where $T_{4,0} = 376°$ and $x_{4,0} = 0$, we have

$$\bar{r}_{4,0} = 0.029$$

and we assume $\bar{r}_P = 0.015$. Then

$$T_{4,1} = 376 + 0.404[\tfrac{1}{4}(197 - 376) + 197 - 2(376) + 400] + 798(0.015)$$

$$= 376 - 83 + 12 = 305°C$$

$$x_{1,1} = 0 + 4.38(0.015) = 0.066$$

From Fig. 13-11 at 305°C and $x = 0.066$ it is evident that the rate is close to zero; thus

$$\bar{r}_P = \frac{0.029 + 0}{2} = 0.015$$

which agrees with the assumed value. Hence at $n = 4$ and $L = 1$ the calculated temperature will be 305°C and the conversion 6.6%.

Since at $n = 5$ the wall is reached, the temperature remains 197°C. The rate is zero at this temperature, and so there will be no conversion due to reaction. Hence $T_{5,1} = 197°C$, and $x_{5,1} = 0$. At higher bed depths the conversion at the wall will not be zero, not because of reaction, but because of diffusion of the product SO_3 from the center of the tube.

The computations have now been made across the radius of the reactor at $L = 1(z = 0.05$ ft). The temperature results are indicated by the six points marked $\times$ in Fig. 13-13. The results computed by this stepwise procedure do not form a smooth curve at low bed depths. It is desirable before proceeding in the next increment to draw a smooth curve for this bed depth, as indicated in Fig. 13-13. The computed values and corresponding smooth conversion curve are shown in Fig. 13-14. The temperatures and conversions read from the smoothed curves, and to be used in the calculations at $L = 2$, are given in Table 13-6.

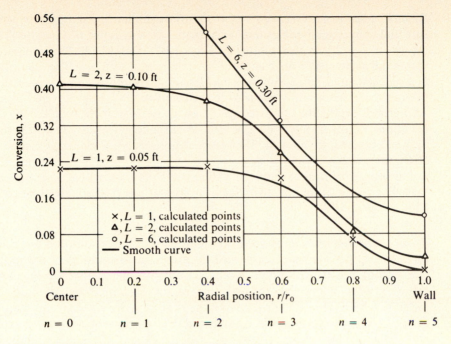

Figure 13-14 Calculated conversions in an SO_2 reactor.

Table 13-6 Temperatures and conversions for SO_2 oxidation by the two-dimensional model

Bed depth, ft	0(center)	0.2	0.4	0.6	0.8	1.0 (wall)
			Radial position			
			Temperature, °C			
$L = 0, z = 0$	400	400	400	400	376	197
$L = 1, z = 0.05$	441	441	437	418	315	197
$L = 2, z = 0.10$	475	471	458	390	298	197
$L = 3, z = 0.15$	496	488	443	378	285	197
$L = 4, z = 0.20$	504	476	437	360	278	197
$L = 5, z = 0.25$	470	466	415	350	269	197
$L = 6, z = 0.30$	451	435	412	334	265	197
			Conversion			
$L = 0, z = 0$	0	0	0	0	0	0
$L = 1, z = 0.05$	0.223	0.223	0.216	0.186	0.066	0
$L = 2, z = 0.10$	0.411	0.402	0.380	0.258	0.090	0.027
$L = 3, z = 0.15$	0.557	0.540	0.464	0.293	0.110	0.053
$L = 4, z = 0.20$	0.658	0.607	0.510	0.311	0.130	0.072
$L = 5, z = 0.25$	0.686	0.638	0.527	0.318	0.150	0.096
$L = 6, z = 0.30$	0.684	0.650	0.525	0.337	0.173	0.122

Results for successive increments $(L = 2$ to 6) By similar calculations temperature and conversion profiles may be obtained for successive bed lengths. The
results at $L = 2$ are also plotted in Figs. 13-13 and 13-14. The computed
values fall more nearly on a smooth curve than those at $L = 1$. The quantities
at $L = 2$ shown in Table 13-6 were read from the smooth curves. These, rather
than the computed points, are used in the calculations for the next bed depth.
At high bed depths the tabulated values are those directly computed from the
equations. A conversion of 68% is reached at the center when $z = 0.30$ ft, and
12% is obtained at the wall. The temperature reaches a maximum value of
504°C at $z = 0.20$ ft and then decreases at higher bed depths because the
radial transfer of heat toward the wall exceeds the heat evolved due to reaction. The temperature and conversion profiles at $z = 0.30$ are also plotted in
Figs. 13-13 and 13-14.

Mean conversion and temperature The bulk mean conversions and temperatures at any bed depth are obtained by graphical integration of the radial
temperature and conversion profiles. The bulk mean temperature is the value
resulting when the stream through the reactor is completely mixed in the
radial direction. Hence the product of the heat capacity and the temperature
at each radial position should be averaged. For an element dr the heat capacity of the flowing stream will be $G(2\pi r \, dr)c_p$. Thus the bulk mean temperature is given by the equation

$$\bar{T}_b = \frac{\int_0^{r_0} G(2\pi r \, dr)c_p T}{\pi G r_0^2 \bar{c}_p} = \frac{2\int_0^{r_0} T c_p r \, dr}{r_0^2 \bar{c}_p}$$

If the variable $n = r/r_0$ is substituted for r, this expression becomes

$$\bar{T}_b = \frac{2}{\bar{c}_p}\int_0^1 T c_p n \, dn \qquad (E)$$

In Eq. (E) c_p is the specific heat at the temperature T and $\bar{c}_p$ is for the bulk
mean temperature $\bar{T}_b$. Equation (E) can be integrated by plotting the product
$T c_p n$ vs. n and evaluating the area under the curve.

Similarly, the bulk mean conversion $\bar{x}_b$ corresponds to complete radial
mixing of the flow through the reactor. The moles of SO_2 converted in an
element of thickness dr are $x(G/\bar{M})y_0 2\pi r \, dr$, where $G/\bar{M}$ represents the total
moles per unit area entering the reactor and y_0 is the mole fraction SO_2 in the
feed.

Integrating over all radial elements gives

$$\frac{\pi G r_0^2 y_0 \bar{x}_b}{\bar{M}} = \int_0^{r_0} x \frac{G}{\bar{M}} y_0 2\pi r \, dr$$

$$\bar{x}_b = \frac{2\int_0^{r_0} xr \, dr}{r_0^2}$$

Table 13-7 Data for calculation of mean conversion at $L = 6$ ($z = 0.30$ ft)

n	x	xn
0	0.684	0
0.2	0.650	0.130
0.4	0.525	0.210
0.5	0.422	0.211
0.6	0.337	0.202
0.8	0.173	0.139
1.0	0.122	0.1222

If r/r_0 is replaced by n, then†

$$\bar{x}_b = 2\int_0^1 xn \, dn \tag{F}$$

From the data in Table 13-6, graphs can be made of $Tc_p n$ and xn vs. n to represent the values of the integrals in Eqs. (E) and (F). Actually, the mean conversion is the quantity of interest. Table 13-7 shows xn at a bed depth of $z = 0.30$ ft. These data are plotted in Fig. 13-15, and the area is

$$\int_0^1 xn \, dx = 0.150$$

† The velocity does not appear in Eqs. (E) and (F) because it is assumed that G is constant across the diameter of the reactor.

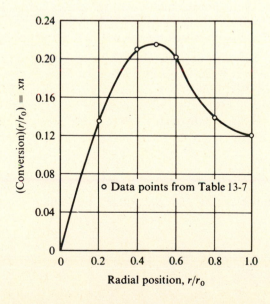

Figure 13-15 Graph for obtaining mean conversion.

Then, from Eq. (F),

$$\bar{x}_b = 2(0.150) = 0.300 \qquad \text{or } 30\% \text{ conversion}$$

Similar evaluations of the mean conversion have been made at the other bed depths. The results are plotted in Fig. 13-12, which also shows the measured conversions and the conversions calculated by the simplified method of Example 13-6. The two-dimensional method has resulted in better agreement than the one-dimensional model, but the computed conversions are still less than the experimental results. Young and Finlayson[†] have suggested that it is not valid to apply Eqs. (13-27) and (13-28) to the experimental conditions of this example. They have predicted conversions that agree well with Fig. 13-12 with a model that includes axial dispersion of heat and mass. Also, Ahmed and Fahien[‡] have obtained good agreement neglecting axial dispersion but with $(D_e)_r$ and $(k_e)_r$ that vary radially.

Some of the deviations in Fig. 13-12 may be due to uncertainties in the radial diffusivity and thermal conductivity. The predicted results are based upon three independent sets of experimental data: global rates of reaction (Fig. 13-11), $(D_e)_r$, and $(k_e)_r$. The effective thermal conductivity and global rate, particularly, must be accurately known if any model is to predict reliable results.

The temperatures shown in Fig. 13-8 as solid curves represent experimental data for the conditions of this example. The computed values for r/r_0 of 0.0 and 0.6 have been taken from Table 13-6 and plotted as dashed lines on the figure for comparison. Referring to the centerline ($r/r_0 = 0$), we see that the computed results are about 10 to 20°C below the experimental values, although the location of the hot spot is predicted accurately. The comparison at $r/r_0 = 0.6$ is not as good.

When more than one reaction occurs the calculation procedures are similar to those illustrated in Example 13-7. A difference equation is written for each component and these equations are solved simultaneously with the difference equation for the conservation of energy. Froment[§] and Carberry and White[¶] have used one- and two-dimensional models to predict conversion and temperatures in a fixed-bed reactor for the oxidation of o-xylene to phthalic anhydride, CO, and CO_2, with a V_2O_5 catalyst. The reaction scheme is

$$o\text{-xylene} \xrightarrow{\text{air}} \text{phthalic anhydride} \xrightarrow{\text{air}} CO$$
$$\Big\downarrow \text{air}$$
$$\quad\quad\quad\quad\quad CO_2$$

Calculations were made by Froment for a reactor containing a bundle of 2500 one-inch tubes packed with catalyst pellets and surrounded by molten salt to

---

† Loc. cit.

‡ M. Ahmed and R. W. Fahien, *Chem. Eng. Sci.*, **35**, 889, 897 (1980).

§ G. F. Froment, *Ind. Eng. Chem.*, **59**, 18 (1967).

¶ J. J. Carberry and D. White, *Ind. Eng. Chem.*, **61**(7), 27 (1969).

absorb the heat of reaction. The results showed that even with tubes of only 1 in. diameter, radial temperature gradients are severe for this extremely exothermic system. Carberry and White considered a reactor of 5 cm ID packed with 0'5-cm catalyst particles operating at a Reynolds number, $d_p G/\mu$, of 184. Here, too, axial and radial gradients of temperature were severe. It was demonstrated that the radial diffusivity had little effect on conversion to phthalic anhydride but that it was sensitive to the value of $(k_e)_r$.

A thorough study of a nonisothermal fixed-bed reactor for the reaction

$$C_6H_6 + 3H_2 \rightarrow C_6H_{12}$$

has been carried out by Otani.† An experimental differential reactor was used to obtain an equation of the Langmuir-Hinshelwood form [e.g., Eq. (9-32)] for the intrinsic rate, and then conversion and temperature measurements were made for cylindrical catalyst pellets (5 × 5 mm), for which the effectiveness factor was about 0.12. Then the *global* rate equation and estimated values of $(k_e)_r$ and $(D_e)_r$ were used to predict temperature and conversion data in the integral reactor tubes. The model was that described by Eqs. (13-27) and (13-28) and used in Example 13-7. The agreement between the predicted and experimental results was good in this case.

13-7 Dynamic Behavior

Thus far we have considered only steady-state operation of fixed-bed reactors. The response to variations in feed composition, temperature, or flow rate is also of significance. The dynamic response of the reactor to these involuntary disturbances determines the control instrumentation to be used. Also, if the system is to be put on closed-loop computer control, a knowledge of the response characteristics is vital for developing a control policy.

The solution of the steady-state form of the mass- and energy-conservation equations for fixed-bed reactors has been found to be complex. When transient conditions are considered the situation becomes more difficult. For the special case of isothermal operation only the mass-conservation equation is required. Then it is possible for many types of kinetics to obtain solutions for conversion at any point in the reactor in response to fluctuations in the feed. However, the nonisothermal case is most important because of the possibility of instabilities, and it is just this problem that is difficult to solve. For this reason effort has been directed more toward finding criteria for predicting when disturbances in the feed would grow and when they die out, instead of trying to solve the whole problem of conversion and temperature at any point in the reactor. We saw in Example 13-5 (Fig. 13-10) that sharp hot spots can develop when an exothermic reaction occurs in a cooled fixed-bed reactor. In that example conditions were such that the temperature rise was moderately large. Given other combinations of the heat of reaction, activation energy, rate equation, and heat-transfer rate to the surroundings, the temperature rise could be more pronounced. In such instances a positive

---

† S. Otani, "Some Practices in Petrochemical Process Development," presented at Sixty-fourth Annual Meeting of AIChE, New Orleans, March, 1969.

fluctuation in feed temperature or composition could cause so large a temperature rise in the reactor that reaction would be complete within a short section of the bed. Such behavior could be undesirable because of catalyst deactivation and reduced selectivity, but most important, the reactor would become uncontrollable. It is clear that a means for predicting the conditions at which this instability could occur would be helpful.

A number of criteria have been proposed for predicting instability,†—‡¶ and a review is available.¶ A simple proposal is that of Wilson.†† This criterion states that instability cannot occur if

$$\frac{E(T_{max} - T_{sur})}{R_g T_{max}^2} < 1 \tag{13-39}$$

where T_{max} and T_{sur} are the maximum and surroundings (cooling-medium) temperatures. For example, using the results of Example 13-5, $E/R_g = 2958°K^{-1}$, at the hot spot we have

$$\frac{E(T_{max} - T_{sur})}{R_g T_{max}^2} = \frac{2,958(527 - 427)}{(527)^2} = 1.06$$

According to this criterion, the reactor operation is at the point of instability, so that a perturbation in the feed could cause an uncontrollable situation.

The calculations of Froment‡‡ for a reactor for the oxidation of *o*-xylene illustrate the effect of a small fluctuation in feed temperature. The results showed that a 3°C rise in feed temperature (from 357 to 360°C) would lead to a continuous increase of the center temperature with reactor length, instead of the stable behavior of a rise and fall in temperature (for example as in Fig. 13-10). Such an unstable situation for this reaction results in a great loss in selectivity; the xylene is converted almost entirely to CO and CO_2 rather than to phthalic anhydride.

13-8 Variations of Fixed-Bed Reactors

Monolith catalyst beds The removal of oxidizable pollutants from air has stimulated the development of *monolithic* catalysts for use in a fixed-bed arrangement. The catalyst support is a partially or wholly continuous phase rather than consisting of individual particles. Many forms have been developed. Such structures are designed to have a high external surface area and to give a small pressure drop for the air flow. In one type, the support is prepared from crimped metal ribbon and arranged in mat-like sheets. In another form the catalyst base is a honeycomb structure. A third type consists of a bank of airfoil-shaped rods. The

---

† J. Beek, in T. B. Drew, J. W. Hoopes, Jr., and Theodore Vermeulen (eds.), "Advances in Chemical Engineering," vol. 3, Academic Press, Inc., New York, 1962.

‡ C. H. Barkelew, *Chem. Eng. Progr.*, *Symp. Ser.*, **55**, 37 (1959).

§ J. B. Agnew and O. E. Potter, *Trans. Inst. Chem. Eng.* **44**, 216 (1966).

¶ V. Hlavacek, *Ind. Eng. Chem.*, **62**(7), 9 (1970).

†† K. B. Wilson, *Trans. Inst. Chem. Eng.*, **24**, 77 (1946).

‡‡ G. F. Froment, *Ind. Eng. Chem.*, **59**, 18 (1967).

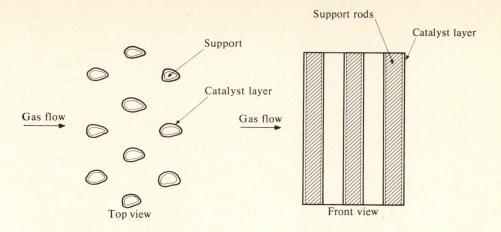

Figure 13-16 Cross-flow monolithic reactor.

air flows around the outer surface of the rods as indicated in Fig. 13-16. Such structures are made of either metal or ceramics.

The chief impetus for development of monolithic catalyst beds has been to reduce automobile exhaust emissions,[†] particularly unburned hydrocarbons and carbon monoxide. The oxidation catalyst, normally a platinum or platinum-palladium, can be electrolytically deposited on the support. Alternately, a thin layer of alumina impregnated with platinum may be deposited on the support. Since the reactions are strongly exothermic and have a sizable activation energy, the reaction rate is high. Hence, the reactive gases diffuse but a short distance into a porous catalyst before reaction is nearly complete. Hence, only a thin layer of catalytic material is usable. On the other hand, in an automobile exhaust line the catalyst bed must be mechanically strong to withstand mechanical and thermal shock and have a low pressure drop. These requirements are more closely satisfied in a monolithic arrangement than in the conventional bed of individual catalyst particles.

Since the catalyst layer is thin, intraparticle temperature and concentration gradients are not significant. Also, the high reaction rates imply that gas-to-solid mass- and heat-transfer resistances may be significant. The continuous nature of the solid phase introduces the possibility of heat transfer through the solid by conduction. Finally, the high temperature level and sharp changes in temperature in the direction of flow, suggest that heat transfer by radiation can be important. For these reasons models for predicting temperature and conversion in monolithic reactors are different from those in conventional fixed beds. Mathematical models based upon these concepts have been discussed.[‡–¶] When all the

---

† F. G. Dwyer, *Cat. Rev.*, **6**(2), 261 (1972).

‡ L. L. Hegedus, *AIChE J.*, **21**, 849 (1975).

§ S. T. Lee and R. Aris, Proceedings of the Fourth International Symposium on Chemical Reaction Engineering, Heidelberg, Section VI, 232 (1976).

¶ V. Hlavacek, Proceedings of the Fourth International Symposium on Chemical Reaction Engineering, Heidelberg, Section VI, 240 (1976).

significant processes are included the solution of the conservation equations and boundary conditions is difficult, even by numerical methods.

We illustrate some of the aspects of the design problem by considering a simple model that ignores radiation. Suppose the catalyst is continuous in the direction of air flow (as indicated in Fig. 13-17; Fig. 13-16 shows a cross-flow arrangement) and suppose that the air moves in plug flow. Suppose also that the oxygen is in excess so that the rate is first order in the concentration of pollutant. Then Eq. (3-1) for a section of reactor of length dz gives, for the gas phase,

$$Q\frac{dC}{dz} + k_m a_L(C - C_s) = 0 \tag{13-40}$$

where Q is the volumetric flow rate, k_m the mass transfer coefficient from gas-to-catalyst surface, and a_L the mass- or heat-transfer area per unit *length* of reactor. For the solid phase, mass conservation of pollutant requires

$$k_m a_L(C - C_s) = k\,\delta C_s \tag{13-41}$$

where k ($k = Ae^{-E/R_g T_s}$) is the reaction-rate constant per unit mass of catalyst (evaluated at T_s) and δ is the mass of catalyst per unit length of reactor.

The energy conservation equation is obtained by applying Eq. (5-1) to the gas and to the catalyst phase. Allowing for axial conduction of heat in the catalyst phase the result is, for the gas,

$$Q\rho c_p\frac{dT}{dz} + h a_L(T - T_s) = 0 \tag{13-42}$$

and, for the catalyst,

$$(\Delta x)a_L\frac{d^2 T_s}{dz^2} + h a_L(T - T_s) + (-\Delta H)k\,\delta C_s = 0 \tag{13-43}$$

where (Δx) is the thickness of the solid support (Fig. 13-17).

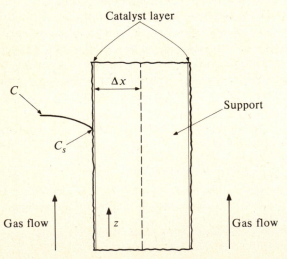

Figure 13-17 Axial section of monolithic catalyst.

If the heat loss from the end of the catalyst at $z = 0$ can be neglected, and the feed concentration and temperature are C_f and T_f, the boundary conditions are

$$C = C_f, T = T_f, \text{ and } dT_s/dz = 0 \qquad \text{at} \qquad z = 0 \qquad (13\text{-}44)$$

$$dT_s/dz = 0 \qquad \text{at} \qquad z = L \qquad (13\text{-}45)$$

The boundary-value problem defined by the four equations (13-40) to (13-43), with the boundary conditions, can be solved for C, C_s, T, and T_s as a function of bed length z. Since k is a nonlinear function of T_s, numerical methods are required.

If the temperature is constant $(T = T_s = \text{a constant})$ only Eqs. (13-40) and (13-41) are required. Now the solution is simple. If Eq. (13-41) is solved for C_s and the result substituted in Eq. (13-40), one obtains

$$\frac{dC}{dz} = -\left(\frac{1}{k\delta} + \frac{1}{k_m a_L}\right)^{-1} \frac{C}{Q}$$

This can be integrated directly to obtain the following solution for the conversion as a function of bed length.

$$1 - x = \frac{C}{C_f} = \exp\left[-\left(\frac{1}{k\delta} + \frac{1}{k_m a_L}\right)^{-1} \frac{z}{Q}\right] \qquad (13\text{-}46)$$

Autothermal reactors† When an exothermic reaction requires a high temperature [examples are the ammonia or methanol synthesis, water-gas shift reaction $(CO + H_2O = H_2 + CO_2)$] the heat of reaction can be used to preheat the feed. This can be accomplished in several ways in fixed-bed reactors. A separate external heat exchanger can be employed to transfer the heat of reaction, residing in the effluent, to the feed stream (Fig. 13-18a). Alternately, the exchanger may be an integral part of the reactor (Fig. 13-18b). Another possibility is to recycle part of the high-temperature effluent, Fig. 13-18c. Such heat-transfer and reaction systems are termed autothermal. Their advantage is that they can be essentially self-sufficient in energy, even though high temperature is required for the reaction to occur at a reasonable rate. An external source of heat is needed during the startup period while the system is coming to thermal equilibrium.

For the operating procedures shown in Fig. 13-18a or 13-18c, the reactor itself would operate adiabatically or with heat transfer to the surroundings. Then the conservation equations developed in Secs. 13-4 to 13-6 are applicable. When the heat exchanger is an integral part of the reactor (Fig. 13-18b) the equation for conservation will be different, but can be derived by applying Eq. (5-1). Conservation equations and their solution for various types of autothermal systems are available for both fixed-bed catalytic reactors‡§ and homogeneous reactors.¶

---

† C. van Heerden, *Ind. Eng. Chem.*, **45**, 1242 (1953).

‡ B. F. Baddour, P. L. T. Brian, B. A. Logeais, and J. P. Eymery, *Chem. Eng. Sci.*, **20**, 281 (1965). These authors used a one-domensional model to predict reactor temperature profiles for an ammonia-synthesis reactor of the type shown in Fig. 13-18a.

§ J. Caha, V. Hlavacek, and M. Kubicek, *Chem. Ing. Techn.*, **45**, 1308 (1973).

¶ T. G. Smith and J. T. Banchero, *J. Heat Transfer*, **95**, 145 (1973).

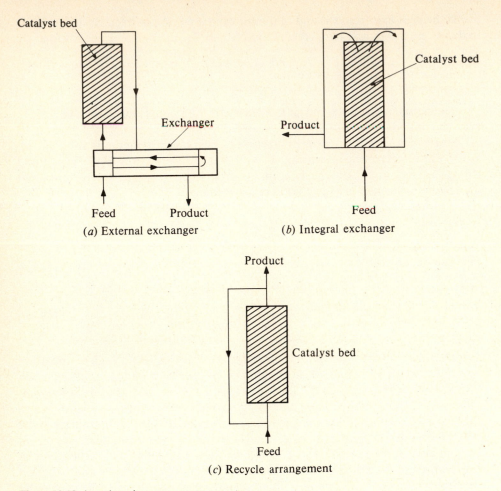

Figure 13-18 Autothermic reactor arrangements.

An interesting feature of autothermal systems is that they may exhibit more than one steady state, much like the phenomena found in Sec. 5-5 for stirred-tank reactors.

13-9 Importance of Transport Processes in Fixed-Bed Reactors

The procedure followed in this chapter for design of *fixed-bed* reactors has presumed that we know the *global* rate. Then only intrareactor transport processes in the radial and axial directions need be considered. As an aid in deciding on an appropriate model for the overall design, it is helpful to assign a relative importance to all the transport processes, including those that affect the global rate.

We divide the processes into three divisions, intrareactor (radial and axial dispersion of heat and mass), interphase (fluid–to–outer-surface of the catalyst particle), and intraparticle. It is impossible to summarize the importance of trans-

Table 13-8 Relative importance of transport effects in fixed-bed reactors

Mass transfer	Heat transfer	Importance of transport process
Intraparticle (effectiveness factor)	Intrareactor [radial direction $(k_e)_r$]	Most severe
Intrareactor [radial direction $(D_e)_r$]	Interphase (h)	Intermediate severity
Interphase (k_m) Intrareactor [axial direction $(D_e)_L$]	Intrareactor [(axial direction $(k_e)_L$] Intraparticle	Least severe

port effects in a way that is applicable to all reaction systems because of differences in such factors as bed length, fluid velocity, temperature level, catalyst particle size. However, for many cases, the relative importance follows the arrangement in Table 13-8. Some of the conclusions suggested in the table have been discussed separately: for example the relatively low importance of interphase mass transfer (Sec. 10-3) and the great importance of radial heat transfer (Sec. 13-6). The table is arranged so as to apply to a nonadiabatic, nonisothermal reactor for which the two-dimensional model (Sec. 13-6) is necessary. For isothermal reactors none of the heat-transfer effects are pertinent, and for adiabatic reactors radial transport is not involved.

One approach to modeling the reactor would be to include only the transport effects of great and intermediate severity. Then we would neglect the effects of intraparticle temperature differences, axial dispersion of heat and mass, and fluid-to-particle (interphase) concentration differences.

FLUIDIZED-BED REACTORS

In Chap. 10 we discussed the effect of increasing velocity (Fig. 10-6) on the movement of gas and particles in a fluidized bed and also heat-transfer rates between fluid and particles (Sec. 10-6). Now we want to consider how to predict the behavior of the fluidized bed as a catalytic reactor; that is, we want to predict the conversion in the effluent from a proposed model for the reactor.

The particles in a fluidized bed are so small that intraparticle concentration and temperature gradients are negligible. As was noted in Sec. 10-6, heat- and mass-transfer rates between fluid and particles are very high so that external temperature and concentration differences are negligible. Thus, the global reaction rate is equal to the intrinsic rate, evaluated at bulk values of the temperature and concentration. This means that the reactor design problem is essentially one of modeling the flow conditions in the reactor as a whole (interactor model).

Most fluidized-bed reactors operate in the bubbling regime as sketched in Fig. 10-6. Under these conditions the upward motion of the "gas bubbles" causes enough mixing in the dense phase that the temperature is nearly uniform in the entire reactor. This effect of the gas bubbles is favorable. However, there is little reaction within the bubbles (because of the low catalyst particle concentration). Hence, the bubbles are unfavorable in that they serve as channels for gas to bypass the catalyst and leave the reactor more or less unreacted. Models have been proposed† for predicting the conversion under these conditions. Such models should, in general, include the extent of reaction in the gas bubbles, in the dense phase, velocities and magnitudes of bubble and dense phases, and extent of mass transfer between the two phases. One of the most attractive proposals is the bubbling-gas model.‡ This approach actually includes a third region, a cloud of particles, surrounding the bubble and within which gas recirculates but does not mix rapidly with the gas in the dense phase. Quantitative treatment for this three-phase model is described in detail by Kunii and Levenspiel. Discussions of the nature of the cloud region and its relation to the bubble and dense phases are available.§¶ Also the model has been carefully compared with other models using experimental data.†† The model includes several parameters: apparent reaction rate constants in the bubble, in the cloud, and in the dense phase, and transport rate constants for the reactant from bubble gas to cloud and from bubble gas to emulsion phase. The useful feature of the model is that these parameters may be estimated solely from the diameter of the gas bubbles. Kunii and Levenspiel‡‡ suggest correlations for obtaining numerical values. One of the requirements for valid models is that they be able to predict a conversion less than that corresponding to complete mixing of the entire reactor contents. If the apparent rate constant in the bubbles is low, because of low particle concentration, and if a significant portion of the total gas flow is in such bubbles, the model successfully predicts these low conversions. Figure 13-19 shows how experimental conversions can be less than those for either an ideal stirred-tank or plug-flow reactor, at higher average gas velocities. The deviation between the fluidized bed and ideal stirred-tank conversions presumably increases with velocity because the extent of bypassing in the gas bubbles increases.

We will not give the quantitative development of the rather complex three-region model. The equations can be illustrated with a two-region ("two-phase") model as applied by Pavlica and Olson.§§ Also these same equations can be used for treating trickle-bed reactors later in this chapter.

An even simpler, "single-phase" model considers the entire reactor volume to

† J. F. Davidson and D. Harrison, "Fluidized Solids," Cambridge University Press, London, 1963; J. R. Grace, AIChE Symposium Series, **67**(116), 159 (1971).

‡ D. Kunii and O. Levenspiel, *Ind. Eng. Chem. Fundam.*, **7**, 446 (1968); *Ind. and Eng. Chem. Proc. Des. Dev.*, **7**, 481 (1968).

§ J. F. Davidson and D. Harrison, loc. cit.

¶ P. N. Rowe and B. A. Partridge, *Proc. Symp. on Interaction Between Fluids and Particles, Inst. Chem. Eng. (London)*, June 1962.

†† Colin Fryer and O. E. Potter, *AIChE J.*, **22**, 38 (1976).

‡‡ Loc. cit.

§§ R. T. Pavlica and J. H. Olson, *Ind. Eng. Chem.*, **62**(12), 45 (1970).

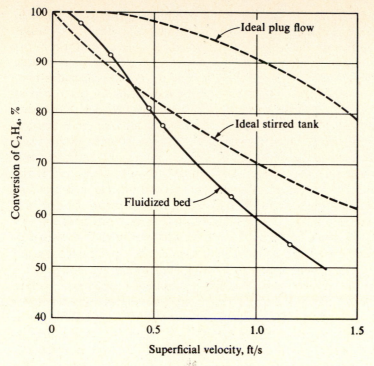

Figure 13-19 Hydrogenation of ethylene in a fluidized-bed reactor [*from W. K. Lewis, E. R. Gilliland, and W. Glass, AIChE J., 5, 419 (1959)*].

have a uniform concentration of catalyst particles. Deviations from plug-flow behavior is accounted for by introducing an axial dispersion term in the mass conservation equation. Then the results are the same in form as those for the isothermal, fixed-bed reactor. Equation (13-14) is the appropriate conservation equation where D_L is an axial dispersion coefficient applicable for a fluidized bed (see Prob. 13-14). The conversion for first-order kinetics is given by Eq. (6-47). A conversion less than that for a stirred tank cannot be obtained with this model.

13-10 Two-Phase Fluidized-Bed Model

Suppose an irreversible catalytic reaction occurs in a fluidized-bed reactor operating in the bubbling-gas regime, as shown in Fig. 13-20. Radial variations in concentrations within the two phases are neglected and isothermal operation is assumed. Suppose that the catalyst particle concentration within the bubbles is so low that reaction in the bubbles can be neglected. Also, assume that the bubbles move in plug flow up through the reactor. Then conservation of mass of reactant requires that the net flow rate of reactant into an element of reactor volume of height Δz be equal to the rate of mass transfer from bubble to dense phase. Application of Eq. (3-1) to the bubble phase gives

$$u_b \frac{dC_b}{dz} + k_m a_v (C_b - C_d) = 0 \tag{13-47}$$

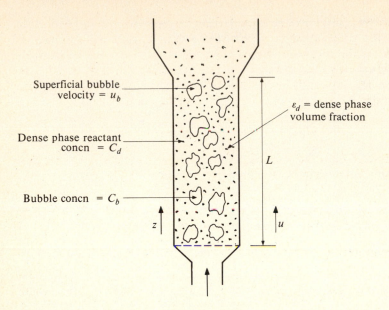

Superficial bubble velocity = u_b

Dense phase reactant concn = C_d

Bubble concn = C_b

ε_d = dense phase volume fraction

L

z

u

Figure 13-20 Two-phase model of fluidized-bed reactor.

where u_b = superficial velocity of the gas bubbles

      (u_b = gas flow rate divided by cross-sectional area of reactor)

   C_b, C_d = concentration of reactant in gas bubbles or in the dense phase

      k_m = mass-transfer coefficient between bubble and dense phases

      a_v = mass-transfer area between bubble and dense phases per unit volume of reactor

Suppose that mixing of the reactant gas in the dense phase can be accounted for with the dispersion model. Then the mass conservation equation for reactant in the dense phase will be similar to Eq. (13-47) but include a term for axial mixing and for reaction. Equation (3-1) under these conditions may be written

$$-u_d\frac{dC_d}{dz} + \varepsilon_d D_L\frac{d^2C_d}{dz^2} - \varepsilon_d\rho_d[kf(C_d)] + k_m a_v(C_b - C_d) = 0 \qquad (13\text{-}48)$$

where $kf(C)$ represents the reaction rate per unit mass of catalyst

      ρ_d is the density of catalyst particles in the dense phase

      u_d = superficial velocity of the gas in the dense phase, and

      ε_d = fraction of reactor cross section occupied by the dense phase (the holdup of the dense phase)

These equations with appropriate boundary conditions describe the concentrations in the two phases as a function of reactor height z. With known values for the parameters $\varepsilon_d, D_L, a_v, k_m$, and the reaction rate, they can be solved for the effluent concentrations C_b and C_d. These concentrations can be combined to give the conversion. However, the experimental evidence suggests that the net upward gas velocity in the dense phase is low and that mixing in this phase is very good. As an approximation, the process may be visualized as the feed stream forming gas bubbles as it enters the reactor and these bubbles flowing upward through the dense phase. This means that the dense region behaves as a well-mixed *batch* phase so that the first two terms in Eq. (13-48) disappear. With this simplification, and for a first-order reaction $[kf(C) = kC]$, Eq. (13-48) reduces to

$$\varepsilon_d \rho_d k C_d = k_m a_v (C_b - C_d) \tag{13-49}$$

Now we can solve Eq. (13-49) for C_d in terms of C_b and substitute this result in Eq. (13-47) for immediate integration. Doing this, Eq. (13-47) becomes

$$u_b \frac{dC_b}{dz} = - \left[\frac{1}{\varepsilon_d k \rho_d} + \frac{1}{k_m a_v} \right]^{-1} C_b \tag{13-50}$$

If the feed concentration is C_f (at $z = 0$), the integrated form of Eq. (13-50) is

$$1 - x = \frac{C}{C_f} = \exp \left[- \left(\frac{1}{\varepsilon_d k \rho_d} + \frac{1}{k_m a_v} \right)^{-1} \frac{z}{u_b} \right] \tag{13-51}$$

Note that this result is similar to Eq. (13-46) for a monolithic reactor. Both reactor models represent the same type of interaction between reaction and mass transport processes.

Example 13-8 The effect of bypassing in a bubbling fluidized bed is determined by the fraction $(1 - \varepsilon_d)$ of the reactor volume that consists of bubbles and by the relative values of the reaction rate and mass transfer rate. The quantities k_m, ε_d, u_b, and a_v all depend upon the bubble diameter. In a particular case suppose that these parameters have the following values, for a first-order reaction operating isothermally in the bubbling regime:

$$\rho_d = 0.01 \text{ g/cm}^3$$
$$(k_m a_v) = 0.60 \text{ s}^{-1}$$
$$k = 50 \text{ cm}^3/(\text{g})(\text{s })$$
$$u_b = \text{velocity of feed} = 10 \text{ cm/s}$$
$$\text{Reactor height, } z = L = 40 \text{ cm}$$
$$\varepsilon_d = 0.80 \text{ (that is, 20\% of the reactor volume is occupied}$$
$$\text{by gas bubbles and 80\% by the dense phase)}$$

A. Calculate the conversion in the effluent.

B. For comparison, calculate the conversion for plug-flow and stirred-tank reactors operating with the same apparent bubble residence time.

SOLUTION

A. The terms in parentheses in Eq. (13-51) may be regarded as time constants that measure the time required for reaction, $1/\varepsilon_d k \rho_d$, and for mass transport between bubble and dense phases, $1/k_m a_v$.

For $\varepsilon_d = 0.80$ Eq. (13-51) gives

$$x = 1 - \exp\left[-\left(\frac{1}{0.80(50)(0.01)} + \frac{1}{0.60}\right)^{-1}\frac{40}{10}\right]$$

$$x = 1 - \exp\left[-4(2.50 + 1.67)^{-1}\right]$$

$$x = 1 - 0.38 = 0.62\ (62\%)$$

The time constant for mass transfer (1.67 s) is only a little less than the time constant for reaction (2.50 s). In this case mass transport from gas bubble to dense phase has an important effect on reactor performance.

B. Equations (4-18) and (4-17) give the conversion for first-order reactions in plug-flow and in stirred-tank reactors. In these equations the residence time is V/Q. Hence, for the same apparent bubble residence time as in the fluidized bed,

$$\frac{V}{Q} = \frac{L}{u_b} = \frac{40}{10} = 4\ \text{s}$$

Also, the rate constant per unit volume in these equations is equal to $k\rho_d = 0.50\ \text{s}^{-1}$. Then Eqs. (4-18) and (4-17) give

$$\text{Plug flow reactor } x = 1 - [\exp - 4(0.5)]$$

$$= 1 - 0.14 = 0.84\ (84\%)$$

$$\text{Stirred tank } x = \frac{0.50(4)}{1 + 0.50(4)} = 0.667\ (67\%)$$

For these conditions we see that bypassing of reactant in the gas bubbles has reduced the conversion below that expected if the whole reactor contents were well mixed (stirred-tank performance).

We have been concerned with but one reaction. If multiple reactions occur, the mass-transfer term in Eq. (13-51) can affect selectivity. The situation is the same as discussed in Sec. 10-5 for the effect of external mass transport on selectivity. If the reactions are consecutive the mass-transfer term $1/k_m a_v$ in Eq. (13-51) has an adverse effect on selectivity. If the reactions are of the parallel type selectivity is unaffected.

13-11 Operating Characteristics

Fluidized beds are particularly suitable when frequent catalyst regeneration is required, or for reactions with a very high heat effect. Often the reactors are vessels of large diameter (10 to 30 ft is not unusual for catalytic cracking units in the petroleum industry). A typical system is shown in Fig. 13-21. The movable catalyst permits continuous regeneration in place. Part of the catalyst is continuously withdrawn from the reactor in line A and flows into the regenerator. The regenerator shown in the figure is another fluidized bed, from which reactivated catalyst is returned to the reactor through line B. It is not necessary to carry out the regener-

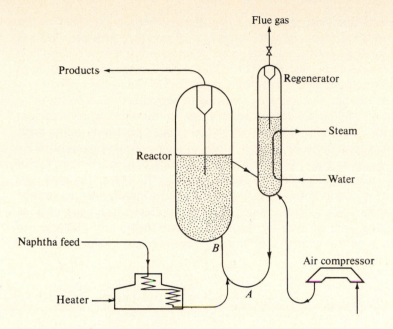

Figure 13-21 Flow diagram for a fluid hydroformer, a fluidized-bed reactor-generator combination (*by permission from Esso Standard Oil Company, Baton-Rouge, La.*).

ation in a fluidized bed, as catalyst could be withdrawn continuously, and reactivated catalyst returned with the regeneration accomplished by any procedure. However, the process is most economical if both reaction and regeneration are carried out in fluidized beds by means of the integral setup illustrated in Fig. 13-21.

An important feature of the fluidized-bed reactor is that it operates at a nearly constant temperature and hence is easy to control. There is no opportunity for hot spots to develop, as in the case of the fixed-bed unit. The fluidized bed does not possess the flexibility of the fixed bed for adding or removing heat. A diluent can be added to control the temperature level, but this may not be desirable for other reasons (requires separation after the reactor, lowers the rate of reaction, increases the size of the equipment). A heat-transfer fluid can be circulated through a jacket around the reactor, but if the reactor is large in diameter, the energy exchange by this method is limited.†

Catalyst loss due to carryover with the gas stream from the reactor and regenerator may be an important problem. Attrition of particles decreases their size to a point where they are no longer fluidized, but move with the gas stream. It has been customary to recover most of these catalyst fines by cyclone separators and electrical precipitation equipment placed in the effluent lines from reactor and regenerator.

---

† A heat-transfer fluid also may be circulated through tubes placed within the fluidized bed.

Deterioration of lines and vessels due to the abrasive action of the sharp, solid particles in the fluidized cracking process has caused concern. This problem has been particularly severe in the small-diameter transfer lines, where the particle velocity is high. These and other matters relating to the commercial operation of fluidized catalytic cracking plants have been discussed in the literature.†

It has been mentioned that catalytic cracking of petroleum gas-oil is the most frequent application of fluidized-reactor technology. In this area the development of synthetic zeolite (molecular-sieve) catalysts has greatly increased the intrinsic rates of the cracking reactions. As a result, much of the reaction may occur in the transfer line (line *AB* is Fig. 13-20) *before* the hydrocarbon enters the reactor proper. Such "transfer-line" reactors are moving-bed units with a very high void fraction. The system consists of two moving phases. Reactor design models can be developed by writing conservation equations for each phase much like was done for the two-phase model of fluid-bed reactors [Eqs. (13-47) and (13-48)]. An important factor in such reactors is the slip velocity, or relative velocity between particles and gas.

SLURRY REACTORS

Slurry reactors are similar to fluidized-bed reactors in that a gas is passed through a reactor containing solid catalyst particles suspended in a fluid. In slurries the catalyst is suspended in a liquid; in fluidized-beds the suspending fluid is the reacting gas itself. The advantages of slurry reactors over fixed beds are similar to those of fluidized beds: more uniform temperature, better temperature control for highly exothermic reactions, and low intraparticle diffusion resistance. For very active catalysts this last factor means that the global rate can be much higher than for fixed-bed reactors. In order to eliminate the retardation of the global rate due to intraparticle diffusion the particles must be very small. This is because diffusivities in *liquid*-filled pores are relatively low, of the order of 10^{-5} cm^2/s in comparison with 10^{-2} cm^2/s for gases. For example, in studies of adsorption of benzene in aqueous slurries of activated carbon‡ it was found that intraparticle diffusion reduced the global rate for carbon particles as small as 160 microns. For slower intrinsic rates (the rate for physical adsorption of benzene was very fast), intraparticle diffusion would be less important.

The most serious disadvantage of slurry reactors is the difficulty in retaining the catalyst in the vessel. Screens and other devices placed in the outlet lines tend to clog or otherwise be unreliable. In some instances,§ the catalyst can be so active that it need not be retained in the reactor. If the production rate of product per unit mass of catalyst is very high, there may be no need to separate the small amount of finely divided catalyst from the effluent. The concentration of catalyst particles in the product is too small to be troublesome.

---

† E. V. Murphree, et al., *Trans. AIChE*, **41**, 19 (1945); A. L. Conn, et al., *Chem. Engr. Progr.*, **46**, 176 (1950).

‡ Takehiko Furusawa and J. M. Smith, *Ind. Eng. Chem. Fundam.*, **12**, 197 (1973).

§ For example, ethylene polymerization.

As pointed out in Chap. 10, when one reactant is a gas and a second is a liquid, reactions catalyzed by a solid must be carried out in a three-phase system. Then, the meaningful comparison of slurry reactors is not with fluid-solid fixed-bed units but with trickle-bed reactors. The latter is the three-phase version of the fixed-bed arrangement. This comparison is taken up later in the chapter after the design of trickle-bed units has been discussed.

13-12 Slurry Reactor Models

In Chap. 10 global rate equations were developed (for the overall transfer of reactant from gas bubble to catalyst surface) in terms of the individual mass-transfer and chemical-reaction steps [Eqs. (10-38) and (10-39)]. The purpose there was to show how the global rate r_v (per unit volume of bubble-free slurry) was affected by such variables as the gas-liquid interfacial area a_g, the liquid-solid interfacial area a_c, and the various mass-transfer coefficients, and the intrinsic rate for the chemical step. Now the objective is to utilize the results for the global rate to predict the performance of the reactor as a whole. For this purpose we need a model to describe the flow and mixing characteristics of the gas, liquid, and solid phases. With three phases there are a large number of possible operating conditions. We first list some of the possibilities and then, as an illustration, quantitatively examine one common form of slurry system.

With respect to reactor modeling we could propose that both gas and liquid could be in plug flow or could be well mixed, while the catalyst phase remains in the tank-type reactor. Alternatively, the liquid phase could remain in the reactor, as in batch operation with continuous flow of gas in and out of the reactor. Regardless of the arrangement, the design procedure is to write mass conservation equations, for the reactant, according to Eq. (3-1), for each reaction.† If the reactant exists in both gas and liquid phases, separate conservation equations may be necessary for each phase. If the global rate is used, these equations will be in terms of bulk concentrations. Hence, the solution gives the relation between extent of reaction and reactor volume, analogous to the results for single-phase reaction systems developed in Chap. 4.

As an example, consider a continuous slurry reactor in which the reactants are both in the gas phase, and the liquid is inert.‡ The purpose of the liquid is simply to suspend the catalyst particles. The overall catalytic reaction is

$$A(g) + B(g) \rightarrow C(g)$$

For a reactor model suppose that the slurry is well agitated so that concentrations in the liquid, both of dissolved reactants and of catalyst particles, are uniform. Suppose also that the gas bubbles are discrete§ and rise in plug flow through the slurry as indicated in Fig. 13-22. We may write a mass conservation equation for

---

† As mentioned in Chaps. 5 and 10, for tank-type reactors mixing results in a uniform tempera-ture unless internals, such as cooling coils, interfere with mixing.

‡ Examples are the hydrogenation of gases with a slurry of nickel particles as catalyst, or oxidation of gases such as SO_2 or H_2S in aqueous slurries of activated carbon particles.

§ That is, they rise individually through the slurry rather than coalesce.

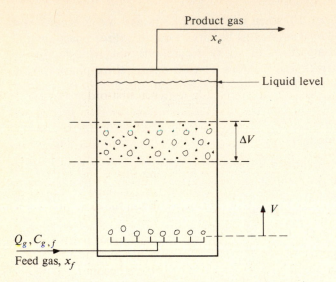

Feed gas, x_f

Figure 13-22 Slurry reactor (batch liquid, continuous-flow gas).

an element of reactor volume, dV, extending across the entire cross section of the slurry. This is because the liquid-phase concentrations are uniform and because we suppose that the gas-flow rate, Q_g is distributed uniformly. For the limiting reactant, at steady state, Eq. (3-1), applied to the gas phase is identical with Eq. (3-14) for a plug-flow reactor. If $\mathbf{r}_v$ is a positive value (rate of disappearance of reactant)

$$-d(Q_g C_g) - \mathbf{r}_v \, dV = 0$$

As noted in Chap. 3, if the conservation equation is to be applicable for a reaction with a change in moles (i.e., variable Q_g), it is advantageous to express $-d(Q_g C_g)$ as $F_A \, dx$. Then we may write

$$\frac{V}{F_A} = \int_{x_f}^{x_e} \frac{dx}{\mathbf{r}_v} \tag{13-52}$$

This is the same as Eq. (3-18) except that V is the volume of bubble-free slurry in the reactor, and $\mathbf{r}_v$ is the global rate of reaction per unit volume of liquid, as defined in Chap. 10. If the reaction is first-order and irreversible, $\mathbf{r}_v$ is given by Eq. (10-38).

Equation (13-52) may be used to calculate the volume of slurry liquid required to obtain a conversion x for a molal feed rate F_A of reactant. For a first-order reaction, and if Henry's law is valid, Eq. (10-39) displays how the various rate constants k_g, k_c, and k, and areas a_g and a_c, influence the process.

If there is no change in moles in the gas phase as a result of reaction, Eq. (13-52) can be easily integrated. Noting that $C_g = C_{gf}(1 - x)$ and $F_A = Q_g C_{gf}$, and using Eq. (10-38) for $\mathbf{r}_v$, Eq. (13-52) becomes, for $x_f = 0$,

$$\frac{V}{Q_g C_{gf}} = \int_0^{x_e} \frac{dx}{k_0 a_c C_{gf}(1 - x)}$$

or

$$\frac{V}{Q_g} = \frac{-1}{k_0 a_c} \ln (1 - x_e) \tag{13-53}$$

which is similar to Eq. (4-18). If the reaction had not been first-order, the simple relation between r_v and the various rate constants, as given by Eqs. (10-38) and (10-39) would not have been valid. Then the global rate would not be a simple function of gas-phase concentration like Eq. (10-38). Numerical methods would have been needed to eliminate the liquid and surface concentrations, C_L and C_s, and express r_v in terms of C_g. The procedure for doing this is the same as illustrated in Example 10-2 for two-phase systems. Once r_v is obtained as a function of C_g, Eq. (13-52) could be integrated numerically to obtain the exit conversion.

Only the gas phase has been considered in our application because both reactants were in the gas phase. If the slurry liquid were reacting, as in hydrogenation of oils, then mass conservation expressions would need to be written for liquid and gas phases. For a batch liquid system the problem becomes a dynamic one because the concentrations in the liquid will change with time. If the process is continuous, and at steady state with respect to the liquid, there would be a combination of plug flow of gas and stirred-tank behavior for the liquid. The method of solution is the same, in principle, for these cases: conservation equations are written for gas and liquid phases and solved for final and/or effluent values (see Prob. 13-20). Quantitative treatments for these forms of slurry reactors are available.† ‡

Examples 13-9 and 13-10 illustrate the calculation of conversion in slurry reactors.

Example 13-9 Ethylene is to be hydrogenated by bubbling mixtures of H_2 and C_2H_4 through a slurry of Raney nickel catalyst particles suspended in toluene. The gas bubbles are formed at the bottom of a tubular reactor and rise in plug flow through the slurry. The slurry is well mixed, so that its properties are the same throughout the tube. A large concentration of small catalyst particles will be used. The temperature and pressure are to be 50°C and 10 atm. At these conditions the overall rate has been shown to be determined by the rate of diffusion of hydrogen from the bubble interface to the bulk liquid.§ This means that r_v is first order in the gas-phase concentration of hydrogen, regardless of the intrinsic kinetics at the catalyst site.

Estimate the volume of bubble-free slurry required to obtain a conversion of 30% for a hydrogen feed rate of 100 ft³/min (at 60°F and 1 atm). By a light-transmission technique, Calderbank measured gas-liquid interfacial areas of 0.94 to 2.09 cm²/cm³ for bubble sizes likely to be encountered in this system. Suppose for this illustration $a_g = 1.0$ cm²/cm³ of bubble-free slurry.

† Hiroo Niiyama and J. M. Smith, *AIChE J.*, **22**, 961 (1976).

‡ Takehiko Furusawa and J. M. Smith, *Ind. Eng. Chem. Fundam.*, **12**, 360 (1973).

§ P. H. Calderbank, F. Evans, R. Farley, G. Jepson, and A. Poll, "Catalysis in Practice," *Symp. Inst. Chem. Engrs. (London)*, 1963.

The Henry's law constant for hydrogen in toluene at 50°C is 9.4 (g mol/cm^3)/(g mol/cm^3), and its diffusivity is 1.1×10^{-4} cm^2/s. The density and viscosity of toluene at 50°C are 0.85 g/cm^3 and 0.45 centipoises, respectively. Equimolal feed rates of ethylene and hydrogen will be used.

SOLUTION For the reaction

$$C_2H_4(g) + H_2(g) \rightarrow C_2H_6(g)$$

the number of moles of each component at a conversion level x is

$$H_2 = F(1 - x)$$
$$C_2H_4 = F(1 - x)$$
$$C_2H_6 = Fx$$
$$\text{Total moles} = F(2 - x)$$

where F is the molal feed rate of H_2 or C_2H_4. Then the concentration of H_2 in the gas bubbles is

$$(C_{H_2})_g = \frac{p_t}{R_g T} y_{H_2} = \frac{p_t}{R_g T} \frac{1 - x}{2 - x} \tag{A}$$

Equation (13-52) is applicable to this reactor with r_v given by Eq. (10-38). Substituting Eqs. (A) for C_g in Eq. (10-38), Eq. (13-52) becomes

$$\frac{V}{F} = \frac{R_g T}{k_o a_c p_t} \int_0^{x_e} \frac{2 - x}{1 - x} \, dx \tag{B}$$

Integrating from zero to the exit conversion gives

$$\frac{V}{F} = \frac{1}{k_o a_c} \frac{R_g T}{p_t} [x_e - \ln(1 - x_e)] \tag{C}$$

The overall rate constant k_o would, in general, be a function of several rate parameters. Since it is known that only the resistance to diffusion of hydrogen from the bubble interface is significant, Eq. (10-39) reduces to

$$\frac{1}{k_o} = \frac{a_c}{a_g} \frac{H}{k_L}$$

or

$$k_o a_c = \frac{a_g k_L}{H} \tag{D}$$

Hence Eq. (C) may be written

$$\frac{V}{F} = \frac{H}{a_g k_L} \frac{R_g T}{p_t} [x_e - \ln(1 - x_e)] \tag{E}$$

Everything in Eq. (E) is known except the mass-transfer coefficient for hydrogen in the liquid. This may be estimated from Eq. (10-46), since this correlation was based on data for gas bubbles rising through a liquid phase. Thus

$$k_L \left[\frac{0.45 \times 10^{-2}}{0.85(1.1 \times 10^{-4})} \right]^{2/3} = 0.31 \left[\frac{(0.85 - 0.8 \times 10^{-3})(0.45 \times 10^{-2})(32.2)}{0.85^2} \right]^{1/3}$$

$$k_L = \frac{0.31}{13.2}(0.55) = 0.013 \text{ cm/s}$$

Substituting this result for k_L and the other numerical value in Eq. (E) gives

$$\frac{V}{F} = \frac{9.4}{1.0(0.013)} \frac{82(273 + 50)}{10} [0.3 - \ln(1 - 0.3)]$$

$$= 1.25 \times 10^6 \text{ cm}^3/(\text{g mol/s})$$

For a hydrogen feed rate of 100 ft³/min, at 60°F and 1 atm, the slurry volume required would be

$$V = \frac{100}{379} \left(\frac{454}{60} \right) (1.25 \times 10^6)(10^{-3}) \left(\frac{1}{28.32} \right) = 88 \text{ ft}^3$$

In this example many important properties of the slurry have been omitted. For example, the questions of bubble diameter and volume fraction of gas in the slurry have been avoided by giving a directly measured value of a_g. Alternately, the interfacial area can be estimated from Eqs. (10-50) and (10-52).

Example 13-10 An aqueous slurry of 3×10^{-5} m (0.03 mm) (diameter) carbon particles is to be used, at 25 C and 1 atm pressure, to remove SO_2 from a gas stream. The gas contains 2.3% SO_2 and 97.7% air, and its rate is 2.75 m³/s. The density ρ_p of the carbon particles is 0.80×10^3 kg/m³ (0.80 g/cm³) and their concentration, m_s will be 70 kg/(m³ of water) or 0.07 g/cm³. For the distributor to be used the gas bubble size will be 3×10^{-3} m (3 mm). The gas holdup, or bubble volume per unit volume of liquid, is estimated to be 0.08. Assume that the gas bubbles are uniformly distributed across the diameter of the reactor and rise individually in plug flow. The slurry will be well mixed. Calculate the volume of bubble-free liquid needed to convert 70% of the SO_2 to H_2SO_4. Base the calculations on the steady-state period before the H_2SO_4 concentration has increased enough to retard significantly the reaction rate.

Use the mass-transfer coefficients found for the same system in Example 10-7. The intrinsic rate at the carbon site under these operating conditions is first-order in oxygen and zero order in SO_2. In terms of a rate, r_v, per unit volume of bubble-free liquid, it is given by[†]

$$(r_v)_{O_2} = \tfrac{1}{2}(r_v)_{SO_2} = (k\eta)a_c C_s \tag{A}$$

---

† Hiroshi Komiyama and J. M. Smith, *AIChE J.*, **21**, 664 (1975).

where $k = 0.0033 \times 10^{-2}$ m/s

$\eta = 0.86\dagger$(for $d_p = 3 \times 10^{-5}$ m)

C_s = concentration of oxygen in the liquid at a carbon site, kg mol/m^3

SOLUTION For 70% conversion, the concentration of SO_2 in the gas will decrease by 2.3 (0.7) = 1.6 mol/(100 mole of gas). By the reaction

$$SO_2 + \tfrac{1}{2}O_2 \to SO_3$$

the oxygen concentration will change by $\tfrac{1}{2}(1.6)$ mol. Hence, the oxygen concentration will fall from 97.7(0.21) = 20.5 % to 20.2 %. This small change will not affect the rate significantly. We may use a constant, arithmetic average value of 20.3%, and in so doing introduce less error than the uncertainty in the various rate constants. The kinetics are independent of the SO_2 concentration. This means that the rate will be the same throughout the reactor. Then Eq. (13-52) can be immediately integrated to give

$$V = F_A \frac{(x_e - 0)}{\mathbf{r}_v} = \frac{F_A x_e}{\mathbf{r}_v} \tag{B}$$

The product $F_A x_e$ is equal to the moles reacted in the entire reactor per unit time. We can express this product as $Q_g C_{g_f} x_e$ and evaluate it in terms of either SO_2 or O_2, as long as $\mathbf{r}_v$ is on the same basis. In terms of SO_2, Eq. (B) becomes

$$V = \frac{Q_g (C_{g_f})_{SO_2} (x_e)_{SO_2}}{(\mathbf{r}_v)_{SO_2}} \tag{C}$$

The required volume is calculable from Eq. (C) once the global rate has been determined.

Since oxygen is but slightly soluble in water at 25°C, the resistance for mass transfer from gas to liquid is determined by the liquid-side coefficient k_L. Also, Henry's law is valid for oxygen in water at 25°C. [H = 35.4(mol/cm^3 of gas)/(mol/cm^3 of liquid)]. Then Eqs. (10-40) and (10-41) give the global rate in terms of the two mass-transfer coefficients k_L, k_c, and the intrinsic rate constant $(k\eta)$. For spherical carbon particles and spherical gas bubbles, using Eqs. (C) and (10-50) of Example 10-7,

$$a_c = \frac{6m_s}{d_p \rho_p} = \frac{6(70)}{3 \times 10^{-5}(0.80 \times 10^3)} = 175 \times 10^2 \text{ m}^2/\text{m}^3$$

$$a_g = \frac{6}{d_b} V_B = \frac{6}{3 \times 10^{-3}}(0.08) = 1.6 \times 10^2 \text{ m}^2/\text{m}^3$$

Then Eq. (10-40) becomes

$$\frac{1}{k_o H} = \frac{175}{1.6}\left(\frac{1}{k_L}\right) + \frac{1}{k_c} + \frac{1}{k\eta} \tag{D}$$

---

† For 0.03-mm particles the effectiveness factor in the liquid-filled pores of activated carbon was found to be 0.86 (loc. cit.)

We could estimate k_L and k_c from Eq. (10-42) and Fig. 10-10. However, conditions are essentially the same as in Example 10-7. In that example, for bubbles of the same size, k_L was found to be 0.08×10^{-2} m/s. The liquid-to-particle coefficient was $k_c = 0.027 \times 10^{-2}$ m/s for 0.542-mm particles. According to Fig. 10-10, the effect of particle size on k_c is very small (the effects of d_p on Re and on Sh about cancel each other). Using these values and the intrinsic rate constant, Eq. (D) gives

$$\frac{1}{k_o H} = \frac{175}{1.6}\left(\frac{10^2}{0.08}\right) + \frac{10^2}{0.027} + \frac{10^2}{0.86(0.0033)}$$

$$= (1370 + 37 + 352) \times 10^2 = 1759 \times 10^2$$

$$k_o H = 5.7 \times 10^{-6} \text{ m/s}$$

The global rate from Eq. (10-41) is

$$(r_v)_{O_2} = k_o a_c H(C_L) = (k_o H)a_c[(C_g)_{O_2}/H]$$

Substituting numerical values,

$$(r_v)_{SO_2} = 2(r_v)_{O_2} = 2(5.7 \times 10^{-6})(175 \times 10^2)\left[\frac{1(0.203) \times 10^3}{82(298)35.4}\right]$$

$$= 4.68 \times 10^{-5} \text{ kg mol/(s)(m}^3 \text{ of liquid)}$$

With this global rate, Eq. (C) gives for the required volume of water

$$V = \frac{2.75\left[\dfrac{1(0.023)}{82(298)} \times 10^3\right]0.7}{4.68 \times 10^{-5}} = 39 \text{ m}^3$$

In this example, mass transfer from gas to liquid has a greater effect on the global rate than either liquid-to-particle mass transfer or the intrinsic rate. The relatively small effect of liquid-to-particle mass transfer is due to the large ratio, a_c/a_g, which is over 100 (i.e., 175/1.6). For lower concentrations of particles this ratio would decrease. Also, larger carbon particles would in-increase the resistance of liquid-to-particle mass transport. Note that the resistance to the reaction at a site also would become relatively larger, because η decreases as d_p increases. For example, it was found[†] that η decreased from 0.86 to 0.098 when the particle size increased from 0.03 to 0.54 mm.

It should not be inferred from Examples 13-9 and 13-10 that diffusion from the bubble to the bulk liquid always controls the global rate. For example, data for different slurry reactions given by Sherwood and Farkas[‡] show other results. For the hydrogenation of α-methyl styrene containing a slurry of palladium black (diameter 55 microns), the global rate was controlled by k_L and k_c, with the significance of k_L decreasing at low catalyst concentrations. For catalyst concentrations below about 0.5 g of catalyst per liter of slurry, and at temperatures less

---

[†] Hiroshi Komiyama and J. M. Smith, loc. cit.
[‡] T. K. Sherwood and E. J. Farkas, *Chem. Eng. Sci.*, **21**, 573 (1966).

than 30°C, the rate of diffusion of dissolved hydrogen to the catalyst particles (that is, k_c) essentially controlled the rate. Results for the hydrogenation of cyclohexene in an aqueous suspension of 30-micron palladium-black particles indicated again that the predominant resistance was the diffusion to catalyst particles. However, the data of Kolbel and Maennig† for the slower hydrogenation of ethylene in a slurry of Raney nickel particles in a paraffin oil show different results. Here the catalyst particles were small enough (5 microns) and their concentration high enough that $k_c a_c$ was very large. Sherwood and Farkas were able to correlate these data by assuming that the chemical step on the catalyst particles controlled the rate.

TRICKLE-BED REACTORS

In Chap. 10 we discussed the characteristics of cocurrent down-flow of gas and liquid over a fixed bed of catalyst particles (see Fig. 10-16). In Secs. 10-10 to 10-12, correlations were presented for mass transfer from gas to liquid and from liquid to solid for the trickling-flow regime (continuous gas phase and liquid rivulets cascading *down* over the particles). Now we wish to use this information, along with intrinsic kinetics and a reactor model, to predict the behavior of trickle-bed reactors. Since the catalyst particle sizes in these reactors are relatively large, intraparticle resistances will be more significant than for slurry reactors. On the other hand, there is no difficulty in retaining the large particles in the bed, in contrast to the situation in slurries.

As in Chap. 10, the treatment will be limited to the trickling flow regime and to isothermal operation. Also, it will be assumed that flowing liquid completely covers the particles so that reaction can occur only by reactant mass transfer through the liquid-particle interface. Isothermal conditions and complete liquid coverage are reasonable assumptions for many trickle-bed processes, except at very low liquid rates. Capillary forces normally draw liquid into the pores of the particles. Therefore, in evaluating intraparticle mass-transfer effects (effectiveness factors) it is satisfactory to use liquid-phase diffusivities.

The main need for three-phase reactors is when one reactant is too volatile to liquefy and a second is too nonvolatile to vaporize. Hence, the main emphasis will be for the case of a gaseous component reacting with a second reactant in the liquid phase; that is for a reaction of the form:

$$aA(g) + B(l) \rightarrow C(g \text{ or } l) + D(g \text{ or } l) \tag{13-54}$$

This general form would include hydrogenation (e.g., hydrodesulfurization of petroleum fractions, hydrogenation of oils) and oxidation (oxidation of pollutants dissolved in liquids) reactions.

---

† H. Kolbel and H. G. Maennig, *Elektrochem.*, **66**, 744 (1962).

13-13 Trickle-Bed Reactor Model

For isothermal operation and uniform distribution of gas and liquid phases across the reactor diameter, there will be no radial gradients of concentration or velocity for the gas or liquid, except on a particle-size scale. Of course there will be concentration and velocity changes near the phase boundaries, as shown in Fig. 10-14. However, the concentrations and velocities will be the same in the *bulk* fluid at any radial position if the flow distribution is uniform.†

With no radial gradients, a one-dimensional, isothermal model is adequate. The mass-conservation equations are of the same form as those developed in Sec. 13-10 for the "two-phase," fluid-bed reactor, except that equations must be written for both reactants A and B. For most applications axial dispersion in the gas phase is negligible, and such dispersion may be unimportant in the liquid.‡ Let us write the conservation equations for plug-flow of gas and use a dispersion model (with axial dispersion coefficient D_L) for the liquid. For reactant A in the gas phase, at steady state, and for a volume element that extends across the reactor (Fig. 13-23),

$$u_g \frac{d(C_A)_g}{dz} + (K_L a_g)_A [(C_A)_g / H_A - (C_A)_L] = 0 \qquad (13\text{-}55)$$

where $(C_A)_g / H_A$ is the liquid-phase concentration in equilibrium with the bulk gas concentration; K_L is an *overall* mass-transfer coefficient between gas and liquid; u_g is the superficial velocity of the gas. Since Henry's law has been assumed to be applicable for A, K_L is related to the individual films coefficients k_L and k_g by the equation:

$$\frac{1}{K_L} = \frac{1}{Hk_g} + \frac{1}{k_L} \qquad (13\text{-}56)$$

As noted in Chap. 10, for many trickle-bed processes, A is slightly soluble in the liquid (H is large) so that $K_L \approx k_L$.

For reactant A in the liquid phase, the conservation equation is

$$D_L \frac{d^2(C_A)_L}{dz^2} - u_L \frac{d(C_A)_L}{dz} + (K_L a_g)_A [(C_A)_g / H_A - (C_A)_L]$$

$$- (k_c a_c)_A [(C_A)_L - (C_A)_s] = 0 \quad (13\text{-}57)$$

---

† In a trickle bed, the gas is distributed uniformly across the diameter, but the liquid tends to flow toward the reactor wall. Approximately, a reactor-to-particle diameter ratio of 18 or more will ensure uniform liquid distribution. Correlations for predicting the flow distribution and the catalyst bed depth required to achieve the equilibrium distribution are available [Mordechay Herskowitz and J. M. Smith, *AIChE J.*, 24, 439 (1978)].

‡ Criteria for the importance of axial dispersion are available, e.g., D. E. Mears, *Chem. Eng. Sci.*, 26, 1361 (1971).

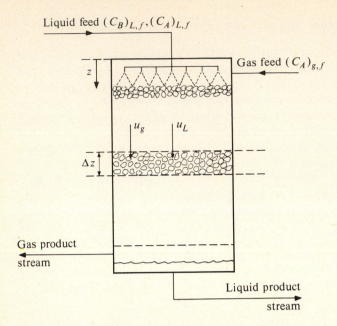

Gas product stream

Liquid product stream

Figure 13-23 Trickle-bed for catalytic reaction $aA(g) + B(l) \rightarrow$ product.

which includes terms for mass transfer from gas-to-liquid and from liquid-to-particle (coefficient $k_c a_c$). The third mass-conservation equation needed is that for reactant B in the liquid phase:

$$(D_L)_B \frac{d^2(C_B)_L}{dz^2} - u_L \frac{d(C_B)_L}{dz} - (k_c a_c)_B[(C_B)_L - (C_B)_s] = 0 \qquad (13\text{-}58)$$

Here it has been assumed that B is nonvolatile. The velocity u_L is the superficial value for the liquid, and $(C)_s$ is the liquid-phase concentration at the outer surface of the catalyst particle. The areas per unit volume of empty reactor a_g and a_c are the same as those used in Secs. 10-10 to 10-12.

The reaction rate expressed in terms of surface concentrations provides the relationship between $(C)_s$ and $(C)_L$. From the definition of the effectiveness factor η in Chap. 11 [Eq. (11-42)], we may express the required equality of mass-transfer and reaction rates, $\mathbf{r}_A$ and $\mathbf{r}_B$ as

$$(k_c a_c)_A[(C_A)_L - (C_A)_s] = \mathbf{r}_A = \rho_B \eta f[(C_A)_s, (C_B)_s] \qquad (13\text{-}59)$$

$$(k_c a_c)_B[(C_B)_L - (C_B)_s] = \mathbf{r}_B = \frac{\mathbf{r}_A}{a} = \frac{\rho_B}{a} \eta f[(C_A)_s, (C_B)_s] \qquad (13\text{-}60)$$

where $f[(C_A)_s, (C_B)_s]$ represents the intrinsic rate of reaction, per unit mass of catalyst, for the disappearance of A, and ρ_B is the bulk density of the catalyst particles in the bed.

With appropriate boundary conditions, the five equations [Eqs. (13-55), (13-57) to (13-60)] can be solved for the five concentrations, $(C_A)_g$, $(C_A)_L$, $(C_B)_L$, $(C_A)_s$, $(C_B)_s$ as a function of reactor bed depth z. It is necessary to know all the mass-transfer coefficients, the intrinsic rate equation, $f[C_A, C_B]$, and the effectiveness factor.

The problem becomes much simpler (an initial-value rather than a boundary-value problem) if axial dispersion in the liquid can be neglected (plug flow of liquid). Then the terms involving the second derivative in Eqs. (13-57) and (13-58) can be omitted. The boundary conditions for the three first-order equations [Eqs. (13-55), (13-57), and (13-58) with dispersion terms omitted] are, at $z = 0$,

$$(C_A)_g = (C_A)_{g,f} \left.\vphantom{\begin{matrix}1\\1\\1\end{matrix}}\right\} \qquad (13\text{-}61)$$

$$(C_A)_L = (C_A)_{L,f} \left.\vphantom{\begin{matrix}1\\1\end{matrix}}\right\} \text{ feed concn} \qquad (13\text{-}62)$$

$$(C_B)_L = (C_B)_{L,f} \qquad (13\text{-}63)$$

Both the axial dispersion and plug-flow models have been solved and compared with experimental data for the oxidation of aqueous solutions of formic† and of acetic acids with air.‡ Plug-flow solutions for design of trickle-bed reactors are illustrated in the examples that follow. When the intrinsic kinetics are first order, analytical solutions for the conversion are possible. Otherwise, the set of differential equations must be solved numerically.

Example 13-11 A hydrocarbon oil is to be desulfurized prior to catalytic cracking. Of the various sulfur compounds present (mercaptans, sulfides, disulfides, etc.) one of the most difficult to desulfurize is thiophene. With a sulfided, cobalt-molybdenum oxide catalyst on alumina, thiophene reacts with hydrogen to form butane and H_2S.

(B)

$$\underset{\displaystyle \underset{S}{\diagdown\diagup}}{\overset{\displaystyle HC-CH}{HC \qquad CH}} \; + \; 4H_2 \; \longrightarrow \; C_4H_{10} + H_2S$$

A trickle-bed reactor is to be designed on the supposition that if the refractory thiophene is reacted, the other sulfur compounds will also have been hydrogenated.

Pure hydrogen and the hydrocarbon liquid will be fed to the top of the catalyst bed, operating at 200°C and 40 atm. Neglect the vaporization of thiophene from the liquid at these conditions.

A. For the first case assume that the reaction rate at the catalyst sites and the mass-transfer rate from liquid to catalyst particle are slow enough that the liquid is saturated with hydrogen throughout the column. While the reaction probably is second order, assume that the thiophene concentration is relatively large with respect to that of hydrogen dissolved in the oil. Then the intrinsic rate will be pseudo-first-order in hydrogen. Also, the intrinsic rate is slow enough that the effectiveness factor is unity. Derive an expression for the

† S. Goto and J. M. Smith, *AIChE J.*, **21**, 706 (1975).
‡ J. Levec and J. M. Smith, *AIChE J.*, **22**, 159 (1976).'

fractional removal (conversion) of thiophene from the oil, assuming plug flow of liquid.

B. For very low thiophene concentrations the intrinsic kinetics will become first order in C_B. Consider the extreme case where the intrinsic rate is independent of hydrogen concentration. What will be the expression for the fractional removal of thiophene?

C. Calculate the catalyst bed depth required, for conditions of parts A and B, to remove 75% of the thiophene. For part A, the feed concentration of thiophene is 1000 ppm and for part B suppose it is 100 ppm. The superficial liquid velocity will be 5.0 cm/s. The first-order rate constants are $k_H = 0.11$ cm^3/(g)(s), and $k_B = 0.07$ cm^3/(g)(s), and the volumetric mass-transfer coefficients from liquid to particle are $(k_c a_c)_{H_2} = 0.50$ s^{-1} and $(k_c a_c)_B = 0.3$ s^{-1} (estimated from correlations in Sec. 10-11), $\rho_B = 0.96$ g/cm^3.

SOLUTION

A. Since the liquid is saturated with hydrogen, and pure hydrogen is used as the gas stream, Eqs. (13-57) and (13-55) are not needed; the concentration of hydrogen in the liquid $(C_{H_2})_L$ is a constant and equal to $(C_{H_2})_g/H_{H_2}$. Further, the intrinsic rate is independent of the thiophene concentration so that the function, $f[(C_A)_s, (C_B)_s]$, is the first-order form:

$$f[(C_A)_s, (C_B)_s] = k_H(C_{H_2})_s \tag{A}$$

Then Eq. (13-59) for hydrogen is

$$(k_c a_c)_{H_2}[(C_{H_2})_L - (C_{H_2})_s] = \rho_B k_H(C_{H_2})_s$$

or

$$(C_{H_2})_s = \frac{(k_c a_c)_{H_2}}{(k_c a_c)_{H_2} + \rho_B k_H} (C_{H_2})_L = \frac{(k_c a_c)_{H_2}}{(k_c a_c)_{H_2} + \rho_B k_H}\left[\frac{(C_{H_2})_g}{H_{H_2}}\right] \tag{B}$$

Hence, Eq. (13-60) for thiophene (B) may be written

$$(k_c a_c)_B[(C_B)_L - (C_B)_s] = \frac{\rho_B}{4}\frac{(k_c a_c)_{H_2} k_H}{(k_c a_c)_{H_2} + \rho_B k_H}\left[\frac{(C_{H_2})_g}{H_{H_2}}\right] \tag{C}$$

Since $(C_{H_2})_g$ is constant, Eq. (C) shows that the rate of mass transfer of thiophene from liquid to particle (and the rate of reaction) is the same throughout the reactor. Equation (C) can be substituted in Eq. (13-58) and the result easily integrated to determine how C_B varies with catalyst bed depth. Thus, for plug flow of liquid, Eq. (13-58) becomes

$$u_L\frac{d(C_B)_L}{dz} + \tfrac{1}{4}(k_H^\circ)\frac{(C_{H_2})_g}{H_{H_2}} = 0 \tag{D}$$

where the overall coefficient is given by

$$\frac{1}{k_H^\circ} = \frac{1}{(k_c a_c)_{H_2}} + \frac{1}{\rho_B k_H} \tag{E}$$

The boundary (feed) condition is $(C_B)_L = (C_B)_{L,f}$ at $z = 0$. Then the integrated form of Eq. (D) is

$$(C_B)_L - (C_B)_{L,f} = -\left(\frac{k_H^\circ z}{4u_L}\right)\frac{(C_{H_2})_g}{H_{H_2}}$$

and the fractional removal of thiophene is

$$x = \frac{(C_B)_{L,f} - (C_B)_L}{(C_B)_{L,f}} = \left(\frac{k_H^\circ z}{4u_L}\right)\frac{(C_{H_2})_g/H_{H_2}}{(C_B)_{L,f}} \tag{F}$$

where z is the catalyst bed depth.

B. In this case the intrinsic rate function (for *thiophene*) becomes

$$f[(C_A)_s, (C_B)_s] = k_B(C_B)_s \tag{G}$$

Now, Eq. (13-59) for hydrogen is not involved. Equation (13-60) may be written

$$(k_c a_c)_B[(C_B)_L - (C_B)_s] = \rho_B k_B(C_B)_s$$

or

$$(C_B)_s = \frac{(k_c a_c)_B}{(k_c a_c)_B + \rho_B k_B}(C_B)_L \tag{H}$$

Substituting this result in Eq. (13-58) gives, for plug flow of liquid,

$$u_L\frac{(dC_B)_L}{dz} + k_B^\circ(C_B)_L = 0 \tag{I}$$

where, now,

$$\frac{1}{k_B^\circ} = \frac{1}{(k_c a_c)_B} + \frac{1}{\rho_B k_B} \tag{J}$$

Integrating Eq. (I) with the boundary condition at $z = 0$ gives

$$x = 1 - \exp\left(-\frac{k_B^\circ z}{u_L}\right) \tag{K}$$

C. For part A, Eq. (F) is applicable. The hydrogen concentration at 40 atm and 200°C is

$$(C_{H_2})_g = \frac{P_{H_2}}{RT} = \frac{40}{82(473)} = 1.03 \times 10^{-3} \text{ g mol/cm}^3$$

From hydrogen solubility data it is estimated that H_{H_2} at 200°C is 50[g mol/(cm^3 gas)]/[g mol/(cm^3 in liquid)]. From the rate constant data and Eq. (E),

$$\frac{1}{k_H^\circ} = \frac{1}{0.50} + \frac{1}{0.96(0.11)} = 11.5 \text{ s}$$

The concentration of thiophene in the feed (1000 ppm) is $(1000/84) \times 10^{-6} = 1.19 \times 10^{-5}$ g mo./cm^3. Substituting these results in Eq. (F) gives

$$z = x \left(\frac{4u_L}{k_H^\circ} \right) \frac{(C_B)_{L,f}}{(C_{H_2})_g / H} = 0.75 \frac{4(5)}{(1/11.5)} \frac{1.19 \times 10^{-5}}{1.03 \times 10^{-3}/50}$$

$$z = 100 \text{ cm}$$

For part B, we first must calculate k_B° for thiophene. From data given, Eq. (J) gives

$$\frac{1}{k_B^\circ} = \frac{1}{0.3} + \frac{1}{0.96(0.07)} = 18.2 \text{ s}$$

Then Eq. (K) is applicable:

$$0.75 = 1 - \exp \left[- \frac{(1/18.2)z}{(5)} \right]$$

Solving for z yields

$$z = 126 \text{ cm}$$

Example 13-12 Reconsider Example 13-11, part A, but do not assume that the liquid is saturated with hydrogen. The hydrogen concentration in the feed liquid is zero, and the gas-to-liquid mass-transfer coefficient is $(K_L a_g)_{H_2} = (k_L a_g)_{H_2} = 0.030 \text{ s}^{-1}$.

Derive an expression for the fractional removal of thiophene. Also calculate the bed depth required for removal of 75% of the thiophene for the conditions of Example 13-11.

SOLUTION Since pure hydrogen is used, Eq. (13-55) is again unnecessary; $(C_{H_2})_g$ is constant and equal to the value in the feed. Equation (13-57) for plug flow of liquid becomes

$$-u_L \frac{d(C_{H_2})_L}{dz} + (K_L a_g)_{H_2} \left[\frac{(C_{H_2})_g}{H_{H_2}} - (C_{H_2})_L \right] - (k_c a_c)_{H_2} [(C_{H_2})_L - (C_{H_2})_s] = 0 \tag{A}$$

For the first-order (in hydrogen) kinetics, Eq. (A) of Example 13-11 is applicable. Then, $(C_{H_2})_s$ from Eq. (13-59) is

$$(C_{H_2})_s = \frac{(k_c a_c)_{H_2}}{(k_c a_c)_{H_2} + \rho_B k_H} (C_{H_2})_L \tag{B}$$

Using this expression for $(C_{H_2})_s$, we may write

$$(C_{H_2})_L - (C_{H_2})_s = \frac{\rho_B k_H}{(k_c a_c)_{H_2} + \rho_B k_H} (C_{H_2})_L \tag{C}$$

Then Eq. (A) may be written solely in terms of $(C_{H_2})_L$ as a variable:

$$-u_L \frac{d(C_{H_2})_L}{dz} + (K_L a_g)_{H_2} \left[\frac{(C_{H_2})_g}{H_{H_2}} - (C_{H_2})_L \right] - k_H^\circ (C_{H_2})_L = 0 \qquad \text{(D)}$$

where k_H° is given by Eq. (E) of Example 13-11.

Equation (D) can be integrated to give the following equation for $(C_{H_2})_L$ as a function of z:

$$(C_{H_2})_L = \frac{(C_{H_2})_g / H_{H_2}}{1 + [k_H^\circ / (K_L a_g)_{H_2}]} \left\{ 1 - \exp\left[-\{(K_L a_g)_{H_2} + k_H^\circ\} \frac{z}{u_L} \right] \right\} \qquad \text{(E)}$$

Next we want to integrate Eq. (13-58) to find the thiophene concentration at any bed depth. To do this the surface concentration in the term in brackets must be eliminated. This is done by using Eq. (13-60). Thus, from Eq. (B) and Eq. (A) of Example (13-11), Eq. (13-60) becomes

$$(k_c a_c)_B [(C_B)_L - (C_B)_s] = \frac{\rho_B k_H}{4} (C_{H_2})_s = \frac{\rho_B k_H}{4} \frac{(k_c a_c)_{H_2}}{(k_c a_c)_{H_2} + \rho_B k_H} (C_{H_2})_L$$

$$= \frac{k_H^\circ}{4} (C_{H_2})_L \qquad \text{(F)}$$

Substituting Eq. (E) for $(C_{H_2})_L$ in Eq. (F),

$$(k_c a_c)_B [(C_B)_L - (C_B)_s] = \frac{k_H^\circ (C_{H_2})_g / H_{H_2}}{4(1 + \{k_H^\circ / (K_L a_g)_{H_2}\})}$$

$$\times \left\{ 1 - \exp\left[-\{(K_L a_g)_{H_2} + k_H^\circ\} \frac{z}{u_L} \right] \right\} \qquad \text{(G)}$$

$$= \alpha[1 - \exp(-\beta z)] \qquad \text{(H)}$$

where α and β are constants defined by comparing Eqs. (G) and (H).

Now we can use Eq. (H) to write Eq. (13-58), for plug flow of liquid, in terms of only $(C_B)_L$:

$$u_L \frac{d(C_B)_L}{dz} + \alpha[1 - \exp(-\beta z)] = 0 \qquad \text{(I)}$$

Integrating, with the boundary condition $(C_B)_L = (C_B)_{L, f}$ at $z = 0$, gives

$$(C_B)_L - (C_B)_{L, f} = -\alpha \cdot \frac{z}{u_L} - \frac{\alpha}{\beta u_L} (e^{-\beta z} - 1)$$

In terms of fractional removal of thiophene this is

$$x = \frac{\alpha}{(C_B)_{L, f}} \left[\left(\frac{z}{u_L} \right) + \frac{1}{\beta u_L (C_B)_{L, f}} (e^{-\beta z} - 1) \right] \qquad \text{(J)}$$

If the expressions for α and β are inserted, Eq. (J) becomes,

$$x = \frac{k_H^\circ [(C_{H_2})_g / H_{H_2} (C_B)_{L, f}]}{4[1 + \{k_H^\circ / (K_L a_g)_{H_2}\}]} \left\{ \frac{z}{u_L} + \frac{\exp[-\{(K_L a_g)_{H_2} + k_H^\circ\}(z/u_L)] - 1}{[(K_L a_g)_{H_2} + k_H^\circ]} \right\} \qquad \text{(K)}$$

From Example 13-11, $k_H^\circ = 1/11.5 = 0.087$ s, $(C_{H_2})_g = 1.03 \times 10^{-3}$ g mol/cm³, and $(C_B)_{L,f} = 1.19 \times 10^{-6}$ g mol/cm³. Substituting these values in Eq. (K),

$$x = 0.75$$

$$0.75 = \frac{0.087[(1.03 \times 10^{-3}/50)1.19 \times 10^{-5}]}{4[1 + (0.087/0.030)]}$$

$$\times \left\{ \frac{z}{5} + \frac{\exp -(0.03 + 0.087)(z/5) - 1}{0.03 + 0.087} \right\}$$

$$77.7 = \frac{z}{5} + 8.55(e^{-0.023z} - 1)$$

Solving this expression for z gives

$$z = 430 \text{ cm}$$

All of the increase from 100 (obtained in Example 13-11) to 430 cm is not due to the mass-transfer resistance between gas and liquid. In Example 13-11 it was supposed that the liquid was always saturated with hydrogen. In the present example, the concentration of hydrogen in the liquid feed was taken as zero. If the boundary condition for Eq. (D) were the saturated condition, $(C_{H_2})_L = (C_{H_2})_{g,f}/H_{H_2}$ at $z = 0$. Then, the bed depth required is $z = 270$ cm. The increase in z from 100 to 270 cm is the result of the mass-transfer resistance from gas to particle. The change from 430 to 270 cm reflects the effect of saturating the hydrocarbon liquid with hydrogen before it enters the reactor.

In the previous two examples, the gas-phase conservation expression [Eq. (13-55)] was not needed, because the gas was pure hydrogen. This is not so when a trickle-bed reactor is used to remove pollutants from a gas stream. Example 13-13 is an illustration of this type of application.

Example 13-13 Sulfur dioxide is to be removed from air in a bed of activated carbon particles at 25°C and 1 atm pressure. The reactor will be of the trickle-bed type in which the gas stream, containing 3% SO_2, 18% O_2, and 79% N_2, and pure water are fed to the top of the reactor. Activated carbon catalyzes the oxidation of SO_2 to SO_3, and the SO_3 dissolves in the water to produce H_2SO_4. It has been suggested† that the intrinsic rate of reaction is controlled by the rate of adsorption of oxygen on the carbon and is independent of the SO_2 concentration.

Derive equations for steady-state operation which give the fraction of the SO_2 from the gas stream removed *by reaction,* as a function of the catalyst bed depth. Assume that liquid and gas streams are in plug flow and that the feed water is in equilibrium with the gas feed with respect to oxygen. To simplify the mass balances written for the *entire* reactor, neglect the difference in oxygen content of the liquid feed and effluent streams.

---

† Hiroshi Komiyama and J. M. Smith, *AIChE J.,* **21**, 664 (1975).

SOLUTION Equations (13-55) and (13-57) with the second derivative omitted, and Eq. (13-59) are applicable when applied to oxygen. For simplicity, we omit the subscript A so that C designates the oxygen concentration. The intrinsic rate function is

$$\mathbf{r}_{O_2} = \rho_B \eta f[(C_A)_s, (C_B)_s] = \rho_B \eta k C_s \tag{A}$$

Then Eq. (13-59) becomes

$$(k_c a_c)(C_L - C_s) = \rho_B \eta k C_s$$

Solving for C_s,

$$C_s = \frac{k_c a_c}{k \eta \rho_B + k_c a_c} C_L = \alpha C_L \tag{B}$$

If this value for C_s is substituted into Eq. (13-57), with the second derivative omitted, there is obtained

$$-u_L \frac{dC_L}{dz} + K_L a_g \left(\frac{C_g}{H} - C_L \right) - k_c a_c (1 - \alpha) C_L = 0 \tag{C}$$

Equation (13-55) may be rearranged to the form

$$\frac{C_g}{H} - C_L = -\left(\frac{u_g}{K_L a_g} \right) \frac{dC_g}{dz} \tag{D}$$

or

$$C_L = \frac{C_g}{H} + \left(\frac{u_g}{K_L a_g} \right) \frac{dC_g}{dz} \tag{E}$$

Differentiation of Eq. (E) yields

$$\frac{dC_L}{dz} = \left(\frac{1}{H} \right) \frac{dC_g}{dz} + \left(\frac{u_g}{K_L a_g} \right) \frac{d^2 C_g}{dz^2} \tag{F}$$

If Eqs. (D), (E), and (F) are substituted in (C), a second-order equation is obtained with C_g as the only dependent variable. The expression may be written:

$$\frac{d^2 C_g}{dz^2} + \beta \frac{dC_g}{dz} + \gamma C_g = 0 \tag{G}$$

where the constant coefficients are

$$\beta = \frac{u_g + u_L/H + k_c a_c (1 - \alpha) u_g / K_L a_g}{u_L u_g / K_L a_g} \tag{H}$$

$$\gamma = \frac{(k_c a_c) K_L a_g (1 - \alpha)}{H u_L u_g} \tag{I}$$

$$\alpha = k_c a_c / (k \eta \rho_B + k_c a_c) \tag{J}$$

The boundary conditions (for oxygen) are

$$\text{at } z = 0, \qquad C_g = C_{g,f} \tag{K}$$

$$C_L = C_{g,f}/H \qquad \text{(equilibrium)} \tag{L}$$

Boundary condition (L) can be written in terms of C_g by applying Eq. (E) at $z = 0$. This gives

$$\text{at } z = 0, \qquad \frac{dC_g}{dz} = 0 \tag{M}$$

Equation (G) with boundary conditions (K) and (M) can be solved by standard methods to give

$$\left(\frac{C_g}{C_{g,f}}\right)_{O_2} = \frac{1}{m_2 - m_1}[m_2 e^{m_1 z} - m_1 e^{m_2 z}] \tag{N}$$

where

$$m_1 = -\frac{\beta}{2} + \frac{1}{2}(\beta^2 - 4\gamma)^{1/2} \tag{P}$$

$$m_2 = -\frac{\beta}{2} - \frac{1}{2}(\beta^2 - 4\gamma)^{1/2} \tag{Q}$$

The reactions involved are

$$SO_2 + \tfrac{1}{2}O_2 \rightarrow SO_3$$

$$SO_3 + H_2O \rightarrow H_2SO_4$$

Neglecting the change in O_2 concentration of the liquid between feed and effluent, the first reaction requires that the SO_2 removed from the gas stream by reaction be twice that of oxygen; that is,

$$(C_{g,f} - C_g)_{SO_2} = 2(C_{g,f} - C_g)_{O_2}$$

Then the fractional removal of SO_2 is

$$\left(\frac{C_{g,f} - C_g}{C_{g,f}}\right)_{SO_2} = 2\frac{(C_{g,f})_{O_2}}{(C_{g,f})_{SO_2}}\left[1 - \left(\frac{C_g}{C_{g,f}}\right)_{O_2}\right] \tag{R}$$

Equation (R) with Eq. (N) for $(C_g/C_{g,f})_{O_2}$ can be used to calculate the fractional removal of SO_2 from the gas by reaction. This does not account for the SO_2 that might be removed in the effluent liquid. Problem 13-23 is a numerical illustration for this type of trickle-bed reactor.

A more realistic approach to the hydrodesulfurization process in Examples 13-11 and 13-12 would have been to use nonlinear kinetics, particularly when the concentrations of thiophene and hydrogen in the liquid are of the same magnitude. However, an analytical solution for the effluent concentrations cannot be obtained. Another example of nonlinear kinetics is the catalytic oxidation of aqueous solutions of formic acid. The kinetics of the reaction

$$HCOOH(aq) + \tfrac{1}{2}O_2(g) \rightarrow H_2O(l) + CO_2$$

using a $CuO \cdot ZnO$ catalyst have been found† to be second order with a rate equation

$$\mathbf{r}_{O_2} = \eta \rho_B k_{O_2}(C_{O_2})_s(C_F)_s$$

where $(C_F)_s$ represents the concentration of formic acid in the liquid at the catalyst site. For this case the five equations (13-55) and (13-57) to (13-60) become, for plug flow of liquid:

$$u_g \frac{d(C_{O_2})_g}{dz} + (K_L a_g)_{O_2}\left[\frac{(C_{O_2})_g}{H_{O_2}} - (C_{O_2})_L\right] = 0 \qquad (13\text{-}64)$$

$$-u_L \frac{d(C_{O_2})_L}{dz} + (K_L a_g)_{O_2}\left[\frac{(C_{O_2})_g}{H_{O_2}} - (C_{O_2})_L\right] - (k_c a_c)_{O_2}[(C_{O_2})_L - (C_{O_2})_s] = 0$$
$$(13\text{-}65)$$

$$-u_L \frac{d(C_F)_L}{dz} - (k_c a_c)_F[(C_p)_L - (C_F)_s] = 0 \qquad (13\text{-}66)$$

$$(k_c a_c)_{O_2}[(C_{O_2})_L - (C_{O_2})_s] = \mathbf{r}_{O_2} = \eta \rho_B k_{O_2}(C_{O_2})_s(C_F)_s \qquad (13\text{-}67)$$

$$(k_c a_c)_F[(C_F)_L - (C_F)_s] = 2\mathbf{r}_{O_2} = 2\eta \rho_B k_{O_2}(C_{O_2})_s(C_F)_s \qquad (13\text{-}68)$$

The required boundary conditions are the feed concentrations; that is, Eqs. (13-61) to (13-63) with $A = O_2$ and $B =$ formic acid.

These equations can be solved numerically in the following way. Starting at $z = 0$ where $(C_F)_L$ and $(C_{O_2})_L$ are known, calculate $(C_F)_s$ and $(C_{O_2})_s$. Then Eqs. (13-64) to (13-66) can be solved for the first increment, for example by the Runge-Kutta method, to give $(C_{O_2})_g$, $(C_{O_2})_L$ and $(C_F)_L$ at the end of the increment chosen for z. Repetition of the process ultimately gives the concentrations at any bed depth.

A complication arises if intraparticle diffusion is not negligible (i.e., $\eta \neq 1.0$). This is because η is a function of the surface concentrations when the intrinsic kinetics are not first order (see Fig. 11-8). This means that η will vary with bed depth. A method of accounting for this variation is available‡ as well as a detailed description of the solution procedure for the concentration profiles in the reactor. For the linear kinetics considered in Example 13-13, this complication did not exist. The effectiveness factor is independent of concentration [Fig. 11-8 or Eq. (11-52)] so that η is a constant throughout the reactor.

We have not discussed countercurrent flow in packed-bed reactors. The equations for predicting the performance for this flow arrangement are essentially the same (only the signs of some of the terms in the conservation equations are different) as those for trickle beds, and the results are similar. A comparison has been made of the performance of slurries, trickle-beds and countercurrent packed beds as examples of 3-phase reactors.§ The three types were compared for the

---

† G. Baldi, S. Goto, C. K. Chow, and J. M. Smith, *Ind. Eng. Chem., Proc. Des. Dev.*, **13**, 447 (1974).
‡ S. Goto and J. M. Smith, *AIChE J.*, **21**, 706 (1975).
§ S. Goto and J. M. Smith, *AIChE J.*, **24**, 286 (1978).

removal of SO_2 from a gas stream. The differences in performance are due primarily to the different values for the mass transfer effects. For the same reactor volume and flow rates, the countercurrent packed bed gave somewhat higher SO_2 removal. For the same mass of catalyst, the slurry reactor gave the greatest removal. Practical factors, such as the problem of retaining small particles in the slurry reactor, were not considered.

OPTIMIZATION

Although we shall not consider the quantitative aspects of optimization, some general comments are necessary in order to give proper emphasis to the material that has been covered. As indicated in Sec. 1-1, the purpose of this text has been to present the concepts necessary to design a reactor. We started with chemical kinetics (Chap. 2) and then discussed physical processes in terms of the process-design features of large-scale reactors, first for homogeneous reactions (Chaps. 3 to 6) and then for heterogeneous catalytic reactions (Chaps. 7 to 13). The general approach was that the reactor form was known, and the objective was to predict the performance for a single set of operating conditions. Nevertheless, aspects of optimum performance were often introduced. As early as Chap. 1, the interrelationship between kinetics and thermodynamics for an exothermic reversible reaction was employed. We saw that the maximum attainable conversion in such reactions decreases as the temperature increases, but the rate of reaction increases. These contrasting effects suggest that improved conversion (per unit volume of reactor) could be obtained by operating the system at different temperatures: first in a high-temperature reactor, where most of the conversion is obtained at a high rate, and then in a second reactor operated at a lower temperature to achieve the higher conversion dictated by thermodynamics. In Chap. 4, Secs. 4-4 and 4-5, conclusions were reached about operating conditions and reactor types for maximum yield of the desired product in multiple-reaction systems. Optimum temperature profiles for exothermic reactions in a tubular reactor were discussed in Sec. 5-7. More on these subjects is available.†

The term "optimizing the performance" does not properly describe the goal, since the ultimate objective concerns economics. However, this term does indicate a dilemma of optimization studies. Rarely can the profit from a chemical reactor be described quantitatively in terms of operating conditions. First, the reactor is probably only one unit of a plant, and the most economical operation of the reactor may conflict with the economy of subsequent separation processes. Hence overall economy may in fact require operating the reactor at nonoptimum conditions. Second, market conditions for the reaction products, even when they are known, are subject to fluctuation. Hence the economics of the entire plant may be time dependent or uncertain. Third, it is difficult to establish valid figures for *all* the costs that accrue to a reactor or a plant. Because of these uncertainties, most

---

† R. Aris, "The Optimal Design of Chemical Reactors," Academic Press, Inc., New York, 1961; K. G. Denbigh, "Chemical Reactor Theory," chap. 5, Cambridge University Press, Cambridge, 1965; H. Kramers and K. R. Westerterp, "Elements of Chemical Reactor Design and Operation," chap. VI, Academic Press, Inc., New York, 1963.

published optimization studies have dealt with conversion and selectivity rather than with profits. While this approach does not take into consideration any of the cost factors, it does provide a constant solution to the *technical* problem. For *economic* studies a unit value must be applied to each product, determined on the basis of operating and initial costs and marketing uncertainties. Such studies require a knowledge of how operating conditions, such as temperatures, pressures, and feed compositions, affect the production rate of the products, and the problem is to determine the particular operating conditions that will give the maximum profit.

It is supposed that the most profitable reactor type has already been chosen, and the question is: What conditions will afford the most profitable operation of this type? Actually, complete optimization would require simultaneous solution for both considerations. Consider a highly exothermic catalytic reaction (such as air oxidation of naphthalene to phthalic anhydride) which must be carried out below an upper temperature limit to prevent undesirable side reactions (oxidation to CO_2 and H_2O). Several choices of reactor exist. A single, large-diameter, adiabatic fixed bed could be used, with an excess of inert diluent added to the feed to absorb the heat of reaction. This would reduce the initial cost of the reactor. Alternately, a large number of parallel small tubes, packed with catalyst pellets and surrounded by a cooling fluid, might be employed; in this case the diluent might be reduced, and the temperature rise would be limited by heat transfer to the cooling fluid. Operating costs might be reduced, but the initial cost would be high for a reactor consisting of hundreds of small tubes manifolded together. Another possibility would be a large-diameter fluidized bed, either with extensive diluent in the feed and no internal cooling tubes, or with less diluent but a bank of tubes in the bed through which a cooling medium flowed. To decide among these types, optimization studies and a comparison of the results for each would be necessary. More study of this broader type of optimization could be rewarding.

Let us return to the problem of optimizing total conversion and selectivity. To begin with, it is important to note the relation between the two quantities. For a single reaction optimum operation corresponds to the maximum production rate of the product per unit mass of catalyst. When two or more reactions are involved the situation is more complicated, because both total conversion and selectivity, that is, the production rate of each product, are likely to be important. Moreover, this importance may depend on factors other than the reactor, specifically the difficulty (cost) of separation and recycling unreacted feed components and the separation of desirable and undesirable products. Sometimes separation is not feasible.

As an example, consider the catalytic reforming of naphtha to gasoline in a fixed bed of platinum catalyst. The aromatization and cracking processes that occur with the naphtha feed (which itself contains many components) lead to a product that contains literally hundreds of individual components. An overall measure of selectivity is usually all that is possible, and this is generally the octane number of product; separation of individual products is not attempted.† The

---

† Automated chromatographs with improved resolution have simplified greatly the analysis of mixtures of a large number of individual components.

profitability of the reactor depends on the total production of reformed product and its selectivity (octane number). As temperature increases, total conversion increases, but octane number decreases (many other operating conditions, of course, affect these two measures of performance). Hence a profit function giving appropriate emphasis to total conversion and octane number must be calculated, and then operating conditions that maximize this function must be determined. A complete mathematical analysis, taking into account all the reactions, is impossible—in this case because all the reactions and their kinetics are not known, and not because of the magnitude of the calculations. In simpler situations involving only a few variables, complete analysis is not difficult. Kramers and Westerterp† consider several cases; for example, maximizing the production rate of B in the reaction $A \rightarrow B \rightarrow C$ in an ideal stirred-tank reactor for various temperature conditions, and maximizing the profit in a tubular-flow reactor where three reactions occur, $A \rightarrow B$, $A \rightarrow C$, and $A \rightarrow D$, and B is the valuable product. For somewhat more complex cases with more variables, solutions are still feasible with machine computation for the extensive calculations. In these cases mathematical concepts such as those embodied in the optimality theory of Bellman‡ or the "method of steepest descent"§ are useful.

In summary, the requirement for economical commercial operation suggests the need for further optimization studies. Mathematical procedures are available for carrying out almost any reactor optimization problem. The limitations appear to reside not so much in optimization methods as in formulation of the profit function and complete knowledge of the technical aspects of the process.

PROBLEMS

13-1 The vapor-phase hydration of ethylene is being carried out in a commercial-scale, fixed-bed reactor using spherical, phosphoric acid-on-kieselguhr catalyst pellets whose diameter is d_p. Ethylene and steam (15 moles of steam per mole of ethylene) enter the reactor at 150°C. The heat of reaction

$$C_2H_4(g) + H_2O(g) \rightarrow C_2H_5OH(g)$$

is essentially constant; $\Delta H_R = -22,000$ cal/g mol.

The system pressure is 40 lb/in.² abs and at these conditions no liquid phase is present. Also no reaction occurs in the absence of the catalyst.

External mass transfer (bulk-gas-to-pellet-surface) and intrapellet diffusion both can affect the performance of the reactor. The intrinsic rate at a catalyst site is first order in ethylene and first order in water vapor. Neglect the reverse reaction.

A. Derive an explicit equation for the conversion of ethylene leaving the reactor in terms of the mass-transfer coefficient, k_m, based upon the outer surface area of the pellet, the effectiveness factor, η, and other quantities as needed. Assume that there is no axial dispersion in the tubular reactor, no radial gradients of concentration or velocity, and assume isothermal conditions. Neglect pressure drop through the catalyst bed.

B. Calculate the conversion of C_2H_4 in the reactor effluent for the following conditions:

---

† H. Kramers and K. R. Westerterp, "Elements of Chemical Reactor Design and Operation," chap. 6, Academic Press, Inc., New York, 1963.

‡ See R. Aris, *Chem. Eng. Sci.*, **13**, 18 (1960).

§ F. Horn and U. Trolten, *Chem. Eng. Techol.*, **32**, 382 (1960).

1. Effective diffusivity of ethylene in the catalyst pellets at 150°C and 40 lb/in.² abs = 1.05 cm²/s
2. Pellet diameter, $d_p = \frac{1}{4}$ in.
3. External mass-transfer coefficient, $k_m = 1.0$ cm/s
4. Intrinsic rate constant for second-order reaction at a catalyst site, $k_2 = 2.13 \times 10^5$ cm⁶/(s)(mol)(g of catalyst)
5. Density of catalyst pellet, $\rho_p = 1.5$ g/cm³ of pellet
6. Density of bed of catalyst pellets, $\rho_B = 1.2$ g/(cm³ of reactor volume)
7. Diameter of reactor, $2R = 6$ in.
8. Length of catalyst bed = 5 ft
9. Total volumetric flow rate entering the reactor, at 150°C and 40 lb/in.² abs, $Q_F = 9.0$ ft³/s

13-2 Suppose it were possible to operate an ethyl benzene–dehydrogenation reactor under approximately isothermal conditions. If the temperature is 650°C, prepare a curve for conversion vs. catalyst-bed depth which extends to the equilibrium conversion. The catalyst to be used is that for which rate data were presented in Example 13-3. Additional data are:

Average pressure = 1.2 atm
Diameter of catalyst tube = 3 ft
Feed rate of ethyl benzene per tube = 8.0 lb mol/h
Feed rate of steam per tube = 225 lb mol/h
Bulk density of catalyst as packed = 90 lb/ft³

Equilibrium-constant data are given in Example 13-3.

13-3 In this case assume that the reactor in Prob. 13-2 operates adiabatically and that the entrance temperature is 650°C. If the heat of reaction is $\Delta H = 60{,}000$ Btu/lb mol, compare the curve for conversion vs. bed depth with that obtained in Prob. 13-2.

13-4 Begley[†] has reported temperature data taken in a bed consisting of $\frac{1}{4} \times \frac{1}{4}$-in. alumina pellets packed in a 2-in. pipe (actual 2.06 in. ID) through which heated air was passed. No reaction occurred. The tube was jacketed with boiling glycol to maintain the tube wall at about 197°C. For a superficial mass velocity (average) of air equal to 300 lb/(h)(ft²) the experimental temperatures at various packed-bed depths are as follows:

Radial position	Experimental temperature, °C				
	0.076 ft	0.171 ft	0.255 ft	0.365 ft	0.495 ft
0.0	378.7	354.7	327.8	299.0	279.3
0.1	377.2	353.7	327.0	298.0	278.9
0.2	374.6	349.9	324.1	294.7	277.0
0.3	369.5	343.9	319.7	289.2	273.2
0.4	360.3	336.3	313.8	282.1	267.6
0.5	347.7	327.4	306.4	274.0	260.8
0.6	331.9	316.1	298.2	265.0	252.7
0.7	313.2	300.7	287.9	254.8	243.8
0.8	291.0	282.8	273.1	242.2	234.5
0.9	256.5	257.9	244.2	224.8	224.6

Calculate the effective thermal conductivity $(k_e)_r$ vs. radial position from these data. Neglect axial dispersion.

---

† J. W. Begley, master's thesis, Purdue University, Purdue, Ind., February, 1951.

13-5 The temperature-profile data in Prob. 13-4 are to be represented by a constant k_e (across the tube diameter) and a wall heat-transfer coefficient h_w. Estimate the values of $(k_e)_r$ and h_w which best fit the temperature data. Note that at the boundary layer between the wall and the central core of the bed the following relation must apply:

$$h_w(T_i - T_w) = -(k_e)_r \left(\frac{\partial T}{\partial r}\right)_i$$

where $(k_e)_r$ is constant for central core of bed

$T_i =$ temperature at the interface between film and central core

$\left(\dfrac{\partial T}{\partial r}\right)_i =$ gradient in central core at the interface between core and wall film

13-6 The rate of the catalytic hydrogenation of carbon dioxide to produce methane is [*Ind. Eng. Chem.*, **47**, 140 (1955)]:

$$\text{Rate} = \frac{k p_{CO_2} p_{H_2}^4}{[1 + K_1 p_{H_2} + K_2 p_{CO_2}]^5}$$

where p_{CO_2} and p_{H_2} are partial pressures in atmospheres.

At a total pressure of 30 atm and 314°C, the values of the constants are

$$k = 7.0 \text{ kg mol of } CH_4/(\text{kg catalyst})(\text{h})(\text{atm})^{-5}$$
$$K_1 = 1.73 \text{ (atm)}^{-1}$$
$$K_2 = 0.30 \text{ (atm)}^{-1}$$

A. For an isothermal tubular-flow, fixed-bed, catalytic reactor with a feed rate of 100 kg mol/h of CO_2 and stoichiometric rate of hydrogen, calculate the mass of catalyst required for 20% conversion of the carbon dioxide. Assume that there are no diffusional or thermal resistances (i.e., the global rate is given by the above equation) and that axial dispersion in the reactor is negligible.

B. Repeat part A neglecting the change in total moles due to reaction.

13-7 In the German phthalic anhydride process† naphthalene is passed over a vanadium pentoxide (on silica gel) catalyst at a temperature of about 350°C. Analysis of the available data indicates that the rate of reaction (pound moles of naphthalene reacted to phthalic anhydride per hour per pound of catalyst) can be represented empirically by the expression

$$r = 305 \times 10^5 p^{0.38} e^{-28,000/R_g T}$$

where p is partial pressure of naphthalene in atmospheres and T is in degrees Kelvin. The reactants consist of 0.10 mole % naphthalene vapor and 99.9% air. Although there will be some complete oxidation to carbon dioxide and water vapor, it is satisfactory to assume that the only reaction (as the temperature is not to exceed 400°C) is

$$C_{10}H_8 + 4.5O_2 \rightarrow C_8H_4O_3 + 2H_2O + 2CO_2$$

The heat of this reaction is $\Delta H = -6300$ Btu/lb of naphthalene, but use a value of -7300 Btu/lb in order to take into account the increase in heat release owing to the small amount of complete oxidation. The properties of the reaction mixture may be taken as equivalent to those for air.

The reactor will be designed to operate at a conversion of 80% and have a production rate of 6000 lb/day of phthalic anhydride. It will be a multitude type (illustrated in Fig. 13-1) with heat-transfer salt circulated through the jacket. The temperature of the entering reactants will be raised to 340°C by preheating, and the circulating heat-transfer salt will maintain the inside of the reactor-tube walls at 340°C.

Determine curves for temperature vs. catalyst-bed depth for tubes of three different sizes: 1.0, 2.0, and 3.0 in. ID. In so doing, ascertain how large the tubes can be without exceeding the maximum

† *FIAT Repts. 984, 649; BIOS Repts. 1597, 957, 753, 666; CIOS Rept. XXVIII 29; XXVII 80, 89.*

permissible temperature of 400°C. The catalyst will consist of 0.20 by 0.20-in. cylinders, and the bulk density of the packed bed may be taken as 50 lb/ft^3 for each size of tube.† The superficial mass velocity of gases through each tube will be 400 lb/(h)(ft^2 tube area). Use the one-dimensional design procedure.

13-8 In order to compare different batches of catalysts for fixed-bed cracking operations it is desired to develop a numerical catalyst activity by comparing each batch with a so-called "standard catalyst" for which the x-vs.-W/F curve is known. If the activity of a batch is defined as the rate of reaction of that batch divided by the rate for the standard catalyst at the same conditions, which of the following two procedures would be the better measure of catalyst activity?

(a) Determine from the curves the values of W/F required to obtain the same conversion x, and call the activity the ratio

$$\frac{(W/F)_{standard}}{(W/F)_{actual}}$$

(b) Determine from the curves the values of x at the same W/F, and define the activity as

$$\frac{x_{actual}}{x_{standard}}$$

Sketch the x-vs.-W/F curve for a standard catalyst and a curve for a catalyst with an activity less than unity.

13-9‡ Design a reactor system to produce styrene by the vapor-phase catalytic dehydrogenation of ethyl benzene. The reaction is endothermic, so that elevated temperatures are necessary to obtain reasonable conversions. The plant capacity is to be 20 tons of crude styrene (styrene, benzene, and toluene) per day. Determine the bulk volume of catalyst and number of tubes in the reactor by the one-dimensional method. Assume that two reactors will be needed for continuous production of 20 tons/day, with one reactor in operation while the catalyst is being regenerated in the other. Also determine the composition of the crude styrene product.

With the catalyst proposed for the plant, three reactions may be significant:

$$C_6H_5C_2H_5 \overset{k_1}{\rightleftharpoons} C_6H_5CH{=}CH_2 + H_2$$

$$C_6H_5C_2H_5 \overset{k_2}{\rightleftharpoons} C_6H_6 + C_2H_4$$

$$H_2 + C_6H_5C_2H_5 \overset{k_3}{\rightleftharpoons} C_6H_5CH_3 + CH_4$$

The mechanism of each reaction follows the stoichiometry indicated by these reactions. The forward rate constants determined for this catalyst by Wenner and Dybdal are

$$\log k_1 = \frac{-11{,}370}{4.575T} + 0.883$$

$$\log k_2 = \frac{-50{,}800}{4.575T} + 9.13$$

$$\log k_3 = \frac{-21{,}800}{4.575T} + 2.78$$

where T is in degrees Kelvin, and

k_1 = lb mol styrene/(h)(atm)(lb catalyst)
k_2 = lb mol benzene/(h)(atm)(lb catalyst)
k_3 = lb mol toluene/(h)(atm)^2(lb catalyst)

† This represents an approximation, since the bulk density will depend to some extent on the tube size, especially in the small-diameter cases.
‡ From an example suggested by R. R. Wenner and F. C. Dybdal, *Chem. Eng. Progr.,* **44,** 275 (1948).

The overall equilibrium constants for the three reversible reactions are as follows:

t, °C	K_1	K_2	K_3
400	1.7×10^{-3}	2.7×10^{-2}	5.6×10^4
500	2.5×10^{-2}	3.1×10^{-1}	1.4×10^4
600	2.3×10^{-1}	2.0	4.4×10^3
700	1.4	8.0	1.8×10^3

The reactor will be heated by flue gas passed at a rate of 6520 lb/(h)(tube) countercurrent (outside the tubes) to the reaction mixture in the tubes. The flue gas leaves the reactor at a temperature of 1600°F. The reactant stream entering the reactor will be entirely ethyl benzene. The reactor tubes are 4.03 in. ID, 4.50 in. OD, and 15 ft long. The feed, 425 lb of ethyl benzene/(h)(tube), enters the tubes at a temperature of 550°C and a pressure of 44 lb/in.^{2} abs; it leaves the reactor at a pressure of 29 lb/in.^{2} abs. The heat-transfer coefficient between the reaction mixture and the flue gas is 9.0 Btu/(h) (ft^2)(°F) (based on outside area) and $\rho_B = 61$ lb/ft^3.

The thermodynamic data are as follows:

$$\text{Average specific heat of reaction mixture} = 0.63 \text{ Btu/(lb)(°F)}$$
$$\text{Average specific heat of flue gas} = 0.28 \text{ Btu/(lb)(°F)}$$
$$\text{Average heat of reaction 1, } \Delta H_1 = 53,600 \text{ Btu/lb mol}$$
$$\text{Average heat of reaction 2, } \Delta H_2 = 43,900 \text{ Btu/lb mol}$$
$$\text{Average heat of reaction 3, } \Delta H_2 = -27,700 \text{ Btu/lb mol}$$

To simplify the calculations assume that the pressure drop is directly proportional to the length of catalyst tube. Point out the possibility for error in this assumption, and describe a more accurate approach.

13-10 A pilot plant for the hydrogenation of nitrobenzene is to be designed on the basis of Wilson's rate data (see Example 13-5). The reactor will consist of a 1-in.-ID tube packed with catalyst. The feed, 2.0 mol % nitrobenzene and 98% hydrogen, will enter at 150°C at a rate of 0.25 lb mol/h. To reduce temperature variations the wall temperature of the tube will be maintained at 150°C by a constant-temperature bath. The heat-transfer coefficient between reaction mixture and wall may be taken equal to 20 Btu/(h)(ft^2)(°F).

Determine the temperature and conversion as a function of catalyst-bed depth for the conversion range 0 to 90%. Convert the rate equation in Example 13-4 to a form in which r_p is expressed as lb mol nitrobenzene reacting/(h)(lb catalyst) by taking the void fraction equal to 0.424 and the bulk density of catalyst as 60 lb/ft^3. The heat of reaction is $-274,000$ Btu/lb mol. The properties of the reaction mixture may be assumed to be the same as those for hydrogen.

13-11 Use the two-dimensional method to compute the conversion for bed depths up to 0.30 ft for the oxidation of sulfur dioxide under conditions similar to those described in Examples 13-6 and 13-7. The same reactor conditions apply, except that the superficial mass velocity in this case is 147 lb/(h)(ft^2), and the temperature profile at the entrance to the reactor is as shown below:

t, °C	352.0	397.5	400.4	401.5	401.2	397.4	361.2	197.0
Radial position	0.797	0.534	0.248	0.023	0.233	0.474	0.819	1.00

The measured conversions are given for comparison:

Catalyst-bed depth, in.	0	0.531	0.875	1.76	4.23	5.68
% SO$_2$ converted	0	26.9	30.7	37.8	41.2	42.1

13-12 A monolithic catalyst for a catalytic muffler is made using a metal support of high thermal conductivity. Suppose that the catalyst temperature T_s is constant along the entire reactor length. The air entering the reactor has a pollutant concentration C_f and temperature T_f. Using the nomenclature of Sec. 13-8, derive equations for the conversion and temperature in the air leaving the reactor. Assume that the rate of oxidation is first order (and irreversible) in pollutant concentration and that there is a large excess of oxygen. Hence, $r = A[\exp{(-E/RT_s)}]C_s$, where C_s is the pollutant concentration in the air at the catalyst surface. Neglect heat loss from the reactor.

Also, derive an equation for the constant surface temperature.

13-13 A fluid-bed reactor has been suggested for the oxidation of ethylene to ethylene oxide. The operating conditions would be 280°C and 1 atm pressure. The feed gas whose composition is $C_2H_4 = 8\%$, $O_2 = 19\%$ and 73% N_2 has a superficial velocity in the reactor of 2 ft/s at 280°C.

With the silver catalyst available the intrinsic reaction rate, g mol of ethylene oxide/(s)(g catalyst), is given by

$$r = 5.0C_{C_2H_4}$$

Two competing reactions occur:

$$C_2H_4 + \tfrac{1}{2}O_2 \rightarrow C_2H_4O$$
$$C_2H_4 + 3O_2 \rightarrow 2CO_2 + 2H_2O$$

The selectivity of ethylene oxide with respect to carbon dioxide is independent of conversion and equal to 1.5.

A. Calculate the conversion as a function of reactor height using the bubbling-gas model. Neglect the catalyst concentration in the bubbles, assume that the dense phase is a well-mixed batch fluid and that the gas bubbles rise in plug flow. Other properties of the fluidized bed are:

Density of catalyst particles in the dense phase $= 0.04$ g/cm^3
Mass-transfer coefficient, bubble to dense phase, $k_m a_v = 0.30$ s^{-1}
Volume fraction occupied by gas bubbles, $1 - \varepsilon_d = 0.10$

B. For comparison, calculate the conversion in a plug-flow reactor and in a stirred-tank reactor for the same bubble-phase residence time.

13-14 An irreversible first-order gaseous reaction $A \rightarrow B$ is carried out in a fluidized-bed reactor at conditions such that the rate constant is $k_1 = 0.076$ ft^3/(s)(lb catalyst). The superficial velocity is 1.0 ft/s and the bulk density of catalyst in the bed is 5.25 lb/ft^3. In this case assume that the entire bed has a uniform particle density ("single phase" rather than bubbling-gas concept) and that the mixing conditions correspond to plug flow modified by axial dispersion. The extent of dispersion can be evaluated from Gilliland and Mason's† equation

$$\frac{u}{D_L} = 2.6\left(\frac{1}{u}\right)^{0.61}$$

where u is superficial velocity, in feet per second, and D_L is axial diffusivity, in square feet per second.

What will be the conversion in the effluent from a reactor with a catalyst-bed depth of 5.0 ft? What would the conversion be if plug-flow behavior were assumed?

13-15 A first-order gaseous reaction $A \rightarrow B$ is carried out in a fluidized bed at 500°F and 2 atm pressure. At this temperature $k_1 = 0.05$ ft^3/(s)(lb catalyst). The bulk density of the catalyst bed is 3 lb/ft^3 and the superficial mass velocity of gas is 0.15 lb/(s)(ft^2). If the bed height is 10 ft, what will be the exit conversion? The molecular weight of component A is 44.

13-16 Repeat Prob. 13-15 for a reversible reaction for which the equilibrium constant is 0.6. Compare results obtained for plug flow of gas through the bed and by using the axial dispersion model (with D_L given by the correlation suggested in Prob. 13-14).

---

† E. R. Gilliland and E. A. Mason, *Ind. Eng. Chem.*, **41**, 1191 (1949).

13-17 A reaction $2A \rightarrow B$ is being studied in a fluidized-bed reactor at atmospheric pressure and 200°F. The global rate of reaction may be approximated by a second-order irreversible equation

$$\mathbf{r}_p = k_2 p_A^2$$

where $k_2 = $ lb mol/(s)(atm²)(lb catalyst). Assume that the "single phase" model described in Prob. 13-14 is applicable. At the operating temperature, $k_2 = 4.0 \times 10^{-6}$. The superficial gas velocity is 1.0 ft/s, and the bulk density of the fluidized catalyst 4.0 lb/ft³. (a) Calculate the conversion of A for plug flow of gas, with bed heights of 5, 10, and 15 ft. (b) Correct for the effect of longitudinal diffusion, using the diffusivity data given in Prob. 13-14.

13-18 In the slurry reactor of Example 13-10, an arithmetic average concentration of oxygen in the gas bubbles was used. Then the reaction rate was constant throughout the reactor. To evaluate the error introduced by this assumption, calculate the volume of bubble-free liquid required if the change in oxygen concentration is taken into account.

13-19 A. Reconsider Example 13-10 for a carbon particle size of $d_p = 0.542$ mm; the mass concentration, m_s, of particles will be the same, 0.070 g/(cm³ of water), and the size of the gas bubbles is unchanged so that k_L is 0.08 cm/s. For the 0.542 mm particles the effectiveness factor is 0.098 (note that for the 0.03 mm particles of Example 13-10, $\eta = 0.86$). What volume of water would be required? All other conditions are the same as in Example 13-10.

B. If m_s were reduced to 0.03 g/(cm³ of water) and the reactor volume was that found in part A, what would be the conversion of SO_2 to H_2SO_4?

13-20 A laboratory-scale, continuous slurry reactor is used for the polymerization of ethylene. A slurry of catalyst in cyclohexane is fed to the reactor at a rate of 10^3 cm³/min and the volume of liquid in the vessel is 10^4 cm³. Pure ethylene gas at a rate of 10^5 cm³/min is bubbled into the bottom of the vessel and is dispersed into bubbles which are uniformly distributed throughout the slurry.

At the operating conditions, the values of the transport coefficients are

$$k_L = 0.07 \text{ cm/s}$$

$$k_c = 0.03 \text{ cm/s}$$

The catalyst particle concentration is 0.10 g/cm³ of 0.10 mm particles (particle density $\rho_p = 1.0$ g/cm³). The bubbles will be about 3 mm in diameter and the bubble volume per unit volume of liquid will be 0.09.

While the kinetics are complex, assume that the rate of disappearance of ethylene is controlled by the first-order reaction

$$\mathbf{r}_{C_2H_4} = k a_c (C_L) \qquad \text{mol } C_2H_4/(s)(cm^3 \text{ of liquid})$$

$$k = 0.01 \text{ cm/s}$$

The effectiveness factor for the catalyst particles is unity. Henry's law constant for ethylene is 5 [mol/(cm³ of gas)]/[mol/(cm³ of liquid)].

Calculate the production rate of polymer in terms of moles of ethylene reacted per second. The entering slurry of cyclohexane and catalyst contains no dissolved ethylene or polymer.

13-21 In Example 13-9 the catalyst concentration was so large that the rate of hydrogenation was determined solely by the mass-transfer rate of hydrogen from the gas bubble to the liquid. Reconsider this example by accounting for the effect of liquid-to-particle mass transfer. For the catalyst particle size of 0.5 mm and for the agitation conditions, Fig. 10-10 gives $k_c = 0.02$ cm/s. The density of the particles is $\rho_p = 0.9$ cm³/s. What would be the minimum particle concentration (m_s) in the slurry if inclusion of liquid-to-particle mass transfer is not to change the overall rate of hydrogenation by more than 5%.

13-22 Reconsider Example 13-12 for a gas feed of 50% H_2 and 50% N_2. Neglect the solubility of nitrogen in the liquid. Note that for this case the mass balance for hydrogen in the gas phase is required.

A. Derive an equation for the fractional removal of thiophene as a function of catalyst bed depth.

B. Calculate the bed depth required for removal of 75% of the thiophene. The superficial gas velocity is 20 cm/s (at 200°C and 40 atm).

13-23 The removal of SO_2 from air is being investigated in a laboratory trickle-bed reactor as described in Example 13-13. Using the results of that example, and the following data, calculate the fractional removal of SO_2 from the air by reaction for a catalyst bed depth of 100 cm:

$$(H)_{O_2} = 5.0[\text{g mol/(cm}^3 \text{ of gas)}]/[(\text{g mol/(cm}^3 \text{ of liquid)}]$$
$$\text{Liquid flow rate} = 1.0 \text{ cm}^3/\text{s}$$
$$\text{Gas flow rate} = 1.0 \text{ cm}^3/\text{s}$$
$$(k_c a_c) = 0.20 \text{ s}^{-1}$$
$$(K_L a_g) = 0.02 \text{ s}^{-1}$$
$$\text{Intrinsic reaction rate constant, } k = 0.01 \text{ cm}^3/(\text{g})(\text{s})$$
$$\text{Catalyst particle effectiveness factor} = 0.1$$
$$\text{Density of catalyst bed, } \rho_B = 1.0 \text{ g/cm}^3$$
$$\text{Cross-sectional area of empty volume} = 5.0 \text{ cm}^2$$

13-24 A laboratory trickle-bed reactor is to be designed for the catalytic oxidation of dilute, aqueous solutions of acetic acid using air. A commercial iron-oxide catalyst, for which the intrinsic reaction rate is given by Eq. (A) of Example 10-10, is to be used. Operating conditions will be 67 atm and 252°C. The reactor, 2.54 cm ID, will be packed with 0.0541 cm catalyst particles. Liquid and gas-flow rates are 0.66 cm³/s and 3.5 cm³/s, measured at reactor temperature and pressure. For these catalyst particles the effectiveness factor is unity. The density, ρ_B, of the particles in the bed is 1.17 g/cm³ and the density of the particles themselves is 2.05 g/cm³. Oxygen is slightly soluble so that it is the limiting reactant in the oxidation:

$$CH_3COOH(aq) + 2O_2(g) \rightarrow 2CO_2(aq) + 2H_2O$$

Henry's law constant for oxygen is 2.78 [(g mol)/(cm³ of gas)]/[(g mol)/(cm³ of liquid)] when the concentration in the gas is reported in terms of a volume measured at 252°C and 5.5 atm. The mass-transfer coefficients may be taken as those found in Example 10-9

$$(K_L a_g)_{O_2} = (k_L a_g)_{O_2} = 0.024 \text{ s}^{-1}$$

$$(k_c a_c)_{O_2} = 2.2 \text{ s}^{-1}$$

$$(k_c a_c)_{HA} = 1.5 \text{ s}^{-1}$$

where HA designates acetic acid.

The feed concentrations in the liquid are 2.40×10^{-7} g mol/cm³ for oxygen and 33.7×10^{-7} g mol/cm³ (about 200 ppm) for acetic acid. The gas feed is saturated with water vapor at 252°C ($p_{H_2O} = 40.8$ atm) so that its oxygen partial pressure is $(67 - 40.8)0.21 = 5.5$ atm. Calculate the conversion of acetic acid as a function of catalyst bed depth. Assume plug flow of liquid.

A numerical solution of the mass-conservation equations and boundary conditions is required. These equations are similar to Eqs. (13-64) to (13-68) and (13-61) to (13-63), except that the intrinsic rate is not second order.

FOURTEEN

FLUID-SOLID NONCATALYTIC REACTIONS

There is an important class of fluid-solid reactions in which the solid is a reactant rather than a catalyst. Illustrations are[†]

$$CaCO_3(s) \rightarrow CaO(s) + CO_2(g) \qquad \text{lime kiln}$$

$$Fe_2O_3(s) + 3C(s) + \tfrac{1}{2}O_2(g) \rightarrow 2Fe(s) + CO_2(g) + 2CO(g) \qquad \text{blast furnace}$$

$$ZnS(s) + \tfrac{3}{2}O_2(g) \rightarrow ZnO(s) + SO_2(g) \qquad \text{roasting of ores}$$

$$CaCO_3(aq) + 2NaR(s) \rightarrow Na_2CO_3(aq) + CaR_2(s) \qquad \text{ion-exchange reaction}$$

$$2C(s) + \tfrac{3}{2}O_2(g) \rightarrow CO_2(g) + CO(g) \qquad \text{catalyst regeneration, combustion of coal, etc.}$$

$$C(s) + H_2O(g) \rightarrow CO(g) + H_2(g) \qquad \text{coal gasification}$$

In the first four cases solid products are formed, so that the original particles are replaced with another solid phase as reaction takes place. In the fifth and sixth reactions the solid phase disappears (except for the small amount of ash formed). Since the amount of reactant surface and its availability (in the first four cases) change with extent of reaction, the global rate changes with time. Calculating reactor performance for heterogeneous noncatalytic reactions involves a combination of transport processes and intrinsic kinetics, with the added factor that the properties of the solid reactant change during the course of reaction. Our

---

[†] For a more complete survey see J. Szekely, J. W. Evans and Hong Yong Sohn "Gas-Solid Reactions", Academic Press, New York, 1976; C. Y. Wen, *Ind. Eng. Chem.*, **60**, 34 (1968).

approach toward design in this chapter is the same as for catalytic reactions (Chap. 13). First, methods of predicting the global rate are discussed (Secs. 14-2 and 14-3). Then models for obtaining the performance of the whole reactor are considered (Secs. 14-4 to 14-6). The global rate is the rate expressed in terms of properties of the bulk fluid. Such rates can be developed by examining the behavior of a *single* particle surrounded by fluid. This single-particle basis is employed in Sec. 14-3. It is particularly appropriate for noncatalytic reactions since in many commercial units the particles move individually and continuously through the reactor (e.g., lime kiln, blast furnace, coal gasification, etc.).

14-1 Design Concepts

Reactor design problems for fluid-solid, noncatalytic reactions are similar to those considered in Chap. 13. The special feature of the noncatalytic case is that the rate of reaction is a function of time. If the particles are in continuous flow, the reactor can operate at steady state. Then a proper model for the reactor as a whole will account for the nature of the flow of the solid and fluid phases. If the particles remain in the reactor, as in a batch-fluidized bed or fixed-bed arrangement, with a continuous flow of fluid, steady-state operation is not possible. The global rate varies with reaction time as well as with position in the reactor. Regeneration of deactivated catalyst in a fixed-bed reactor is an example of this type of process.

Models used to represent the continuous flow of particles and fluid can take various forms. For reactions producing a solid product "transfer line" reactors (see Sec. 13-11) may be used. For an approximate treatment for this type, both particles and fluid could be assumed to move through the reactor in plug flow. In a more elaborate model derivations from plug flow in the fluid could be accounted for with a dispersion term in the mass conservation expression. Deviations from plug flow of particles can be treated in terms of a residence-time distribution function. Such a distribution is likely if the particles in the feed do not all have the same size. The residence time concept is advantageous for fluid–solid–noncatalytic reactors, because the flow is completely segregated. That is, the solid particles do not coalesce. The conversion of solid reactant can be evaluated by the method outlined in Sec. 6-8 [Eq. (6-41)]. To use this method the residence-time distribution needs to be determined under actual operating conditions. Also the conversion-vs.-time relation must be established for a single particle. This relation can be found from the global rate as discussed in Sec. 14-3.

In some applications there is a large excess of reactant in the fluid, or the fluid is well mixed, so that the fluid reactant concentration is the same throughout the reactor. Then a model for flow of the fluid phase is not pertinent. All that is necessary to obtain the average conversion of the solid reactant is the conversion-vs.-time relation and the residence-time distribution for the particles.

When the particles are in a fixed bed, the models developed in Chap. 13 are applicable. Similarly, for batch-fluidized beds, the models discussed in Sec. 13-10, can be used. However, in both cases the global rate and behavior of the reactor as a whole becomes a function of time on stream.

Some of the foregoing concepts are developed and applied in Secs. 14-4 to 14-6.

SINGLE PARTICLE BEHAVIOR

14-2 Kinetics and Mass Transfer

When a component in a fluid reacts with a solid, the sequence of steps will be similar to that for fluid-solid catalytic reactions. Reaction at an active site on the solid must occur by adsorption of fluid reactant at the site, followed by surface reaction involving the adsorbed molecule. For the fluid molecule to reach the active site it must first be transported to the outer surface and then diffuse into the particle to an active site. Hence, for an *irreversible* reaction, the global rate will be determined by the four steps:

1. Mass transfer from bulk fluid to the outer surface of the particle
2. Intraparticle diffusion into the particle
3. Adsorption at an active site of solid reactant
4. Intrinsic reaction at the site

Combination of steps 3 and 4 gives the rate in terms of the concentration of reactant in the fluid at an interior site. In Chap. 9, methods of formulating such rates were considered for catalytic reactions, and similar considerations apply for noncatalytic reactions. Hence, we will not be concerned with the detailed mechanism of the steps at the site. It is noted, however, that in writing Eq. (9-1) [or (7-13)], it was assumed that the rate of adsorption of fluid reactant was proportional to the concentration, $\bar{C}_m - \bar{C}$, of vacant sites. This same assumption will be made in this chapter, and, in addition, it is supposed that the number of sites is proportional to the surface area of the solid reactant.

A key parameter in formulating the rate of reaction for a single particle is how the active sites and surface areas for adsorption are distributed. The porosity of the unreacted particle has a major effect on this parameter. Several models have been proposed.† Consider a general form of reaction between gas A and solid reactant B which produces one solid product (F):

$$A(g) + bB(s) \rightarrow E(g) + F(s)$$

The effect of particle porosity is evident by considering three cases:

Shrinking core If reactant B is nonporous, the reaction will occur at its outer surface. This surface recedes with extent of reaction (and time), as shown in Fig. 14-1a. As reaction occurs, a layer of product F builds up around the unreacted core of reactant. A porous particle might also behave in this way if the resistance to reaction is much less than the resistance to diffusion of fluid reactant in the pores of the particle. The key factor in this model is that the reaction always occurs at a surface boundary, that is, at the interface between unreacted core and surrounding solid product.

---

† An extensive review of models is available: C. Y. Wen, *Ind. Eng. Chem.,* **60,** 34 (1968).

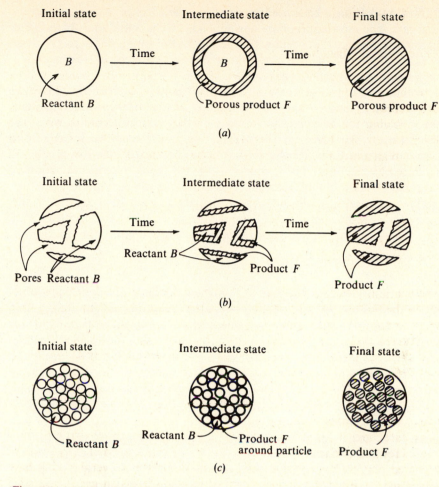

Figure 14-1 Models for gas-solid noncatalytic reactions of the type $A(g) + bB(s) \rightarrow E(g) + F(s)$. (a) Shrinking-core model. (b) Highly porous reactant. (c) Porous reactant pellet composed of non-porous particles.

Highly porous reactant (no pore-diffusion resistance) Suppose the solid reactant is so porous that the fluid reactant can reach all parts of the solid without diffusion resistance; this is depicted in Fig. 14-1b. Now the rate per particle will vary as the surface of solid reactant changes with time and the layer of solid product accumulates. The key factor here is that the concentration of reactant in the fluid phase is the same at any location within the particle.

Porous reactant (*intermediate* pore-diffusion resistance) An example of this case would be a solid reactant formed by compressing nonporous particles into a porous pellet, as shown in Fig. 14-1c. The pores surrounding the particles are supposed to be small enough that the fluid reactant concentration decreases significantly toward the center of the pellet.

All three models have been used to represent the effect of mass transport on the global rate. The most appropriate one depends on the initial form of the solid reactant and the changes that occur with reaction. In the remainder of this chapter the shrinking-core model will be used. It is amenable to analytical treatment and represents many real systems rather well.

The intermediate porosity case is more realistic. It has been developed in detail as the grain model[†] and also described as the particle-pellet model.[‡] However, solutions for the conversion are more involved since conservation equations for both the particles and the pellet (Fig. 14-1c) must be solved. Both isothermal and nonisothermal behavior has been considered with this model[‡]

Our next objective is to derive equations for the global rate as a function of time and concentration of fluid reactant. This is done in the next section, but first it is helpful to consider briefly how rates of reaction may be measured in the laboratory. One method is to record the weight of one or more particles as a function of time. Figure 14-2 shows an apparatus used[§] to study the reaction:

$$UO_2(s) + 4HF(g) \rightarrow 2H_2O(g) + UF_4(s)$$

A single UO_2 pellet (initial diameter = 2 cm) was weighed at various times by attaching a balance to the wire holding the pellet. From the weight change, the conversion of UO_2 to UF_4 could be calculated from the stoichiometry of the reaction. The stirrer was rotated about the suspended pellet in order to obtain a uniform gas composition (ideal stirred-tank performance). The resultant conversion-vs.-time data represent the integration of the global rate equation from the start of the experiment ($t = 0$). To evaluate a suitable rate equation, various possibilities can be integrated and compared with the data. Alternately, the curve of conversion vs. time can be differentiated to find the rate at any time. For this system the UO_2 pellet was nearly nonporous, corresponding to Fig. 14-1a. The controlling step in the global rate changes with time. At low times the thickness of the $UF_4(s)$ layer is low so that the intrinsic processes at a site (steps 3 and 4 in prior list) determine the global rate. At long times the product layer will be relatively large and step 2 will likely be controlling. By suitable adjustment of the stirrer speed the external mass transport resistance (step 1) could be made negligible. Despite the stoichiometry of this reaction, the intrinsic rate has been found[¶] to be first order in $HF(g)$ concentration. Possibly, the adsorption of HF (step 3 of prior list) determines the surface rate.

More complete data can be obtained by measuring the composition of the effluent gas. This permits a mass balance to test the accuracy of the data. With a single pellet or small mass of particles this is difficult, because the quantity of gaseous products may be insufficient to measure accurately. The apparatus of

† J. Szekely, J. W. Evans, and H. Y. Sohn, op. cit.

‡ A. Calvelo and J. M. Smith, Proceedings of CHEMECA '70, paper 3.1, Butterworths (Australia) August, 1971.

§ E. C. Costa and J. M. Smith, *AIChE J.*, **17**, 947 (1971).

¶ L. Tomlinson, S. A. Morrow, and S. Graves, *Trans. Faraday Soc.*, **57**, 1008 (1961).

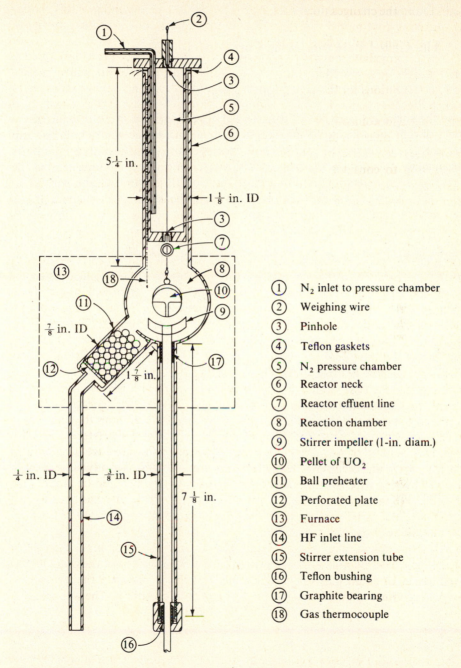

1 N_2 inlet to pressure chamber
2 Weighing wire
3 Pinhole
4 Teflon gaskets
5 N_2 pressure chamber
6 Reactor neck
7 Reactor effuent line
8 Reaction chamber
9 Stirrer impeller (1-in. diam.)
10 Pellet of UO_2
11 Ball preheater
12 Perforated plate
13 Furnace
14 HF inlet line
15 Stirrer extension tube
16 Teflon bushing
17 Graphite bearing
18 Gas thermocouple

Figure 14-2 Stirred-tank single-pellet reactor for hydrofluorination of uranium dioxide.

Fig. 14-2 is a form of thermal gravimetric analysis (TGA) equipment. Such equipment is available as a complete unit and has been used for kinetic studies of solid-noncatalytic reactions.†

14-3 The Global Rate (Shrinking-Core Model)

Suppose that the reaction

$$A(g) + bB(s) \rightarrow E(g) + F(s)$$

obeys the shrinking-core model (Fig. 14-1a), where solid reactant B is initially a sphere of radius r_s. The solid sphere is in contact with gas A, whose bulk concentration is C_b. Consider the case where the temperature is uniform throughout the heterogeneous region. As reaction occurs, a layer of product F forms around the unreacted core of reactant B. For further reaction this layer must be porous, so that reaction occurs by diffusion of A through the layer of F to react at the interface between F and unreacted core. This situation is shown in Fig. 14-3,

---

† M. Suzuki, D. M. Misic, D. M. Koyama, and K. Kawazoe, *Chem. Eng. Sci.*, **33**, 271 (1978).

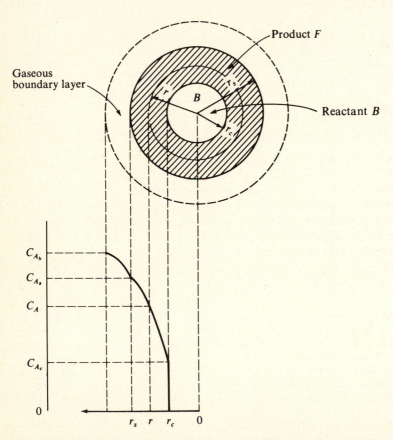

Figure 14-3 Concentration profile in a spherical pellet (shrinking-core model).

where the concentrations of A are labeled at various locations. The shape of the concentration profile from bulk gas to reacting surface is also indicated. It is assumed that the pellet retains its spherical shape during reaction. It is also assumed, for convenience, that the densities of the porous product and the reactant B are the same, so that the total radius of the pellet does not change with time and there is no gaseous region between the pellet and the product layer F.†

One more restriction must be placed on the system before a simple mathematical analysis is feasible: the rate of movement of the reaction interface at r_c, that is, dr_c/dt, is small with respect to the velocity of diffusion of A through the product layer. The requirement for this *pseudo*-steady-state concept has been carefully developed,‡ but approximately stated, it is valid if the density of the gas in the pores of the product layer is small with respect to the density of solid reactant B. This is usually the case.

Granting pseudo-steady-state conditions, the three rates—diffusion of A through the boundary layer, diffusion through the layer of product, and reaction at the interface—are identical. By equating the expressions for each of these processes, the concentration $(C_A)_c$ can be written in terms of the known $(C_A)_b$ and the radius of the unreacted core r_c. The three rate equations, expressed as moles of A disappearing per unit time per particle, are

$$-\frac{dN_A}{dt} = 4\pi r_s^2 k_m[(C_A)_b - (C_A)_s] \qquad \text{external diffusion} \qquad (14\text{-}1)$$

$$-\frac{dN_A}{dt} = 4\pi r_c^2 D_e\left(\frac{dC_A}{dr}\right)_{r=r_c} \qquad \text{diffusion through product} \qquad (14\text{-}2)$$

$$-\frac{dN_A}{dt} = 4\pi r_c^2 k(C_A)_c \qquad \text{reaction at } r_c \qquad (14\text{-}3)$$

In Eq. (14-1) k_m is the external mass-transfer coefficient defined in Chap. 10 [Eq. (10-1)]. The rate through the product layer, Eq. (14-2), is evaluated at $r = r_c$; D_e is the effective diffusivity of A through this porous layer. In writing Eq. (14-3) the chemical reaction at r_c is assumed to be first order in A and irreversible. It is also taken as directly proportional to the outer surface area of unreacted core B.§

To evaluate the gradient in Eq. (14-2) consider the diffusion of A through the layer of F. With the pseudo-steady-state assumption this process can be evaluated independently of the change in r_c. Consider a small element of thickness Δr at

† It is interesting to note that K. G. Denbigh and G. S. G. Beveridge [*Trans. Inst. Chem. Engrs.*, **40**, 23 (1962)], in studying the oxidation of ZnS at high temperatures, observed a gaseous region between solid ZnS and solid ZnO. This was due to vaporization of some of the ZnS.

‡ K. B. Bischoff, *Chem. Eng. Sci.*, **18**, 711 (1963).

§ The number of active sites per unit surface of B will presumably be constant, and the rate proportional to the total number of sites. Hence the rate should be proportional to the surface area of the unreacted core. A feature of the shrinking-core model is that this area is known and is equal to $4\pi r_c^2$ for a spherical core. This is not a realistic area for reaction in real situations, but it is the characteristic of the model that permits mathematical analysis of the process. The actual area, even for nonporous particles, would not be a plane surface but would be much larger due to small-scale indentations in the surface.

location r in the product layer (Fig. 14-3). The steady-state mass conservation expression for A in this layer is

$$-\left(\pi r^2 D_e \frac{dC_A}{dr}\right)_r - \left(\pi r^2 D_e \frac{dC_A}{dr}\right)_{r+\Delta r} = 0 \tag{14-4}$$

Taking the limit as $\Delta r \to 0$ gives

$$\frac{d}{dr}\left(r^2 D_e \frac{dC_A}{dr}\right) = 0 \tag{14-5}$$

If this expression is integrated twice, with the boundary conditions

$$C_A = \begin{cases} (C_A)_s & \text{at } r = r_s \\ (C_A)_c & \text{at } r = r_c \end{cases}$$

used to evaluate the integration constants, the result is

$$C_A - (C_A)_c = [(C_A)_s - (C_A)_c]\frac{1 - r_c/r}{1 - r_c/r_s} \tag{14-6}$$

This expression for the concentration profile may be differentiated with respect to r and then evaluated at $r = r_c$ to give

$$\left(\frac{dC_A}{dr}\right)_{r=r_c} = \frac{(C_A)_s - (C_A)_c}{r_c(1 - r_c/r_s)} \tag{14-7}$$

Substituting Eq. (14-7) in Eq. (14-2) yields

$$-\frac{dN_A}{dt} = 4\pi r_c D_e \frac{(C_A)_s - (C_A)_c}{1 - r_c/r_s} \tag{14-8}$$

Now $(C_A)_s$ and dN_A/dt can be eliminated from Eqs. (14-1), (14-3), and (14-8) to give $(C_A)_c$ in terms of $(C_A)_b$ and r_c. The result is

$$(C_A)_c = \frac{(C_A)_b}{1 + (r_c^2/r_s^2)(k/k_m) + (kr_c/D_e)(1 - r_c/r_s)} \tag{14-9}$$

Using Eq. (14-9) for $(C_A)_c$ in Eq. (14-3) gives the global rate in terms of $(C_A)_b$ and r_c,

$$\text{Rate per particle} = -\frac{dN_A}{dt} = \frac{4\pi r_c^2 (C_A)_b k}{1 + (r_c^2/r_s^2)(k/k_m) + (kr_c/D_e)(1 - r_c/r_s)} \tag{14-10}$$

This rate may be expressed in terms of the rate per unit volume of reactor by multiplying $-dN_A/dt$ by the particle density (particles per unit volume).

Since r_c is a variable, Eq. (14-10) is not useful for reactor design until we express r_c as a function of time. According to the spherical geometry of the particle, the rate of reaction of B (moles per unit time per particle) may be written as

$$\frac{dN_B}{dt} = \frac{\rho_B}{M_B}\frac{d}{dt}\left(\frac{4}{3}\pi r_c^3\right) = \frac{4\pi r_c^2 \rho_B}{M_B}\frac{dr_c}{dt} \tag{14-11}$$

where ρ_B represents the density of solid reactant B.

From the stoichiometry of the reaction,

$$\frac{dN_A}{dt} = \frac{1}{b}\frac{dN_B}{dt} = \frac{4\pi r_c^2 \rho_B}{bM_B}\frac{dr_c}{dt} \tag{14-12}$$

Combining this result with Eq. (14-3) for dN_A/dt gives

$$\frac{dr_c}{dt} = -\frac{bM_B k}{\rho_B}(C_A)_c \tag{14-13}$$

Finally, substituting Eq. (14-9) for $(C_A)_c$ in Eq. (14-13) provides a differential equation whose solution gives $r_c = f(t)$,

$$-\frac{dr_c}{dt} = \frac{bM_B k(C_A)_b/\rho_B}{1 + (r_c^2/r_s^2)(k/k_m) + (kr_c/D_e)(1 - r_c/r_s)} \tag{14-14}$$

Equation (14-14) can be integrated to give r_c as a function of $(C_A)_b$ and t. Substituting this expression for r_c in Eq. (14-10) gives the desired expression for the global rate in terms of $(C_A)_b$ and time, t. However, before Eq. (14-14) can be integrated, it is necessary to know how $(C_A)_b$ varies with time; that is, what are the values of $(C_A)_b$ that the particle is exposed to as it moves through the reactor or as the fluid flows past (for the fixed-bed case)? This is determined by the model used to represent the flow of particles and of fluid through the reactor.

REACTOR MODELS

The simplest flow situation for the fluid is for constant $(C_A)_b$ throughout the reactor. This occurs when the fluid is well mixed, or when there is a large excess of reactant A, regardless of the degree of mixing. We first develop conversions-vs.-time relations for *single* particles (Sec. 14-4), and then use these relations for predicting the performance of reactors where $(C_A)_b$ is constant (Sec. 14-5). When $(C_A)_b$ varies with position in the reactor, a mass conservation equation for A in the fluid phase must be combined with Eq. (14-14) before r_c, and the conversion, can be obtained. Reactors corresponding to this situation are considered in Sec. 14-6.

14-4 Conversion vs. Time for Single Particles (Constant Fluid Concentration)

If $(C_A)_b$ is constant, Eq. (14-14) is easily integrated. Starting with $r_c = r_s$ at $t = 0$,

$$-\frac{bM_B k(C_A)_b}{\rho_B}\int_0^t dt = \int_{r_s}^{r_c}\left[1 + \frac{r_c^2}{r_s^2}\frac{k}{k_m} + \frac{kr_c}{D_e}\left(1 - \frac{r_c}{r_s}\right)\right]dr_c \tag{14-15}$$

It is convenient to express the result in terms of a dimensionless time,

$$t^* = \frac{bM_B k(C_A)_b}{\rho_B r_s}t \tag{14-16}$$

and two groups relating the diffusion and reaction resistances,

$$Y_1 = \frac{D_e}{k_m r_s} = \frac{\text{external diffusion resistance}}{\text{diffusion resistance in product layer}} \quad (14\text{-}17)$$

$$Y_2 = \frac{k r_s}{D_e} = \frac{\text{diffusion resistance in product layer}}{\text{reaction resistance at } r_c} \quad (14\text{-}18)$$

In terms of these paramters Eq. (14-15) can be integrated to give

$$t^* = \left(1 - \frac{r_c}{r_s}\right)\left\{1 + \frac{Y_1 Y_2}{3}\left[\left(\frac{r_c}{r_s}\right)^2 + \frac{r_c}{r_s} + 1\right] + \frac{Y_2}{6}\left[\left(\frac{r_c}{r_s} + 1\right) - 2\left(\frac{r_c}{r_s}\right)^2\right]\right\} \quad (14\text{-}19)$$

Equation (14-19) could, in principle, be solved for r_c, and this result substituted in Eq. (14-10), to obtain the global rate. Practically, this is not possible because Eq. (14-19) is explicit in t. However, this is not necessary when $(C_A)_b$ is constant. The conversion of reactant B can be evaluated from r_c without using the global rate. Thus x_B is related to the radius of the unreacted core by the expression

$$x_B = \frac{\text{initial mass} - \text{mass at } t}{\text{initial mass}} = \frac{\frac{4}{3}\pi r_s^3 \rho_B - \frac{4}{3}\pi r_c^3 \rho_B}{\frac{4}{3}\pi r_s^3 \rho_B}$$

or

$$x_B = 1 - \left(\frac{r_c}{r_s}\right)^3 \quad (14\text{-}20)$$

Eliminating r_c/r_s between Eqs. (14-19) and (14-20) gives the desired relationship between conversion and time.

$$t^* = [1 - (1 - x_B)^{1/3}]\left\{1 + \frac{Y_1 Y_2}{3}[(1 - x_B)^{2/3} + (1 - x_B)^{1/3} + 1]\right.$$

$$\left. + \frac{Y_2}{6}[(1 - x_B)^{1/3} + 1 - 2(1 - x_B)^{2/3}]\right\} \quad (14\text{-}19a)$$

Equation (14-19a) establishes the conversion as a function of time for single particles in a situation where $(C_A)_b$ is constant. Constant total radius r_s has been assumed, as well as an irreversible first-order reaction in A. Solutions when these assumptions are not made are also possible,† but the results are more complicated. Isothermal conditions have also been chosen. This restriction can be eliminated by writing expressions similar to Eqs. (14-1) to (14-3) in order to account for the effect of the heat of reaction and the effect of temperature on the rate. The nonisothermal case also has been solved for the shrinking core model.†

When all three resistances are not significant Eq. (14-19) is less complicated. Two of these simpler results are considered in the following paragraphs.

Chemical reaction controlling If the gas-phase velocity relative to that of the solid particle is high, as in a fixed-bed reactor, external-diffusion resistance may be negligible. Also, for a highly porous product layer and for low conversions, diffu-

---

† J. Shen and J. M. Smith, *Ind. Eng. Chem., Fund. Quart.*, **4**, 293 (1965).

sion resistance through the product may be small. Under such conditions the intrinsic reaction step at r_c will determine the rate, and $Y_2 \rightarrow 0$. Then Eq. (14-19) reduces to

$$t^* = 1 - \frac{r_c}{r_s}$$

or with Eqs. (14-16) and (14-20)

$$t = \frac{\rho_B r_s}{b M_B k (C_A)_b}\left(1 - \frac{r_c}{r_s}\right) = \frac{\rho_B r_s}{b M_B k (C_A)_b}[1 - (1 - x_B)^{1/3}] \qquad (14\text{-}21)$$

The time for complete conversion ($x_B = 1$ and $r_c = 0$) is

$$t_{x_B=1} = \frac{\rho_B r_s}{b M_B k (C_A)_b} \qquad (14\text{-}22)$$

Diffusion through product controlling For rapid chemical reaction at the interface and a low D_e, diffusion through the product layer may determine the rate, even at small conversions. If this is the case, $Y_1 = 0$, Y_2 is large, and Eq. (14-19) becomes

$$t^* = \left(1 - \frac{r_c}{r_s}\right)\frac{Y_2}{6}\left[\frac{r_c}{r_s} + 1 - 2\left(\frac{r_c}{r_s}\right)^2\right] \qquad (14\text{-}23)$$

or, from the definitions of t^* and Y_2,

$$t = \frac{\rho_B r_s^2}{6 D_e b M_B (C_A)_b}\left[1 - 3\left(\frac{r_c}{r_s}\right)^2 + 2\left(\frac{r_c}{r_s}\right)^3\right] \qquad (14\text{-}24)$$

In this case the time for complete conversion of B depends on D_e and is given by

$$t_{x_B=1} = \frac{\rho_B r_s^2}{6 D_e b M_B (C_A)_b} \qquad (14\text{-}25)$$

Weisz and Goodwin† have used Eq. (14-24) successfully to explain how the extent of carbon removal on a deactivated catalyst varies with time. In this case the reactions are

1. $C(s) + O_2(g) \rightarrow CO_2(g)$
2. $C(s) + \frac{1}{2}O_2(g) \rightarrow CO(g)$

and so no solid product is formed. However, the carbon has been deposited throughout the pores of a porous catalyst pellet (for example, petroleum cracking catalysts become deactivated by such carbon deposition). If the rate of the chemical reaction is rapid with respect to the rate of diffusion of oxygen into the pellet, the carbon will be burned off the pellet according to the shrinking-core model. Since such oxidation reactions are normally carried out at a high temperature, they are intrinsically fast, and the shrinking-core assumption is often a reasonable

† P. B. Weisz and R. D. Goodwin, *J. Catalysis,* **2**, 397 (1963); *J. Catalysis,* **6**, 227 (1966).

one. The shrinking-core concept of reaction at a sharp interface has been applied to several other systems, including hydrofluorination of UO_2 and poison deposition on porous catalysts.†

14-5 Conversion in Reactors with Constant Fluid Composition

After the t-vs.-x_B relation has been determined for single particles, the conversion in the reactor can be found from the residence time (θ), or reaction time (t).‡ If all the particles have the same residence time, the result is given immediately by substituting the time value in Eq. (14-19a), or in the appropriate simplified forms: Eq. (14-20) with either (14-21) or (14-24). If the particles have a distribution $J(\theta)$ of residence times, the average conversion can be obtained from $J(\theta)$ and the conversion-vs.-time relation, using Eq. (6-41).

Example 14-1 illustrates the application of these equations for a constant residence, or reaction, time. These conditions would occur, for example, in a transport reactor (Fig. 14-5, cocurrent flow of both fluid and particles) or in a fixed-bed reactor when there was a large excess of reactant A. A distribution of residence times could occur for several reasons when there is a continuous flow of solids. For example, if there is a distribution of particle sizes in a transport reactor the particle velocities may be different, giving rise to a significant variation in residence times. This case is considered in Example 14-2. Another possibility is a *continuous* fluidized-bed reactor, for example, a regenerator for burning carbon from spent cracking catalyst (Fig. 13-21).

Example 14-1 The reduction of FeS_2 particles,

$$FeS_2(s) + H_2(g) \rightarrow FeS(s) + H_2S(g)$$

has been studied under conditions where the concentration of hydrogen in the gas phase was essentially constant.§ Hydrogen at a high flow rate was passed at atmospheric pressure through beds of FeS_2 particles. Under these conditions all the particles are exposed to the same concentration of hydrogen. Also the residence time, which in this case is the reaction time, is the same for all particles. The results indicated a first-order (with respect to hydrogen) reaction. Measurements were made at 450°, 477°, and 495°C, and an activation energy of 30,000 cal/g mol (1.26×10^5 kJ/kg mol) was proposed. The experimental data for conversion of FeS_2 vs. time are shown in Fig. 14-4.

Determine whether the shrinkage-core model will fit these data and evaluate the rate constant (the frequency factor in the rate equation) and an

---

† P. B. Weisz and C. D. Prater, "Advances in Catalysis," vol. VI, p. 143, Academic Press, Inc., New York, 1954; E. C. Costa and J. M. Smith, *Proc. Fourth European Symp. Chem. Reaction Eng.*, Brussels, Sept. 9–11, 1968; J. J. Carberry and R. L. Gorring, *J. Catalysis*, **5**, 529 (1966).

‡ If there is a continuous flow of particles through the reactor, the particle residence time, θ, is appropriate. If the particles remain in the reactor, as in a fixed-bed, the proper time quantity is the time of reaction, t.

§ G. M. Schwab and J. Philinis, *J. Am. Chem. Soc.*, **69**, 2588 (1947).

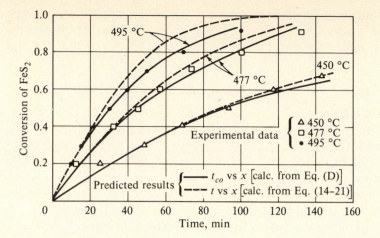

Figure 14-4 Conversion-vs.-time results for hydrogenation of FeS_2.

effective diffusivity D_e. The particles were granular and varied in size from 0.01 to 0.1 mm, but assume that a spherical particle with an average radius of 0.035 mm can represent the mixture.

SOLUTION Since the gas-flow rate is high, the external-diffusion resistance should be small, and Y_1 will approach zero. Furthermore, at low conversions the product layer of FeS will be thin, so that the chemical reaction at the interface may control the rate. This is more likely to be a good assumption at the lowest temperature, where the reaction rate is smallest. Under these conditions Eq. (14-21) is applicable, and an appropriate way to initiate the analysis is to apply this equation to the data at 450°C (723 K). Substituting numerical values $\left(\rho_{FeS_2} = 5.0 \text{ g/cm}^3 \text{ or } 5.0 \times 10^3 \text{ kg/mol}^3\right)$, we have

$$t = \frac{5.0(0.0035)}{1(120)(C_A)_b k}[1 - (1 - x_{FeS_2})^{1/3}] \tag{A}$$

From the ideal-gas law at 450°C,

$$(C_A)_b = \frac{p}{R_g T} = \frac{1}{82(273 + 450)}$$

$$= 1.69 \times 10^{-5} \text{ g mol/cm}^3 \ (1.69 \times 10^{-2} \text{ kg mol/m}^3)$$

Hence Eq. (A) for 450°C is

$$t = \frac{8.6}{k}[1 - (1 - x_{FeS_2})^{1/3}] \tag{B}$$

The data points in Fig. 14-4 can be used with Eq. (B) to evaluate k. It is found that $k = 0.019$ cm/min, or 3.2×10^{-4} cm/s (3.2×10^{-6} m/s), gives a curve in good agreement with the data. This is the dashed curve for 450°C in Fig. 14-4.

According to the Arrhenius equation,

$$k = Ae^{-E/R_gT}$$

Hence

$$A = \frac{3.2 \times 10^{-4}}{e^{-30,000/(1.98)(723)}} = 3.8 \times 10^5 \text{ cm/s } (3.8 \times 10^3 \text{ m/s})$$

Then the rate constant for any temperature is given by

$$k = 3.8 \times 10^5 e^{-30,000/R_gT} \text{ (cm/s)} \qquad (C)$$

By calculating k from Eq. (C) and using it in Eq. (B) we can calculate t-vs.-x curves for 477 and 495°C. These are the dashed lines in Fig. 14-4. At these higher temperatures the predicted conversions are higher than the experimental values. Particularly at the higher conversions, there is considerable deviation. These results suggest that the diffusion resistance through the product layer (FeS) may not be negligible. In fact, the value of Y_2 necessary for agreement with the observed results can be evaluated by applying Eq. (14-19) to this case. If we still assume external diffusion to be negligible ($Y_1 = 0$), Eq. (14-19) reduces to the form

$$t_{co} = \frac{\rho_B r_s (1 - r_c/r_s)}{bM_B k(C_A)_b} \left\{ 1 + \frac{Y_2}{6} \left[\frac{r_c}{r_s} + 1 - 2\left(\frac{r_c}{r_s}\right)^2 \right] \right\} \qquad (D)$$

where the subscript on t indicates that t_{co} is corrected for diffusion resistance in the product layer. Comparison with Eq. (14-21) shows that the term in braces in Eq. (D) is a correction factor for this diffusion resistance; that is,

$$\frac{t_{co}}{t} = 1 + \frac{Y_2}{6} \left[\frac{r_c}{r_s} + 1 - 2\left(\frac{r_c}{r_s}\right)^2 \right] \qquad (E)$$

or, in terms of conversion, using Eq. (14-20),

$$\frac{t_{co}}{t} = 1 + \frac{Y_2}{6} [1 + (1 - x)^{1/3} - 2(1 - x)^{2/3}] \qquad (F)$$

The t values (without subscripts) in Eq. (E) refer to those calculated from Eq. (14-21) and correspond to the dashed lines in Fig. 14-4. Then Y_2 was evaluated at 477°C by finding what value gives the best agreement of t_{co} with the experimental data, making the calculations with Eq. (F). The solid line for 477°C (Fig. 14-4) shows the t_{co} curve for $Y_2 = 0.66$. There appears to be some deviation at the highest conversion, but the agreement for other x values is good. The effective diffusivity is then readily obtainable from the definition of Y_2 and Eq. (C),

$$D_e = \frac{kr_s}{Y_2} = \frac{3.8 \times 10^5 e^{-30,000/R_g(273+477)}(0.0035)}{0.66}$$

$$= 3.6 \times 10^{-6} \text{ cm}^2/\text{s } (3.6 \times 10^{-10} \text{ m}^2/\text{s})$$

This result can be checked by calculating a corrected curve at 495°C. If the diffusivity is assumed to be constant, Y_2 for 495°C will be

$$Y_2 = \frac{3.8 \times 10^5 e^{-30,000/R_g(273+495)}(0.0035)}{3.6 \times 10^{-6}} = 1.0$$

The solid line for 495°C in Fig. 14-4, which agrees well with the data points, represents the values of t_{co} calculated from Eq. (D) with $Y_2 = 1.0$.

The comparison between solid and dashed lines at any temperature shows the effect of introducing the resistance to diffusion through the product layer. The deviation between the lines increases as the temperature increases, since the resistance to reaction decreases. It should be emphasized that the calculations for this example involve several approximations and thus do not represent a thorough check of the shrinking-core model. However, the experimental data do show that it would be unsatisfactory to use a model which did not allow for either a change in reaction area or diffusion through the product layer. Such a model would give a rate independent of time, so that the predicted conversion-time relationship would be a straight line.

Suppose that the particles have a residence time distribution $J_p(\theta)$. Since the particle flow is segregated, the average conversion is given by Eq. (6-41):

$$\bar{x}_B = \int_0^\infty x_B(\theta)\, dJ_p(\theta) \tag{14-26}$$

This expression would be applicable, for example, to a fluidized bed regenerator used to burn off carbon from deactivated catalyst particles. The conversion-residence time relation would need to be known from an expression such as Eq. (14-19a). Also, the residence-time distribution would have to be measured for the particular reactor at the appropriate operating conditions. This might be done by introducing a step function of marked particles and measuring these marked particles in the effluent, as described in Chap. 6. If the particles were well mixed, $J_p(\theta)$ would be given by Eq. (6-13). Numerical solution of Eq. (14-26) is required because Eq. (14-19a) is not explicit in x_B, and because $J_p(\theta)$ would be in numerical form rather than as an equation.

Sometimes $J_p(\theta)$ is determined solely by the particle-size distribution in the feed. For example, this simplification might be approached in a cocurrent, upflow transport reactor (Fig. 14-5). Under these conditions Eq. (14-26) can be written in terms of the particle-size distribution. Thus

$$\bar{x}_B = \int_0^1 x_B(r_s)\, dw_i \tag{14-27}$$

where w_i is the mass fraction of particles with radii between zero and r_s, and $x_B(r_s)$ is the conversion in the residence time corresponding to a particle size r_s. Equation (14-19a) is such a relationship. Since particle sizes are normally known not as

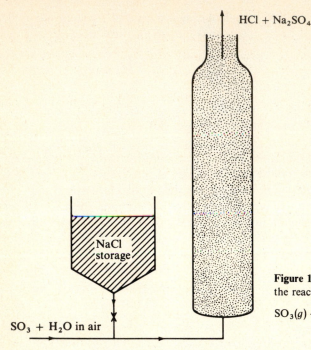

HCl + Na₂SO₄

NaCl storage

SO₃ + H₂O in air

Figure 14-5 Upflow transport reactor for the reaction

$$SO_3(g) + H_2O(g) + 2NaCl(s) \rightarrow$$

$$Na_2SO_4(s) + 2HCl(g).$$

a continuous function but as fractions within two finite sizes (by sieve analysis), it is more appropriate to write Eq. (14-27) as a summation:

$$\bar{x}_B = \sum_{i=1}^{n} x_i(r_{s_i})(\Delta w_i) \tag{14-28}$$

where (Δw_i) is the mass fraction of particles with radii between r_s and $r_s + \Delta r_s$.

Example 14-2 An upflow transport reactor (Fig. 14-5) is to be designed for producing HCl and Na₂SO₄ according to the reaction

$$2NaCl(s) + H_2O(g) + SO_3(g) \rightarrow Na_2SO_4(s) + 2HCl(g)$$

Suppose that the reaction is first order in SO_3 and that the shrinking-core model is applicable. The temperature is 900°F and uniform. Also, an excess of SO_3 and H_2O are used, so that the gas composition is uniform throughout the reactor. The solids flow rate is to be 1000 lb/min into a cylindrical reactor 2 ft in diameter and 30 ft long. The density of the salt particles is 2.1 g/cm³. The particle-size distribution of the feed (assume it does not change in the reactor) is given in Table 14-1. In tests with a batch laboratory reactor with the same gas composition, it was found that the conversion of a sample of 88- to 105-micron particle was 85% in 10 s. Measurements with other sizes of particles indicated that the chemical reaction controlled the rate. RTD studies were carried out in a laboratory reactor with the same flow conditions as expected in the large reactor. The results given in the last column of Table 14-1, and indicated that the residence time was determined by the initial particle size. The *mean residence time* is defined as the volume of solids in the reactor divided by the volumetric feed rate of solids. The fraction of the reactor volume occupied by solids, as determined by measuring the solids holdup in the laboratory flow reactor, was 0.01. It is expected that the porosity will be the same in the large reactor.

Calculate the average conversion of NaCl to Na_2SO_4 in the solid phase in the exit stream.

SOLUTION For chemical reaction controlling the rate, Eq. (14-21) and (14-20) determine $x_B(r_s)$. From the laboratory measurements with 88- to 105-micron particles (average $r_s = 48.2 \times 10^{-4}$ cm), Eq. (14-21) gives

$$\frac{t}{r_s} = \frac{\rho_B}{bM_Bk(C_A)_b}[1 - (1 - 0.85)^{1/3}] = 0.997 \frac{\rho_B}{bM_Bk(C_A)_b}$$

or

$$\frac{\rho_B}{bM_Bk(C_A)_b} = \frac{10}{0.997(48.2 \times 10^{-4})} = 2.08 \times 10^3 \text{ s/cm}$$

The average conversion can be calculated from Eq. (14-28), but first it is necessary to evaluate x_i for each fraction of the particles. To do this we must obtain the residence time for each fraction from the given distribution. From the definition of $\bar{\theta}$,

$$\bar{\theta} = \frac{(\pi 2^2/4)(30)(0.01)}{1,000/[2.1(62.4)(60)]} = 7.1 \text{ s}$$

The residence time of the 53- to 63-micron group of particles is $\theta = 7.1(0.6) = 4.3$ s. The conversion of this group (average $r_s = 29 \times 10^{-4}$ cm) can be calculated from Eq. (14-21), which may be written

$$(1 - x)^{1/3} = 1 - \frac{t}{r_s}\frac{bM_Bk(C_A)_b}{\rho_B}$$

$$1 - x = \left(1 - \frac{4.3}{29 \times 10^{-4}}\frac{1}{2.08 \times 10^3}\right)^3$$

$$x = 1 - (1 - 0.71)^3 = 1 - 0.0246 = 0.975$$

Table 14-1

Sieve no.	Particle diameter, microns	Weight fraction, w_i	Residence time, $\theta/\bar{\theta}$
250–270	53–63	0.10	0.60
200–250	63–74	0.20	0.85
170–200	74–88	0.35	0.99
150–170	88–105	0.25	1.15
115–150	105–125	0.10	1.30

Although it does not occur in this instance, a negative value for the term in parentheses would mean that complete conversion of the particles of the smallest-size group was obtained in less than their residence time of 4.3 s. Actually, the conversion in this group will be 97.5%.

For the group of largest particles, $\theta = 7.1(1.30) = 9.3$ s. Hence the conversion for these particles will be

$$1 - x = \left(1 - \frac{9.3}{57.5 \times 10^{-4}} \frac{1}{2.08 \times 10^3}\right)^3 = (1 - 0.78)^3$$

$$x = 1 - 0.01 = 0.99$$

The conversions for the intermediate-sized groups, calculated in the same way, are given in Table 14-2 (for column height of 30 ft). The interesting feature of these results is that the effects of residence time and particle size tend to balance each other. The large particles, which might be expected to have a lower conversion, have a longer time to react, so that approximately the same conversion is obtained for them as for the smaller particles. This counterbalancing effect is due to the upflow in the reactor. If the particles were in downflow (due to gravity) and the gas passed upward, the smaller sizes would have a longer residence time than the large ones. The effects of residence time and particle size would then supplement each other, leading to a relatively lower conversion for the larger particles and a relatively higher conversion for the smaller particles.

The average conversion, from Eq. (14-28), is

$$\bar{x} = \sum x_i(r_{si})\,\Delta w_i = 0.10(0.975) + 0.20(0.996) + 0.35(0.995)$$
$$+ 0.25(0.993) + 0.10(0.99)$$
$$= 0.991 \qquad \text{or } 99.1\%$$

If nearly complete conversion of NaCl to HCl is desired, a 30-ft reactor is needed. If, however, the conversion need not be more than, say, 75%, a shorter column could be used. If it is assumed that the RTD would be the same in a shorter vessel, the average conversion in it could be evaluated by first obtain-

Table 14-2

Particle diameter, microns	Weight fraction, w_i	Conversion	
		30-ft reactor	15-ft reactor
53–63	0.10	0.975	0.73
63–74	0.20	0.996	0.81
74–88	0.35	0.995	0.81
88–105	0.25	0.993	0.80
105–125	0.10	0.99	0.77

ing a new $\bar{\theta}$. On this basis let us estimate the conversion obtainable in a 15-ft reactor. The average residence time would be

$$\bar{\theta} = 7.1\left(\frac{15}{30}\right) = 3.6 \text{ s}$$

For the group of smallest particles,

$$(1 - x)^{1/3} = 1 - \frac{t}{r_s}\frac{1}{2.08 \times 10^3} = 1 - \frac{3.6(0.6)}{29 \times 10^{-4}}\frac{1}{2.08 \times 10^3}$$

$$1 - x = (1 - 0.355)^3$$

$$x = 0.732$$

The x_i values for the other size groups are shown in the last column of Table 14-2. The average conversion, obtained using these results in Eq. (14-28), is 0.80. A 15-ft reactor would be more than adequate for a 75% conversion.

14-6 Conversion for Variable Fluid Composition

Unless there is a large excess, the concentration of fluid reactant will not be uniform in either fixed or moving beds (transport reactor) of particles. We will analyze both types in a simple way by assuming plug flow of fluid in the fixed-bed, and plug flow of both fluid and particles in the moving-bed reactor.

Fixed-bed reactor Consider first a fixed bed in which the solid particles constitute or contain one of the reactants, while a second reactant is in the fluid phase. In the general case the reactant concentration in the bulk fluid will decrease along the reactor length z. Since the rate at any location is a function of time, the variation of concentration with length will be a function of time; that is, $C_b = f(t, z)$. Thus the process does not operate at steady state. Such behavior is characteristic of several practical processes, such as ion-exchange reactors, regeneration of fouled catalysts by combustion with air, and adsorption from gas or liquid streams. Figure 14-6 shows concentration profiles for the catalyst-regeneration case. Hot air is passed over the bed of particles which have been uniformly deactivated by the deposition of carbon. The oxygen cleans the catalyst by reacting with the deposited carbon to give CO and CO_2, which go out with the air. The upper part of Fig. 14-6 shows how C_b varies with reactor length at various times after first introducing air to the bed. The O_2 concentration in the feed is $(C_b)_f$. The vertical dashed line at $z = L$ represents the exit of the reactor. If the O_2 concentrations at this point for various times are plotted, the *breakthrough* curve shown in Fig. 14-6b is obtained. The lower graph represents the fraction of the original carbon on the catalyst that has been removed. Similar curves describe the behavior of ion-exchange reactors. For example, for a water-softening process, C_b would represent the concentration of calcium or magnesium ions. The breakthrough curve (Fig. 14-6b) indicates the length of time that the bed could be operated before a given concentration of calcium or magnesium would appear in

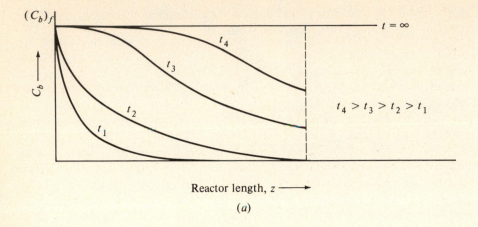

Reactor length, $z \longrightarrow$

(a)

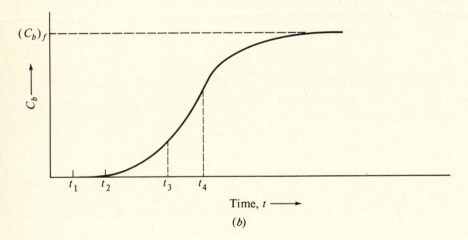

Time, $t \longrightarrow$

(b)

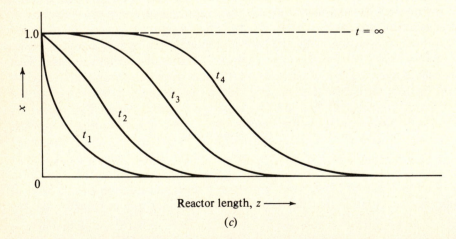

Reactor length, $z \longrightarrow$

(c)

Figure 14-6 Concentration profiles in fixed-bed reactor for catalyst regeneration: (a) reactant (O_2) concentration vs. bed depth; (b) breakthrough curve $(O_2$ concentration vs. time); (c) conversion of solid reactant (carbon).

the water leaving the exchanger. Figure 14-6c would correspond to the fraction of the sodium form of the resin that has been converted.

In Chap. 13 the effects of mass and energy transfer on the design of catalytic *fixed-bed* reactors were considered, and several design procedures of varying degrees of complexity were discussed. The same problems, all related to the flow pattern of the fluid in the bed, exist for noncatalytic reactions except that the global rate is a function of time. Methods† have been developed for predicting the profiles illustrated in Fig. 14-6. Some methods include the effects of axial dispersion in the bed, external mass transfer between fluid and particle, intraparticle diffusion, and chemical reaction, and nonisothermal conditions.‡ Results are usually obtainable only by numerical methods, since simultaneous solution of several partial-differential equations is required. It is not appropriate to go into detail about these solutions, but we can illustrate the concepts involved by employing a simple model. We assume isothermal operation, plug flow of fluid, and that the global rate is given by the shrinking-core model. The latter assumption means that Eqs. (14-10) and (14-14) are applicable.

Suppose the reaction is the usual one

$$A(g) + bB(s) \to E(g) + F(s),$$

and that we write a mass conservation equation for A in the fluid phase. This expression will be the same as Eq. (13-17), except that a term is added for the rate of accumulation of A in the fluid:

$$-u\frac{\partial(C_A)_b}{dz} = \mathbf{r}_P\rho_B + \varepsilon_B\frac{\partial(C_A)_b}{\partial t} \tag{14-29}$$

where ρ_B = density of bed of particles

$\mathbf{r}_P\rho_B$ = global rate of disappearance of A, per unit volume of reactor

u = superficial velocity in the direction of flow

Equation (14-10) gives the rate per particle. For spherical particles packed with a void fraction ε_B, the number of particles per unit volume is $3\varepsilon_B/4\pi r_s^3$, so that

$$\mathbf{r}_P\rho_B = \frac{3\varepsilon_B k(r_c/r_s)^2(C_A)_b}{r_s[1 + (r_c^2/r_s^2)(k/k_m) + (kr_c/D_e)(1 - r_c/r_s)]} \tag{14-30}$$

Then Eq. (14-29) becomes

$$-u\frac{\partial(C_A)_b}{\partial z} = \frac{3\varepsilon_B k(r_c/r_s)^2(C_A)_b}{r_s[1 + (r_c^2/r_s^2)(k/k_m) + (kr_c/D_e)(1 - r_c/r_s)]} + \varepsilon_B\frac{\partial(C_A)_b}{\partial t} \tag{14-31}$$

This equation expresses $(C_A)_b$ in terms of the radius r_c of the unreacted core, which

† J. B. Rosen, *J. Chem. Phys.*, **20**, 387 (1952); *Ind. Eng. Chem.*, **46**, 1590 (1954); H. C. Thomas, *J. Am. Chem. Soc.*, **66**, 1664 (1944); N. K. Heister and T. Vermeulen, *Chem. Eng. Progr.*, **48**, 505 (1952); S. Masamune and J. M. Smith, *AIChE J.*, **10**, 246 (1964); *AIChE J.*, **11**, 34 (1965); *Ind. Eng. Chem., Fund. Quart.*, **3**, 179 (1964).

‡ M. Sagara, S. Masamune, and J. M. Smith, *AIChE J.*, **13**, 1226 (1967).

is a function of time as described by Eq. (14-14). Equations (14-31) and (14-14) establish $C_{Ab}(t, z)$ and $r_c(t, z)$. The boundary conditions are

$$(C_A)_b = (C_A)_0 \qquad \text{at } z = 0 \qquad \text{for } t \geq 0 \qquad (14\text{-}32)$$

$$r_c = r_s \qquad \text{at } t = 0 \qquad \text{for } z \geq 0 \qquad (14\text{-}33)$$

These equations may be solved numerically to give curves of the form shown in Fig. 14-6. The resultant $r_c(t, z)$ can be converted to $x(t, z)$ by means of Eq. (14-20).

Note that if $(C_A)_b$ is essentially constant, corresponding to a large excess of reactant A, Eq. (14-31) disappears, and the solution of Eq. (14-14) is that given in Sec. 14-4; that is, Eq. (14-19), or (14-19a) in terms of conversion.

Moving-bed reactor A second kind of reactor in which C_b may vary with position is the *moving-bed* type. Examples are inclined or vertical kilns and transport reactors with cocurrent or counterflow of fluid and solid phases (Fig. 14-7). In such cases steady state is achieved in that the properties at any location in the bed do not change with time. The behavior of this type of reactor can be described in a relatively simple way, provided solid and fluid phases are in plug flow. Under these conditions the residence time is the same for all particles. It is related to an element Δz of reactor length by the expression

$$\Delta t = \Delta \theta = \frac{(\text{vol. of particles})}{(\text{vol. flow rate of particles})} = \frac{A_c(\Delta z)(\varepsilon_s)}{(\text{mass flow rate})/\rho_s} \qquad (14\text{-}34)$$

$$\Delta t = \frac{\varepsilon_s \rho_s}{G_s} \Delta z$$

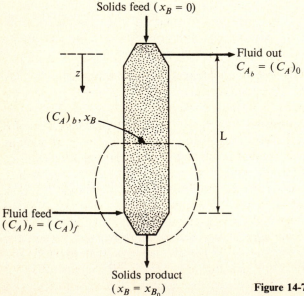

Solids feed ($x_B = 0$)

Fluid out
$C_{A_b} = (C_A)_0$

$(C_A)_b, x_B$

L

Fluid feed
$(C_A)_b = (C_A)_f$

Solids product
($x_B = x_{B_0}$)

Figure 14-7 Counterflow moving-bed reactor.

or

$$dt = \frac{\varepsilon_s \rho_s}{G_s} dz \tag{14-35}$$

where z = reactor length (from entrance of solids)
  ε_s = fractional holdup of solids (fraction of reactor volume occupied by solids
  G_s = superficial mass velocity of solids at z
  ρ_s or ρ_B = density of particles (assume constant)
  A_c = cross-sectional area of column

Substitution of this expression for dt in Eq. (14-14) gives the change in r_c with length of reactor z, provided that the shrinking core concept is valid. The result is

$$\frac{dr_c}{dz} = - \frac{\varepsilon_s b M_B k (C_A)_b}{G_s[1 + (r_c^2/r_s^2)(k/k_m) + (k r_c/D_e)(1 - r_c/r_s)]} \tag{14-36}$$

The radius r_c can be written in terms of conversion through Eq. (14-20). In this form Eq. (14-36) is

$$\frac{dx_B}{dz} = \frac{3\varepsilon_s b M_B k (C_A)_b (1 - x_B)^{2/3}}{G_s r_s \{1 + (k/k_m)(1 - x_B)^{2/3} + (k r_s/D_e)(1 - x_B)^{1/3}[1 - (1 - x_B)^{1/3}]\}} \tag{14-37}$$

Before Eq. (14-37) can be integrated, $(C_A)_b$ must be expressed in terms of x_B. A mass balance around a section of reactor provides this relation. For a counterflow reactor a balance of A around the bottom section, as shown in Fig. 14-7, yields

$$u[(C_A)_f - (C_A)_b] = \frac{G_s}{b M_B} (x_{B_o} - x_B) \tag{14-38}$$

or

$$(C_A)_b = (C_A)_f - \frac{G_s}{u b M_B} (x_{B_o} - x_B) \tag{14-39}$$

Here u is the superficial axial velocity of fluid (constant across the reactor for plug flow), x_{B_o} is the conversion of B in the solids leaving the reactor, and $(C_A)_f$ is the concentration of A in the feed. We can now substitute Eq. (14-39) in Eq. (14-37) and integrate the resulting expression from $x_B = 0$ to obtain the conversion of B for any reactor length. If the process is isothermal and if one mole of gaseous product E is produced for each mole of reactant gas A, the gas velocity u will be constant (except for the effect of change in pressure) through the reactor. However, G_s will, in general, vary since the molecular weight of product F will be different from that of reactant B. However, G_s can be expressed in terms of the mass velocity of solids at the feed and the conversion x_B (see Prob. 14-10). A trial procedure is required if the conversion for a given length is needed, since the desired exit conversion occurs in Eq. (14-39). A value for x_{B_o} may be assumed and the x-vs.-z profile evaluated. When the conversion at $z = L$ agrees with the assumed value, no further trials are necessary. If the length of reactor to achieve a certain conversion is required, the trial procedure is not necessary (Example 14-3).

Example 14-3 A counterflow, moving-bed reactor with downflow of solids and upflow of gas, as in Fig. 14-7, is to be designed for reducing FeS_2 pellets to FeS (see Example 14-1). The reactor will operate at 1 atm pressure and isothermally at 495°C. The reactor diameter is to be 0.90 m with a gas flow rate of 2.6 m³/s of a mixture of 70% CO_2 and 30% H_2 (the CO_2 is inert). Specially prepared, relatively large pellets 0.01 m in diameter (equivalent spheres) will be fed to the top of the reactor at a rate of 0.5 kg/s.

The FeS_2 pellets have a high reactivity and form a very porous layer of FeS during reaction. The rate constant is a hundredfold larger than that given by Eq. (C) of Example 14-1. For the large pores in the FeS product layer the effective diffusivity will be 3.6×10^{-7} m²/s.

Neglect external diffusion resistance, and also any change in the total pellet diameter as reaction proceeds. The solid fraction, ε_s, in the bed will be 0.5. Assume that the pellet density is constant and equal to that of FeS_2, that is, 5.0×10^3 kg/m³. Since the molecular weight of FeS is not greatly different from that of FeS_2, take G_s as constant and equal to G_s for the feed ($x_B = 0$).

Calculate the reactor length required for a conversion of 70% of the FeS_2 to FeS.

SOLUTION Assume that both streams are in plug flow and that the shrinking core model for the global rate is applicable. Then Eq. (14-37) is valid with A designating hydrogen and B designating FeS_2.

The feed concentration of hydrogen is

$$(C_{H_2})_f = \frac{P_{H_2}}{R_g T} = \frac{1(0.30)}{82(495 + 273)} = 4.76 \times 10^{-6} \text{ g mol/cm}^3$$

$$= 4.76 \times 10^{-3} \text{ kg mol/m}^3$$

The gas velocity and solids mass velocity at the feed will be

$$u = \frac{2.6}{(\pi/4)(0.9)^2} = 4.09 \text{ m/s}$$

and

$$G_s = \frac{0.5}{(\pi/4)(0.9)^2} = 0.786 \text{ kg/(m)}^2(\text{s})$$

Since $b = 1$, $M_B = 120$ for FeS_2, and $x_{Bo} = 0.7$, Eq. (14-39) becomes

$$(C_{H_2})_b = 4.76 \times 10^{-3} - \frac{0.786}{4.09(1)120}(0.7 - x_B)$$

$$= 4.76 \times 10^{-3} - 1.61 \times 10^{-3}(0.7 - x_B) \tag{A}$$

From Eq. (C) of Example 14-1,

$$k = 100(3.8 \times 10^5) \exp[-30,000/R_g(768)], \text{ cm/s}$$

$$= 1.08 \times 10^{-1} \text{ cm/s, or } 1.08 \times 10^{-3} \text{ m/s}$$

$$kr_s/D_e = \frac{1.08 \times 10^{-3}(0.01/2)}{3.6 \times 10^{-7}} = 15$$

Since external diffusion is negligible, k_m is very large, and the term involving k_m in Eq. (14-37) approaches zero. Then, substituting Eq. (A) for $(C_{H_2})_b$ and other numerical values in this equation gives

$$\frac{dx_B}{dz} = \frac{3(0.5)(1)(120)1.08 \times 10^{-3}}{0.786(0.01/2)} \frac{[4.76 - 1.61(0.7 - x_B)]10^{-3}(1 - x_B)^{2/3}}{1 + 15(1 - x_B)^{1/3}[1 - (1 - x_B)^{1/3}]}$$

$$\frac{dx_B}{dz} = f(x_B) = 49.5 \times 10^{-3} \frac{[4.76 - 1.61(0.7 - x_B)](1 - x_B)^{2/3}}{1 + 15(1 - x_B)^{1/3}[1 - (1 - x_B)^{1/3}]} \tag{B}$$

At the top of the column, $x_B = 0$ and Eq. (B) yields

$$\frac{dx_B}{dz} = 49.5 \times 10^{-3} \frac{4.76 - 1.61(0.7)}{1} = 0.18 \text{ m}^{-1}$$

If this rate of increase in conversion with reactor length were the same throughout the reactor, a column of only $0.7/0.18 = 3.9$ meters length would be needed for 70% conversion. However dx_B/dz will decrease with z. As a layer of FeS builds up on the particles, the diffusion resistance to hydrogen transfer through the layer increases. This reduces the global rate of reaction. Thus, at the bottom of the bed, $x_B = 0.7$, Eq. (B) gives

$$\frac{dx_B}{dz} = 49.5 \times 10^{-3} \frac{4.76(0.3)^{2/3}}{1 + 15(0.3)^{1/3}[1 - (0.3)^{1/3}]} = 0.024 \text{ m}^{-1}$$

To find the required length of reactor we can write Eq. (B) in integrated form as

$$\int_0^L dz = \int_0^{0.7} \frac{dx_B}{f(x_B)}$$

or

$$L = \int_0^{0.7} \frac{dx_B}{f(x_B)} \tag{C}$$

where $f(x_B)$ is defined in Eq. (B). Numerical integration gives

$$L = 14 \text{ m}$$

PROBLEMS

14-1 A gas-solid noncatalytic reaction of the type discussed in Sec. 14-3 is investigated by measuring the time required for complete conversion of solid B as a function of particle diameter. The results are as follows:

Particle diameter mm	0.063	0.125	0.250
Time for complete conversion, min	5.0	10.0	20.0

If the diffusion resistance in the gas phase around the particle is negligible, what mechanism controls the rate of reaction?

14-2 In measurements for a reaction of the type discussed in Sec. 14-3, the time required for equal conversions is found to be directly proportional to particle diameter at low conversion but becomes proportional to the square of the particle size as the conversion becomes larger. What can be said about the mechanism that controls the rate? Again, diffusion resistance in the gas surrounding the particles is negligible.

14-3 The reaction described in Prob. 14-1 is to be carried out by passing the reactant gas crosswise over a moving grate carrying the solid particles. The velocity of the grate is such that the particles are exposed to the gas stream for 9 min. If the distribution of particle sizes is as given below, what will be the average conversion leaving the reactor? What will be the conversion for the 0.063-mm particles?

Particle diameter, mm	0.063	0.125	0.250	0.500
Wt %	25	35	35	5

The gas composition does not change significantly during flow across the grate.

14-4 A solid feed of $\frac{1}{4}$-in. spherical particles of pure B is to be reacted in a rotary kiln (such as a lime or cement kiln). The gas A in contact with the solids is of uniform composition. The whole process is isothermal. The reaction is first order with respect to A, irreversible, and follows the stoichiometry

$$A(g) + B(s) \rightarrow C(s) + D(g)$$

The substance B is nonporous, but product C forms a porous layer around the unreacted core of B as the reaction proceeds. In the kiln the solids will move in plug flow from one end of the kiln to the other at a velocity of 0.1 in./s. It is desired to design a kiln to obtain 90% conversion of B.

Small-scale studies of the reaction indicate that diffusion resistance between particle surface and the gas is negligible. In a batch agitated reactor operated at the same temperature and gas composition, the following data were obtained:

A conversion of 87.5% in 1 h with $\frac{1}{8}$-in. particles
A conversion of 65.7% in 1 h and 24 min with $\frac{1}{4}$-in. particles

(a) Calculate the length required for the kiln. (b) In the future it may be necessary to handle a feed of B which consists of 20 wt % $\frac{1}{8}$-in. particles, 50 wt % $\frac{1}{4}$-in. particles, and 30 wt % $\frac{3}{8}$-in. particles. Calculate the average conversion in the product from this mixed feed, using the reactor designed in part (a).

14-5 A plant produces HCl and Na_2SO_4 from salt and sulfuric acid in a transport reactor (Fig. 14-5). The reactor operates at about 900°F, so that the NaCl and Na_2SO_4 are solids, HCl is a gas and the H_2SO_4 exists as gaseous H_2O and SO_3 (see Example 14-2). The H_2O is present in great excess. At normal conditions the residence time of the particles ($r_s = 0.05$ cm) in the reactor is 10 s, and the conversion of NaCl is 100%. A new supply of salt is obtained which has particles twice the diameter of the normal material. If velocity of solids and gases and all other operating conditions are held constant, what would the residence time have to be to obtain complete conversion of the new salt particles?

There is a negligible concentration difference of SO_3 between bulk gas and the surface of the particles. The particles remain spherical and of constant diameter, regardless of extent of conversion to Na_2SO_4. Unlike the situation in Example 14-2, it is expected that the resistance to reaction at the salt interface, while significant, is not controlling for either size of salt particles. The chemical reaction is first order in SO_3. The diffusivity of SO_3 through the $NaSO_4$ layer is 0.01 cm^2/s, and the first-order rate constant for the reaction is 0.5 cm/s. The whole reactor is isothermal.

14-6 The reduction of FeS_2 to FeS is carried out in a tubular reactor with upflow of hydrogen and downflow of solids. The reactor will operate at 495°C and 1 atm with pure hydrogen. For these conditions gas-phase diffusion resistance is negligible. The diffusivity of hydrogen in the product layer and the rate constant for the reaction at the FeS_2 surface have the values established in Example 14-1.

The sizes and RTDs for the particles in the reactor are as follows:

Particle radius, mm	0.05	0.10	0.15	0.20
Weight fraction	0.10	0.30	0.40	0.20
Residence time, $\theta/\bar{\theta}$	1.40	1.10	0.95	0.75

If the mole fraction of hydrogen in the gas is assumed to be constant and unity, what will be the average conversion of FeS_2 to FeS in a reactor with a mean residence time of 60 min?

14-7 A. Reconsider Example 14-3 for a gas feed of 2.6 m^3/s of *pure* hydrogen. What conversion of FeS_2 could be achieved in a 14-m-long reactor? All other conditions are the same as in Example 14-3.

B. If a 14-m reactor were used and a conversion of 70% of FeS_2 was adequate, how much could the solids rate be increased if the gas feed was pure hydrogen instead of a 30% H_2-70% CO_2 mixture?

14-8 A. What is the residence time of the particles in the reactor in Example 14-3?

B. What would be the residence time in part B of Problem 14-7?

14-9 In the quantitative material and examples of this chapter, one of the products of the reaction was always a solid. Consider a coal gasification reaction of the type

$$C(s) + H_2O(g) \rightarrow CO(g) + H_2(g)$$

where only gaseous products are produced. This type of reaction could also represent the steam regeneration of spent activated carbon, if $C(s)$ is replaced by the organic substance adsorbed in the activated carbon.

A. For the coal gasification reaction derive an expression analogous to Eq. (14-19a) showing how the conversion of C varies with reaction time.

B. What forms of the equation derived in part A are applicable if
(1) external mass transfer controls the global rate;
(2) external mass transfer resistance is negligible.

14-10 Consider a counterflow, moving-bed coal-gasification reactor (Fig. 14-7). Assume that the only reaction is that given in Prob. 14-9. Derive a differential equation, analogous to Eq. (14-37), relating the extent of gasification to the reactor length. In contrast to Example 14-3, the mass velocity G_s of solids could change markedly along the reactor length. Allow for this variation by expressing G_s in terms of the solids feed rate and the conversion of coal. Note also that in integrating the differential equation, the gas velocity u could vary significantly along the reactor since two moles of gas are produced for each mole of steam consumed.

SUBJECT INDEX